TRANSACTIONS

OF THE

AMERICAN PHILOSOPHICAL SOCIETY

HELD AT PHILADELPHIA
FOR PROMOTING USEFUL KNOWLEDGE

NEW SERIES—VOLUME XXXIV
1944

THE AMERICAN PHILOSOPHICAL SOCIETY
INDEPENDENCE SQUARE
PHILADELPHIA 6

1945

LANCASTER PRESS, INC., LANCASTER, PA.

CONTENTS OF VOLUME XXXIV

PART I

PART II

PART III

TRANSACTIONS

OF THE

AMERICAN PHILOSOPHICAL SOCIETY

HELD AT PHILADELPHIA
FOR PROMOTING USEFUL KNOWLEDGE

NEW SERIES—VOLUME XXXIV, PART I

THE VELOCITY OF LIGHT

N. ERNEST DORSEY
National Bureau of Standards

PHILADELPHIA 6
THE AMERICAN PHILOSOPHICAL SOCIETY
INDEPENDENCE SQUARE
OCTOBER, 1944

LANCASTER PRESS, INC., LANCASTER, PA.

THE VELOCITY OF LIGHT *

N. Ernest Dorsey

CONTENTS

* A contribution from the National Bureau of Standards.

INTRODUCTION

OBJECT

As is well known to those acquainted with the several determinations of the velocity of light, the definitive values successively reported—those values which the several observers give as defining or summing up the result of the experimental work being reported—have, in general, decreased monotonously from Cornu's 300.4 megameters per second [1] in 1874 to Anderson's 299.776 in 1940, the monotony being severely broken by the presence of Perrotin and Prim's 299.90 of 1902, between the adjacent values by Michelson—299.853 in 1882 and 299.802 (first published as 299.820) in 1924. In how far is either this drift or its interruption of physical significance? That is in dispute, some holding one view, and others the opposite. In this paper an answer to that question is sought.

The earlier view, still held by most experienced experimental physicists, is that the drift is of no physical significance, and that the break in it is to be sought in the low precision of Cornu's and of Perrotin and Prim's work, and perhaps in some common systematic error, those two determinations having been made by the same method (Fizeau's, not used in any of the others) and largely by the use of the same apparatus, and carried out in the same manner. Cornu continually advised regarding the technique of the later work. The two determinations differed mainly in only three particulars: the observer, the path, and the lenses used. In each work, the spread of the individual determinations was great, even the nominally identical determinations for the same order of eclipse frequently having a spread of more than 2 percent. In view of such a spread, those holding the opinion now being considered feel that the difference of only 0.2 percent between the definitive values published by these observers and by Anderson is of little, if any, significance.

As for the drift itself, they see it, prior to any critical study of the several reports, as probably, in large part, a psychological phenomenon facilitated by the low precision of the earlier work, which not only obscured the effect of systematic errors but made impossible the experimental discovery of such errors unless their effects were great. They see it as probably arising in large part from two all but universally acting causes: (1) the observer's exaggerated opinion of the accuracy of his own work, and (2) his inability to avoid being influenced in some measure by his preconceived opinion as to what he should find.

As will be seen later, Cornu admitted that from the start he favored a high value, one above 300 megameters per second. Although the agreement of the result of his preliminary determination with that of Foucault partially reconciled him to a lower value, the higher value that he obtained in his later and more accurate work was so pleasing that he reverted to his original opinion, and went back to the records of his earlier work, hunting for a plausible reason for discrediting it. He found such a reason and accepted it, although he did not show that it was efficacious. Thus was established as probable a value greater than 300 megameters per second, which was bolstered by certain astronomical data.

To say that an observer's results are influenced by his preconceived opinion does not in the least imply that those results were not obtained and published in entire good faith. It is merely a recognition of the fact that it seems more profitable to seek for error when a result seems to be erroneous, than when it seems to be approximately correct. Thus reasons are found for discarding or modifying results that do violence to the preconceived opinion, while those that accord with it go untested. An observer who thinks that he knows approximately what he should find labors under a severe handicap. His result is almost certain to err in such a direction as to approach the expected value.

The size of this unconsciously introduced error is, obviously, severely limited by the experimenter's data, by the spread of his values. The smaller the spread, the smaller, in general, will be this error. The size will be much affected also by the circumstances of the work, and by the strength of the bias. If the work is strictly exploratory, its primary purpose being to find whether the procedure followed is at all workable, then only a low accuracy will be expected, and there will be no serious attempt to explain departures from the expected, even though the departures be great. Consequently, this error of bias may be entirely absent from such results. But if the worker is striving for accuracy, then departures from the expected will appear to him serious; and the stronger the bias, the more serious will they seem. He will seek to explain them; and that seeking will tend, in the manner already stated, to introduce an error. An error arising in this way will seldom be negligible, but in no case should one expect it to be great, the work being done in good faith.

The later view, apparently first published by Gheury de Bray [1] * in 1926, is that both the drift and its break are of prime physical significance, indicating that the velocity of light is subject to secular variations, presumably arising from changes in the space or medium in which the earth finds itself from time to time.

In his first papers [1] Gheury de Bray contented himself with a rectilinear relation, but remarked that

[1] For the sake of brevity, the word "second" is used throughout this paper to denote the mean solar second, unless the contrary is clearly indicated.

* Figures in brackets indicate the literature references at the end of this paper.

the data could be more easily explained by assuming that the velocity oscillates with the time. Now, the three values published around 1880 by Newcomb and by Michelson, obtained with much shorter paths than were the others available at the time of Gheury de Bray's paper, lie below the best linear representation of the others. Consequently, he represented the data by two lines: one for "short paths," the other for "long paths," Cornu's value, which he takes as 299.9 megameters per second (Helmert's [2] correction), instead of Cornu's 300.4, being included in both graphs. He thus implies that the secular decrease in the velocity of light is greater when the measurement is made over a short path than when over a long one. Such a dependence on path length is hard to understand. This same idea will be found also in articles he published in 1931, 1932, and 1936 [3]. But he seems to have concluded, even in 1927 [1], that measurements over the shorter paths are the less accurate, and to have stressed more particularly the graph for the longer ones.

Later, F. K. Edmondson [4] proposed a sinusoidal variation, which he showed could be so adjusted as to pass through Perrotin and Prim's value while fitting all the other values satisfactorily, Cornu's value being again taken as 299.9 megameters per second. This proposal was gladly accepted by Gheury de Bray [5].

In the years since Gheury de Bray's paper of 1926, these suggestions, that the velocity of light is subject to a secular variation, have called forth many papers [6]. Not a few of their authors seem to be very favorably impressed by the idea of a secular variation, some seem to be favorable to it, but unwilling to commit themselves, and some are strongly critical. The criticisms have been along two lines: (1) Such a variation would under certain plausible assumptions demand a secular variation in other quantities; and no such variation has been observed. (2) The discrepancy between the earlier definitive values and the more recent ones is seldom more than a little greater than the admitted uncertainties in the earlier values, and consequently is of no significance.

To the last, Gheury de Bray objects that it does no more than replace one strange phenomenon by another, the strange apparent decrease in the velocity, by the existence of some mysterious cause that makes each successive definitive value exceed the true by a smaller amount than did its predecessor [7]. In this he seems to be right.

The trouble is this: The secular variationists and those to whose criticisms Gheury de Bray specifically objects, seem each to forget that the published definitive values, with their accompanying limits of uncertainty, are not experimental data, but merely the authors' inferences from such data. Inferences are always subject to question; they may be criticized, reexamined, and revised at any time. When uncritically accepted, they form an exceedingly weak foundation for a revolutionary suggestion; in fact, the suggestion then rests solely on authority.

True, each worker has much information not available to others to guide him in drawing his inference; and in many cases, such as those involving a comparison of several determinations of the same quantity, and the derivation from them of a definitive value, it is very common and quite proper to accept the worker's inference, modified as may be by considerations of published criticisms by others working in the same field and of the general impression obtained by a rather casual inspection of his report. This is thoroughly justified by the principle of economy of effort. It assumes that each of those working in the same field will have examined critically all such previous work and will have brought to light errors and omissions not recognized by those earlier workers. Unfortunately, there has been published very little independent critical discussion of the several determinations of the velocity of light. In general, each worker seems to have confined his attention to the particular determination on which he was then engaged, accepting the published definitive values of earlier work without other criticism than one of arbitrary weighting.

Be that as it may, when it is a question of basing a revolutionary suggestion on work reported by others, it behooves one to examine each piece of that work carefully in order to see whether it is sound enough to carry the superstructure that is to be placed on it. An ill-founded revolutionary suggestion may cause many to waste much time and effort.

And such a revolutionary suggestion offered without that examination cannot be satisfactorily attacked by merely pointing out that only slight changes in the admitted uncertainties of the measurements will render the suggestion unnecessary, especially if those changes must exhibit some kind of regularity. The most that can be accomplished by such criticism is to show the weakness of the foundation on which the suggestion rests, to show that the suggestion is unproven; whereas the critic presumably wishes to show that there is no basis at all for the suggestion. For that, such a detailed study of the several reports as will presently be described is necessary.

For these reasons it is every author's duty to publish amply sufficient primary data and information to enable a reader to form a just and independent estimate of the confidence that may be placed in the inferences that the author has drawn therefrom. If he does not, he is false to both his reader and himself, and his inferences should carry little weight, no matter how great his reputation may be; for even the greatest is not infallible.

Indeed, values reported without such satisfactorily supporting evidence have no objective value whatever, no matter how accurate they may happen to be. They rest solely on the authority of the reporter, who is never infallible. The acceptance in scientific matters of conclusions resting on authority alone has for long been, quite properly, considered as thoroughly "unscientific."

In order to evaluate satisfactorily the strength of the foundation on which the suggestion of such a secular variation rests, it is necessary to go behind the inferences of the several experimenters and to see how far those inferences are justified by the experimental work. That demands in each case a study of the method employed, of the means adopted for the realization of the method, of the systematic errors that might be expected to affect the results, of the author's diligence in searching out and eliminating such errors, of the degree of concordance of the observations, and of the procedure by which the author derived his definitive value from the experimental data. In every case it is the objective value of the work that is to be independently appraised.

The present paper is a report of such a study. Of course, the conclusions reached are themselves inferences; this time, mine. An endeavor has been made to show the grounds on which they rest. Some readers may agree with me; others will dissent. It is to be hoped that the latter will publish clear and specific statements of their reasons for disagreeing, to the end that the true status of the subject may be established.

It was initially intended to carry the study in each case only so far as is required to substantiate or to disprove the objective value of the data on which the suggestion of a secular variation rests. But as the study progressed, it became plain that with only a little more labor it would be possible to arrive at an objective estimate of the accuracy that might validly be ascribed to the work covered by each of the reports; and from those estimates to derive a definitive value for the velocity of light—the best value that can be inferred from the collective reports. It seeming worth while, that extra work has been done.

Obviously, one who has never participated in such measurements is not qualified to estimate the actual magnitude of those systematic errors for which no pertinent information is given in the paper under study; neither should his failure to recognize some important sources of systematic error cause any surprise. All he can hope to do is to call attention to such sources of systematic error as are mentioned in the paper or occur to him, and to seek carefully for such errors, omissions, and improper procedures as may be revealed by the report being studied, hoping that he himself may not fall into more grievous ones.

PLAN

The plan that is followed is this: Certain remarks concerning the theory of errors, the method of least squares, averaging, and absolute physical measurements are given first, so as to avoid digressions later.

Then the determinations by Fizeau and by Foucault are considered, and those by Cornu and by Perrotin and Prim, by the Fizeau method of the toothed wheel, are each critically studied. This is followed by a similar study of the determinations by Newcomb, by Michelson, and by Michelson, Pease, and Pearson, all of whom used the Foucault method of rotating mirror. Then the work of Mittelstaedt, of Anderson, and of Hüttel, each of whom used Kerr cells, is briefly considered. The work of Young and Forbes, who used the toothed-wheel method, is not considered, it being generally admitted that their work is seriously in error, and is reported unsatisfactorily.

Then come two appendixes: one dealing with the theories of the methods and their inherent difficulties, and the other with vibrations maintained by periodic impulses.

All that is contained in the remarks and in the appendixes is involved somewhere in the discussion of the reports, and although none of it is particularly new, some of it seems to have escaped the attention of one or more of those who have been engaged in the determination of the velocity of light.

REMARKS

THEORY OF ERRORS

The mean of a family of measurements—of a number of measurements of a given quantity carried out by the same apparatus, procedure, and observer—approaches a definite value as the number of measurements is indefinitely increased. Otherwise, they could not properly be called measurements of a given quantity. In the theory of errors, this limiting mean is frequently called the "true" value, although it bears no necessary relation to the true quaesitum, to the actual value of the quantity that the observer desires to measure. This has often confused the unwary. Let us call it the limiting mean.

Let e denote the amount by which a given member of the family departs from the limiting mean, and let e_q denote that value which in the indefinitely extended family is surpassed by half of the e's; that is, it is an even chance that a given member of such an extended family departs from the limiting mean by as much as e_q.

The quantity e_q, the quartile error, commonly called the probable error of a single observation, will in this study be called the technical[2] probable error of a single member of such a family.

It is obvious that the mean of all the e's approaches a definite limit as the number of members in the family increases indefinitely; and the same is true of any power of e. Denote this limiting mean of e by η, and

[2] The addition of this qualifier is to indicate that the term "probable error," which has several connotations, is used in the strictly technical sense in which it is employed in the theory of errors. In this study that qualifier will always be used when the term is to be so understood.

that of e^2 by σ^2; that is,

$$\eta \underset{n\to\infty}{\equiv} (\Sigma e)/n \qquad \text{and} \qquad \sigma^2 \underset{n\to\infty}{\equiv} (\Sigma e^2)/n,$$

n being the number of members in the family, and Σ indicating the sum of all n values of the symbol following it. The second of these quantities, σ, is known as the standard deviation of the members of the family.

Each of the quantities e_q, η, and σ is determined solely by the characteristics of the family; none depends in any way on the number of members involved in a given study. If the distribution of the e's is "normal," as may be assumed in most problems of physics, these quantities are related as shown in eq 1.

$$\left.\begin{aligned}\eta^2 &= 2\sigma^2/\pi \\ e_q &= 0.6745\ \sigma = 0.8454\ \eta.\end{aligned}\right\} \tag{1}$$

Now consider ν groups of n members each, all belonging to the same family, n having a fixed finite value. Let the mean value of a group be denoted by a. These a's may be regarded as defining a new family. The limiting mean of this new family, as ν increases indefinitely, will, obviously, be the same as the limiting mean of the original family. Let e_n denote the departure of an a from that mean, and let e_{qn} be that value which is exceeded by half the e_n's when ν is increased indefinitely. Then e_{qn} is the technical probable error of an a; that is, of the mean of a group of n members belonging to the original family. With reference to the new family, the quantities

$$\eta_n \underset{\nu\to\infty}{\equiv} (\Sigma e_n)/\nu \qquad \text{and} \qquad (\sigma_n)^2 \underset{\nu\to\infty}{\equiv} (\Sigma e_n^2)/\nu$$

play exactly the same parts as do η and σ^2 in the original one.

Since the new family is built from the original one, its properties can be expressed in terms of those of the original, together with the common number of original members represented by each of its own.

$$\eta_n^2 = \eta^2/n; \quad \sigma_n^2 = \sigma^2/n \tag{2}$$

$$e_{qn} = 0.6745\ \sigma_n = 0.8454\ \eta_n = e_q/\sqrt{n} \tag{3}$$

All the expressions so far considered involve departures from the limiting mean. But in practice one does not have an unlimited number of members to work with. He does not know the limiting mean, nor σ, nor η. Hence it becomes necessary to infer the values of η and σ from the deviations δ of the individual members from the mean of a limited number n of members of the family. That can be done if n is so great that the group is a fair sample of the family, and if the distribution of the errors is what is called "normal." The last can be safely assumed in most measurements in physics. The relations are as follows, the new quantities being $\delta_m \equiv (\Sigma\delta)/n$, and $(\delta_{m2})^2 \equiv (\Sigma\delta^2)/n$; δ being in every case the deviation of an individual member from the mean of n members:

$$(\delta_{m2})^2 = (\delta_m)^2\pi/2 \qquad \text{or} \qquad \delta_{m2} = 1.25\ \delta_m \tag{4}$$

$$\eta = \delta_m\{n/(n-1)\}^{\frac{1}{2}} \tag{5}$$

$$\sigma = \delta_{m2}\{n/(n-1)\}^{\frac{1}{2}} = 1.25\ \delta_m\{n/(n-1)\}^{\frac{1}{2}} \tag{6}$$

$$e_q = 0.6745\ \delta_{m2}\{n/(n-1)\}^{\frac{1}{2}} = 0.8454\ \delta_m\{n/(n-1)\}^{\frac{1}{2}} \tag{7}$$

$$e_{qn} = 0.6745\ \delta_{m2}/(n-1)^{\frac{1}{2}} = 0.8454\ \delta_m/(n-1)^{\frac{1}{2}}. \tag{8}$$

As before, e_q is the technical probable error of a single member of the family, and e_{qn} is that of the mean of n members. The mean deviation (η_n) of the average of n members from the limiting mean is

$$\eta_n = \delta_m/n^{\frac{1}{2}}. \tag{9}$$

It should be noticed that all these quantities are simply related to δ_m, the mean deviation of the individual members from the mean of a group that truly represents the family. Each of these equations, except eq. 4, 8, and 9, contains the factor $\{n/(n-1)\}^{\frac{1}{2}}$, which never exceeds 1.414 ($n=2$) and is only 1.06 when $n=10$.

In actual practice, the group is often far too small to be a fair sample of its family.[3] Then relations 4 to 8 are no longer correct. Nevertheless, they are the best we have, and the approximations they afford are amply sufficient for many purposes; but it would be useless to carry out the computations to more than one, or possibly two, significant digits.

It should be noticed that the technical probable error either of a single measurement or of the mean of a group of n measurements indicates merely the closeness with which that measurement or mean probably approaches the limiting mean. It tells nothing whatever about the actual quaesitum, and so it is of very minor interest to the experimental physicist engaged in absolute measurements.

To him its main interest is threefold:

(*a*) It tells him when it has become profitless to take additional routine observations; but in most cases other and more important considerations set another limit.

(*b*) It may enable him to state positively that a systematic error affects one or both of two rival families of measurements.

(*c*) It, as applied to a relatively small number of observations, enables him to state positively that systematic errors smaller than a certain amount cannot with certainty be detected experimentally with the apparatus and procedures employed in obtaining those measurements.

The last is, for him, by far the most valuable property of the technical probable error. But in practice

[3] A group must contain at least 100 members before one can determine whether it is a fair sample. See page 89 of W. E. Deming's exceedingly interesting and valuable paper [8].

he seldom thinks of it in that connection. By what seems to be a kind of intuition, he recognizes rough numerical relations between the minimum detectable error and the mean deviation of the several determinations from their mean. And he studies those deviations without thinking about the technical probable error. Actually, the relations he uses are practically those that may be derived in the following manner from the technical probable error.

The argument runs as follows: If the means of two groups of measurements do not differ by at least the sum of their technical probable errors, then the existing difference is not sufficient to justify the assumption that they do not belong to the same statistical family. Consequently, if the only basic difference between the groups were the presence in one of a systematic error that was absent from the other, then the presence of that error could not be certainly established from the difference, unless it amounted to at least the sum of the two technical probable errors. Conversely, it cannot be proved that the measurements are not affected by such an error.

Consequently, the result obtained, in whatever manner, from such measurements—from those having this value of δ_m—is necessarily dubious by twice the technical probable error of the mean of a group of the size used in the search for systematic errors.

For practical reasons, only small groups of measurements can be so used. But the number of members in a single family of routine measurements can be made as great, and consequently the technical probable error of the mean of that family can be made as small, as one may desire. But the smallness of that technical probable error does not affect in the least the dubiety arising from the discordance δ_m of the individual measurements.

Throughout this paper, the term "systematic error" is used to cover all those errors which cannot be regarded as fortuitous, as partaking of the nature of chance. They are characteristics of the system involved in the work; they may arise from errors in theory or in standards, from imperfections in the apparatus or in the observer, from false assumptions, etc. To them, the statistical theory of errors does not apply. They are frequently called "constant errors," and very often they are constant throughout a given set of determinations, but such constancy need not obtain. For example, if the value found by a certain measurement depends upon the humidity of the air, which the experimenter fails to record, thinking that it is of no consequence, then the measures will be affected by a systematic error which will, in general, vary throughout the day, and especially from day to day.

SUMMING AND AVERAGING

Any set of numbers may be weighted as desired, and summed and averaged, and the result can be carried out to as many digits as one may wish. The procedures are simple, exact, and not open to any question or criticism. They are purely arithmetical.

But if the numbers represent physical quantities, then questions arise concerning both the validity of averaging and the number of digits that have a physical significance.

(1) It is sometimes forgotten that the averaging of a set of values, even of the same kind, may be a physically invalid procedure. That is, that the average may not deserve greater confidence as an estimate of the quaesitum than do the individual values.

For example, consider a series of sets of determinations, each set being affected by a systematic error peculiar to it; that error being constant throughout any given set, but varying from set to set. Superposed on that error are fluctuating errors of various kinds. These last are minimized, set by set, by averaging the determinations composing a set. This averaging is entirely proper. But it leaves one with a series of values that differ, one from another, on account of the presence of systematic errors peculiar to each. In general, the averaging of such a series of values will be quite invalid; in general, the average will not deserve more confidence than do the individual values. The only cases in which it will be justifiable when the values differ by more than can be accounted for by the irregularities inherent in each of the several sets, are three: those in which it is definitely known—or perhaps is very highly probable—that the variation in the systematic error from one value to another either is (*a*) strictly fortuitous, in which case the fluctuating part of the error is minimized by the averaging, or (*b*) arises from the error fluctuating between equal and fixed positive and negative values, the number of positive values being essentially equal to the number of negative ones, or (*c*) arises from the error varying progressively from a positive value to a negative one as certain uncontrolled conditions change, and those conditions are known to vary in such a way that each negative error will in the long run be matched by an equal positive one.

Only when one knows a great deal about the systematic error can one be sure that any of these conditions are satisfied. And when he knows that much, he can often arrange to eliminate, or to evaluate, the error; and he should do so.

The cases that most frequently give trouble are those in which the data give evidence of the presence of a systematic error, but the experimenter does not know its source, and those in which another studying the data finds evidence of a systematic error that was overlooked by the experimenter. In such cases one may not know how the error varies with the conditions. If it makes all the values too great, then the smaller ones will be better than the average. Or the reverse may be true. Or the error may be present in some and absent from others; then averaging will not improve things.

Under such conditions it is quite improper to present the average as being superior to the individual values.

One is never justified in merely guessing that averaging will minimize or eliminate the effect of a systematic error. He must know it, must know that under the actually existing conditions the error is so minimized or eliminated.

In the absence of such knowledge, the proper brief summation of the work would seem to consist in a giving of the extreme values with a statement that at least some of the values seem to be affected by a systematic error of unknown origin. To this might well be added the experimenter's opinion, and if he wishes, the arithmetical average, with a clear statement of its questionable value. To give merely the average tends to mislead the reader, to blind him to the presence of systematic errors. The reader must always be on guard, as it is not very uncommon for a writer to average his results quite invalidly, either because he has not awaked to the fact that averaging may be invalid or because he has failed to recognize the evidence for the existence of systematic error.

(2) The number of digits that are of physical significance in the sum and in the average must be carefully considered. If δ_m is the mean deviation of the individual members from their average, then the mean deviation η_n of the average of n members from the limiting average as n is indefinitely increased is (eq. 9) $\eta_n=\delta_m/n^{\frac{1}{2}}$, the error to fear in the sum of n is $\delta_s=n\eta_n=\delta_m n^{\frac{1}{2}}$, and the technical probable error of the average of n is (eq. 8) $e_{qn}=0.8454\ \delta_m/(n-1)^{\frac{1}{2}}$. Whence one obtains the ratios given in table 1, from which it may be seen that only under exceptional conditions should the average be carried farther to the right than one place beyond that in which lies the first digit of δ_m, all places beyond that being physically meaningless; and the sum should generally stop one place to the left of that in which the first digit of δ_m lies.

TABLE 1

RATIOS OF CERTAIN QUANTITIES FOR GROUPS OF n MEMBERS TO THE MEAN DEVIATION (δ_m) OF THE INDIVIDUAL MEMBERS OF THE FAMILY

η_n = mean deviation of the averages of groups of n from the limiting mean of the family; e_{qn} = technical probable error of the average of n members; δ_s = error to fear in the sum of n members.

n	5	10	25	50	75	100	400
η_n/δ_m	0.447	0.316	0.200	0.141	0.115	0.100	0.050
e_{qn}/δ_m	0.423	0.282	0.173	0.121	0.098	0.085	0.042
δ_s/δ_m	2.24	3.16	5.00	7.07	8.66	10.00	20.00

All of this tacitly assumes that the groups of n members are each a fair sample of the statistical family to which they belong. And this, in turn, implies that the mean of a group is not seriously affected by ignoring one of its members. That condition is often not fulfilled. Among the n members there may be a few for which δ is 10 or 100 times as great as δ_m. Each such case must be individually considered; and the question arises whether the abnormal measures should be given lower weights than the others, or perhaps be omitted. But in physical measurements, each of those procedures is recognized as dangerous when its sole basis is that of discrepancy. They should be used only with extreme caution. If all the members be retained with equal weight, and if in the greatest of those δ's the first digit counting from the left occurs in the rth decimal place, then in general that place in the sum is uncertain, and the following ones are physically meaningless. And in no case should a digit to the right of the $(r+1)$th place in the sum be considered as physically significant.

The number of physically significant digits in the average is determined by its percentile uncertainty, which is the same as that of the sum.

LEAST SQUARES

It frequently happens that the observed value v is a known linear function of several unknown constants, the values of the coefficients being known from the conditions of the observations. Then, for each observed value one has an "equation of condition" of the form

$$a_1x+b_1y+\cdots+f_1t=v_1, \tag{10}$$

where a_1, b_1, $\cdots f_1$, and v_1 are known, and x, y, $\cdots t$ are unknown constants, the same for every observed value.

In order to take advantage of the "smoothing" that is secured by averaging, many more observed values v are determined than there are unknown constants in the equations of condition. The problem then arises: How shall this large family of equations be solved so as to obtain the "best" values for the few constants that they contain?

If the observations contained no errors, the family would contain no more independent equations than there are unknown constants, and the solution would involve no difficulty. But each observed value is in error by an unknown quantity δ. Hence the equation of condition may be replaced by one of the form

$$a_1x+b_1y+\cdots+f_1t-v_1=\delta_1. \tag{11}$$

In 1806 Legendre suggested that the best solution of the equations of condition will be that for which x, y, $\cdots$, t are so determined that the sum of all the δ^2's is a minimum (hence the term "least squares"); and he showed how that condition can be secured. His suggestion has been quite generally accepted by physicists as satisfactory, and his procedure has been commonly followed.

Forming and adding the squares of the δ's, and writing $[aa]$, $[ab]$, $\cdots$ for the sum of the squares of the a's, of the products of the a's and b's having the same subscript, etc., one obtains

$$[aa]x^2+2[ab]xy+\cdots+2[af]xt-2[av]x+[bb]y^2 \\ +\cdots+2[bf]yt-2[bv]y+\cdots+[vv]=[\delta^2].$$

The conditions for minimum, called "normal equations," are as follows, one for each unknown:

$$\left.\begin{aligned} \frac{1}{2}\frac{d[\delta^2]}{dx}&=[aa]x+[ab]y+\cdots+[af]t-[av]=0 \\ \frac{1}{2}\frac{d[\delta^2]}{dy}&=[ab]x+[bb]y+\cdots+[bf]t-[bv]=0 \\ &\cdots\cdots\cdots\cdots \\ &\cdots\cdots\cdots\cdots \end{aligned}\right\} \quad (12)$$

and the value of $[\delta^2]$ when these equations are satisfied is

$$[\delta^2]_{min}=[vv]-[av]x-[bv]y\cdots-[fv]t. \quad (13)$$

On comparing eq. 12 with eq. 10, it is seen that the several normal equations are derived from the equations of condition by the following simple process:

Multiply each equation of condition by the coefficient of x in it, and form the sum of all. Multiply each equation of condition by the coefficient of y in it, and form the sum of all. Likewise, for each of the other unknown constants.

If the equations have different weights, w_1, w_2, $\cdots$, then by definition each equation is to be taken as many times as is indicated by its w; and the corresponding condition for the best solution is that $[w\delta^2]$ shall be a minimum. This changes the normal equation from (eq. 12) into (eq. 14).

$$\left.\begin{aligned} [waa]x+[wab]y+\cdots+[waf]t-[wav]=0 \\ [wab]x+[wbb]y+\cdots+[wbf]t-[wbv]=0 \\ \cdots\cdots\cdots\cdots \\ \cdots\cdots\cdots\cdots \end{aligned}\right\}. \quad (14)$$

It should be noticed that the w's enter to the first power only. It is not proper, as some have done, to multiply eq. 10 by its weight w_1 and to call the resulting equation an "equation of condition"; for that leads to the making of $[w^2\delta^2]$ a minimum, which in turn leads to normal equations containing the second powers of w. It weights the equations by w^2, not by w.

The solution of a set of n normal equations in n unknowns may be tedious, but can be obtained by any one of several well-known procedures. In certain cases, however, the elegant solution by means of determinants is to be preferred, as it makes the value of each unknown stand strictly on its own feet, unaffected by uncertainties in the values of the others. The determinant of the coefficients of normal equations being symmetric with reference to the principal diagonal, the number of independent minors to be computed is only $n(n+1)/2$ instead of n^2. Furthermore, the coefficients of the unknowns are frequently numbers that are exactly known, the only numbers affected with experimental errors being the constant terms. This permits one to concentrate his attention upon the numerator determinants, those involving the constant terms, when he seeks to determine the dubiety inherent in the determination of the unknowns from the available experimental data.

It should be remembered that if each of a series of observations is uncertain by x percent, then every sum of those observations is also uncertain by x percent, and no product nor sum of products of those observations by other numbers can be less uncertain than x percent, no matter how accurately the value of the other member of the product may be known. Consequently, if in the solving of the normal equations containing those observations there is involved a difference of two such numbers that are equal to within x percent, that difference has no physical significance, and the same is true of the numerical value formally found for that unknown. The proper procedure in such a case is to ignore that unknown, reducing by one the number of the normal equations.

In the foregoing discussion, attention has been given solely to the fluctuating deviation δ of the individual values of the derived quantity from its "best" value, that best being defined as the one that makes $[\delta^2]$ a minimum.

In certain cases, that δ arises almost, or quite, exclusively from fluctuating experimental errors ϵ in a single term of the equation of condition. In that case, the "best" value will be that corresponding to those values of the constants that make $[\epsilon^2]$ a minimum.

For example, in the present study an equation of condition of the form of eq. 15 will be met with.

$$x+y/a_1=v_1, \quad (15)$$

where the values of a_1 and v_1 are known, x and y are the constants to be determined, and y is affected by a fluctuating experimental error, ϵ.

In complete analogy with Legendre's suggestion, the "best" values for x and y will be those that make $[\epsilon^2]$ a minimum. But the value of ϵ is given by eq. 16.

$$\epsilon_1=a_1\{v_1-x-y/a_1\}\equiv -a_1\delta_1, \quad (16)$$

where in accordance with eq. 11, $x+y/a_1-v_1=\delta_1$. Obviously, $[\epsilon^2]=[a^2\delta^2]$.

That is, the minimizing of $[\epsilon^2]$ is exactly the same as the minimizing of $[a^2\delta^2]$, which is the same as the minimizing of the sum of the squares of δ when each equation of condition is given the weight a^2.

If each v is the mean of p determinations, all for the same value of a, then to each equation of condition must be assigned a weight equal to its p; and the quantity to be minimized is $[p\epsilon^2]=[pa^2\delta^2]$. These define two new quantities, ϵ_{m2} and δ_{m2}, such that

$$\left.\begin{aligned} (\epsilon_{m2})^2[p]=[p\epsilon^2] \\ (\delta_{m2})^2[pa^2]=[pa^2\delta^2] \end{aligned}\right\}. \quad (17)$$

Each of these root-mean-squared deviations is of interest; ϵ_{m2} fixes the precision with which the value of y can be determined, and δ_{m2} fixes the precision of x.

The general normal equations are

$$\left.\begin{array}{l}[pa]x+[p]y=[pav]\\ [pa^2]x+[pa]y=[pa^2v]\end{array}\right\}, \tag{18}$$

and the corresponding value of the minimum sum is given by eq. 19,

$$[p\epsilon^2]_{min}=[pa^2\delta^2]_{min}=[pa^2v^2]-\{x[pa^2v]+y[pav]\}. \tag{19}$$

Although in the preceding it was said that each v is the mean of p determinations, it will be noticed that in eq. 18 each product containing v also contains p, and each enters to the first power only. Hence the values of those sums of products, and hence of x and y, are exactly the same as if the individual values of the v's had been used instead of their means. But in computing the minimum value of $[p\epsilon^2]=[pa^2\delta^2]$, eq. 19, it does make a difference whether one uses the individual determinations or the mean of p determinations, since in $[pa^2v^2]$ the p and v are of different powers. The difference is that between the mean of the squares and the square of the mean.

In eq. 19, the individual values of v should be used. If only means are available, then the computed values will be in error.

ABSOLUTE MEASUREMENTS

By an absolute measurement of a physical quantity, such as the velocity of light, is meant the determination of the value of that quantity in terms of the significant fundamental units of length, mass, time, etc., and of those constant parameters that characterize the accepted system of theoretical equations that connect the several pertinent quantities.

Quaesitum

The quaesitum of the investigation is the actual value of the quantity. The particular value yielded by a given apparatus, procedure, and observer is of no interest in itself, but only in connection with such a study as will enable one to say with some certainty that the value so found does not depart from the quaesitum by more than a certain stated amount. No investigation can establish a unique value for the quaesitum, but merely a range of values centered upon a unique value. The quaesitum may lie anywhere within that range, but the wiser and more careful the experimenter's search for systematic errors, and the more completely he has eliminated them, the less likely is it to lie near the limits of the range. The wider the range, the less becomes the physical significance of the particular value on which the range is centered.

Definitive Value

The term "definitive value" is used in two distinct, though related, senses. (*a*) In a narrower, particular sense, it denotes the value that is believed to lie as near the quaesitum as any that can be legitimately derived from the observations taken in the course of the work being reported. It is the ultimate or definitive value to which that work itself leads. It is often called the "final" value of the work. (*b*) In a broader, general sense, it denotes the value that is believed to lie as near the quaesitum as any that can be derived from a consideration of all the determinations that have been made, and of all other available pertinent information. Whenever not otherwise indicated by the context or a modifier, it is in this broader sense that the term is to be understood.

Every report of measurements of a physical quantity should state clearly the particular definitive value to which those measurements lead. It may also give the broader definitive value based on everything that is known. But the two should not be confused, as unfortunately they often are.

Dubiety

The determination of the range is of an importance that is secondary only to that of its center. No absolute measurement has been completed until values have been established for both of those quantities. The determination of the range necessarily involves an element of judgment, and the limits cannot be set with precision. Nevertheless, it is possible to assign a lower limit; and although no fixed upper limit can be assigned, it is possible to say that if suitable care and diligence had been employed, it is not likely that the range exceeds a certain specified value.

In order to distinguish this range from the numerous kinds of "errors" that abound, its half will in this study be called the "dubiety" of the value found. If that value be denoted by V, and the dubiety by D, then the quaesitum will likely lie within the range $(V-D)$ to $(V+D)$. By this, one means that nothing has come to light in the course of the work to indicate that the quaesitum lies outside that range.

The dubiety is made up of three distinct additive terms to which it is convenient to give descriptive names. They are as follows:

Mensural dubiety arises from the uncertainties in the several primary measurements and in the elimination of known systematic errors. It is common practice to take the arithmetical sum of the effects of these individual uncertainties as an upper limit for the mensural dubiety.

Discordance dubiety arises from the fact that the discordance in the individual determinations limits the smallness of a systematic error that can be experimentally detected. The result cannot be less dubious than the size of the largest systematic error that can escape detection. This term of the dubiety is generally the most important by far, and the least understood and least appreciated by those who are not experimentalists.

Deficiency dubiety arises from the determinations being too few; in particular, finite in number. It is equal to the technical probable error of the result. This term, much honored by those not skilled in experimentation, is always smaller than the discordance dubiety and frequently is negligible in comparison therewith.

Of these three terms, the second alone needs to be especially considered here. In searching for systematic errors, the logical procedure is to make a series of measurements, then to change something and to make another series, and to compare the means of the two groups. This will be repeated as often as may seem necessary. None of the series can be long, for an extended delay offers opportunity for unanticipated changes to occur. If the two means being compared do not differ by more than the sum of their technical probable errors, their difference is of no physical significance—it proves nothing. Hence, the presence of a systematic error that does not exceed the sum of the technical probable errors of the two groups of observations used in the search cannot be established without great difficulty, if at all. That sets a minimum limit for the discordance dubiety.

From eq. 8 one finds the following values (eq. 20) for this minimum discordance dubiety when the test groups contain n measurements each, and δ_m is the mean deviation of the measurements from their mean. (It is better to determine δ_m from as many suitable measurements as are available, than merely from the small number in the compared groups.) The approximate values in the third column of eq. 20 are amply accurate for the present purposes.

n	Min. discord. dub.		
3	$1.2\ \delta_m$	$1.2\ \delta_m$	
5	$0.85\ \delta_m$	δ_m	
10	$0.56\ \delta_m$	$\delta_m/2$	(20)
25	$0.34\ \delta_m$	$\delta_m/3$	
50	$0.24\ \delta_m$	$\delta_m/4$	
100	$0.17\ \delta_m$	$\delta_m/6$	

If the tests consist in nothing more than the comparison of a single determination with the mean of all, then a difference of $2\ \delta_m$ will be of very doubtful significance. Under those conditions it would seem that a conservative estimate of the minimum dubiety would be that indicated by the equation

$$\text{Min. discord. dub., one vs. mean,} = 2.5\delta_m. \quad (21)$$

In all these cases, the minimum discordance dubiety is of the same order of magnitude as the mean deviation, δ_m, lying between $\delta_m/6$ and $2.5\delta_m$.

Obviously, no one should claim a discordance dubiety that is smaller than the smallest systematic error that he might certainly have detected by the tests he made. But there may be reasons that seem to him sound for believing that the actual dubiety is smaller than that. In such case he may, and generally should, state his belief and the reasons therefor; but the statement should never be of such a kind as to lead the reader to confuse the writer's estimate with the minimum discordance dubiety as just defined.

In studying another's work in which there is no clear indication of the size of the groups used in a test, it will usually be conservative to assume that the groups did not contain over 25 determinations each, thus taking $\delta_m/3$ as the minimum discordance dubiety (see eq. 20). This is for the work itself.

But on comparing a series of determinations made by different persons with significantly different apparatus and procedures, it may be found that the several members of the series agree more closely than their individual dubieties would lead one to expect. Then if the differences in apparatus and procedure are sufficiently fundamental, one might be justified in thinking it very improbable that the quaesitum lies far outside the range of the means of the several members of the series. And from the whole he might infer a smaller range of possible values than that demanded by the dubieties of the several determinations.

Although no fixed upper limit can be set, if the search for systematic errors has been careful and comprehensive, it seems unlikely that the net systematic error would have remained undetected if it had been as great as δ_m. And if several distinctly different procedures and instrumental equipments have been used, a correspondingly smaller upper limit might be set. But if there has been no search, or merely a perfunctory one, then so far as the work itself is concerned, there is no upper limit, and the work is worthless. However, certain general considerations resting on other investigations may indicate a possible limit, and so in part salvage it.

No one is really interested in how near the quaesitum the definitive value may possibly lie, for he knows that by chance the two may coincide even though the work be very poorly done. But one does keenly desire to know how far the two are likely to differ—how dubious the definitive value may be. And it is the plain duty of the experimenter not merely to show that his definitive value may be that of the quaesitum, but to prove that it is unlikely to depart from the quaesitum by more than a certain stated amount. In order to obtain the information needed to meet that demand, the careful experienced investigator will proceed somewhat as follows.

Procedure

Before one undertakes an absolute measurement in physics, he will make a careful theoretical study of the problem, including, among other things, methods of attack, sources of errors and how they can be avoided or eliminated, and types of computation. On the basis of that study, the apparatus will be constructed and set up. Only then does the investigation itself begin.

Working standards of the absolute units required must be carefully compared with primary standards. This will ordinarily be done at some standardizing laboratory, which will certify those working standards as being correct under certain specified conditions to within, say, a in 10^n. That value is accepted by the experimenter and sets the top limit to the known accuracy attainable in the work. If, for example, the absolute measurement attempted were simply a length, and the working standard were certified as correct to 3 in 10^5, then the absolute measurement (which determines merely the ratio of the measured length to that of the working standard) could under no condition give the value of the quaesitum to a known accuracy that exceeds 3 in 10^5. No matter how small the technical probable error of the measurements might be, the dubiety of the result cannot be less than 3 in 10^5. Indeed, the dubiety of the value found for the quaesitum will in general be distinctly greater than that, on account of errors inherent in the absolute measurement itself.

The experimenter will measure each of the involved quantities in terms of the appropriate working standard, taking pains to observe as well as may be the conditions laid down by the standardizing laboratory, and to determine carefully whatever is necessary to correct for the actual deviations from those conditions. He will do this repeatedly, and he will also measure them under deliberately different conditions, so as to obtain a check on the accuracy with which he can correct for departures from the specified conditions.

Having found that the apparatus seems to be working properly, he will change, one by one, and by known amounts, each of the adjustments, and will note how each change affects his observations. If possible, he will carry each maladjustment to a point where it produces an easily measurable change in his observations; and if maladjustments in both directions (positive and negative) are possible, he will similarly study each. Thus he will find how important the several adjustments are, the accuracy with which they must be made, and perhaps how to detect each maladjustment experimentally and to correct for the error that it produces.

Readjusting the apparatus, he will proceed to change, one by one, every condition he can think of that seems by any chance likely to affect his result, and some that do not, in every case pushing the change well beyond any that seems at all likely to occur accidentally.

There still remains the possibility of systematic errors arising from unsuspected causes, from secular variation in laboratory conditions (temperature, humidity, light, vibration, etc.), possibly from solar, lunar, or atmospheric effects, etc. So the observer will take long series of observations, extending over weeks, months, or years, noting carefully everything that seems either pertinent in itself or of assistance in fixing the attendant conditions. These will be worked up, day by day, carefully compared with one another, and probably plotted in such a way as to show clearly any change that might appear. From time to time changes will appear, and will be studied.

Thus the experimenter presently will feel justified in saying that he feels, or believes, or is of the opinion, that his own work indicates that the quaesitum does not depart from his own definitive value by more than so-and-so, meaning thereby, since he makes no claim to omniscience, that he has found no reason for believing that the departure exceeds that amount.

That is exactly what he means. He does not mean, as some have suggested, that he is of the opinion that the chances are only one in a hundred, or in a thousand, or in some other number n, that the quaesitum's departure from his definitive value exceeds that amount. He, differing from those others, feels that it would be foolish for him to make such a statement, that it could be nothing more than a gambler's guess. For how can one say, without stultifying himself, that the chance is one in n that the error produced in his result by an entirely unknown, and possibly non-existent, cause exceeds so-and-so, n being a definite specified number? And what can the word "chance" mean in that connection? Quantitative "chance" has significance only in relation to a family of events, and its value for a given event depends upon the characteristics of the family as well as upon that of the event itself. But as regards the uneliminated systematic errors, his observations define no family. He has nothing from which to compute a chance. All he can validly do is to express an opinion; and that opinion can validly relate only to certain theoretical considerations and to the magnitude of the errors that might have escaped his attention, not to any chance that his result might be in error by a given amount.

In every report, such an opinion of the limits within which the quaesitum is believed to lie, based solely on the work being reported, should be given. But in addition to that, previous measurements of the same quantity, when available, will usually be compared with those being reported, for one or more of the following purposes: supporting the author's value; setting other limits for the range within which the author thinks the quaesitum lies; deriving a general definitive value. But even in these cases only the same kind of opinion can be expressed, the number of absolute determinations that have been made of any given physical quantity being far too small to define a statistical family.

The futility of attempting to be more exact, of claiming that the chances are, say, even that the value found departs from the quaesitum by exactly x percent may be illustrated by Kelvin's computation of the age of the earth. He concluded, from a consideration of what was then known and believed, that the length of the period during which the earth had existed in essentially its present state lay between twenty and forty million years. His work was entirely correct, but his conclusion was vitiated by systematic errors that were totally unsuspected at that time, an important one being the presence of radioactive elements. As a result, the earth's age is now believed to be scores of times as great as Kelvin's estimate. There was, and is, no possible way by which the "chance" of such an error could have been estimated.

The experimenter's opinion must rest on evidence, if it is to have any weight. And the only evidence available comes from theory, the series of observations made in the course of the work, and the diligence with which errors were sought. These, and in particular the discordance of the observations and the diligence of the search, are what must be depended upon. Dependence on theory is weak, for the actual conditions never accord exactly with those assumed in the theoretical work.

He knows that it is impossible to avoid systematic errors, that even when he has done his best, his result is still haunted by the ghosts of such errors. His whole problem has been to seek such errors out, and to eliminate them when found; and he believes that in his long search any existing combination of them that would have produced an effect greater than the limit he sets would have been found. But he would be the first to admit that he may be wrong, that his result might be affected by a much larger error arising in such a way that, in spite of the many changes made in the course of the work, it remained essentially unchanged; but he thinks that contingency is highly unlikely. However, he is not entitled to that opinion unless he has carried out the indicated search, for in no other way can a foundation be found on which to base an opinion.

In the absence of such a search, the worker can do no more than hope that all is going well. The fact that he sees no reason for suspecting the presence of an unknown systematic error is of no importance at all, no matter who the observer is. The really troublesome errors are exactly those that are not suspected. The suspected ones can usually be to some extent eliminated.

In brief, the careful experimenter engaged in absolute measurement in physics is an extreme Baconian. He refuses to trust implicitly in theoretical conclusions as to how his apparatus should behave, for he knows that such ideal conditions as are assumed in the theory are never actually secured; and he insists on testing experimentally everything that he can. It is that careful, thoroughgoing, experimental testing that distinguishes true absolute measurement from the mere piling up of a long series of routine observations.

Report

The work should be fully reported, so that the reader may know what was done, may have the means for forming an independent judgment of the work and for checking possible errors and omissions, and may have the worker's experience to build upon in case he himself should undertake a similar piece of work. The last is certainly a very important function of such a report, and should never be ignored.

The report should, of course, give a clear indication of the care with which search was made for sources of error, and of the thought that was given to it. Otherwise, one has no choice but to conclude either that no search was made, or that the author attached no special importance to it. In either case, the work is of little, if any, objective value; its acceptance can rest only on authority, on subjective grounds.

Data should be reported as fully as may be. But in every series of observations some are erratic, especially at the start. How should they be treated? Those that occur in the body of the work should certainly be reported as fully as if they were not erratic, and if the cause of the trouble is known, that should be explained.

Those that occur peculiarly at the beginning of the series, arising mainly from maladjustment and inexperience, furnish very valuable information regarding details of adjustment and manipulation that had escaped the foresight of the worker, and that might, therefore, readily escape the attention of the reader and of subsequent workers. In certain cases they give valuable information about unsuspected sources of error. For such reasons, they should never be completely omitted. They need not always be given in full, but they should be given to such an extent and in such detail as will show the reader what they were like and how they were related to the pertinent conditions, and should be accompanied by such explanatory text as will show him how they were regarded by the worker, and how he contrived to remove the disturbing conditions.

In brief, the report should give the reader a perfectly candid account of the work, with such descriptions and explanations as may be necessary to convey the worker's own understanding and interpretation of it. Anything short of that is unfair to the writer as well as to the reader. Every indication that significant information has been omitted reduces the reader's confidence in the work.

It is the unquestioned privilege of the worker to say where the boundary lies between preliminary or trial determinations, made primarily for studying and adjusting the apparatus and procedures, and those that were expected to be correct. But he should give good reasons for placing that boundary where he does; and those preliminary determinations should be reported to the extent already indicated.

Furthermore, it is scarcely fair, to any one concerned, to describe a series of determinations as "preliminary," thus implying, in accordance with common usage, that they are open to question, that they are merely preparatory for something better, and then, later on, to include that same series in the list of good, acceptable determinations. To do so, both confuses the reader and suggests to him that the use of the adjective "preliminary" may have been merely a face-saving device intended to justify the ignoring of that series in case it should be found to disagree uncomfortably with later ones.

FIZEAU'S WORK

In 1849 Fizeau [9] reported the result of the first successful direct determination of the velocity of light. In that work he used what is now known as the Fizeau method (see Appendix A), in which an outgoing beam of light is chopped by a toothed wheel into a series of segments, which are returned by a distant mirror along the same path, and are viewed between the teeth of the wheel. For certain speeds of the wheel the returning light segments will be blocked (eclipsed) by the teeth; for others they will be transmitted.

He determined certain speeds for which the returned light was eclipsed. From those speeds and the distance, the velocity of light can be computed.

His optical system consisted of two telescopes, one at each end of the path of the light, so adjusted that their optic axes coincided. Then in the center of the focal plane of each there is formed an image of the objective of the other. The teeth of the rotating wheel cut across the optic axis in one of these focal planes, and in the other was a mirror with its reflecting surface perpendicular to, and centered on, that axis. The returned light was viewed by reflection from an inclined plate of unsilvered plane glass.

The wheel had 720 teeth, and was driven by clockwork made by Froment. One telescope was at Suresnes, the other at Montmartre, the distance between their focal planes being 8.633 km.

Details of the experimental work are not given in this article, and the promised later one [4] has not been found.

He stated, as the mean of 28 determinations, that the velocity of light is 70,948 leagues of 25 to the degree per second ("70,948 lieues de 25 au degré"), which value differs little from that accepted by astronomers. Since the meter was supposed to be a ten-millionth part of a meridional quadrant, a league of 25 to the degree may, for present purposes, be taken as equal to 40/9 km., whence one obtains for Fizeau's value

Velocity of light = 315 megameters per second.

[4] He wrote: "J'aurai l'honneur de soumettre au jugement de l'Académie un Mémoire détaillé lorsque toutes les circonstances de l'expérience auront pu être étudiées d'une manière plus complète."

The meagerness of the report makes it impossible to estimate the dubiety of this value, but it may be expected to be great, this being but a first attempt to use this method.

FOUCAULT'S WORK

The second successful direct determination of the velocity of light was made by Foucault [10] in 1862 by means of the rotating mirror method—now called the Foucault method. In that method (see Appendix A) the light from a small fixed source is reflected from a rotating mirror through a converging system to a distant fixed concave mirror that returns it along its outgoing path to the rotating mirror. Diffraction effects excepted, the returned light forms an exact image of the source; and that image coincides exactly with the source if the mirror is at rest, but is deflected from the source if the mirror is rotating. From that angular deflection, the speed of the mirror, and the length of the light-path between the mirrors, the velocity of light can be computed.

Foucault called attention [11] to two difficulties inherent in the method: (1) Owing to diffraction, especially by the small rotating mirror, the returned image is unavoidably impaired in sharpness and altered in contour. (2) Unless prevented by some special device, the intensity of the returned image will decrease rapidly as the distance between the mirrors is increased. And he devised means for reducing their evil effects.

For minimizing the evil effects of diffraction, he used for a source a grid of equal and equidistant parallel lines of light, 10 to a millimeter. Then, although the image of each line would be damaged by diffraction, that of the grid would still consist of a series of lines alternately of maximum and minimum intensity, the distance between the maxima being the same as that between the lines of light in the grid itself. But it seems to the present writer probable that the altering of the contours might cause those maxima to be all similarly displaced with reference to the positions they would have occupied had there been no such alteration. Of that, Foucault said nothing; and he may have worked in such a way as to eliminate any error so arising.

In order to reduce the loss in intensity that accompanies an increase in the length of path when the simple arrangement shown in (*b*) of figure 20 (Appendix) is used (the intensity of the image in that case varies as $1/D^3$), he proposed that a chain of an odd number of concave mirrors [5] be used; and in his determination he actually used a chain of five. The chain is to be arranged in this manner: The first concave mirror, M_1, is to be placed beyond the objective, L (fig. 20*b*), and at such a distance that L forms on its surface an image of the grid, S, exactly as in that

[5] He first [11] suggested an equivalent chain of convex lenses.

figure, which is a one-mirror chain. When other mirrors are to be added to the chain, M_1 is so oriented that the light from it misses L, and strikes the second concave mirror M_2 which is so placed that M_1 forms on its surface an image of the rotating mirror R; M_2 passes the light to M_3, forming on its surface an image of the image of the grid formed on M_1; and so on, an image of the grid being formed on the surface of each odd-numbered concave, and one of the rotating mirror on each even-numbered one. The axis of the last mirror M_{2n+1} of the chain is made to coincide with that of M_{2n}, so that the beam of light is sent back along its outgoing path to the rotating mirror.

It is obvious that if the curvatures and apertures of the several concave mirrors are suitably adjusted, then, after leaving the first M_1, the beam of light will be continually confined to the chain; it will not be subject to the inverse-square law, and it will never sweep off the surface of a mirror. There will be no further loss of light from those effects, no matter how large a number n may be—no matter how much the length of the path has been increased by the use of the chain. Obviously, there will be other losses, arising from absorption and scattering by the air, from imperfect reflection, etc., but the relatively enormous effect produced by increasing the length $O'O'''$ in the simple arrangement of (*b*) in figure 20 is by this means avoided.

Michelson seems to have failed to appreciate the light-saving property of Foucault's chain of mirrors, the very property that Foucault was seeking; for he has written [12] that he thinks that Foucault's limiting of the distance to 20 m. (chain of five mirrors) may have been necessary on account

> of the streak of light caused by the direct reflection from the revolving mirror, which in Foucault's experiments was doubtless superposed on the former [the returned image]. The intensity of the return image varies inversely as the cube of the distance, while that of the streak remains constant.

With Foucault's chain of mirrors the intensity of the image does not vary as the inverse cube of the distance.

Foucault's rotating mirror was a plane glass disk 14 mm. in diameter, silvered on one side and blackened on the other. It was mounted in a ring forming an integral part of the (vertical) axis of a turbine, of 24 vanes, driven by compressed air. The bearings were lubricated by a continuous stream of oil flowing under a constant head. The air was delivered by a precision blower and regulator, by which the total pressure of 30 cm.-H_2O was kept constant within 0.2 mm. (7 in 10,000).

The speed of the mirror was determined stroboscopically in terms of that of a toothed disk driven by precision clockwork at the rate of 2 turns per second. He stated that with a mirror speed of 400 turns per second, the mirror and disk would keep in step within 1 part in 10,000 for minutes at a time [10, second part]. He actually used a speed of 500 turns per second [13, pp. 219–226].

By using a chain of five concave mirrors, he obtained a folded path of 20 m., from rotating mirror to concave mirror at the far end of the chain, entirely within a room.

The deflection of the returned image was at first measured by means of the screw of the eyepiece micrometer of the observing microscope, the pitch of the screw being inferred from measurements by it of 10, later of 7, spaces of the grid, which had been made with great care by Froment, and which served as standard of length. The screw was thus calibrated for each determination, but whether always the same portion of the screw was used, is not stated. However, he has stated [10, second part] that, finding that the screw was not as good as he had thought, he discarded all of those observations, and took others in which the distance r from the grid to the mirror (about 1 m.) was always adjusted until the deflection was exactly 7 divisions of the grid (0.7 mm.). In a sample set of 20 determinations of that distance the extreme range was 5 mm., and the mean deviation from the mean was only 1 mm. (1 part in 1,000). But nothing has been found that tells either how he determined when the deflection was exactly 7 divisions, or with what precision. If he set the micrometer line on the seventh division from the center of the image of the grid as seen with the mirror at rest, and then, with the mirror rotating, adjusted the distance r until the center of the image of the grid was exactly on the micrometer line, the result would involve the error with which a single setting of the micrometer line to a line of the image could be made. Since one division of the micrometer head corresponded to about 0.01 mm., it seems likely that a single setting on a line of the image would be uncertain by at least 2 or 3 microns, which amounts to an uncertainty of 3 or 4 parts in 1,000. To that must be added such uncertainty as existed in the spacing of the lines on the grid.

Nothing is said regarding any checking of the accuracy of the grid or of the speed of the disk of the stroboscope, or any search for periodicities in the speed of the mirror. There is nothing that will enable one to estimate the minimum dubiety of his reported result:

Velocity of light in air = 298 megameters per second

His estimated uncertainty is 0.5 megam./sec.; it seems to be decidedly too small.

He stated that the value generally accepted at that time for the velocity was 308 megam./sec., which he was convinced is too great.

CORNU'S WORK

CORNU'S WORK OF 1872

In 1872 Cornu [14] made a series of measurements of the velocity of light by means of Fizeau's toothed-wheel method [6] over a path 10.310 km. long. The apparatus was crude; the precision was low. Two driving mechanisms were used: (1) a weight-driven motor built by Froment was used for 86 of the determinations; and (2) the spring-driven mechanism of a commercial clock was used for the remaining 572. Both directions of rotation of the wheel were used. With the wheel running with a slight acceleration, he determined the velocity ω_{e+} at which the returned star vanished into the faintly luminous background and that ω_{r+} at which it reappeared again; similarly, with a slight deceleration, he determined ω_{e-} and ω_{r-}. From the average of these four, he computed the velocity of light. It may be shown (see Appendix A) that under certain conditions, which he seems to have assumed were realized, that value will be freed from the larger observational errors of a systematic nature.

By using both directions of rotation and averaging the two sets of results, the effect of any fixed lateral displacement of the returned image from its ideal position was eliminated.

His unit of time was one-half of the 2-second interval of a clock gaining 8 to 12 seconds per day. He thought that the errors in reading the intervals of time did not exceed 0.5 percent; [7] and that the distance was certainly known with an accuracy of 1 in 1,000. The last can probably be accepted without question, but as the time intervals recorded in his table range from 0.4 to 6.2 seconds, averaging about 3 seconds, it seems probable that his measurements of those intervals were more uncertain than he supposed.

His published values of the velocity V of light in air are given in table 2, and in table 3 are his weighted means arranged by order n of eclipse. From these weighted means he inferred that $V = 298.4$ megameters per second in air (298.5 in vacuum), with an uncertainty equal to or less than 1 in 300, greater or smaller.[8]

[6] For a discussion of this method and of the more important correction terms and errors, see Appendix A.

[7] In his later *Mémoire* [15] he says that the glass scale of diverging lines that was used in this work for subdividing the 2-second interval read directly to 0.1 sec. and to 0.01 or 0.02 sec. by estimation. "Elle permet de subdiviser l'intervalle de deux secondes en 1/10 de seconde, et d'estimer le 1/50 ou le 1/100; mais il se prête difficilement à une approximation supérieure" (p. A.152).

[8] "Dans l'hypothèse la plus défavorable . . . l'approximation de la valeur précédent serait encore égale ou inférieur à 1/300 en plus ou en moins" (p. 178).

But in his later *Mémoire* [15] (1876) he wrote of this result: "J'avais cru pouvoir conclure à une approximation de 1/300, probablement par défaut" (p. A.298), and referred to page 178 of this paper; and earlier in the *Mémoire*: "Dont l'erreur probable est notablement inférieure à 1/100" (p. A.3).

May we not see in this an illustration of how one's opinion may affect his conclusion? Being confident that Fizeau's method

TABLE 2

CORNU'S DETERMINATIONS OF 1872 (IN AIR)

$q = 2n - 1$, n = order of eclipse, N = number of determinations, S = quality of the observations, V = derived value of velocity of light in air, $\delta = V - V_{ml}$, where V_{ml} is the weighted mean of the group, the weight is taken as the product of S multiplied by the number of determinations. The direction of rotation of the wheel is indicated by R (to right) and L (to left).

In the summary, V_m is the mean of V for the two directions of rotation; weighted mean is $\Sigma qSNV$ divided by ΣqSN, without distinction between R and L; average mean deviation is the average of the arithmetical values of the δ's.

Unit of V and δ = 1 megameter/second; weight = $S(N)$

Date (1872)	q	N	S	Rotation R V	R δ	Rotation L V	L δ
8/10	3	5	2			298.6	−4.9
8/23		6	2	299.7	−0.8		
8/23		6	2			307.2	+3.7
8/29		7	1			304.3	+0.8
8/30		1	2	305.2	+4.7		
Weighted mean				300.5		303.5	
Mean deviation					2.8		3.1
8/7	5	22	3			298.1	+0.7
8/7		46	3			298.1	+0.7
8/10		8	2			299.0	+1.6
8/10		13	3			292.0	−5.4
8/10		5	1			295.6	−1.8
8/23		8	2	302.2	+3.1		
8/23		6	2			298.6	+1.2
8/26		2	3			302.4	+5.0
8/27		4	3	301.6	+2.5		
8/29		11	2			300.2	+2.8
8/29		7	1	304.4	+5.3		
8/30		11	4	296.5	−2.6		
8/30		12	4			296.6	−0.8
Weighted mean				299.1		297.4	
Mean deviation					3.4		2.2
8/10	7	4	1			290.9	−6.7
8/10		17	2			295.1	−2.5
8/10		20	3			294.8	−2.8
8/10		4	1			295.1	−2.5
8/23		5	3			296.7	−0.9
8/23		10	2	299.8	+0.4		
8/23		10	2			296.6	−1.0
8/26		5	3			296.4	−1.2
8/26		24	3			296.1	−1.5
8/26		33	3	299.5	+0.1		
8/27		12	3	299.9	+0.5		
8/27		24	3			298.4	+0.8
8/27		4	1			302.2	+4.6
8/29		17	2			302.6	+5.0
8/29		13	1	300.6	+1.2		
8/30		15	4	299.6	+0.2		
8/30		13	4			299.9	+2.3
8/31		1	4			305.3	+7.7
8/31		5	3	295.8	−3.6		
8/31		4	3			301.9	+4.3
Weighted mean				299.4		297.6	
Mean deviation					1.0		3.1

TABLE 2—*Continued*

Date (1872)	q	N	S	Rotation R V	Rotation R δ	Rotation L V	Rotation L δ
8/10	9	17	1			295.9	−2.8
8/10		19	2			300.8	+2.1
8/10		30	3			297.4	−1.3
8/23		9	3			297.2	−1.5
8/23		5	2	297.7	−1.2		
8/23		8	2			293.4	−5.3
8/26		18	3			297.1	−1.6
8/26		40	3	299.4	+0.5		
8/27		21	3	297.9	−1.0		
8/27		7	3			304.6	+5.9
8/27		3	1	302.7	+3.8		
8/31		7	2			298.8	+0.1
8/31		7	2	302.4	+3.5		
8/31		8	3	297.2	−1.7		
8/31		8	3			305.8	−7.1
Weighted mean				298.9		298.7	
Mean deviation					2.0		3.1
8/10	11	4	3			290.4	−7.6
8/23		21	3			299.5	+1.5
8/31		3	3	292.9			
Weighted mean				292.9		298.0	
Mean deviation							4.5
8/27	13	5	1	301.2			
8/27		15	1			300.1	

SUMMARY

q	R	L	V_m
3	300.5	303.5	302.0
5	299.1	297.4	298.2
7	299.4	297.6	298.5
9	298.9	298.7	298.8
11	292.9	298.0	295.4
13	301.2	300.1	300.6
Mean	298.7	299.2	298.9
Weighted mean			298.4
Average mean deviation			2.7

was superior to Foucault's, Cornu undertook this work with the opinion that the velocity of light exceeded 300 megameters per second, as indicated by Fizeau's determination and by certain astronomical data which he accepted. But the value he found, agreeing closely with Foucault's, and lying below 300 by more than his admitted uncertainty, convinced him, distinctly against his will, that Foucault's much lower value was the better (see following footnote); and when he published this paper, he was confident that his own work, though of low accuracy, indicated that the range in which the quaesitum lies was centered on 298.5 megameters per second. But the results of his next determination, to which he ascribed a much higher accuracy, plainly indicated to him a center of over 300, one nearer to what he had previously expected to find. So he reconsidered the present work, searching for some overlooked cause that would have made its result too small. Finding one qualitatively of the right kind, he assumed, without any quantitative data to justify the assumption, that the effect of that cause was to make the derived value of the velocity too small by one-third of one percent. That makes the two determinations agree sufficiently well. There is nothing to indicate that he would have restudied this work had the two determinations agreed initially, nor that he made any search for an overlooked cause that would have produced the opposite effect.

Also, as will be seen in the study of his later work, he there discredited a theoretically valid correction that would have reduced the value of the derived velocity, and introduced another (actually invalid) that would increase it; thus getting a value nearer that which he expected.

In none of this is it implied that the search and alterations were made in other than good faith. It was merely a question of searching and altering in the apparently more profitable direction, in that in which he thought the error lay.

TABLE 3

CORNU'S WEIGHTED MEANS, 1872

These means differ from V_m of table 2 simply because no distinction has here been made between the R and L values.

His weighted mean is 298.4 megam./sec. in air (298.5 in vacuum). In the last line is given the percentile excess of each of the means in air above 298.4. Order = order of eclipse. The several mean velocities are those he gives in a similar table.

Unit of velocity = 1 megameter/second

Order	2	3	4	5	6	7
Mean velocity	302.5	297.7	298.2	298.8	297.5	300.5
Number of observations[a]	25	155	240	207	28	20
Departure (%)	+1.4	−0.2	−0.1	+0.1	−0.3	+0.7

[a] These values, taken from his long table, total 675, whereas he states that there were 658.

That estimate of the dubiety seems to rest in part on the fact that the mean departure from 298.4 of the values for orders 3 to 6, as given in table 3, is 0.5 megam./sec., or 1 in 600. Rounding this to 1 in 500 and adding it to the 1 in 1,000 for the uncertainty in the length, gives about 1 in 300—actually a little less than that. But this takes no account of any systematic error other than those that might vary with the order so rapidly as to become prominent when the number of the order is doubled. In order to cover others, one must (see eq. 20) increase the dubiety by at least one-third of the average mean deviation of the several determinations that are nominally identical, which amounts to 0.9 megam./sec. (see table 2).

Thus one finds that the dubiety of his result is at least 1.9 megam./sec., say an even 2; and that one may conclude that his observations indicate that the velocity of light in vacuum may lie within the range

296.5 to 300.5 megameters per second.

But in his *Mémoire* [15, p. A. 298] he gives reasons for thinking that the center of the range should be higher than the 298.5 derived in this paper.

It is interesting to notice that he admits that at the beginning of the work he doubted the correctness of

Foucault's low value (298 megam./sec.), and that the unexpected agreement of his own value with Foucault's largely removed his earlier doubts.[9]

CORNU'S REPORT OF 1874

THE REPORT

On December 14, 1874, Cornu [16] presented to the French Academy a preliminary report on the measurements completed the preceding September and reported in detail in his *Mémoire* [15] of 1876, to be presently studied in detail.

Presumably this incomplete report, containing only partially studied data, was prepared for the use of astronomers in preparation for their observations of the transit of Venus on December 9, as mentioned in the beginning of Cornu's *Mémoire*. Its publication, as in most cases of preliminary reports containing only partially studied and incomplete data, was most unfortunate, and can be excused only on the ground of some such pressing demand for an estimate of some kind prior to a fixed and uncontrollable date, and the attendant desirability of a permanent record of the estimate.

In this report, mean values according to order of eclipse and covering 504 determinations are given; as compared with 546 unrectified (526 rectified) determinations used in the *Mémoire*. Many of these means differ markedly from the corresponding ones given in the *Mémoire*, and there seems to be no way to find out how the differences arose. He gave as the weighted mean of these: velocity of light = 300.33 megam./sec. in air, or 300.40 megam./sec. in vacuum.

After the *Mémoire* itself had been published, this preliminary report and all conclusions based on it should have been ignored, except as they directly related to astronomical or other work done in the interim. That was not done. The note by Helmert based on it, and now to be considered, is still being quoted.

HELMERT'S DISCUSSION

On examining Cornu's preliminary report [16] of the work fully reported in his *Mémoire* [15] of 1876, Helmert [2] observed that, as compared with their mean, the values corresponding to low speeds of the wheel were on the whole too great, and those to high speeds were too small. This suggested that the reported values were probably affected by a systematic error of the form $V_{\text{reported}} = V_{\text{true}} + (H/q_n)$ (*cf.* eq. 103, where $q_n = 2n-1$). On that assumption, he found $V = 299.99$ megam./sec. and $H = 7.1$. But he admits that prior to a thorough study of the details of the observations, which are not given in this brief report, one cannot be sure how the values should be corrected.[10] He also calls attention to the fact that the introduction of the H-term alone is not enough to eliminate the irregularities, but that such failure is not surprising, since Cornu used several wheels and did not distribute their data uniformly over all the values of q_n. From an inspection of a graph of the reported values as a function of q_n, he concluded that one could not tell the actual amount of the correction that should be applied to the reported value, but that the uncertainty was such that for a preliminary estimate of the velocity of light one might take the mean of Foucault's 298 and this 299.99, or 299 megam./sec.

Even today this "corrected Cornu value" or "Helmert-Cornu" value is quoted, but, strange to say, the value given is not Helmert's computed value (299.99), but 299.9. In most cases, that value is given instead of Cornu's definitive one (300.4), although Helmert did not have access to Cornu's *Mémoire* at the time he prepared his note, and Cornu firmly refused [17] to accept Helmert's conclusion. Cornu's criticism of Helmert's treatment of his data rests, strange to say, on exactly those disabilities of the data that were recognized by Helmert.

It will be noticed that Helmert's formula is exactly that found in the present study of Cornu's *Mémoire* to be applicable. The value he found for V is somewhat higher than that here found for all wheels treated together, and his H is lower than the 10 to 17 here found.

In view of the nature of the data on which it rests, one must conclude that this value of Helmert's has acquired an importance far beyond its deserts. Although the type of correction that he applied was entirely correct, the value he computed was never of more than temporary interest. It should never have been used beyond the short interim between its publication and that of Cornu's *Mémoire*. After that had appeared, those who approved of Helmert's suggestion should have tested it against the voluminous information given in the *Mémoire;* and if they found it applicable, they should have applied it to those data. That would have been laborious, but it would have prevented serious mistakes, much misunderstanding, and perhaps the publishing of other incorrect values.

CORNU'S REPORT OF 1876

INTRODUCTION

The work published in this extensive *Mémoire* [15] was decided upon by the Council of the Observatory of Paris at its session on April 2, 1874, on the proposal

[9] "Je ne chercherai pas à dissimuler que le nombre trouvé par Foucault me paraissait suspect" (p. 139).

"On verra que le résultat auquel je suis arrivé diffère à peine de celui de Foucault. Cette coincidence, à laquelle j'étais loin de m'attendre, je dois l'avouer, dissipe en grande partie les doutes que j'avais conçus sur la validité de la méthode du miroir tournant, et donne, je crois, une grande probabilité d'exactitude à un résult obtenu par deux méthodes si différentes" (p. 140).

[10] "In welcher Weise die Beobachtungen zu reduciren sind, um der Wahrheit näher zu kommen, lässt sich ohne eingehende Studien aller Beobachtungsdetails gar nicht sagen."

of LeVerrier, director of the Observatory, and Fizeau, and was entrusted to A. Cornu. Its purpose was to obtain a value for the velocity of light that might be used by astronomers in connection with observations to be made on the transit of Venus on December 9, 1874. The accuracy desired was 1 in 1,000. It will be noticed that only eight months were allowed for the construction of much apparatus, its installation and testing, the making of the observations, and an at least partial reduction of them. In the opinion of the writer, that time was much too short, hardly more than would have been needed for a satisfactory study of the behavior of the apparatus.

The *Mémoire* contains a very detailed account of the work, including an elaborate discussion of correction terms and their elimination, and a discussion of the data.

METHOD AND APPARATUS

Fizeau's toothed-wheel method [11] was used. The sending telescope had an objective of 38-cm. aperture, focal length 890 cm., and was mounted in the Observatory. The returning collimator, objective of 15-cm. aperture and focal length about 200 cm., was mounted on the tower of Montlhéry. The distance D from the toothed wheel to the mirror of the collimator was 22.910 km., based on old surveys, of which the monuments were still standing. The weight-driven, friction-brake-controlled mechanism for driving the wheel, the wheels, the chronograph, and the auxiliary oscillators for subdividing the mean sidereal second of the observatory clock, were all made especially for this work.

The diameter of the utilized portion of the light that is focused on the wheel is fixed by D and the dimensions just given, and is approximately given by eq. 22 (see eq. 72)

$$\text{Diameter} = d_c \cdot f_s / D, \qquad (22)$$

where d_c is diameter of the collimator lens, f_s is focal length of the sending lens, and D is distance from wheel to collimator mirror; D is supposed to be very great as compared with the focal lengths of the lenses. From that equation one finds that this diameter is 0.059 mm., which is also the diameter of the returned star, exclusive of diffraction effects. One half of that, or 0.029 mm., is the greatest amount by which a point of light can be displaced from the line joining the centers of the lenses, and still have its light returned by the collimator.

The wheel could be rotated in either direction. Large variations in speed were made by changing the driving weight; small ones, by manual adjustment of an ivory-shod brake acting on one of the arbors. At each 40th, or at each 400th, turn of the wheel, as the experimenter might desire, an electric circuit was automatically closed, making a record on a chronograph sheet.

Four smoked wheels, of aluminium foil 1/10 to 1/15 mm. thick, were used. Their significant constants are given in table 4.

TABLE 4

CONSTANTS OF CORNU'S TOOTHED WHEELS

Number = number of teeth; shape = shape of teeth; angle = angle between centers of teeth; distance = distance between centers of teeth.

Unit of diameter and distance = 1 mm.

Diameter.......	30	35	40	45
Number........	144	150	180	200
Shape..........	Pointed	Pointed	Square	Pointed
Angle..........	2.5°	2.4°	2.0°	1.8°
Distance.......	0.65_4	0.73_4	0.69_8	0.70_6
Thickness......	One-tenth to one-fifteenth millimeter			

The chronograph drum, 50 cm. long and 95 cm. in circumference, was turned by a weight-driven motor controlled by a centrifugal wing governor. It turned once in about 51 sec., 1 sec. corresponding to about 1.85 cm. Two interchangeable drums were used. The carriage, carrying four styluses, advanced about 15 mm. per turn of the drum. The styluses were arranged in two banks, and adjusted so that all four points lay near together and approximately along a generating line of the drum, the two outermost points being those of the upper bank. Taking the points in regular sequence, they recorded (1) seconds from the observatory clock, (2) tenth-seconds from an auxiliary vibrator, (3) signals from the revolution counter attached to the wheel, and (4) signals made by the observer. Records were made on smoked drawing paper, and then fixed. The drawing paper was "grand aigle" cut to 50 by 100 cm., allowing a 5-cm. overlap. It was moistened on the less glazed side, allowed to stand, then moistened again, except at the overlap, and wrapped around the drum, and at the overlap the two layers were stuck together with thick mucilage. After the paper had thoroughly dried, it was smoked.

The chronograph records were read by means of a microscope with a ruled micrometer in the eyepiece. The microscope was so constructed and mounted that its magnification could be progressively varied without disturbing the focusing of the record upon the micrometer. Hence, 10 divisions of the micrometer could easily be made to correspond exactly with the distance between adjacent time-records of the auxiliary vibrator, which in this case was equivalent to 0.1 sec. Thus times could, by estimation, be read to 0.001 sec., provided that the vibrator could be trusted to that extent.

The auxiliary vibrator, which furnished the unit of time that was actually used, was a 1/20-sec. oscillator

[11] For a discussion of this method and of the more important correction terms and errors, see Appendix A.

that closed an electric circuit once every complete vibration. It was electrically driven by a half-second pendulum that closed a circuit when passing through the center of its arc; and that was in turn driven by the observatory clock, which closed a circuit every second. The observatory clock recorded sidereal time, and the proper factor for converting sidereal to mean solar time was introduced into the computational formula. The vibrator and the half-second pendulum were each individually adjusted as closely as might be to its nominal value while oscillating freely, while undriven.

PROCEDURE

In the ideal case, one would determine the speed of the wheel at which the returned star is exactly eclipsed, and from that would compute V the velocity of light by means of the simple formula 75, here repeated as eq. 23

$$V=4DNm/(2n-1), \tag{23}$$

where N is the number of teeth of the wheel, m is the number of turns of the wheel per second, and n is the order of the eclipse.

But that is not practical. The best one can do is to determine the speed of the wheel at which the returned star vanishes into the always slightly luminous background, or that at which it reappears. One of these speeds is greater than that corresponding to a true eclipse, the other is smaller; the same is true of the two values V_e and V_r of the velocity of light, each computed as though the speed of the wheel were that corresponding to a true eclipse. Furthermore, the returned star is not a point, as in the ideal case, but has a finite diameter.

It may be shown (see Appendix A) that if the wheel and its motion are ideally perfect and the star has a finite diameter, then the relations of V to V_e and V_r, corresponding respectively to a disappearance and the following reappearance when the wheel has a constant positive acceleration, α_+, are of the form of eq. 24 (see eq. 101).

$$\left.\begin{aligned} V_{e+}&=V+\frac{\Delta'-(H_e-B_e)}{q}+\frac{L_e\alpha_+}{q}+\frac{S}{q}\\ V_{r+}&=V-\frac{\Delta'-(H_r+B_r)}{q}+\frac{L_r\alpha_+}{q}+\frac{S}{q}\end{aligned}\right\} \tag{24}$$

and if the wheel has a negative acceleration, $-\alpha_-$, they are of the form of eq. 25 (see eq. 101).

$$\left.\begin{aligned} V_{e-}&=V-\frac{\Delta'-(H_e-B_e)}{q}-\frac{L_e\alpha_-}{q}+\frac{S}{q}\\ V_{r-}&=V+\frac{\Delta'-(H_r+B_r)}{q}-\frac{L_r\alpha_-}{q}+\frac{S}{q}\end{aligned}\right\}. \tag{25}$$

In these equations, Δ' is determined by the diameter of the returned star and by the amount by which the pertinent breadth of an interdental space exceeds that of a tooth, H by the brightness of the star at the instant of disappearance or of reappearance, B by that part of the observer's hesitancy in deciding whether or not the star is really seen, which depends on the rate of change in its apparent intensity, L depends upon all the constant delays between the occurrence of the observed phenomenon and the record on the chronograph sheet, S upon the lateral displacement of the star from the position of its actual source, and $q=2n-1$, n being the order of the eclipse.

These same equations hold whenever the observation lies in a normal region,[12] no matter how the conditions otherwise depart from ideality, but the values of the coefficients vary with that departure.

If both S and the α's are negligible, then the mean of V_{e+} and V_{r+} (call it V_{er+}) is

$$V_{er+}=V-[(H_e-B_e)-(H_r+B_r)]/2\,q,$$

which will be V if $H_e-B_e=H_r+B_r$. Similarly, the mean of V_{e-} and V_{r-} is

$$V_{er-}=V+[(H_e-B_e)-(H_r+B_r)]/2\,q.$$

And the mean of V_{er+} and V_{er-} is $V_{er\pm}=V$. Cornu called each of the means, V_{er+} and V_{er-}, a *double observation;* and the mean of the two, he called a *crossed double observation.*

Cornu's avowed procedure was to take observations in normal regions only, and to eliminate the several correction terms by crossing the double observations, eliminating the term in S by taking observations with both directions of rotation of the wheel, and assuming that the terms in α will be sufficiently well eliminated by the averaging, since all the α's were small. Believing that the values of α would be still smaller if he reversed their sign between an eclipse and a reappearance, so that V_{e+} would be followed by V_{r-} and V_{e-} by a V_{r+}, he took such readings also, determining the double observations $V_{e+r-}=(V_{e+}+V_{r-})/2$, and $V_{e-r+}=(V_{e-}+V_{r+})/2$, and the crossed double observation $V_{e\pm r\mp}\equiv(V_{e+r-}+V_{e-r+})/2$. His notation for these several double observations was as follows: $V\equiv V_{er+}$, $v\equiv V_{er-}$, $U\equiv V_{e-r+}$, $u\equiv V_{e+r-}$; to these he added a fifth type, designated by w, to cover those double observations for which the speed of the wheel as derived from the chronograph record remained unchanged throughout the interval between the eclipse and the reappearance of the star. Omitting type w, which is plainly abnormal, his double observations fall into eight classes, one for each direction of rotation of the wheel for each of the four types, V, v, U, u.

OBSERVATIONS AND DUBIETY

Cornu's individual values for each class are given in table 5 for the 200-tooth wheel; and in table 6 are

[12] For the distinction between a "normal" and a "critical" region, see Appendix A.

TABLE 5

Cornu's Values (in Air) for the 200-tooth Wheel

Each value was obtained from a "double observation." For those in columns L the wheel was rotating counterclockwise (top moving to left); in columns R, clockwise. n is the order of the eclipse; $q=2n-1$; δ is the excess of a value above the mean of the set; $V\equiv V_{er+}$, $v\equiv V_{er-}$, $U\equiv V_{e-r+}$, and $u\equiv V_{e+r-}$ indicate the four types of double observations; δ_m is average, irrespective of sign, of all the δ's in the column; δ_{m2} is the square root of the mean δ^2. If the distribution of the δ's were "normal" and if the values in the column were a fair sample of the family, then $\delta_{m2}/\delta_m=1.25$ (see eq. 4).

Unit of V and of $\delta=1$ megameter/second

n	q	$V\equiv V_{er+}$				$v\equiv V_{er-}$				$U\equiv V_{e-r+}$				$u\equiv V_{e+r-}$			
		L	δ	R	δ	L	δ	R	δ	L	δ	R	δ	L	δ	R	δ
8	15	296.7	−0.1			304.1	+2.9			300.5	0.0						
		297.0	+0.2			304.8	+3.6										
						300.4	−0.8										
						295.6	−5.6										
Mean		296.8				301.2				[a]300.5							
9	17	294.9	−1.8	298.0	−2.3	304.2	+3.3	301.0	+0.2			296.6	0.0	298.2	+1.4	296.2	−3.2
		298.0	+1.3	302.6	+2.3	307.5	+6.6	300.5	−0.3					295.5	−1.3	295.0	−4.4
		297.3	+0.6			298.0	−2.9									301.9	+2.5
						298.6	−2.3									302.1	+2.7
						296.0	+4.9									302.0	+2.6
Mean		296.7		300.3		300.9		300.8				296.6		296.8		299.4	
9	17	304.2	+3.3	298.7	−2.0	302.6	+1.5	302.1	+2.3	299.2	−1.9	300.2	+0.1	301.0	+0.9	288.0?	−6.7
		299.0	−1.9	303.8	+3.1	304.6	+3.5	298.4	−1.4	303.4	+2.3	294.7	−5.4	300.5	+0.4	299.1	+4.4
		300.7	−0.2	302.4	+1.7	298.5	−2.6	305.5	+5.7	300.8	−0.3	303.4	+3.3	299.3	−0.8	297.1	+2.4
		302.5	+1.6	297.0	−3.7	298.7	−2.4	299.1	−0.7			299.7	−0.4	300.4	+0.3		
		299.8	−1.1	301.1	+0.4	298.4	−2.7	300.4	+0.6			301.0	+0.9	299.9	−0.2		
		303.3	+2.4	293.4	−7.3	300.0	−1.1	297.7	−2.1			301.5	+1.4	303.6	+3.5		
		300.1	−0.8	301.9	+1.2	297.1	−4.0	299.4	−0.4					297.2	−2.9		
		298.2	−2.7	300.6	−0.1	301.1	0.0	290.0	−9.8					300.2	+0.1		
		302.4	+1.5	303.7	+3.0	302.5	+1.4	300.3	+0.5					298.0	−2.1		
		300.2	−0.7	304.7	+4.0	300.2	−0.9	303.4	+2.6					301.4	+1.3		
		299.5	−1.4	302.7	+2.0	304.9	+3.8	298.6	−1.2					298.7	−1.4		
				301.0	+0.3	301.6	+0.5	302.4	+2.6					302.0	+1.9		
				300.2	−0.5	303.6	+2.5	299.7	−0.1					298.9	−1.2		
				296.0	−4.7	299.9	−1.2										
				302.1	+1.4	300.6	−0.5										
				299.7	−1.0	298.6	−2.5										
				302.0	+1.3	305.7	+4.6										
				301.3	+0.6												
Mean		300.9		300.7		301.1		299.8		301.1		300.1		300.1		294.7	
10	19	301.6	+2.2	300.9	−0.1	300.9	−0.3	300.4	+0.4	294.3	−5.7	301.1	+0.3	304.7	+5.9	296.1	−3.9
		295.6	−3.8	301.0	+0.0	299.2	−2.0	302.2	+2.2	303.6	+3.6	398.1	−2.7	293.3	−5.5	298.0	−2.0
		299.7	+0.3	298.0	−3.0	300.3	−0.9	294.5	−5.5	301.5	+1.5	303.0	+2.2	298.5	−0.3	299.7	−0.3
		302.8	+3.4	298.4	−2.6	304.6	+3.4	299.7	−0.3	298.7	−1.3	299.3	−1.5			301.5	+1.5
		300.7	+1.3	301.7	+0.7			300.9	+0.9	303.1	+3.1	300.6	−0.2			297.3	−2.7
		297.8	−1.6	299.9	−1.1			300.6	+0.6	303.7	+3.7	301.3	+0.5			299.8	−0.2
		297.9	−1.5	300.3	−0.7			301.2	+1.2	295.0	−5.0	300.1	−0.7			302.7	+2.7
				299.4	−1.6			299.8	−0.2			302.2	+1.4			299.3	−0.7
				302.0	+1.0			299.8	−0.2			301.6	+0.8			297.8	−2.2
				304.2	+3.2			301.9	+1.9							298.7	−1.3
				300.5	−0.5			300.6	+0.6							299.7	−0.3
				299.7	−1.3			301.3	+1.3							301.6	+1.6
				303.2	+2.2			300.4	+0.4							301.2	+1.2
				300.5	−0.5			298.6	−1.4							306.5	+6.5
				306.0	+5.0			297.6	−2.4							299.2	−0.8
								301.4	+1.4							301.1	+1.1
								299.1	−0.9								
								301.8	+1.8								
								298.2	−1.8								
Mean		299.4		301.0		301.2		300.0		300.0		300.8		298.8		300.0	

[a] There being no u to match it, this U was ignored by Cornu.

TABLE 5—*Continued*

n	q	$V \equiv V_{er+}$				$v \equiv V_{er-}$				$U \equiv V_{e-r+}$				$u \equiv V_{e+r-}$			
		L	δ	R	δ	L	δ	R	δ	L	δ	R	δ	L	δ	R	δ
11	21	298.1	−1.9	298.9	−1.9	300.1	+1.8	300.1	−0.6	298.6	−2.2	300.1	+0.2	298.9	−0.4	301.6	−0.2
		298.7	−1.3	297.6	−3.2	299.0	+0.7	302.7	+2.0	302.4	+1.6	301.5	+1.6	301.6	+2.3	297.5	−4.3
		300.2	+0.2	301.8	+1.0	297.7	−0.6	301.7	+1.0	299.5	−1.3	299.8	−0.1	298.6	−0.7	305.4	+3.6
		300.4	+0.4	300.0	−0.8	296.5	−1.8	300.8	+0.1	301.5	+0.7	296.1	−3.8	300.0	+0.7	299.5	−2.3
		299.5	−0.5	301.0	+0.2	298.1	−0.2	298.5	−2.2	302.4	+1.6	301.1	+1.2	297.3	−2.0	303.5	+1.7
		299.7	−0.3	302.7	+1.9			300.2	−0.5	297.6	−3.2	299.3	−0.6			305.6	+3.8
		301.7	+1.7	302.8	+2.0			301.9	+1.2	305.1	+4.3	300.4	+0.5			299.2	−2.6
		300.3	+0.3	301.8	+1.0			298.9	−1.8	302.1	+1.3	299.9	0.0				
		299.4	−0.6	300.8	0.0			300.6	−0.1	299.0	−1.8	298.9	−1.0				
		298.5	−1.5					305.4	+4.7	300.3	−0.5	300.9	+1.0				
		300.0	+0.9					297.2	−3.5			301.2	+1.3				
		302.4	+2.4					300.0	−0.7								
								303.4	+2.7								
								298.8	−1.9								
Mean		300.0		300.8		298.3		300.7		300.8		299.9		299.3		301.8	
δ_m			1.3_6		1.7_8		2.2_9		1.6_4		2.2_3		1.2_3		1.6_3		2.4_3
δ_{m2}			1.6_7		2.3_4		2.8_0		2.4_1		2.7_0		1.7_6		2.2_6		2.9_0
δ_{m2}/δ_m			1.2_3		1.3_1		1.2_2		1.4_7		1.2_1		1.4_3		1.3_8		1.1_9
Number			35		44		35		48		21		27		23		31

Averages of the preceding means, L and R, and their averages

n	V_{er+}	V_{er-}	$V_{er\pm}$	V_{e-r+}	V_{e+r-}	$V_{e\mp r\pm}$
8	296.8	301.2	299.0	300.5		
9	298.5	300.8	299.6	296.6	298.1	297.4
9	300.8	300.4	300.6	300.6	297.4	299.0
10	300.2	300.6	300.4	300.4	299.4	299.9
11	300.4	299.5	300.0	300.4	300.6	300.5

given his averages, by order of eclipse, for each class and each wheel. It will be noticed that, in general, the number of double observations for a given class and order is greater for the 200-tooth wheel than for any of the others. That is the reason it has been chosen for the display of individual values in table 5.

In table 5 it will be noticed that the 264 deviations δ from the means of nominally homogeneous groups of values range from −9.8 to +6.6, giving for the mean deviation $\delta_m = 1.8$ megameters per second. Consequently, it would have been (eq. 20) practically impossible for him to have detected the presence of a systematic error that did not significantly exceed 0.6 megam./sec. Since many series contain very few observations, the discordance dubiety of the final result, that arising from the mere discordance of the observations, is surely greater than 0.6 megam./sec., which is twice the limit he admits.

It will also be noticed that there is no clear evidence of any difference between the L and the R values (left and right rotation), or between the four types of double observations (V, v, U, and u), or even between the several orders, of which only 8, 9, 10, and 11 are represented. Such systematic differences as surely exist are quite completely masked by the erratic ones.

In table 6 are given the means, by orders, of each class for each wheel, each followed in parentheses by the number of double observations involved in it. The values in the L class of the V group for the 150-tooth wheel decrease, in general, as n increases, but marked irregularities occur. To all the others, practically everything already said about the values for the 200-tooth wheel applies.

The values are very poorly distributed. Many classes contain only 1, 2, or 3 values; values for one direction of rotation are frequently not matched by any for the other; only for the 150-tooth wheel are the values distributed over a sufficient range of orders to justify an attempt to determine the value of the correction term involving $(H \pm B)$, and then one must assume that the term involving the acceleration has been automatically eliminated by the averaging of the data.

There seems to be no reason for thinking that the dubiety of the final result can be less than the minimum (0.6 megam./sec.) inferred from a consideration

TABLE 6

CORNU'S MEANS (IN AIR) FOR EACH WHEEL

Each entry is the mean of all the double observations of that type, and is followed, in parentheses, by the number of double observations that is involved.

Unit of $V = 1$ megameter/second

	$V \equiv V_{er+}$		$v \equiv V_{er-}$		$U \equiv V_{e-r+}$		$u \equiv V_{e+r-}$	
q	L	R	L	R	L	R	L	R
	150-tooth wheel							
7	307.3(7)	297.3(1)	298.1(1)	289.6(2)	300.4(3)	299.8(1)	303.4(2)	
9	305.3(6)	302.0(7)	299.6(7)	297.5(3)	296.9(1)	300.4(3)	296.2(2)	299.8(2)
11	305.5(6)	298.9(5)	295.8(2)	300.9(8)		[a]307.1(2)		
13	301.1(3)	305.4(2)	307.0(1)	301.0(2)	301.9(2)		299.5(3)	
15	307.3(1)	301.5(3)	297.6(1)	298.6(3)			[a]304.7(1)	[a]302.0(3)
17	302.3(3)	304.7(2)	297.4(5)			[a]305.3(1)		
23		304.9(2)		296.1(1)				[a]300.9(1)
27	302.3(1)	304.0(2)		299.3(2)	300.9(1)	299.3(2)	298.4(1)	
31		299.3(3)	301.9(1)					
33	300.4(6)	298.0(1)	298.9(8)	297.9(3)	303.2(1)		301.9(2)	298.4(2)
35	302.2(4)	301.3(4)	300.6(3)	298.3(5)	300.2(4)	299.1(1)		298.7(3)
37		301.4(4)	297.7(1)			[a]299.2(2)		
41	298.5(3)	300.2(2)	301.1(8)	298.1(2)	300.8(4)	301.8(1)	299.4(4)	300.3(3)
	200-tooth wheel							
15	296.8(2)		301.2(4)		[a]300.5(1)			
17	296.7(3)	300.3(2)	300.9(5)	300.8(2)		296.6(1)	296.8(2)	299.4(5)
17	300.9(11)	300.7(18)	301.1(17)	299.8(13)	301.1(3)	300.1(6)	300.1(13)	294.7(3)
19	299.4(7)	301.0(15)	301.2(4)	300.0(19)	300.0(7)	300.8(9)	298.8(3)	300.0(16)
21	300.0(12)	300.8(9)	298.3(5)	300.7(14)	300.8(10)	299.9(11)	299.3(5)	301.8(7)
	144-tooth wheel							
27	302.9(3)		297.6(6)		[a]301.5(1)			
29	303.0(13)	302.2(12)	297.9(18)	298.1(9)	301.6(2)	302.4(3)	298.5(4)	299.2(4)
31	300.6(1)		297.2(2)					
	180-tooth wheel							
25	300.8(5)		300.5(3)		300.5(1)		299.0(1)	

[a] There being no match (U or u) to this value, Cornu ignored it.

of the discordance of the values found for the 200-tooth wheel.

SOURCES OF ERROR

But this is not all. Although Cornu continually emphasized the necessity of avoiding the critical region, and implied that he had done so, he has offered no evidence to show that he did succeed in so doing. Neither has he given a satisfactory experimental study of his 1/20-second oscillator, although his time measurements rest solely on that. Both must be carefully considered. That will be done, and it will be found that there are good reasons for thinking that all his observations lay in the critical region, and that his oscillator did not function in the way he assumed.

Critical Region

Throughout his *Mémoire*, Cornu implies[13] that his observations were so taken as to avoid the critical region, possibly excepting a single situation to be mentioned presently. But no more positive statement that they were so taken has been found than those

[13] "Il est bon de remarque que le mode d'observations doubles employé, permettant de se tenir loin des *phases critiques*, est affranchi en principe de ce genre d'erreur. En effet, la phase utilisée du phénomène est telle, qu'aux environs de la vitesse observée les variations d'intensité sont proportionelles aux vitesses de la roue dentée" (p. A. 273).

"D'abord, ainsi que je l'ai longuement démontré, la méthode d'observation double, en permettant d'éviter les phases critiques, annule l'influence de l'inegalité des dents" (p. A. 287).

in the quotations just given; and he gives no evidence to prove it.

Now the actual wheel may be regarded as derived from an ideal one by the making of additions, positive or negative, to the edges of the teeth of that wheel; and such additions may also simulate the effect of eccentricity and of periodic irregularities in the speed of the wheel. In Appendix A, it is shown that if several of those additions are as broad as a quarter of the used interdental gap, g, then practically every observation will have to lie in a critical region. Also, if many of the additions are half as wide as an open space in the ideal equivalent sectored disk when set for the observation, then that setting of the actual wheel will lie in a critical region. (As shown in Appendix A, the appearance to the observer is as if he were viewing a steadily shining star through a rotating sectored disk composed of the toothed wheel and an angularly displaced phantom of itself. That equivalent sectored disk is the one here referred to.)

There are no actual data for the width of that space when so set. But the apparent brightness of the returned star, when the speed of the wheel was midway between those that corresponded to consecutive eclipses, must have been quite significantly greater than when the star vanished into the background. If it had been only 10 times as great, which seems to be a conservative guess, then the width of the open space in the disk when the observation was made would have been 0.1 g, and the allowable additions to the teeth would not have exceeded 0.05 g. From table 4 it is seen that, for the several wheels, g lay between 0.33 and 0.37 mm. Hence the irregular additions to the sides of the teeth should not have exceeded 16.5 to 18.5 microns, if the observations were to lie in a normal region.

That is, for the metal wheels themselves.

But the wheels were smoked in order to reduce the illumination of the field by light reflected from them; and they were smoked repeatedly. Furthermore, Cornu stated[14] that the thickness of the smoke made a relatively great change in the breadth of the teeth, and that grains of the soot were continually being rearranged and thrown off by the motion of the wheel.

From all of which it seems certain that many, probably all, of his observations lay in critical regions, for a heavily smoked edge that does not present irregularities of over 18 microns is a rarity. And the smoking of numerous edges so that the thickness of the smoke layer on each shall be the same within 18 microns would surely be difficult and require far more care than is indicated in the report. In fact, the report does not indicate that any particular care was taken.

Oscillator

As Cornu devoted many pages to a theoretical study of an oscillator driven by periodically applied impulses, and to a very queer and unsatisfactory discussion of certain of the chronograph records made by the oscillator in the course of his observations for determining the velocity of light, it seems well to indicate in some detail what he did, before going on to an independent study of the oscillator; especially since the writer cannot accept Cornu's conclusions.

Cornu's study of the oscillator.—Cornu considered first a vibrating body acted upon by periodically applied impulses of constant strength, the period of the impulses being nearly the same as the free period of the body; and he found that the number of vibrations made by the body in a given time, when so driven, is the same as the number of the impulses. All of which is well known. From this he concluded that each swing of the half-second pendulum driven by the clock will be made in exactly one-half of a clock second, and that each oscillation of the 1/20-second oscillator will be made in exactly one-tenth of a swing of the half-second pendulum, and consequently in exactly one-twentieth of a clock second.

Those conclusions are, of course, wrong. The 1/20-second oscillator, receiving an impulse every half-second, would make four complete vibrations with its own free period, and during the fifth, its phase would be so adjusted by the impulse as to conform properly with that of the half-second pendulum. (See Appendix B.) This, as well as his treatment, assumes that the intervals between successive impulses from the half-second pendulum were exactly equal. The same reasoning applies to the half-second pendulum when driven by the clock.

Consequently, the accuracy with which the oscillator recorded tenth-seconds depended upon the closeness of the "tuning" both of the half-second pendulum and of the oscillator, and upon the equality of the intervals between impulses.

Although it seems that Cornu did not in the least doubt his conclusions, he thought it desirable to give in his report data that would bear them out. So, after the completion of his observations on the velocity

[14] "Il y a enfin une circonstance particulière qui favorise singulièrement l'élimination de l'influence fâcheuse des inégalités de la denture, dans le cas même où elle aurait pu devenir appréciable: c'est l'opération fréquente de l'enfumage des dents. En effet, la denture se trouve recouverte d'une couche notable de noir de fumée, qui modifie la largeur des dents dans une proportion relative assez grande et, par suite, la loi des inégalités produites lors de la taille des dents; on peut même dire que ces inégalités varient sans cesse, car des grains de noir de fumée sont inévitablement déplacés ou projetés de temps à autre sous l'action de la force centrifuge et des vibrations des engrenages. C'est donc en réalité avec une roue dentée sans cesse variable un point de vue optique qu'on opère. Cette variation continuelle est éminement favorable à l'élimination des erreurs causées par la forme de la denture, puisque les choses se passent comme si l'on opérait successivement avec un nombre considérable de roues différentes. C'est probablement à cette cause que l'on doit la concordance si satisfaisante des déterminations obtenues avec les diverses roues dentées" (p. A. 287). His argument may be questioned, but what he says about the wheel is all that is of present interest.

TABLE 7

CORNU'S MEASUREMENTS OF 10-SECOND INTERVALS IN TERMS OF THE COMPLETE VIBRATION OF HIS 1/20-SECOND OSCILLATOR

In the first column are given the numbers assigned by Cornu to the several measured intervals; in the second is given the number of nominal 0.1 seconds in each of the 10-second intervals; δ_1 is the amount by which the corresponding number exceeds 100; $Mean_{10}$ = mean of the values in column 2 taken in consecutive groups of 10; last group contains 5 only; δ_{10} is amount by which the adjacent $Mean_{10}$ exceeds 100; δ_m is average of δ_1 irrespective of sign.

Unit of "Interval" and of δ = 0.1 nominal second

No.	Interval	$100\delta_1$	No.	Internal	$100\delta_1$
1	99.94	− 6	41	100.20	+20
2	100.02	+ 3	42	99.77	−23
3	100.06	+ 6	43	100.06	+ 6
4	99.77	−23	44	99.81	−19
5	100.09	+ 9	45	100.05	+ 5
6	100.00	0	46	100.14	+14
7	99.84	−16	47	99.89	−11
8	99.88	−12	48	99.90	−10
9	99.80	−20	49	99.98	− 2
10	100.10	+10	50	100.12	+12
$Mean_{10}$	99.951		$Mean_{10}$	99.992	
δ_{10}	−0.049		δ_{10}	−0.008	
11	100.05	+ 5	51	100.31	+31
12	99.87	−13	52	99.99	− 1
13	99.96	− 4	53	99.95	− 5
14	99.88	−12	54	100.03	+ 3
15	99.97	− 3	55	100.03	+ 3
16	100.02	+ 2	56	100.08	+ 8
17	100.07	+ 7	57	99.85	−15
18	100.06	+ 6	58	100.10	+10
19	100.21	+21	59	100.07	+ 7
20	99.80	−20	60	100.02	+ 2
$Mean_{10}$	99.989		$Mean_{10}$	100.043	
δ_{10}	−0.011		δ_{10}	+0.043	
21	99.99	− 1	61	100.03	+ 3
22	99.88	−12	62	99.85	−15
23	99.77	−23	0	99.88	−12
24	99.73	−27	5+	99.71	−29
25	99.96	− 4	10+	100.17	+17
26	99.97	− 3	15+	100.24	+24
27	100.08	+ 8	20+	100.03	+ 3
28	100.02	+ 2	25+	99.95	− 5
29	100.06	+ 6	30+	100.01	+ 1
30	99.92	− 8	35+	99.94	− 6
$Mean_{10}$	99.938		$Mean_{10}$	99.981	
δ_{10}	−0.062		δ_{10}	−0.019	
31	99.80	−20	40+	100.16	+16
32	99.81	−19	45+	100.09	+ 9
33	99.95	− 5	50+	100.03	+ 3
34	100.00	0	55+	100.28	+28
35	100.05	+ 5	60+	100.15	+15
36	100.04	+ 4			
37	99.84	−11	$Mean_5$	100.142	
38	100.00	0	δ_5	+0.142	
39	99.89	−11			
40	99.93	− 7	$Mean_{all}$	99.98_7	
			$100\delta_m$		$10._2$
$Mean_{10}$	99.931				
δ_{10}	−0.069				

of light, he made 63 measurements of 10-second intervals in terms of his oscillator unit, using the chronograph records made in the course of his observations. Those records were always read to the nearest millisecond. The values reported, together with 12 others, are given in table 7, and shown in figure 1. They exhibit wide variations; an extreme range of 60 milliseconds. His mean of the 63, averaged first in groups of 10 (one of 3), and then the group-means averaged, is 99.972 intervals per 10 seconds.

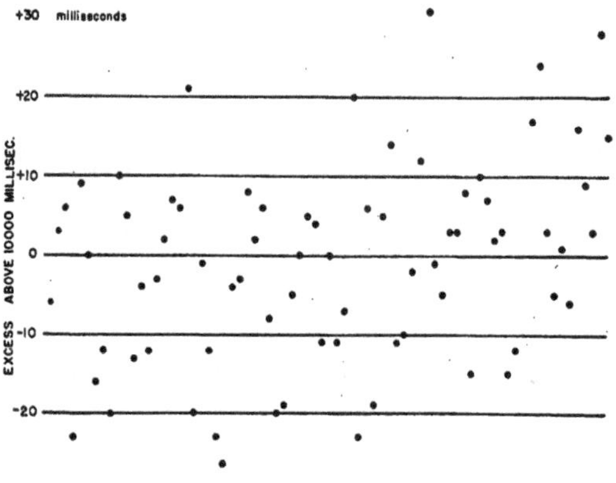

FIG. 1.—Display of Cornu's 75 measurements of 10-second intervals in terms of his 1/20-second oscillator.

Ordinates are the values of the excess of the measured interval above 10 000 nominal milliseconds (of the 100 δ_1 of table 7). The abscissas have neither significance nor function other than to spread the points.

Looking for an explanation of the difference between this value and the even 100 that he expected, he in some way arrived at the idea that the measurements should have been uniformly distributed over the entire length of a chronograph sheet, whereas in measuring the 63 he had (quite properly) excluded the portions where the ends of the sheet overlap. As he remarked, the fixed ends of the flexible tips of the two styluses—that for the clock and that for the oscillator—lie at different levels above the sheet; consequently, an increase in thickness of the paper will cause a relative displacement of the tips upon the paper.[15] So he measured 12 additional 10-second intervals, such that in each case one end of the interval lay upon the double thickness of paper. These ends were distributed over the entire length (5 cm.) of the overlapped ends. (A 10-second interval corresponded to a length of about 18.5 cm.)

These, the last 12 in table 7, averaged higher than the others. He then assembled the 75 measurements

[15] From his plate VI, figure 3, and on the assumption that the tips of the styluses bend as if they were rigid and pivoted where they are attached to the stiff rods, it appears that a 1-mm. change in thickness would have caused a differential shift of about 0.020 sec. in the positions of the two tips on the paper.

in 5 groups, according to the position each occupied with reference to the end of the sheet, and found 100.00 for the average of the 5 group averages. He concluded that on the average the interval between consecutive signals was exactly a tenth of a sidereal second, and that the accidental variations were due to general causes inherent in chronographs with electrical recording devices.[16] Of course they averaged properly, but that is not sufficient; and as will be presently shown, the irregularities did not arise primarily from the chronograph and the electrical recording, but from the oscillator itself.

Obviously, the reason he gave for taking the additional 12 measurements is thoroughly unsound. The relative shift of the tips of the styluses occurred only as they passed from one thickness of paper to two, or the reverse. So long as the same thickness was under each tip and at each end of the adjacent intervals to be compared, the actual thickness was of no consequence. The change on passing from one thickness to two is to be classed as an instrumental error, and should have been obviated by ignoring all intervals that included such transition.

The procedure by which he attempted to derive from these measurements an estimate of how much his definitive result was affected by such uncertainties in the determination of the time, is also queer. For each of the 5 groups into which he arranged his 75 measurements he determined the mean square of the deviations of the individual measurements from the group mean, and found for the mean of those 5 mean squares the value $\epsilon^2 = 178.012$ (milliseconds)^2. Whence he found 9.24 milliseconds for the technical probable error of an interval of time as so derived. Taking the number of individual determinations of the velocity of light as 630,[17] he concluded that, so far as the oscillator's chronograph record was concerned, the technical probable error of his definitive value for the velocity is $9.24/(630)^{1/2} = 0.37$ millisecond per 10-second interval, or a relative error of 0.000037, which is entirely negligible as compared with the precision (1/1000) to which he aspired.[18]

He then endeavored to convince the reader that in an actual determination of the velocity of light the "relative probable error" arising from such uncertainties will be of the same order as that (0.000037) just computed for 630 measurements of 10 seconds each. His argument runs thus: Each determination involved the reading, in terms of the records of the oscillator, of the times of occurrence of three, four, five, and sometimes of six signals made by the revolution counter. These were spread over an interval of from 4 to 10 seconds. Although the speed of the wheel was not constant, its variation throughout the interval considered was slight and need not be considered in this study; it will be entirely satisfactory to ignore the middle one of the three times (t_0, t_1, t_2) used in a determination, since t_1 serves merely to determine the rate of variation; and to consider the mean speed over the double interval from t_0 to t_2, an interval of 2 to 4 seconds. Furthermore, as he used the method of double-observation, each determination of the velocity of light rests upon the mean of the speeds over two such intervals. Thus the conditions are very near those characterizing his 75 ten-second intervals, and one may take 0.000037 as being near the relative technical probable error of his definitive value for the velocity of light, as affected by the determination of the time. The adopted method for determining the speed is therefore practically perfect.[19]

---

[16] "En résumé, on peut avoir toute confiance dans les indications du chronographe; les signaux consécutifs représentent exactement, en moyenne, des dixièmes de seconde sidérale. Quant aux erreurs accidentelles qui peuvent se rencontrer, elles sont dues à des causes générales indépendantes du mode particulier de synchronisme auquel il doit sa propriété fondamentale, et qui egiraient au même degré sur des chronographes quelconques entretenus électriquement" (p. A. 219).

[17] He records 624 sets of observations, of which he utilizes only 546 (see p. A. 266).

[18] "L'erreur probable relative serait 0.000037; comparée à la limite 0.001 que nous somme imposée, on voit qu'elle serait vingt-sept fois moindre" (p. A. 224).

[19] "Nous pouvons en conclure aussi l'erreur probable de la moyenne des 630 déterminations de la vitesse de la lumière, du moins en ce qui concerne l'influence de l'enregistreur. Si ces mesures consistaient dans l'évaluation d'une vitesse uniforme se réduisant au relevé d'un seul intervalle voisin de 10 secondes, l'erreur probable sur la moyenne des résultats s'obtiendrait en divisant l'erreur probable d'une mesure de $10^s = 0^s00924$ par $\sqrt{630}$ ou 25.1, ce qui donnerait 0.00037. L'erreur probable relative serait 0.000037; comparée à la limite 0.001 que nous nous somme imposée, on voit qu'elle serait vingt-sept fois moindre.

"Il est facile de voir que cette erreur probable sera dans le cas réel à peu près du même ordre. En effet, les déterminations comprennent le relevé de 3, 4, 5 ou même de 6 signaux sur un intervalle variant de 4 à 10 secondes.

"Sans entrer à ce sujet dans une discussion approfondie qui nous mènerait un peu loin, on peut dire que, en moyenne, si l'erreur relative augmente par suite de la diminution de l'intervalle à mesurer, elle diminue probablement un peu par l'influence des mesures intermédiaires, et qu'on peut compter sur une compensation approximative.

"D'autre part, si ce n'est pas une vitesse rigoureusement uniforme qu'on determine, la variation de la vitesse dans l'intervalle considéré est si faible qu'on peut la négliger pour l'évaluation approximative de l'erreur. En effet, il suffirait, sur les trois signaux employés pour l'interpolation, de faire abstraction du signal intermédiaire qui ne sert qu'à mettre en évidence l'accélération du mouvement et de raisonner sur la vitesse moyenne. Chaque determination étant double est ainsi déterminée par 4 lectures indépendantes formant 2 groupes comprenant 2 intervalles de 2 à 4 secondes dont on prend la moyenne. On est ainsi ramené à des conditions bien voisines de celles qui ont été prises comme point de départ et condut aux mêmes conclusions. On peut donc admettre, avec beaucoup de vraisemblance, que la moyenne des 630 determinations de la vitesse de la lumière, qui vont être discutées, ne sera affectée, du fait de l'emploi de l'enregistreur comme intermédiaire de mesures, que d'une ***erreur probable relative*** voisine de 0.000037, negligeable par conséquent vis-à-vis de la limite 0.001 que nous nous sommes imposée. Elle

To the casual reader, all this is very seductive unless he recalls that 36 pages earlier in the report (p. A. 188) Cornu has given an illustration of how he actually used the observations in computing the velocity of light. On referring to that, he finds that Cornu did not determine the speed from the double interval t_0 to t_2 of from 2 to 4 seconds, but from a 1-second interval. And from the times recorded in his long table (pp. A. 194–210) it may be seen that in Cornu's actual computations the time interval used varied from 0.2 to 2.1 seconds, there being 42 that were less than 0.6 second and only 9 that exceeded 2 seconds; the mean of all being 1.18 seconds. Furthermore, the complete over-all interval for the three to six records for which he read the times, which interval was divided into two parts, one for each phase of the eclipse, if the number of records covered exceeded three, never amounted to 10 seconds; for the 584 determinations that were completely reduced they averaged only 3.62 seconds, and were distributed as shown below:

Interval...	0–1	1–2	2–3	3–4	4–5	5–6	6–7	7–8	8–9 sec.
Number...	4	53	156	155	134	51	22	8	1

Whence it is evident that Cornu's treatment of this portion of the subject is misleading, and his conclusion is wrong.

If, according to Cornu, the deviations of the measurements of the 10-second intervals be regarded as strictly fortuitous, then the average error in the determination of an interval of time from the chronograph record of the oscillator is about 10 milliseconds (see table 7), which is about 0.93 percent of the average interval (1.18 sec.) used in computing the speed of the disk. Hence the corresponding approximate technical probable error of the mean of 630 such determinations is $0.8453(0.93)/(630)^{1/2}=0.03$ percent, over eight times as great as Cornu's value. It is, however, amply small if the assumptions on which it is based are valid.

But this is not all. The computed speed depends not only on a difference between two readings of the time, but also upon the difference of two such differences, the speed being not uniform, but accelerated. In Cornu's illustrative computation (p. A. 188), the speed is inversely proportional to $\mu''+\delta''$, where μ'' is a difference and δ'' depends upon a second difference, in time. In that case, μ'' was 1.049 sec., and the second difference $(\mu_2-\mu_1)$ was +0.011 sec., which led to $\delta''=0.0071$ sec., or 0.7 percent of μ''. Had $\mu_2-\mu_1$ been only 0.001 sec. (the average deviation of his readings for 10-sec. intervals was 0.010 sec.), then δ'' would have been only 0.0006 sec., and the computed value of the speed would have been over 6 parts in 1,000 greater than that he computed. Of such uncertainties, Cornu says nothing.

But none of this has touched the fundamental question as to why the values in table 7 vary so greatly. It has merely been assumed that they are fortuitous. The only suggestions offered by Cornu are these: (1) uncertainty in setting the microscope on the trace made by the stylus on the smoked paper; (2) irregularities in the movement of the chronograph drum; (3) variations in the batteries and in the electric resistances of the contacts; and (4) irregularities in the chronograph sheet, such as inequalities in grain or in thickness of the paper (see pp. A. 215–216). The effects produced by all except the extra thickness of the overlapped ends of the sheets will be fortuitous, and he seems to think that he has brought that into the same category by taking the last 12 of his 75 measurements of 10-second intervals. It has already been pointed out that his reasoning regarding those 12 is unsound. The effect of the overlap is probably negligible. And it is hard to see how any probable irregularities in the motion of a large chronograph drum turned by a weight-driven motor controlled by a centrifugal wing governor and running in clean, well-lubricated bearings can give rise to such variations in the comparative readings of two adjacent traces as are shown in table 7.

New study of the oscillator.—But there are other factors that might very definitely have affected the quantities measured, and that are in part amenable to experimental investigation; such as, irregularities (1) in the signals from the standard clock, (2) in the signals from the 1/20-second oscillator, (3) in the period of that oscillator, (4) in the period of the half-second pendulum, (5) in the intervals between consecutive closings of the electric circuit by the standard clock.

It is well known that irregularities of types (1) and (2) may on occasion cause differences of some thousandths of a second. A careful study of the magnitude of the changes that were produced in the measurements by known changes in the conditions of the electric contacts would have been of much value in fixing the extent of the variations that might be expected from those irregularities.

As to the free period of the half-second pendulum, we are told no more than that, by comparing the chronograph record of the half-second pendulum, swinging freely for some tens of oscillations, with that of the standard clock, it was possible to adjust its period very approximately to the desired value, for example, to 1 in 1,000.[20] The adjustment of the elec-

pourrait être deux ou trois fois plus grande sans cesser d'être négligeable.

"Le mode d'enregistrement est donc *pratiquement* parfait, et les erreurs qu'il reste à éliminer ne dépendent plus que de la difficulté propre aux observations" (pp. A. 224–225). No later profound discussion of this subject has been found; his reference seems to be to his general discussion of the uncertainties and errors in his computed values for the velocity of light.

[20] "On arrive à un synchronisme très-approximatif, par exemple à une oscillation sur mille. A ce degré, le réglage est suffisant" (p. A. 145).

tric contact, closed by this pendulum, to the center of the arc was judged by an ear estimate of the equality of the intervals between the sounds emitted by a telegraph sounder placed in circuit with the contact. No estimate of the accuracy of this adjustment is given.

The 1/20-second oscillator was adjusted in the same manner, its chronograph record, while oscillating freely, being compared with that of the standard clock. By working systematically, one soon reaches a very approximate adjustment. ("Un essai méthodique . . . permet d'arriver rapidement à un réglage très-approximatif.") No estimate of the approximation is given.

One is told that this oscillator closes an electric circuit at each double oscillation by making contact with a platinum wire carried at the end of a spring, the position of contact being adjustable by means of a screw pressing against the spring near its base (p. A. 147). This is all one is told about the contact and the tuning of this oscillator.

Although Cornu discussed at some length the case of a body oscillating under the action of regularly recurring impulses, he seems to have completely failed to recognize the fundamental physical difference between that case and the one in which the body is driven by a simple harmonic force, and to have thought that in the first case the motion is continuously as strictly simple harmonic as it is in the second. Only on the assumption of such confusion can one understand the very meager information given regarding the tuning of the pendulum and oscillator, his satisfaction with approximate tuning, the total absence of any indication of a serious study of the behavior of these fundamentally important pieces of apparatus, and his uniform assumption that the period of the oscillator is exactly 0.1 sec.

Of the uniformity of the intervals between successive closings of the circuit by the standard clock, one is told incidentally that the pendulum has a one-second swing and that the signals are not produced under exactly the same conditions except on alternate seconds;[21] and in the detailed account of how the work was done one is told that the signals from the vibrator were counted so as to verify their coincidences 10 by 10, or rather 20 by 20, with the signals from the clock, because the period of that is really 2 seconds;[22] and still later, at the beginning of Cornu's discussion of the measurement of 10-second intervals in terms of the oscillator signals (those in table 7), one is told that a simple inspection of the chronograph records is enough to enable one to sort out the odd and even seconds, which are not absolutely equal.[23] This seems to be all that he tells one about this vitally important item.

Obviously, if consecutive seconds of the standard clock are unequal, the half-second pendulum driven by them can never reach the steady state considered both in his treatment and in Appendix B, but will have its phase changed every second swing, advanced at one time, retarded the next. Similarly, the oscillator, driven by it, can never reach a steady state, but will have its phase changed every half second (every five signals), now advanced, and now retarded, the change being especially pronounced at each impulse from the standard clock. Under such conditions, the signals from the oscillator cannot possibly mark equal intervals of time for more than 0.4 second, at the most, unless there is some other departure from the ideal conditions that tends to smooth out irregularities.

One may reply that, although this is true, it is probable that the adjacent intervals between successive signals from the clock were actually so nearly alike that the irregularities caused by their difference were negligible. Fortunately, Cornu has, quite inadvertently, furnished means by which this can be tested.

In figure 9 on his plate VI he has given a facsimile of the chronograph record for observation no. 143. The engraving is excellent, and the record can be read with precision. Changes in the paper arising from its age and variations in its moisture content may well cause the distances measured on it to differ from those measured on the original sheet, but such changes will have little effect upon the relative positions of signals in adjacent tracings; and those relative positions are the items of importance. That facsimile has been measured by means of a micrometer slide with a traveling microscope. One division on the micrometer head corresponded to a motion of 5 microns, but the distances were read only to the nearest 0.01 mm., which corresponded to about 0.0005 second; in general, repeated settings agreed within that amount.

Before considering the measurements, it is desirable to notice that this facsimile shows that the contact closed by the oscillator remained closed for about 1/20 second, for one-half of a complete vibration. As the circuit was closed, the armature of the recording mechanism was jerked against its magnet-stop, the flexible stylus overshot its mark, and settled back with highly damped vibrations; a little later, as the circuit was opened, the armature was jerked against its back stop, the stylus overshot, and settled back to its new position. The chronograph record of the two halves

[21] "La période d'oscillation du balancier de l'horloge qui produit les signaux est de deux secondes, et les signaux ne se trouvent rigoureusement tracés dans les mêmes conditions que de deux en deux secondes" (p. A. 152).

[22] "Afin de vérifier leurs coincidences de 10 en 10, ou plutôt de 20 en 20 avec les battements de l'horloge (car la période des battements de l'horloge est réellement 2 secondes)" (p. A. 174).

[23] "Effectivement, à la simple inspection des tracés graphiques, on reconnaît que les signaux du chronographe [oscillator] coincident avec une grande régularité, de 20 en 20, avec ceux de la seconde; la coincidence est assez caractéristique pour faire distinguer à première vue les secondes, d'ordre pair ou impaire, dont l'egalité n'est pas absolue" (p. A. 211).

of the cycle are remarkably similar, and very closely of the same length. It is thus evident that for half its period the vibrator was subjected to the rubbing of the contact and to the pressure of its spring. This is a significant departure from the ideal case, and may perhaps have led to an appreciable variation of the period with the amplitude.

The times α corresponding to the make-circuit signals from the standard clock, and those β corresponding to the break-circuit ones, each as measured from an arbitrary origin and in terms of a complete period of the vibrator (nominally 0.1 second), were found to be as given in table 8.

TABLE 8

RESULTS OF MEASUREMENT OF FACSIMILE OF CHRONOGRAPH RECORD FOR CORNU'S OBSERVATION NO. 143

α and β are the times, measured from an arbitrary origin, corresponding to make-circuit records of the clock and to break-circuit records, respectively.
Unit of α and β = 1 complete period of oscillator (approximately 0.1 sec.)

Clock second	0	1	2	2	3	3	4	5
α.....		8.22	18.76	18.77	28.25	28.25	38.74	48.22
β......	−0.01	9.52	19.98	19.98	29.54	29.50	39.98	49.48
$\beta-\alpha$..		1.30	1.22	1.21	1.29	1.25	1.24	1.26

Intervals	1 to 3	3 to 5	0 to 2	2 to 4	0 to 1	1 to 2	2 to 3	3 to 4	4 to 5
$\Delta\alpha$....	20.03	19.97		19.98		10.54	9.49	10.49	9.48
$\Delta\beta$....	20.00	19.96	19.99	20.00	9.53	10.46	9.54	10.46	9.50
Mean..	19.99				9.53	10.50	9.52	10.48	9.49

From these observations it is evident that any 2-second interval is equivalent to exactly 20 periods of the oscillator, within the limits of precision of the measurements; and that the even-second intervals (1–2, 3–4) are longer than the odd-second ones (0–1, 2–3, 4–5) by 10.49−9.51=0.98 period, essentially 0.1 second.

Hence alternate intervals between impulses given by the clock differ by essentially one complete period of the oscillator. This cannot fail to cause irregularities in the oscillations of both the half-second pendulum and the oscillator. Neither one can ever settle down to a really steady state.

Additional information regarding the behavior of the oscillator may be obtained from a study of the distances between its successive signals. The measurements are given in table 9. Even a casual examination of the differences shows that those of the second decade exceed those of the first and third. This is shown more clearly in figure 2, where these differences are plotted, each against the arbitrary ordinal number of the signal bounding it on the left. These ordinal numbers are those assigned by Cornu, and were so chosen that zero corresponds to a signal that nearly coincided with a break-circuit signal from the clock, which clock signal was also numbered zero.

TABLE 9

MEASUREMENT OF THE PUBLISHED CHRONOGRAPH RECORD OF THE 1/20-SECOND OSCILLATOR

A traveling microscope, moved along a millimeter scale by means of a micrometer screw, was set on the successive break-circuit signals of the oscillator. The scale being but 50 mm. long, the instrument had to be moved twice in order to cover the entire record with the desirable overlaps. All three sets of readings are tabulated.

No. = number of signal; Read. = reading (mm.) of the micrometer; Diff. = difference between successive readings = distance between adjacent signals. The time interval between adjacent signals being very nearly 0.1 sec., 0.01 mm. corresponds to about 0.0005_4 sec. = 0.5_4 milliseconds.

The mean differences are plotted in fig. 2.

Unit = 1 mm. = 54 milliseconds, approx.

No.	Read.	Diff.	Read.	Diff.	Read.	Diff.	Mean Diff.
1	5.55	1.78					1.78
	7.33	1.81					1.81
	9.14	1.79					1.79
	10.93	1.82					1.82
5	12.75	1.80					1.80
	14.55	1.80					1.80
	16.35	1.78					1.78
	18.13	1.79					1.79
	19.92	1.86					1.86
10	21.78	1.91					1.91
	23.69	1.86					1.86
	25.55	1.86					1.86
	27.41	1.88					1.88
	29.29	1.86					1.86
15	31.15	1.90					1.90
	33.05	1.89					1.89
	34.94	1.86					1.86
	36.80	1.89					1.89
	38.69	1.88					1.88
20	40.57	1.78					1.78
	42.35	1.82					1.82
	44.17	1.80					1.80
	45.97	1.78	0.23	1.79			1.78
	47.75	1.82	2.02	1.82			1.82
25	49.57		3.84	1.81			1.81
			5.65	1.81			1.81
			7.46	1.81			1.81
			9.27	1.81			1.81
			11.08	1.85			1.85
30			12.93	1.89			1.89
			14.82	1.86			1.86
			16.68	1.88			1.88
			18.56	1.90			1.90
			20.46	1.84			1.84
35			22.30	1.89			1.89
			24.19	1.87			1.87
			26.06	1.85			1.85
			27.91	1.89			1.89
			29.80	1.88			1.88
40			31.68	1.82	25.76	1.83	1.82
			33.50	1.79	27.59	1.79	1.79
			35.29	1.80	29.38	1.80	1.80
			37.09	1.83	31.18	1.82	1.82
			38.92	1.76	33.00	1.78	1.77
45			40.68	1.81	34.78	1.81	1.81
			42.49	1.80	36.59	1.80	1.80
			44.29	1.80	38.39	1.80	1.80
			46.09	1.79	40.19	1.78	1.78
			47.88		41.97	1.85	1.85
50					43.82	1.83?	1.83?
					45.65?		

It is at once obvious that the differences fall into two main groups. One averages about 1.80 and the other about 1.87_5. The groups alternate; each persists for 10 signals, beginning at or near one of the signals 9, 19, 29, 39. The general pattern of each group remains the same, but differs from that of the other.

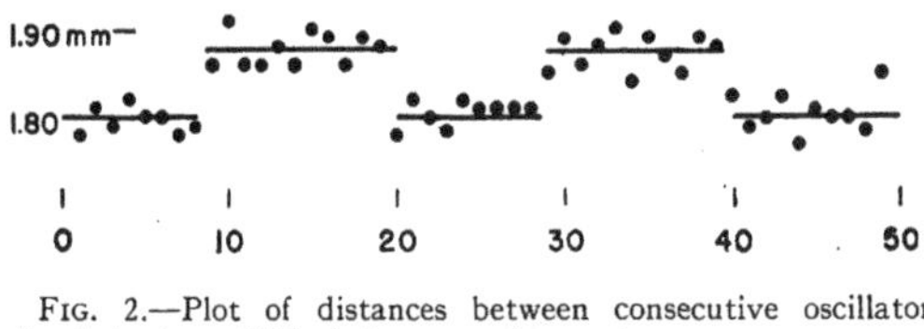

FIG. 2.—Plot of distances between consecutive oscillator signals in the published chronograph record.

Abscissas are the ordinal numbers of the signals at the left boundary of the interval. Ordinates show the distance (mm., 1 mm. = 54 milliseconds, approximately) between consecutive signals, one made at each complete period of the oscillator, at nominally 0.1-second intervals. They are the values in the column "Mean diff." of table 9.

In view of the relatively great difference (0.1 sec.) between the adjacent intervals defined by the clock, none of this is surprising, unless it be the absence of any clear tendency for the oscillator to return to a single fixed frequency for two or three signals just preceding the impulses from the half-second pendulum. But that may result from the amplitude corresponding to one group having been greater than that for the other, and from a variation of the period with the amplitude. Furthermore, it is probable that the phase adjustment to the clock impulse was much more nearly complete for the half-second pendulum than it was for the vibrator; that would result in the next impulse from the half-second pendulum producing upon the vibrator the same kind of readjustment as that produced a half-second earlier by the clock. The data are too scanty to justify more definite statements. But they clearly show that in this particular case the oscillator had two distinct average periods differing by about 4 percent (i. e., by 0.004 sec.), one corresponding to the odd-numbered seconds, and the other to the even-numbered ones.

Consequently, when these changes in the frequency of the oscillator are ignored, as they are in Cornu's work, variations of 4 percent, due to these changes alone, may be expected in the individual determinations of the velocity of light. Obviously, the effect of this error upon the mean of a large number of determinations distributed in no systematic manner over the two distinct groups of frequencies shown in figure 2 will be much smaller, but scarcely smaller than 0.1 percent, even if the number be so great as 630. Although the distribution of the errors arising from this irregularity of the oscillator will be unsystematic, it will not be Gaussian.

Whence it is concluded that the use of the oscillator introduced irregularities amounting at times to 4 percent in a double observation, and perhaps of 0.1 percent in the definitive value for the velocity of light.

Approximations

Cornu assumed that the terms involving the acceleration α in eq. 24 and 25 would for each order be, for all practical purposes, averaged out of the crossed double observations: $(V_{er}+V_{er-})/2 \equiv V_{er\pm}$ and $(V_{e-r+}+V_{e+r-})/2 \equiv V_{e\mp r\pm}$. He also assumed that the displacement term S, whatever its source, would be eliminated by averaging, without separation, the observations for each direction of rotation of the wheel; even though there might be only one observation for one direction and seven for the other (see table 6).

Although he writes of determining the exact law according to which the wheel turns ("la loi complète du mouvement," pp. A.5, A.68), he in fact merely assumed that the speed was uniformly accelerated throughout the interval used in determining it; and he did this even when the accelerations in adjacent intervals had opposite signs, as in his classes

$$U=(V_{e-}+V_{r+})/2 \equiv V_{e-r+}$$

and

$$u=(V_{e+}+V_{r-})/2 \equiv V_{e+r-}.$$

The assignment of an observation to one of the four groups—$V=V_{er+}$, $v=V_{er-}$, $U=V_{e-r+}$, and $u=V_{e+r-}$—was not determined at the time of observation, but solely from the sign of the acceleration of the wheel, as derived from the chronograph record (see p. A.192), which in half the cases rests upon differences that do not exceed 0.01 second. But his measurements of 10-second intervals (table 7), which, involving intervals of an even number of seconds, are the most favorable he could have chosen, have an average uncertainty of 0.01 second. Hence in half the cases, his classification is open to great doubt.

He recognized some uncertainty, but only when the time-differences did not exceed 0.003 second, and he endeavored to show that it is of little importance. But his argument is scarcely satisfactory, especially when one recalls how few determinations go into any one average.

Neither was the order of the eclipse determined at the time of observation, but $q_n \equiv 2n-1$ was taken as the nearest odd integer to the quotient of $4DNm$ divided by 300,000 km./sec. (see eq. 23). That is entirely satisfactory if n is not great and if m is not seriously in error, but the last is not always true.

CORNU'S TREATMENT AND DISCUSSION OF HIS DATA

Treatment of Data

Cornu averaged all the double observations for a given group and order, irrespective of the direction of rotation of the wheel, and then "crossed" them, thus getting the unrectified values in table 10. (They may be obtained directly from those in table 6.) Assigning to each the weight pq^2, where p is the total number of double observations involved, and $q=2n-1$, n being the order of the eclipse and reappearance, he obtained the indicated weighted mean. For that mean, each of the sums $\Sigma p\epsilon^2$ and $\Sigma pq^2\delta^2$ is a minimum (see eq. 15–17), where ϵ and δ are defined by the expression $\epsilon = q[V - V_n - H/q] = q\delta$, V being the derived value of the velocity, V_n the weighted mean of the observed values for order n, and H is a constant independent of the order. The assumed expression corresponds to case *b* of figure 17 (see Appendix A).

Had his observations completely avoided the critical region, and had they been in each case sufficiently numerous to give a reliable mean, then H would have been zero, and $V_{er\pm}$ would have equaled $V_{e\mp r\pm}$, each being the correct value for the velocity of light. Their difference for each order is given in the table, and is quite significant, averaging 1.3 megam./sec.

As his observations surely lay in the critical region, and as there are uncertainties in the computed speeds and in the group assignment of the double observations, especially of V_{e-r+} and V_{e+r-}, this difference is not surprising. But it does show again the uncertainty that resides in Cornu's values.

It will be noticed that Cornu has carried the values to five apparently significant figures, and the weighted mean to six. But the fourth digit being uncertain by several units, the fifth and sixth have no physical significance.

The preceding refers to Cornu's unrectified values ("valeurs non rectifiées"). He then went over his individual observations, throwing away those that differed much from the mean, and changing other values from one of the four groups to another so as

TABLE 10

CORNU'S AVERAGES (IN AIR) THAT SERVE AS FOUNDATION FOR HIS DEFINITIVE VALUE FOR THE VELOCITY OF LIGHT

$(V+v)/2$ and $(U+u)/2$ are Cornu's notations for the crossed double observations denoted in Appendix A by $V_{er\pm}$ and $V_{e\mp r\pm}$, respectively. Here each value is the mean of all such determinations for the indicated value of $q=2n-1$; p = number of double observations involved. These values have been copied directly from Cornu's memoir, pages A.266 and A.268.

The theory on which he relied, required $V_{er\pm}=V_{e\mp r\pm}$; in the columns "Diff." are given the several differences, $V_{e\mp r\pm}-V_{er\pm}$. At the foot of the table are given certain sums and averages, and Cronu's weighted means.

Unit of $V=1$ megameter/second

		UNRECTIFIED					RECTIFIED				
Wheel	q	$V_{er\pm}$		$V_{e\mp r\pm}$			$V_{er\pm}$		$V_{e\mp r\pm}$		
		$\frac{1}{2}(V+v)$	p	$\frac{1}{2}(U+u)$	p	Diff.	$\frac{1}{2}(V+v)$	p	$\frac{1}{2}(U+u)$	p	Diff.
150	7	299.28	11	301.79	6	+2.51	300.24	9	299.07	5	−1.17
	9	301.26	23	298.78	8	−2.48	300.79	23	299.21	7	−1.58
	11	300.05	22				300.84	20			
	13	302.93	8	300.69	5	−2.24	301.06	6	300.69	5	−0.37
	15	300.65	8				300.65	8			
	17	300.35	10				299.93	8			
	23	300.50	3				300.50	3			
	27	301.35	5	299.12	4	−2.23	301.35	5	299.12	4	−2.23
	31	300.62	4				300.62	4			
	33	299.33	18	301.69	5	+2.36	299.82	17	301.09	6	+1.27
	35	300.47	16	299.31	8	−1.16	300.47	16	299.31	8	−1.16
	37	299.55	5				299.55	7			
	41	299.87	15	300.38	12	+0.51	300.22	18	299.89	9	−0.33
200	15	299.04	6				300.10	7			
	17	299.50	12	297.69	8	−1.81	299.82	12			
	17	300.65	59	299.76	25	−0.89	300.47	56	300.56	27	+0.09
	19	300.29	45	300.14	35	−0.15	300.58	47	299.81	33	−0.77
	21	300.21	40	300.54	33	+0.33	299.94	38	300.57	35	+0.63
180	25	300.65	8	299.85	2	−0.80	300.28	6	300.43	4	+0.15
144	27	300.25	9				300.10	7			
	29	300.27	52	300.44	13	+0.17	300.24	51	300.22	12	−0.02
	31	298.88	3				298.88	3			
Sum		6605.95	382	3900.18	164	17.64	6606.45	371	3599.97	155	9.77
Average		300.27	17.3_6	300.01	12.6_2	1.36	300.29	16.8_6	300.00	12.9_2	0.81
Wt. mean		300.175		300.168			300.225		300.122		

to improve the consistency of the values lying in a given group. Thus he got other means which he called rectified values ("valeurs rectifiées"). These are given in the right-hand half of table 10. The justification of such a manipulation of experimental data is open to question, but it will be noticed that, although some individual values have been changed by more than 1 megam./sec., both the averages and the weighted means have remained essentially unchanged.

Cornu then, in search for the best value to give as definitive, tried other kinds of averaging. The natural one to use, if the precision justifies it, is that (see eq. 101) required on the assumption that the mean value V_n computed for order n is related to the true velocity V in accordance with the equation

$$V_n = V + (H+\epsilon)/q, \tag{26}$$

where H is the same for all orders, ϵ is a fluctuating error, and $q=2n-1$, the α and S terms having been eliminated by averaging. (This H stands for one of the quantities $\pm[\Delta' - (H_e - B_e)]$ and $\pm[\Delta' - (H_r + B_r)]$ of eq. 101, depending upon which of the four classes of observations is being considered.)

But he was not satisfied with that formula. He introduced here, for the first time seriously, the idea that the vibration of the wheel and its mechanism might throw an observation into a critical region, and he undertook to derive an analytical expression to cover that case. His procedure is so queer that a rather detailed consideration of it seems desirable.

Erroneous K-formula

In his attempt to derive an expression for the effect of vibrations of the wheel and its mechanism, he proceeded thus (pp. A.273–A.274): If there were no vibration, the angular velocity, ω, of the wheel when there is an eclipse will, by eq. 74, be $\omega_n = (2n-1)\pi/N_T \equiv q_n\pi/N_T$. If there is vibration, in particular, if there are periodic inequalities in the angular velocity, then the eclipse setting will be $\omega_n + \delta\omega$. If the mechanism is well made, $\delta\omega$ will in general be small and may be developed in a series of ascending powers of ω_n, thus

$$\delta\omega = a\omega_n + b\omega_n^2 + c\omega_n^3 + \cdots \tag{27}$$

There is no constant term, because for $\omega_n = 0$ there is evidently no error. Hence the value of the angular velocity corresponding to the actual setting will be

$$\omega = \omega_n(1 + a + b\omega_n + c\omega_n^2 + \cdots), \tag{28}$$

and the false velocity of light V_f computed from it will be related to the true V as shown in eq. 29

$$V_f = V(1 + a + b\omega_n + c\omega_n^2 + \cdots) \tag{29}$$

or

$$V_f = V\left(1 + a + bq_n\frac{\pi}{N_T} + cq_n^2\left(\frac{\pi}{N_T}\right)^2 + \cdots\right). \tag{30}$$

Since, by hypothesis, $\delta\omega$ arises solely from the mechanism, it is probable that the form of the function representing it is independent of the direction of rotation of the wheel. Consequently, if one averages a pair of observations differing only in the direction of rotation, the average will contain only even powers of ω_n; the a and c terms will not appear. Hence, by averaging determinations that differ only in the direction of rotation of the wheel, one obtains for the mean V_{fm} the value

$$V_{fm} = V\left(1 + bq_n\frac{\pi}{N_T} + \cdots\right) \tag{31}$$

or

$$V_{fm} = V + Kq_n \tag{32}$$

or the more general form

$$V_{fm} = V + Kq_n + \epsilon/q_n. \tag{33}$$

This treatment is unsatisfactory in several respects. First, $\delta\omega$ being entirely unknown, there is no ground for assuming that it can be developed in the series shown in eq. 27; and even if it could be, there is no ground for assuming that the first terms of the series would be the overpowering ones (ω_n is not infinitesimal, but is large). One must first show that $\delta\omega$ can be developed in a convergent series of the form of eq. 27, and then show that in that series the sum of all the terms of higher power is negligible in comparison with the sum of those of power lower than themselves. No attempt was made to prove either of these propositions; it is probable that neither is true. The value of $\delta\omega$ and its dependence on ω_n will surely be much influenced by the various natural periods of the system.

Secondly, averaging two determinations differing only in the direction of rotation will not eliminate the a term, because the quantity that enters into each V_f is not $\delta\omega$ alone, but the absolute value of $\omega_n + \delta\omega$; that is, of eq. 28.

Thirdly, since $\delta\omega$ arises from those irregularities that characterize the critical state, and they vary with the order of the eclipse, the coefficients in eq. 28 are not constants, but are necessarily functions of n. Hence K is also a function of q_n; it is not a constant.

Consequently, one is forced to conclude that Cornu's treatment of this problem reduces to a mere assumption that the effect of vibrations is to increase the observed angular velocity for an eclipse of order n from ω_n to $\omega_n(1+b\omega_n)$, where b is independent of n. No valid grounds were advanced for that assumption; and it seems to be quite wrong.

Cornu's Discussion of Data

In addition to the weighted mean given in table 10, and the values derived from eq. 26 and 33, he tried other kinds of weighting. The more important of the values so derived are given in table 11. They correspond to the following four hypotheses concerning the values of V_n ($= V_{er\pm}$, or $V_{e\mp r\pm}$).

I. V_n contains no systematic error: $V_n = V + (\epsilon/q)_n$.

II. V_n contains a systematic error arising from the speed of the wheel having been $\omega_n + C$ instead of ω_n, where C has the same value for every order: $V_n = V + (H+\epsilon_n)/q_n$ (see eq. 26).

III. V_n contains a systematic error arising from the speed of the wheel having been $\omega_n + b\omega_n^2$ instead of ω_n: $V_n = V + Kq + \epsilon_n/q_n$ (see eq. 33).

IV. V_n contains the same type of systematic error

TABLE 11

CORNU'S SEVERAL COMPUTED VALUES FOR THE VELOCITY (V) OF LIGHT IN AIR

Each value has been derived, by means of the indicated equation, from the average values of $(V_n + v_n)/2 \equiv V_{er\pm}$ or of $(U+u)/2 \equiv V_{e\mp r\pm}$ that are given in table 10, the values of the appropriate constants (V, H, K) being so determined as to make the sum ($\Sigma\epsilon_n^2$) of the squares of the fluctuating errors a minimum. The four equations are as follows:

(I) $V_n = V + \epsilon_n/q_n$ (II) $V_n = V + (H+\epsilon_n)/q_n$ (III) $V_n = V + Kq_n + \epsilon_n/q_n$ (IV) $V_n = V + Kq_n + \epsilon_n$

where V_n is the observed value $V_{er\pm}$ or $V_{e\mp r\pm}$ for order n, V is the computed value (V_v or V_u), and $q_n = 2n-1$. The residual (δ) equals ϵ_n/q_n for I, II, and III, and equals ϵ_n for IV: hence Cornu's $m^2 \equiv \Sigma pq^2\delta^2/\Sigma pq^2$ for I, II, and III is equivalent to $\Sigma p\epsilon^2/\Sigma pq^2$; p is the number of double determinations involved in V_n.

Except as indicated, all values have been copied directly from Cornu's paper (pages A.270–A.286).

Unit of V, H, and K = 1 megameter/second

No.	Eq.	$(V_n+v_n)/2 \equiv V_{er\pm}$				$(U_n+u_n)/2 \equiv V_{e\mp r\pm}$					No.
		V_v	H	1000 K	m^2	V_u	H	1000 K	m^2	$(V_v+V_u)/2$	
		Unrectified: All wheels									
1	I	300.175			0.2625	300.168			0.5004	[a]300.172	1
2	II	299.714	+11.5		0.23025	300.512	−8.6		0.4864	[a]300.113	2
3	III	301.031		−30.692	0.2159	299.764		+14.366	0.4838	300.398	3
4	IV	300.795		−21.99		299.779		+13.57		300.287	4
		Unrectified: 150-tooth wheel									
5	I	300.066			0.4267	300.196			0.6519	[b]300.131	5
6	II	299.635	+12.31		0.3674	300.284	−2.87		0.6497	[a]299.960	6
7	III	301.182		−33.81	0.3418	299.612		+16.18	0.6391	300.397	7
8	IV	300.930		−25.42		300.175		− 0.55		[c]300.558	8
		Unrectified: 200-tooth wheel									
9	I	300.296				300.069				300.183	9
		Rectified: All wheels									
10	I	300.225			0.1310	300.122			0.2797	[a]300.174	10
11	II	299.921	+ 7.7		0.1167	299.872	+6.2		0.2714	[a]299.896	11
12	III	300.593		−12.394	0.1148	300.601		+17.375	0.2810	[c]300.597	12
13	IV	300.742		−18.48		300.095		+ 2.43		300.419	13
		Rectified: 150-tooth wheel									
14	I	300.227			0.1720	299.890			0.4274	[c]300.058	14
15	II	299.951	+ 8.17		0.14	299.979	−2.88		0.4253	[a]299.965	15
16	III	300.886		−19.52	0.1462	299.838		+ 1.47	0.3923	[c]300.362	16
17	IV	300.885		−19.37		299.403		+14.36		300.144	17
		Rectified: 200-tooth wheel									
18	I	300.290				300.310				300.300	18

[a] Cornu does not give these means.

[b] Cornu gives 300.181, apparently a misprint.

[c] For these, Cornu gives the following slightly different values: 300.560, 300,596, 300.059, and 300.357.

as III, but the fluctuating error in the speed is proportional to the speed, $\epsilon = \epsilon'\omega_n$: $V_n = V + Kq_n + \epsilon_n'$. In all cases the adjustment is such as to minimize the sum $\Sigma\epsilon^2$.

It has just been shown that the formulas with the K coefficient are unjustified; and of the last hypothesis (IV) Cornu did no more than to suggest that it might be well to try it.

CORNU'S DEFINITIVE VALUE

Cornu derived his definitive value in this manner. From the values in table 11 he chose nos. 3 (300.398), 12 (300.597), and 9 (300.183), nos. 3 and 12 depending on his erroneous formula (III) and being derived from the mean values for all the wheels, no. 3 being for the unrectified values, and no. 12 for the rectified. On the other hand, no. 9 depends on formula I and the unrectified values for a single wheel, the 200-tooth one. He averaged the mean of nos. 3 and 12 (300.497) and no. 9, getting 300.340 megam./sec. for his definitive value for the velocity of light in air. Adding to this 0.080, to correct to vacuum, and rounding to four digits, he gets

Velocity of light in vacuum = 300.4 megameters per second

But from the 36 values in table 11, why did he choose those particular three? And why did he combine them in that particular way, giving to no. 9 twice the weight he gave to each of the other two?

No satisfactory answer has been found. But it seems probable that he was in part influenced by the uniformity of the following means of means, which he gives at various places (pp. A.279–A.286). The primary means may be found in table 11, as indicated by the attached number.

		Mean
Unrectified, all wheels, eq. III, No. 3	300.398	
Rectified, all wheels, eq. III, No. 12	300.597	
		300.497
Unrectified, 150-tooth wheel, eq. III, No. 7	300.397	
Rectified, 150-tooth wheel, eq. III, No. 16	300.357[24]	
		300.377
Unrectified, all wheels, eq. IV, No. 4	300.287	
Rectified, all wheels, eq. IV, No. 13	300.419	
		300.353
Unrectified, 150-tooth wheel, eq. IV, No. 8	300.558	
Rectified, 150-tooth wheel, eq. IV, No. 17	300.144	
		300.351

But why did he average the rectified and the unrectified? Averaging the bad with the better does not improve the better. But without such averaging there is no such uniformity.

Furthermore, values based on formula II, which has a firm theoretical foundation, have been completely ignored, not only in these means of means, but also in the derivation of his definitive values. Those values for the mean of V_v and V_u range from 299.9 to 300.1, a smaller range than that for any of the other formulas. Why were values based on formula II ignored?

[24] So given by him; proper average is 300.362.

The values and equations given in table 12, and Cornu's remarks about them, are of some interest in connection with that question. Notice first the equations, which are not given in Cornu's memoir. It will be seen that every value of H and of K depends upon the difference of two M's, the coefficients of both H and K being exact numbers that may validly be carried out to as many digits as may be desired. Those differences for equations II and III, which are the only ones of special interest, are as given below.

	Unrectified		*Rectified*	
Formula II . . . . H	0.138	0.025	0.084	0.024
Formula III . . . . K	0.075	0.022	0.039	0.002

It will be noticed that in every case the difference that determines H exceeds the one that determines K. Nevertheless, after having given the solution for II, but not the equations, Cornu remarked that the values so determined for V and H are very uncertain, and that the mean squared deviation is almost the same as for the simple weighted mean (formula I). Whence he concluded that there is very little chance of formula II representing the systematic error.[25] Whereas after having given the solutions for III, but not the equations, he remarked that the values for V and K are well determined, but the mean squared deviation is only slightly smaller than that for the simple weighted mean (formula I); and that the mean of the two V's for the unrectified values is almost the same as that for the rectified, and each is much greater than the simple weighted mean.[26]

It is not at all evident why he should have considered that the equations for determining K are any more satisfactory and compatible than those for H; indeed, the reverse seems to be true. Nor is it evident how he can derive much more satisfaction from a comparison of the values of the mean squared deviations for III with those for I, than from a similar comparison of II with I (see table 11, columns m^2).

There is no evident justification for those contrasting statements tending to discredit formula II as compared with formula III.

[25] "Comme dans le cas de l'étude de l'ensemble des résultats (278), les systèmes d'équations sont presque incompatibles, de sorte que les paramètres H et V sont très-incertains.

"En outre, le carré moyen de l'écart avec la formula est resté, à fort peu près, le même que celui qui correspond à la moyenne principale; cette forme d'équations a donc très-peu de chances de représenter l'erreur systématique . . . " (p. A. 283).

[26] "Dans le cas présent, les coefficients des équations de condition sont bien déterminés; mais la valeur du carré de l'écarte moyen avec la formula n'a que peu diminué sur celle relative à la moyenne principale. On remarquera que les moyennes des paramètres V, qui sont censés affranchis de l'erreur systématique, sont presque identiques et plus fortes que les moyennes principales M_2" (p. A. 284).

TABLE 12

CONSTANTS AND COMPUTATIONS FOR CORNU'S 150-TOOTH WHEEL

The values of the constants, Σp to M_3, have been copied directly from Cornu's memoir. The assumed equations II, III, and IV, connecting V_n and V, are those similarly numbered in table 11, and the values of V and either H or K have been obtained by solving the appropriate pair of the following equations:

$$\begin{array}{lll} \text{II.} & V+H(\Sigma pq)/(\Sigma pq^2)=M_2; & V+H(\Sigma p)/(\Sigma pq)=M_1 \\ \text{III.} & V+K(\Sigma pq^3)/(\Sigma pq^2)=M_2; & V+K(\Sigma pq^4)/(\Sigma pq^3)=M_3 \\ \text{IV.} & V+K(\Sigma pq)/(\Sigma p)=M_0; & V+K(\Sigma pq^2)/(\Sigma pq)=M_1 \end{array}$$

where $M_0=(\Sigma pV_n)/(\Sigma p)$, $M_1=(\Sigma pqV_n)/(\Sigma pq)$, $M_2=(\Sigma pq^2V_n)/(\Sigma pq^2)$, $M_3=(\Sigma pq^3V_n)/(\Sigma pq^3)$, V_n is the observed average value of $V_{er\pm}$ or of $V_{e\mp r\pm}$ (in Cornu's notation, of $(V+v)/2$ or of $(U+u)/2$) for order n, and p is the sum of the number of double observations involved in V_{er+} and V_{er-}, or in V_{e-r+} and V_{e+r-}, belonging to a given value of $q=2n-1$.

Unit of $V=1$ megameter/second, in air

	UNRECTIFIED		RECTIFIED	
	$(V+v)/2=V_{er\pm}$	$(U+u)/2=V_{e\mp r\pm}$	$(V+v)/2=V_{er\pm}$	$(U+u)/2=V_{e\mp r\pm}$
Σp	148	48	144	44
Σpq	3 202	1 224	3 270	1 118
Σpq^2	91 444	40 120	96 880	36 036
Σpq^3	3 017 554	1 447 344	3 272 118	1 275 446
Σpq^4	106 272 724	54 179 200	117.054 160	46 878 876
M_0	300.380	300.175	300.445	299.768
M_1	300.204	300.171	300.311	299.866
M_2	300.066	300.196	300.227	299.890
M_3	299.991	300.218	300.188	299.892
Eq. II	$V+0.03501\ H=300.066$	$V+0.03051\ H=300.196$	$V+0.03375\ H=300.277$	$V+0.03102\ H=299.890$
	$V+0.04622\ H=300.204$	$V+0.03922\ H=300.171$	$V+0.04404\ H=300.311$	$V+0.03936\ H=299.866$
	$0.01121\ H=\ 0.138$	$-0.00871\ H=\ 0.025$	$0.01029\ H=\ 0.084$	$-0.00834\ H=\ 0.024$
	$\therefore H=+12.31,\ V=299.635$	$\therefore H=-2.87,\ V=300.284$	$\therefore H=+8.17,\ V=299.951$	$\therefore H=-2.88,\ V=299.979$
Eq. III	$V+32.999\ K=300.066$	$V+36.0754\ K=300.196$	$V+33.775\ K=300.227$	$V+35.394\ K=299.890$
	$V+35.218\ K=299.991$	$V+37.4335\ K=300.218$	$V+35.773\ K=300.188$	$V+36.755\ K=299.892$
	$-2.219\ K=\ 0.075$	$1.3581\ K=\ 0.022$	$-1.998\ K=\ 0.039$	$1.361\ K=\ 0.002$
	$\therefore K=-0.0338,\ V=301.182$	$\therefore K=+0.01620,\ V=299.612$	$\therefore K=-0.01952,\ V=300.886$	$\therefore K=+0.00147,\ V=299.838$
Eq. IV	$V+21.635\ K=300.380$	$V+25.500\ K=300.175$	$V+22.709\ K=300.445$	$V+25.409\ K=299.768$
	$V+28.558\ K=300.204$	$V+32.778\ K=300.171$	$V+29.627\ K=300.311$	$V+32.233\ K=299.866$
	$-6.923\ K=\ 0.176$	$-7.278\ K=\ 0.004$	$-6.918\ K=\ 0.134$	$6.824\ K=\ 0.098$
	$\therefore K=-0.02542,\ V=300.930$	$\therefore K=-0.00055,\ V=$[a]$300.189$	$\therefore K=-0.01937,\ V=300.885$	$\therefore K=+0.01436,\ V=299.403$

[a] Cornu gives this as 300.175.

But it is certainly true that the actual values so determined for H, and still more so for K, are of little value, resting as they do on differences between the M's that exceed a tenth of a unit in only a single case, and that for H. Since the M's involve V_n as a factor, and the tenth's place of V_n is uncertain, that place is uncertain in the M's also; the two following places given by Cornu are of no physical significance whatever.

Hence, if Cornu regarded all four formulas represented in table 11 as potentially valid, the only conclusion that he would have been justified in drawing from those results is that the value of the velocity of light in air probably lies somewhere around, probably within, the range of the extreme values (299.6 and 301.2), and that differences of that amount might arise from the presence of systematic errors that are so small as to be almost completely obscured by the fluctuating errors that affect the observations.

But he, like many others, seems to have failed to realize that differences between values that have each been derived by a potentially valid type of combination of the individual observations, and that each yield essentially the same mean squared deviation, are themselves of no significance whatever. They do no more than show that the several derived values are all inherently uncertain by an amount that is at least comparable to those differences. Of course invalid combinatory procedures should never be used.

Coming back to the question of why he discarded the potentially valid H-formula in favor of the K-formula derived late in the reduction of the observations, and which has here been shown to rest on unsupported, and probably false, assumptions, could one not consider it possible that he was influenced (1) by his belief—and that of his contemporary astronomers—that the velocity was greater than 300, and (2) by the fact that the value 300.33, in air, was given in his preliminary report of December 1874? On finding that the weighted mean of all his values (300.17, see table 11, nos. 1 and 10) fell below that, he sought for an overlooked systematic error that would account for it. The H-formula led to a still smaller value, the change, however, being of no real significance. The effect of the vibrations of the wheel and its mechanism occurred to him as a possible explanation. And for that he derived the K-formula, which, as he explicitly stated[27] in a quotation already given, leads to values much higher than the weighted mean. This met his requirement; and that seemed to him to show that the application of the K-formula was not only justified, but necessary.

Having decided that the K-formula properly applies, he found (table 11) that when the data for all the wheels were used he obtained with the unrectified values 300.398 (no. 3) and with the rectified 300.597 (no. 12). Each is higher than the value given in his preliminary report. Their mean is 300.497. But for the 200-tooth wheel the number of double observations for a given class and order exceeded, in general, that for any of the other wheels, hence the observations with it should be given more weight than is given to the others (actually that was done in the computations involving all wheels, but overlook that for the present). Unfortunately, the observations with the 200-tooth wheel were concentrated in the middle orders and so did not lend themselves readily to the K-formula. Consequently, their simple weighted mean was taken. For the unrectified values, this gave 300.183 (no. 9), which, when averaged with the 300.497, gave 300.340, practically the same (300.33) as was given in his preliminary report. That value was accepted as his definitive one.

If such were the mental processes by which he was led to the chosen definitive value, this instance is an excellent illustration of how an experimenter may be influenced by his preconceived opinions. Such bias is always to be feared, and is rarely entirely absent when an experimenter has a preconceived opinion about what he should find.

However that may be, since Cornu's definitive value rests on a repudiation of the potentially valid H-formula and on an acceptance of the invalid K one, it is necessary to discard it. His definitive value is entirely untrustworthy.

NEW DEFINITIVE VALUE

Since Cornu's definitive value is untrustworthy, it becomes necessary to derive another from his observations.

If none of his observations had lain in the critical region, and if errors introduced by the irregular functioning of the oscillator had introduced no systematic error, then the crossed double observations would probably have satisfied formula I of table 11. As those conditions were not fulfilled, it is not possible to forecast whether formula I should be fulfilled or not, but it is more likely that the data should satisfy formula II (see Appendix A), especially those for the uncrossed double observations.

From table 11 it will be seen that the extreme values found by Cornu from the crossed double observations, when adjusted to formulas I and II, are 299.6 and 300.5 megam./sec., in air; and the extremes of the corresponding values of $(V_v+V_u)/2$ are 299.9 and 300.3 (for formula I, 300.1 and 300.3; for formula II, 299.9 and 300.1). Although m^2 is always smaller for formula II than for formula I, the difference is so slight that one might doubt whether it is of physical significance.

It appeared that some light might be thrown on that question by a study of the results of similar least-squares adjustments of the simple double observations, both when all the observations are used and when one uses only those orders for which there are at least four or five double observations; similarly for the crossed double observations, and for observations of all types treated collectively without distinction of type. Such adjustments, except for the crossed observations, $V_{e\mp r\pm}$, have been made for the 150-tooth wheel, for which the observations extend over the widest range of orders; and the collective treatment has been applied also to all observations for all wheels. (The observations for $V_{e\mp r\pm}$ being relatively few and unsatisfactory, it seemed profitless to consider them in this study.) Furthermore, the values of the root-mean-square errors and deviations (ϵ_{m2} and δ_{m2}) have been computed for each of two cases: (1) when the mean for the order is regarded as a determination of weight p, p being the number of double observations in the mean, and (2) when each double observation is treated individually.

The results of these computations are given in table 13. For each kind of combination of the data, either two or four sets of computed values are given. In every case the first set corresponds to all the available observations, and the last to only those for which there are at least five (in two cases four) double observations of the same type and order. The values of V for these last, being considered the better, have been placed in a separate column, but the average value is essentially the same as when all the values are used.

On examining the values of ϵ_{m2} and of δ_{m2}, it will be seen that in only a single case (V_{e+r-}, $\Sigma p_n=19$) does the value for formula II ($H\neq 0$) exceed that for formula I; and for the separately considered types of uncrossed observations they are, with the same exception, all smaller for formula II than for formula I. The difference is always small, so small that, if one were concerned with a single instance, it would be of doubtful significance, but when it runs consistently in one direction, as here, it is highly probable that the sign of the difference is significant. It may, therefore, be concluded that formula II is preferable to formula I.

But even when the means by orders are treated as though they were suitably weighted individual values,

[27] ". . . Les moyennes des paramètres V . . . sont . . . plus fortes que les moyennes principales M_2" (p. A.284).

TABLE 13

RESULTS OF NEW ADJUSTMENTS OF CORNU'S OBSERVATIONS

Adjustments are made to one or other of the formulas (I) $V_n = V + \epsilon_n/q_n$ and (II) $V_n = V + (H + \epsilon_n)/q_n$, where V_n is the value for a double observation of order n, ϵ_n is an erratic fluctuating error in the speed, $q_n = 2n - 1$, and H is a constant independent of n; $\delta_n = \epsilon_n/q_n$; $(\epsilon_{m2})^2 = (\Sigma p_n \epsilon_n^2)/(\Sigma p_n)$, and $(\delta_{m2})^2 = (\Sigma p_n q_n^2 \delta_n^2)/(\Sigma p_n q_n^2)$ when V and H have been so chosen as to make them each a minimum. Two sets of values of ϵ_{m2} and of δ_{m2} have been computed; for one the observations have been considered individually, for the other the mean for each order has been considered as a single observation of weight p, p = number of double observations involved in the mean. In the column "wheel," the "150" indicates that only observations with the 150-tooth wheel were used.

Unit of V = 1 megameter/second in air

Group	Wheel	Σp_n	p_n		Formula II									Formula I					
								Means			Individuals					Means		Individuals	
			Min.	Max.	V	V	H	ϵ_{m2}	ϵ_{m2}/H	δ_{m2}	ϵ_{m2}	ϵ_{m2}/H	δ_{m2}	V	V	ϵ_{m2}	δ_{m2}	ϵ_{m2}	δ_{m2}
V_{er+}	150	78	2	13	299.4		46	30	0.65	1.26	63	1.4	2.7	301.1		38	1.6	67	2.8
V_{er+}	150	62	5	13		299.2	45	23	0.51	1.02	59	1.3	2.6		300.9	33	1.5	64	2.9
V_{er-}	150	70	1	11	300.3		−28	28	1.00	1.09	74	2.6	2.8	299.4		31	1.2	75	2.9
V_{er-}	150	50	5	11		300.3	−26	22	0.85	0.78	78	3.0	2.8		299.5	25	0.9	79	2.8
V_{e-r+}	150	29	1	5	300.0		16	34	2.1	1.24	48	3.0	1.7	300.6		35	1.3	48	1.8
V_{e-r+}	150	18	4	5		300.8	−10	14	1.4	0.49	41	4.1	1.4		300.5	15	0.5	41	1.4
V_{e+r-}	150	29	1	7	299.3		14	23	1.6	0.83	46	3.3	1.6	299.8		24	0.9	47	1.6
V_{e+r-}	150	19	4	7		299.4	18	25	1.4	0.84	59	3.3	1.9		300.0	21	0.7	57	1.9
$V_{er\pm}$	150	148	3	[a]23	299.6		12	15	1.2	0.60				300.1		16	0.6		
$V_{er\pm}$	150	148	1	[b]11	299.7		12	15	1.2	0.60				300.1		16	0.6		
$V_{cr\pm}$	150	104	5	[b]11		299.7	9	11	1.2	0.43					300.0	12	0.5		
V_{er+}, V_{er-}, V_{e-r+}, V_{e+r-}	150	203	4	31	299.7		14				68	4.9	2.6	300.2				68	2.6
V_{er+}, V_{er-}, V_{e-r+}, V_{e+r-}	150	144	17	31		299.8	8				68	8.5	2.5		300.0			68	2.5
Everything...	All	609	4	88	299.4		17				69	4.1	2.9	300.0				69	2.9
Everything...	All	382	3	[a]59	299.7		12	12	1.0	0.50				300.2		12	0.5		
Everything...	All	382	1	[b]29	299.8		11	10	0.9	0.45				300.2		11	0.5		
Everything...	All	312	5	[b]29		299.8	10	8	0.8	0.34					300.2	9	0.4		
Mean of all					299.6_9	299.8_6								300.1_7	300.1_6				
Mean of last 6 lines					299.6_8	299.8_0								300.1_5	300.1_0				

[a] These are the weights Cornu assigned to the values of $V_{er\pm}$ corresponding to the several orders; each is the sum of the number of determinations in the groups V_{er+} and V_{er-}.

[b] These are the numbers of determinations in the smaller of the two groups V_{er+} and V_{er-}, and are the weights here given to the corresponding values of $V_{er\pm}$.

the error ϵ_{m2} essentially equals or exceeds H. Consequently, the actual value of H cannot be satisfactorily determined.

The best value one can derive from the observations seems, from these data, to be 299.8 megam./sec. in air, with a possible range of ±0.2.

But in the study of the data in table 5 it was found that the discordance dubiety is at least 0.6 megam./sec. The irregularity of the oscillator introduces uncertainties also, but they may perhaps have been sufficiently included in the discordance dubiety. Hence, taking 0.080 for the correction to vacuum, one may conclude that the

Velocity of light in a vacuum = 299.9 megameters per second
Dubiety at least........... = ±0.6 " " "

That is, the velocity of light in a vacuum seems to lie near or within the range 299.3 to 300.5 megam./sec.

If the computed value is not seriously affected by systematic errors, then the velocity probably lies nearer 300 than either 299 or 301, but the data do not justify one in assuming the absence of such errors.

PERROTIN AND PRIM'S REPORT OF 1908

SUMMARY

In 1908, thirty-two years after the publication of Cornu's memoir, the report of the determination of the velocity of light by Joseph Perrotin (1845–1904), director of the observatory at Nice, and his assistant, Prim, was published in the *Annales* of that observatory [18]. Fizeau's method[28] was used. Cornu was much interested in the work and constantly advised regarding it.[29] The work extended from 1898 to 1902, the year

[28] For a discussion of the method, correction terms, and errors, see Appendix A.

[29] "J'ajoute immédiatement que nous avons eu la bonne fortune d'entreprendre et de poursuivre ce travail sous la haute direction du savant Physicien [Cornu] qui dès le début, n'a pas hésité à venir installer lui-même sur le Mont-Gros les appareils dont il s'était servi lors des expériences que je viens de rappeler et qu'il n'a cessé de nous prodiguer ses plus précieux conseils" [19].

"Cornu, qui s'intéressait vivement à nos opérations et fut constamment pour nous un conseil éclairé" [18, p. A.6]. See also [20].

of Cornu's death. Perrotin died in 1904, before the final calculations had been completed.

Observations were made over two paths: Nice to the village of La Gaude, a distance of 11.8622 km.; and Nice to Mont Vinaigre, a distance of 45.9507 km. For the shorter path, the sending lens had a diameter of 16.0 cm. and a focal length of 215 cm., and the collimator lens had a diameter of 6.5 cm. and focal length of 80 cm. For the longer distance the corresponding quantities were 76 cm., focus 18 meters, for the sending lens; and 38 cm., focus 7 meters, for the collimator lens. Hence, the diameter of the utilized portion of the light focused on the wheel was (see eq. 72) 0.011 mm. for the LaGaude series, and 0.149 mm. for the Mont Vinaigre. Exclusive of diffraction effects, these were also the diameters of the returned stars; and they are twice the distances that a point of light could have been displaced from the line joining the centers of the lenses and still have its light returned by the collimator.

The authors refer readers to Cornu's memoir [15] for everything concerning the description and adjustment of the apparatus, the discussion of most formulas and of sources of errors, and of the means for eliminating the effects of those errors. Cornu's illuminator, mechanism for driving the toothed wheel, chronograph, subsidiary pendulums for subdividing the time (including his 1/20-second oscillator), and microscope with variable magnification, were all used in this work. Only one toothed wheel was used. It was of aluminium, 35.5 mm. in diameter, 0.8 mm. thick, and had 150 teeth, each an isosceles triangle of 0.7-mm. base and 2-mm. height. It seems to have differed from Cornu's 150-tooth wheel only in the thickness of the metal. Hence, the distance (see table 4) between centers of adjacent teeth was about 0.74 mm., width of interdental gap = 0.37 mm., which is only about 2½ times the diameter of the effective source of light for the longer path.

Although nothing is said about smoking the wheel, it seems likely that it was smoked in the same manner as were Cornu's; and consequently, as in Cornu's work, it is quite likely that the observations lay in the critical region, so that a complete automatic elimination of the various errors considered by Cornu was impossible except as the result of fortunate chance. The subject is not discussed.

Nothing is said about the equality, or inequality, of successive intervals between the impulses delivered by the master clock to the intermediate pendulum, and there is no indication that the functioning of the 1/20-second oscillator was studied. Such silence implies that Cornu's erroneous opinions of thirty years before were accepted; if so, serious errors may exist in the timings, as has been shown in the discussion of Cornu's work.

Although systematic errors and their elimination are mentioned, there is nothing in the report to show that they received any experimental study. Supreme reliance seems to have been placed on Cornu's theoretical discussion of the obvious errors, which discussion rested upon the assumption, probably invalid, that the observations did not lie in critical regions. They did not use, or even mention, Cornu's erroneous K-formula. But they did derive certain new formulas that are in error. Those formulas will be discussed in the proper place.

They reduced their observations in two ways: (1) by so averaging as to eliminate all except the small L-terms (eq. 102), and (2) by a least-squares solution of their equivalents of equations 101. The averages used were $V_e\pm$ and $V_r\pm$, which are superior to Cornu's double observations, which have to be "crossed."

The results obtained in the first way were discarded as unsatisfactory. Those obtained over the shorter path, the La Gaude ones, were thought to be too few to justify a least-squares treatment; and so were completely discarded. Their definitive value rests solely on the Mont Vinaigre determinations, as reduced by a least-squares solution of their equivalents of equations 101. Their definitive value for the velocity of light in a vacuum is given as 299901±84 km./sec. Since the formulas used in the computation were erroneous, as will presently be shown, this value is totally unworthy of confidence.

DATA AND DISCUSSION

Even a casual examination of the data shows that the dubiety of their value is much greater than that indicated by the ±84. Illustrative determinations by Perrotin for two orders are given in table 14; values, by orders, published before the work was completed, are compared with the corresponding final ones in table 15; and all the final means by orders, and combinations of those means, are given in table 16.

It will be noticed in table 14 that the mean deviation of the individual determinations from the mean of a group of one kind averages over 2 megam./sec. If all the observations in a group corresponded to the same nominal experimental conditions, and if groups of 25 had been used in the search for systematic error, then, with a mean deviation of that amount, the dubiety of any result derived from them would, on account of this discordance alone, amount to at least 0.7 megam./sec. (eq. 20). But the report states that observations were taken for each direction of rotation of the wheel, and it does not indicate which is which. If the two sets differed significantly, then the dubiety just stated would be too great. Actually, the two sets should have agreed very closely, if the apparatus was properly adjusted; but there is nothing in the report that will enable one to tell whether they did or not.

But in table 14 it will be seen that for order 15 the mean value of $V_e\pm$ derived from the column exceeds

TABLE 14

ILLUSTRATIVE DETERMINATIONS BY PERROTIN

The values for the velocity of light in air, V_{e+}, V_{e-}, V_{r+}, V_{r-}, have been taken directly from Perrotin and Prim's Tableau II for the Mont Vinaigre observations and deviations have been determined from the values and averages as given in that table. The subscript e indicates that the value refers to an eclipse, r to a reappearance; the sign + or − following an e or r indicates the sign of the acceleration of the wheel. If V_{ne+} denotes the mean of the V_{e+} values for order n, then $\delta_{e+} = V_{e+} - V_{ne+}$; similarly for the others.
Values for each direction of rotation of the wheel are included in each V-column, but there is no notation for distinguishing between them.

Unit of V and δ = 1 megameter/second

n	V_{e+}	V_{e-}	$V_{e\pm}$	δ_{e+}	δ_{e-}	V_{r+}	V_{r-}	$V_{r\pm}$	δ_{r+}	δ_{r-}
15	303.89	295.22	299.56	+3.10	−3.41	315.31	287.25	301.28	+6.28	−3.96
	295.72	297.86	296.79	−5.07	−0.77	300.05	290.97	295,51	−8.98	−0.24
	305.56	296.86	301.21	+4.77	−1.77	310.49	291.66	301.08	+1.46	+0.45
	300.94	298.88	299.91	+0.15	+0.25	307.62	294.01	300.82	−1.41	+2.80
	301.03	300.78	300.90	+0.24	+2.15	310.23	291.43	300.83	+1.20	+0.22
	303.27	297.36	300.32	+2.48	−1.27	311.82	289.36	300.59	+2.79	−1.85
	299.30	302.78	301.04	−1.49	+4.15	306.07	289.81	297.94	−2.96	−1.40
	300.54	300.04	300.29	−0.25	+1.41	305.56	284.40	294.98	−3.47	−6.81
	305.82	307.87	306.84	+5.03	+9.24	311.55	304.29	307.92	+2.52	+13.08
	298.57	299.55	299.06	−2.22	+0.92	306.36	285.73	296.04	−2.67	−5.48
	294.01	297.11	295.56	−6.78	−1.52	314.23	296.15	305.19	+5.20	+4.94
		300.78			+2.15		291.20			−0.01
		299.30			+0.67		294.48			+3.27
		294.93			−3.70		290.49			−0.72
		302.27			+3.64		298.32			+7.11
		298.00			−0.63		291.09			−0.12
		298.09			−0.54		293.30			+2.09
		300.54			+1.91		288.21			−3.00
		297.39			−1.24		293.06			+1.85
		296.39			−2.24		291.20			−0.01
		299.30			+0.67		290.50			−0.71
		297.11			−1.52		290.97			−0.24
		300.54			+1.91		284.62			−6.59
		300.54			+1.91		292.12			+0.91
		299.06			+0.43		290.96			−0.25
		294.01			−4.62		285.06			−6.15
		298.09			−0.54		290.50			−0.71
		296.70			−1.93		294.00			+2.79
		292.83			−5.80		289.95			−1.26
Mean	300.79	298.63	300.13	2.87	2.17	309.03	291.21	300.20	3.54	2.72
	$V_{e\pm}$ = 299.71					$V_{r\pm}$ = 300.12				
16	303.28	299.64	301.46	+1.63	+1.05	310.24	292.72	301.48	+1.89	+1.03
	299.46	297.31	298.38	−2.19	−1.28	304.72	289.99	297.36	−3.63	−1.70
	300.86	302.55	301.70	−0.79	+3.96	309.19	292.97	301.08	+0.84	+1.28
	301.47	298.09	299.78	−0.18	−0.50	306.90	288.76	297.83	−1.45	−2.93
	304.98	296.00	300.49	+3.33	−2.59	306.35	289.74	298.04	−2.00	−1.95
	302.29	297.81	300.05	+0.64	−0.78	307.63	292.47	300.05	−0.72	+0.78
	302.01	298.32	300.16	+0.36	−0.27	305.53	294.72	300.12	−2.82	+3.03
	300.63	299.36	300.00	−1.02	+0.77	308.85	292.95	300.90	+0.50	+1.26
	300.69	299.90	300.30	−0.96	+1.31	303.09	287.77	295.43	−5.26	−3.92
	303.08	298.84	300.96	+1.43	+0.25	308.55	294.47	301.51	+0.20	+2.78
	299.85	296.25	298.05	−1.80	−2.34	309.70	289.26	299.48	+1.35	−2.43
	302.53	299.62	301.08	+0.88	+1.03	306.62	292.68	299.65	−1.73	+0.99
	303.08	302.02	302.55	+1.43	+3.43	308.55	297.54	303.04	+0.20	+5.85
	300.95	296.75	298.85	−0.70	−1.84	308.55	290.71	299.63	+0.20	−0.98
	303.62	296.24	299.93	+1.97	−2.35	313.93	290.21	302.07	+5.58	−1.48
	300.68	297.02	298.85	−0.97	−1.57	307.16	288.99	298.08	−1.19	−2.70
	300.42	302.54	301.48	−1.23	+3.95	308.20	292.70	300.45	−0.15	+1.01
	299.89	301.74	300.82	−1.76	+3.15	316.55	290.95	303.75	+8.20	−0.74
		293.58			−5.01		291.70			+0.01
		300.68			+2.09		294.47			+2.78
		301.84			+3.25		296.33			+4.64
		299.63			+1.04		294.21			+2.52
		294.97			−3.62		288.75			−2.94
		295.48			−3.11		285.61			−6.08
Mean	301.65	298.59	300.27	1.29	2.11	308.35	291.69	300.00	2.11	2.43
	$V_{e\pm}$ = 300.12					$V_{r\pm}$ = 300.02				

TABLE 15

Comparison of Perrotin and Prim's Preliminary and Final Values for the Velocity of Light

The preliminary values [19, 20] are referred to a vacuum; the final ones [18] to air. To the precision of the values here given, the difference is without significance. In no case does a final value rest on more than twice as many observations as does the preliminary one.

Unit of V = 1 megameter/second

La Gaude

	Perrotin			Prim		
n	Prelim. Vac.	Final Air	Diff. P−F	Prelim. Vac.	Final Air	Diff. P−F
4				298.24	299.22	−0.98
5	300.13	300.46	−0.33	297.88	297.56	+0.32
6				300.56	301.14	−0.58
7	300.02	300.55	−0.53	299.90	300.01	−0.11
8	300.09	300.74	−0.65	299.91	299.89	+0.02
9	299.79	299.60	+0.19	299.88	299.80	+0.08
10	299.35	298.84	+0.51	299.93	299.59	+0.34
		Mean	0.44		Mean	0.35

Mont Vinaigre: Perrotin

n	Prelim. Vac.	Final Air	Diff. P−F	n	Prelim. Vac.	Final Air	Diff. P−F
16	300.52	300.07	+0.45	25	300.03	299.96	+0.07
17	299.72	299.40	+0.32	26	299.89	299.95	−0.06
18	299.60	299.95	−0.35	27	300.24	299.65	+0.59
19	300.31	299.77	+0.54	28	299.72	299.57	+0.15
20	300.13	299.96	+0.17	29	300.38	300.13	+0.25
21	299.55	299.22	+0.33	30	300.52	300.11	+0.41
22	299.88	299.97	−0.09	31	299.73	299.83	−0.10
23	299.58	299.43	+0.15	32	299.50	299.56	−0.06
24	299.86	299.51	+0.35				
						Mean	0.26

that derived from the means of the columns V_{e+} and V_{e-} by 0.4 megam./sec.; and similar, but smaller, differences exist for other such combinations. And in tables 15 and 16 it will be seen that nominally equivalent means differ on the average by over 0.3 megam./sec. Which again shows that it would have been practically impossible for Perrotin and Prim to have experimentally detected with certainty a systematic error that did not exceed 0.6 megam./sec.

When all this is taken into consideration, and when it is noticed that deviations of several megameters per second are not uncommon in tables 14 and 16, it seems conservative to take 0.6 megam./sec. as the least allowable value for the discordance dubiety.

The value of that dubiety is entirely independent of the procedure used for deriving the velocity from the data, and it does not in the least depend upon the total number of observations taken, nor upon the technical probable error of the definitive value.

FIRST METHOD OF REDUCTION

In table 16 are given all the means, by orders, of the separate direct determinations, the averages $V_{e}\pm$ and $V_{r}\pm$, $V_{er}\pm$ (the mean of $V_{e}\pm$ and $V_{r}\pm$), the deviation δ of each $V_{er}\pm$ from the average, the root-mean-squared errors ϵ_{m2} and δ_{m2}, as defined by eq. 17, and Perrotin and Prim's "mean error" E, which is defined by means of the equation $E^2=(\Sigma w\delta_1^2)/(\nu-1)\Sigma w$, where w is the weight attached to the mean for the order (here $w=pq^2$), δ_1 is the deviation of that mean from the weighted mean for all orders, and ν is the number of orders involved.

These values of $V_{er}\pm$ are those obtained by their first, or "combined," method of reduction; and their weighted mean, each order being given the weight pq_n^2, is the accepted ultimate value for each series.

But it is obvious from table 16, and even more so from figure 3, in which $V_{er}\pm$ is plotted against $1/q_n$, that these means vary over such a wide range that no kind of average over all values of n can fairly be regarded as less dubious than several units in the fourth digit, and figure 3 shows that the Mont Vinaigre values of $V_{er}\pm$ are affected by a large systematic error.

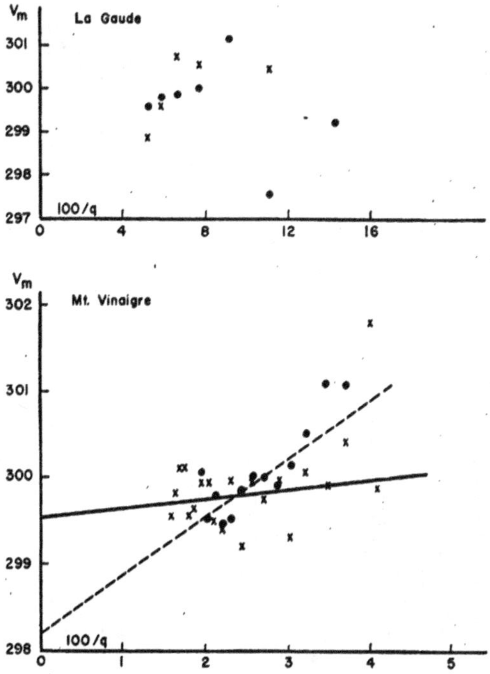

Fig. 3.—Plot of V_m against $100/q$.

The values of $V_m\equiv(V_{e\pm}+V_{r\pm})/2\equiv V_{er\pm}$ have been taken from table 16; crosses represent Perrotin's values, dots Prim's. The order of the eclipse, or reappearance, being n, $q\equiv 2n-1$. The lines through the Mont Vinaigre values have been determined by least squares; their equations are as follows: Perrotin's $V_m=299.5+10.7/q$; Prim's $V_m=298.2+67.8/q$. The solid line refers to Perrotin's values; the dashed one, to Prim's. The unit of V is 1 megameter per second.

TABLE 16

Mean Values (in Air) from Perrotin and Prim's Tableau III

n = order of eclipse; $q=2n-1$; V_e and V_r are, respectively, the means of p observed values of the velocity of light in air for an eclipse and a reappearance of the indicated order; signs + and − in subscript are the signs of the acceleration of the wheel; $V_{e\pm}=(V_{e+}+V_{e-})/2$; $V_{r\pm}=(V_{r+}+V_{r-})/2$; $V_{er\pm}=(V_{e\pm}+V_{r\pm})/2$; δ = deviation of $V_{er\pm}$ from its average; E = Perrotin and Prim's "mean error," defined by $E^2=(\Sigma w\delta_1^2)/(\nu-1)\Sigma w$, where w is weight (here pq^2) assigned to the δ_1, and ν = number of orders involved in the mean of the $V_{er\pm}$ column, and δ_1 is the deviation of $V_{er\pm}$ from its weighted mean, $(\Sigma pq^2V)/(\Sigma pq^2)$; ϵ_{m2} and δ_{m2} are the root-mean-squared deviations defined by eq. 17; H is a constant independent of n; and ϵ_n is an erratic fluctuating error. The values of δ, ϵ_{m2}, δ_{m2}, the "average," the values of $(\Sigma pV)/(\Sigma p)$, and of V and H as defined by $(V_{er\pm})_n=V+(H+\epsilon_n)/q_n$, have all been determined for this study; the other values have been copied directly from the report, those for E being obtained from page A.27.

Unit of V = 1 megameter/second, in air

n	q	V_{e+}	p	V_{e-}	p	$V_{e\pm}$	V_{r+}	p	V_{r-}	p	$V_{r\pm}$	$V_{er\pm}$	p	δ	E	ϵ_{m2}	δ_{m2}
							Nice — Mont Vinaigre: Perrotin										
12	23	300.62	6	300.34	3	300.48	307.84	6	290.67	3	299.26	299.87	9	0.03			
13	25	302.62	11	299.30	13	300.96	313.58	11	291.67	13	302.63	301.80	24	1.90			
14	27	302.70	10	299.48	17	301.09	309.10	10	290.38	17	299.74	300.42	27	0.52			
15	29	300.79	11	298.63	29	299.71	309.03	11	291.21	29	300.12	299.92	40	0.02			
16	31	301.65	18	298.59	24	300.12	308.35	18	291.69	24	300.02	300.07	42	0.17			
17	33	300.04	21	298.97	39	299.51	306.28	21	291.93	39	299.11	299.31	60	0.59			
18	35	301.22	19	299.10	22	300.16	306.54	19	292.93	22	299.74	299.95	41	0.05			
19	37	300.85	20	299.48	33	300.17	306.27	20	292.46	33	299.37	299.77	53	0.13			
20	39	300.91	38	299.42	47	300.17	306.63	38	292.87	47	299.75	299.96	85	0.06			
21	41	299.61	32	299.45	40	299.53	303.94	32	293.88	40	298.91	299.22	72	0.68			
22	43	300.35	25	299.48	52	299.92	306.27	25	293.77	52	300.02	299.97	77	0.07			
23	45	300.57	35	299.14	62	299.86	305.22	35	292.77	62	299.00	299.43	97	0.47			
24	47	300.04	40	299.48	81	299.76	304.85	40	293.67	81	299.26	299.51	121	0.39			
25	49	300.24	44	300.12	63	300.18	305.28	44	294.20	63	299.74	299.96	107	0.06			
26	51	299.38	32	300.61	73	300.00	304.57	32	295.23	73	299.90	299.95	105	0.05			
27	53	300.01	36	300.12	45	300.07	304.29	36	294.17	45	299.23	299.65	81	0.25			
28	55	299.25	26	299.26	47	299.26	305.23	26	294.53	47	299.88	299.57	73	0.33			
29	57	300.46	23	299.74	39	300.10	305.47	23	294.84	39	300.16	300.13	62	0.23			
30	59	299.40	31	300.49	29	299.95	304.86	31	295.65	29	300.26	300.11	60	0.21			
31	61	299.71	34	300.40	43	300.06	304.57	34	294.62	43	299.60	299.83	77	0.07			
32	63	299.73	59	300.14	80	299.94	303.43	59	294.91	80	299.17	299.56	139	0.34			
		Average				300.05	Average				299.74	299.90		0.32			
		$(\Sigma pV)/(\Sigma p)$				299.96	$(\Sigma pV)/(\Sigma p)$				299.62	299.79					
		$(\Sigma pq^2V)/(\Sigma pq^2)$				299.93_4	$(\Sigma pq^2V)/(\Sigma pq^2)$				299.58_1	299.75_9			0.065	12.77	0.26_5
							$(V_{er\pm})_n=V+(H+\epsilon_n)/q_n$ gives $V=299.54$ $H=10.7_0$									13.6	0.28_3
							Nice — Mont Vinaigre: Prim										
14	27	302.25	9	299.00	5	300.63	310.90	9	292.13	5	301.52	301.07	14	0.99			
15	29	301.92	6	299.75	7	300.84	308.81	6	293.87	7	301.34	301.09	13	1.01			
16	31	301.33	37	300.47	32	300.90	307.44	37	292.80	32	300.12	300.51	69	0.43			
17	33	300.20	46	299.86	44	300.03	306.44	46	294.11	44	300.28	300.15	90	0.07			
18	35	300.70	38	299.08	32	299.89	306.93	38	292.94	32	299.94	299.92	70	0.16			
19	37	300.68	62	300.31	64	300.50	305.75	62	293.27	64	299.51	300.01	126	0.07			
20	39	300.30	73	300.08	73	300.19	305.52	73	294.22	73	299.87	300.03	146	0.05			
21	41	299.69	62	299.86	67	299.78	305.47	62	294.36	67	299.92	299.85	129	0.23			
22	43	298.98	51	299.69	59	299.34	304.40	51	295.04	59	299.72	299.53	110	0.55			
23	45	299.63	47	299.57	41	299.60	303.92	47	294.71	41	299.32	299.46	88	0.62			
24	47	299.80	43	300.30	49	300.05	303.71	43	295.38	49	299.55	299.80	92	0.28			
25	49	299.46	21	299.54	25	299.50	303.72	21	295.37	25	299.55	299.53	46	0.55			
26	51	299.77	8	300.88	12	300.33	302.61	8	296.98	12	299.80	300.07	20	0.01			
		Average				300.12	Average				300.03	300.08		0.38			
		$(\Sigma pV)/(\Sigma p)$				300.02	$(\Sigma pV)/(\Sigma p)$				299.88	299.90					
		$(\Sigma pq^2V)/(\Sigma pq^2)$				299.94_9	$(\Sigma pq^2V)/(\Sigma pq^2)$				299.75_5	299.85_2			0.086	11.92	0.297
							$(V_{er\pm})_n=V+(H+\epsilon_n)/q_n$ gives $V=298.18$ $H=67.8$									7.42	0.185

TABLE 16—*Continued*

n	q	V_{e+}	p	V_{e-}	p	$V_{e\pm}$	V_{r+}	p	V_{r-}	p	$V_{r\pm}$	$V_{er\pm}$	p	δ	E	ϵ_{m2}	δ_{m2}
							Nice — La Gaude: Perrotin										
5	9	301.34	10	300.33	9	300.84	313.64	10	286.50	9	300.07	300.46	19	0.42			
7	13	299.49	49	302.65	39	301.07	315.48	49	284.57	39	300.03	300.55	88	0.51			
8	15	300.20	155	301.63	141	300.92	313.79	155	287.31	141	300.55	300.74	296	0.70			
9	17	298.90	99	301.07	100	299.99	301.54	99	287.88	100	299.21	299.60	199	0.44			
10	19	301.64	26	300.42	26	301.03	306.76	26	286.53	26	296.65	298.84	52	1.20			
		Average				300.77	Average				299.26	300.04		0.65			
		$(\Sigma pV)/(\Sigma p)$				300.66	$(\Sigma pV)/(\Sigma p)$				299.75	300.21					
		$(\Sigma pq^2V)/(\Sigma pq^2)$				300.61_0	$(\Sigma pq^2V)/(\Sigma pq^2)$				299.55_3	300.08_4			0.342	10.68	0.684
							Nice — La Gaude: Prim										
4	7	290.10	31	305.52	17	297.81	324.19	31	277.09	17	300.64	299.22	48	0.38			
5	9	298.07	18	300.10	9	299.09	317.96	18	274.08	9	296.02	297.56	27	2.04			
6	11	303.73	70	299.97	30	301.85	319.40	70	281.43	30	300.42	301.14	100	1.54			
7	13	299.51	149	300.71	99	300.11	315.83	149	283.96	99	299.90	300.01	248	0.41			
8	15	298.89	163	301.24	147	300.07	313.82	163	285.57	147	299.70	299.89	310	0.29			
9	17	298.53	93	301.05	84	299.79	312.81	93	286.79	84	299.80	299.80	177	0.20			
10	19	297.61	9	300.30	14	298.96	312.04	9	288.38	14	300.21	299.59	23	0.01			
		Average				299.67	Average				299.53	299.60		0.70			
		$(\Sigma pV)/(\Sigma p)$				300.05	$(\Sigma pV)/(\Sigma p)$				299.80	299.93					
		$(\Sigma pq^2V)/(\Sigma pq^2)$				300.02_9	$(\Sigma pq^2V)/(\Sigma pq^2)$				299.81_0	299.92_4			0.172	5.98	0.422

Nevertheless, and although ultimately ignored in the derivation of the definitive value, those means, carried out to six nominally significant digits, are published in the report. That tends to confuse the unwary reader.

SECOND METHOD OF REDUCTION

ERRORS IN EQUATIONS

Turn now to the second method used for reducing the observations. That involved the solution by least squares of a long set of equations, one for each order and each class, the average for the order being used. These equations for a single observation of order n for each of the four classes are given by Perrotin and Prim (p. A.20) as follows, except that their V_e has been replaced by the suitable one of the corresponding symbols (V_{e+}, V_{e-}, V_{r+}, V_{r-}) used in this study, and $(\mu_1-\mu_0)/(\tau_1-\tau_0)$ by R.

$$\begin{aligned} V-V_{e+}-zq^{-1}+yq^{-1}+qRx/M&=0,\\ V-V_{e-}-zq^{-1}-yq^{-1}+qRx/M&=0,\\ V-V_{r+}+zq^{-1}+y'q^{-1}+qRx'/M&=0,\\ V-V_{r-}+zq^{-1}-y'q^{-1}+qRx'/M&=0, \end{aligned} \qquad (34)$$

where V is the true velocity of light in air,

$$\begin{aligned} x&=V^2(r+c+\epsilon+\eta_0)/4DN,\\ x'&=-V^2(r'+c+\epsilon+\eta_0')/4DN,\\ y&=(2DN\eta/\pi)\lvert d\omega'/dI\rvert,\\ y'&=(2DN\eta'/\pi)\lvert d\omega''/dI\rvert,\\ z&=2(k-0.5-d)V, \end{aligned}$$

M = number of turns of the toothed wheel between two consecutive chronograph records by the counter; in notation of eq. 84, $k\equiv w/(w+g)$, $d\equiv d'/(w+g)$, d' = diameter of effective source.

In the notation of this report $x=-L_e'$, $x'=-L_r'$, $y=B_e$, $y'=B_r$, and $z=-\Delta'$ (see eq. 107, 100, and 85). On replacing the x, x', y, y', and z by these values and comparing the resulting equations with eq. 108, it will be seen that Perrotin and Prim have neglected the terms in H and in S, and that there are errors in sign in their equations.

The neglect of the S term is justified by their use of averages covering both directions of rotation of the wheel (p. A.8); and the neglect of H is of no practical importance, since it combines directly with B, and only the combination of the two can be determined experimentally.

The errors in the signs are, however, serious. They seem to have arisen thus: After having obtained on pages A.13 and A.14 the correct expressions for Θ_{e+} and Θ_{r+} of eq. 88—viz., in their notation,

$$(2n-1)\Theta_{e+}/\Theta_n=2(n-k+d)$$

and

$$(2n-1)\Theta_{r+}/\Theta_n=2(n-1+k-d),$$

where $k(w+g)=w$ and $d(w+g)$ is the diameter of the effective source of light—on page A.19 they write their eq. 2 for eclipses as follows:

$$\delta V=2(n-k+d)\,\frac{1}{M}\left(\frac{\mu_1-\mu_0}{\tau_1-\tau_0}\right)\pm\frac{y}{2(n-k+d)}\begin{cases}+\text{vitesses croissantes.}\\-\text{vitesses decroissantes.}\end{cases}$$

Here the $2(n-k+d)$ comes from the expression for Θ_{e+}. The correct expression for decreasing speeds differs from that for increasing ones not only in the sign of the y term, but also in the substitution of Θ_{e-} for Θ_{e+}; and by eq. 88, $\Theta_{e-}=\Theta_{r+}$. Hence the $2(n-k+d)$ must be replaced by $2(n-1+k-d)$ when the velocity is decreasing. Similar remarks apply to their eq. 3 on page A.20. They have failed to recognize that with decreasing speed the eclipse of the returned star is by the trailing edge of the tooth, whereas with increasing speed it is by the leading edge. Similar remarks apply to a reappearance. This error accounts for the erroneous sign of the z term in each of their equations for decreasing speed.

When these errors are corrected, the z and y terms change sign together; they merge to form for an eclipse or for a reappearance a single unknown constant that, contrary to their equations, cannot be separated by any least-squares adjustment.

In addition to these errors in their basic equations, another occurs in the setting-up of their "equations of condition" (tableau V), from which are derived the normal equations for determining the least-squares values of the constants. Entirely correctly they say that each observational equation representing the average of p determinations for a single order n must be given the weight pq_n^2 ($q_n \equiv 2n-1$). But they have erroneously assumed that this means that each "equation of condition" is pq^2 times the corresponding observational equation (compare their tableaux V and VI). That, however, leads to the occurrence of the square of pq^2 in the normal equations, instead of its first power (see eq. 14 and accompanying text).

THEIR DEFINITIVE VALUE UNTRUSTWORTHY

As a result of these errors in Perrotin and Prim's equations, the results derived by their use are untrustworthy. And it is on them alone that the published definitive value rests. Consequently, no confidence whatever can be placed in that definitive value.[30]

It thus becomes necessary to recompute the value, using correct equations.

RECOMPUTATION

Classification of Data

Before going on to find the best value one can get from the Mont Vinaigre observations, it is well to recall a somewhat casual remark to be found in their report (p. A.22), and to consider its implications.

When an eclipse and its following reappearance are observed, there are two criteria for determining whether they refer to an increasing, or to a decreasing, speed: (*a*) the sign of the acceleration α; and (*b*) the relative sizes of the values of the computed velocity of light ($V_{e+}<V_{r+}$; $V_{e-}>V_{r-}$). The two should agree. The authors do not mention the first, but without comment adopt the second.[31] This is somewhat surprising, when one recalls that a determination of the corresponding α is essential to the determination of the velocity of the wheel at the time of observation, thus making the sign of α at once available for classifying the value; and when he also recalls that a knowledge of α, as to sign as well as to magnitude, is essential to the evaluation of the "hesitation" term (that in x or x') in their basic equations.

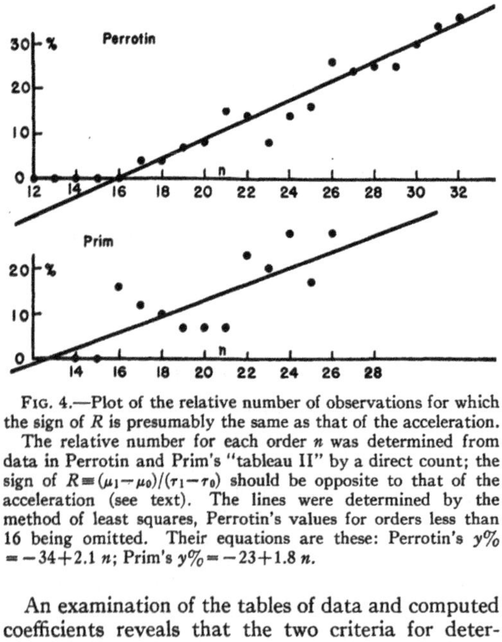

FIG. 4.—Plot of the relative number of observations for which the sign of R is presumably the same as that of the acceleration. The relative number for each order n was determined from data in Perrotin and Prim's "tableau II" by a direct count; the sign of $R \equiv (\mu_1-\mu_0)/(\tau_1-\tau_0)$ should be opposite to that of the acceleration (see text). The lines were determined by the method of least squares, Perrotin's values for orders less than 16 being omitted. Their equations are these: Perrotin's $y\% = -34+2.1\,n$; Prim's $y\% = -23+1.8\,n$.

An examination of the tables of data and computed coefficients reveals that the two criteria for determining whether the speed of the wheel is increasing or decreasing are frequently contradictory. The authors have chosen the second for classifying their V's, thus implying that the sign of their computed α is not trustworthy; but in determining the coefficients of the x and x' terms in their equation, they have placed implicit trust in the sign as well as in the magnitude of the computed α, which for each order is proportional to $-R/M$. Nothing has been found in their text to explain or to justify this strange procedure, and

[30] It does not necessarily follow that their definitive value is greatly in error. The terms involved may be of negligible size; or there may by chance be a compensation of errors; or errors in equations and computations may just offset the observational errors. For example, in the course of this study a stupid arithmetical blunder led to a result that did not differ by 1 in 30,000 from what is now the preferred value for the velocity of light. But when the blunder was corrected, the value found was several parts in 3,000 different therefrom.

[31] "Les valeurs V_1 et V_2 relatives à une disparition et à une réapparition consécutives déterminent le sens croissant ou décroissant de la vitesse de la roue dentée par la condition $V_2>V_1$ (vitesse croissante), $V_2<V_1$ (vitesse décroissante)" (p. A.22).

nothing is said about the frequent inconsistencies between the two criteria.

By a direct count of the abnormal signs in their tableau II— of the signs that disagree with the ordering of the V's in accordance with the second criterion— the results shown in figure 4 were obtained. At low speeds all signs are normal, but at higher speeds the number of abnormal signs increases at a rate of about 2 percent per unit increase in the order of the eclipse. The rate of increase is essentially the same for each observer. Although the data are very rough, there seems no reasonable ground for doubting that the number of such abnormalities increases markedly with the speed, if that exceeds a certain value.

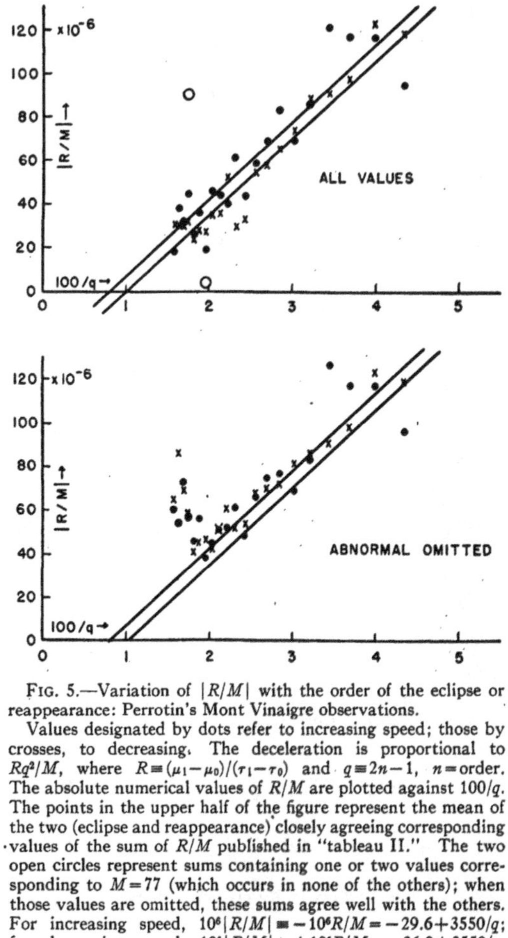

FIG. 5.—Variation of $|R/M|$ with the order of the eclipse or reappearance: Perrotin's Mont Vinaigre observations.

Values designated by dots refer to increasing speed; those by crosses, to decreasing. The deceleration is proportional to Rq^2/M, where $R \equiv (\mu_1-\mu_0)/(\tau_1-\tau_0)$ and $q \equiv 2n-1$, n=order. The absolute numerical values of R/M are plotted against $100/q$. The points in the upper half of the figure represent the mean of the two (eclipse and reappearance) closely agreeing corresponding values of the sum of R/M published in "tableau II." The two open circles represent sums containing one or two values corresponding to $M=77$ (which occurs in none of the others); when those values are omitted, these sums agree well with the others. For increasing speed, $10^6|R/M| \equiv -10^6R/M = -29.6+3550/q$; for decreasing speed, $10^6|R/M| \equiv +10^6R/M = -36.0+3550/q$. Adjustment by least squares.

The lower half of the figure differs from the upper only in the omission of all values for which the sign of R is abnormal. The lines from the upper half have been redrawn here for reference.

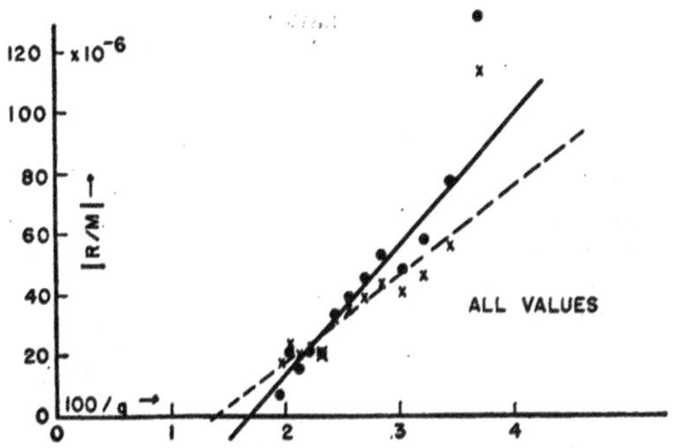

FIG. 6.—Variation of $|R/M|$ with the order of the eclipse or reappearance: Prim's Mont Vinaigre observations.

This figure differs from the upper half of figure 5 only in the use of Prim's data instead of Perrotin's. For increasing speed, $10^6|R/M| \equiv -10^6R/M = -74+4300/q$; for decreasing speed, $10^6|R/M| \equiv +10^6R/M = -42+2900/q$.

The first impression one derives from this is that the abnormalities probably arise from the vibration of the machine, which of course increases, in general, as the speed rises, and may perhaps cause eclipses and reappearances to occur at improper times. Were that the correct explanation, their strange procedure, to which attention has been called, would be quite unjustifiable.

But there is another possibility. If the 1/20-second oscillator, in terms of whose period all time intervals were measured, was subject to the same type of irregularity as has been found to have characterized it when used by Cornu, and if by some chance the manually controlled acceleration of the wheel varied in a regular manner with the speed and in such a way that the numerical value[32] of $R/M \equiv (\mu_1-\mu_0)/(\tau_1-\tau_0)M$ decreased as the speed increased, then when the speed shall have reached a certain value, the irregularity in the performance of the oscillator will be just sufficient to reduce to zero some of the values of R/M; at higher speeds, some of those values will be carried beyond zero, the number so carried increasing with the speed, but never exceeding 50 percent of the whole.

If this be the true explanation, then their strange procedure ceases to be strange, and becomes correct. The classification of the V's should be by the second criterion, and in determining the coefficient of the hesitation term all values of R/M, each with its own sign, whether normal or abnormal, should be given equal weight, because in the long run, when many readings are concerned, the average of the time intervals as measured by the oscillator would presumably be correct.

It would have been easy in the course of the work to have determined unambiguously by experimental

[32] $\alpha = -(\pi V^2 q_n^2/8D^2N^2)\cdot(R/M)$ (see eq. 129).

tests which of these explanations is correct. It is a great pity that such a study of the apparatus was not made.

Study of Acceleration of Wheel

The best that one can do now is to see whether the recorded data give any indication of a marked decrease in the numerical value of R/M as the speed increases. The available data are very rough; the recorded values of R rarely contain more than two significant digits. For each order, the authors give in tableau II the algebraic sum of all the values of R/M. The mean of these sums, for the same order, for V_{e+} and V_{r+} as well as those for V_{e-} and V_{r-}, has been divided by the number of observations involved. This gives in each case the mean value of R/M. In figures 5 and 6 the numerical values of these means are plotted against q^{-1} in the section marked "all values."

These plots show most plainly that, contrary to what one would expect of a manually controlled speed, the control being guided solely by a visual observation of the changing apparent brightness of the returned star, the numerical value of R/M does decrease rapidly as the speed increases; and more surprising still, the decrease is approximately linear in the reciprocal of q.

All of this points to a highly standardized procedure in adjusting and controlling the speed. That procedure should have been carefully described, for every standardized procedure is a potential source of hidden systematic error.

For Perrotin's observations there were also determined the mean values of only those values of R/M that have normal signs. Those means, necessarily greater than the former, containing all values, have been plotted in the lower part of figure 5, in that marked "abnormal omitted." The sharp rise as q^{-1} decreases below 0.02 is in strong contrast with the continued smooth decrease when all values are retained.

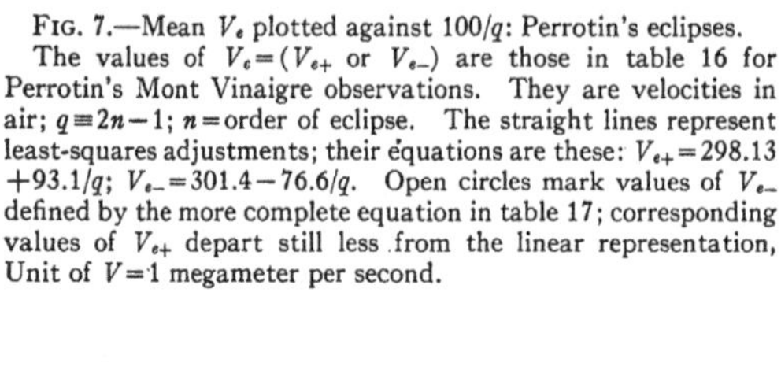

FIG. 7.—Mean V_e plotted against $100/q$: Perrotin's eclipses.

The values of $V_e=(V_{e+}$ or $V_{e-})$ are those in table 16 for Perrotin's Mont Vinaigre observations. They are velocities in air; $q\equiv 2n-1$; n = order of eclipse. The straight lines represent least-squares adjustments; their equations are these: $V_{e+}=298.13+93.1/q$; $V_{e-}=301.4-76.6/q$. Open circles mark values of V_{e-} defined by the more complete equation in table 17; corresponding values of V_{e+} depart still less from the linear representation, Unit of $V=1$ megameter per second.

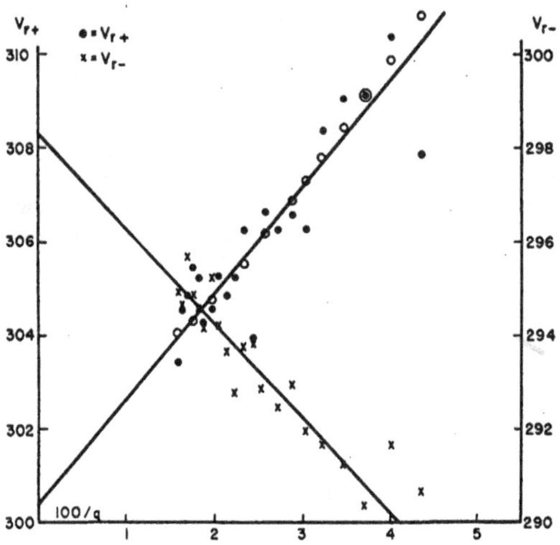

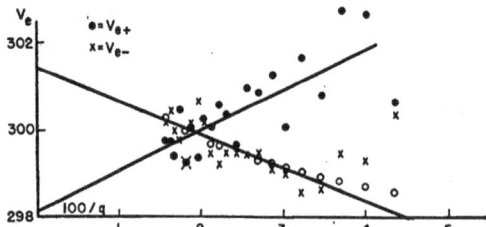

FIG. 8.—Mean V_r plotted against $100/q$: Perrotin's reappearances.

This figure differs from figure 7 only in that the plotted values are V_{r+} and V_{r-} instead of V_{e+} and V_{e-}, that the scale of V_{r-} is at the right and is displaced with reference to that of V_{r+}, and that the open circles mark values of V_{r+} defined by the more complete equation of table 17—the corresponding values for V_{r-} depart still less from the linear representation. The equations of the lines are these: $V_{r+}=300.34+227.7/q$; $V_{r-}=298.29-203.7/q$. Unit of $V=1$ megameter per second.

This suggests that the abnormal values should be included; that is, that the irregularity in sign of R/M arises from irregularities in the functioning of the oscillator.

Whence it seems likely, but by no means certain, that the discrepancies between the two criteria for classifying the V's arose from irregularities in the performance of the oscillator, and that the apparently strange procedure followed by the authors was justified.

Since $\alpha=-(\pi V^2 q_n^2/8D^2N^2)(R/M)$, by eq. 129, one infers from figures 5 and 6 that the accelerations for Perrotin's observations were as given by eq. 35

$$\alpha_+=[-29.6q^2+3550q]\frac{\pi V^2(10^{-6})}{8D^2N^2}\text{ radians/sec.}^2$$
$$\alpha_-=[+36.0q^2-3550q]\frac{\pi V^2(10^{-6})}{8D^2N^2}\text{ radians/sec.}^2 \qquad (35)$$

and for Prim's as given by eq. 36

$$\alpha_+=[-74q^2+4300q]\frac{\pi V^2(10^{-6})}{8D^2N^2}\text{ radians/sec.}^2$$
$$\alpha_-=[+42q^2-2900q]\frac{\pi V^2(10^{-6})}{8D^2N^2}\text{ radians/sec.}^2 \qquad (36)$$

Study of Data

Return now to the Mont Vinaigre observations, all of which, averaged by orders, are given in table 16 and displayed in figures 7, 8, and 9. It is obvious from the figures that each of the four sets by each observer can be represented by a right line just about as closely as by any other curve. The equations of the right lines of best fit, as determined by least squares, are given in the legends. When these equations are compared with the theoretical ones (eq. 101) from which the S terms have been eliminated and the L terms, presumably small, omitted, a striking contrast is revealed (eq. 37).

$$\left.\begin{array}{ll}
\text{Perrotin} & \text{Theoretical} \\
V_{e+}=298.1+\ 93.1q^{-1} & V+(\Delta'-H_e+B_e)q^{-1} \\
V_{e-}=301.4-\ 76.6q^{-1} & V-(\Delta'-H_e+B_e)q^{-1} \\
V_{r+}=300.3+277.7q^{-1} & V-(\Delta'-H_r-B_r)q^{-1} \\
V_{r-}=298.3-203.7q^{-1} & V+(\Delta'-H_r-B_r)q^{-1} \\
\text{Prim} & \text{Theoretical} \\
V_{e+}=295.6+181\ q^{-1} & V+(\Delta'-H_e+B_e)q^{-1} \\
V_{e-}=301.4-58.8\ q^{-1} & V-(\Delta'-H_e+B_e)q^{-1} \\
V_{r+}=295.5+383\ q^{-1} & V-(\Delta'-H_r-B_r)q^{-1} \\
V_{r-}=299.0-187\ q^{-1} & V+(\Delta'-H_r-B_r)q^{-1}
\end{array}\right\} \quad (37)$$

Whereas for each of the eclipse equations the q^{-1} term in the observational equation has the same sign as in the theoretical one, the same is not true for either of the reappearance equations. In the former, $(\Delta'+B_e)$ is the controlling quantity; in the latter, (H_r+B_r).

That is, in approaching an eclipse, the fading star was followed into the region that would have appeared dark had the star been a mere point, but the star could not be detected on reappearance until its brightness had exceeded that corresponding to the effect of the star's finite size.

That, initially surprising, situation comes from Perrotin and Prim's having used a large collimator lens and a sending lens of very long focus. That caused (eq. 72) the star to have a diameter of 0.15 mm., exclusive of diffraction effects, which is about 0.4 of the interdental gap of the wheel, at midheight of the teeth.[33] By eq. 100 and 84, $\Delta' \equiv V\Delta = V(2d+g-w)/(w+g)$, where d is the diameter of the star, w is the pertinent breadth of the tooth, and g that of the interdental gap. Putting $g=w=0.37$ mm., and $d=0.15$ mm., one finds $\Delta'=0.41V=122$; whereas Perrotin's values for $(\Delta'-H_e+B_e)$ are 93.1 and 76.6. There seems to be no inherent inconsistency between these values.

Similarly, if in Prim's observations the average value of w had been 0.29 mm., making $g=0.45$ mm., then $\Delta'=187$; whereas Prim's values for $(\Delta'-H_e+B_e)$ are 181 and 59. Again there is no inherent inconsistency.

[33] Since the wheel was 35.5 mm. in diameter and had 150 triangular teeth, the width of an interdental gap at midheight of a tooth was about 0.37 mm.

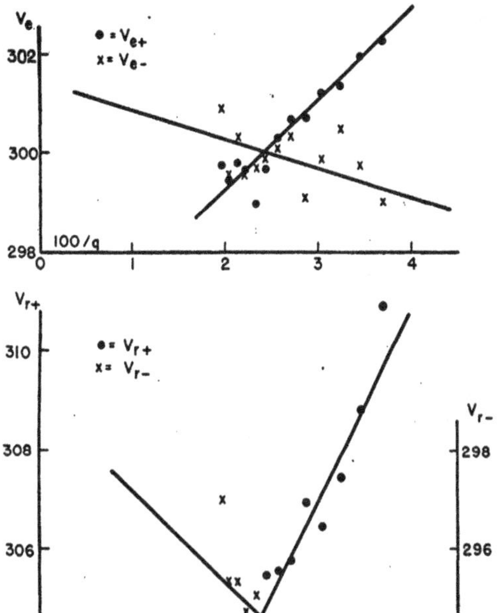

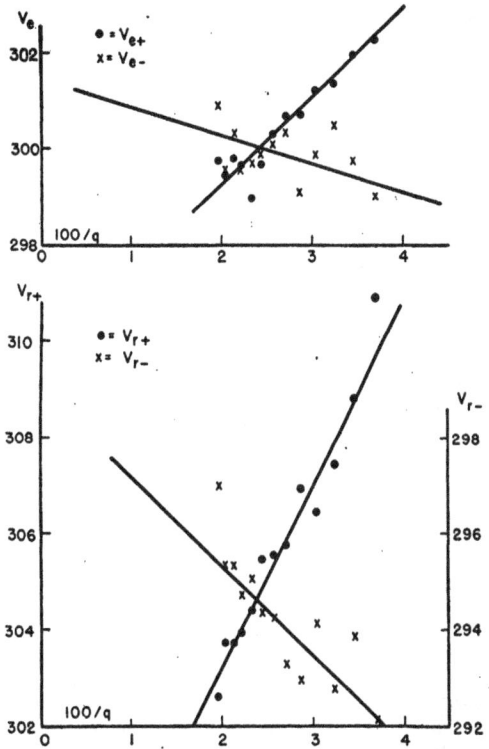

FIG. 9.—Mean V_n plotted against $100/q$: Prim's observations. The values of $V_n=(V_{e+}, V_{e-}, V_{r+},$ or $V_{r-})$ are those in table 16 for Prim's Mont Vinaigre observations. They are velocities in air. This figure corresponds to figures 7 and 8. The equations of the lines are these: $V_{e+}=295.6+181/q$; $V_{e-}=301.4-58.8/q$; $V_{r+}=295.5+383/q$; $V_{r-}=299.0-187/q$. Unit of $V=1$ megameter per second.

Furthermore, the elevation of the wheel was adjustable, so that the star could be placed at such a depth in the gap as the observer might find most satisfactory for easy observation. There is no reason for supposing that it was placed at exactly midheight, or at exactly $w=0.29$ mm. And it is probable that the adjustment was changed from time to time.

Referring again to the empirical equations (eq. 37), one sees that the derived values of V differ widely.

Although few who are experienced in such work would expect to be able to get from the data any values significantly superior to those given by the right lines of figures 7, 8, and 9, it seemed well to carry through a least-squares solution for each of the four classes, using the correct theoretical equation (eq. 101) containing all the terms considered by Perrotin and Prim in their attempted solution. Following them, it will be assumed that the S terms have been eliminated by averaging. Each α will be replaced by the proper empirical formula (eq. 35, 36) that has been found to represent it.

Each of those formulas may, for convenience, be put in the form

$$L\alpha = C'q^2 + Cq. \tag{38}$$

Then the general equation (eq. 101) for V_{e+} takes the form

$$V_{e+} = V + \frac{(\Delta' - H_e + B_e)}{q_n} + C'_{e+}q_n + C_{e+}. \tag{39}$$

The C_{e+}, being independent of q_n, fuses with V to form a single term

$$V' = V + C_{e+}. \tag{40}$$

Hence eq. 39 takes the form

$$V_{e+} = V' + k_e/q_n + C'_{e+}q_n, \tag{41}$$

where

$$k_e \equiv (\Delta' - H_e + B_e). \tag{42}$$

Similarly for the other equations 101.

The results of least-squares computations for each of the four classes of Perrotin's observations[34] are given in full in table 17—once, for all of his observations, and again, for all except the observations for the highest order ($n = 32$).

Values defined by the constants so obtained, when all the observations are used, are shown by open circles in figures 7 and 8. Their departures from the straight lines being negligible in comparison with similar departures by the observed values, it is obvious that the observed values do not suffice to determine with certainty which representation is to be preferred. Furthermore, six of the values so determined for C' contain not more than one digit that has any physical significance, and both of the remaining two, which may possibly contain two digits of significance, arise from computations in which observations of order 32 were ignored; and all the values are positive, whereas eq. 35 shows that those for positive acceleration are essentially negative.

All of which shows that the discordance in the observed values is so great that the value of C' cannot be satisfactorily determined.

Systematic Error

It will be noticed (table 17) that the value of V for $e+$ is not very different from that for $r-$; and similarly for $e-$ and $r+$. But the two pairs differ by a percent or more.

[34] In view of the low accuracy of the work, it has seemed unnecessary to carry out a similar computation and study of Prim's observations, which are less numerous than Perrotin's and seemingly (fig. 4) less satisfactory.

TABLE 17

RESULTS OF LEAST-SQUARE ADJUSTMENTS OF PERROTIN'S MONT VINAIGRE OBSERVATIONS TO THE GENERAL EQUATION

$$V_{obs} = V + (k + \epsilon)q^{-1} + C'q + aC'$$

V = velocity of light, ϵ is a fluctuating experimental error; $V + aC'$, k, and C' are determined by least-squares; a is determined from the data of figure 5; ϵ_{m2} and δ_{m2} are the root-mean-square errors defined by eq. 17.

Unit of V and δ_{m2} = 1 megameter/second

V_{obs}	V	k*	C'	aC'	ϵ_{m2}	δ_{m2}
			All observations			
V_{e+}	$298._5$	$+12_1.$	$+0.01_4$	$-1._7$	$0_{22}.$	$0._4$
V_{r+}	$301._3$	$+29_4.$	$+0.03_3$	$-4._0$	$0_{40}.$	$0._9$
V_{e-}	$301._6$	$-2_0.$	$+0.02_8$	$-2._8$	$0_{19}.$	$0._4$
V_{r-}	$298._3$	$-1_{96}.$	$+0.00_4$	$-0._4$	$0_{24}.$	$0._5$
			Observations of $n = 32$ ignored			
V_{e+}	$298._1$	$+10_4.$	$+0.00_3$	$-0._3$	$0_{20}.$	$0._4$
V_{r+}	$305._4$	$+45_6.$	$+0.13_8$	$-16._6$	$0_{37}.$	$0._8$
V_{e-}	$302._0$	$+0_{15}.$	$+0.051$	$-5._0$	$0_{20}.$	$0._4$
V_{r-}	$298._8$	$-1_{49}.$	$+0.03_4$	$-3._3$	$0_{19}.$	$0._4$

* In solving by determinants, the value of each quantity is obtained as a fraction, of which the denominator is exactly known, but the numerator consists of three terms, each involving an experimentally determined factor. Consequently, the numerator is not of physical significance beyond a certain number of digits. For V_{e+} all observations, the numerator for k is $487._{757} + 447.3_{66} - 961._{300} = 3._{823}$, and $k = 12_{1.6}$; although the several terms are perhaps known to one part in 500, scarcely more than one digit of k is of physical significance. Similarly in other cases.

Now it may be seen that when the observations lie in the critical region, as they almost surely do in this work, then it may happen that the k of eq. 42 has the form $k = k_0 + k_1 q$ (see fig. 17 and adjacent text). In that case the k_1 adds to the V, producing a constant error, exactly as in the case when $L\alpha/q$ is of that form (eq. 38). If the k_1 were positive and of nearly the same size for both the e and the r classes of observations, then the computed values of V would be related in exactly the way observed. It was hoped that the presence of such an error could be certainly accounted for either by an eccentricity of the wheel or by a periodic component in its speed, the presence of each of which is to be expected. But it seems that such irregularities cause H to decrease, and B to increase, as q increases (see Appendix A). Hence, the only general conclusion that can be drawn about the variation of k with q is that k is unlikely to be independent of q when such periodic errors exist.

No source of systematic error that will certainly give rise to the relations noted has been identified.

CONCLUSION

On referring again to table 17, it will be seen that not only do the values of V as derived from all observations vary over a range of 1 percent, but the omission of the observations of a single order changes three of those values by 4 or 5 units in the fourth digit, and the fourth by 10 times as much, even though over 500 determinations are involved in each case. This again confirms our earlier conclusion that the dubiety arising from the discordance alone amounts to at least 0.6 megam./sec.

Consequently, and in view of the undoubted presence of systematic errors, no statement more specific than the following is justified by the data:

It is probable that the velocity of light lies between 298 and 302 megameters per second, and it may be closer to their mean (300) than to either 299 or 301; but the obvious presence of systematic errors of unidentified origin throws serious doubt on the validity of taking the mean as the best representation of the whole. The dubiety arising from discordance alone is at least 0.6 megam./sec.

NEWCOMB'S WORK, 1880–82

SUMMARY

The report [21] of Simon Newcomb's measurements of the velocity of light during the years 1880–82 was published in 1891. In it we are told that he had been considering the subject since 1867, that in March 1879 Congress made an appropriation for the work, and that he was charged with the duty of doing it. In the meantime it became known that A. A. Michelson was also preparing to carry out such measurements, "but before the reliability of Mr. Michelson's work had been established, the preparations for the present determination had been so far advanced that it was not deemed advisable to make any change in them on account of what Mr. Michelson had done" (p. 120). At Newcomb's request, Michelson was detailed to assist him in this work, and did so during a portion of the first series of observations, until September 1880, when he went to the Case Institute, Cleveland, Ohio.

Newcomb used Foucault's method (see Appendix A), in which a pencil of light is reflected from a rotating mirror to a distant fixed mirror which returns it; if during the interval of time τ required for the light to travel the distance $2D$, to the fixed mirror and back to the rotating one, the mirror has turned through an angle θ, then the returned pencil after reflection from the rotating mirror will make an angle 2θ with its initial direction. The distance $2D$, the angle 2θ, and the angular velocity ω of the mirror are measured. From them the velocity of light V is determined by means of the equation

$$V=2D\omega/\theta. \tag{43}$$

Two distances were used, and numerous speeds m of the mirror; their approximate values and those of the corresponding rates of sweep s of the light across the distant mirror (table 38) were as follows: $D=2.55$ km., $m=114$ to 254 turns/sec., $s=0.012V$ to $0.027V$; $D=3.72$ km., $m=176$ to 268 turns/sec., $s=0.027V$ to $0.042V$.

The measurement of D with suitable accuracy involves no great difficulty; it was done by the U. S. Coast and Geodetic Survey. The angular velocity ω was determined from a chronograph record of every 28 revolutions of the mirror, the time record being given by a rated chronometer.

As would have been expected, Newcomb gave much careful attention to the designing of the optical portions of the apparatus and of the means for measuring the angle 2θ. He called his apparatus a "phototachometer."

The rotating mirror was a square steel prism, all four faces of which were nickel-plated and polished. It was 85 mm. long and 37.5 mm. square, and was rotated about its long axis by means of air blasts directed against the vanes of two fan wheels, one attached rigidly to either end of the prism. Each wheel had 12 vanes. Although nothing is said about it in the text, it appears from figure 5 of plate VI that the axial planes passing through the edges of the prism about coincided with four of the vanes of the lower wheel, but lay about midway between vanes of the upper one. The prism rotated inside a metal housing with two opposite open windows.

Swinging about an axis that coincided with the axis of rotation of the mirror was a stiff frame that carried the observing telescope and, at is farther end, a pair of microscopes for reading the divided arc over which it swung (radius 2.4 meters, p. 127). That arc, attached to the central base by a rigid frame anchored at each end to brick piers, rested on the stone cap of one of the piers, to which one end of the arc was firmly bolted.

Immediately above the objective end of the observing telescope was that of the similar sending one, which was, however, bent at a right angle as near as possible to its objective, was anchored to the pier, and was supported at its other end by a third pier. At the far end of the sending telescope was an adjustable slit which, illuminated by sunlight reflected from a heliostat, served as the source of light to be observed.

Each objective was something over 4 cm. in diameter, was about 8 cm. from the housing of the rotating mirror, and looked directly at the mirror through the front opening in the housing. Light from the slit was reflected from the upper half of a face of the rotating mirror to the distant mirror, and on its return was reflected from the lower half of the same face into the receiving telescope.

His procedure was this: The receiving telescope being set in a suitable position, the speed of the mirror was adjusted until the returned image was on the cross hair of the telescope, and was then held as constant as possible until a suitably long chronograph record had been secured. The receiving telescope was then

TABLE 18

NEWCOMB'S OBSERVED TRANSIT INTERVALS

τ is the observed nominal time required for light to travel the distance $2D$, where D = distance from the rotating mirror to the distant fixed one; $\delta = \tau - \tau_m$, τ_m being the average of all the τ's belonging to the series except those marked (?), to which Newcomb assigned zero weight. The mean magnitude of δ, irrespective of sign, is given at the foot of the column on the line marked "mean."

Unit of τ and of δ = 1 mean solar second

Series 1. $2D = 5.10190$ km.

Date	$10^6\tau$	$10^9\delta$
1880		
6/28	17.036	+8
6/29	32	+4
6/29	31	+3
6/29	31	+3
6/30	34	+6
6/30	40	+12
6/30	39	+11
7/3	43	+15
7/3	27	−1
7/9	29	+1
8/9	28	0
8/9	28	0
8/9	22	−6
8/9	25	−3
8/9	27	−1
8/9	23	−5
8/10	29	+1
8/13	33	+5
8/13	24	−4
8/13	26	−2
8/13	22	−6
8/13	26	−2
8/13	27	−1
8/16	27	−1
8/16	28	0
8/16	23	−5
8/16	15	−13
8/16	32	+4
8/16	27	−1
8/17	28	0
8/17	18	−10
8/17	30	+2
8/17	30	+2
8/17	31	+3
8/17	27	−1
8/21	74?	
8/21	25	−3
8/21	28	0
8/21	33	+5
8/21	30	+2
8/21	26	−2
8/24	26	−2
8/24	20	−8
8/24	27	−1
8/24	30	+2
8/24	26	−2
8/24	25	−3
8/25	26	−2
8/25	31	+3
8/25	21	−7
8/25	32	+4
8/25	26	−2
8/25	22	−6
9/3	25	−3
9/3	25	−3

Series 1. $2D = 5.10190$ km.

Date	$10^6\tau$	$10^9\delta$
9/3	17.027	−1
9/3	27	−1
9/3	24	−4
9/3	31	+3
9/4	21	−7
9/4	27	−1
9/4	27	−1
9/4	20	−8
9/4	32	+4
9/4	31	+3
9/4	30	+2
9/4	23	−5
9/4	31	+3
9/4	28	0
9/10	27	−1
9/10	27	−1
9/10	27	−1
9/10	22	−6
9/10	28	0
9/10	29	+1
9/10	27	−1
9/10	31	+3
9/10	31	+3
9/10	23	−5
9/10	28	0
9/11	34	+6
9/11	36	+8
9/11	33	+5
9/11	00?	
9/11	31	+3
9/11	30	+2
9/11	30	+2
9/11	31	+3
9/11	35	+7
9/11	28	0
9/11	29	+1
9/11	31	+3
9/11	33	+5
9/11	36	+8
9/11	35	+7
9/13	42	+14
9/13	38	+10
9/13	35	+7
9/13	32	+4
9/15	33	+5
9/15	28	0
9/15	30	+2
9/15	27	−1
9/15	31	+3
9/17	30	+2
9/17	29	+1
9/17	26	−2
9/17	28	0
9/17	30	+2
9/17	28	0
9/17	28	0

TABLE 18—*Continued*

Series 1. $2D = 5.10190$ km.

Date	$10^6\tau$	$10^9\delta$
9/18	17.029	+1
9/18	27	−1
9/18	29	+1
9/18	31	+3
9/18	30	+2
9/18	28	0
9/18	30	+2
9/18	25	−3
9/18	27	−1
9/20	32	+4
9/20	30	+2
9/20	28	0
9/20	27	−1
9/20	30	+2
9/20	33	+5
9/20	27	−1
9/20	27	−1
9/20	31	+3
1881		
3/25	26	−2
3/25	26	−2
3/25	22	−6
3/25	28	0
3/28	87?	
3/28	27	−1
3/28	26	−2
3/28	26	−2
4/7	31	+3
4/7	23	−5
4/7	26	−2
4/7	28	0
4/7	27	−1
4/7	28	0
4/15	27	−1
4/15	28	0
4/15	28	0
4/15	26	−2
4/15	27	−1
4/15	27	−1
4/15	29	+1
Mean	17.028_4	3.0

Mean $|\delta| \div$ mean τ
$= 17_6$ in 10^6

$300 \times 17_6 \times 10^{-6}$
$= 0.05_3$ megam./sec.

$2D \div$ mean τ
$= 299.61_1$ megam./sec.

Series 2. $2D = 7.44242$ km.

Date	$10^6\tau$	$10^9\delta$
1881		
8/8	24.834	−2
8/8	36	0
8/8	40	+4
8/8	40	+4
8/8	33	−3
8/8	34	−2

Series 2. $2D = 7.44242$ km.

Date	$10^6\tau$	$10^9\delta$
8/8	24.837	+1
8/10	31	−5
8/10	34	−2
8/10	29	−7
8/10	31	−5
8/10	35	−1
8/10	30	−6
8/10	29	−7
8/10	35	−1
9/12	35	−1
9/12	54	+18
9/13	45	+9
9/13	34	−2
9/13	49	+13
9/13	21	−15
9/13	35	−1
9/13	53	+17
9/19	32	−4
9/19	39	+3
9/19	38	+2
9/19	24	−12
9/19	25	−11
9/19	31	−5
9/19	27	−9
9/24	00	−36
9/24	54	+18
9/24	38	+2
9/24	42	+6
9/24	39	+3
9/24	62	+26
9/24	36	0
9/24	37	+1
9/24	50	+14
Mean	24.836_1	$7._1$

Mean $|\delta| \div$ mean τ
$= 28_6$ in 10^6

$300 \times 28_6 \times 10^{-6}$
$= 0.08_6$ megam./sec.

$2D \div$ mean τ
$= 299.66_1$ megam./sec.

Series 3. $2D = 7.44242$ km.

Date	$10^6\tau$	$10^9\delta$
1882		
7/24	24.828	+1
7/24	26	−1
7/24	33	+6
7/26	24	−3
7/26	24.834	+7
7/26	24.756?	
7/26	24.827	0
7/31	16	−11
7/31	24.840	+13
7/31	24.798	−29
7/31	24.829	+2
8/9	22	−5
8/9	24	−3

TABLE 18—*Continued*

Series 3. $2D=7.44242$ km.			Series 3. $2D=7.44242$ km.		
Date	$10^6\tau$	$10^9\delta$	Date	$10^6\tau$	$10^9\delta$
8/9	24.821	− 6	9/1	24.827	0
8/10	25	− 2	9/1	31	+ 4
8/10	30	+ 3	9/1	27	0
8/10	23	− 4	9/1	26	− 1
8/10	29	+ 2	9/1	33	+ 6
8/10	31	+ 4	9/1	26	− 1
8/10	19	− 8	9/2	32	+ 5
8/10	24	− 3	9/2	32	+ 5
8/10	20	− 7	9/2	24	− 3
8/10	36	+ 9	9/2	39	+12
8/11	32	+ 5	9/2	28	+ 1
8/11	36	+ 9	9/2	24	− 3
8/11	28	+ 1	9/2	25	− 2
8/11	25	− 2	9/2	32	+ 5
8/11	21	− 6	9/2	25	− 2
8/11	28	+ 1	9/5	29	+ 2
8/25	29	+ 2	9/5	27	0
8/25	37	+10	9/5	28	+ 1
8/25	25	− 2	9/5	29	+ 2
8/25	28	+ 1	9/5	16	−11
8/29	26	− 1	9/5	23	− 4
8/29	30	+ 3			
8/29	32	+ 5	Mean	24.827_3	4.3
8/29	36	+ 9			
8/29	26	− 1			
8/29	30	+ 4			
8/30	22	− 5			
8/30	36	+ 9			
8/30	23	− 4			
9/1	27	0			
9/1	27	0			
9/1	28	+ 1			

Mean $|\delta| \div$ mean τ $=17_3$ in 10^6

$300\times 17_3\times 10^{-6}$ $= 0.05_2$ megam./sec.

$2D\div$ mean τ $=299.78_1$ megam./sec.

moved to a position about equally far to the other side of the position of no deflection, and the speed of the motor in the reverse direction was determined in the same manner. The angular distance between those two positions of the receiving telescope is the value of 2θ that corresponds to the algebraic difference in the two angular velocities of the mirror (the arithmetic sum of the two speeds, one in each direction).

In this way he obviated the necessity for setting the observing telescope on the undeflected image, a setting that involves serious difficulties on account of the breadth of the image of the slit at the distant mirror (p. 192); and it would seem that this procedure also eliminated any error that might otherwise have arisen from a lack of central symmetry in the diffraction pattern, a source of error emphasized by Cornu in his report to the Paris Congress of 1900 [17].

Great pains were taken in determining the angular motion of the telescope that corresponded to one division of the graduated arc, and although the several determinations differed more than Newcomb had expected, it is probable that the average obtained is amply correct. A calibration of the arc and of the several scales used in determining the value of its division would have been desirable, but Newcomb thought it unnecessary, thought that the divisions in each case could be assumed to be sufficiently uniform; and the concordance of the data in each of his series of determinations seems to bear this out.

THREE SERIES OF OBSERVATIONS

Newcomb made three distinct series of determinations, before each of which the pivots of the mirror were examined and reground by the makers (p. 192): series 1, in which he was for a time assisted by Michelson, was between Fort Myer and the Naval Observatory, $2D=5.10190$ km.; series 2, between Fort Myer and the Washington Monument, $2D=7.44242$ km.; series 3, in which he was assisted by Holcombe, between the same stations as series 2, but after the apparatus had been changed so that either the observing or the sending telescope could be placed above the other.

The observed time of transit τ for each determination is given in table 18, together with its deviation from the mean of the series. The three mean deviations, being 17_6, 28_6, and 17_3 parts in a million, corresponding to 0.05_3, 0.08_6, and 0.05_2 megameters per second in the velocity, are only about 0.03 as great as those of Perrotin and Prim. With this degree of concordance, Newcomb might (eq. 20), by suitable experiments, have detected the presence of systematic errors that amounted to no more than a few units in the fifth digit of the velocity, but he could do no better. That is the lowest discordance dubiety that should be attributed to his results.

It will be noticed that the mean deviation for series 2 is 60 percent greater than that for either of the other two. During this series, certain abnormalities were noticed for the first time (p. 168). There seemed to be slight relative displacements of portions of the image received from different faces of the mirror; and on September 12, 1881, the image was definitely split into two pairs of parts, so arranged as to indicate that the splitting arose from an axial vibration of the mirror, the period being half the time of rotation (p. 185). This vibration did not produce a sensible effect "until the mirror attained a certain speed, which limit of speed, however, was very variable." During the rest of this short series, he tried to keep the speed below that critical limit, but with indifferent success. After the observations on September 13, the mirror was sent to the maker to be balanced. He reported it to be sensibly out of balance, and the pivot to be not perfectly round. After its return, more trouble of the same kind was experienced, and on September 24 "the pivot of the mirrors suddenly cohered to its conical cap, and the mirror was sent to the makers for another thorough overhauling of its pivots." At the same time other portions of the apparatus were also sent to the maker for such alterations as would permit an interchange in the positions

TABLE 19

DAILY MEANS OF TRANSIT INTERVALS AND OF VALUES DERIVED FOR VELOCITY OF LIGHT IN AIR

The values of τ and their average for each series have been taken directly from Newcomb's report [21, pp. 193–194]. The individual values of V have been computed by the writer, but the average value for each series is that derived by Newcomb from his corresponding average value of τ; $\delta = V -$Average of series. In series 3 the relative positions of the two telescopes are indicated by the letters R and D, D presumably indicating that the sending telescope was above the observing one, as in series 1 and 2.

Unit of $\tau = 1$ mean solar second; of $V = 1$ megameter/second, in air

Series 1. $2D = 5.10190$ km.				
Date	$10^6\tau$	wt.	V	δ
1880 6/28	17.036	1	299.48	−0.14
6/29	31	3	57	−0.05
6/30	37	3	46[a]	−0.16
7/3	32	2	.55	−0.07
7/9	29	1	.60	−0.02
8/9	25	5	.67[b]	+0.05
8/10	29	0	.60	−0.02
8/13	26	5	.65	+0.03
8/16	25	5	.67[b]	+0.05
8/17	27	5	.64	+0.02
8/21	29	3	.60	−0.02
8/24	26	5	.65	+0.03
8/25	26	5	.65	+0.03
9/3	26	5	.65	+0.03
9/4	27	7	.64	+0.02
9/10	27	7	.64	+0.02
9/11	32	7	.55	−0.07
9/13	36	3	.48	−0.14
9/15	30	4	.58	−0.04
9/17	28	3	.62	0
9/18	29	6	.60	−0.02
9/20	29	5	.60	−0.02
1881 3/25	26	2	.65	+0.03
3/28	26	2	.65	+0.03
4/7	28	4	.62	0
4/15	27	6	.64	+0.02
Average	17.028_2		299.61_5	

Series 2. $2D = 7.44242$ km.				
Date	$10^6\tau$	wt.	V	δ
1881 8/8	24.836	4	299.66	−0.02
8/10	32	4	71	+0.03
8/12	31	1	72[b]	+0.04
8/13	37	3	.65	−0.03
8/19	31	4	.72[b]	+0.04
8/24	40	2	.61[a]	−0.07
Average	24.834_4		299.68_2	

Series 3. $2D = 7.44242$ km.				
Date	$10^6\tau$	wt.	V	δ
1882 7/24	24.828	4 R	299.76	−0.01
7/26	28	3 R	.76	−0.01
7/31	19	2 D	.87[b]	+0.10
8/9	22	2 R	.83	+0.06
8/10	28	5 D	.76	−0.01
8/10	25	5 R	.80	+0.03
8/11	28	6 R	.76	−0.01
8/25	29	4 D	.75	−0.02
8/29	31	6 R	.72[a]	−0.05
8/30	27	4 R	.77	0
9/1	28	8 ?	.76	−0.01
9/2	29	9 ?	.75	−0.02
9/5	26	6 D	.78	+0.01
Average	24.827_6		299.76_6	

[a] Smallest in the series.
[b] Largest in the series.

of the two telescopes, so that the light could be either sent out from the upper half of the rotating mirror and received on the lower half, or vice versa.

ELASTIC VIBRATIONS SUSPECTED

The reason for the latter change was this: Newcomb believed that the vibrations that caused the multiple images observed on September 12, and subsequently, were elastic torsional vibrations of the mirror, the top of the mirror being twisted about its axis with reference to the bottom. If the period of these vibrations were half the time for one rotation of the mirror, the image would be split as observed on September 12; but if the period were one quarter the time for one rotation, then the image would give no indication of the vibration, although the determination would be affected by a systematic error. However, if the two telescopes be interchanged, the sign of this systematic error will be changed. By comparing the results corresponding to the two positions of the telescopes, one could obtain an estimate of the size of that error; and by averaging the two, the error could be partially, or wholly, eliminated.

After these changes had been made, the mirror rebalanced, and pivots reground, the observations of series 3 were made. They give no certain evidence that interchanging the telescopes made any difference in the result obtained. As Newcomb puts it: "The difference between the two classes of results is too small for taking account of." It will be noticed (table 19) that observations with the telescopes in the position designated by D were taken on only four days, and that the smallest value of τ in the entire series is that for one of those four. Furthermore, the weighted means of the two sets of τ differ by 0.001_1 μsec., which corresponds to 0.01_4 megam./sec. difference in V, a difference, as already seen, that is smaller than any that could be surely established.

FIRST TWO SERIES DISCARDED

Newcomb, therefore, concluded that in series 3 there were present no vibrations of the kind considered, and that the value of the velocity derived from that series was correct, those from the other two series being vitiated by the presence of such vibrations, which could not have been detected by the procedure followed unless their period had happened to be a multiple, higher than unity, of the time for a quarter revolution, as in some of the observations in series 2.

HIS DEFINITIVE VALUE

Consequently he based his definitive value exclusively on series 3, as follows:

Observed in air	299.76_6	megameters per second
Correction for curvature[35]	$+0.01_2$	"
Reduction to vacuum	$+0.08_2$	"
Velocity in vacum	299.86_0	"

Of this he writes (p. 201):

> If we estimated the probable error of this result from the discordance of the separate measures, it would be less than 10 kilometers. But we can have no ground for assigning any definite numerical value to the probable error, owing to the possibility of constant errors. Indeed, judges may not be wanting to maintain that the results of all three series of observations should have been taken into account, on the ground that we cannot be sure of having eliminated all systematic errors from any of them. On this hypothesis we might fairly assign the respective weights 2, 3, and 6 to the three series. This would give:
>
> Velocity in air 299,728
> Velocity in vacuum 299,810
>
> The probable error might then be estimated at 40 or 50 kilometers.

It will be noticed that he here uses the term "probable error" in two distinct senses: (1) in its technical sense, as used in the theory of chance errors and computed "from the discordance of the separate measures"; (2) in the sense of likely uncertainty from all causes—which cannot be computed, but must be simply estimated. In this second sense it seems to be essentially equivalent to what in this study has been called the dubiety of the result; it seems to be intended to do no more than mark the limits of the region within which the true value is believed to lie. Such confusion in the use of the term "probable error" is not uncommon.

On page 202 occurs the sentence: "Making a liberal allowance for probable error, I think we may conclude as the most probable result—

"Velocity of light in vacuo = 299,860 ± 30 km."

Although this ±30 is not infrequently regarded as the technical probable error of his result, the context shows that it is nothing of the kind. And that is confirmed by the wording of the sentence; in the technical probable error there is no place for a "liberal allowance." He is here using the term in the sense of dubiety; and he arrives at the given value of the dubiety by considering all available determinations that he thinks reliable. It is not an individual characteristic of his own work.

One studying the various determinations of the velocity of light needs to be continually on guard if he would avoid being misled by the confusion of two distinct sets of numerical data. First, those derived directly and solely from the experimenter's observations; second, the experimenter's inference from all available sources. Each is valuable, but the two should not be confused. One making an independent appraisal of the work should consider the first only, especially as regards the uncertainty of the experimental work.

CRITICISMS

But this allowance of ±30 km./sec. is surely too small. Not only is it about as small an error as could have been detected, the disordance of the separate determinations being what they were, but it is only a little greater than twice the difference between the means of the two classes D and R of values of series 3, which difference he described as "too small for taking account of."

ELASTIC VIBRATIONS INSUFFICIENT

But there is a far more serious criticism to be considered. He has nowhere offered any evidence that elastic vibrations of the kind he considered ever existed in any of his work. He merely assumed them; and when he had the apparatus changed in such a way as to permit their detection, if present, he failed to find them.

Is it probable that he ever had such elastic vibrations? Consider the results of series 1 and series 3, for the velocity in air. The first is 299.61_5; the second is 299.76_6—a difference of 50 parts in 100,000. The double angle measured in series 1 ranged from 10,500″ to 22,300″, the average being 15,000″, of which 50 parts in 100,000 amounts to 7.5″. That is twice the amount by which the assumed vibration must change the angle between the axial planes through the center of the upper half of a mirror face and that through the center of the lower half of the same face, as one passes from the first half of a determination (rotation positive) to the second (rotation negative). Assume that the actual twist in each half of a determination was half of that, one being positive and the other negative. Then the corresponding twist of the extreme top of the mirror with reference to the extreme bottom was $3.75'' = 18 \times 10^{-6}$ radian. If, while one end of a square prism of length l and side a is held fast, an axial torque T applied to the other end twists it through an angle β, then the amount of that torque is given by the equation

$$T = 0.843\left(\frac{8a^4\mu\beta}{3l}\right) = 2.24\mu\frac{a^4\beta}{l}, \qquad (44)$$

[35] Faces of the rotating mirror were not perfectly plane.

where μ is the coefficient of rigidity of the material. For the mirror, $l=85$ mm., $a=37.5$ mm., and μ may be taken as 8,400 kg./mm.^{2}; hence for $\beta=3.75''$ the required torque is 7,900 kg. mm. Hence, if the torque were produced by two equal forces F lying along the top edges of opposite faces of the mirror, and oppositely directed, each of those forces would be

$$F=210 \text{ kg.}=463 \text{ lbs.} \qquad (45)$$

If these forces were applied to the sides of the pivot instead of to the faces of the mirror, they would have to be far greater, amounting to nearly 1.75 tons weight each, if the diameter of the pivot were 5 mm.; and the diameter was probably smaller than that. In view of the greatness of this force required for a static twist, it seems that no such elastic vibration great enough to have caused the observed difference between the results of those two series could possibly have existed.

PERIODIC TERMS IN SPEED OF MIRROR

Nevertheless he was surely correct in attributing the observed splitting of the image to a vibration of the mirror. But, although he considered the effect of dissipative forces in reducing the speed between impulses to the fan wheels, he overlooked the fact that the mirror might vibrate as a whole about its state of dynamic equilibrium. Indeed, he seems to have been entirely unaware of that very common phenomenon, of which a common illustration of today is the "hunting" of a motor-generator when the load is suddenly changed. In the simplest case of this kind, the angular position φ of the perpendicular to any one face of his mirror would be given by the equation

$$\varphi=\varphi_0+\omega t+A \sin (bt+d), \qquad (46)$$

where A is the amplitude of the vibration, $b/2\pi$ is its frequency, d determines its phase with reference to the origins from which φ and t are measured, t is the time that has elapsed since $\varphi=\varphi_0+A \sin d$, and ω is the mean angular velocity. In every actual case there is also a damping coefficient.

In the motor-generator illustration the value of b is determined by the dynamical system itself, by what may be called its free period. The vibration is gradually destroyed by dissipative forces unless it is maintained by some outside action. That may be done by an applied periodic force of the same $b/2\pi$ frequency.

Similar, but not necessarily simple harmonic, vibrations may be set up by any periodically applied impulse, exactly as in the case of a pendulum. This is considered more fully in Appendix B. The fundamental frequency of the impressed vibration is that of the impulse.

It should be noticed that if the period of such a vibration were equal to the time for the mirror to make a quarter revolution, or were an aliquot part of that time, then the contribution by the vibration to the velocity of the mirror at the instant the image appeared in the telescope would be the same for every face, and the appearance in the telescope would be exactly as though there were no vibration. Interchanging the telescopes would produce no change in the apparent velocity of light. No test reported by Newcomb could possibly have detected the presence of such a vibration.

But if the period were an integral multiple (greater than unity) of the time for a quarter revolution, then the phase of the vibration at the time a face reflects the light into the telescope would vary from face to face, and the image would appear to be split into two or more parts, displaced in the direction of the rotation. In particular, if the period were twice the time for a quarter revolution, the image would be doubled, and if the axis of rotation were not exactly parallel to the faces, the image would be split in the way observed by Newcomb on September 12, 1881 (footnote, p. 185).

It will be remembered that on the next day (Sept. 13) the mirror was removed and sent to the maker, who reported that it was out of balance, and that the pivots were not round. As is well known, either of those defects may cause such a vibration as that just considered. Consequently, the observed splitting of the image can be satisfactorily explained as arising from vibrations of the prism as a rigid whole, being caused by the pounding of the pivots in their bearings.

But were there present other forces that could have given rise to such vibrations? And in particular, were there forces having a period equal to the time required for the mirror to make a quarter revolution, or to an aliquot part of that time? The mirror was driven by airblasts striking the vanes of two fan wheels of 12 vanes each, the wheels being relatively displaced through half the angle between adjacent vanes. Consequently, the mirror was subjected to impulses of frequencies 24, 12, 8, 6, 4, 5, 2, and 1 per rotation of the mirror. Of these, those of 24, 12, 8, and 4 had periods that were aliquot parts of the time for the mirror to make a quarter revolution, and would tend to cause vibrations that could not possibly have been detected by any test made by Newcomb. Furthermore, the mirror rotated in a cylindrical housing with two opposite open windows, the angular openings of the windows being less than 90°. Each time a corner of the mirror entered the angle defined by a window, it was subjected to an aerodynamic impulse; it was also subjected to an aerodynamic impulse each time it left that angle. These impulses would have been in opposite directions, but there is no reason for assuming that they would have been equal. Furthermore, the two did not occur at the same time, with reference either to a single corner or to a pair of corners. Consequently, it is to be expected that these impulses would have set up vibrations having a period one quarter as great as the time for one revolution—vibrations that would not have been detected.

There are, therefore, good grounds for thinking that such vibrations did actually exist. But the report contains no data from which their magnitudes can be independently determined. Nevertheless, if one assumes that the difference (50 parts in 100,000) between the results of series 1 and series 3 arose from the presence of a simple harmonic vibration having a period of a quarter of a revolution ($b=4\omega$), then one can get an idea of the order of magnitude of the amplitude of that vibration, and of the forces involved.

For that period, eq. 46 becomes

$$\varphi=\varphi_0+\omega t+A\sin(4\omega t+d), \qquad (47)$$

and the angular speed at the time t is

$$\dot{\varphi}=\omega[1+4A\cos(4\omega t+d)]. \qquad (48)$$

For the special case in which the time of reflection by a given face is $t=2n\pi-d/4\omega$, n being an integer, the instantaneous speed is

$$\dot{\varphi}=\omega[1+4A]. \qquad (49)$$

And if A is to account for the discrepancy between series 1 and series 3, the result of series 3 being assumed to be correct, it must have the value given in the equation

$$\begin{aligned} 4A &= 50\times10^{-5}\text{ radians,}\\ A &= 125\text{ microradians}\\ &= 25.8''. \end{aligned} \qquad (50)$$

Even a very much greater amplitude would not seem unreasonable.

In order to get an idea of the magnitude of the forces involved, one may proceed thus: The kinetic energy E of a rotating body being $\frac{1}{2}I\omega^2$ where I is the moment of inertia of the body, and ω is its angular velocity, the increase in E when ω is changed by $\delta\omega=k\omega$ is $kI\omega^2$. The constant torque T that will produce that change while the body rotates through the angle ψ is given by the equation

$$T\psi=kI\omega^2. \qquad (51)$$

For a square prism of length l, side a, and density ρ,

$$I=(1/6)\rho la^4. \qquad (52)$$

For Newcomb's mirror, $l=8.5$ cm., $a=3.75$ cm., and ρ may be taken as 7.7 g./cm.^{3}. Hence

$$I=2160\text{ g.cm.}^2 \qquad (53)$$

Putting this in eq. 51 and replacing k and ω by their values (50/100,000, and 1070 radians/sec.), gives

$$T\psi=1.235\times10^6\text{ dyne.cm.radian.} \qquad (54)$$

Since $b=4\omega$, the displacement due to the vibration alone passes from zero to its maximum while the body is making 1/16 of a revolution. Hence if the 50 in 100,000 is the maximum effect that can be produced by that vibration, then $\psi=\pi/8$ radians and

$$T=3.14\times10^6\text{ dyne.cm.}=32\text{ kg.mm.} \qquad (55)$$

This is less than 0.5 percent of that (7,900 kg.mm.) found for Newcomb's assumed elastic vibrations. If it be represented by a couple with its plane normal to the axis of rotation, the forces lying in the opposite faces of the mirror, then each force of the couple will be only

$$F=850\text{ g.}=1.87\text{ lb.} \qquad (56)$$

And if the forces act on opposite sides of a pivot 5 mm. in diameter, each will be only 6.4 kg. = 14 lbs.

Although these calculations show that the torques required for setting up vibrations of the mirror as a whole are very minute as compared with those required for setting up the elastic vibrations assumed by Newcomb, it is quite obvious that they give no information regarding the forces actually acting on Newcomb's mirror. The problem has been much oversimplified. The vibration was not simple harmonic, but contained components corresponding to at least the eight frequencies already mentioned. Since the vibrations were built up by the action of periodically applied impulsive torques, those torques might have been much smaller than the values just computed, and still have given rise to vibrations of much greater amplitudes than that used in the computation. Furthermore, Newcomb computed the velocity of light from the change in the deflection of the returned light when the direction of rotation of the mirror was changed, its speed being nearly the same in each direction. But the relation of the vanes of the fan wheels to the air blasts and that of the corners of the mirrors to the windows in the housing were not the same for a positive direction of rotation as for a negative one. Consequently, both the amplitude and the phase of a vibration would, in general, have differed in the two cases. The absence of pertinent data makes it unprofitable to carry the discussion further. It's a pity that the fan-wheels did not have, say, 13 vanes, and that the windows were not made larger and provided with adjustable shutters, so that their effects could have been studied experimentally.

CONCLUSIONS

It seems certain that the discrepancy between the results of series 1 and series 3 cannot be accounted for by elastic vibrations of the kind assumed by Newcomb, but may be accounted for by vibrations of the mirror as a whole, about its condition of dynamic stability. From the construction of the apparatus, vibrations of the latter type were to have been expected.

Whence it is concluded that Newcomb erred in selecting the result of series 3 as the proper representation of the outcome of his work. All that he was justified in saying was that his results for the velocity of light in vacuo ranged from 299.71 to 299.86 megam./sec., and were obviously affected by systematic errors of unknown sign and magnitude.

The presence of such systematic errors makes it improper to present any kind of average of the values found in the three series as being more reliable than the individual value given by any one series. It also removes every ground on which one can validly base an estimate of the range within which the quaesitum lies.

It is a great pity that Newcomb did not return to the Naval Observatory station and make a fourth series of observations, using the modified instrument. That would surely have shown him the presence of a systematic error that he had not considered.

AN UNEXPLAINED VARIABILITY

One other variability in the data should probably be mentioned. Newcomb expressed regret that it was not practicable with his apparatus to set his telescope accurately on the undeflected return light; consequently he reported no such settings. However, from the data given in his tables it is possible to determine the setting β_0 that corresponded to the midposition of the telescope between the extremes corresponding to equal positive and negative rotations of the mirror. That may be called the apparent setting for the undeflected light. It would be the true setting if the mirror were without vibration or if the vibration increased each speed by the same amount at the times during which the light was going to the distant mirror and returning. When these values of β_0 were computed, it was found that usually throughout any one day, and sometimes for several consecutive days, they remained the same within a few seconds of arc. But frequently between consecutive days, and very rarely between observations on the same day, there were wide variations, the variations between days amounting at times to 100″ to 600″. The value found for the velocity of light seems, in general, to have been unaffected by these changes. They scarcely arose from the displacement of the graduated arc; for that was bolted to the pier; and if it had not been rigidly fixed, one would expect to find marked changes whenever the telescope was moved. They might have arisen from changes in the position of the sending telescope or some change in the illumination of the slit, but that telescope seems to have been anchored firmly to a pier. A minute displacement of the distant mirror might have been the cause, or a variation in the amplitude and phase of the vibration of the mirror; but the last would have had to be of a very special kind, as it did not change the value of the computed velocity of light. None of these possibilities were studied. Indeed, there is no indication that Newcomb was aware of these variations.

NEWCOMB'S SUGGESTED IMPROVEMENTS

In chapter VIII of his report, Newcomb offers suggestions for improvements. Among them is the suggestion that such a prismatic mirror be used that the returned light is reflected from a face adjacent to that which reflected the outgoing light. He suggested a pentagonal mirror. He also proposed the use of such great distances that during the passage of the light to the distant mirror and back, the mirror would turn by an angle that is nearly equal to that between the normals to adjacent faces. Each of these improvements was utilized by Michelson many years later.

Newcomb's paper should be carefully studied by any one planning to undertake such work.

Although in his preface Newcomb wrote that he "would be happy to co-operate with any physicist who may desire to utilize it [his phototachometer] for further researches," it seems that no one except D. B. Brace [22] has accepted the offer, and his work could not be carried to completion.

MICHELSON'S WORK

INTRODUCTION

Michelson's preliminary determination of the velocity of light in 1878 and his more precise one of the following year, both by Foucault's method, slightly antedated Newcomb's, and were entirely independent of it. They were the first measurements that were in any way precise, and seem to have been privately initiated and carried out by Michelson himself. In that particular they are unique, all others, including Newcomb's, having been sponsored and assisted in some way by an institution. Too much credit cannot be given him for his initiative and boldness in attacking such a problem unaided, and in securing privately the funds needed for the more precise work—especially when one recalls that this was before there was more than sporadic attempts to carry out basic experiments in physics in this country.

Most unfortunately, his reports on the velocity of light, late as well as early, are marred by ambiguity in expression, are deficient in essential details, and give no evidence of any serious search for systematic errors. Illustrations of these imperfections will appear in the discussions of his several reports.

MICHELSON'S WORK OF 1878

The object of Michelson's work of 1878 [23], by Foucault's method (see Appendix A), was to show that it was possible to obtain a much greater displacement of the image than that obtained by Foucault.

This was secured by using a greater distance (500 ft.) between the mirrors, by focusing the slit upon the surface of the distant plane mirror by means of a long-focus lens, and by placing the rotating mirror far (30 ft.) from the slit and between that and the lens. The mirror was a circular disk of glass, about an inch in diameter, silvered on one side. It was driven by an air-blast directed against the mirror itself, and attained a speed of 130 turns per second. The cost was $10. Stroboscopic observations were used in controlling the speed, and for determining its amount. He reported, without further detail, the "ten independent observations" of V, the velocity of light in air, given in table 20.

TABLE 20

MICHELSON'S RESULTS OF 1878

Unit of V = 1 mi./sec. = 1.6093 km./sec.

V	$V-186\ 510$
186 720	+ 210
188 820	+2310
186 330	− 180
185 330	−1180
187 900	+1390
184 500	−2010
185 000	−1510
186 770	+ 260
185 800	− 710
187 900	+1390
186 508[a] =300 147 km./sec.	1115 =1794 km./sec.

[a] So printed; should be 186 507.

As the differences, which are here added, show, the value of the fourth digit is very uncertain and any value assigned to the fifth is entirely devoid of physical significance. If he had not made the tenth determination, the mean would have been 186,350, which would have been discordant with the value (186,600) he gives for Cornu's result,[36] with which he compares his own with much satisfaction. The giving of a value, other than zero, to the fifth digit is misleading; it tempts the reader to ascribe to the result a higher precision than the data justify.

Where this work is summarized in his paper of 1880, the average is given as 186,500±300 mi./sec. "or 300,140 kilometers per second," five significant digits again being given. That ±300 mi./sec. is the technical probable error of the mean, computed on the assumption that those 10 values form a fair sample of the statistical family to which they belong. That assumption is certainly not fulfilled; the ±300 mi./sec. is without physical significance.

Since there was no search for systematic errors, this being merely an exploratory determination, the data do not justify a statement more exact than this: The observed values range from 185,000 to 189,000 mi./sec. (297 to 304 megam./sec.), and seem to indicate that the correct value probably lies nearer to 186,500 than to either 186,000 or 187,000 mi./sec. (nearer to 300 than to either 299 or 301 megam./sec.).

[36] Cornu gave his result as 300,400±300 km./sec., which equals 186,660±190 mi./sec., or rounded off, 186,700±200.

As illustrations of ambiguous statements that might easily mislead the reader into supposing that the work was done under better conditions than actually existed, attention may be called to the following:

Immediately after the short paragraph in which he states that the mirror was driven by an air blast directed against it, and in which a diagrammatic illustration of the arrangement of the blast is given, occurs the paragraph: "This crude piece of apparatus is now supplanted by a turbine wheel which insures a steadier and more uniform motion."

And the concluding paragraph is this: "In conclusion, I take this opportunity of tendering thanks to Mr. A. G. Heminway, of New York, for contributing $2000 for the purpose of carrying out these experiments."

On reference to the summary of this work as given in a later report [24, p. 115], it becomes evident that each of these quotations refers strictly to work that was yet to be done; they have no relation to the work he was reporting, although the reader might very excusably infer that they did.

MICHELSON'S WORK OF 1879

INTRODUCTION

During the last half of 1878 and the first part of 1879 new instruments were constructed and plans were made for repeating the work over a much longer path, using again Foucault's method (see Appendix A). A report of that work, which was done at the Naval Academy, Annapolis, where Michelson was an instructor, was submitted to the Secretary of the Navy, and by him was referred to the Nautical Almanac Office, of which Simon Newcomb was Superintendent. Newcomb states, in the introductory note to Michelson's published report, [24] that at his suggestion "the paper was reconstructed with a fuller general discussion of the processes, and with the omission of some of the details of individual experiments." There is in the archives of the Naval Observatory a manuscript report which may be the one originally submitted to the Secretary of the Navy. An examination of that manuscript, through the courtesy of the present Superintendent, Captain J. F. Helweg, U. S. N. (Ret.), yielded nothing new of significant importance. Besides a rearrangement of the material, the main difference between that and the published paper consists in the elimination of such things as the logarithmic computation of each day's result, a better summarizing of the data, and the correction of errors that in the manuscript were covered by lists of errata. The manuscript was not compared, page by page, with the published report, but was merely read carefully after the report had been studied, keeping in mind certain desired information that was not given in the report.

A report of this work was presented before the American Association for the Advancement of Science and published in its *Proceedings* [25]. It is essentially the same as that [24] now to be studied, differing from it only in a few minor details. What purports to be an abstract of the report presented before the Association, prepared by the author, may be found in the *American Journal of Science* [26]. But the table of the 100 values for the velocity in air that is given there was later replaced by a "corrected" one prepared by the author. That was inserted in the volume that I examined. Todd's discussion [27] of this work is based on the values given in the "corrected" table. No explanation of the correction has been found. Furthermore, both the table and its correction differ from the table published in each of the two detailed reports. In the table first published the values given for the velocity of light in air are from 140 to 250 km./sec. smaller, and those in its correction are a flat 20 km./sec. smaller, than the values in the table in the detailed report. No explanation of these discrepancies has been found. It will here be assumed that all the values given in the abstract are erroneous.

Michelson's apparatus and procedure differed from Newcomb's in six main particulars. (1) Instead of using two telescopes, one to send the light and the other to receive it, he used a single long-focus lens placed between the rotating mirror and the distant fixed one. Hence the portion of the rotating mirror that reflected the incoming light was at essentially the same level as that which reflected the outgoing, and the axis of rotation had to be slightly inclined to the vertical so as to keep the rotating mirror from flashing light directly into the eyes of the observer. (2) The mirror was a circular disk of glass, 1.25 inches in diameter and 0.2 inch thick, silvered on one side and mounted in a metal ring that formed an integral portion of the spinning axle. It spun in a rectangular frame that held the bearing sockets. (3) The mirror was driven by a type of air turbine having six outlets and attached to the axle of the mirror. (4) The mirror rotated in one direction only. (5) The speed of the mirror was stroboscopically determined by means of an electrically driven tuning fork, which was compared by means of beats with a standard fork mounted on a resonator and vibrating freely. (6) The angle through which the returned light was deflected was determined from its tangent, which was measured by means of an eyepiece moved by a micrometer screw, the line from the center of the face of the mirror to one end of the distance traversed being perpendicular to the screw and of a measured length. The angle was about 45′, whereas in Newcomb's work it ranged from 1.6° to 4.16°, and the angle actually measured was twice that. In Michelson's work, the eyepiece had to be set once on the slit that defines the source of light, and again on the returned image of the slit—two qualitatively different objects. The slit was rigidly clamped to the frame of the micrometer screw that moved the eyepiece; and the micrometer stand was on the wooden table on which stood the electrically driven fork.

The approximate values of the instrumental constants were as follows: distance between mirrors $D=0.605$ km.; lens, not achromatic, 8 in. (20.3 cm.) in diameter, focal length 150 ft. (45.7 m.); fixed mirror was flat, 7 in. (17.8 cm.) in diameter; usual speed of rotating mirror, $m=258$ turns/sec.; distance from rotating mirror to micrometer, $r=28.2$ or 33.3 ft. (8.6 or 10.1 m.); rate of sweep of light across distant mirror (table 38, see Appendix), $s=0.0013\ V$ or $0.0015\ V$.

Seeing conditions were suitable for measurements for only two hours of the day—the hour after sunrise and that before sunset.

DISCUSSION

General

Michelson measured all distances himself in terms of scales which he compared with certain standards. He also calibrated the micrometer screw, his calibration being checked satisfactorily by Professor A. M. Mayer, of the Stevens Institute, Hoboken, N. J. Unfortunately, he does not give sufficient data, regarding either the standards used, or his individual measurements, or his method of applying the several necessary corrections, to enable one to form an independent estimate of the reliability of the work. Possibly his estimate of accuracy is correct, but that he was unfamiliar with such work is obvious from the error he made in applying a temperature correction to his final result. In some way he arrived at the idea that the tangent of the angle of deflection should be corrected for the thermal expansion of the brass revolution counter attached to the ways of the micrometer slide (p. 141), and he applied a correction of 12 km./sec. to his computed value of the velocity of light. But before his next paper appeared he had learned of the error, and then corrected it [28, p. 243, bottom].

He gives as the distance from the rotating mirror to the distant fixed one 1,986.23 ft. (p. 128), with an error not exceeding 4 in 100,000 (p. 140); that is, $D=1{,}986.23$ within ± 0.08 ft. (605.40 m. within ± 2 cm.).

Only two other distances have to be measured: (1) the distance from the center of the face of the mirror to the cross hair of the eyepiece when that is at one end of the distance it traverses in measuring the deflection of the light, and the micrometer is so oriented about a vertical axis that the screw is perpendicular to the line joining the cross hair to the axis of rotation; and (2) the distance traversed by the eyepiece in going from the slit to its returned deflected image.

The first he called the "radius" and denoted by r. Owing to unspecified causes, the optical system did not remain in adjustment. Before each hour of runs, whether morning or afternoon, it was necessary to readjust the moving mirror by sliding it about on its pier, and tipping it slightly forward or backward until the light returned by the distant mirror was found to strike it properly (p. 122). This changed both D and r. The first change was perhaps of no importance; nothing is said about it. But r was remeasured before each set of runs. That would seem to be a very unsatisfactory feature of the work.

Measurement of Radius

The precision with which the velocity of light can be determined cannot exceed that with which r can be measured. Nevertheless, the only information given about that important measurement is this:

> The distance between the front face of the revolving mirror and the cross-hair of the eye-piece was then measured by stretching from the one to the other a steel tape, making the drop of the catenary about an inch, as then the error caused by the stretch of the tape and that due to the curve just counterbalance each other (p. 124).

On its face, the procedure seems very crude. He gives no illustration of how closely successive measurements agreed, but in his table of data he records those distances to the nearest 0.001 ft. (0.3 mm.). How was the drop in the catenary determined? How were the distances from the cross hair and from the face of the mirror to given divisions of the scale determined? Were the graduations of the scale calibrated? How was it determined that such a catenary gave the stated compensation? No information is given regarding these important questions.

It is possible, however, to get a rough idea as to the degree of compensation that might be expected, but it can only be rough; for what does "about an inch" mean? Also, was this "a steel tape" the same as "the steel tape" of the next page, which was studied and used in measuring D? Assume that it was. That tape was 100 ft. long, and under a tension of 10 pounds it stretched 0.0167 ft. (p. 128). That is, a tension of 10 pounds increased its length by 167 parts in a million. Hence its cross-sectional area A is given by eq. 57, where E is the Young's modulus of the tape.

$$A = 10^7/167E. \tag{57}$$

For steel tapes E is about 28×10^6 lb./in.^{2}, hence $A = 0.0021$ in.^{2} This is of the right order of magnitude. Since the density of steel is about 7.80 g./cm.^{3} $= 0.2818$ lb./in.^{3}, a tape for which $A = 0.0021$ in.^{2} weighs 0.0071 lb./ft. The length S along the curve of a catenary with a horizontal chord of length L, the lowest point being a distance B below the chord, and the weight of the catenary being w per unit of length, is given by the equation

$$S = L\left[1 + \frac{1}{3!}\left(\frac{L}{2a}\right)^2 + \frac{1}{5!}\left(\frac{L}{2a}\right)^4 + \cdots\right], \tag{58}$$

where

$$a = \frac{L^2}{8B} = \frac{T_0}{w}, \tag{59}$$

T_0 being the tension of the catenary at its lowest point. For present purposes, the tension may be regarded as the same throughout the entire length.

When a length L of the tape considered is subjected to a tension T_0, its length becomes

$$L' = L[1 + 167T_0 \times 10^{-7}]. \tag{60}$$

Hence, to a close approximation,

$$L' - S = L[167T_0 \times 10^{-7} - L^2/24a^2], \tag{61}$$

which may be put in either of the following forms.

$$\frac{L' - S}{L} = \frac{167}{8} \cdot \frac{wL^2}{B}(10^{-7}) - \frac{8B^2}{3L^2}, \tag{62}$$

$$\frac{L' - S}{L} = 167T_0(10^{-7}) - \frac{w^2L^2}{24T_0^2}. \tag{63}$$

Putting into these expressions the value already found for w (0.0071 lb./ft.), and taking $L = 30$ ft., which is near the average of the values of r actually used, one finds that $(L' - S)/L = 0$ when $T_0 = 4.83$ lb. and $D = 0.165$ ft. $= 1.98$ in. This can scarcely be regarded as about an inch. But how far can B

TABLE 21

EFFECT OF SAG OF TAPE

L = length of horizontal chord of the catenary
S = length of the catenary cut off by the chord
B = distance from chord to bottom of the catenary
L' = length under tension of an unstretched length L of the tape, the tension being that at the bottom of the catenary

Units: first line, 1 inch; second and fourth lines, 1 foot

B	1	1.25	1.5	1.75	2	2.25	2.5	2.75	3
B	1/12	5/48	1/8	7/48	1/6	3/16	5/24	11/48	1/4
$10^5(L'-S)/L$	+13.9	+9.6	+6.0	+2.9	−0.2	−3.3	−6.5	−9.7	−13.2
$10^3(L'-S)_{L=30}$	+ 4.2	+2.9	+1.8	+1.0	−0.1	−1.0	−1.9	−2.9	− 4.0

depart from that value without introducing a significant error? That can be readily determined by giving to B in eq. 62 a series of values. In that way the data in table 21 have been obtained. Since the measured value of r is reported to the nearest 0.001 ft., the drop B in the catenary should have lain within the range 1.75 to 2.25 in. If it actually were 1 in. and no corrections were applied for stretch, then, *from that cause alone*, the reported values of r and the values of the velocity of light computed therefrom will each be too small by 1.4 in 10,000, which corresponds to 40 km./sec. in the velocity.

That is, the use of the indefinite expression, "about an inch," combined with the omission of all other information concerning the measurement of r, raises serious doubts as to the confidence that should be accorded to those measurements. And that in turn impairs one's confidence in the value derived for the velocity of light, the impairment amounting to parts in 10,000.

But this is not all. On page 124, farther down the page than one would expect to find it, is a reference to a footnote that reads thus:

> The deflection being measured by its tangent, it was necessary that the scale should be at right angles to the radius (the radius drawn from the mirror to one or the other end of that part of the scale which represents this tangent). This was done by setting the eye-piece approximately to the expected deflection, and turning the whole micrometer about a vertical axis till the cross-hair bisected the circular field of light reflected from the revolving mirror. The axis of the eye-piece being at right-angles to the scale, the latter would be at right angles to the radius drawn to the cross-hair.

Obviously, this adjustment must have been made before the radius was measured; and in a preceding paragraph, already quoted, it is stated that the radius was measured from the mirror to the cross hair. Hence there would seem to be no doubt that the radius actually measured was the long leg of the right triangle defined by the line through the cross hair and parallel to the screw of the micrometer and the lines joining the center of the face of the mirror with the slit and with the cross hair respectively. And the values for the velocity were computed accordingly. But in his next paper [28] he writes:

> In the previous work two errors were committed: 1st, in neglecting to make allowance for the fact that in measuring r the hypothenuse of a triangle was measured instead of the base; 2nd, the correction on page 141 for φ should be omitted (p. 243, bottom).

The second of these corrections is the thermal expansion one that has already been discussed; it is the first that is of interest now. The phrase "of a triangle" is very indefinite, but there seems to be no doubt that the triangle referred to is that just defined. And for two reasons: first, there seems to be no other triangle involved; and second, the total correction applied being a reduction of 34 km./sec., of which 12 km./sec. is accounted for by the second correction, the triangle correction amounts to 7.3 in 100,000, and the ratio of the tangent of his angle (2695", see p. 134) to the sine exceeds unity by 8.6 in 100,000. The difference between the 7.3 and the 8.6 is too small to be of any physical importance, and may well have arisen from arithmetical errors. Hence, in his correction he seems to say that the measurement of r was not along the long leg of the triangle here defined, but along its hypotenuse. In view of the positive statements already quoted from the paper now under consideration, how could such a mistake possibly have happened? Those statements and his later correction seem to be irreconcilably contradictory.

Measurement of Displacement of Image

The other factor involved in the determination of the tangent of the deflection is the displacement of the returned image of the slit that serves as the source of light. It is based upon measurements by the micrometer, and exceeds the distance (BH, fig. 10) so measured by the amount of AB, where AMH is the angular deflection, H is the cross hair when set on the returned image, HBA is parallel to the screw of the micrometer, which, by adjustment, is perpendicular to MH, and B is the position of the cross hair when set on the slit S; BS is parallel to MH. It may easily be shown that $AB/BH = BS/(MH-BS)$. The report contains no mention of this correction AB. If one may infer that such omission indicates that it was not applied, then the recorded displacement is too small, and the velocity of light computed from it is too great by the fraction AB/BH of itself. With the exception of a few determinations near the end of the published table of data, BH is approximately either 112.6 or 133.2 mm., and is recorded to the nearest 0.01 mm.; and the corresponding values of MH are approximately 28.2 and 33.3 ft. (8.60 and 10.15 m.). Hence, if AB is not to exceed 0.01 mm.,

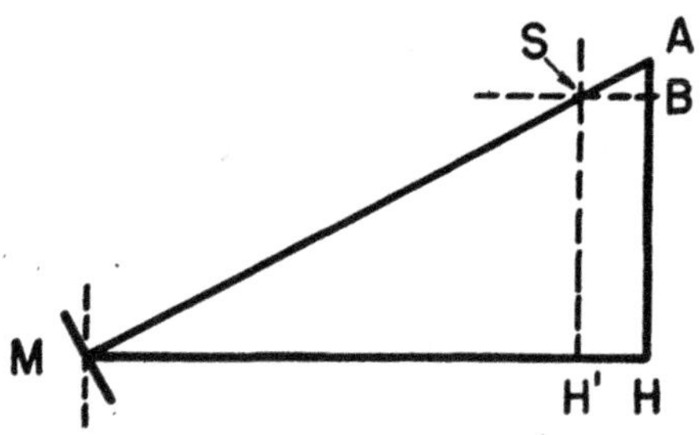

FIG. 10.—Measurement of displacement of image (Michelson). M is rotating mirror; H is cross hair moved by micrometer screw along HA, which is perpendicular to HM; S is slit that serves as source of light; BS is parallel to HM; $\tan AMH = AH/MH \neq BH/MH$.

corresponding to about 25 km./sec. in the velocity, then BS must not exceed 3/4 mm. The only information given about BS is that the cross hair is "in nearly the same plane as the face of the slit" (p. 120).

If, however, the measured r was not MH but MS (as may be intimated by the "correction" published in 1883, which has been considered in the preceding discussion of the measurement of r), then HB is the correct measure for the sine of SMH.

If, on the other hand, that "correction" is intended to take account of his having initially overlooked the AB correction, and consists merely in assuming that the distance MS is equal to MH, then it assumes that the length BS is given by equations 64 and 65

$$(MS)^2=(MH-BS)^2+(BH)^2, \qquad (64)$$

$$MS=MH. \qquad (65)$$

Hence, approximately,

$$BS=(BH)^2/2(MH). \qquad (66)$$

Therefore, if $BH=112.6$ mm. and $MH=8.6$ m., then $BS=0.74$ mm.; and if $BH=133.2$ mm. and $MH=10.15$ m., then $BS=0.87$ mm. If BS actually had these values, then AB was close to 0.01 mm., amounting to about one unit in the last digit recorded in the micrometer measures.

As has already been seen, the contradictions between the wording of his statement of the nature of the "correction" and that of his statements in the original report involve the entire subject in doubt.

Frequency of Fork

The standard fork, with which was compared the electrically driven fork used for controlling the speed of the mirror by stroboscopic means, was carefully studied by Michelson and Professor A. M. Mayer, of Stevens Institute, Hoboken, after the completion of the observations for determining the velocity of light. The last of those observations is dated July 2, 1879, and the first set of data for the fork is dated July 4. Two series of determinations of the frequency are recorded, their averages being 256.072 and 256.068 vib./sec. at 65° F. (18.3° C.).

This agreement is most excellent. But there are two statements that are, at least at first sight, somewhat disturbing. On page 128 occurs the sentence: "The fork was armed with a tip of copper foil, which was lost during the experiments and replaced by one of platinum having the same weight, 4.6 mgrs." And on page 132 it is stated that work on the fork prior to July 4 "was omitted on account of various inaccuracies and want of practice, which made the separate results differ widely from each other"; and what seems to be the result of that earlier work is given as "256.180" vib./sec., presumably at 65° F. No other information regarding that earlier work is given.

From the precision with which the weight of the tip is specified, the reader is likely to infer that the removal of the tip would change the frequency quite significantly. In which case a number of questions arise. How much difference did the loss of the tip make in the frequency? Was the copper tip lost before, during, or after the measurements on the velocity of light? If before or during those measurements, was it at once replaced by the platinum tip? Was the copper tip weighed before the work was started, or was its weight estimated from its remembered size? Is the loss of the tip in any way associated with the fact that the determinations of V prior to to June 14 are preponderantly higher than the mean (see table 22)? No answer to any of these questions has been found in the report. If the fork was a heavy one (his fig. 9 indicates that it was a König fork), the loss of the tip may have made no significant difference in its pitch. But in any case the reader should have been given some indication of the amount by which the loss of the tip affected the frequency.

The relatively great discrepancy between the earlier and the later determinations of the frequency of the fork (0.110 vib./sec., corresponding to a difference of 129 km./sec. in the velocity) is disturbing, because it suggests that something untoward may have happened to the fork. Had details of the earlier work been reported, one might have been able to explain the difference otherwise, but in the absence of all information except the author's confession of inexperience, the possibility of change cannot be lightly brushed aside.

DATA

Published Values

The data on which Michelson based his value of the velocity of light consist of exactly 100 sets of observations taken between June 5 and July 2, 1879, inclusive (pp. 135–138). They were preceded by 30 sets taken between April 2 and June 5, of which the report tells only that the separate results differed widely "on account of various inaccuracies and want of practice" (p. 116).

A "set" of observations consisted of the mean of 10 consecutive settings of the micrometer, together with the values of such other quantities as are required for a determination. All those other quantities were supposed to remain unchanged while the 10 settings were being made.

The electric fork was compared by beats with the standard "two or three times before every set of observations" (p. 124), and the temperature was read. These served to determine the frequency of the electric fork. Otherwise the several sets of a session were merely repeated measurements of the same thing. (Actually, the recorded number of beats is generally the same for every set of a given hour-long session.)

The values of V corresponding to the 100 reported sets of observations are given in table 22, together with certain means and deviations. The means have a range of 90 km./sec. For the first set, an electric lamp was used for illuminating the slit; for all the others, the sun. Each set of June 7 and 9 carries the same remark:[37] "Frame inclined at various angles," but no indication of the size of the angle is given. (From the text (p. 144), it is inferred that this inclination was a change in the azimuth of the frame

[37] Obviously, the change in inclination was made between sets, not during a set.

TABLE 22

MICHELSON'S DETERMINATIONS OF 1879 (IN AIR)

Thermometer reading $=T$°F.: V = velocity of light in air, values copied directly from his table; V_s = mean of V for a single session, morning (A) or afternoon (P); δ_1 and δ_a = excess of V and V_s, respectively, above 299.85; $\delta_s = V - V_s$ = deviation from mean of session.
Before each session, the rotating mirror was readjusted and the radius (r) was measured. Such readjustment and remeasurement presumably occurred between the several measurements of June 7 and 9 when the azimuth of the frame of the mirror was changed.

Unit of V, V_s, δ_1, and δ_s = 1 megameter/second; temperature $= T$ °F.

T	Date	V	V_s	$100\delta_1$		$100\delta_a$	$100\delta_s$	T	Date	V	V_s	$100\delta_1$	$100\delta_a$	$100\delta_s$
76	6-5-79	299.85			0			77	17 P	299.80		− 5		− 6
			299.85			0		77		.88		+ 3		+ 2
72	[a]7 P	.74			−11			77		.88		+ 3		+ 2
72		299.90			+ 5			77		.88		+ 3		+ 2
72		300.07			+22			77		.86		+ 1		0
72		299.93			+ 8						299.86		+ 1	
72		.85			0			58	18 A	.72		−13		+ 3
			.94			+ 9		58		.72		−13		+ 3
83	[a]9 P	.95			+10			59		.62		−23		− 7
83		.98			+13						.69		−16	
83		.98			+13			75	18 P	.86		+ 1		− 7
83		299.88			+ 3			75		.97		+12		+ 4
83		300.00			+15			75		.95		+10		+ 2
			.96			+11					.93		+ 8	
90	10 P	299.98		+13			+ 2	60	20 A	.88		+ 3		0
90		.93		+ 8			− 3	61		.91		+ 6		+ 3
			.96			+11		62		.85		0		− 3
71	12 A	.65		−20			− 9	63		.87		+ 2		− 1
71		.76		− 9			+ 2				.88		+ 3	
71		299.81		− 4			+ 7	78	20 P	.84		− 1		0
			.74			−11		79		.84		− 1		0
72	13 A	300.00		+15			+ 1	80		.85		0		+ 1
72		300.00		+15			+ 1	79		.84		− 1		0
72		299.96		+11			− 3	79		.84		− 1		0
			.99			+14		79		.84		− 1		0
79	13 P	.96		+11			+ 4				.84		− 1	
79		.96		+11			+ 4	61	21 A	.89		+ 4		+ 6
79		.94		+ 9			+ 2	62		.81		− 4		− 2
79		.96		+11			+ 4	63		.81		− 4		− 2
79		.94		+ 9			+ 2	64		.82		− 3		− 1
79		.88		+ 3			− 4	65		.80		− 5		− 3
79		.80		− 5			−12				.83		− 2	
			.92			+ 7		80	21 P	.77		− 8		+ 1
64	14 A	.85		0			− 1	81		.76		− 9		0
64		.88		+ 3			+ 2	82		.74		−11		− 2
65		.90		+ 5			+ 4	82		.75		−10		− 1
66		.84		− 1			− 2	81		.76		− 9		0
67		.83		− 2			− 3				.76		− 9	
			.86			+ 1		89	23 P	.91		+ 6		+ 2
84	14 P	.79		− 6			− 5	89		.92		+ 7		+ 3
85		.81		− 4			− 3	90		.89		+ 4		0
84		.88		+ 3			+ 4	90		.86		+ 1		− 3
84		.88		+ 3			+ 4	90		.88		+ 3		− 1
84		.83		− 2			− 1				.89		+ 4	
			.84			− 1		72	24 A	.72		−13		− 9
62	17 A	.80		− 5			+ 2	73		.84		− 1		+ 3
63		.79		− 6			+ 1	74		.85		0		+ 4
64		299.76		− 9			− 2	75		.85		0		+ 4
			299.78			− 7		76		299.78		− 7		− 3
											299.81		− 4	

[a] Azimuth of frame of mirror was changed from time to time by unstated amounts.

TABLE 22—*Continued*

T	Date	V	V_s	$100\delta_1$			$100\delta_a$	$100\delta_s$
86	26 P	299.89		+ 4				+ 3
86		.84		− 1				− 2
			299.86				+ 1	
73	27 A	.78		− 7				− 1
74		.81		− 4				+ 2
75		.76		− 9				− 3
75		.81		− 4				+ 2
76		.79		− 6				0
76		.81		− 4				+ 2
			.79				− 6	
85	[b]30 P	.82				− 3		
86		.85				0		
86		.87				+ 2		
86		.87				+ 2		
			.85				0	
83	[b]7-1 P	.81				− 4		
84		.74				−11		
86		.81[c]				− 4		
86		.94				+ 9		
			.82				− 3	
86	[d]2 P	.95				+10		
86		.80				− 5		
86		.81				− 4		
85		299.87				+ 2		
			299.86				+ 1	
	Average	299.85	299.85	5.9	9.1	4.7	5.5	2.7

Mean of all (100)	299.85_2
Mean of 6.10 to 6.27, (77)	$.84_4$
Mean of A's, of the 77	$.82_0$
Mean of P's, of the 77	$.86_6$
Mean before 6.14	299.91_0

[b] Mirror inverted in its bearings.
[c] So printed, but the reported data differ but little from those for the following value, and actually lead to $V=299.93$.
[d] Speed varied from set to set.

that carried the bearings of the rotating mirror.) For the sets of June 30 and July 1, the mirror was inverted in its bearings; and for those of July 2 the speed of the mirror differed from set to set. For all the others (77 in all), the speed was very nearly the same, close to 257.5 turns per second, and presumably the azimuth of the frame was essentially unchanged, and the mirror was in its normal position in its bearings.

Discordance Dubiety

From table 22 it may be seen that the average deviations, δ_1 and δ_a respectively, of V and of its session average V_s from the grand mean (299.85) are essentially the same ($\delta_1=59$ and $\delta_a=55$ km./sec.); whereas that (δ_s) of V from V_s is only 27 km./sec. These fix the minimum sizes of the discordance dubieties—of the systematic errors that could have been certainly detected under various conditions (see eq. 20).

Since there were never more than seven sets in a single session, the discordance dubiety with reference to tests that could have been made during a single session is at least (1.2)(27) = 32 km./sec. Likewise, the minimum discordance dubiety of a test involving six sessions, three to a group, is 66 km./sec., and that of one based on the comparison of a single V or V_s with the grand mean is (eq. 21) $2.5\delta_1$ or $2.5\delta_a$, respectively; i.e. 148 or 138 km./sec. Or in round numbers, the minimum discordance dubiety of a test based on a single session is 30 km./sec., on six sessions is 70 km./sec., and on a comparison of a single V or V_s with the grand mean is 140 km./sec.

Tests covering the effects of four changes have been reported: (1) change in azimuth of frame, (2) change in observer, (3) inverting of mirror, and (4) change in speed. The third involved two sessions, the others only one each, and in all cases the V or V_s was compared with the grand mean. Hence, the discordance dubiety of these tests is at least 140 km./sec., or possibly 100 for the third.

Since his own values of δ_1 for the routine observations run as high as 230 km./sec., many being over 130, the fact that for the tests the deviations do not exceed 220 for (1), 60 for (2), 110 for (3), and 100 for (4) means no more than that with those few observations he was unable to detect the effects of those changes. And the limit of that ability is, as has been seen, 140 km./sec., or possibly 100 km./sec. in the case of the third.

Whence it may be concluded that the discordance dubiety of his grand mean is at least 140 km./sec.

Inconclusive Tests

Furthermore, certain of the reported tests are inconclusive, either in themselves or from imperfections in the report.

Uniformity of speed of mirror.—One is told that the test covering the effect of change in azimuth of the frame was made for the purpose of seeing whether there was sufficient variation in the speed of the mirror to affect the result. This is explained as follows on page 144, under the head "Periodic variation in friction":

> If the speed of rotation varied in the same manner in each revolution of the mirror, the chances would be that, at the particular time when the reflection took place, the speed would not be the same as the average speed found by the calculation. Such a periodic variation could only be caused by the influence of the frame or the pivots. For instance, the frame would be closer to the ring which holds the mirror twice in every revolution than at other times, and it would be more difficult for the mirror to turn here than at a position 90° from this. Or else there might be a certain position, due to want of trueness of shape of the sockets, which would cause a variation of friction at certain parts of the revolution.
>
> To ascertain if there were any such variations, the position of the frame was changed in azimuth in several experiments. The results were unchanged showing that any such variation was too small to affect the result.

This indicates that what was being looked for was a slight reduction in the speed while the mirror is turning through a certain small angle, the speed elsewhere being essentially the mean speed. If no change in the result is produced when the frame is rotated through a reasonably small angle in both directions from the position used in the work, then there is no error due to such reduction. Holding this idea, which seems to have been the same as Newcomb's with respect to the slowing of his mirror between impulses on the fan wheels, he seems to have thought it sufficient to report the results without giving any information regarding the several angles through which the frame was turned. Although the omission of information about the angles is unfortunate, it would make no great difference, if the variation in speed were of the kind he assumed.

But, as was pointed out in the study of Newcomb's work, that is neither the only nor the most probable type of variation. It is far more likely that the aerodynamic action of the frame and of the turbine, which had six openings, set up vibrations of the mirror about its stable state of kinetic equilibrium. And the reported observations are not at all satisfactory as a search for that, they are not sufficiently numerous, and a knowledge of the actual position of the frame for each set is essential to their proper interpretation. If the effect of the vibrations had been a maximum when the frame was in the position used in the measurements, then a small displacement one way or the other would have had no observable effect, even though the effect of the vibration on the deviation of the ray of light were great. The few observations that are reported, and the way they are reported, are not sufficient to throw any light at all on this important question.

Here again one wishes information about the unreported first 30 sets of observations. They probably served as a guide in the adjustment of the apparatus. A satisfactory report of them might have thrown some light on the present subject.

Bias.—Since the decade means of the micrometer settings made during a single session seldom differed by more than a few hundredths of a millimeter, it is not surprising to find that the mean value of δ_s (table 22) is only half that of δ_1, if one recalls the difficulty in avoiding bias in such settings, especially when, as here, the head of the micrometer has a handle to assist one's muscular remembrance.

Michelson cites the close agreement of the first three observations of June 14, P. M., with those of June 17, P. M., and with the mean of all observations as proof that bias did not enter, the micrometer settings of each of those groups having been made by a different observer, other than himself. But it is not stated that the micrometer was distinctly out of setting when those observers took charge. If it was not, then the tests are of little value.

Effect of change in speed.—The four sets on July 2 were each made at a different speed of the mirror, which was erect in all cases. The speeds, in the order used, were approximately 193, 128.6, 96.5, and 64.3 turns per second, being respectively 6, 4, 3, and 2 times one-eighth of the speed (257.3) used in most of the work. In addition to the observations being, as previously seen, too few to establish the absence of effects that do not exceed 140 km./sec., they are of a kind that will not reveal the presence of a systematic error arising from the presence of a harmonic of one-eighth of the speed usually used.

Other Uncertainties

Temperature.—It is somewhat disturbing to find that the morning observations average 46 km./sec. smaller than the evening ones [38] (table 22). This suggests an uncertainty in the temperature of the standard fork. Although the temperature is reported for each set, one is not told where the thermometer was placed (in the Association report [25] a thermometer is shown lying on the table near the standard König fork; it is not shown in the corresponding figure in the report under study), and there is no indication of any attempt to control the temperature of the fork. It appears that the temperature of the air, as indicated by the thermometer, was assumed to be identical with that of the fork; and the pitch of that was assumed accordingly. But immediately after sunrise the air temperature rises rapidly; and just before sunset, it falls. And after a change in air temperature, the temperature of the massive fork may be expected to lag behind that of a good thermometer. Consequently, one would expect that the fork would be cooler than the thermometer in the morning, and warmer in the evening. Such a differential error of 3.3° F. will account for the difference in the mean V's, since the pitch of the fork decreased 47 parts in a million per 1° F. increase in temperature. Whether on the average the morning and evening errors were equal, as well as opposite, is an open question.

Break in values.—It is surprising to find such an excess of large positive values of δ preceding June 14 (table 22). The average V of those 26 sets is 299.91; even if the first 11 be omitted, the average is 299.90. True, there are among them a few large negative values of δ; nevertheless, one can scarcely overlook the possibility that something untoward happened between June 13 and June 14. Nothing is said about it in the report.

---

[38] Michelson does not give these averages, but he does give, without comment, the average for the extreme temperatures, 299.91 (90° F) and 299.80 (58–62° F). They differ by 110 km./sec.

DEFINITIVE VALUE

Michelson's

Michelson's estimate of the total error that could have affected each of his directly measured quantities was as follows:

Distance D	4 in 10^5
Radius r	4 in 10^5
Deviation d	5 in 10^5
Speed m	2 in 10^5
Total	15 in 10^5

To this he added the technical probable error (2 in 10^5) as computed from the discordance of his several determinations, obtaining, for the total, 17 in 100,000 or 51 km./sec. as fixing the extreme limits within which he expected the true velocity of light to lie.

By the use of these estimates he arrives at his definitive value in this way:

Mean of all values in air	299,852 km./sec.
Correction for temperature	12 km./sec.
Correction to vacuum	80 km./sec.
Uncertainty, as estimated	±51 km./sec.
Velocity in vacuum	299,944±51 km./sec.

(This ±51 km./sec. is *not* the technical probable error, but is his estimate of the dubiety of his result; it defines the range within which he believed the actual value of the velocity to lie.)

With reference to this value he writes:

> It remains to notice the remarkable coincidence of the result of these experiments with that obtained by Cornu by the method of the "toothed wheel." Cornu's result was 300,400 kilometers, or as interpreted by Helmert 299,990 kilometers. That of these experiments is 299,940 kilometers.

In his preliminary work of 1878 he took Cornu's value as 186,600 mi./sec. (=300,300 km./sec.) and found close agreement with his 186,508 mi./sec.; here he takes Cornu's value as 299,990 km./sec., and finds close agreement with his present result.

The value (299,944) just given is revised in his next paper [19, p. 243, bottom] by deleting the false correction for temperature, and by changing the interpretation of r from the long leg of a right triangle to the hypotenuse, as already explained. The sum of those two additive corrections he gives as −34 km./sec. Thus he obtains the value

Velocity of light in vacuo 299,910±50 km./sec., quite properly ignoring the value of the sixth digit.

Criticisms

But, as already remarked, this change in the interpretation of r cannot be accepted unreservedly. Furthermore, in arriving at his ±50 km./sec. he has considered only uncertainties in the direct measurements of the four quantities D, r, d, and the mean value of m, and it has been seen that the discordance of his determinations is such that by the procedure he followed he could not have detected experimentally a systematic error that did not exceed 140 km./sec. Hence, the dubiety of his result is not 50 km./sec., but at least 190 km./sec., and the best he would have been justified in claiming for the work would have been this:

Velocity of light in a vacuum	299.9 megam./sec.
Dubiety at least	±0.2 megam./sec.

That is, the velocity of light in a vacuum may lie between 299.7 and 300.1 megam./sec. The report, however, gives no indication that any search was made for systematic errors other than the four already mentioned.

Now it is certain that the speed of the mirror could not have been strictly uniform and that departures from uniformity might introduce error; it is probable that there were errors in the assumed temperature of the fork; and it is not at all certain that there were not other sources of systematic error.

OTHER POTENTIAL SOURCES OF ERROR

Near the close of his report, he shows that if the returned pencil of light is merely the outgoing one reversed, then no drag of the pencil by the rotating mirror will affect the result, the drag being presumably the same for the outgoing as for the returning light.

He also shows that, with the same assumption regarding the mere reversal of the pencil, no distortion of the mirror will change laterally the position of the center of the returned image, although the image will, in general, be broadened. Similarly for any imperfection of the lens.

But the returned pencil is not merely the outgoing one reversed. It is much modified by diffraction, the distant mirror subtending a very small angle. Indeed, the returned image is largely a diffraction pattern. It is important to know whether the center of intensity of that pattern can be laterally displaced either by rotation or distortion of the mirror, or by imperfections in the lens, or by modifications of the edges of the distant mirror. Of these, nothing is said.

He concluded, from Foucault's value being correct to within 1 percent, that his own result is not affected by any retardation on reflection by more than 3 in 100,000; i.e., by not more than 9 km./sec.

Although the ideal adjustment sought (p. 117) required the slit to be focused on the face of the distant mirror, nowhere in the report has there been found an account of how that adjustment was secured, nor has there been found even a statement that it was secured. And there is no discussion of either the presence or the absence of errors arising from a maladjustment. It will be recalled that Newcomb states that with his apparatus, using a much longer light path, the image at the mirror was wider than the mirror itself. He focused the light by adjusting the slit until it lay in the plane of the image of the distant mirror.

MICHELSON'S WORK OF 1882

In September, 1880, having accepted an appointment in the Case Institute, Michelson moved to Cleveland, terminating his short association with Newcomb in the latter's measurement of the velocity of light. Michelson's last recorded observation with Newcomb's apparatus was on September 13, 1880, before the first series of measurements—those between Fort Myer and the Naval Observatory—had been completed.

When those observations were worked up, they were found to yield a value (299,697) quite different from that (299,944) derived by Michelson from his observations of 1879. So Newcomb asked him to make a new determination at Cleveland, and says, in the introductory note to Michelson's report [28] of the new work, that "no instructions or suggestions were sent him except such as related to the investigation of possible sources of error in the application of his method." It is this new work, also by Foucault's method (see Appendix), that is now to be considered.

The approximate values of the instrumental constants were as follows: $D=0.625$ km.; lens 8 in. (20.3 cm.) in diameter, focal length 150 ft. (45.7 m.); fixed mirror was slightly concave, diameter 15 in. (38.1 cm.); usual speed of rotating mirror, $m=258$ turns/sec.; distance from rotating mirror to micrometer, $r=33.3$ ft. (10.15 m.); rate of sweep of light across the distant mirror, $s=0.0015\,V$.

This new work differs from that of 1879 in no essential feature except geographical location. The length of the light path, measured by John Eisenmann and himself, is given as 2,049.355 ft. (=0.624645 km.) —the old one was 1986.23 ft.; the same micrometer, rotating mirror, lens, and apparatus for the air blast driving the mirror, were used in both investigations; and the general optical arrangements were the same. The distant fixed mirror, 15 inches in diameter, was, however, nearly twice as large as the old one (7 inches), and "was slightly concave."

In the previous work the micrometer (then on a table, now on a brick pier) was always so oriented that when the eyepiece was set at the expected reading for the deflected image it looked directly at the center of the mirror. In the present work the orientation was always adjusted so that when the eyepiece was set for a deflection of 138 mm. it looked at the mirror; and a suitable correction was applied for the difference between that and the actual deflection.

This time he states that "the 'radius' was measured by finding the distance from the surface of the mirror to the slit, and therefore the *sine* of the deflection was measured instead of the tangent." He used the same tape as before, but it was supported throughout its length, and was used without tension. One division of the tape was placed in coincidence with a mark on the frame of the mirror; the tape passed about an inch below the center of the slit, "and divisions and tenths [were] read off." The scratch on the frame of the mirror was 0.0050 ft. from the surface of the mirror. This is all that is told of the measurement. No specimen set of individual measures is given.

He used his previous calibration of the micrometer screw, but this time he compared the pitch of the screw with the total length of the portion of the steel tape that was used in measuring r. An auxiliary scale on brass served as intermediary. Assuming that his previous determination gave the correct value (0.996307 mm.) for the mean pitch over the first 140 turns, this comparison showed that 0.00328081 nominal foot of the tape has a length of 1 mm. Since 1 mm. is actually equivalent to 0.00328083 ft., it would seem that all the comparisons are satisfactory. But he gives no details of the comparisons. In the earlier work he, seemingly, had merely assumed that the ratio of the whole length of the tape to that used for measuring r was correctly given by the subdividing marks on the tape. The comparisons mentioned in the present report seem to justify that assumption.

The inclination α of the plane of rotation of the mirror was determined from the trace, on a vertical wall, of sunlight reflected from the mirror. No numerical value of α is reported; in his earlier work it was about 1°.

For the stroboscopic control of the speed he used an electrically driven fork making 128 vib./sec. (in previous work, 256), and compared it by beats with another of nearly the same frequency, vibrating freely. Such comparisons, when combined with a direct comparison of the electric fork with a clock, determine the frequency of the standard free fork. Comparison with the clock was by observation of a Geissler tube flashed each second and reflected from a mirror attached to the electric fork. Thus he determined that the frequency of the standard was 128 vib./sec. at 71° F. and decreased by 0.0079 vib./sec. for each degree Fahrenheit increase above that temperature. Observations were extended from 54° F. to 73.5° F.

Whence it seems that, although certain details absent from the earlier report are given in this, and although the procedure for measuring r is superior to that in the earlier work, this is essentially merely a continuation of that, and is subject to essentially the same uncertainties. One would expect the result to be essentially the same for each.

In table 23 are given the values of the several significant quantities. They are plainly means; he gives no individual values. In column "no." are given the "number of observations," which is not more specifically defined. Whether those numbers indicate the number of "sets" as defined in the preceding

TABLE 23

MICHELSON'S DETERMINATIONS OF 1882 (IN AIR)

Except as indicated below, these tabulated values have been copied directly from Michelson's table. t °F = temperature of room, No. = number of observations (interpretation is doubtful, see text), wt. = weight assigned (see text), S = source of light (s = sun, e = electric light), v = distinctness of image or visibility, r = distance from slit to face of rotating mirror, d = displacement of image from slit, m = number of turns of mirror per second, Δ = difference between greatest and least values of d, φ is angular displacement of returned light, V = derived velocity of light in air, $\delta = V - 299.77$.

Michelson tabulated the values of V to six digits, but as V is inversely proportional to φ, and that is given (and used) to only five digits, the sixth digit of V is physically meaningless, as is also shown by the values of δ. For that reason the sixth digit is not given in this table. Its presence in the report tends to give the reader a false impression as to the precision of the results. The values of δ and of the unweighted mean have been determined for this study.

Unit of r = 1 ft.; of d = 1 mm.; of m = 1 turn per sec.; of φ = 1″; of V = 1 megameter/second.

Date	t	No.	wt.	S	v	r	d	100 Δ	m	φ	V	100 δ
10–12–1882	75.0	40	7	s	3	33.350	137.920	15	258.254	2788.7	299.88	+11
12	75.0	*	5	s	3	33.350	137.742	*	257.871	2785.1	.82	+ 5
12	75.0	*	5	s	3	33.350	138.233	*	258.754	2795.0	.78	+ 1
10–14–1882	71.0	56	3	s	2	33.350	137.933	27	258.214	89.0	.80	+ 3
10–16–1882	73.2	25	5	e	2	33.351	137.900	21	.042	88.2	.68	− 9
18	61.5	65	4	e	3	.356	137.917	19	.058	88.1	.71	− 6
19	56.0	19	6	s	3	.354	138.037	17	.212	90.7	.61	−16
19	54.7	10	2	e	3	.356	138.067	17	.258	91.3	299.60	−17
20	58.0	22	3	s	3	.355	137.774	25	.082	85.2	300.05	+28
21	64.3	68	9	s	2	.355	137.887	20	.072	87.6	299.78	+ 1
24	56.8	20	1	s	1	.355	138.010	25	.128	90.1	.58	−19
25	59.0	10	10	s	3	.356	137.897	9	.094	87.7	.80	+ 3
25	59.0	30	2	e	2	.356	137.905	26	.094	87.9	.77	0
26	59.0	10	1	s	1	.355	137.873	35	258.078	87.3	.82	+ 5
31	73.0	15	8	s	3	.355	137.754	12	257.814	84.9	.77	0
31	73.0	11	2	e	2	.355	137.787	22	257.814	85.6	.70	− 7
11– 4–1882	53.0	30	2	s	3	.360	103.572	20	193.634	2093.0	.57	−20
8	56.0	20	6	s	3	.357	.470	12	.581	91.2	.75	− 2
8	56.0	46	10	e	3	.357	.472	11	.581	91.2	.75	− 2
11	70.5	20	10	e	3	.357	.352	9	.390	88.8	.80	+ 3
11	70.5	20	6	e	3	.357	68.907	10	128.927	1392.3	.85	+ 8
14	40.5	6	7	s	2	.362	69.070	7	129.196	95.4	.81	+ 4
14	40.5	20	4	e	2	.362	69.091	11	129.196	95.8	299.72	− 5
	Mean	27									299.76	7.6
									Weighted mean		299.77	

* In the report this place contains merely a dot.

report, or merely the number of repetitions of the setting of the micrometer, or something else, is not known. All that is said regarding the weights assigned is this: "The weights . . . are deduced from the formula $w = 1/E^2$." But the basis from which E is measured is not stated. However, the difference between his weighted mean and the unweighted one is only 0.01 megam./sec., which is of no significance.

The average value Δ_m of Δ, and its ratio to d, for each of the three values of d used in the work, are as follows:

d	138	103	69 mm.
Δ_m	0.21	0.13	0.09 mm.
$1000\Delta_m/d$	1.5	1.3	1.3

That is, the average extreme range of the individual values of d is about 1.4 in 1000. Hence the value of the last digit (0.001 mm.) in each of the tabulated values of d is without physical significance. An uncertainty of only 0.01 mm. in the largest d (138 mm.) causes one of 20 km./sec. in V. The mean value of δ being 76 km./sec., the discordance dubiety in V, even if groups of 25 means of 27 determinations each had been used in a thorough search for systematic errors, would have been at least $0.076/3 = 0.025$ megam./sec. (see eq. 20), the average number of determinations per δ being 27.

This also shows that Michelson was entirely unjustified in giving V to six significant digits; doing so tends to mislead the reader.

If the search for systematic errors had consisted in the comparison of the mean of 27 determinations with the grand mean, 27 determinations being the average number involved in each of the tabulated values of V, then the discordance dubiety would have been at least $2.5 \times 76 = 190$ km./sec. (see eq. 21). It will be noticed that values of δ amounting to 160, 170, 190, 200, and 280 km./sec. occur in the table.

It will be noticed that the mean deviation from his weighted mean is 0.07_6 megam./sec., as compared with 0.06 in the earlier work; but the former refers to values that are averages of 6 to 65 determinations (average 27), whereas the latter refers to single determinations.

He takes the weighted mean, carried to six digits, as the definitive value to be derived from this work, giving

Velocity in air	299,771 km./sec.
Reduction to vacuum	82 km./sec.
Velocity in vacuum	299,853±60 km./sec.

This ±60 km./sec. is not the technical probable error, but, like the ±51 of the earlier work, is his estimate of the dubiety; it sets the range within which he believed the actual value of the velocity to lie. He does not say why he has here taken the dubiety 9 km./sec. greater than before.

But it has been seen that the discordance dubiety would have been at least 190 km./sec. if he had sought diligently for systematic errors by comparing the means of groups with the grand mean, even when each group contained at least 27 determinations. In that case his actual dubiety would have been 190+60=250 km./sec.; and he would not have been justified in claiming more for the work than this:

Velocity of light in a vacuum	299.85 megam./sec.
Dubiety at least	0.25 megam./sec.

That is, the velocity of light in a vacuum may lie between 299.6 and 300.1 megameters per second, essentially the same as for the earlier work.

MICHELSON'S WORK OF 1924

In 1924 appeared the report [29, 30] of Michelson's first set of measurements of the velocity of light between Mount Wilson and Mount San Antonio, a distance of 22 miles. They were made under the auspices of the Carnegie Institution of Washington; Foucault's method (see Appendix A) was used.

The approximate values of the instrumental constants were as follows: $D=35.4$ km.; two concave mirrors of 24-in. (61-cm.) aperture, 30-ft. (9.14-m.) focus; the source of light was at the focus of one, and a small concave mirror at that of the other; rotating mirror was an 8-sided glass prism, speed $m=528$ turns/sec.; distance from rotating mirror to micrometer seems to have been $r=25$ cm. (see p. 258, text is not clear); rate of sweep of light over distant 24-in. concave mirror (see Appendix A, table 38), $s=0.021\,V$.

The work seems to have been done at night. The source of light was a Sperry arc focused on a slit. Two large concave mirrors, each of 24-inch aperture and 30-foot focus, replaced the long-focus lens and plane distant mirror previously used. The adjustment was such that the mirror at the observing station sent a parallel beam of light to the distant one, which formed an image of the slit on the surface of a small concave reflector at its focus. That returned the light to the large mirror and back to the observing station, where both outgoing and returning light were reflected from the octagonal rotating mirror. By the use of a system of right-angle prisms, the two beams were so directed that they were reflected from diametrically opposite faces of the octagon. No further details regarding the adjustments are given.

Except that mirrors were used, instead of lenses, the general optical set-up, similar to that shown in figure 11, was essentially that proposed by Cornu [17] in 1900 for use with the Foucault method, which proposal appears to have been forgotten by Michelson[39]; as also the fact that this set-up is merely a particular case of that developed and used by Foucault in his determination.

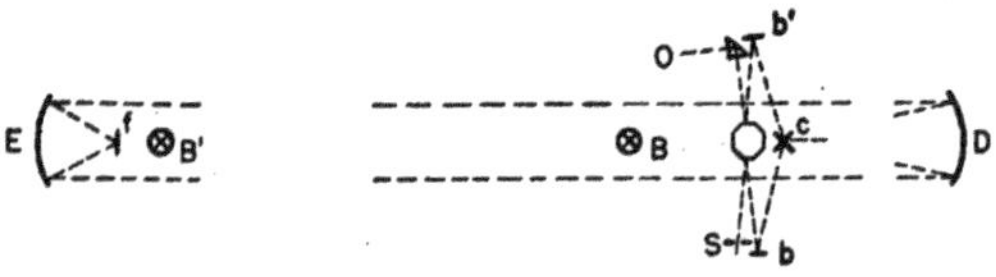

FIG. 11.—Schematic representation of optical system used in Michelson's work of 1926.

Except for scale, this is essentially a copy of the figure given in Michelson's report [31] of 1927. Light from the slit *S* is reflected from the octagonal mirror to the flat *b*, to one of the flats *c*, to the concave mirror *D* which sends a beam of parallel rays to the concave mirror *E*, which forms an image of the slit on the mirror *f*, which returns it to *D*, to the other of the flats *c*, to the flat *b'*, to the opposite face of the octagon, to a right-angle prism which reflects it to an eyepiece at *O*. *B* and *B'* are bench marks, between which the distance was measured by the U. S. Coast and Geodetic Survey. The report [29] of 1924 states that *D* and *E* were each of 24-inch aperture and 30-foot focus, and that *f* was "a small concave reflector"; that [31] of 1927 says nothing about any of them, but the figure, here reproduced, shows *f* as a flat, and indicates that the focal length of *E* was much shorter than that of *D*. The last may be a draftsman's error.

The distance was measured by the United States Coast and Geodetic Survey, and it is stated that "it is estimated that the result is accurate to within one part in two million." That statement is incorrect, as any experienced experimental physicist should know, and as may be seen by reference to Major William Bowie's report of the work, published as Appendix III to Michelson's later paper [31]. There, Bowie states (p. 20): "The Bureau of Standards certified that the lengths of the tapes were correct within 1 part in 300,000 and that the probable error [in the determinations of their lengths] did not exceed 1 part in 2,000,000." Hence the known accuracy of the measured distance between the mountains could not possibly have exceeded about 1 in 300,000, even if there had been no errors of any kind in the field work.

Michelson has confused "probable error"—the technical probable error as derived from the discordance of individual determinations—with accuracy.

[39] "This is the arrangement also used in the method of Fizeau and Cornu, but so far as I know it was not supposed to be applicable to the method of the revolving mirror" (note 1, p. 259).

In the present case he may have been somewhat confused by Bowie's statement (p. 16) that

> The methods adopted for the field measurements and the office computations were such as to assure the attainment of an accuracy, for the straight-line distance between Mount Wilson and San Antonio Peak, corresponding to a probable error of about 1 part in 2,000,000, derived from field measurements and observations alone, and to an actual error surely less than 1 part in 300,000. It is the feeling of those who have been engaged in the work that the actual error is somewhere between 1 part in 500,000 and 1 part in 1,000,000.

All of this obviously refers solely to what was done by the Survey. Those who did that work felt that in the doing of it they had not introduced an error that was as great as 1 part in 500,000. To that uncertainty of possibly 1 in 500,000 must be added the uncertainty of 1 in 300,000 in the standardization of the tapes, making the total uncertainty in the distance between the peaks about 1 in 190,000.

As compared with other uncertainties inherent in the determination of the velocity, this is amply accurate, although it is ten times the uncertainty stated in Michelson's report.

The airline distance between the two Coast Survey markers is given by Bowie as 35,385.53 meters. To that must be added a suitable correction at each end in order to obtain the length D of the light-path from the rotating mirror to the distant fixed one.

Michelson computed the value of D in this manner:

Distance between C. G. S. marks.........	35,385.50 m.
Provisional distance from each mark to focus of its mirror = 12 ft.; 24 ft. = ...	7.32 m.[40]
Focal length each mirror = 30 ft.; 120 ft...	36.58 m.[40]
Correction[41].........................	−3.2 m.
	D = 35,426.18 m.

Beginning with this work, he used regular prismatic mirrors, and used them in the way proposed by Newcomb in the report [21] of his own observations of 1880–81. Although Newcomb is not mentioned in Michelson's reports of his work with prismatic mirrors, an acknowledgment of Newcomb's suggestions may be found in Michelson's *Studies in Optics* (University of Chicago Press, 1927) and in his article on the velocity of light in *Encyclopaedia Britannica* (ed. 14, **23**, 1929).

In this work Michelson used an octagonal mirror, which, as one may find in the report next to be considered, was of glass. It's size is not stated. It was driven by an air blast, the pressure near the nozzle being about 40 cm. of mercury; and the speed was regulated by a valve controlling a counter-blast (p. 259). Photographs of this "small octagon" are shown in figure 2 of his report [31] of 1927. There it seems at first glance to have had but a single nozzle, but apparently a second is almost hidden behind the mirror. That figure shows that the blast was directed against the vanes of an open fan wheel. It is difficult to determine the number of vanes, but there seem to be eight; possibly there were only six. The text gives no information about it.

The speed of the mirror was controlled and determined by stroboscopic means, as in the past. The stroboscope was an electrically driven fork, of 132.25 vib./sec., that carried a mirror. It was compared with a free auxiliary pendulum, which was in turn compared with an invar gravity pendulum rated and loaned by the Coast and Geodetic Survey. The comparison of the fork with the free auxiliary pendulum was by means of sparks from an induction coil whose primary circuit was made and broken by the passage through a mercury drop of a platinum point attached to the pendulum. The same device was used for comparing the free pendulum with the invar one, the sparks in that case being viewed by reflection in a mirror attached to the invar pendulum.

No detail of any of the work is given, nor is any specimen set of observations. There is nothing that will enable the reader to determine either the correctness of the procedure actually followed or the reproducibility of any of the necessary types of observations.

The formulas given for the reduction of the observations are these

$$V = 16mD/(1-\beta),\quad \beta \equiv (\alpha_1-\alpha_2)/\pi,\quad N = N_1/P_1 + C^{-1}, \tag{67}$$

where P_1 is the period of the free auxiliary pendulum, $N_1(=133)$ is the nearest whole number of vibrations of the fork in the interval corresponding to one swing of the pendulum, C is the number of seconds required for the image of the spark to trace one complete cycle, as seen in the mirror attached to the fork, N is the true number of vibrations of the fork per second, α_1 and α_2 are the angular displacements of the reflected returned light as measured in the same direction from an arbitrary zero, one for either of the two directions of rotation of the mirror, $m(=4N)$ = number of rotations of the mirror per second, D is half the distance traversed by the light in going from the rotating mirror to the distant fixed one and returning to the face of the rotating mirror that reflects it to the eyepiece, V is the velocity of light in air, β is defined by the equation already given, which implies that the α's are to be measured in the direction that corresponds to what is called the negative direction of rotation of the mirror. These seem to be the proper definitions of the symbols, several of which are not at all clearly defined in the report.

From those formulas and the relation $m=4N$ one finds

$$V = 64DK, \tag{68}$$

[40] He combined these two, getting 144 ft., which he called 44 m., thus getting D = 35,426.3. The difference between the two values is of no importance.

[41] No explanation of this "correction" has been found.

TABLE 24

MICHELSON'S DATA OF 1924

The first six columns have been copied directly from his report; the next two show that in computing K he used the formula $K=N/(1+\beta)$, whereas his published formula is $K=N/(1-\beta)$. This may result from a change in convention regarding the measuring of the α's; δ=deviation from the mean.

The lower portion of the table shows that the use of a more accurate value of N_1/P_1 results in an increase of 0.01 vib./sec. in K. For explanations of symbols, see text.

Unit of $P_1=1$ sec.; of N_1/P_1, $1/C$, K, and $N/(1-\beta)=1$ vib./sec.

Date	P_1	N/P_1[a]	$1/C$	$\beta=\frac{\alpha_1-\alpha_2}{\pi}$	K	$N/(1-\beta)$	$N/(1+\beta)$	$100\,\delta$
8-4-1924	1.00630	132.16	+0.07	+0.00020	132.20	132.26	132.20	0
5	630	.16	−0.03	− 60	.21	.05	.21	+1
7	622	.17	.04	54	.20	.06	.20	0
8	628	.16	.01	70	.24	.06	.24	+4
9	633	.16	.01	40	.20	.10	.20	0
9	633	.16	.01	50	.22	.08	.22	+2
10	635	.16	.00	20	.19	.13	.19	−1
10	635	.16	−0.05	−0.00030	.15	.07	.15	−5
				Mean	132.20±0.006			1.6
	Mean, omitting observations of 8.10				132.21			

Date	P_1	$133/P_1$	$1/C$	β	$K=N/(1+\beta)$	$100\,\delta$
8-4-1924	1.00630	132.167	+0.07	+0.00020	132.21	0
5	630	.167	−0.03	− 60	.22	+1
7	622	.178	.04	54	.21	0
8	628	.170	.01	70	.25	+4
9	633	.164	.01	40	.20	−1
9	633	.164	.01	50	.22	+1
10	635	.161	.00	20	.19	−2
10	635	.161	.05	−0.00030	.15	−6
				Mean	132.21	1.9
	Mean, omitting observations of 8.10				132.22	

For $K=132.20$, $V=299.73_5$ megam./sec. in air
For $K=132.21$, $V=299.75_7$ megam./sec. in air
For $K=132.22$, $V=299.78_0$ megam./sec. in air

[a] So published, but the N should be N_1.

where

$$K\equiv N/(1-\beta)=\left(\frac{N_1}{P_1}+\frac{1}{C}\right)(1+\beta).$$

The only experimental data given in the report, other than the computation of D, already considered, are those in table 24. By trial, it has been found that the tabulated values of K require one to interpret the N as N_1 (probably a typographic error), and to change the sign of $(\alpha_1-\alpha_2)/\pi$. Whether the last means that the K's, and consequently the concluded values for V, are in error, or that the convention followed in measuring the α's differed from that assumed in defining β, there is no way to tell. But the values of K obtained by using the signs as published have a range of about 0.21 vib./sec., whereas those with the signs reversed have a range of only 0.09 vib./sec.

Furthermore, as shown in the lower section of the table, the first four of the published values of N_1/P_1 and of K should each be 0.01 vib./sec. greater than those published. Also, had the observations of August 10 not been taken, his mean value of K would have been greater by 0.01 vib./sec.

Hence, it is obvious that the value of K is dubious by at least 0.02 vib./sec. Furthermore, the mean value of δ being 0.016 vib./sec., the discordance dubiety will be at least $1.2\times0.016=0.019$ vib./sec., essentially 0.02 vib./sec., if tests had been made between groups of 3 values each (see eq. 20); and $2.5\times0.016=0.040$ vib./sec., if tests had been limited to a comparison of single values with the mean (see eq. 21). Whence it seems conservative to regard the discordance dubiety as being at least 0.02 vib./sec., which corresponds to 45 km./sec.

Accepting the value $K=132.20$ vib./sec., Michelson published

$$V=299{,}735 \text{ km./sec. in air}$$
$$=299{,}820 \text{ km./sec. in vacuo.}$$

He regarded this determination as "provisional."

But the vacuum correction here applied (85 km./sec.) is entirely too great for the altitude at which the work was done. This error was corrected in his next report [31] where the vacuum correction is given as 67 km./sec., making

$$V = 299{,}802 \text{ km./sec.}$$

which Michelson thought was "probably correct to within one part in ten thousand." If the value was considered doubtful by 3 in the fifth digit, the sixth digit is obviously without physical significance, and no value other than zero should have been given it.

But it should be remembered that the value used for D contains an unexplained "correction" amounting to 3.2 m. (9 in 100,000) and involves two "provisional" distances of 12 ft. each. If these represent Michelson's uncertainty of 30 km./sec., then one must add to that at least another 40 km./sec. on account of the discordance of the values of K, as already shown.

Hence the best he would have been justified in claiming for the work is

Velocity of light in a vacuum.......	299.80 megam./sec.
Dubiety at least....................	±0.07 megam./sec.

That is, the velocity of light in a vacuum may lie between 299.73 and 299.87 megameters per second.

MICHELSON'S REPORT OF 1927

Michelson's report [31] of 1927 covers a continuation of the work with regular prismatic mirrors that was reported in 1924, and sums up the observations made in 1924–26 inclusive.

The approximate values of the instrumental constants were as follows: two concave mirrors, presumably those used in the work of 1924, for each of which the aperture was 24 in. (61 cm.) and the focal length was 30 ft. (9.14 m.), were used; the image of the source of light in the rotating mirror was at the focus of one of these mirrors, and a small, presumably concave, mirror was at that of the other; the other constants were as follows:

Date	Rotating Mirror	D	m	r	s
1924	G8	35.42618 km.	528 turns/sec.	25 cm.(?)	$0.021V$
1925	G8	35.42515	528	?	—
1926	G8	35.425	528	?	—
1926	G12	35.4245	352	53	$0.030V$
1926	G16	35.4245	264	55	$0.023V$
1926	G16	?	264	?	—
1926	S12	?	352	?	—
1926	S8	?	528	?	—

Question marks indicate that the values are not given in the report; G8 indicates an 8-sided glass prism, and S12, a 12-sided steel one, etc.; each value of D contains all the reported digits; r is the length of the optical path of the returned light from the face of the rotating mirror to the fiducial mark in the eyepiece; the rate s at which the light sweeps across the distant 24-in. mirror has been computed by formula (table 38, see Appendix), and is expressed as a fraction of V, the velocity of light.[42]

But he first computed the correction from air to vacuum for a pressure of 625 mm.-Hg and 20° C., finding it to be 67 km./sec. This same correction (67 km./sec.) was applied to all determinations over the line from Mount Wilson to Mount San Antonio. In a footnote to the correction he remarks: "The correction should be applied to the individual observations; but the result is not appreciably altered by taking the mean values given above." No indication of the temperature and pressure actually existing at the time a particular determination was made is contained in the report. The reader is given no means for estimating the amount by which the proper correction for a given determination may differ from this 67 km./sec.[43]

[42] Landenburg's statement [32], that the rate of sweep was nearly 1/3 the velocity of light, seems to be entirely wrong. Possibly he considered case a instead of c (see fig. 20, Appendix), and lost a factor 2.

[43] Through the courtesy of Major E. H. Bowie, in charge of the Weather Bureau at San Francisco, the following estimates, prepared by Mr. L. G. Gray, Meteorologist in that office, have been made available to the writer.

ESTIMATED TEMPERATURES FROM SUN-DOWN TO MIDNIGHT ALONG A HORIZONTAL LINE FROM MOUNT WILSON OBSERVATORY TO MOUNT SAN ANTONIO

1926	Highest	Mean Maximum	Mean Minimum
June	80° F. (26.7° C.)	77.0° F. (25.0° C.)	64.0° F. (17.8° C.)
July	83° F. (28.3° C.)	77.5° F. (25.3° C.)	66.0° F. (18.9° C.)
Aug.	78° F. (25.6° C.)	75.5° F. (24.2° C.)	64.0° F. (17.8° C.)
Sept. 1–15	76° F. (24.4° C.)	72° F. (22.2° C.)	60.0° F. (15.6° C.)

1926	Lowest	Average
June	55° F. (12.8° C.)	70.5° F. (21.4° C.)
July	54° F. (12.2° C.)	71.7° F. (22.1° C.)
Aug.	57° F. (13.9° C.)	69.7° F. (21.0° C.)
Sept. 1–15	51° F. (10.6° C.)	66.0° F. (18.9° C.)

From sun-down to midnight the temperature probably fell by about 11° to 14° F. (6.1° to 7.8° C.).

ATMOSPHERIC PRESSURE AT 5 P.M. REDUCED TO SEA-LEVEL

1926	Highest	Lowest	Average
June	29.95 in. (760.7 mm.)	29.70 in. (754.4 mm.)	29.79 in. (756.7 mm.)
July	29.96 in. (761.0 mm.)	29.60 in. (751.8 mm.)	29.80 in. (756.9 mm.)
Aug.	29.91 in. (759.7 mm.)	29.70 in. (754.4 mm.)	29.81 in. (757.2 mm.)
Sept. 1–15	29.86 in. (758.5 mm.)	29.69 in. (754.1 mm.)	29.78 in. (756.4 mm.)

Between sun-down and midnight the pressure would be about 0.03 to 0.05 in. (0.8 to 1.3 mm.) greater than at 5 p.m. The corresponding pressures at the altitude of the Mount Wilson Observatory were about 4.85 in. (123.2 mm.) less than these tabulated values.

From these estimates it seems that the average temperature was probably about 21° C., and the average pressure about 635 mm.; whereas Michelson used 20° C. and 625 mm.-Hg. The former lead to a vacuum correction of 68 km./sec.; the latter to 67 km./sec. (In each case the index of refraction is taken as 1.0002765 at 15° C. and 760 mm.-Hg, corresponding to $\lambda = 5900$A.) The difference is small, but not negligible if the sixth digit is to be retained.

But in July the temperature ranged from 28.3° C. to 12.2° C., and the pressure from 638.8 to 629.6 mm.-Hg, corresponding to a range of 6 km./sec. in the correction for the temperature, and of 1 km./sec. for the pressure. It is thus evident that variations up to 7 km./sec. in individual determinations, and several kilo-

He then remarks that the correction (85 km./sec.) used in the 1924 report was erroneous. It should have been this 67 km./sec. Consequently, the value given by that work for the velocity of light in a vacuum is 299,802 km./sec., not the 299,820 km./sec. published in the several papers of 1924.

But all such observations measure the "group velocity." Hence the correction to vacuum should be based on the group index, and not, as Michelson assumed, on the ordinary index. This long-known fact has recently been recalled by Anderson [33]. For air and $\lambda = 5900$A, the group index is 3 percent greater than the ordinary one;[44] hence the proper correction to vacuum is 69 km./sec. instead of Michelson's 67.

In 1925 he made a second series of measurements with the same glass octagonal mirror as was used the preceding year. The only difference seems to have been the use of a fork of 528 vib./sec. "driven by a vacuum-tube circuit," and the direct comparison of the fork with the Coast and Geodetic Survey pendulum, instead of an indirect one involving the use of an auxiliary pendulum. He states that this new drive gave "a far more nearly constant" rate. Details of the circuit are not given.

No illustrative specimen of any of the types of observations involved in a determination is given, nor any other detail, but merely the value used for the length of the light path and the means for each of 10 series of observations. The values of the velocity so found are given in table 25.

He regarded both these and the determinations of 1924 as "preliminary." "The definitive measurements were begun in June 1926 and continued until the middle of September."

For these definitive measurements, the rotating mirror and the subsidiary reflectors for directing the light to and from it were so adjusted (see fig. 11) that the light was incident almost normally upon diametrically opposite faces of the prismatic rotating mirror. The old "small" glass octagon, used in 1924 and 1925, and four new prismatic mirrors were used. The new mirrors were as follows:

12-face, glass, 6.25 cm. in diameter, speed 350 rev./sec.
16-face, glass, 7.5 cm. in diameter, speed 264 rev./sec.
8-face, nickel-steel
12-face, nickel-steel

Neither the size nor the speed of either of the steel mirrors is stated, but the speed was presumably the same as for the glass mirrors having the same number of faces; viz. such that a face will be almost exactly replaced by the next in the interval required for the light to go to the distant mirror and return.

Each of the glass mirrors was driven by an air blast impinging on the vanes of an open fan wheel. There were two nozzles, so directed that the mirror could be rotated in either direction. The number of vanes is not stated, and it is difficult to tell from the illustration; there seem to have been eight, but there may have been only six.

The steel mirrors were also driven by air, but for them the air issuing from four co-operating nozzles, arranged at intervals of 90°, impinged on buckets cut in the edges of a wheel attached to the axle. Each mirror had two oppositely directed bucket wheels and nozzles; so that it could be rotated in either direction. The number of buckets on a wheel is not stated, but the one (octagon) shown in his figure 3 seems to have had 24.

The lengths of the several elements making up the distance $D = 35.4245$ km. that the light travels in going from the rotating mirror to the distant mirror that returns it, as well as their sum, are given in Appendix I of his paper. Two sets of comparisons of the C. G. S. pendulum with the observatory clock, one on July 1 and the other on August 13, agreeing to less than 1 in 100,000 after being corrected for the rate of the clock, are given in Appendix II; but in the next report to be considered [36] it is said that consistent readings, with the C. G. S. pendulum then used, could not be obtained until the pendulum case had been inclosed in a constant-temperature one. And Major William Bowie's report on the measurement of the distance between the C. G. S. markers on the two mountains is given in Appendix III. From the last, it may be seen that Major Bowie considers that the total uncertainty in that distance does not exceed about 1 in 190,000 (see the present study of Michelson's report of 1924).

No other details of the work are given. Only series of averages and sets of series are reported. Consequently, there is no foundation on which the reader can base an objective evaluation of the published result.

However, the deviations δ of the several reported values, presumably averages, from the mean of the set including them are such (see table 25) that it would have been impossible to have been sure that a similar value obtained under intentionally changed condition

meters per second in group means may have resulted solely from atmospheric variations.

When the precision of the measurements is otherwise such that retention of the sixth digit in the velocity is justified, the mean temperature and pressure along the entire path should be known for each determination, and the corresponding correction should be applied. But whether it is practicable to determine, with satisfactory accuracy, that mean temperature over such a long line seems open to question.

[44] The commonly accepted relation $U = V - \lambda dV/d\lambda$ for deriving the group velocity U from the wave velocity V is the best available, but Ehrenfest [34] has pointed out that its current justification leaves much to be desired. The only direct evidence for its correctness in the case of light is Michelson's measurement [28] in 1885 of the ratio of the velocity of light in carbon disulphide to that in air. A repetition of that work with the much better facilities now available, and an extension of it to other substances, are much to be desired. A physical explanation of the difference between U and V has been offered by Osborne Reynolds [35].

TABLE 25

MICHELSON'S VALUES BY PRISMATIC MIRRORS: 1924–26

For the 1924 series, Michelson gives but one value of V—299.735 for the mean K (see the study of his report of 1924). The values for the eight individual values of K have been computed for the present paper. All other values of V have been directly copied from his report. At the bottom of each column of V's for 1926, are given both his weighted mean (W. M.), and the unweighted mean (M), wt. is the weight he assigned to the adjacent value of V; δ is the excess of the corresponding V above the M for its column, and at the bottom of each column of $1000\,\delta$ is the mean value irrespective of sign.

V = velocity of light in air; its unit is 1 megameter/second

Octagon, glass; D=35.4263(?) 1924			Octagon, glass; D=35.4251 1925		Octagon, glass; D=35.425 1926			12-face, glass; D=35.4245 1926		
K	V	$1000\,\delta$	V	$1000\,\delta$	wt.	V	$1000\,\delta$	wt.	V	$1000\,\delta$
132.20	299.735	− 3	299.695	+ 6	2	299.747	− 2	1	299.736	+ 4
.21	.757	+ 19	.651	−38	2	.747	− 2	3	.745	+13
.20	.735	− 3	.671	−18	3	.738	−11	3	.733	+ 1
.24	.826	+ 88	.677	−12	3	.762	+13	3	.730	− 2
.20	.735	− 3	.722	+33	3	.729	−20	1	.700	−32
.22	.780	+ 42	.695	+ 6	3	.759	+10	5	.727	− 5
.19	.712	− 26	.725	+36	1	.792	+43	5	.718	−14
132.15	299.621	−117	.686	− 3	4	.794	− 5	5	.727	− 5
			.707	+18	4	.741	− 8	1	.757	+25
132.20	299.738	38	299.662	−27	4	.747	− 2	2	.766	+34
					4	.744	− 5	2	.748	+16
			299.689	20	4	299.741	− 8	5	.724	− 8
								5	.742	+10
					W.M.	299.746	11	5	.718	−14
					M	299.749		5	299.715	−17
								W.M.	299.729	13
								M	299.732	

16-face, glass; D=35.4245 1926			16-face, glass; D=? 1926			12-face, steel; D=? 1926			Octagon, steel; D=? 1926		
wt.	V	$1000\,\delta$	wt.	V	$1000\,\delta$	wt.	V	$1000\,\delta$	wt.	V	$1000\,\delta$
1	299.727	−12	1	299.766	+38	2	299.712	−16	3	299.730	+ 2
2	.766	+27	5	.721	− 7	4	.730	+ 2	3	.721	− 7
2	.748	+ 9	5	.727	− 1	4	.730	+ 2	3	.733	+ 5
5	.707	−32	1	.733	+ 5	5	.727	− 1	5	.718	−10
5	.727	−12	5	.709	−19	5	.730	+ 2	3	.723	− 5
4	.737	− 2	5	.724	− 4	5	.739	+11	3	.744	+16
3	.769	+30	2	.709	−19	2	.718	−10	3	.733	+ 5
2	.724	−15	2	.706	−22	2	.727	− 1	5	.730	+ 2
2	.763	+24	3	.739	+11	3	.748	+20	5	.724	− 4
4	.715	−24	3	.742	+14	3	.724	− 4	5	.724	− 4
4	.730	− 9	3	.718	−10	3	299.718	−10	5	299.730	+ 2
2	.727	−12	1	.712	−16						
2	.742	+ 3	1	299.763	+35	W.M.	299.729	7	W.M.	299.728	6
3	.742	+ 3				M	299.728		M	299.728	
3	299.760	+21	W.M.	299.722	16						
			M	299.728							
W.M.	299.736	16									
M	299.739										

departed significantly from the others if it did not depart from the mean of the set by much more than the technical probable error e_q of a single value of the set. In fact, the minimum discordance dubiety for such a test is about 2.5δ (see eq. 21). These values are given below. The smallest of these discordance

Mirror	Glass	Glass	Glass	Glass	Glass	Glass	Steel	Steel
Faces	8	8	8	12	16	16	12	8
Year	1924	1925	1926	1926	1926	1926	1926	1926
δ	38	20	11	13	16	16	7	6 km./sec.
e_q	34	18	10	11	14	14	6	5 km./sec.
2.5δ	95	50	28	32	40	40	18	15 km./sec.

dubieties (2.5δ) amounts to 15 km./sec. Consequently, a systematic error could not have been detected by such a test, even in the most favorable case, unless it was at least as great as 15 km./sec. And the report does not even state that any test for systematic error was made.

The reported sets of average values for the velocity of light are given in table 25, together with the deviations from the mean. It will be noticed that the average deviation for the steel mirrors is only about half as great as that for the 1926 determinations with the glass ones, and the latter is less than half that for the glass octagon in 1924. Even if the large fluctuations in 1924 be ascribed to fluctuations in the frequency of the fork, which need not necessarily be the correct explanation, no such explanation can be accepted for the difference between the steel and the glass mirrors in 1926. That difference requires serious consideration, but is not mentioned in the report.

Furthermore, the means of the two sets of values for the 16-face glass mirror differ by almost 3/4 of the mean deviation of the values in either set. That needs consideration, but none is given it.

When the several means are assembled, as in table 26, other interesting relations appear.

TABLE 26

SUMMARY OF MICHELSON'S RESULTS WITH PRISMATIC MIRRORS, 1924–26: VELOCITY OF LIGHT IN AIR

In column WM_1 are given Michelson's weighted means of the eight sets of observations given in table 25, the set of 1924 with the eight-sided glass prism being indicated by G8 1924; similarly for the others. In M_1 are the unweighted means of the same sets, each followed by δ_m, the mean deviation of the individual values from M_1. In WM_2 are Michelson's definitive values for each of the mirrors; in M_2 are the corresponding unweighted means; $\delta' = M_1 - 299730$. (Michelson added a flat 67 km./sec. correction to vacuum to each of the WM_2 values before summarizing and averaging).

Unit of WM, M, $\delta = 1$ km./sec.

	WM_1	M_1	δ_m	WM_2	M_2	δ'
G8 1924	299 735	299 738	38			+ 8
G8 1925	689	689	20	299 730	299 725	−41
G8 1926	746	749	11			+19
S8 1926	728	728	6	728	728	− 2
G12 1926	729	732	13	729	732	+ 2
S12 1926	729	728	7	729	728	− 2
G16 1926	736	739	16			+ 9
G16 1926	299 722	299 728	16	299 729	299 734	− 2
Mean	299 729	299 727		299 729	299 729	
Mean:						
Omitting G8	299 731			299 730		
Omitting G8, 1st G16	729					
Omitting G8, 2nd G16	732					
Omitting G8, 1925	735					
Mean 1926	734 Definitive measurements.					

Except for the flat 67 km./sec. correction to vacuum, the values in column WM_2 are those given by Michelson immediately below his table VIII, and of which he wrote: "When grouped in series of observations with the five mirrors the results show a much more striking agreement." The agreement is indeed "striking." But if, as in column M_2, they had not been weighted, the agreement would have been far less striking. Furthermore, the mean for G8 includes both the very abnormally low value of 1925 and the value of 1924, which rests on a distance D that involves two "provisional" lengths and an unexplained "correction" of 3.2 m. (see the present study of his 1924 report); also the mean deviations of the components of these values are 2 and over 3 times as great as the average mean deviation for the 1926 values; and besides, the 1924 and 1925 values are explicitly described as "preliminary" (p. 3). Whence, the propriety of including them is open to serious question.

If these "preliminary" values be omitted, the value in WM_2 for G8 becomes 299,746, which is 17 km./sec. greater than the mean of the others; and in M_2 it becomes 299,749. The "striking" agreement has now completely vanished. It arose solely from a happy weighting and a combining of very discordant values.

The unweighted mean of all the sets of "definitive measurements" (1926) is 299,734, which is 5 km./sec. greater than the value given by Michelson as definitive. It is only by ignoring all determinations with G8, or by including the "preliminary" ones of 1924 and 1925, that a value fairly concordant with the mean of the others can be obtained.

Finally, an omission comes to light when one compares the values given in his *Studies in Optics* (1927: 136) with those given in this report of 1927. There, he states that eight determinations made in 1924 gave for the velocity of light in air the value 299,735, another series in 1925 gave 299,690, and a third series, in which the vibration of the fork was maintained by an "audion circuit," gave 299,704. When these are given the respective weights 1, 2, and 4, their weighted mean is 299,704, and the corresponding velocity in vacuo is 299,771. This "should be considered as provisional." A little farther along, on page 137, it is stated:

> Observations with the same layout were resumed in the summer of 1926, but with an assortment of revolving mirrors. The first of these was the same small octagonal glass mirror used in the preceding work. The result obtained this year was $V = 299{,}813$ [299,746 in air]. Giving this a weight 2 and the result of the preceding work weight 1 gives 299,799 for the weighted mean [in a vacuum].

On comparing these values with those in table 26, which contains in column WM_1 Michelson's weighted means of the several sets given in table 25, which contains all the values given in his 1927 report, it will be seen that the first, second, and fourth (299,735, 299,690, and 299,746) are the same as the first three in that table, except for a unit in the last place of the second. But the third, 299,704, also made with the small glass octagon in 1925, does not appear in his report, although in *Studies in Optics* it is rated as twice as good as the 299,689, which does appear. Furthermore, this 299,704 value also appears in the article on the velocity of light published over Michelson's name in edition 14 of the *Encyclopaedia Britanica* (**23**: 34–38, 1929). But this apparently better value was omitted from the report.

Michelson gets his definitive value by adding 67 to the mean of column WM_1 of table 26, and writes it thus: 299,796±4.

The significance of the ±4 is not stated. But, as already pointed out, an uncertainty of 1 or 2 km./sec. arises from uncertainties in the temperature and pressure of the air; another of about 1.5 comes from an uncertainty of 1 in 190,000 in the distance; a third is the discordance dubiety which exceeds 15; and a fourth, 5, comes from uncertainties as to how the several sets should be combined. The total uncertainty exceeds 20 km./sec. And the correction for the "group" velocity increases Michelson's value by 2 km./sec. Whence it seems that the best he would have been justified in claiming for the 1924–26 series of determinations is this:

Velocity of light in a vacuum..........	299,798 km./sec.
Dubiety at least.....................	±20 km./sec.

That is, the velocity of light in a vacuum may, but does not necessarily, lie between 299.78 and 299.82 megameters per second.

MICHELSON, PEASE, AND PEARSON'S REPORT OF 1935

The report [36] of Michelson, Pease, and Pearson covers the measurements made between February 19, 1931, and February 27, 1933, inclusive, the total light path being either 8 or 10 miles long and lying throughout nearly its entire length in an exhausted tube about a mile long, the additional length being obtained by multiple reflections within the tube. The work was proposed and planned by Michelson, who, however, died on May 9, 1931, when only 36 of the 233 series of observations had been completed.

In that it gives far more details, this report is much more satisfactory than the others of this series, but a number of things that are essential to a clear understanding and independent evaluation of the work are lacking. Some of these will be mentioned in what follows.

The pressure in the pipe, which contained 4 fixed, but adjustable, mirrors, varied from 0.5 to 5.5 mm. of mercury; correction was made for the presence of this air, assumed to be of atmospheric composition. Light from the rotating mirror entered the pipe through a slightly inclined, plane parallel plate of glass, and after traveling 8 or 10 miles, depending on the adjustment of the mirrors, it emerged through the same plate, to strike the rotating mirror again, and to be reflected by it to the eyepiece.

In that a single converging system was used to focus the light from the rotating mirror on the distant plane mirror, the optical set-up resembled that in Michelson's work of 1879 [24] and of 1882 [28], but the lens then used was now replaced by a concave mirror, which formed on the distant flat an image of the illuminated slit.

The rotating mirror was a regular glass prism of 32 faces. The prism was 0.25 inch long and 1.5 inches along the diagonals of its cross-section; its angles were "correct to 1.0″ and its surfaces flat to 0.1 wave." "The mounting is one of those used in the Mount Wilson-San Antonio experiments having compressed-air turbine drive capable of rotation in either direction." Nothing more is told about the mounting or the motor, but it seems from plate II that the mirror was not spanned by a rectangular frame to carry the upper bearing, as in figure 2 of the preceding report [31], but that nearly half of that frame had been cut away, leaving merely an inverted L.

The approximate values of the instrumental constants were as follows: the flats were 22 in. (55.9 cm.) in diameter; the concave mirror was 40 in. (101.6 cm.) in diameter, 49.28 ft. (15.02 m.) focus; the rotating mirror was a 32-face prism of glass; the slit was 0.003 in. (0.075 mm.) wide; distance from rotating mirror to micrometer, $r = 11.8$ in. (30 cm.). Two distances were used: $D = 7.99987$ km., $m = 585$ turns/sec., rate of sweep of light across distant mirror (table 38), $s = 0.004V$; and $D = 6.40559$ km., $m = 730$ turns/sec., $s = 0.004V$.

The speed of the mirror was again controlled and determined by a stroboscopic comparison with a tuning fork adjusted to the desired frequency, the final adjustment being "made by adding a small lump of universal wax to each prong." No information is given as to the means used for driving the fork. The period of the fork was determined by stroboscopic comparisons with a free pendulum formerly used for gravity determinations by the United States Coast and Geodetic Survey. This pendulum swung in a heavy bronze box that was exhausted to a low pressure; corrections were made for the temperature and the amplitude. The period of the pendulum was stroboscopically compared with a timepiece, which was itself compared with time signals from Arlington, Va. This was done "several times during 1931 and before and after each experiment in 1932–1933." "Consistent readings could not be obtained with the pendulum in 1931, but its inclosure in a constant-temperature case in 1932–1933 eliminated this difficulty" (p. 39). No such difficulty is mentioned in the report of 1927 [31], although it seems that the same type of pendulum was used.

The speed of rotation of the mirror was such that during the time taken by the light to go to the distant mirror and return, the mirror turned almost exactly 1/32 of a revolution, thus replacing each face by the following one. The small difference from this 1/32 was determined from the deflection of the returned image, as measured by the micrometer screw that moved the observing eyepiece. Actually, it was not the deflection itself that was measured, but the doubled deflection produced by reversing the direction of rotation of the mirror. It is a pity that no single deflections are reported, since they might have given valuable information regarding the behavior of the rotating mirror. In particular, they would have shown whether the deflection was the same in both directions. True, the doubled deflection in the single published sample set of observations was only 0.00571 in. (0.145 mm.); so that the deflection in one direction was, in that case, quite small (0.072 mm.). Nevertheless a difference of 0.01 mm. in it corresponds to a difference in the computed velocity of light of about 26 km./sec., and should have been easily detectable if the accuracy was such as to justify the retention of the sixth digit of V.

The light from the slit was reflected from the upper half of a face of the rotating mirror, and the returned light from the lower half of the following face. Reversing this procedure was found to produce no difference in the result.

"The slit, condensing lens, rotating mirror, air controls, etc., were mounted outside the tube. In 1931, they were on a metal table bolted to the cement floor" (p. 30). (A few lines farther along, one finds a statement that seems to say that this "floor" was in reality a "massive concrete pier 3 feet thick, whose top lay flush with the floor.") After 1931, "the slit, prism [for directing the light], rotating mirror, and observing eyepiece were mounted on a heavy cast-iron base, fastened to a solid concrete pier," which was itself fastened to the thick pier previously mentioned. All of this seems obviously to have been intended to provide solidity and to obviate vibrations of the several parts. But vibrations have a habit of traveling along and through piers and tables, and of building up surprisingly in parts that have the proper natural frequencies. The numerous rods and clamps shown in plate II would seem to provide many opportunities for such building up of vibrations, especially those of high frequency, which would be the most troublesome, especially if they were aliquot parts of the number of turns per second of the mirror (585, 730, or 732).

A very careful search for such vibrations and for effects produced by them should have been made. But all that is published about it is this: For the first 25 series the directing prism

> was mounted directly on the rotating-mirror support, but for the remaining work it was mounted on a shelf attached to the table, thus eliminating displacements due to any possible turbine reaction. Later experiments showed that these displacements were negligible (p. 33.).
>
> Attempts to explain these variations in velocity [of light] as a result of instrumental effects have not thus far been successful (p. 28).

One would like to know the nature and extent of the tests that were made.

The source of light was an arc lamp placed outside the observing room, and focused on the adjustable slit. The slit was about 0.003 in. (0.075 mm.) wide during the observations, which were usually made during the early hours of the night, but some were around midnight and 3 A.M.

The length of the light path $2D$ was determined by the experimenters themselves in terms of tapes and a baseline a mile long laid out about 10 feet to the west of the pipe, marked and measured by the United States Coast and Geodetic Survey. The average of three series of determinations of its length by the Survey is given as 1,594,265.8 mm. The number of digits in this value is obviously excessive. The length cannot be known to a higher accuracy than that of the tapes that were used in measuring it, and it is improbable that they were certified to a higher accuracy than those used in measuring the distance from Mount Wilson to Mount San Antonio; viz. 1 in 300,000 (see Appendix III of the report [31] of 1927 and the present discussion of the report [29] of 1924). Hence the 8 has no physical significance whatever, and the preceding 5 is uncertain by 5; i. e., the uncertainty in this length amounts to at least 5 mm. It is interesting to notice that the three sets of measurements made by the Survey have a range of only 13 mm.

The face of each of the two flats between which the multiple reflections occurred was essentially normal to the baseline and near one of its terminal marks. The longitudinal distance of each of those planes from its neighboring mark was determined by a 12-ft. straight edge inserted through an opening in the tube. Other linear distances, none over 14 m., were measured by means of steel scales and stretched steel tapes. Few details of the work and no specimen observation, from which one might estimate the uncertainties, are given. One is told that a scale and a trammel bar were clamped to the tape, but nothing is said of the effect of this upon the reading of the tape—upon the nominal value of the distance measured. The published means of each of the component lengths are given to the nearest tenth of a millimeter, and for each of the three periods of continuous work—1931, 1932, and 1932–1933. The total light path $2D$, as so determined for each period and distance, but rounded off to the nearest millimeter, is as follows:

	1931	1932	1932–33
10-mile;	$2D = 15.999744$	———	——— km.
8-mile;	$2D = 12.811183$	12.811208	12.811223 km.

These values include correction for the effective vacuum length of the light path through the air outside the tube and through the glass window. Anderson has suggested [33] that an additional correction should be applied to take care of the difference between the phase velocity and the group velocity in the glass window. But the window was only 2 cm. thick (total path 4 cm.), whereas the total light path was 13 to 16 km.; hence, if the glass had been completely ignored, the result would not have been changed by more than 5 parts in a million. Correction for the residual air in the tube was applied to each determination of the velocity.

By a "set" of observations is meant (usually) 5 groups of micrometer settings, 5 settings to a group.

The groups correspond alternately to left and to right rotation of the mirror. Several sets (average is about 4), covering an interval of about an hour, are combined to form a "series"; 233 series of observations were made.

In the sample set of observations given in their table IV, the sign of d, and consequently of a, is at variance with that demanded by the definition and value of f. This probably resulted from a change in convention; it may be that the definition of f assumed that the micrometer readings increased in a direction opposite to what they actually did. The report does not contain such information as will enable the reader to decide.[45]

During the first 25 series only two groups of settings were taken for each set, one for left rotation, the other for right; but a slight drift in the values was noticed. For that reason, later sets consisted of an odd number of groups, which were so combined as to eliminate the drift. From five groups, three different combinations that eliminate the drift may be obtained. For that reason, they regarded such sets as equivalent to three determinations of the velocity, and the entire 233 series, comprising about 1,110 sets, as equivalent to 2,885 determinations.

No explanation of the drift is offered. The drift shown by the single published set of observations (table 27) rests almost entirely on the difference (0.00058 inch) in the averages of the two groups corresponding to rotation in the direction designated as R. As the micrometer reads directly to only 0.001 in., and as the mean deviation of the five readings of each group from their mean is 0.00027 and 0.00025 in., respectively, the presence of any real drift in that sample is open to question. It will be noticed that even when the drift has been "eliminated" by combining the groups three by three, the resulting values for the double deflection d still vary from 0.00602 to 0.00541 in. This change of 0.00061 inch corresponds to a change of 6.6 in 100,000 in V; i. e., to 20 km./sec. They use the average (0.00571 in.) of the three differences obtained by combining the groups three by three. That is the best one can do; but this question obtrudes itself: What would have been found had two or four more groups of observations been taken? Would the value of d have become still smaller?

With only this one set of observations as a guide, it is impossible to answer these questions, but it seems to the present writer probable that there was no true drift and that it was impossible to determine the actual value of the double deflection d to an accuracy exceeding 0.0003 or 0.0004 inch (say 0.01 mm.), which corresponds to 10 or 13 km./sec. in the velocity of light. This seems to be borne out by the fact that the residuals of the sets from the mean for their series run as high as 60 km./sec., and their average deviation is about 11 km./sec. (see their tables VI and VII).

They summarize their results in table VII, the contents of which are here reproduced as table 28. The means of the four groups into which the results have been distributed range from 299.770 to 299.780, averaging 299,774 km./sec. The average deviation of the "series" members of each group from their mean ranges from 8 to 12, averaging 10 km./sec. for the whole; and of the "set" from their "series" ranges from 9 to 12, averaging 11 km./sec. Of the 2,885 determinations, 1,095 lie within the range 299,770 to 299,780 km./sec., leaving 1,790 outside that range. The curve showing the frequency distribution of the several determinations is given in the report. It is

[45] Serious confusion results from the authors' having used N in two distinct senses: (1) to denote the nearest whole number of vibrations of the fork per second (p. 41), and (2) to denote twice that number (p. 46). The second results in an erroneous relation in the last line on page 46, where it is stated that $b=a/N=n/365\times2$. The "$\times2$" should be deleted. The authors actually require an additional symbol, say m, to denote the multiplicity of the stroboscopic image. Then the denominators of the expressions for T on page 41 will contain m as a factor; the title of their table III will be "Values of $32mND$," and the second column will be split into two, one headed N and the other m; in their table IV, "$N=365\times2$" will be replaced by $N=365$, $m=2$; in the last line of page 48 the "$\times2$" will not occur; and on p. 49 "$32ND$" will become $32mND$.

TABLE 27

SAMPLE SET OF MICROMETER SETTINGS (JUNE 30, 1932)

L = rotation to left (counterclockwise); R = rotation to right; δ = deviation from mean.

Unit is 0.001 in. (0.0254 mm.)

Rotation	L	δ	R	δ	L	δ	R	δ	L	δ
Settings	13.6	0.08	19.2	0.54	14.0	0.24	18.8	0.36	13.2	0.54
	13.8	.12	19.6	.14	14.2	.44	18.9	.26	14.2	.46
	13.7	.02	20.0	.26	14.0	.24	19.2	.04	13.5	.24
	13.4	.28	19.9	.16	13.1	.66	19.5	.34	13.6	.14
	13.9	0.22	20.0	0.26	13.5	0.26	19.4	0.24	14.2	0.46
Mean	13.68	0.14	19.74	0.27	13.76	0.37	19.16	0.25	13.74	0.37
Mean of adjacent two			13.72		19.45		13.75			
Difference			6.02		5.69		5.41			

TABLE 28

MICHELSON, PEASE, AND PEARSON'S SUMMARY OF THEIR VALUES FOR THE VELOCITY OF LIGHT IN A VACUUM

From their table VII. Number = number of separate determinations, as defined in the text. A. D. = average deviation of the "set" values from the mean value for the "series" = average of the corresponding values of A. D. in their table VI; δ = mean deviation of the "series" values from their mean (computed for this study):

Unit of V = 1 km./sec.

Series	Date	Number	Mean V	A. D.	δ
1 to 54	Feb. 19 to July 14, 1931	493	299 770	12	12
55 to 110	Mar. 3 to May 13, 1932	753.5	780	11	12
111 to 158	May 13 to Aug. 4, 1932	742	771	9	8
159 to 233	Dec. 3, 1932 to Feb. 27, 1933	897	775	11	10
		2885.5	299 774	11	10

plainly unsymmetrical, there being an excess of small values. Their table shows that there are 1,025 values below 299,770, and only 765 above 299,780. For three or four short periods, the values found remained persistently abnormal; the most striking instance is the 10 consecutive series in the interval from March 26 to April 3, 1931. Those 10 values lie within the range 299,728 and 299,757, averaging 299,746 km./sec., 28 km./sec. lower than the mean of all.

The authors endeavored to establish a connection between their several determinations and the variations in the tidal forces, but without conspicuous success. Values much less discordant than these must be obtained before such a study can be profitably undertaken.

Near the end of the report they write:

> A vibration of the mirror system with a period equal to a fraction of that of the rotating mirror conceivably may have produced the rapid fluctuations observed in the individual readings. Further experiments on a more stable terrain, with improved self-recording apparatus, carried on continuously over an extended period of time will be necessary to clear up the problem.

But still there is no indication of an awareness that the speed of the mirror itself may vary periodically, and none that it is essential that a thorough experimental study of the apparatus itself be made. As in the other papers already studied, the measured and averaged quantities are carried out to an excessive number of apparently significant digits. (A change of 1 km./sec. corresponds to a change in the micrometer reading—rotation left to rotation right—of only 0.000031 in.; and the micrometer reads directly to only 0.001 in.).

The authors give no estimate of the accuracy of their mean value, which is this:

Velocity of light in a vacuum..........	299,774 km./sec.
Average deviation of a "set"..........	11 km./sec.

With that average deviation, the discordance dubiety would have been at least 27 km./sec. (see eq. 21), if the tests had consisted solely of comparisons of single "sets" with the grand mean, and at least 4 km./sec. (see eq. 20), if they had consisted of comparisons of pairs of groups of 25 sets each. Although the report indicates that tests were made, it gives no information regarding them. Perhaps it will be safe to say that the observations indicate that the velocity of light in a vacuum lies between 299,764 and 299,784 km./sec., and that if the mean is unaffected by systematic errors that could have been detected by a comparison of pairs of groups of 25 sets each, then the velocity probably lies between 299,770 and 299,778 km./sec.

MICHELSON'S LISTS OF VALUES AND ESTIMATES

In addition to his own numerous experimental determinations of the velocity of light, Michelson has published from time to time lists of the values found by others, and what he thought from time to time was the best estimate one could make of the velocity, those estimates resting in general on the weighted means of certain selected determinations.

As pointed out by Gheury de Bray [37], errors and inconsistencies occur in both the quoted and the estimated values. Furthermore, others have sometimes taken these estimates for experimental determinations, and it is not always at once obvious whether a given value refers to the velocity in air, or *in vacuo*, the two occasionally occurring in close proximity and without a plain distinction. As a result, confusion has occurred not infrequently.

For such reasons, it has seemed desirable to collect those values in a single table (table 29), arranging them in the chronological order in which they were published, and to call attention to some of their noteworthy peculiarities.

It will be noticed that the combined result of his determinations 4 and 5 is given as 299.895 in D and once in G; a second time in G it appears as 299.880; and in the column of estimates it appears twice as 299.882, which is the plain mean of 4 and 5.

Newcomb's lower value (g 299.810) is used in the estimates of group C and in the quotation of the better values in G, but the high one (f 299.860) is given in another place in G, and in D as the better.

In E he combines 7, 8, and 9 to get 299.797 for the glass octagon; in G he combines 7, 8, and a third value, which is not included in his report, to get a provisional 299.771; and then combines that value with 9 to get 299.799, all for the same octagon. Obviously, the value found with this octagon is uncertain by at least 1 in 10,000; consequently in the following group of values the use of 299.797 for it is quite misleading.

In C, c_1 "Cornu (discussed by Listing)" is given as 299.990, but in G it is given as 299.950, which value also occurs as h in D.

TABLE 29

VALUES FOR THE VELOCITY OF LIGHT IN VACUUM, AS DERIVED BY MICHELSON FROM HIS OWN OBSERVATIONS OR QUOTED BY HIM, OR ESTIMATED BY HIM

The numbers under "Ref." refer to the bibliography; his several experimental values are indicated in columns "No." by arabic numerals; groups of quoted values or of estimates, by capital letters; individual determinations by others, by small letters. Where he has assigned weights to the several determinations of a series, these are given under the symbol "wt." in a column preceding the values to which they apply. The designations Glass 8, or G8; Glass 12, or G12; etc., indicate that the mirror was a regular glass prism of 8, 12, etc. sides. Similarly Steel 8, or S8; and Steel 12, or S12, indicate that it was a regular steel prism of 8 and of 12 sides. A. D. = average deviation from the mean.

Unit of V = 1 megameter/second; velocity in vacuum except as indicated

Ref.	No.	Date	Determinations		Quoted Values			Estimates			
			V	Remarks	No.	Source	*V*	No.	Source	*V*	Remarks
23	A	1878			a	Foucault*	298				
					b	Cornu*	300.4				
23, 24	1	1878	300.14†								
24	2	1879	?	30 Unreported							
24	3	1879	299.944 ±0.051								
24	B	1879			b	Cornu	300.4				
					c	Cornu interp. by Helmert	299.990				
28	4	1882	299.853 ±0.060								
28	5	(1879)	299.910 ±0.050	Corrected No. 3							
38	C	1902				Quotes from Newcomb:					
					a	Foucault 1862	298	b	Cornu	300.400	
					d	Cornu 1874	298.5	4, 5	Michelson	299.882	
					b	Cornu 1878	300.4	g	Newcomb	299.810	
					c_1	Cornu (Listing)‡	299.99		Mean	300.030	
					e	Young and Forbes (1880–81)	301.382		A. D.	0.250	
					5	Michelson (No. 5)	299.910				
					4	Michelson (No. 4)	299.853	c_1	Cornu	299.990	
					f	Newcomb (Selected)	299.860	4, 5	Michelson	299.882	
					g	Newcomb (all)	299.810	g	Newcomb	299.810	
						Mean (true = 299.745)	299.644		Mean	299.890	Preferred
						A. D. (true = 0.664)	0.600		A. D.	0.060	
29, 30	6	1924	299.820 ±0.006	Glass 8							
							wt.				
30	D				h	Cornu	1 299.950§				
					i	Perrotin	1 299.900				
					4, 5	Michelson	2 299.895‖				
					f	Newcomb	3 299.860				
					6	Michelson	3 299.820				
31	7	(1924)	299.802	Corrected No. 6							
31	8	1925	299.756	Glass 8							
31	9	1926	.813	Glass 8						wt.	
31	E							7	Michelson	1 299.802	G8
								8	Michelson	2 .756	G8
								9	Michelson	5 .813	G8
									Mean	299.797	
31	10	1926	299.796	Glass 12							
31	11	1926	.803	Glass 16							
31	12	1926	.789	Glass 16							
31	13	1926	.796	Steel 12						wt.	
31	14	1926	.795	Steel 8				7	Michelson	1 299.802	G8
31	F	1927						8	Michelson	1 .756	G8
								9	Michelson	3 .813	G8
								14	Michelson	5 .795	S8

* Michelson gives Foucault's value as 185 200 mi./sec.; Cornu's as 186 600 mi./sec.

† Michelson [23] gives his own value as 186 508 mi./sec.; and in reference [24] as 186 500 mi./sec. or 300 140 km./sec. (p. 115).

‡ "Discussed by Listing." This c_1 is identical with c, which was derived by Helmert from the data in Cornu's preliminary report [16]. The name "Listing" is believed to be erroneous; Listing seems [40] to have accepted Cornu's 300.4 without question.

§ Perhaps this is intended for c, 299.990.

‖ How does he get this 299.895? In C he gave it as 299.882.

TABLE 29—*Continued*

Ref.	No.	Date	Determinations		Quoted Values			Estimates			
			V	Remarks	No.	Source	V	No.	Source	V	Remarks
								10	Michelson 3	.796	G12
								13	Michelson 5	.796	S12
								11	Michelson 5	.803	G16
								12	Michelson 5	.789	G16
									Mean	299.796 ±0.004	
								7, 8, 9	Michelson	299.797	G8
								14	Michelson	.795	S8
								10	Michelson	.796	G12
								13	Michelson	.796	S12
								11, 12	Michelson	.796	G16
12	G	1927							Velocity in air (p. 136):		
					b	Cornu (p. 124)	300.400		wt.		
					c_1	Cornu (Listing) (p. 124)	299.950§	7	Michelson 1	299.735	
					i	Perrotin (p. 124)	299.900	8	Michelson 2	.690	
					a	Foucault (p. 127)	298.000	—	Michelson 4	.704¶	
					4, 5	Michelson (p. 129)	299.895		Mean	299.704	
					f	Newcomb (p. 129)	299.860		Cor. to vac.	0.067	
									Mean in vac.	299.771**	Provisional
									Preceding 1	299.771	
								9	Michelson 2	.813	
									Mean	299.799††	
								7, 8, 9	Michelson	299.797	G8
								14	Michelson	.795	S8
								10	Michelson	.796	G12
								13	Michelson	.796	S12
								11, 12	Michelson	.796	G16
									Mean	299.796 ±0.001	
12	G	1927				The more reliable results:					
					c	Cornu	299.990				
					i	Perrotin	.900				
					4, 5	Michelson	.880‡‡				
					g	Newcomb	.810				
					j	Michelson	.800§§				
39	H	1929			5	Michelson	299.910	7, 8, 9	Michelson	299.797	G8
					k	Newcomb (1880–81)	.709	14	Michelson	.795	S8
					l	Newcomb (1881)	.776	10	Michelson	.796	G12
					f	Newcomb (1882)	.860	13	Michelson	.796	S12
					4	Michelson	.853	11, 12	Michelson	.796	G16
						Velocity in air:			Mean	299.796 ±0.001	
					7	Michelson (1924)	299.735				
					8	Michelson (1925)	.690				
					—	Michelson (1925)	.704¶				
						Mean, in vacuum	299.771**				
					9	Michelson	.813				

¶ This 1925 value with the glass octagon is not given in his report [31] of the 1924–26 work.

** This "provisional" average (299.771) has not been found except in G and H. It involves a series of determinations that does not appear in his report of the work.

†† This average, involving the preceding one, has not been found elsewhere.

‡‡ This value appears in C correctly, as 299.882; in D and in G (p. 129) as 299.895; and here in G, as 299.880.

§§ This value 299.800 may be supposed to represent the preceding mean (299.796±0.001) of Michelson Nos. 7 to 14, or the 299.799 obtained by combining Michelson No. 9 with the weighted mean of Michelson Nos. 7, 8, and –.

TABLE 29—*Continued*

Ref.	No.	Date	Determinations		Quoted Values			Estimates			
			V	Remarks	No.	Source	V	No.	Source	V	Remarks
36	15	1931	299.770								
36	16	1932	.780								
36	17	1932	.771								
36	18	1932–1933	.775								
		Mean	299.774								
		A. D.	0.011	(see table 28)							

CONCLUDING REMARKS

The present study and comparison of the papers by Michelson on the velocity of light reveal certain striking peculiarities and similarities.

1. Although each series of determinations has yielded a value that differs from each of the others, Michelson has made no attempt in his reports, or elsewhere, so far as I know, to account for those differences.

2. Not one of his reports contains sufficient detailed information to enable a reader to form an independent and objective evaluation of the result. Whatever value he may attach to it, is purely subjective, resting solely on his confidence in Michelson.

3. The amount of detailed information that is given in a report decreases continuously from the report of the work of 1879 to the report of 1927, but the decrease is not a mere avoidance of repetition.

4. When details are given, they have to do with the simplest of the measurements, those open to the least question. Of the more recondite measurements, those involving real difficulty and where mistaken procedures would not be especially surprising, little or nothing is said.

For example, take his report of the work of 1879. There, in the simple straightaway measurement along and on level ground one is told even the kind of markers that were used at the ends of a tape-length. But when it comes to determining, from that length on the ground, the distance between specific points on the tops of two piers 11 feet high, situated near the ends of the measured line, nothing is said about how it was done. Merely the result is given. And of the determination of the distance from the face of the rotating mirror to the hair-line of the micrometer—a determination that involves real difficulties and that is of importance equal to that of the distance between the mirrors—no more is said than that it was determined by a tape so stretched that the drop in the catenary was "about an inch." Not even a specimen set of those measurements is given, nor anything else that will indicate their reproducibility. But values of that distance are tabulated to 0.001 ft. (0.3 mm.).

In the 1927 report there are given Major William Bowie's report on the measurement of the distance between the Coast and Geodetic Survey's markers at the two stations, and two sets of data for determining the period of the C. G. S. pendulum with which their tuning fork was compared, which is a relatively simple operation. But no detail whatever is given of any of the other measurements involved, and nothing that will enable one to form an idea of their reproducibility.

The last report (1935), written after Michelson's death, is much more satisfactory than that of 1927, but nevertheless is deficient in important details.

5. Emphasis has been continually placed on the measurement of the distance between the mirrors, and on the determination of the mean speed of rotation of the mirror [46]—two measures that can be made rather easily. But except for a few observations in the report [24] of the work of 1879, no attention has been given to possible irregularities in the speed of the mirror, and no emphasis has been placed on the difficulties that are involved in measuring the equivalent angular deflection of the returned beam. Indeed, a casual reader might well be excused for inferring from the wording of the reports that because in the later work the micrometer has to measure only the small amount by which the equivalent deflection differs from twice the angle of the prismatic mirror, therefore that measure is easily made to the required precision. Such, of course, is not the case. The total equivalent deflection being the same, a given error in the angular equivalent of the micrometer reading is just as important when the mirror is prismatic as when it is a simple plane. If the mirror turns through the angle θ while the light is going and returning, then the angular equivalent of the micrometer measurement must be exact within $\theta/150{,}000$ if the velocity derived from it

[46] The reason given for undertaking the last series of observations (report of 1935) is this:

"The measurements involve two distinct elements: first, the time; and second, the distance. It was estimated that with a rated tuning fork and stroboscopic methods the time of rotation of the mirror could be measured to one part in a million. . . . In the 1924–1926 experiments the determination of the distance required the measurement of a long base line and an extended triangulation from this base into the mountains. . . . It was felt that the direct measurement of a short base line, without subsequent triangulation, might yield an even higher order of accuracy," and the use of an exhausted tube would eliminate the necessity of applying a correction for the air. (See [36], pp. 26–27.)

is to be correct within 1 km./sec. If φ is the angle between adjacent faces of the prism, and one face is almost exactly replaced by the next while the light is going the distance $2D$, then θ is very nearly φ, and the error δ in the angular equivalent of the micrometer reading that will produce an error of 1 km./sec. in the velocity of light will be as given in table 30.

TABLE 30

REQUIRED PRECISION OF SETTING (FOUCAULT'S METHOD)

"Faces" = number of effective faces of the rotating mirror; θ = angle turned by mirror while light is going and returning; δ = angular equivalent of that error in the micrometer reading which will produce an error of 1 km./sec. in the deduced velocity of light; "Year" = year in which Michelson used the indicated type of mirror.

Faces	1	8	12	16	32
θ........	1348″	45°	30°	22.5°	11.25°
δ........	0.009″	1.08″	0.72″	0.54″	0.27″
Year.....	1879	1924–26			1931–32

Such small δ's are not easily measured. Newcomb has stated [21, pp. 122–123] that "the astronomical limit of accuracy may be considered as an important fraction of a second of arc." Furthermore, a lateral displacement of the apparent center of the image by an amount δ from its ideal position will produce in V just as great an error as an error of δ in the reading. There is nothing in the reports that suggests that any attempt has been made to show the absence of such displacement.

As for irregularities in the speed of the mirror, it seems almost certain that the only irregularity thought of and looked for in the 1879 work was a slight temporary slowing of the mirror while it was passing through a certain fixed angle. The reported observations were, as has been seen, not sufficient to do more than show that there was no marked change in the neighborhood of the angle at which reflection occurred. They were entirely insufficient to establish the absence of a vibration of the mirror about its state of uniform speed. But that is a type of disturbance that is to be expected, that may be easily overlooked, and that may produce serious systematic errors. There is nothing in the reports to indicate that the possibility of such a disturbance was ever considered.

6. If a vibrational motion about the axis of rotation had been actually superposed upon the motion of uniform angular velocity of the mirror, as is to be expected, what observable effects would have been produced? Would those effects have led to differences of the kind observed when Michelson's several results are compared?

Such vibration may produce one or more of several effects. In general, it will cause a broadening and blurring of the image, and the broadening may be asymmetric; it may give rise to a multiplicity of images; those components that have periods equal to, or a submultiple of, the time taken for a face of the mirror to be exactly replaced by the following one will, in general, give rise to a steady image displaced from the position it would have occupied had there been no vibration.

With the single exception of multiplicity of images, none of these effects are of a kind to attract the attention of the observer. He will become aware of them only when he suitably changes the experimental conditions and compares the results. They must be looked for. Nevertheless, both the asymmetric broadening and the displacement of the image introduce systematic error.

If a disk mirror be so adjusted that small changes in the azimuth of its mounting produce no appreciable effect upon the derived velocity of light, then the effect of the vibration is at its greatest. This seems to have been the case in Michelson's experiments of 1879 and 1880. Also, if a prismatic mirror rotates at such a speed that one face is exactly replaced by the next in the time taken for the light to go and return, then the vibration will produce no steady displacement of the image. This was closely the case in the determinations of 1924. And the derived value for the velocity of light dropped from 299.85 in 1879 and 1880 to 299.80 in 1924; just as one would expect if vibrations had been present.

But even with prismatic mirrors used at that speed of rotation, errors from asymmetry and broadening remain. These, and consequently the steadiness of the image and the reproducibility of the results, will depend upon the amplitudes of the several components of the vibration. And other things being the same, the amplitudes of those vibrations will be the smaller the greater the moment of inertia of the mirror. On referring to table 25, it will be seen that the mean deviations δ for the heavy steel mirrors are much smaller than those for the glass ones; and of the latter, those for the smallest (the glass octagon) are distinctly greater than those for the others. Again the change is in the direction that one would expect.

The variations in Michelson's data from report to report are of the kind that would have been caused by such vibrations as are here considered.

It may be asked: How did it happen that his determination of 1882 agrees so closely with that of 1879? It may have been mere chance. But it may have occurred somewhat as follows.

In 1879 he made 30 sets of unpublished determinations before beginning the 100 sets that he reported. It is reasonable to suppose that from those 30 sets he obtained information that guided him in the ultimate adjustment of the apparatus. And from the fact that the first few of the reported sets are intended to convince the reader that changing the azimuth of the frame of the mirror produced no change in the result, one may infer that part of the omitted 30 had to do with that question. Although one cannot speak with certainty about what was actually done, is it not plausible to assume that from those 30 sets he found that when the frame of the mirror had a certain azimuth with the line joining the mirrors, then the result obtained was independent of small displacements from that azimuth, and concluded that that unique result was the correct value for the velocity of light? The argument is, of course, fallacious, but he was only twenty-seven years old, was only recently graduated from the Naval Academy, and was not an experienced investigator. His only previous published investigation was the brief report [23] of his preliminary work of 1878, presented to the American Association for the Advancement of Science.

Having arrived at a criterion for the adjustment of the mirror, it would have been quite natural for him to apply the same criterion in the adjustment of the same mirror, driven in the same way, and used for measuring the velocity of light over a path of essentially the same length, at Cleveland. And consequently he would then obtain essentially the same, equally erroneous, result. There is nothing occult or peculiar about the agreement of the two results. They should have been expected to agree. The second determination is, in a very real sense, merely a continuation of the first.

As to why the chosen criterion was that which corresponded to a maximum, instead of to a minimum, value, it is profitless to speculate.

7. Numerical values are usually given with an excess of apparently significant digits.

8. In the report of the work of 1879 it is stated that 30 sets of determinations were omitted; no adequate explanation is given. And it has been seen that the result of a series of measurements made in 1925 is given in *Studies in Optics* and in the article on the velocity of light in edition 14 of *Encyclopaedia Britannica*, but is not mentioned in the report of 1927, covering the determinations of 1924–26.

9. Both metric and customary units are used in the same report, often without a clear designation of which applies to the number given. And although the mathematical relations involved are identical in every case, the symbols used and the form in which the relations are expressed vary from report to report, unnecessarily increasing the labor of one trying to intercompare them.

10. In 1900 Cornu [17] pointed out that the rapid sweep of the light across the distant mirror in Foucault's method might cause an error in the result. This, with other objections raised at the same time, was considered by H. A. Lorentz [41], who wrote: "Dans ces circonstances, il est difficile de se former une idée exacte de la propagation des ondes qui forment cette image." But by means of certain assumptions he arrived at the conclusion that Newcomb's observations, with sweeps of 1 to 4 percent of the velocity of light, are probably not in error on account of that motion by more than one part in 10,000, or 30 km./sec. And of this Michelson [38] has written: "It seems to me that M. Lorentz has satisfactorily answered M. Cornu's questions."

Michelson gives no consideration to this question in any of his reports. It is not mentioned in his reports of 1924 and 1927, although in that work the rates of sweep were of the same order as in Newcomb's, ranging from 2 to 3 percent of the velocity of light. While seeming to claim that his value is in error by no more than a few kilometers per second, he ignores this potential source of error, rated by Lorentz as being possibly of the order of 30 km./sec.

11. In none of his reports on the velocity of light prior to 1935 does one find any indication of a thorough experimental study of his apparatus and procedure.

KAROLUS AND MITTELSTAEDT'S REPORT OF 1929

The investigation published by Mittelstaedt [42] in 1929, of which a preliminary report was given by Karolus and Mittelstaedt [43] in 1928, is an outgrowth of an attempt begun by Karolus in 1925 to replace the toothed wheel of Fizeau's investigation by an inertialess, electrical interrupter of the light. Its purpose was primarily to determine what precision might be expected of such a method, and how it should be modified in order to secure greater accuracy.[47] The improvements proposed are not of immediate interest.

The method is a compensation one, based upon the action of the Kerr electro-optic cell (see Appendix A). If a plane polarized pencil of light be passed through an activated Kerr cell placed with its electric field at 45° to the plane of polarization, the emerging light will be elliptically polarized. If that light be passed at once through a second Kerr cell placed so that its field is perpendicular to that of the first and is activated by the same alternating potential, then the second cell will tend to remove the ellipticity imposed by the first, and if its dimensions are suitably chosen, it will restore the original state of plane polarization. But if, on account of the time taken for the light to pass from the first cell to the second, the phase of the field at the time of passage is not the same for each cell, then the light emerging from the second cell will be elliptically polarized. When elliptically polarized, the light will pass a Nicol so oriented as to block the initial plane polarized pencil.

Here is a layout that closely reproduces the conditions of Fizeau's method, and that may be similarly used for measuring the velocity of light. It differs from Fizeau's method in that the light between the cells does not consist of a series of discontinuous groups

[47] "Die vorliegenden Messungen, die sich zum Ziel gesetzt hatten, festzustellen, welche Genauigkeit sich mit dieser Methode erreichen lässt, erlauben nun eine weitgehende Aussage über die Massnahmen, die man auf Grund der gewonnenen Erfahrungen ergreifen muss, um die bisherigen Messungen auf einen höheren Grad von Genuigkeit zu bringen." [42, pp. 310–311.]

of waves, but of an uninterrupted pencil of light, of which the polarization varies continuously and periodically from plane to elliptical and to plane again, but of which the intensity does not undergo periodic fluctuations.

Whether the velocity observed under such conditions is the "group velocity" or the "phase velocity" seems not to have been discussed. To the present writer it seems probable that it is the phase velocity.

A long light path between the two cells was obtained by multiple reflections between plane mirrors placed about 40 meters apart. The procedure was to set the mirrors so as to give a path of the desired length, and then to determine the frequency that gives an eclipse of a known order.

The theory, construction and arrangement of the apparatus, sources of error, and measurements of lengths and of frequencies, are all given careful consideration; and for each series of observations a curve is given showing the frequency-distribution of the observations as a function of the frequency of the field applied to the Kerr cells.

The path was either 250 or 333 m. long. It was carefully measured, and a correction of 101 mm. for glass, nitrobenzene, and air traversed by the light, was added in order to obtain the equivalent path in vacuo. The equivalent length of the 250 m. path was, from various causes, uncertain by ±1 cm., and the 330 m. one by ±1.2 cm. Roughly, the uncertainty was 4 in 100,000.

The frequency was uncertain by ±200 cycles per second, which for the lowest frequency employed (3.6 megacycles/sec.) amounted to 5.6 parts in 100,000; for the highest (7.2 megacycles/sec.), to 2.8.

Hence the total mensural dubiety of a determination is about 7 or 10 in 100,000; i. e., 20 or 30 km./sec.

Observations were by eye.

TABLE 31

DETERMINATIONS BY KAROLUS AND MITTELSTAEDT

V = velocity of light in a vacuum; n = order of eclipse; ν = frequency; No. = number of determinations; δ = deviation of V from weighted mean.
Unit of path = 1 m., of ν = 1 cycle/sec., of V and δ = 1 km./sec.

Path	n	ν	No.	V	δ
250.053±0.010	3	3 596 570	108	299 778	0
250.044±0.010	4	4 795 700	295	299 784	+ 6
332.813±0.012	4	3 603 130	130	299 791	+13
332.813±0.012	5	4 503 436	117	299 761	−17
332.813±0.012	8	7 205 614	125	299 760	−18
Weighted mean				299 778	11
Estimated dubiety				±20	

The result of each of his five sets of observations, covering a total of 775 individual determinations, is given in table 31. He takes as his definitive value:

Velocity of light in vacuo = 299,778±20 km./sec., to which Anderson [33, p. 196] would add 6 km./sec. to correct for the difference between the "group velocity" and the "phase velocity," giving for the velocity of light in vacuo, 299,784±20 km./sec. But, as already stated, it seems to the present writer likely that the velocity observed was the "phase velocity," in which case Anderson's correction would not apply. In any case, it is here a minor matter, in view of the dubiety being given as 20 km./sec.

Although the mensural dubiety lay between 20 and 30 km./sec. and the discordance dubiety was, perhaps, at least 4($=\delta_m/3$, see eq. 20), making a total dubiety of 24 to 34 km./sec., the evidence of care and attention to detail is such that the present writer is inclined to accept the author's estimate of 20 as probably great enough.

It should be noticed that the ±20 is not the technical probable error, but is the estimated dubiety.

ANDERSON'S REPORT OF 1937

In 1936 W. C. Anderson [44] carried out at Harvard University a series of measurements of the velocity of light. He used a Kerr electro-optic cell, as had Mittelstaedt eight years previously, but his arrangement and procedure were superior to those used in the earlier work.

He used a single Kerr cell placed between crossed polarizers, so oriented that their planes of polarization were at 45° to the direction of the electrical field of the cell. The cell was subjected to a biasing constant voltage, in addition to a radiofrequency one, so that when a pencil of light was passed through the system, it emerged as a plane polarized pencil of light in which the intensity, as observed at any point fixed with reference to the apparatus, varied almost sinusoidally. This pencil was split into two portions by a half mirror, and the two portions, after passing over paths of different lengths, each entered the same photo-electric cell. If the intensities of the two portions are in the same phase when they enter the cell, there will be co-operation, and a photo-electric current having the same high frequency as the intensities, which is the same as the radiofrequency applied to the cell, will be excited. If the intensities are in exactly opposite phases, there will be opposition, and little or no high-frequency photo-electric current.

The photo-electric current is fed into a circuit tuned to the same frequency, and is suitably amplified.

If the length of path of one of the portions into which the light is split by the half-mirror be progressively changed while that of the other is kept fixed, the amplified photo-electric current will be observed to pass through a series of maxima and minima. The distance between consecutive minima is equal to the velocity of light divided by the radiofrequency applied to the cell. And the difference in the lengths of the two light-paths when there is a minimum is an integral multiple of half the distance between two minima.

Hence the velocity V of light in air is given by the equation

$$V=2fs/n \tag{69}$$

where s is the difference in the lengths of the two paths, f is the frequency of the alternating field applied to the cell, and n is an integer, the order of the eclipse.

The theory, apparatus and its adjustments, sources of error, and measurement of lengths and of frequencies, are all given careful consideration, but additional information about the actual measurements, and an indication of the actual frequency and path-difference used for each set of values, would have been welcome.

The results are presented in the form of a table, reproduced here as table 32. As may be seen by comparison with the preliminary abstract of the report, what is marked "weight" is actually the number of determinations made on the indicated date. The "mean path difference" is given as 15,934.78±0.60 cm., and the mean(?) frequency is 14,105,120±160 cycles/sec.

TABLE 32

ANDERSON'S VALUES OF 1936

His table I with certain additions. V_0 = velocity of light in air +81 km./sec. = velocity in vacuum − correction for difference in air between "group velocity" and "phase velocity." The ±15 km./sec. is the dubiety of the mean, as estimated by Anderson; it is not the technical probable error.

Unit of V_0 = 1 km./sec.

Date (1936)	Weight	Daily mean, V_0	Average deviation from daily mean	Deviation of daily mean from final mean
June 22	52	299 773	9	+9
June 23	66	757	12	−7
Oct. 15	32	765	16	+1
Oct. 23	123	754	11	−10
Oct. 24	17	772	20	+8
Oct. 30	22	773	16	+9
Oct. 31	3	769	5	+5
Nov. 2	78	772	16	+8
Nov. 6	5	778	4	+14
Nov. 7	107	761	10	−3
Nov. 14	132	770	8	+6
Dec. 5	14	748	16	−16
	651	299 764	12	8
Correction for "group velocity"		7		
Velocity of light		299 771±15 km./sec., in vacuum		

His definitive result is given as 299,683 km./sec. in air, to which he adds 81 km./sec. for the effect of the air, obtaining 299,764±15 km./sec. for the velocity in a vacuum. To this he added later [33] a further correction of 7 km./sec. for the difference between the observed "group velocity" and the true "phase velocity," thus getting

Velocity of light in a vacuum	299,771 km./sec.
Dubiety	±15 km./sec.

It should be noticed that his ±15 km./sec. is not the technical probable error of the mean, but is his "estimated error," i. e., the dubiety. The discordance dubiety (groups of 25) is at least as great as 4 km./sec., (see eq. 20).

Since the deviation of the daily mean from the final one does not exceed the mean deviation from the daily mean, it is obvious that throughout the series, extending over five months, there occurred no change due to variations in external and uncontrolled conditions that affected the value by a certainly detectable amount, say, by so much as 12 km./sec.

ANDERSON'S REPORT OF 1941

Anderson's report [33] of 1941 covers a continuation of the work reported [44] in 1937, but with many improvements in apparatus and procedure. It was undertaken for the purpose of obtaining data of such precision that objective answers can be made to the questions raised respectively by Gheury de Bray [1] and by Pease and Pearson [36] regarding a possible secular variation in the velocity of light, and a possible effect of lunar forces upon the observed velocity of light. The results obtained were of high accuracy and gave no indication of any such changes, but the estimated dubiety was not reduced below that of his previous work.

The principle employed and the general plan of the work was the same as before. But the photo-electric cell previously used was replaced by a photosensitive electron multiplier tube of eleven stages, a more intense source of light was used, the variation in the photo-electric current as the difference in the lengths of the two paths was changed through the position of minimum current was recorded photographically and automatically, without interference by the experimenter, and the means for determining the difference in length of the two paths were superior to those previously used. A careful discussion of the whole is given. Anderson was of the opinion that the limit of accuracy was set by the functioning of the multiplier tube.

The difference in the lengths of the two paths was about 171.815 m., and was known with an uncertainty of about 1 in 100,000; the frequency used was 19.2 megacycles/sec., and was known to within 1 in a million, or better.

His results are given in a table, here reproduced in table 33. His final value, taking into consideration the fact that the observed velocity was the "group velocity" in air is

Velocity of light in a vacuum	299,776 km./sec.
Dubiety (estimated)	±14 km./sec.

The ±14 is the estimated dubiety, not the technical probable error. How it was got is not stated. It is

TABLE 33

ANDERSON'S VALUES OF 1939–40

Essentially his table I. The correction to a vacuum includes the difference between the "group velocity" (observed) in air and the phase velocity. The ±14 is Anderson's estimate of the dubiety of his final value; it is not the technical probable error of that value. "Daily" and "Final" indicate the mean from which the deviations are measured. The deviations of the individual observations are measured from the daily mean, and of the daily mean from the final mean.

Unit of velocity and of deviations = 1 km./sec.

Date	Weight	Mean velocity in vacuum	Average deviation from mean	
			Daily	Final
1939				
May 21........	20	299 774	7	2
Nov. 8........	17	775	9	1
Nov. 13........	35	759	10	17
Nov. 15........	79	772	9	4
Nov. 16........	140	780	8	4
Nov. 27........	103	781	13	5
1940				
Jan. 10........	39	774	5	2
Jan. 23........	46	774	3	2
Mar. 4........	30	757	3	19
Mar. 7........	56	745	9	31
Mar. 8........	257	754	3	22
Mar. 11........	147	749	9	27
Apr. 4........	348	808	9	32
Apr. 5........	122	774	10	2
Apr. 8........	125	769	7	7
Apr. 9........	197	771	7	5
June 15........	322	801	19	25
June 16........	94	775	1.4	1
June 21........	148	768	18	8
July 1........	293	789	12	13
July 7........	147	741	3	35
July 8........	130	758	9	18
	2895	299 776	9	14
Estimated dubiety.........		±14		

here assumed to include the discordance dubiety, which for groups of 25 is at least 3 km./sec. (see eq. 20).

The precision and accuracy realized are essentially the same as in the earlier work; and as there, there is no certain indication of any effect arising from variations in uncontrolled conditions external to the apparatus.

HÜTTEL'S REPORT OF 1940

In 1940 A. Hüttel [45] reported measurements of the velocity of light made by himself at Leipzig, under the direction of A. Karolus. It was in a quite real sense a continuation of the work by Karolus and Mittelstaedt, about a decade earlier, but with greatly modified apparatus and procedure.

As in the earlier work, a Kerr cell was used to modulate the beam of light. But in this work the cell, biased by a constant field, was placed between a pair of crossed Nicols, forming thus an electro-optical shutter. Thus the intensity of the light in the beam beyond the shutter varied sinusoidally about a fixed value; whereas in the earlier work there was no second Nicol, and the beam was of uniform intensity and elliptically polarized, the ellipticity varying sinusoidally. In that work the beam of light, after traveling a certain path, was received by a second Kerr cell followed by a Nicol, each crossed with reference to the one at the other end of the beam. In this it was received by a vacuum photo-electric cell activated by an alternating voltage of the same frequency as that applied to the Kerr cell.

Whereas in the present case the observed velocity was that of a group of waves, that in the earlier one was probably the phase velocity.

Here the intensity of the source was adjusted so that the maximum intensity of the photo-electric current was of a predetermined value, the same for each of a pair of settings, and the difference in the lengths of the paths was so determined that the photo-electric current for each setting was the same, and close to that corresponding to the zero value, on the rising side, of the approximately sinusoidal component of the variation of the photocurrent with the path length. Thus the difference in path length corresponded to an integral number of complete oscillations of the field applied to the two cells, and each setting was where the photocurrent was most sensitive to a change in setting.

The difference $2D$ in the distances traveled by the light for the two settings of the mirror was varied from 48.5 to 118.4 m. and was uncertain by about 3 mm. (the author says 0.004 percent of 80 m.).

The frequency was controlled by quartz oscillators and is said to have been uncertain by 0.007 percent.

Various potential sources of error seem to have been carefully studied, experimentally as well as theoretically. But the report is deficient in illustrative experimental data and in certain information essential to an objective appraisal of the work. For example, one is not told how the essential length-measurement was made, and nothing at all is said about correcting for the refraction of the air. To the present writer it seems probable that a refraction correction of some kind was applied to each of the several length-measurements, as in the work of Karolus and Mittelstaedt, but in the absence of definite information on the subject, this opinion is open to question.

His results are given in table 34, his definitive value being this:

Velocity of light in a vacuum(?).......	299,768 km./sec.
Dubiety (estimated)..................	±10 km./sec.

It should be noticed that his ±10 is not the technical probable error, but, as here stated, is his estimate of the dubiety. How it was obtained is not stated. But it is close to what one would infer from the discordancies and from the uncertainties in the measurements of length and frequency, taking into considera-

TABLE 34

HÜTTEL'S DETERMINATIONS OF THE VELOCITY OF LIGHT, 1940

V = velocity of light, presumably in a vacuum, but not so stated in the report; No. = number of determinations in the set, n = "order" = number of waves of frequency ν in the distance 2 D; ν = frequency of field applied to the two cells; δ = average deviation of the single determinations from the mean of the set.

Unit of V and δ = 1 km./sec.; of ν = 1 cycle/sec.

No.	n	V	ν	δ
20	2	299 772	5 058 560	17
40	1	762	5 058 560	10
10	2	785	5 058 560	18
10	1	758	5 058 560	11
13	2	750	8 299 889	5
10	3	770	12 646 400	5
28	3	776	12 646 400	9
4	2	775	12 353 472	14
Mean		299 768		11

tion the number of distinct lengths and frequencies used.

SUMMARY AND CONCLUSIONS

After remarks on several subjects related to a discussion of the reports of measurements of the velocity of light, comes a detailed study of each of those reports, beginning with Cornu's work of 1872. Then, following this summary and conclusion, come two appendixes treating more fully certain problems involved in the work.

INDIVIDUAL REPORTS

The detailed study of the reports published since 1862 has led to the following conclusions.

Cornu's work of 1872 (method of Fizeau) indicated that V lies within the range 296.5 to 300.5 megam./sec., centered on 298.5.

Cornu's report of 1874 (method of Fizeau), discussed by Helmert in 1876, was a strictly preliminary report of the work published in 1876. Both it and Helmert's discussion of it ceased to be of other than historical interest as soon as Cornu's complete and final report was published; and they should not thereafter have been considered in any attempt to arrive at the probable value of V.

Cornu's report of 1876 (method of Fizeau) covers all the work of which a portion was reported in his preliminary report of 1874, and completely supersedes that report. The value he reported rests strictly on the following assumptions:

(*a*) That all his observations lay in a normal region; that is, for the speed corresponding to any given observation, each side of each tooth of the Fizeau wheel played exactly the part it would have played had the teeth been all of equal size and uniformly spaced, had there been no eccentricity of the wheel, and had the speed been constant or uniformly accelerated or decelerated.

(*b*) That his 1/20-second oscillator recorded at intervals of exactly 0.1 sec., and that the time intervals could be read correctly to the nearest 0.001 sec.

(*c*) That his K-equation applied to the data, but that the H-equation did not.

(*d*) That his selection and averaging of the computed values were the proper ones.

It has been found that not one of those assumptions is valid. His data seem to indicate that V lies in the range 299.3 to 300.5 megam./sec., centered on 299.9.

Perrotin and Prim's report of 1908 (method of Fizeau) is little more than a repetition of Cornu's work with a longer path and other lenses. Most of the apparatus and procedures were those used by Cornu. The observations were reduced by two methods: one similar to Cornu's, and a second by least squares. The result by the first method was discarded. It has been found that the result by the second method cannot be accepted because the equations on which it rests are incorrect. Furthermore, assumptions (*a*) and (*b*) of Cornu's work are likewise made in this, and are similarly invalid. And it has been found that the acceleration of the wheel varied with the order of the eclipse in such a way as to introduce a constant error that need not be the same for deceleration as for acceleration.

The data indicate that V is likely to lie between 298 and 302 megam./sec., and may lie between 299 and 301. But the presence of systematic errors throws doubt on the validity of any kind of averaging of the values found for the several types of observations.

Newcomb's observations of 1880–82 (method of Foucault) comprise three series. During the second series, which was very short, trouble arose from a lack of balance and from damaged pivots. Newcomb thought that the trouble observed arose from an elastic torsional twist of the prismatic mirror about its axis; and before the next series was begun he had the apparatus changed so that such a twist could be definitely measured and eliminated, if it existed. Finding no indication of it in the third series, he inferred that the first, in which it could not have been detected, was vitiated by it, and discarded that series.

It has been found that such an elastic twisting as Newcomb assumed could not possibly have been great enough to have produced an observable effect. Consequently his discarding of series 1 was unjustified.

But it is certain that the mirror must have vibrated as a whole about its stable state of uniform speed, and the amplitude of such a vibration need be but small in order to account for the observed trouble and for the difference between the series. No tests mentioned in the report would have revealed such a vibration. Hence each series is presumably affected by a systematic error of unknown size and sign. That makes it improper to present any average of the series as being more nearly correct than any of the individual series. Consequently, no range can be set. All that can be done is to present the result of each series (299.71 and 299.86), neglecting the second, which was short and known to be affected by trouble of some sort, and to state that each is presumably in error.

Michelson's determinations have all been made by Foucault's method, the speed of the mirror being stroboscopically controlled and determined. The determinations of 1878, 1879, and 1882 were made with disk mirrors rotating about a diameter. In all cases the motion of the mirror must have had a periodic component, but the reports give no indication of any tests that could have revealed its existence. Consequently, as in Newcomb's work, it is to be expected that the result in each case will be affected by a systematic error. Since the work of 1882 was in reality merely a continuation of that of 1879 at a different geographical location, it is to be expected that the two will be affected by essentially the same errors, and will yield essentially the same result. The reports are very deficient in essential detail.

The work of 1878 was strictly preliminary; the results ranged from 297 to 304 megam./sec. Those of 1879 indicated the range: 299.7 to 300.1; and of 1882, 299.6 to 300.1 megam./sec.

His reports of 1924 and 1927 covered work with prismatic mirrors, as advised by Newcomb in 1882. The effect of the periodic component of the speed of the mirror will be small, perhaps zero, in this work. But, again, the reports give no indication of any search for systematic error, and are very lacking in essential detail. The report of 1924 indicates the range 299.73 to 299.87; and that of 1927, 299.77_8 to 299.81_8 megam./sec. In each case the existence of systematic errors is probable.

Michelson, Pease, and Pearson's report of 1935 covers work with a 32-face prismatic mirror; most of the path traversed by the light was in an exhausted pipe. The report is more detailed than the preceding, but contains little indication of serious search for systematic errors arising from the apparatus. If there were no systematic error, the value of V would lie in the range 299.76_4 to 299.78_4 centered on 299.77_4 megam./sec.

Karolus and Mittelstaedt's reports of 1928 and 1929 cover the same measurements, in which a Kerr cell was used. The work seems to have been carefully done and was well reported. It indicates the range 299.75_8 to 299.79_8, centered on 299.77_8 megam./sec.

Anderson's reports of 1937 and 1941 (Kerr cell used) contain each a careful study of the various elements and sources of error that are involved in the work. His observations of 1937 indicate the range 299.75_6 to 299.78_6, centered on 299.77_1 megam./sec.; and those of 1941, with an improved layout, 299.76_2 to 299.79_0, centered on 299.77_6.

Hüttel's report of 1940 (Kerr cell used) also is detailed in many particulars, but it is lacking in such illustrative data as will enable the reader to form an objective estimate of the worth of his result. And he fails to say anything about correcting his result for the refraction of the air. If it may be assumed that he has properly made the several necessary measurements, and has properly applied the correction for the refraction of the air, his observations indicate a range of 299.75_8 to 299.77_8, centered on 299.76_8 megam./sec.

These several values are assembled in table 35, and all except the strictly preliminary ones are plotted in figure 12.

TABLE 35

SUMMARY OF THE VALUES FOR THE VELOCITY OF LIGHT IN A VACUUM THAT ARE INDICATED BY THE SEVERAL SETS OF OBSERVATIONS MADE SINCE 1862

Unit of V = 1 megameter/second, in vacuum

	Observer and date	Range for V	Center V	Remarks
1	Cornu, 1872	296.5 –300.5	298.5	Preliminary
2	Cornu, 1876	299.3 –300.5	[a]299.9	Uncertain
3	Perrotin and Prim, 1908	299 301	300	Systematic error
4	Newcomb, 1880–82	299.71 299.86	299.78	Systematic error
5	Michelson, 1878	297 304	300	Preliminary
6	Michelson, 1879	299.7 300.1	299.9	Systematic error
7	Michelson, 1882	299.6 300.1	299.85	Systematic error
8	Michelson, 1924	299.73 299.87	299.80	Report deficient
9	Michelson, 1927	299.77_8 299.81_8	299.79_8	Report deficient
10	Michelson, Pease, and Pearson, 1935	299.76_4–299.78_4	299.77_4	Little study of apparatus
11	Karolus and Mittelstaedt, 1929	299.75_8–299.79_8	299.77_8	Apparatus studied
12	Anderson, 1937	$.75_6$– $.78_6$	$.77_1$	Apparatus studied
13	Anderson, 1941	$.76_2$– $.79_0$	$.77_6$	Apparatus studied
14	Hüttel, 1940	$.75_8$– $.77_8$	$.76_8$	Apparatus studied
		Mean of last five	299.77_3	

[a] The range is so great that this center value, 299.9, means nothing more than that the likely value is nearer 300 than either 299 or 301.

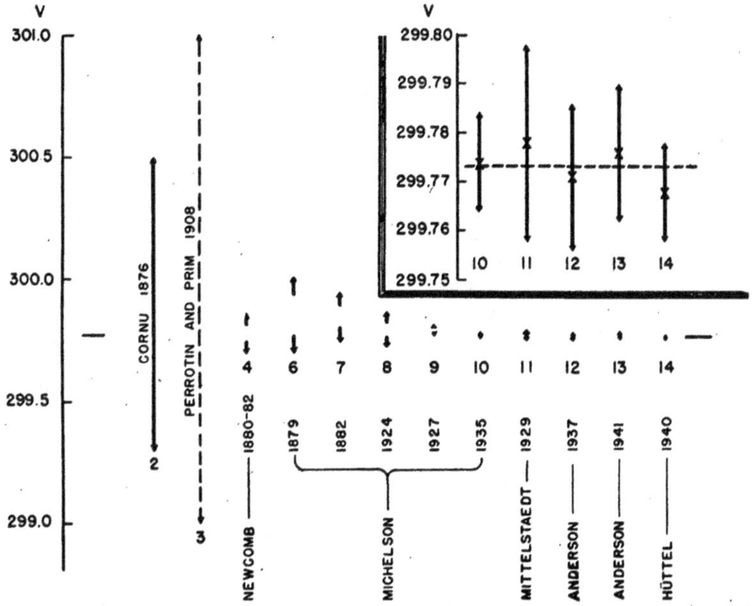

FIG. 12.—Plot of the ranges within which the several determinations indicate that the correct value for the velocity of light lies.

The values, referred to a vacuum, have been taken from table 35, and are designated by the same numbers as there. They are not in chronological order. For determinations 3 to 9, which are presumably affected by systematic errors and are not reported in sufficient detail, no range can be indicated; merely extreme values are given. The position of the mean of the last five sets is indicated by two dashes, one near either side of the figure. In the upper right corner, determinations 10 to 14 are plotted on a more open scale, their midpoints being marked by crosses, and a dashed line indicates the mean of the five. Unit of V is 1 megameter per second.

At best, a determination defines only a range within which it is likely that the correct value lies; and the significance of the center of that range decreases rapidly as the breadth of the range increases. When there are good reasons for thinking it probable that the results are affected by systematic errors that may vary in an unknown manner from one determination to another, then the center of the apparent range cannot validly be regarded as the value that is most likely to be correct. The correct value may lie entirely outside the apparent range. For such reasons, where systematic errors are present and where the report is too deficient to enable one to form an opinion regarding such errors, the limits of the apparent range have not been connected by dashes in table 35, nor by solid lines in figure 12.

NO SECULAR VARIATION

With the exception of no. 9 (table 35 and figure 12), all the apparent ranges include the value 299.77_3, which is the mean of the center values of the last five determinations; and no. 9 almost includes it. Hence, in view of the uncertainty of the significance of the center values in the first 9 determinations, it is obvious that the data give no indication of any secular change in the velocity of light.

THE VELOCITY OF LIGHT

Determinations prior to 1928 seem to be of historical interest only. The centers of the ranges found since that date range from $299{,}76_8$ to $299{,}77_8$ km./sec., and average $299{,}77_3$ km./sec. It thus seems to the writer probable that the best value that can be derived from the data now available is

Velocity of light in a vacuum.......... $299{,}77_3$ km./sec.
Dubiety, perhaps, but probably less than, $\pm 1_0$ km./sec.

If the centers of the five ranges determined since 1928 may be taken as the best representations of those determinations, then limits may be set to the errors that may be introduced by two of the outstanding potential disturbances: viz., the delay at reflection, and the rapid sweep of the light over the distant mirror or lens; the second is characteristic of the Foucault method.

DELAY AT REFLECTION

It there be a delay at reflection, then the velocity derived on the assumption of no delay, as in all of this work, will be too small whenever it is based either on the full-length path or on the difference in the lengths of two paths, the longer one involving more reflections than does the shorter. The only work that does not involve one or other of these conditions is that of Hüttel. Although he used the difference in the lengths of two paths, the same number of reflections was involved in each. Hence his result is unaffected by any such delay. On the other hand, the result obtained by Karolus and Mittelstaedt is based on the full-length path. Two such paths were used. One involved five reflections; and the other, seven. And Anderson based his result on the difference in the lengths of two paths, the longer path involving three more reflections than the shorter. Nevertheless, Hüttel's value is the smallest, and Anderson's are intermediate. All of which indicates a negative delay, which seems to be impossible. Hence, one is forced to conclude that the actual delay at reflection is so small as to be completely overshadowed by the existing fluctuating errors.

Since the mean of the individual spreads of these four determinations is 30 km./sec., it seems unlikely that the delay at reflection can be as great as that which would correspond to Hüttel's value increased by 15 km./sec. (making $299{,}78_3$ km./sec.) and Karolus and Mittelstaedt's smaller value, for the 250-m. path (five reflections), decreased by 15 km./sec. (making $299{,}76_3$ km./sec.). That is, the delay resulting from five reflections is not likely to cause an error of 20 km./sec. in the velocity as measured over a 250-m. path. If the delay per reflection were x sec., the effect of five reflections in a 250-m. path would cause the velocity to be too small by the amount $\delta V = 5xV^2/0.250 = 1.8x \times 10^{12}$ km./sec. (see eq. 70). If this is unlikely to be as great as 20 km./sec., x is unlikely to be as great as 11×10^{-12} sec., which corresponds to the time required for the passage of about 5500 light waves.

EFFECT OF RATE OF SWEEP

The rate at which the light swept across the distant mirror in the Michelson, Pease, and Pearson work was $0.004\,V = 1{,}200$ km./sec.; whereas the rate of sweep was zero in the last four determinations in table 35. Consequently, if the Michelson, Pease, and Pearson value is not seriously in error, one may conclude that a sweep of not more than 0.4 percent of the velocity of light produces an effect that is too small to be detected with the precision yet attained in this work. The effect seems to be certainly less than 10 km./sec., and may be zero. Ten kilometers per second is a third of the limit suggested by Lorentz as possibly applicable when the sweep is eight times as rapid.

APPENDIX A

EXPERIMENTAL METHODS FOR DETERMINING THE VELOCITY OF LIGHT

GENERAL REMARKS

Every experimental method for measuring the velocity of light involves a measurement of the time taken for light to pass over a measured path. Since one must work with a train of waves, it is necessary to place identifying marks upon the train so as to be sure that the instants defining the terminals of the time interval measured correspond to the passages of the same point upon the train. The several methods differ in the way the train is marked, and in the nature of the train while traversing the measured path.

Four distinct methods have been used: (1) Fizeau's method, in which the train is chopped by a toothed wheel into a series of discrete groups of waves; (2) Foucault's method, in which a rotating mirror reflects along the path a train of finite length; (3) Karolus and Mittelstaedt's method in which a Kerr electro-optic cell impresses upon a train, unlimited in length and of constant intensity, a sinusoidally varying elliptical polarization, and a second Kerr cell indicates the ellipticity of the received light; and (4) Anderson's method, in which a Kerr electro-optic shutter imposes upon the train an intensity that varies sinusoidally about a fixed mean value. This same method was used by Hüttel.

In methods 1, 2, and 4 the observed velocity is that of a group of waves lying between regions of darkness. (In method 4 these groups were superposed on a train of constant intensity.) Those methods measure the "group velocity."

But in method 3 the marking of the train is not done by introducing regions of darkness, but by changing the polarization of the light. For that reason it seems to the present writer probable that in method 3 it is the phase velocity that is measured, not the "group velocity." Anderson has, however, suggested the opposite.

In the first two methods the returned image of the source is directly observed, and the result obtained depends upon the lateral position of the center of brightness of that image. On account of the great distance of the returning mirror from the observer, the returned image is mainly a diffraction pattern. Can maladjustment or imperfections in the optical parts, particularly near their edges, introduce a lateral asymmetry in the image, and thus affect the result? Although this question was raised by Cornu [17] in his report to the Paris Conference of 1900, no serious consideration of it has been found in any report of an experimental determination of the velocity.

In all these methods the time interval that is measured includes a possible delay in the reflection of light. Let x denote that delay at each reflection, n = number of reflections that occur during the measured time interval, D = length of the path that is traversed twice during that interval, once in each direction, then the true velocity V will exceed that computed on the assumption that there is no delay, the excess δV being that given by the equation

$$\delta V = nxV^2/2D, \qquad (70)$$

the same unit of length being used in both V and D.

The possible effect of this delay upon the experimentally determined velocity has been considered in only the following two of the reports. Using an equation equivalent to eq. 70, Cornu pointed out in his memoir [15] of 1876 (p. 309) that this delay must have produced in Foucault's work a relative error that was about 7,000 times as great as that produced in his own. Hence if this effect were appreciable in his work, for which an accuracy of only 0.1 percent was claimed, then Foucault should have obtained a most fantastic result.

Following him, Michelson wrote, in the report [24] of his 1879 work, as follows (p. 142); "In my own experiments the same reasoning shows that if this possible error made a difference of 1 percent in Foucault's work (and his result is correct within that amount), then the error would be but .00003 part."

Each of these conclusions is entirely correct, but Michelson's parenthetical clause, implying that this error in Foucault's case could not have exceeded 1 percent, is misleading. It totally ignores the possibility that other errors may have partially masked the one under consideration. In fact, nothing is known about the errors that inhere in Foucault's work.

From a comparison of the recent determinations of the velocity—those made within the last 13 years—it seems that the delay may be zero, and that it is surely less than 12 $\mu\mu$sec. (see preceding section).

Fizeau's method involves many observational difficulties, but the physical principles underlying it are relatively simple.

Foucault's method involves fewer observational difficulties, but some of the physical principles underlying it are obscure, as was pointed out by Cornu in his report [17] to the Paris Conference of 1900. Among those, he emphasized three, which may be presented in the form of questions: (*a*) Are the laws of reflection the same for a rapidly rotating mirror as for a fixed one? (*b*) Does the intense disturbance of the air in the neighborhood of the rotating mirror affect the result? (*c*) When a narrow beam of light sweeps over a fixed mirror with a speed that is not negligible as compared with the speed of light, is it reflected in the same manner as is a stationary beam?

The following year H. A. Lorentz [41] considered these questions, concluding that the answer to (*a*) is "Yes"; and that to (*b*) is "No." There is no carrying along of the light by the mirror; and the effect of the air disturbance is negligible.

Question (*c*) is more difficult to answer. In Newcomb's determination the beam swept over the fixed mirror with speeds of 1.2 to 4.2 percent of that of light, averaging about 3 percent. Lorentz stated that under such conditions it is difficult to form an exact idea of how the waves that form the returned image are propagated. But by making plausible assumptions he arrived at certain limits, and concluded that Newcomb's determination is probably not in error from this cause by more than 1 part in 10,000; that is, by not more than 30 km./sec.

It should be noticed that this estimated limiting uncertainty is as great as the entire dubiety that Newcomb had allowed for his determination.

Of these conclusions by Lorentz, Michelson wrote [38]: "It seems to me that M. Lorentz has satisfactorily answered M. Cornu's questions."

Since then, those questions have never been discussed. Although in Michelson's work [31] in 1924–27 the beam of light swept the distant mirror with a speed that exceeded 2 percent of the velocity of light (about two-thirds of that in Newcomb's work), there is nothing in the report to indicate that any consideration was given to the possibility that such a rapid sweep might introduce a serious systematic error.

FIZEAU'S METHOD (TOOTHED WHEEL)

GENERAL THEORY. TEETH AND MOTION UNIFORM

In Fizeau's method a narrow pencil of light, passing normally between the teeth of a rapidly rotating wheel, is chopped by that into a series of groups of waves which pass through a suitable optical system to a distant mirror, and are returned by that to their source. If, while the light is going and returning, the wheel has turned just enough for a tooth to occupy the position of the gap through which the outgoing light passed, the returning light will be eclipsed. By the order of an eclipse is meant the number of teeth that have successively blocked the gap while the light was going and returning.

The optical arrangement employed both by Cornu and by Perrotin and Prim was as shown diagrammatically in figure 13. The image of a small intense source of light is formed at *c* in the plane *W* of the wheel by the lens *L*, after reflection from the transparent plate *G*, which directs the axis of the beam along the axis *cd* of the optical system. So much of this image on the wheel as is actually used will hereafter be called the source of the light forming the returned star. The lenses *S* and *S′* are so adjusted that an image of the wheel is formed on the central plane of the collimator lens *S′* of aperture *bb′*. *M* is a concave mirror of radius of curvature equal to the focal length of *S′*; it is so placed that an image of the sending lens *S* will be formed on the sur-

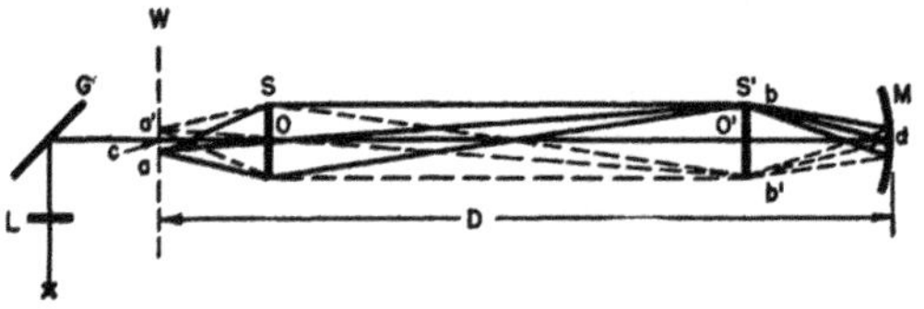

FIG. 13.—Schematic representation of optical system used by Cornu.

Light from a small source is focused on the side of the toothed edge of the wheel *W* by the lens *L*, after reflection from the transparent plate *G*. That image is focused by lens *S* on the midplane of lens *S′*, which forms on *M* an image of the image formed by *L* on *W*. *M* is a concave mirror with its center of curvature at *O′*; it returns the light to a focus on *W*. Light from the point *c* on the line *OO′* is returned to *c*; that from another point *a* is returned to a point *a′* such that $ca = ca'$. No light coming from a point more distant from *c* than that where *bO* intersects *W* can be returned, *b* being the extreme edge of the aperture of *S′*.

face of *M*. Then the light from some point *a* will be brought to a focus at *b*, and after reflection from *M* it will be focused at *b′* and proceed to *a′* on the side of the axis *cd* opposite to *a*. The returned star is completely inverted with reference to the axis *cd*. Light proceeding from a point more distant from *c* than *a* will entirely miss the lens *S′* and will not be returned. The only portion of the image focused on *W* that is actually used is that lying in the circle of diameter *aa′*. That is the actual source of light, and except for diffraction effects, *aa′* is likewise the diameter of the returned star.

The value of *aa′* is given by the equation

$$aa'/bb' = Oc/OO'. \qquad (71)$$

bb′ is the diameter d_c of the collimator lens, *Oc* is the focal length f_s of the sending lens, and $OO' = D - f_s - f_c$, where *D* is the distance from the wheel to the distant mirror, and f_c is the focal length of the collimator lens. Since *D* is exceedingly great as compared with $f_s + f_c$, eq. 71 may be written in the form

$$aa' = d_c f_s / D, \text{ approximately.} \qquad (72)$$

The greater the diameter of the collimator lens, and the longer the focus of the sending lens, the greater is the diameter of the source and the star.

The lenses need not be exactly perpendicular to *OO′*, nor exactly parallel to one another; *OO′* need not be the principal axis of the lenses, but it is the axis of inversion of the star. Neither is it necessary for the mirror *M* to have the curvature specified; only a very small portion of its surface is actually used, and that may be plane without seriously affecting the brightness of the star. The position of the returned star is uninfluenced by the curvature of M. But it is exceedingly important that the reflecting surface of M should lie in the plane of the image of S; otherwise, the returned star will not be focused in the plane of W. The following notation will be used:

D = distance cd from the wheel to the reflector of the collimator
$m = \omega/2\pi$ = number of turns of the wheel per unit of time
N = number of teeth in the circumference of the wheel
n = order of the eclipse. If the returned star is eclipsed by the n'th tooth that passed in view after its light had passed the wheel, the eclipse is said to be of the n'th order.
$q_n = 2n-1$
t = time
V = velocity of light
$\alpha = d\omega/dt$ = angular acceleration of the wheel
Θ = angular displacement of the wheel with reference to its phantom, in the direction of rotation. If the source of light is a point at c (fig. 13), $\Theta = \Psi$.
Ψ = angle through which the wheel turns in the time interval τ
τ = time taken for light to go from the wheel to the mirror and back again
ω = angular speed of the wheel

Subscripts:

n indicates the order of the eclipse.
$e+$ and $e-$ indicate an eclipse when ω is increasing (+) and decreasing (−), respectively.
$r+$ and $r-$ indicate a reappearance when ω is increasing (+) and decreasing (−), respectively.

With this notation,

$$\tau = 2D/V, \quad \Psi = \tau\omega, \quad V = 2D\omega/\Psi, \quad \omega = 2\pi m \qquad (73)$$

and, as will presently be seen, when there is an eclipse, equations 74 and 75 hold good in the ideal case:

$$\Psi_n = \pi q_n/N = \tau\omega_n, \qquad (74)$$

$$V = 2D\omega_n/\Psi_n = 4DNm/q_n. \qquad (75)$$

Each element of light passing through an interdental gap in the wheel returns τ seconds later to illuminate the rear face of the wheel at a point Ψ radians back of the point through which it initially passed. If that point were at c (fig. 13), then $\Theta = \Psi$; if it were displaced from c by the angle ϵ' in the direction of rotation of the wheel, then, owing to the inversion of the returned image, $\Theta = \Psi + \epsilon'$. The angle ϵ' is necessarily small, never exceeding $aa'/2r$, where r is the radius of the wheel and aa' is given by eq. 72.

Thus there is traced on the back of the wheel a series of luminous arcs, each determined by the gap that is Ψ radians in advance of it, and by the diameter of the utilized effective source of light. The portions of those luminous arcs that fall upon the gaps pass through, and are seen by the observer; those that fall on the teeth do not. The appearance to the observer is exactly the same as that of a steadily shining star observed through a rotating sectored disk formed of two coaxial and coplanar toothed wheels relatively displaced by an angle Θ—one, the actual wheel; the other, the phantom wheel defined by the arcs of light traced on the back of the actual wheel. This way of looking at the problem was proposed by Cornu.

From the way the phantom wheel arises, it is obvious that it is not the image of the actual wheel as formed by the optical system. The two are entirely distinct, and have few features in common.

The angle Θ will be called the "setting" of the equivalent sectored disk. If the effective source of light and the returned star were each a true point, then the phantom would be a replica of the actual wheel, and the setting of the disk would be $\Theta = \Psi + \epsilon'$, ϵ' having the value already defined. If the effective source is not a true point, then $\Theta = \Psi + \epsilon$, where ϵ is determined by the size and shape of the effective source and by its position with reference to c (fig. 13). As long as the speed ω of the wheel remains constant, so does Ψ, and therefore the setting Θ of the disk. As ω increases, so does Θ, the two being related as shown by the equation

$$\Theta = \Psi + \epsilon = \omega\tau + \epsilon. \qquad (76)$$

Point Source on Axis

If the source and star were each a true point situated at c (fig. 13), and if the teeth and gaps had a common width and were uniformly distributed around the circumference of the wheel, then there would be an eclipse (the opening in the sectored disk would be zero) whenever Θ has one of the values of Θ_n defined by the equation

$$\Theta_n = (2n-1)\pi/N = \pi q_n/N. \qquad (77)$$

And whenever $\Theta = \Theta_n \pm \delta\Theta$, $\delta\Theta \overline{\leqq} \pi/N$, the width of each opening in the disk would be $\delta\Theta$. Furthermore, under the stated conditions, $\Theta_n = \Psi_n = \tau\omega_n$, and by eq. 76 the numerical value of $\delta\Theta$ would be given by the equation

$$\delta\Theta = |\tau\,\delta\omega|, \qquad (78)$$

where $\delta\omega = \omega - \omega_n$, and $|\tau\,\delta\omega|$ indicates the absolute value of $\tau\,\delta\omega$.

If the gaps were wider than the teeth, there would never be a total eclipse, but the opening in the disk, and consequently the apparent brightness of the star, would decrease to a persisting minimum, and would then increase to a momentary maximum, and repeat.

If the gaps were narrower than the teeth, there would be a persisting eclipse.

Consider the sequence of events when the gaps are narrower than the teeth, first with the speed of the wheel increasing, then with it decreasing. Starting with the pre-eclipse stage and with slowly increasing speed, the spaces in the sectored disk, each bounded by the trailing edge of a tooth of the phantom and by the leading edge of a tooth of the wheel, slowly decrease, becoming zero when $\Theta=\Theta_{e+}$, such that the light just ceases to get by the leading edges of the teeth of the wheel, those edges then coinciding with the trailing edges of the teeth of the phantom. This eclipse persists until $\Theta=\Theta_{r+}$, such that light just reappears at the trailing edges of the teeth of the wheel, and the spaces in the disk are now bounded by the leading edges of the teeth of the phantom and by the trailing edges of the teeth of the wheel. As Θ increases beyond Θ_{r+}, the brightness steadily increases to a maximum, after which the pre-eclipse stage of the next order is entered.

If, after Θ has exceeded Θ_{r+}, the speed be slowly decreased, all the preceding steps will be retraced in the reverse direction. The brightness will decrease, becoming zero, the star eclipsed, when $\Theta=\Theta_{e-}=\Theta_{r+}$; will remain zero until $\Theta=\Theta_{r-}=\Theta_{e+}$, when the star will reappear. The value of Θ for an eclipse with increasing ω (Θ_{e+}) is exactly that for a reappearance with decreasing ω (Θ_{r-}); and that for an eclipse with decreasing ω (Θ_{e-}) is exactly that for a reappearance with increasing ω (Θ_{r+}): $\Theta_{e+}=\Theta_{r-}$, light at leading edge of tooth of wheel; $\Theta_{e-}=\Theta_{r+}$, light at trailing edge of tooth of wheel.

If the gaps are wider than the teeth, but not so much wider that the minimum brightness is as great as the least that can be detected, the idea of eclipses and reappearances may be retained, as well as the definitions just used; viz., with increasing speed, an eclipse occurs when the star just ceases to get by the leading edges of the teeth of the wheel, and a reappearance occurs when it just succeeds in getting by the trailing edges of those teeth; with decreasing speed, it is the trailing edges that just stops the star at an eclipse, and the leading ones that it just escapes at its reappearance. Exactly the same relations exist between the Θ's as when the teeth are wider than the gaps.

If the teeth and gaps are of the same width, all four Θ's are equal; each equal to $\Theta_n=q_n\pi/N$.

The observer attempts to find the value of ω that corresponds to Θ_n, but actually finds that corresponding to a different value Θ. Hence, he computes the value V_c instead of the true value V.

$$V_c=\frac{2D\omega}{\Theta_n}; \qquad V=\frac{2D\omega}{\Theta};$$

$$\therefore \quad V_c=V\Theta/\Theta_n. \tag{79}$$

Or he may for a certain value of Θ determine, not ω, but the incorrect value $\omega'=\omega+\delta\omega$. Then

$$\Theta_n V_c=2D(\omega+\delta\omega); \qquad \Theta V=2D\omega;$$

$$\therefore \quad \Theta_n V_c=V\Theta+2D.\delta\omega \tag{80}$$

or

$$V_c=V\frac{\Theta+\tau.\delta\omega}{\Theta_n} \tag{81}$$

That is, an error in ω has the effect of increasing Θ by $\tau.\delta\omega$, $\delta\omega$ being the amount by which the false value of ω exceeds the ω that corresponds to Θ.

Hence in this case also the formal equation $V_c=V\Theta/\Theta_n$ applies. The experimenter's problem is to determine the value of Θ/Θ_n; that is, to determine the value of Θ corresponding to his observation.

Factors Affecting Setting of Disk

Among the factors causing Θ to differ from Θ_n even when the size and spacing of the teeth are strictly uniform and the motion of the wheel is ideal, are the following:

1. Breadth of tooth not equal to that of gap.
2. Finite area of the effective source of light and of the returned star.
3. Brightness of the background, which forces the observer to determine Θ in terms of the vanishing of the star into the background, and of its reappearance therefrom.
4. The preceding effect is complicated by the fact that an observer can follow a star fading into a background to a lower brightness than that at which he can detect its emergence therefrom.
5. Delay in recording the observation. This is composed of various factors. Cornu, and Perrotin and Prim, considered four: (*a*) Persistence of vision r; (*b*) hesitation in deciding whether the star has vanished (or reappeared) $\eta_0+\eta/|dI/dt|$, I = apparent intensity of the star; (*c*) delay between the decision and the manual act that records it; and (*d*) electrical and mechanical delays. Cornu wrote the hesitation in the form just given, for this reason: If the brightness of the star is changing rapidly, the hesitation will be slight, but it will probably never be zero.
6. Displacement of the center of the returned star from that of the effective source of its light. Such a displacement may be caused by optical imperfections, maladjustment, diffraction.

The first two of these factors may conveniently be considered together. Let d be the diameter of the effective source of light, r be the distance of its center from the axis of rotation of the wheel, g be the width of a gap and w that of a tooth, each measured along the circumference of radius r. The returned star is completely inverted with reference to the optical axis through the center of its source. With the rim of the wheel moving to the right, the source of light will be just beginning to be uncovered when the wheel is in the position shown at A in figure 14, and the first light to pass (at A) will return to the point A', diffraction effects being ignored. If with increasing speed there

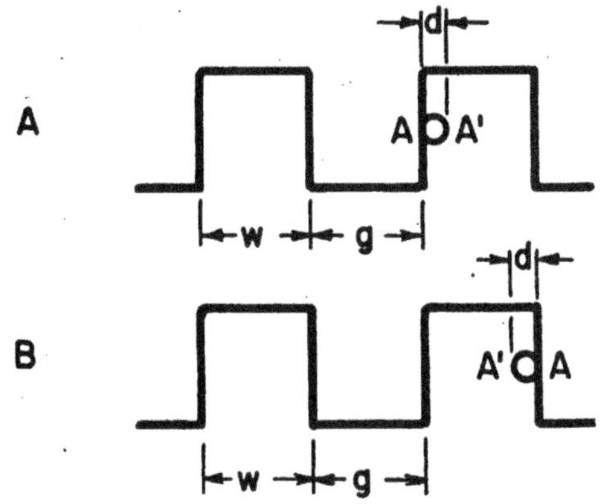

FIG. 14.—Illustrating an eclipse and a reappearance when the source has a finite diameter d.

is to be an eclipse, then the leading edge of a tooth must be at A' when the first light returns. Hence, so far as this factor is concerned, the angle Θ_{e+} must be given by eq. 82.

$$r\Theta_{e+}=n(w+g)-w+d. \quad (82)$$

But $w+g=2\pi r/N$; and $2w=(w+g)+(w-g)$. Hence

$$\Theta_{e+}=(2n-1)\pi/N+(2d+g-w)/2r. \quad (83)$$

Writing

$$\Delta.(w+g)=2d+g-w \quad (84)$$

and remembering that $(2n-1)\pi/N=\Theta_n$, one obtains

$$\Theta_{e+}=\Theta_n+\pi\Delta/N. \quad (85)$$

With rotation in the same direction as before, but with decreasing speed, the light from the source will have just ceased to pass (the leading edge of) a tooth when the wheel is in the position shown at B in figure 14. The light that last passed at A returns to A', and if it is to be eclipsed, the trailing edge of a tooth must be at A' when it returns. Hence

$$r\Theta_{e-}=n(w+g)-g-d; \quad (86)$$

$$\therefore \quad \Theta_{e-}=\Theta_n-\pi\Delta/N. \quad (87)$$

Whence

$$\left.\begin{array}{l}\Theta_{e+}=\Theta_{r-}=\Theta_n+\pi\Delta/N,\\ \Theta_{e-}=\Theta_{r+}=\Theta_n-\pi\Delta/N,\end{array}\right\} \quad (88)$$

where $\Delta\equiv(2d+g-w)/(w+g)$.

The third and fourth factors causing Θ to differ from Θ_n may likewise be considered together. If the brightness of the background and the sensitivity of the observing eye remain unchanged, then it may be expected that the actual brightness of the star when it vanishes will be the same whether the speed be increasing or diminishing. That is, it may be expected that $\Theta_n-\Theta_{e+}=\Theta_{e-}-\Theta_n=h_e$.

Similiarly for its brightness on reappearance; that is, $\Theta_{r+}-\Theta_n=\Theta_n-\Theta_{r-}=h_r$. But h_r will probably exceed h_e; both are essentially positive. Hence, so far as these causes are concerned

$$\left.\begin{array}{ll}\Theta_{e+}=\Theta_n-h_e, & \Theta_{r+}=\Theta_n+h_r,\\ \Theta_{e-}=\Theta_n+h_e, & \Theta_{r-}=\Theta_n-h_r.\end{array}\right\} \quad (89)$$

In considering the fifth factor, it must be recognized that the value both of the lag r, due to the persistence of vision, and of the hesitation coefficient η_0 may not be the same for an eclipse as for a reappearance, but there is no reason for thinking that the effects of causes (c) and (d) differ for the two cases. However that may be, all four terms are mere lags in time and cannot be separated one from another without a very special study; and there is no indication that such a study has been made by those who have used this method. Hence it is desirable to group them all under a single symbol l, the value of which is essentially positive, does not depend upon either the speed or the acceleration of the wheel, but may depend upon whether an eclipse or a reappearance is being observed. Denote these two values by l_e and l_r, respectively. These symbols cover all kinds of fixed lags.

As a result of these fixed lags, the speed of the wheel as determined from the record will not be that ω at which the observation was made, but will be

$$\left.\begin{array}{l}\omega'=\omega+l_e\alpha_e \text{ for eclipses,}\\ \omega'=\omega+l_r\alpha_r \text{ for reappearances,}\end{array}\right\} \quad (90)$$

α being the angular acceleration of the wheel; α_e may, but need not, be equal to α_r.

There remains to be considered the term $\eta/|dI/dt|$. The value of η, always positive, for an eclipse observation may differ from that for a reappearance, but there seems to be no reason for expecting it to depend on the sign of the acceleration of the wheel.

It may be easily seen that the sign of dI/dt is such that $\eta/|dI/dt|$ takes the forms $-\eta_e/(dI/dt)_e$ and $+\eta_r/(dI/dt)_r$. Hence the velocity of the wheel, as determined from the record, will not be ω, but ω', where

$$\omega'_e=\omega-\eta_e\alpha_e/(dI/dt)_e \text{ and } \omega'_r=\omega+\eta_r\alpha_r/(dI/dt)_r. \quad (91)$$

Replacing dI/dt by its equivalent, $\alpha(dI/d\omega)$, leads to the equation

$$\left.\begin{array}{l}\omega'_e=\omega-\eta_e/(dI/d\omega)_e, \text{ eclipse.}\\ \omega'_r=\omega+\eta_r/(dI/d\omega)_r, \text{ reappearance.}\end{array}\right\} \quad (92)$$

In the ideal case, and also in many others, the numerical value of $dI/d\omega$ is a constant, independent of the speed and acceleration of the wheel, and of whether an eclipse or a reappearance is being observed, it being no more than the variation of I with the setting of the equivalent sectored disk.

Write

$$b \equiv |d\omega/dI|. \tag{93}$$

The sign of $dI/d\omega$ varies with the type of observation as indicated below

$$\left.\begin{aligned} &\left(\frac{dI}{d\omega}\right)_{e+} \text{negative;} \quad &\left(\frac{dI}{d\omega}\right)_{r+} \text{positive;} \\ &\left(\frac{dI}{d\omega}\right)_{e-} \text{positive;} \quad &\left(\frac{dI}{d\omega}\right)_{r-} \text{negative.} \end{aligned}\right\} \tag{94}$$

Hence the effect of all the lags is to make the derived speed ω' differ from the true ω as shown in the equation

$$\left.\begin{aligned} \omega'_{e+} &= \omega + l_e\alpha_{e+} + \eta_e b, \\ \omega'_{e-} &= \omega + l_e\alpha_{e-} - \eta_e b, \\ \omega'_{r+} &= \omega + l_r\alpha_{r+} + \eta_r b, \\ \omega'_{r-} &= \omega + l_r\alpha_{r-} - \eta_r b. \end{aligned}\right\} \tag{95}$$

If Θ is the setting corresponding to ω, then the effective value of Θ when the derived speed is ω' will be

$$\left.\begin{aligned} \Theta'_{e+} &= \Theta_{e+} + l_e\tau\alpha_{e+} + \eta_e\tau b, \\ \Theta'_{e-} &= \Theta_{e-} + l_e\tau\alpha_{e-} - \eta_e\tau b, \\ \Theta'_{r+} &= \Theta_{r+} + l_r\tau\alpha_{r+} + \eta_r\tau b, \\ \Theta'_{r-} &= \Theta_{r-} + l_r\tau\alpha_{r-} - \eta_r\tau b. \end{aligned}\right\} \tag{96}$$

The sixth factor, arising from a shift of the center of the returned star from the center of its actual source, has the effect of increasing every Θ by a fixed amount s equal to the least angle ($\pm$) through which the wheel must turn in the direction of its rotation in order that a specified radius may sweep from the center of the effective source to the center of the returned star. The sign and magnitude of s are exactly the same for every value of Θ and every rate of change in the speed of the wheel, but the sign of s changes when the direction of rotation is changed.

Assembling the effects of these six factors, one obtains the following general equations for a wheel with uniformly distributed teeth of a common size.

$$\left.\begin{aligned} \Theta_{e+} &= \Theta_n + \frac{\pi\Delta}{N} - h_e + \eta_e\tau b + l_e\tau\alpha_{e+} + s, \\ \Theta_{e-} &= \Theta_n - \frac{\pi\Delta}{N} + h_e - \eta_e\tau b + l_e\tau\alpha_{e-} + s, \\ \Theta_{r+} &= \Theta_n - \frac{\pi\Delta}{N} + h_r + \eta_r\tau b + l_r\tau\alpha_{r+} + s, \\ \Theta_{r-} &= \Theta_n + \frac{\pi\Delta}{N} - h_r - \eta_r\tau b + l_r\tau\alpha_{r-} + s. \end{aligned}\right\} \tag{97}$$

Remembering (eq. 74) that $\Theta_n = \pi q_n/N$, it is obvious from the preceding that

$$\left.\begin{aligned} \Theta_{e+}/\Theta_n &= 1 + \frac{\Delta}{q_n} - \frac{N}{\pi}\left\{\frac{h_e - \eta_e\tau b}{q_n}\right\} + \frac{l_e\tau N\alpha_{e+}}{\pi q_n} + \frac{sN}{\pi q_n}, \\ \Theta_{e-}/\Theta_n &= 1 - \frac{\Delta}{q_n} + \frac{N}{\pi}\left\{\frac{h_e - \eta_e\tau b}{q_n}\right\} + \frac{l_e\tau N\alpha_{e-}}{\pi q_n} + \frac{sN}{\pi q_n}, \\ \Theta_{r+}/\Theta_n &= 1 - \frac{\Delta}{q_n} + \frac{N}{\pi}\left\{\frac{h_r + \eta_r\tau b}{q_n}\right\} + \frac{l_r\tau N\alpha_{r+}}{\pi q_n} + \frac{sN}{\pi q_n}, \\ \Theta_{r-}/\Theta_n &= 1 + \frac{\Delta}{q_n} - \frac{N}{\pi}\left\{\frac{h_r + \eta_r\tau b}{q_n}\right\} + \frac{l_r\tau N\alpha_{r-}}{\pi q_n} + \frac{sN}{\pi q_n}, \end{aligned}\right\} \tag{98}$$

and

$$V_{e+} = V + \frac{V\Delta}{q_n} - \frac{NV}{\pi}\left\{\frac{h_e - \eta_e\tau b}{q_n}\right\} + \frac{l_e\tau NV\alpha_{e+}}{\pi q_n} + \frac{sNV}{\pi q_n}, \tag{99}$$

and similarly for the others.

Writing

$$\left.\begin{aligned} \Delta' &= V\Delta, \\ B &= \frac{NV\eta\tau b}{\pi}, \\ H &= NVh/\pi, \\ L &= \frac{N}{\pi}l\tau V = 2NDl/\pi, \\ S &= \frac{NVs}{\pi}, \end{aligned}\right\} \tag{100}$$

and applying the proper suffixes, eq. 99 becomes eq. 101.

$$\left.\begin{aligned} V_{e+} &= V + (\Delta' - H_e + B_e)/q_n + L_e\alpha_{e+}/q_n + S/q_n, \\ V_{e-} &= V - (\Delta' - H_e + B_e)/q_n + L_e\alpha_{e-}/q_n + S/q_n, \\ V_{r+} &= V - (\Delta' - H_r - B_r)/q_n + L_r\alpha_{r+}/q_n + S/q_n, \\ V_{r-} &= V + (\Delta' - H_r - B_r)/q_n + L_r\alpha_{r-}/q_n + S/q_n. \end{aligned}\right\} \tag{101}$$

Cornu, and Perrotin and Prim, averaged these in various ways, obtaining what Cornu called "double observations." Denote such averages as follows:

$$\begin{aligned} V_{e\pm} &\equiv (V_{e+} + V_{e-})/2; \\ V_{r\pm} &\equiv (V_{r+} + V_{r-})/2; \\ V_{er+} &\equiv (V_{e+} + V_{r+})/2 \equiv \text{Cornu's } V; \\ V_{er-} &\equiv (V_{e-} + V_{r-})/2 \equiv \text{Cornu's } v; \\ V_{e-r+} &\equiv (V_{e-} + V_{r+})/2 \equiv \text{Cornu's } U; \\ V_{e+r-} &\equiv (V_{e+} + V_{r-})/2 \equiv \text{Cornu's } u. \end{aligned}$$

Cornu's "crossed double observations" were the following means of his double observations:

$$\begin{aligned} V_{er\pm} &\equiv (V_{er+} + V_{er-})/2 \equiv \text{Cornu's } (V+v)/2; \\ V_{e\mp r\pm} &\equiv (V_{e-r+} + V_{e+r-})/2 \equiv \text{Cornu's } (U+u)/2. \end{aligned}$$

The expressions for these several averages are given in eq. 102.

$$\left.\begin{aligned}
V_{e\pm} &= V+L_e(\alpha_{e+}+\alpha_{e-})/2q_n+S/q_n,\\
V_{r\pm} &= V+L_r(\alpha_{r+}+\alpha_{r-})/2q_n+S/q_n,\\
V_{er+} &= V-\{(H_e-B_e)-(H_r+B_r)\}/2q_n\\
&\quad +(L_e\alpha_{e+}+L_r\alpha_{r+})/2q_n+S/q_n\\
&\qquad \equiv \text{Cornu's } V,\\
V_{er-} &= V+\{(H_e-B_e)-(H_r+B_r)\}/2q_n\\
&\quad +(L_e\alpha_{e-}+L_r\alpha_{r-})/2q_n+S/q_n\\
&\qquad \equiv \text{Cornu's } v,\\
V_{e-r+} &= V-\{2\Delta'-(H_e-B_e)-(H_r+B_r)\}/2q_n\\
&\quad +(L_e\alpha_{e-}+L_r\alpha_{r+})/2q_n+S/q_n\\
&\qquad \equiv \text{Cornu's } U,\\
V_{e+r-} &= V+\{2\Delta'-(H_e-B_e)-(H_r+B_r)\}/2q_n\\
&\quad +(L_e\alpha_{e+}+L_r\alpha_{r-})/2q_n+S/q_n\\
&\qquad \equiv \text{Cornu's } u,\\
V_{er\pm} &= V+\{L_e(\alpha_{e+}+\alpha_{e-})\\
&\qquad +L_r(\alpha_{r+}+\alpha_{r-})\}/4q_n+S/q_n,\\
V_{e\mp r\pm} &= V+\{L_e(\alpha_{e+}+\alpha_{e-})\\
&\qquad +L_r(\alpha_{r+}+\alpha_{r-})\}/4q_n+S/q_n.
\end{aligned}\right\} \tag{102}$$

It should be noticed that for each of Cornu's four types of double observations the expression is of the same form, eq. 103.

$$V,\ v,\ U,\ u = V+(H+A+S)/q_n, \tag{103}$$

where S has the same value in all four types, and its sign changes with the direction of rotation of the wheel; A depends on the accelerations, and varies from type to type; and H has one or other of two values, each of which occurs once with each sign. Furthermore, for each of his crossed double observations, the expression is exactly the same, and of the form of eq. 104.

$$V_{er\pm} = V_{e\mp r\pm} = V+(A'+S)/q_n. \tag{104}$$

In every case S is the same for every order, and so is H when the observations lie in normal regions. But A and A' will, in general, vary from order to order, and if they are of the form $A_0+A_1q_n$, they will give rise to a systematic error, making the computed velocity too great by the amount A_1.

If in each case there are the same number of observations for each direction of rotation of the wheel, then the mean will be independent of S. The accelerations being always small, the difference between any two α's having the same sign is still smaller, and may be negligible.

In the notation of Cornu and of Perrotin and Prim, their $(\mu_1-\mu_0)/(\tau_1-\tau_0)$ being replaced by R, the value of α is given by eq. 105 (see eq. 129).

$$\alpha = -\pi V^2 q_n^2 R/8D^2N^2M \tag{105}$$

and

$$L\alpha/q_n = -lV^2Rq_n/4DNM \equiv -L'Rq_n/M, \tag{106}$$

where

$$L' \equiv lV^2/4DN. \tag{107}$$

On replacing $L\alpha/q_n$ in eq. 101 by its equivalent from eq. 106, one obtains eq. 108.

$$\left.\begin{aligned}
V'_{e+} &= V+(\Delta'-H_e+B_e)/q_n\\
&\qquad -L'_eR_{e+}q_n/M_{e+}+S/q_n,\\
V_{e-} &= V-(\Delta'-H_e+B_e)/q_n\\
&\qquad -L'_eR_{e-}q_n/M_{e-}+S/q_n,\\
V_{r+} &= V-(\Delta'-H_r-B_r)/q_n\\
&\qquad -L'_rR_{r+}q_n/M_{r+}+S/q_n,\\
V_{r-} &= V+(\Delta'-H_r-B_r)/q_n\\
&\qquad -L'_rR_{r-}q_n/M_{r-}+S/q_n.
\end{aligned}\right\} \tag{108}$$

Since the speed of the wheel corresponding to V_{r+} is necessarily greater than that for the immediately preceding V_{e+}, $V_{e+} < V_{r+}$. Similarly, $V_{e-} > V_{r-}$.

MECHANICAL IRREGULARITIES

General Treatment

In the ideal case, every opening in the equivalent sectored disk, formed by the wheel and its phantom, is of the same size: for the conditions $r+$ and $e-$, its leading side is bounded by the trailing edge of a tooth of the wheel and its trailing side by the leading edge of a tooth of the phantom, and for conditions $r-$ and $e+$ the reverse is true (*cf.* fig. 15). The symbols $r+$, $r-$, $e+$, and $e-$ have the same significance as before; $r+$ = reappearance with increasing speed.

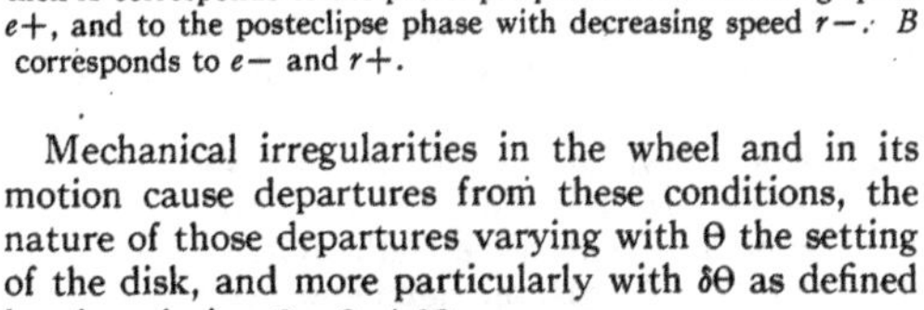

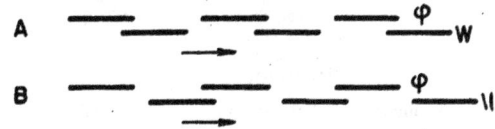

FIG. 15.—Idealized section perpendicular to planes of wheel W and phantom φ and through the teeth.
When the motion of the teeth is in the direction of the arrow, then A corresponds to the preeclipse phase with increasing speed $e+$, and to the posteclipse phase with decreasing speed $r-$. B corresponds to $e-$ and $r+$.

Mechanical irregularities in the wheel and in its motion cause departures from these conditions, the nature of those departures varying with Θ the setting of the disk, and more particularly with $\delta\Theta$ as defined by the relation $\Theta = \Theta_n + \delta\Theta$.

If the irregularities are so small that for a given setting Θ the openings in the equivalent sectored disk are of the same number and each is bounded by the same tooth-edges as in the ideal case, then the setting may be said to lie in a *normal region*. In such case the irregularities do no more than change the sizes of the openings, without either closing some of the ideal openings or making openings that do not exist in the ideal case.

If, however, the irregularities are so great that those conditions are not both fulfilled, then, following Cornu, the setting may be said to lie in a *critical region*.

When the setting lies in a critical region, the functioning of the edges of the teeth of the phantom interferes in some way with the normal functioning of those of the wheel; and vice versa.

As shown in figure 16, six suitably displaced edges, each displacement amounting to one quarter of the width of the ideal interdental gap, are sufficient to make the critical region embrace every possible value of Θ for a given eclipse region. And if the angular value of several displacements amounts to at least one half of $\delta\Theta$, then $\Theta \equiv \Theta_n + \delta\Theta$ will lie in the critical region.

FIG. 16.—Showing an effect of irregularities in the teeth.

This is an idealized section perpendicular to planes of wheel W and phantom φ and through the teeth, represented by lines. The wheel has been generated from the ideal one, in which all gaps and teeth have the same breadth, by adding to, or cutting from, one or both edges of certain of those ideal teeth a fixed amount equal to 1/4 the breadth of the ideal tooth. The breadth and position of the central tooth of W is that of the ideal. As W moves to the right with reference to φ, each of the edges bounding the A gap plays the same part as if the teeth were of normal breadth, but those of B do not. There the edge of the middle tooth of φ usurps the function of that of the middle one of W, and the edge of the right tooth of W usurps that of the right tooth of φ. This persists until W has moved half the width of the ideal tooth (or gap). Then the gap at A is just closed, and that at B is 3/4 the breadth of an ideal tooth. Whenever there is such interference, or the closing of a gap that would be open were the wheel ideal, or the opening of one where there would ideally be none, the setting is said to lie in a critical region. In the case here illustrated, every possible setting for this order lies in a critical region.

Consequently, if there are many displacements that are as great as a quarter of the gap width, then it is practically certain that every observation will lie in a critical region; the same is true if there are many displacements of an angular value at least equal to $0.5\ \delta\Theta$. The latter is the essential criterion, but the former may be valuable when the value of $\delta\Theta$ is not known.

From all of which it might be concluded that the way to minimize the effects of such irregularities of fixed amount is to use wide teeth and gaps, and large values of $\delta\Theta$. That is, the number of teeth to a given wheel should be as small as is practical. But that ignores the fact that $dI/d\Theta$ is proportional to the number of teeth, I being the apparent brightness of the returned star. Hence, decreasing the number of teeth decreases the precision with which the settings can be made. Cornu was so impressed with the latter that he ignored the former, using wheels with many small teeth. A compromise is necessary.

From the way in which the several correction terms in eq. 98 to eq. 102 arise, it is obvious that the irregularities now being considered affect the values of only Δ, b, and the h's. Terms involving them are the only ones that need to be considered.

When the setting is in a normal region, each edge of every tooth serves as a boundary of an open space in the equivalent sectored disk. Hence, in such a region, these irregularities will not change the value of $b \equiv |d\omega/dI|$, but they will change the values of Δ, h_e, and h_r. Those changes, however, will each be independent of the way in which the edges happen to be matched. The values of Δ, h_e, and h_r will each be independent of the order of the eclipse and reappearance. All that the irregularities have done is to change the magnitudes of those quantities, which in any case have to be determined experimentally. All of this is true whether the irregularities arise from irregularities in the wheel or in its motion.

Whenever the observations lie in a normal region, the mean of V_{e+} and V_{e-} will be independent of the values of Δ, h, and b; and the same will be true of the mean of V_{r+} and V_{r-}. These means were determined by Perrotin and Prim. Also, if h_e were equal to h_r and η_e were equal to η_r, then Cornu's "*double observations*" (the mean of V_{e+} and V_{r+}, and the mean of V_{e-} and V_{r-}) would each be independent of those quantities. But $h_e \neq h_r$ and $\eta_e \neq \eta_r$; consequently the double observations are not so independent, but their mean is. Cornu called this double averaging the method of "*doubled and crossed*" observations.

But if the setting is in a critical region, then the value both of b and of h, but not of Δ, will depend on how the edges of the teeth happen to be associated as boundaries of the open spaces in the equivalent sectored disk. Hence, those values will vary with the order n; and for a given order, the value of h_{e+} will in general differ from that of h_{e-}, similarly for h_{r+} and h_{r-}; and b_{e+}, b_{e-}, b_{r+}, and b_{r-} may differ one from another. Furthermore, no one of those quantities need have the same value for rotations in opposite directions.

Hence, when observations lie in critical regions, the term $h + \eta\tau b$ cannot, in general, be eliminated by any combination of them.

But in certain cases the value of $h + \eta\tau b$ may be approximately represented as the sum of a function of $q = 2n - 1$ and a fluctuating term: $h + \eta\tau b = f(q) + \epsilon$. For example, consider the observations $e+$ for the eclipses with accelerating speed, and imagine the corresponding values of $h + \eta\tau b$ plotted against q_n. It is obvious that the numerous possibilities include the four represented in figure 17. In (*a*), $f(q) = 0$ and $h + \eta\tau b$ merely merges with other erratic errors; in (*b*), $f(q)$ is a constant and the term in $h + \eta\tau b$ not only contributes to the erratic errors, but has a value in its own right; in (*c*) and (*d*), $f(q)$ is linear in q and not only contributes to the erratic errors and has a value in its own right, but contributes a term that is independent of q. That constant term merges with V in eq. 99, producing a constant systematic error that may be either positive or negative.

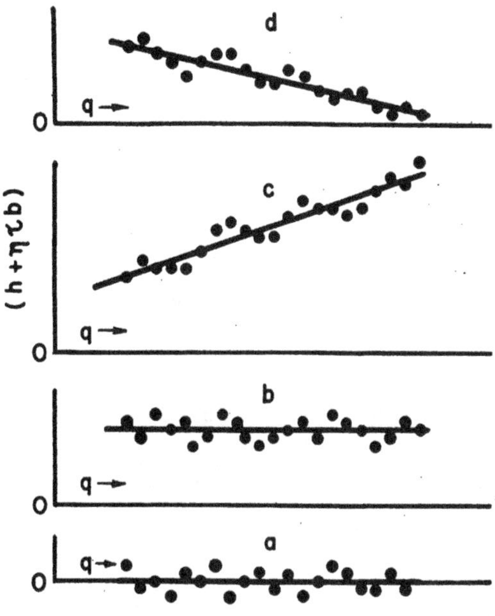

FIG. 17.—Four possible types of variation of $h+\eta\tau b$ with q when the setting lies in a critical region.

The case represented by (*a*) is possible only if the average value of $h+\eta\tau b$ is zero, the average being extended over all the orders studied. The most probable case is that represented by (*b*), or its negative, in which that average has a finite value. Cases (*c*) and (*d*) can arise only when the irregularities follow a definite law.

In the critical region $f(q)$ will not, in general, be the same for any two of the four classes of observations $e+$, $e-$, $r+$, $r-$.

There are, however, three special and important cases in which positive statements can be made regarding the behavior of h and b when Θ lies in a critical region. Namely, when the critical region arises solely (*a*) from a periodic error in the placing of the teeth, or (*b*) from an eccentricity of the wheel, or (*c*) from the speed of the wheel having a periodic component, the period in all three cases being that of one circumference. Other periods might be considered, but this is the important one.

These cases should always be expected to be present unless special precautions have been taken to avoid them. The first is considered in the following section where it is shown that when Θ lies in a critical region so caused, then the numerical value of h decreases as q increases, and that of b increases. Hence, in the absence of information regarding the relative magnitudes of h and $\eta\tau b$ and the extent of the periodic error, it is impossible to say how $h+\eta\tau b$ varies with q; but vary, it most likely does.

Identically the same mathematical treatment applies to the other two cases.

Periodic Variation in the Distribution of the Teeth

Consider a thin wheel of N teeth, each bounded laterally by radial lines, number the edges of those teeth consecutively, from 1 to $2N$, in such a way that the higher-numbered edge of any tooth carries an even number. Each tooth edge of the phantom Φ of that wheel will carry the same number as the corresponding edge of the wheel W. (Such expressions as "edge of W" and "edge ν of Φ" are to be understood as being equivalent to "edge of a tooth of the wheel" and "edge numbered ν of a tooth of the phantom.")

If the wheel were uniformly cut, so that every tooth and every interdental gap had the same angular breadth ($\beta=\pi/N$), then whenever edge $2N$ of W coincided with any edge of Φ, every edge of W would coincide with an edge of Φ.

But, if each edge were displaced from the place $x\beta$ it occupied in that uniformly cut wheel by a small amount $\gamma \sin x\beta$ ($\gamma<\beta/2$), then neither the teeth nor the interdental spaces would be of the same uniform size. Corresponding edges of Φ would be similarly displaced. Let the angle $x\beta$ be measured from edge $2N$ in the direction of increasing numbers. If edge $2N$ of W coincides with edge $2N$ of Φ, each edge of W will coincide with an edge of Φ. But if edge $2N$ of W coincides with an edge of Φ other than $2N$, then all edges of W will not coincide severally with all edges of Φ.

The problem is to find how the edges are related in that case; and in particular to find the size of the total angular opening in the equivalent sectored disk composed of Φ and W, when the angular displacement Θ of W with reference to Φ is either equal to, or near, the value $\Theta=(2n-1)\beta\equiv\Theta_n$, Θ being measured in the direction in which the numbers assigned to the edges increase. The size of that opening determines the settings of the disk at which the star vanishes and reappears, respectively, and is, therefore, directly involved in the determination of the coefficient h occurring in eq. 99, by which the velocity of light is computed from the experimental data. And the b term in that formula varies inversely as the rate at which the size of the opening varies with Θ. If the observations all lay in normal regions, each of those coefficients would be independent of the order of the eclipse, and could be eliminated by averaging suitable observations, as previously shown. But, as shown in the preceding section, if the settings lie in critical regions, both h and b will, in general, vary with the order, and cannot be eliminated by such averaging, the residual error depending on the way the sum $h+\eta\tau b$ varies with the order. It is that variation that is of present interest.

Were $\gamma=0$—the wheel uniformly cut—then when $\Theta=\Theta_n$, edges 2ν and $(2\nu+1)$ of W would, respectively, coincide with edges $(2\nu+2n-1)$ and $(2\nu+2n)$ of Φ, ν having any integral value from 1 to N, inclusive. There would be no opening in the sectored disk; there would be an eclipse of order n. The edges that coincide under those conditions may be called "matched" edges for order n.

When γ is finite and n is neither zero nor an integral multiple of N, matched edges will not, in general, coincide when $\Theta=\Theta_n$. In some cases the W-tooth will overlap the Φ-one; in others a gap will appear between the edges. From the way in which the edges are numbered and the direction in which Θ is measured, it is evident that whenever the displacement of a Φ-edge, in the direction of the numbering of the edges, exceeds that of its matched W-edge, then there will be a gap if the matched W-edge has an even number, and an overlap if its number is odd. If the displacement of Φ is in the opposite direction, the reverse will be true.

The widths of these gaps are given by eq. 109 and eq. 110 where g_e is the width of an even gap (gap at an even-numbered edge of W), and g_o is that of an odd one.

$$g_e=\gamma[\sin (2\nu+2n-1)\beta-\sin 2\nu\beta]$$
$$=2\gamma \sin \frac{(2n-1)\beta}{2}\cdot\cos (2\nu+n-\tfrac{1}{2})\beta, \qquad (109)$$

$$g_o=\gamma[\sin (2\nu+1)\beta-\sin (2\nu+2n)\beta]$$
$$=-2\gamma \sin \frac{(2n-1)\beta}{2}\cdot\cos (2\nu+n+\tfrac{1}{2})\beta. \qquad (110)$$

For g_e, the number of the W-edge is 2ν; for g_o, it is $2\nu+1$. Negative values of g_e and g_o indicate an overlap of the teeth, instead of a gap.

In each case the cosine factor defines a diameter of the wheel such that when gaps of that type occurring on one side of it are open, corresponding ones on the other side are closed. At each end of that diameter the value of the cosine factor is zero. Furthermore, since the angle of the cosine term in g_o exceeds that in g_e by an amount β, which is the angular separation of neighboring edges, the two diameters coincide. Call that diameter the base diameter.

The width (positive or negative) of any potential gap at the W-edge distant $\pm\alpha$ from either end of the base diameter will be equal to $2\gamma\{\sin (2n-1)\beta/2\}\sin \alpha$. Thus, the potential gaps fall into groups of four, as indicated in table 36, where C or O indicates that the gap is closed or open, respectively; and o or e indicates that the pertinent W-edge is odd or even, respectively. In each group of four corresponding to the same value of α, two of the gaps are open, and two are closed.

For convenience, the gaps in one quadrant are supposed to be numbered regularly from 1 to $N/2$ (to $(N-1)/2$ if N is odd), starting from one end of the base diameter; and l is used as the general symbol for the number of a gap.

As N has been even in all past work, only that case will be considered here.

If N is even, the total angular opening G_n in the sectored disk when $\Theta=\Theta_n$ is given [48] by eq. 111.

$$G_n=2[g_1+g_2+\cdots+g_{N/2}]$$
$$=4\gamma \sin \frac{(2n-1)\beta}{2}\cdot\left[\sin\frac{\beta}{2}+\sin\frac{3\beta}{2}+\sin\frac{5\beta}{2}+\cdots+\sin\frac{(N-1)\beta}{2}\right]$$
$$=2\gamma\left(\sin\frac{(2n-1)\beta}{2}\right)\bigg/\sin\frac{\beta}{2}. \qquad (111)$$

And the width g_{nl} of gap l when $\Theta=\Theta_n$ is given by eq. 112.

$$g_{nl}=2\gamma \sin\frac{(2n-1)\beta}{2}\cdot\sin\frac{(2l-1)\beta}{2}$$
$$=G_n \sin\frac{\beta}{2}\cdot\sin\frac{(2l-1)\beta}{2}$$
$$=\tfrac{1}{2}G_n\{\cos (l-1)\beta-\cos l\beta\}. \qquad (112)$$

The width $g_{n\,\max}$ of the widest gap when $\Theta=\Theta_n$, N being even, is that for which $l=N/2$. Its value is

$$g_{n\,\max}=\tfrac{1}{2}G_n \sin \beta. \qquad (113)$$

Consider now the way the size of the total opening in the sectored disk varies as Θ departs slightly from Θ_n, the difference $\Theta'\equiv\Theta-\Theta_n$ being smaller than β.

On referring to table 36 it will be seen that in each of the groups of four potential gaps having the same α, one of the two open gaps is associated with an even edge of W, and the other with an odd edge. The same is true of the two closed ones. Furthermore, from the way the edges of the teeth are numbered it is obvious that as Θ increases (as W turns in the direction of increasing numbers), the widths of the potential gaps associated with the even edges of W will decrease, and those with the odd will increase, each by the same amount. Hence as Θ is gradually increased beyond Θ_n, one of the open gaps in the l-quartet gradually closes, and the other opens wider, the sum of their openings remaining $2g_{nl}$ until $\Theta-\Theta_n\equiv\Theta'=g_{nl}$. For that value of Θ, one of the initially open l-gaps is just closed, the opening of the other is $2g_{nl}$, one of the initially closed l-gaps is just about to open, and at the fourth l-gap the matched teeth overlap by the amount $2g_{nl}$. As Θ' increases beyond the amount g_{nl}, the sum of the openings of the four l-gaps

[48] If $S\equiv\sin mx+\sin (m+\delta)x+\sin (m+2\delta)x+\cdots+\sin nx$, $n\equiv m+k\delta$, k being an integer, multiply through by $2\cos \delta x$, replace the products of the form $2\sin (m+\kappa\delta)x\cdot\cos \delta x$ by $[\cos (m+\kappa\delta-\delta)x+\cos (m+\kappa\delta+\delta)x]$, and simplify, finding

$$2S\sin\frac{\delta x}{2}=\cos\left(m-\frac{\delta}{2}\right)x-\cos\left(n+\frac{\delta}{2}\right)x.$$

TABLE 36

STATUS OF POTENTIAL GAPS IN SECTORED DISK WHEN $\Theta=\Theta_n$, $\gamma\neq 0$

N = number of teeth in the wheel; A and B indicate the two ends of the base diameter; α = angular position of that tooth-edge of W that is associated with the potential gap; it is measured from an end of the base diameter and in the direction indicated by the signs prefixed to A and B, the positive direction being that in which the edges are numbered (a tooth is bounded by an odd and the next higher even-numbered edge); O_e and O_o indicate that the gap is open and that it is, respectively, associated with an even and an odd edge of W, similarly for C_e and C_o, except that the C indicates that the gap is closed; l is the serial number of the gap. When N is odd, one end (A) of the base diameter coincides with an even edge of W, and the other with an odd edge, the four gaps ($\alpha=0$) reduce to two pairs of coinciding gaps of width zero.

N is even						N is odd					
α	$-A$	$+A$	$-B$	$+B$	l	α	$-A$	$+A$	$-B$	$+B$	l
	Status of gap						Status of gap				
$\beta/2$	O_e	O_o	C_e	C_o	1	0	C_e	O_e	C_o	O_o	1
$3\beta/2$	C_o	C_e	O_o	O_e	2	β	O_o	C_o	O_e	C_e	2
$5\beta/2$	O_e	O_o	C_e	C_o	3	2β	C_e	O_e	C_o	O_o	3
$7\beta/2$	C_o	C_e	O_o	O_e	4	3β	O_o	C_o	O_e	C_e	4
.	.	.	.	.	.	.	.	.	.	.	.
.	.	.	.	.	.	.	.	.	.	.	.
.	.	.	.	.	.	.	.	.	.	.	.
$(N-1)\beta/2$	{ O_e	O_o	C_e	C_o	$N/2$ odd	$(N-1)\beta$	{ O_o	C_o	O_e	C_e	$(N-1)/2$ odd
	C_o	C_e	O_o	O_e	$N/2$ even		C_e	O_e	C_o	O_o	$(N-1)/2$ even

becomes $2g_{nl}+2(\Theta'-g_{nl})=2\Theta'$, exactly what it would have been had $\gamma=0$, had there been no periodic error.

If λ is the value of l defined by $g_{nl}\leqq\theta'\leqq g_{n(l+1)}$, which then becomes $g_{n\lambda}\leqq\Theta'\leqq g_{n(\lambda+1)}$, it is evident that the total opening in the sectored disk is $G_{n\theta}$, as given by the equation

$$\begin{aligned}G_{n\theta}&=G_n-2(g_1+g_2+\cdots+g_\lambda)+2\lambda\theta'\\&=2(g_{\lambda+1}+g_{\lambda+2}+\cdots+g_{N/2})+2\lambda\Theta'\\&=G_n\cos\lambda\beta+2\lambda\Theta'\end{aligned}\qquad(114)$$

and if $\Theta'=g_{n\lambda}$, this becomes

$$G_{n\theta\lambda}=G_n\{\lambda\cos(\lambda-1)\beta-(\lambda-1)\cos\lambda\beta\}.\qquad(115)$$

If $\Theta'=g_{n\ max}=\frac{1}{2}G_n\sin\beta$, then $\lambda=N/2$ and $G_{n\theta\lambda}=N\Theta'$; this particular value for $G_{n\theta\lambda}$ for order n will be denoted by $G_{n\ N/2}$.

Whence Θ lies in a normal region if $\Theta-\Theta_n\equiv\Theta'\geqq g_{n\ max}$, and in a critical region if $\Theta'<g_{n\ max}$.

If G_r is the smallest total opening in the sectored disk through which the observer can detect the returned star, then there will be no apparent eclipse unless $G_r>G_n$.

Now it has been seen that Θ will not lie in a critical region unless $\Theta'<g_{n\ max}$, for which value the total opening in the disk is $G_{n\ N/2}=\frac{1}{2}NG_n\sin\beta$. Hence if the setting for an apparent eclipse of order n lies in a critical region, $G_r<G_{n\ N/2}$.

Hence, if an apparent eclipse of the star is to be possible when Θ lies in a critical region, eq. 116 must obtain.

$$G_n<G_r<G_{n\ N/2}=\tfrac{1}{2}NG_n\sin(\pi/N).\qquad(116)$$

Hence the greatest range of orders (m to n) over which there are apparent eclipses while Θ lies in a critical region and G_r remains unchanged, is given by the equation

$$G_n/G_m=\tfrac{1}{2}N\sin\beta=\tfrac{1}{2}N\sin(\pi/N).\qquad(117)$$

For a given N, a given low order m, and a given minimum brightness of the returned star (i. e., a given G_r), the value of Θ corresponding to G_r and lying in a critical region when the order of the eclipse is m—for these conditions, there can be in a critical region no apparent eclipse of order greater than the n defined by eq. 117. And for a fixed n, Θ will lie in a normal region if m is less than that defined by eq. 117.

When there are such periodic errors as are here considered, their amplitude being independent of the order of the eclipse, then for some orders, values of Θ corresponding to G_r will lie in critical regions, and for some they will not.

The setting at which a reappearance of the returned star occurs is given by $G_{n\theta}=G_r$, which by eq. 114 is equivalent to eq. 118, G_r being a constant.

$$G_r=G_n\cos\lambda\beta+2\lambda\Theta'.\qquad(118)$$

As n increases, so does G_n; consequently, Θ' must decrease if λ remains constant. But a decrease in Θ' necessarily results in a decrease in λ, and so does an increase in G_n. The presence of these two contrary effects makes it difficult to determine by mere inspection of eq. 118 whether or not Θ' does actually decrease as n increases.

But, by the definition of λ ($g_{n\lambda}\leqq\Theta'\leqq g_{n(\lambda+1)}$), G_r lies between $G_{n\theta\lambda}$ and $G_{n\theta(\lambda+1)}$, the last being the greater. Hence λ is the largest integer for which $G_{n\theta\lambda}<G_r$. The value of $G_{n\theta\lambda}/G_n$ in terms of λ is given by eq. 115, and displayed in figure 18 for the case $N=150$. From that figure the proper value for λ can be determined for $N=150$ when G_r/G_n is known; and then Θ' can be determined from eq. 118 for the same conditions.

It was thus found that Θ' does decrease as n increases, as shown in table 37 and figure 19. Hence the h term in eq. 99 decreases as n increases.

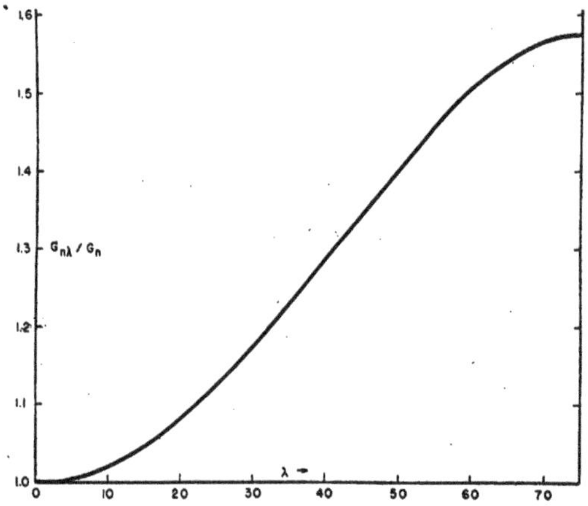

FIG. 18.—Plot of $G_{n\theta\lambda}/G_n$ as a function of λ, for a wheel of 150 teeth affected by a sinusoidal error.

G_n = total angular opening in the equivalent sectored disk when $\Theta = \Theta_n$ (speed adjusted for true eclipse of order n when distribution of teeth is ideally uniform) and the distribution of the edges of the teeth is affected by a sinusoidal error (period = one revolution of the wheel); $G_{n\theta\lambda}$ = total opening when $\Theta = \Theta_n + \Theta'$, Θ' being the angular width of the λ'th gap when $\Theta = \Theta_n$, λ being counted from one end of the (base) diameter that marks the transition of the widths of the gaps associated with the even (or with the odd) numbered edges of the teeth of the wheel from positive to negative values. $G_{n\theta\lambda}/G_n = \lambda \cos(\lambda-1)\beta + (\lambda-1)\cos\lambda\beta$ (see eq. 115); $\beta = 1.2°$.

But the b term must also be considered. That is proportional to the reciprocal of $|dI/d\Theta|$, the rate at which the apparent brightness of the returned star changes as Θ is increased. From eq. 114 it is obvious that $|dI/d\Theta| \propto \lambda$. Hence b, varying as $1/\lambda$, increases as the order increases, as shown in table 37 and figure 19.

Thus in the expression $h+\eta\tau b$ plotted in figure 17, the h term decreases as n increases, and the b term increases. Consequently, as the actual values of h, b, and η are not known, it is impossible to determine how that expression as a whole varies with the order of the eclipse. It might either increase or decrease. In either case, the variation would introduce an error in the computed value for the velocity of light. It seems scarcely likely that the opposite variations of h and b will exactly balance.

All of the preceding has to do with a reappearance with increasing speed. Other cases may be similarly treated, but it seems profitless to do it now.

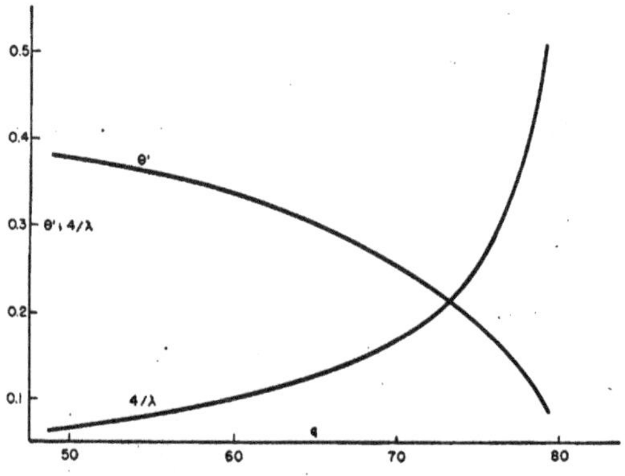

FIG. 19.—Plot of Θ' and of $4/\lambda$ against q, for the wheel considered in table 37.

$\Theta_n+\Theta'$ is the angular displacement (in degrees) of the wheel with reference to its phantom when the total angular opening in the equivalent sectored disk is 57°; λ is the number of the widest gap, counted as in figure 18, that does not exceed 57° when $\Theta'=0$; $q \equiv 2n-1$; n is the order of the eclipse, or reappearance.

COMPUTATION OF SPEED AND ACCELERATION OF WHEEL

The formulas used by Cornu, and also by Perrotin and Prim, for computing the speed and the acceleration of the toothed wheel, the acceleration being assumed to be constant, may be derived as follows:

TABLE 37

CERTAIN COMPUTED DATA FOR A WHEEL OF 150 TEETH AFFECTED BY A PERIODIC ERROR OF THE KIND CONSIDERED IN THE TEXT

The amplitude (γ) of the periodic error is taken as 0.4°, β being 1.2°. It is assumed that the returned star does not appear until the total opening in the equivalent sectored disk amounts to 57°; i. e. $G_r = 57°$. Since this value satisfies the relation $G_{40} < G_r < (G_{25})_{N/2} = \frac{1}{2}NG_{25}\sin\beta$, every setting throughout this range of orders (25 to 40) lies in a critical region when the total opening in the disk is 57°. But orders 25 and 40 lie very near the limits within which the conditions of eq. 116 are fulfilled. In figure 19, Θ' and $4/\lambda$ are each plotted against q_n, λ being determined by the relations $g_{n\lambda} \gtreqless \Theta' \gtreqless g_{n(\lambda+1)}$, where Θ' is such that $G_{n\theta} = G_r$, $G_{n\theta}$ being defined by eq. 114.

n	Order of eclipse	25	30	34	35	37	38	39	40
q_n	$\equiv 2n-1$	49	59	67	69	73	75	77	79
Θ_n	$\equiv (2n-1)\beta$	58.8°	70.8°	80.4°	82.8°	87.6°	90.0°	92.4°	94.8°
$g_{n\ max}$	Widest gap ($\Theta=\Theta_n$)	0.3927°	0.4634°	0.5163°	0.5290°	0.5537°	0.5657°	0.5774°	0.5889°
G_n	Total opening ($\Theta=\Theta_n$)	37.50°	44.26°	49.30°	50.52°	52.88°	54.02°	55.14°	56.24°
$G_{nN/2}$	Total opening ($\Theta=\Theta_n+g_{n\ max}$)	58.91°	69.51°	77.45°	79.35°	83.06°	84.85°	86.61°	88.33°
G_r/G_n	When $G_r=57°$	1.520	1.288	1.156	1.128	1.078	1.055	1.034	1.014
λ	When $G_r=57°$	62	40	28	25	19	15	12	8
$g_{n\lambda}$	When $G_r=57°$	0.3771°	0.3410°	0.2810°	0.2602°	0.2089°	0.1691°	0.1141°	0.0922°
Θ'	When $G_r=57°$	0.3784°	0.3424°	0.2844°	0.2649°	0.2172°	0.1874°	0.1478°	0.0969°
$4/\lambda$	When $G_r=57°$	0.0645	0.1000	0.1428	0.1600	0.2105	0.2667	0.3333	0.5000

Let M = number of turns of the wheel during the interval between two consecutive signals by the revolution counter;
m turns per second be the speed of the wheel;
a turns per second per second be the acceleration, assumed to be constant.
α = angular acceleration $= 2\pi a$;
t_0, t_1, t_2 be the times of 3 consecutive signals by the revolution counter;
t' be the time between t_0 and t_2 at which the values of m and a are desired.
$\mu_0 \equiv t_1 - t_0$; $\mu_1 \equiv t_2 - t_1$;
$\tau_0 \equiv (t_0 + t_1)/2$; $\tau_1 \equiv (t_1 + t_2)/2$;
m_0, m_1, m' = speed at the times τ_0, τ_1, and t', respectively;
$\mu' \equiv M/m'$;
D = distance from wheel to collimator mirror;
N = number of teeth;
n = order of eclipse;
$q_n = 2n - 1$;
V = velocity of light;
$R \equiv (\mu_1 - \mu_0)/(\tau_1 - \tau_0)$.

Then, since a is, by hypothesis, constant,

$$m_0 = \frac{M}{\mu_0}, \qquad m_1 = \frac{M}{\mu_1}, \tag{119}$$

$$a = (m_1 - m_0)/(\tau_1 - \tau_0), \tag{120}$$

and the speed m' at time t' is

$$m' = m_0 + (t' - \tau_0)a \equiv M/\mu', \tag{121}$$

where μ' is to be determined.

$$\frac{m'}{M} \equiv \frac{1}{\mu'} = \frac{1}{\mu_0} + \frac{t' - \tau_0}{\tau_1 - \tau_0}\left(\frac{1}{\mu_1} - \frac{1}{\mu_0}\right)$$

$$= \frac{1}{\mu_0}\left[1 + \frac{t' - \tau_0}{\tau_1 - \tau_0}\left(\frac{\mu_0 - \mu_1}{\mu_1}\right)\right] \tag{122}$$

$$= \frac{1}{\mu_0 + \dfrac{t' - \tau_0}{\tau_1 - \tau_0} \cdot \dfrac{(\mu_1 - \mu_0)\mu_0}{\mu_1}}, \text{ approximately.} \tag{123}$$

If $\mu_1 - \mu_0$ is such a small quantity that its square may be neglected, μ_0/μ_1 may here be replaced by unity, giving

$$\mu' = \mu_0 + (t' - \tau_0)(\mu_1 - \mu_0)/(\tau_1 - \tau_0) \tag{124}$$

and $m' = M/\mu'$. This is Cornu's formula.

Similarly,

$$a = -\frac{(\mu_1 - \mu_0)M}{(\tau_1 - \tau_0)\mu_0\mu_1}. \tag{125}$$

But by eq. 75, at an eclipse of order n, $m = Vq_n/4DN = M/\mu$; also, μ_0, μ_1, and μ' differ but slightly, a being small. Hence, quite approximately

$$\frac{1}{\mu_0\mu_1} = \left(\frac{1}{\mu'}\right)^2 = V^2q_n^2/16D^2N^2M^2$$

and

$$a = -\frac{M}{(\mu')^2} \cdot \frac{\mu_1 - \mu_0}{\tau_1 - \tau_0} \tag{126}$$

or

$$a = -\frac{V^2q_n^2}{16D^2N^2M} \cdot \frac{\mu_1 - \mu_0}{\tau_1 - \tau_0}. \tag{127}$$

Writing $R \equiv (\mu_1 - \mu_0)/(\tau_1 - \tau_0)$, this becomes

$$a = -\frac{V^2R}{16D^2N^2M} \cdot q_n^2. \tag{128}$$

The angular acceleration, $\alpha = 2\pi a$, is

$$\alpha = -\frac{\pi V^2R}{8D^2N^2M} \cdot q_n^2. \tag{129}$$

The formula used by Perrotin and Prim for the acceleration is the angular equivalent of eq. 126.

FOUCAULT'S METHOD (ROTATING MIRROR)

In Foucault's method for measuring the velocity of light, light proceeding from a reticle or a narrow slit is reflected by a rapidly rotating mirror, and passes through a suitable optical system to a distant stationary mirror that returns it to the rotating mirror, which then reflects it to the observing device. The angle θ through which the mirror turns while the light is going and returning is used to measure the time τ required for the light to travel from the rotating mirror to the distant stationary one, and back again.

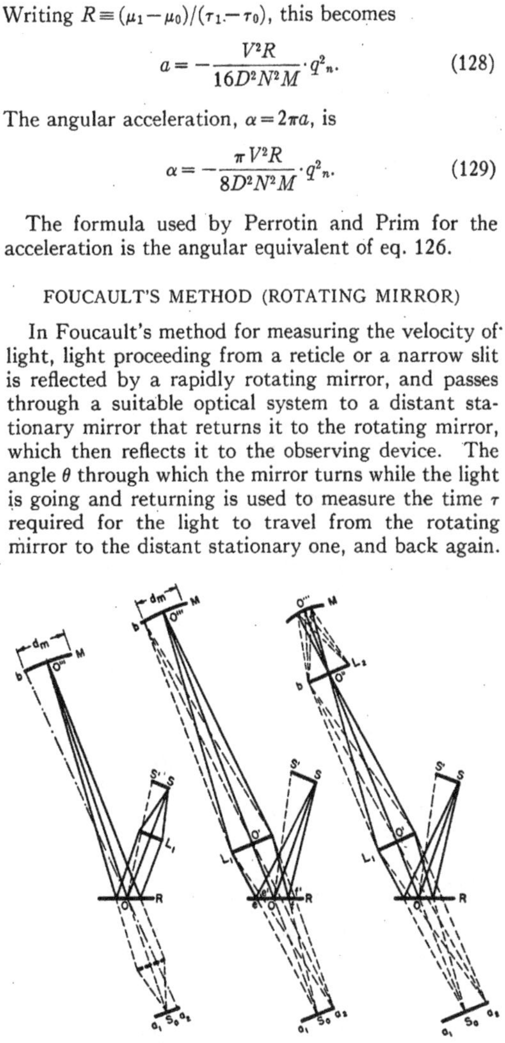

FIG. 20.—Types of optical systems used in the Foucault method for measuring the velocity of light.

S is the source of light; L_1 and L_2 are converging lenses; R is the rotating mirror; M is the distant fixed mirror; S_0 is the image of S in R. See text for more information. In every case a lens may be replaced by a suitable concave mirror properly placed, perhaps with the addition of a redirecting plane mirror.

Three distinct optical systems, shown diagrammatically in figure 20, have been used in connection with the rotating mirror. With any of them, additional fixed mirrors may be introduced for the purpose of lengthening the path traveled by the light, and of conserving the intensity of the returned image. As the sole purpose of the lenses is to focus and direct the light, they may be replaced by concave mirrors suitably placed. There is no need to give special attention to that modification.

In each case the light from S is reflected by the rotating mirror R to the fixed mirror M, distant $D=OM$ from R, which returns it to a focus at S if R is at rest, or to some such point as S' if R is rotating. If R is a simple plane mirror, the angle SOS' is twice the angle θ through which R has turned while the light was going from R to M and back again. If R is a prism of n faces and is turning at such a rate that each face is almost exactly replaced by the following one while the light is going and returning, and if the light is reflected either from that following face or from the face that is diametrically opposite it, then the angle turned through by the mirror while the light was going and returning will be $(2\pi/n)\pm\frac{1}{2}\angle SOS'$, the sign being determined by the direction both of the rotation of the mirror and of the deflection of the returned image.

In systems (a) and (b), the lens L_1 focuses S on M; and M is preferably concave, the radius of curvature being OO''' for (a), and $O'O'''$ for (b). But since OO''' and $O'O'''$ are great, M may be plane without much loss of light. In system (c), the preferred adjustment is that used by Cornu in his work by the Fizeau method, in which S is focused by L_1 on the midplane of L_2, and L_2 focuses L_1 on M, the radius of curvature of M being $O''O'''$.

In no case can light from S be returned by M unless the image of S in R lies between a_1 and a_2, where a_1 and a_2 lie on the lines drawn (a) from the sides of M through O, as bOa_2; (b) from the sides of M through O', the center of L_1, as $bO'a_2$; and (c) from the sides of L_2 through O', as $bO'a_2$.

As the mirror turns, the image of S in R moves from a_1 to a_2, and that formed by L_1 sweeps rapidly across M in systems (a) and (b), and across L_2 in system (c). If s is this rate of sweep, φ the angle through which the mirror turns while S_0 passes from a_1 to a_2, and θ the angle through which it turns during the time τ required for the light to go to M and return, then the values of a_1a_2, φ, s, and θ for each of the three systems are given by the expressions in table 38. The expressions for system (a) are exact; those for (b) and (c) assume that the focal lengths of the lenses are negligible in comparison with D. The symbols not yet defined have the following significance: r = distance OS, d_M = horizontal breadth of mirror M, D = distance OM, F_1 = focal length of L_1, d_{L2} = diameter of L_2,

TABLE 38

FORMULAS FOR SOME QUANTITIES INVOLVED IN THE FOUCAULT METHOD

See text for explanation of symbols.

System	(a)	(b)	(c)
a_1a_2..	rd_M/D	F_1d_M/D	F_1d_{L2}/D
φ....	$d_M/2D$	$F_1d_M/2rD$	$F_1d_{L2}/2rD$
s.....	$4\pi mD$	$4\pi rmD/F_1$	$4\pi rmD/F_1$
θ.....	$4\pi mD/V$	$4\pi mD/V$	$4\pi mD/V$
θ.....	$8\pi mD^2\varphi/Vd_M$	$8\pi mD^2r\varphi/F_1Vd_M$	$8\pi mD^2r\varphi/F_1Vd_{L2}$

m = number of turns of R per second, V = velocity of light.

In the simple form of system (a) shown in the figure, the angle SOS' must be small. But if M be slightly tipped so that the returned light will pass, say, above the outgoing light so as to strike R above that, and missing L_1, pass through a second similar lens that can be rotated about a line passing through O perpendicular to the plane of the paper, then the angle SOS' can be made as great as desired. That is what Newcomb did.

This use of two telescopes enables one to get a dark field in which to view the returned star, and permits both the axis of rotation of the mirror and that around which the observing telescope swings to be vertical, without blinding the observer with the directly reflected light.

In systems (b) and (c), if the reflection of the returned star is not from the opposite side of R, the axis of rotation of R must be slightly tipped so as to prevent directly reflected light from reaching the eyes of the observer. In these systems, the portion of the pencil of light from S that is actually returned varies as its image moves from a_1 to a_2. For example, in (b), when the image is at S_0, the light that strikes the portion ef of R is returned; whereas when it is at a_2, the portion returned is $e'f'$. It seems possible that this might under certain conditions give rise to a systematic error. An experimental search for such an error by those using these systems is much to be desired. But no report of such a search has been found.

In Michelson's [29, 31] use of method (c) with concave mirrors replacing the lenses, S was at the principal focus of L_1, and L_2 formed on M an image of S. With that adjustment, portions of the beam of parallel rays between L_1 and L_2 would miss L_2, and not be returned. For example, when the image of S in R is at a_2, then one-half of the light from a_2 that strikes L_1, will miss L_2. On the other hand, some of the light from points more distant than a_2 from S_0 will strike L_2, and be utilized if M is sufficiently broad. Nevertheless, it is not obvious that Michelson's adjustment is in any way superior to the more elegant one used by Cornu in his work with the Fizeau method.

The time τ required for the light to travel the distance $2D$ is computed from the angular deflection 2θ of the returned light and the angular velocity ω of the mirror, by means of eq. 130.

$$\tau = \theta/\omega. \tag{130}$$

The proper value of ω to use in that expression is the actual mean angular velocity of the mirror during that particular very short interval τ. That may depart widely from the average velocity over an interval of some seconds. The sole requirement for steadiness of the returned star, so far as that is determined by the motion of the mirror, is that the mirror shall have that same particular velocity for each of the successive intervals of duration τ during which the observed light is traveling between the mirrors. If that relation is not fulfilled, the value of θ, and so the position of the returned star, will vary from one reflection to another, giving rise either to a blurred and broadened or to a multiple image. The broadening may be asymmetric.

Since it is obvious that any system possessing inertia and having in the domain considered a condition of stable equilibrium will, in general, vibrate about that condition unless it is protected from all disturbing forces, it is to be expected that the rotating mirror will as a whole execute vibrations about its stable condition of uniform angular velocity. It is to be expected that the angular position of the mirror will be given by an expression of the form

$$\theta = A + \omega t + a_1 \sin \nu_1 t + a_2 \sin \nu_2 t + \cdots + b_1 \cos \nu_1 t + b_2 \cos \nu_2 t + \cdots \tag{131}$$

and it becomes of prime importance to determine by how much the periodic terms may affect the final result.

Strange as it may seem, not one report of any determination of the velocity of light by Foucault's method gives any indication that the author was aware of the probable presence of such terms; and consequently none give any indication that he took any step designed either to eliminate them or to eliminate their effect by a suitable observational procedure.

The only type of possible irregularity in the motion of the mirror as a whole that was considered in those reports was a dead-beat one arising from a temporary retardation of the motion at fixed angular positions of the mirror; and that was considered in but two reports: Newcomb's and the preceding one of Michelson's. Newcomb considered only the dissipational loss of momentum between the impacts of the driving air blasts on the vanes of the fan-wheel attached to the mirror; and he concluded that it was, for his purposes, negligible. Michelson considered the loss of momentum due to localized retarding forces, either in the bearings or from the aerodynamic reaction of the frame of his motor, and from the fact that changing the azimuth of the frame of his motor by certain unstated amounts did not produce in his result a change that exceeded the usual discordances, he concluded that such retarding forces as existed were not great enough to affect significantly the value he found for the velocity of light. Michelson's statement is far from clear, but this interpretation seems to be the only one that accords with the manner in which he has presented the data. If, however, he did have in mind such vibrations as are here considered, then the data he presented are totally useless for establishing his conclusion. This subject has not been mentioned in any of his later reports.

Beginning in 1924, Michelson used prismatic mirrors, as proposed by Newcomb in 1882. The avowed advantages of using a prismatic mirror were but two: (1) it permitted the use of a large θ without increasing the difficulties in measuring it; and (2) undesired light could more readily be kept from the eye of the observer.

But along with these avowed advantages went unheralded another of very great importance: The almost complete elimination of the effects of vibrations having frequencies that are integral multiples of that with which any face of the prism is exactly replaced by the next following one. It is exactly those vibrations that may otherwise be expected to produce the greatest error in the result. The values obtained with the prisms are all fairly concordant, and markedly different from the earlier ones.

If the time τ taken for the light to go and return were exactly that required for a face of the prism to be exactly replaced by the next, then vibrations of the frequencies mentioned would each be always in exactly the same phase when the light returned as it was when it started, and so far as they are concerned the mean angular velocity of the mirror during that interval τ would be exactly the same as the mean for a complete rotation.

As that condition is departed from, effects of vibrations of those frequencies may be expected to appear. They are of two kinds: a shift, and an asymmetrical broadening of the image. These need not be the same for both directions of rotation. Hence, although the substitution of a prism for a disk greatly reduced the evil effects of these vibrations, it did not remove all necessity for searching for and eliminating them. So far as one can judge from the reports, such a search has not been made.

Vibrations of other frequencies may give rise to multiple images or may merely broaden and perhaps distort the image, and these effects may change slowly if there are frequencies that are very near some of those specified, but not identical with them.

KERR-CELL METHODS

Methods involving the use of a Kerr electro-optical cell for modulating the beam of light have been used by Karolus and Mittelstaedt, Anderson, and Hüttel.

These methods have much in common with the Fizeau method, but as frequencies of millions of cycles per second can be used, the light path can be so shortened that the entire apparatus can be set up in a long room. This greatly facilitates the adjustment of the apparatus and the determination of the temperature and pressure of the air throughout the path, and improves the steadiness of the observed light.

The procedure in these methods is to determine the length of path traveled by the light during a known number of cycles of the field applied to the cell. This has been done in either of two ways: (1) Either the path or the frequency was adjusted, the other remaining fixed, until the phase of the cycle was the same at each end of the path. (2) The length of the path was so changed that the change in the time taken to run it corresponded exactly to an integral number of cycles. In the first case, the entire length of the path is involved in the determination; in the second, only the difference in the two lengths.

Observations may be made by eye or by means of a photo-electric cell.

In the work of Karolus and Mittelstaedt the modulation did not change the intensity of the beam, but merely the ellipticity of its polarization. In that case it seems probable that the velocity measured is the "phase velocity." In the other works mentioned, the modulation varied the intensity of the light. In them, the velocity measured would seem to be the "group velocity."

APPENDIX B

MOTION MAINTAINED BY PERIODIC IMPULSES

When a vibrating body is driven by a series of periodically repeated impulses, its motion between impulses is, of course, solely under the action of its own forces; that is, in accordance with its own free period. The impulses do no more than change instantaneously its energy and phase.

When a steady state has been set up, the energy added by an impulse just equals that dissipated between impulses, and the shift of phase produced by the impulse is just sufficient to bring the phase back to what it was immediately after the preceding impulse. If the free period of the body is smaller than the period of the impulse, then just before an impulse the position of the body will be further advanced than it was just after the preceding impulse, and the coming impulse will retard it, so as to bring it back to that position.

In the steady state, each impulse produces the same change in phase, that change being equal and opposite to the change, with reference to the impulses, that accumulates between impulses when the body swings freely.

PERIOD OF IMPULSES NEARLY AN INTEGRAL MULTIPLE OF FREE PERIOD

If the interval θ between successive impulses differs but little from an integral multiple k of the free period T of the driven body, then the motion of that body during the interval $t=n\theta$ to $t=(n+1)\theta$, n being a positive integer, is given by equation 132,

$$x_n = Ae^{-\alpha(t-n\theta)}\cos\frac{2\pi}{T}(t+\delta-n\eta), \qquad (132)$$

where $\eta \equiv \theta - kT$. This equation applies from just after the impulse at $t=n\theta$ to just before that at $t=(n+1)\theta$.

Of course, the last value of x_n, that at $t=(n+1)\theta$, is equal to the first one of x_{n+1} at the same instant; and the last value of dx_n/dt falls short of the first of dx_{n+1}/dt by a fixed amount v, the amount by which each impulse increases the velocity of the driven body. By means of these relations one can readily derive the relations connecting the successive values of the amplitude and of the phase, for each of the intervals of duration θ, starting from any values of A and δ desired. It will be found that, ultimately, A and δ steadily approach the constant values defined by equations 133, 134, and 135.

$$A = \frac{vT}{2\pi R - (\alpha T/\pi R)e^{-\alpha\theta}\sin(2\pi\eta/T)}, \qquad (133)$$

$$\delta = \frac{T}{2\pi}\tan^{-1}\left(\cot\frac{2\pi\eta}{T} - e^{\alpha\theta}\csc\frac{2\pi\eta}{T}\right), \qquad (134)$$

where

$$R^2 \equiv 1 + e^{-2\alpha\theta} - 2e^{-\alpha\theta}\cos(2\pi\eta/T). \qquad (135)$$

If there were no damping ($\alpha=0$), then

$$A_0 = vT/4\pi \sin(\pi\eta/T) \qquad \text{and} \qquad \delta_0 = (T/4) - \eta.$$

It should be noticed that after the steady state has been reached, the time intervals between successive occurrences of the same phase of x, say that corresponding to $x=0$, will be exactly T for each of the first $k-1$ occurrences following an impulse, and will be $T+\eta$ for the next one. Then, by the definition of k and η, the cycle repeats. In each k intervals there are $k-1$ of length T, and one of length $T+\eta$. Also, the sum of any k consecutive intervals is exactly θ; in that sense, and only in that sense, are the vibrations synchronous with the impulses. The time corresponding to the interval between $m+\epsilon=\mu k+a$ successive recurrences of the same phase, ϵ and a/k being positive proper fractions and μ an integer, is either $\mu\theta+aT+\eta$ or $\mu\theta+aT$, depending upon whether or not an impulse occurred within a recurrences after the beginning of the interval. A knowledge not only of θ, but also of both T and the positions of the ends of the interval with reference to the times of impulses, is essential to the determination of a time interval from a record of the vibrations of a body driven by the impulses.

PERIOD OF IMPULSES MUCH SMALLER THAN FREE PERIOD

Closely related to the preceding case, though contrasting with it, is that in which θ is much smaller than T. It is easy to see what happens then. Suppose the body is a very long heavy pendulum initially at rest. The first impulse displaces it slightly. The pendulum is so sluggish that it has not returned to its equilibrium position when the next impulse is delivered; that increases its displacement, and so on, until presently the force of restitution, proportional to the displacement, is such that in the interval θ the pendulum goes from where it was struck to the end of its swing and returns to exactly the point at which it was struck. A steady state has then been reached, and the pendulum will keep moving back and forth over this arc so long as the impulses continue to be applied, the period of a complete cycle being θ.

If there were no damping, the motion would be exactly the same as if on its return swing the pendulum bob struck normally a perfectly reflecting surface. That would exactly reverse its velocity, and the pendulum would retrace its path. It would continually trace and retrace the portion of its path between its extreme elongation and the mirror, and the time required to traverse that path would be the same for each direction.

But if there were damping, the time required for the pendulum's return from its extreme elongation to a given point would be less than that taken in going from that point to that elongation; and if a steady state is to be maintained, additional energy would have to be added in order to replace that dissipated.

The necessary equations can be set up by the procedure outlined in the preceding case, but those for the steady state can be readily written down if one remembers that the motion is that of the freely oscillating pendulum, and that the time required to run the arc in one direction differs from that in the other.

If the impulses are applied at the instants defined by $t=n\theta$, n being an integer, then in the interval between $t=n\theta$ and $t=(n+1)\theta$ the displacement of the body from its equilibrium position is given by eq. 136,

$$x_n=Ae^{-\alpha\tau}\cos\frac{2\pi\tau}{T}, \tag{136}$$

where $\tau\equiv t-(n+1/2)\theta-\eta$, T = free period of the body, $(\frac{1}{2}\theta+\eta)$ being the outgoing time, and $(\frac{1}{2}\theta-\eta)$ the incoming one. The values of η, A, and the amplitude A_0 of the oscillations, each in terms of θ, the constants of the body, and the velocity v imparted to the body by each impact, are given by equations 137, 138, and 139.

$$\tan\frac{2\pi\eta}{T}=\cot\frac{\pi\theta}{T}\tanh\frac{\alpha\theta}{2}, \tag{137}$$

$$A=\frac{vT}{4\pi}\cdot\frac{e^{-\alpha\eta}}{\left[4\sinh^2\left(\frac{\alpha\theta}{2}\right)+2\sin^2\left(\frac{\pi\theta}{T}\right)\right]^{\frac{1}{2}}}, \tag{138}$$

$$2A_0=A\left[1-e^{\alpha(\frac{1}{2}\theta+\eta)}\cos\frac{2\pi}{T}(\tfrac{1}{2}\theta+\eta)\right]. \tag{139}$$

As before, the transformations are tedious, but not difficult.

It is obvious that the motion of the body is not a damped simple harmonic function of the time, of amplitude A_0 and period θ, although it can be represented by a Fourier's series with that period for its fundamental.

REFERENCES

1 Gheury de Bray, M. E. J.: *Bull. soc. astron. France* (=*Astronomie*) **40**: 113, 1926; **41**: 380–382, 504–509, 1927. *Astron. Nachr.* **230** (5520): 449–454, 1927. In the second article he states that he had been bringing this secular variation to the attention of the editor of *Nature* since December, 1924; but nothing from him regarding the velocity of light has been found in *Nature* between 1923 and 1927.

2 Helmert: *Astron. Nachr.* **87** (2072): 123–124, 1876.

3 Gheury de Bray, M. E. J.: *Ciel et terre* **47**: 110–124, 1931. *Nature* **129**: 573, 1932. *Isis* (Belgium) **25**: 437–448, 1936.

4 Edmondson, F. K.: *Nature* **133**: 759–760, 1934.

5 Gheury de Bray, M. E. J.: *Astron. Nachr.* **252** (6038): 235–236, 1934. *Nature* **133**: 948–949, 1934.

6 Gheury de Bray, M. E. J.: *Astron. Nachr.* **230** (5520): 450–454, 1927. *Bull. soc. astron. France* (=*Astronomie*) **41**: 504–509, 1927. *Ciel et terre* **47**: 110–124, 1931. *Nature* **127**: 522, 739–740, 892, 1931; **133**: 464, 948–949, 1934; **144**: 285, 945, 1939. *Isis* (Belgium) **25**: 437–448, 1936. Editorial and other notes referring to his suggestion: *Ciel et terre* **43**: 189, 222, 1927. *Nature* **120**: 594, 1927; **129**: 573, 1932; **138**: 681, 1936.

Salet, P.: *Bull. soc. astron. France* (=*Astronomie*) **41**: 206, 1927.

Maurer, H.: *Physik. Zeitschr.* **30**: 464, 1929.

Mittelstaedt, O.: *Physik. Zeitschr.* **30**: 165–167, 1929.

Vrkljan, V. S.: *Zeitschr. f. Physik* **63**: 688–691, 1930. *Nature* **127**: 892, 1931; **128**: 269–270, 1931.

Takéuchi, T.: *Zeitschr. f. Physik* **69**: 857–858, 1931. *Proc. Phys. Math. Soc. Japan* **13**: 178, 1931.

Kennedy, R. J.: *Nature* **130**: 277, 1932. *Phys. Rev.*, ser. 2, **47**: 533–535, 1935.

Wilson, O. C.: *Nature* **130**: 25, 1932.

Edmondson, F. K.: *Nature* **133**: 759–760, 1934.

Gramatzki, H. J.: *Zeitschr. f. Astrophysik* **8**: 87–95, 1934.

Birge, R. T.: *Nature* **134**: 771–772, 1934.

Machiels, A.: *Zeitschr. f. Astrophysik* **9**: 329–330, 1935.

Omer, G. C., Jr.: *Astrophys. Jour.* **84**: 477–478, 1936. *Nature* **138**: 587, 1936.

Kitchener: *Nature* **144**: 945, 1939.

Smith, A. B.: *Science*, n.s., **93**: 475, 1941.

Anderson, W. C.: *Jour. Optical Soc. America* **31**: 187–197, 1941.

7 Gheury de Bray, M. E. J.: *Ciel et terre* **47**: 110–124, 1931.

8 Deming, W. E.: *Jour. Washington Acad. Sci.* **31**: 85–93, 1941.

9 Fizeau, H.: *Compt. rend.* **29**: 90–92, 132, 1849.

10 Foucault, L.: *Compt. rend.* **55**: 501–503, 792–796, 1862.

11 Foucault, L.: *Compt. rend.* **30**: 551–560, 1850. *Annales de chimie et physique*, ser. 3, **41**: 129–164, 1854.

12 Michelson, A. A.: Studies in optics. Chicago: 120–138, 1927.

13 Foucault, L.: Recueil des travaux scientifiques de Léon Foucault. Paris: 173–226, 517–518, 546–548, 1878. Here are given a detailed description of the apparatus, with drawings (pp. 546–548), and certain numerical data (pp. 219–226) that have not been found elsewhere.

14 Cornu, A.: *Jour. de l'École polytechnique* **27** (44): 133–180, 1874.

15 Cornu, A.: *Annales de l'Observatoire Paris, Mém.* **13**: A.1–A.315, 1876.

16 Cornu, A.: *Compt. rend.* **79**: 1361–1365, 1874.

17 Cornu, A.: *Rapp. Cong. internat. de physique (Paris)* **2**: 225–246, 1900.

18 Perrotin, J., and Prim: *Annales de l'Observatoire Nice* **11**: A.1–A.98, 1908.

19 Perrotin, J.: *Compt. rend.* **131**: 731–734, 1900.

20 Perrotin, J.: *Compt. rend.* **135**: 881–884, 1902.

21 Newcomb, Simon: *Astron. Papers Am. Ephemeris* **2**: 107–230, 1891.

22 Brace, D. B.: *Science*, n.s., **16**: 81–94, 1902.

23 Michelson, A. A.: *Proc. Am. Assoc. Adv. Sci.* **27**: 71–77, 1878. See also *Astron. Papers Am. Ephemeris* **1**: 109–145, 1880. *Am. Jour. Sci.*, ser. 3, **15**: 394–395, 1878.

24 Michelson, A. A.: *Astron. Papers Am. Ephemeris* **1**: 109–145, 1880.

25 Michelson, A. A.: *Proc. Am. Assoc. Adv. Sci.* **28**: 124–160, 1879.

26 Michelson, A. A.: *Am. Jour. Sci.*, ser. 3, **18**: 390–393, 1879.

27 Todd, D. P.: *Am. Jour. Sci.*, ser. 3, **19**: 59–64, 1880.

28 Michelson, A. A.: *Astron. Papers Am. Ephemeris* **2**: 231–258, 1891.

29 Michelson, A. A.: *Astrophys. Jour.* **60**: 256–261, 1924.

30 Michelson, A. A.: *Jour. Franklin Inst.* **198**: 627–628, 1924. *Nature* **114**: 831, 1924.

31 Michelson, A. A.: *Astrophys. Jour.* **65**: 1–22, 1927.

32 Ladenburg, R.: *Handb. d. Exper. Physik* (Wien-Harms) **18**: 1–36, 1928.

33 Anderson, W. C.: *Jour. Optical Soc. America* **31**: 187–197, 1941.

34 Ehrenfest, P.: *Annalen d. Physik*, ser. 4, **33**: 1571–1576, 1910.

35 Reynolds, O.: *Nature* **16**: 343–344, 1877.

36 Michelson, A. A., F. G. Pease, and F. Pearson: *Astrophys. Jour.* **82**: 26–61, 1935.

37 Gheury de Bray, M. E. J.: *Ciel et terre* **45**: 82–88, 1929; **46**: 216–223, 1930. *Isis* **25**: 437–448, 1936.

38 Michelson, A. A.: *Phil. Mag.*, ser. 6, **3**: 330–337, 1902.

39 Michelson, A. A.: *Encyc. Brit.*, ed. 14, **23**: 34–38, 1929.

40 Listing, J. B.: *Astron. Nachr.* **93** (2232): 367–376, 1878.

41 Lorentz, H. A.: *Arch. Néerlandaises des Sci.*, ser. 2, **6**: 303–318, 1901.

42 Mittelstaedt, O.: *Annalen d. Physik*, ser. 5, **2**: 285–312, 1929.

43 Karolus, A., and O. Mittelstaedt: *Physik. Zeitschr.* **29**: 698–702, 1928.

44 Anderson, W. C.: *Rev. Sci. Inst.* **8**: 239–247, 1937.

45 Hüttel, A.: *Annalen d. Physik*, ser. 5, **37**: 365–402, 1940.

INDEX *

* If a name appears only in a foot-note or a reference, the appropriate reference symbol is attached to it; e.g., Birge, R. T. [6]. Other numbers refer to pages. The entire group of pages in a study of an author's work is given, even though the author's name does not appear on all of them. A *d* indicates that a definition is given on that page.

INDEX OF EQUATIONS

TRANSACTIONS

OF THE

AMERICAN PHILOSOPHICAL SOCIETY

HELD AT PHILADELPHIA

FOR PROMOTING USEFUL KNOWLEDGE

NEW SERIES—VOLUME XXXIV, PART II

THE ARC SPECTRUM OF IRON (Fe I)

PART I. ANALYSIS OF THE SPECTRUM

Based on the work of many investigators and including unpublished studies by MIGUEL A. CATALÁN

HENRY NORRIS RUSSELL and CHARLOTTE E. MOORE

Princeton University Observatory

PART II. THE ZEEMAN EFFECT

DOROTHY W. WEEKS

Wilson College

THE AMERICAN PHILOSOPHICAL SOCIETY

INDEPENDENCE SQUARE

PHILADELPHIA 6

DECEMBER, 1944

LANCASTER PRESS, INC., LANCASTER, PA.

PREFACE

The present work deals with a spectrum which has been the object of active study for more than seventy years, and carries one phase of the investigation—the term-analysis of the structure—about as far as existing material appears to permit. The spectrum, though complicated, is found to be highly orderly, and almost all the principal terms predicted by theory are now known. Present methods of observation in the laboratory, however, do not succeed in bringing out the spectrum as completely as it is exhibited in the sun, and future observation, especially for the discovery of better laboratory sources, is still highly promising.

The writers take pleasure in expressing their gratitude to many colleagues who have generously made new and unpublished material available. Special thanks are due to Doctor Miguel A. Catalán for the communication of many spectral terms; to Messrs. Arthur S. King and Harold D. Babcock for laboratory and solar spectroscopic material obtained at Mount Wilson; and to Professor George R. Harrison for observations of the Zeeman effect secured at the Massachusetts Institute of Technology.

HENRY NORRIS RUSSELL
CHARLOTTE E. MOORE
DOROTHY W. WEEKS

THE ARC SPECTRUM OF IRON (Fe I)

PART I

ANALYSIS OF THE SPECTRUM

HENRY NORRIS RUSSELL AND CHARLOTTE E. MOORE

I. INTRODUCTION

(1) GENERAL CONSIDERATIONS

The neutral atom of iron provides the classical example of a complex atomic spectrum. The ubiquity of the metal, and the richness and ease of production of its spectrum, have led to the general adoption of its lines as standards,[1] so that their wave-lengths are now more accurately known than those for any other element for which data have been published. Full data are available on temperature classification [2] and Zeeman effect, and a good beginning has been made [3] in the determination of the line intensities, transition-probabilities, and f-values. The spectrum is therefore of unusual interest both to the practical spectroscopist and to the theoretical physicist, and most of all to the astrophysicist. The high abundance of iron causes even its faint arc lines to appear in the cooler stars, and the stronger ones in those as hot as Sirius. Many lines predicted by the analysis of Fe I, and not yet observed in the laboratory, agree so closely, in wave-length and estimated intensity, with unidentified lines in the solar spectrum, that there can be no doubt of their presence there (see § 22).

The great number of iron lines and the wide range in intensity and excitation among them make them especially valuable for the determination of curves of growth, and for the investigation of stellar atmospheres.

A complete term-analysis of this spectrum is therefore of prime importance to astronomers and physicists alike. Practically complete analyses of Fe II [4] and Fe III [5] have already been published. One of comparable extent for Fe I has been delayed partly by the richness of the spectrum but more by political conditions (§ 2). It is at last possible to present it here.

(2) PREVIOUS INVESTIGATIONS

Shortly after Catalán's discovery of multiplets, twenty of them were identified in Fe I by F. M. Walters, Jr.[6] in 1923. In 1924 he published a list of 52 multiplets involving 26 spectral terms [7] of multiplicities 3, 5, and 7—though the last were not connected with the others. This was accomplished almost at the same time by the independent work of Laporte,[8] who classified about 600 lines and derived an ionization potential of 8.15 volts—which he revised to 8.06 in 1926 [9]—adding a few more terms. These results were shown by Hund [10] to be in complete agreement with his general theory of the structure of atomic spectra. Additional terms, almost all discovered by Walters, and communicated by him to colleagues, were announced by Meggers [11] and by Moore and Russell.[12]

Notable extensions of the analysis were accomplished (again independently in the main) by Burns and Walters [13] and by Catalán.[14] The latter accounted for 2350 lines by combinations among 304 energy levels, identified 51 terms, and determined the ionization potential as 7.83 volts. Many fairly strong lines remained unclassified, and 135 of the energy levels were not grouped into terms.

The only published addition to this appears to be Green's identification [15] of a few high terms, but it was generally known that Dr. Catalán was continuing the analysis. In 1936, in response to a request from the writers for his new terms for use in the "Multiplet Table," he generously sent a list of the low even terms, and stated that the remaining terms would be sent as soon as he could copy them. This list included singlet terms, making this the first spectrum (and up to the present, the only one) in which terms of four different multiplicities are known.

---

[1] *Trans. Internat. Astron. Union* **3**: 86, 1928; **4**: 60, 1932; **5**: 84, 1935; **6**: 79, 1938.

[2] King, A. S., *Mount Wilson Contr.* No. 66, 1913; No. 247, 1922; and No. 496, 1934; *Astrophys. Jour.* **37**: 239, 1913; **56**: 318, 1922; **80**: 124, 1934.

[3] King, R. B., and A. S. King, *Mount Wilson Contr.* No. 528, 1935; and No. 581, 1938; *Astrophys. Jour.* **82**: 377, 1935; **87**: 24, 1938.

[4] Dobbie, *Annals Solar Physics Observ.* (Cambridge) **5**, pt. 1, 1938.

[5] Edlén and Swings, *Astrophys. Jour.* **95**: 532, 1942.

[6] *Jour. Washington Acad. Sci.* **13**: 243, 1923.

[7] *Jour. Optical Soc. America* **8**: 245, 1924.

[8] *Zeitschr. f. Physik* **23**, 135, 1924; **26**: 1, 1924.

[9] *Proc. Nat. Acad. Sci.* **12**: 496, 1926.

[10] *Linienspektren*: 163. Berlin, Springer, 1927.

[11] *Astrophys. Jour.* **60**: 60, 1924.

[12] *Mount Wilson Contr.* No. 365, 1928; *Astrophys. Jour.* **68**: 151, 1928.

[13] *Publ. Allegheny Observ.* **6**: 159, 1929.

[14] *Anales Soc. Española Fisica y Quimica* **28**: 1239, 1930.

[15] *Phys. Rev.* **55**: 1209, 1939.

Shortly afterward, the Spanish War broke out. Communications were interrupted, and it was not known whether Dr. Catalán's laboratory and papers were accessible to him—or still in existence. To aid in continuing his work, a line-list of Fe I, containing all data available to the writers and unpublished Zeeman data by Babcock, was prepared at Princeton and sent to him. More than four years later Dr. Catalán sent a list of spectral terms and Zeeman g-values containing a great amount of new material. Many of the g-values appear to have been derived from Babcock's data. Communication is still very difficult, and he has been unable to send his long list of classified lines. This has had to be reconstructed from the term-values.

Additional data became available in 1941 when Professor G. R. Harrison, of the Massachusetts Institute of Technology, entrusted for discussion to Professor Dorothy W. Weeks, of Wilson College, a complete set of records of the Zeeman effect of iron, made on his automatic measuring machine, from spectrograms taken with the great Bitter magnet. This material is much more extensive and more fully resolved than any which had previously existed. Miss Weeks' results, including patterns for 1038 lines and the determination of g-values for 392 energy levels, are reported by her in Part II of this memoir.

For this work, a comprehensive analysis of the spectrum was prerequisite, and Miss Weeks and the writers were able to collaborate to great mutual advantage. The Zeeman patterns for previously unclassified lines were of great value in suggesting identifications, and the g-values which were derived for many known levels were often conclusive, either in confirming previous term-assignments or in suggesting new ones. At the same time the writers reviewed the existing analysis of the spectrum and were able to find a number of new levels and terms. Many of these were identified as terms predicted by Hund's theory, and nothing was found inconsistent with it.

The progress of the analysis of this spectrum has been curiously uneven. The numbers of energy levels detected and published in successive intervals are given in table 1. An attempt has been made in table A (p. 134) to assign each separate energy level to the investigator who first presented good evidence for its existence (irrespective of later assignment of a term-classification).

A list of the total would be misleading, since Walters and Laporte in 1924 worked and published independently and almost simultaneously, and Catalán in 1930 did not have access to the work of Burns and Walters in 1929. The principal contributors to the published analyses have been Walters, Catalán, and Laporte. The most important unpublished contribution is Catalán's. The analysis which is here presented is thus the co-operative work of many investigators, and the work of the writers has been largely editorial. Dr. Catalán would appear as a joint author of this work were it not for the difficulties of communication which have prevented the continuous exchange of results which the writers would otherwise have desired. As things are, they have been obliged to take the responsibility for the final assignment of term designations, the configuration analysis, and the rejection of doubtful levels. A few of Catalán's designations have been altered, mainly on the basis of new and more precise Zeeman data.

The present work is therefore one more example of the cordial international co-operation among spectroscopists which has been so largely responsible for the rapid advance of the analysis of complex spectra. It was not terminated until it was obvious that the law of diminishing returns was in active operation and that very little more in the way of discovery of additional energy levels would result from extensive effort.

The present lists contain 4860 classified lines, arising from combinations among 464 energy levels. Thanks to the complete Zeeman data, all but 19 of these have been grouped into 146 terms, which combine to give 1342 multiplets.

In so complicated a spectrum as this there can be no hope of classifying all the faint lines. There is good reason, however, to believe that the analysis is substantially complete. Almost all the stronger lines have been classified. Taking as a criterion the lines recorded by King in his temperature classification, which number 1753 and lie between 10469 and 2298A, we find that 46 remain unclassified, of which one is of intensity 20 and the next 8 on a scale in which the strongest lines reach 1500. The configuration analysis in §§ 10–14 shows also that almost all the terms which theory predicts as likely to give even moderately prominent lines have been identified.

Though this familiar spectrum has been so much observed, it is probable that a very large number of additional arc lines of iron can be detected when better sources for faint lines are devised (§ 25).

TABLE 1

PROGRESS OF ANALYSIS OF THE SPECTRUM

Dates	Low Even	High Even	Odd	Total
1923–24	16	20	94	130
1924–28	18	12	16	46
1929–30	3	62	57	122
1930–44			17	17
Total published	37	94	184	315
Catalán unpublished	12	9	68	89
Present writers	13	19	28	60
Grand total	62	122	280	464

II. OBSERVATIONAL DATA

(3) WAVE-LENGTHS

A complete and accurate list of wave-lengths and wave-numbers is the prime necessity for an analysis. There is a wealth of material for Fe I, but this is very far from homogeneous. The principal lines have been measured with very high precision—especially the International Secondary and Tertiary Standards,[16] which depend on three or more independent sets of interferometer measures. Many other lines have been measured by interference methods, and a great many more with gratings of adequate dispersion. There remain, however, many faint lines which have never been accurately measured in the laboratory—some only to 0.1A. Many of these agree with the predictions of the term-analysis within their limits of error and are undoubtedly due to iron.

A detailed and critical list of wave-lengths, based on homogeneous material and standards, is much to be desired, but there is no hope of this during the war.

The wave-lengths which appear in table B (p. 139) represent, therefore, a compilation from all available sources. Reasonable care has been taken to adopt determinations from those sources which, in general, were judged to be most reliable, but it should be emphasized that the list cannot be considered a definitive one. Wave-lengths of extreme precision are essential in the first attempts to break into complex spectra like those of some of the rare earths, which contain few outstanding lines; but the extension of an analysis in which many accurate values of terms and term-differences have already been found makes less severe demands. The existing material has proved generally adequate for the purpose, although the probable error of the worst measure in the table is at least twenty times that of the best.

The long list of sources at the end of table B is arranged roughly in order of accuracy—beginning with the standards, and ending with the rough measures just mentioned. The detailed order of the reference letters has, however, been adopted partly as a matter of convenience (e. g., good measures in the infra-red come early in the list); and the writers explicitly disclaim the assumption that an earlier letter in the alphabet always, or even usually, indicates that they regard the measures referred to as better than some others which may come a few places lower.

The numerous "predicted" lines of Fe I which have been taken from The Revised Rowland Table[17] are discussed in §§ 18–23 and listed in table C (p. 170) along with many others taken from Babcock's extension in the infra-red.[18]

[16] *Trans. Internat. Astron. Union* **3**: 86, 1928; **4**: 60, 1932; **5**: 84, 1935; **6**: 79, 1938.

[17] *Publ. Carnegie Inst. Washington* No. 396, 1928.

[18] Unpublished material.

(4) INTENSITIES

The recorded intensities of the lines in the great majority of spectra are in a state of primeval chaos. Different observers have used radically different scales. The older estimates were usually made on a scale from 1 for the weakest lines to 10 for the strongest. In some recent work (e. g., on the rare earths) a far more open scale has been adopted—from 1 to 1000 or even to 10000. Even the last probably falls short of representing the actual range.

In a few cases where an experienced observer, who has a fairly stable scale of estimation, has photographed the whole range of the spectrum, and made his estimates approximately homogeneous by comparison of overlapping plates, the recorded intensities give a good idea of the relative strengths of the lines on the photographs. Allowances must be made for the varying sensitiveness of the plates and for self-reversal and masking of weak lines by strong neighbors, but the experienced reader can obtain a fairly reliable idea of the actual intensity relations.

In the present case, however, one has to deal with a hodge-podge of estimates on all sorts of scales. The attempt to reduce these to an even roughly homogeneous system is hopeless. An experienced observer making rapid eye estimates on a good set of plates could produce a much better list of intensities.

The intensities found in table B have been taken from what seemed to be as good a source as any (not always the same as for the wave-length). The estimates by King, Meggers, and Kiess, which are on the open scale,[19] are given without parentheses; those of other observers, which are almost all on the narrow scale, in parentheses.

In the course of the analysis, where all known multiplets were written out in detail, the writers obtained a rough idea of the meaning of the estimates of different workers, and they believe that errors in the multiplet analysis resulting from the raggedness of the intensity estimates have been avoided.

All users of table B should, however, be explicitly warned that the tabulated intensities afford only a rough general indication, and should not be used for any quantitative purpose without special study.

(5) TEMPERATURE CLASSIFICATION

These data, which are also of primary importance in the analysis, are taken entirely from King's work.[20] The differences in openness of the intensity scale noted above have very little effect on the temperature classification. The tendency toward assignment of a higher temperature class in the ultra-violet is well known.

[19] Even here several different scales are involved; e. g., King's estimates in his first two papers differ systematically where they overlap, and his later scale in the deep-red is much more open.

[20] *Mount Wilson Contr.* No. 66, 1913; No. 247, 1922; No. 496, 1934; *Astrophys. Jour.* **37**, 239, 1913; **56**: 318, 1922; **80**: 124, 1934.

(6) ZEEMAN EFFECT

The observations and results are discussed by Miss Weeks in Part II.

III. THE ANALYSIS OF THE SPECTRUM

(7) ENERGY LEVELS. ACCURACY OF DETERMINATIONS

The main outlines of the structure of the spectrum have been known since 1924. Burns and Walters in 1929[21] determined the relative positions of 199 energy levels to 0.001 cm^{-1} with the aid of lines measured with the interferometer. These values have been adopted without change except for those levels whose values have been recommended by the International Astronomical Union,[22] although some of the standard wave-lengths upon which they depend have been slightly altered. The remaining 265 levels—most of which were discovered by Catalán—have been determined by the usual process of approximation, finding values for each new level from all the transitions which connect it with previously determined levels, working backward from these to derive improved values for such of the first group as have not already been taken as final, and so on. These values are given to 0.01 cm^{-1}.

The number of combinations from which a level has been determined ranges from more than thirty to one or two in a few instances. In the latter case, the levels were not accepted as real unless their existence was confirmed by position in multiplets, Zeeman effect, or both. Most of them are leading components of terms which have high J-values and give strong lines.

A complete study of the accuracy of the tabular level values would involve one of the accuracy of the various wave-lengths, and is not here attempted. A general idea may be obtained from the residuals of the individual determinations of each level from the mean. For 1118 determinations of 155 odd levels the mean residual, regardless of sign, is ± 0.036 cm^{-1}, and for 325 of 48 high even levels, ± 0.045 cm^{-1}. If these residuals arose solely from the errors of the levels which are being determined, the correct values of the average error could be found by dividing them by $\sqrt{(n-1)/n}$, where n is the number of observations contained in the mean. The average value of the divisor for the individual odd levels is 0.884, and for the even ones 0.873, giving ± 0.041 and ± 0.052 as the average error of a determination, or a general mean of ± 0.043 cm^{-1}. This should be increased to allow for the error of the other level-values from which these were derived, but a large proportion of the others were accurate interferometer values, so that the estimate of ± 0.05 cm^{-1} as the average error of a determination of a level from a single line may be adopted. Results derived from lines in the red will be more accurate, and from ultra-violet lines less so—assuming equal accuracy of wave-lengths. This refinement may await a definitive list of iron wave-lengths. Meanwhile, it may safely be assumed that the level-values given to two places in table A have average errors ranging from ± 0.02 to ± 0.04 cm^{-1}. Twenty-three cases in which the uncertainty is greater are marked by colons. These include:

(*a*) 15 values based on a single well-measured line. (The reality of the level is usually confirmed by other poorly measured lines.)

(*b*) 7 values in which the observations are discordant, with average residual exceeding 0.08 cm^{-1}.

(*c*) one value based on four lines measured to ± 0.1A ($u^5P_3^\circ$). For these levels the uncertainty may reach ± 0.1 cm^{-1}.

Table A (p. 134) contains those energy levels (464 in number) which have been finally accepted as real. It is arranged in order of term-types, the even terms preceding the odd; singlets, triplets; etc.

The first column gives the adopted electron configurations (when assignable with reasonable probability—see §§ 8–15); the second, the term-designations; the third, the energy levels above the ground state a^5D_4; and the fourth, the differences between successive components of a multiple term—positive when, as usual, the order is "inverted." The fifth column gives references to the author who appears to have been the first to present good evidence of the existence of each energy level, and the last, the definitive g-values determined by Miss Weeks, which are identical with the "corrected" values of table E (p. 203).

The unclassified even levels are listed after the classified even levels and similarly for the odd. The notation is that generally adopted.

IV. SPECTRAL STRUCTURE. IONIZATION POTENTIAL

(8) THE LOW LEVELS

The theoretical interpretation of the spectrum given in Hund's classic monograph has been fully confirmed by all subsequent work. The principal low terms arise from the even electronic configurations $3d^6\,4s^2$ and $3d^7\,4s$, and are thoroughly intermingled. These combine with very numerous odd terms coming from $3d^6\,4s\,4p$ and $3d^7\,4p$; and these again with terms of high even configurations, in which 4p is replaced by 5s or 4d.

The low even terms which may theoretically be anticipated are as follows:

Configuration	Quintets	Triplets	Singlets
d^6s^2	D	P D F G H (P F)	S D F G I (S D G)
d^7s	P F	P F, P D F G H (D)	P D F G H (D)
d^8		(P) F	(S D G)

[21] *Publ. Allegheny Observ.* **6**: 159, 1929. See also **8**: 39, 1931.

[22] *Trans. Internat. Astron. Union* **4**: 65, 1932.

The terms inclosed in parenthesis may be expected to be so high as to be almost unobservable.

The quintet terms and the lowest triplet, a^3F, were assigned to these configurations by Laporte in 1924. The assignment of the others on the basis of combination intensities would not be easy, as many cross-connections are strong; but all ambiguity is removed by comparison with the low terms in Fe II and Fe III, which have been thoroughly investigated by Dobbie[23] and by Edlén and Swings.[24] In the latter, the terms derived from d^6 lie much lower than the others, and have been identified with certainty. Their relative positions and those of certain levels of Fe I are given in table 2, the levels being measured from the lowest, a^5D_4, in both cases. The parallelism of the two sets of values is remarkable. The terms a^1I, b^1G, and

TABLE 2

Fe III, d^6			Fe I, d^6s^2			I–III	I–0.96III
Term	J	Level	Term	J	Level		
a^5D	4	0	a^5D	4	0	0	0
	3	436		3	416	− 20	− 3
	2	739		2	704	− 35	− 6
	1	932		1	888	− 44	− 7
	0	1027		0	978	− 49	− 8
a^3P	2	19405	a^3P	2	18378	−1027	−249
	1	20688		1	19552	−1136	−308
	0	21208		0	20038	−1170	−322
a^3H	6	20051	a^3H	6	19390	− 661	+141
	5	20301		5	19621	− 680	+132
	4	20482		4	19788	− 694	+125
a^3F	4	21462	b^3F	4	20641	− 821	+ 37
	3	21700		3	20875	− 825	+ 43
	2	21857		2	21039	− 818	+ 56
a^3G	5	24559	b^3G	5	23784	− 775	+207
	4	24941		4	24119	− 822	+176
	3	25142		3	24339	− 803	+203
a^3D	3	30858	b^3D	3	29372	−1486	−252
	2	30716		2	29357	−1359	−130
	1	30726		1	29320	−1406	−177
a^1I	6	30356	a^1I	6	29313	−1043	+171
a^1G	4	30886	b^1G	4	29799	−1087	+147
a^1S	0	34812					
a^1D	2	35804	b^1D	2	34637	−1167	+265
a^1F	3	42897					
b^3F	4	50276					
	3	50295					
	2	50185					
b^3P	2	50412					
	1	49577					
	0	49148					

[23] *Proc. Roy. Soc.* A **151**: 703, 1935; *Annals Solar Physics Observ.* (Cambridge) **5**, pt. 1, 1938.

[24] *Astrophys. Jour.* **95**: 532, 1942.

TABLE 3

Fe II d^7			Fe I d^7s							
Term	J	Level	Term	J	Level	I–II	Term	J	Level	I–II
a^4F	4½	1873	a^5F	5	6928	5055	a^3F	4	11976	10103
	1½	3118		1	8155	5037		2	12969	9851
a^4P	2½	13474	a^5P	3	17550	4076	b^3P	2	22838	9364
	½	13905		1	17927	4022		0	23052	9147
a^2G	4½	15845	a^3G	6	21716	5871	a^1G	4	24575	8730
	3½	16369		4	22249	5880				
a^2P	1½	18361	c^3P	2	24336	5975	a^1P	1	27543	9182
	½	18887		0	25091	6204				
a^2H	5½	20340	b^3H	6	26106	5766	a^1H	5	28820	8480
	4½	20806		4	26628	5822				
a^2D	2½	20517	a^3D	3	26225	5708	a^1D	2	28605	8088
	1½	21308		1	26406	5098				
b^2F	3½	31999	d^3F	4	37046	5047	a^1F	3	40534	8535
	2½	31812		2	36940	5128				
d^2D	2½	48039								
	1½	47675								

b^1D in Fe I were discovered with the aid of this relation, the search being fairly easy, with their relative positions so closely predicted. The ^1S and ^1F terms were not found despite careful searching.

The comparison between d^7 of Fe II and d^7s of Fe I involves two terms in the latter for each one in the former, with multiplicities greater and less by one. The results are shown in table 3. For brevity, only the components of highest and lowest J are tabulated. Here again the agreement, both in position and in separation, is conclusive. Dobbie assigns the configuration d^7 in Fe II to the term c^2F at 44915 and d^6s to b^2F, but decisive evidence for interchanging the two appears upon comparing d^6 of Fe III with d^6s of Fe II,

TABLE 4

Fe III d^6	Fe II d^6s						II–III	Fe I d^6s^2
Desig	Desig	Level	II–III	Desig	Level	II–III.	Wtd. Mean	Desig
a^5D	a^6D	0	0	a^4D	7955	7955	3182	a^5D
a^3P	b^4P	20831	1426	b^2P	25787	6382	3078	a^3P
a^3H	a^4H	21252	1201	b^2H	26170	6119	2840	a^3H
a^3F	b^4F	22637	1175	a^2F	27315	5853	2734	b^3F
a^3G	a^4G	25429	870	b^2G	30389	5830	2523	b^3G
a^3D	b^4D	31483	625	b^2D	36253	5395	2215	b^3D
a^1I	a^2I	32876	2520					a^1I
a^1G	c^2G	33467	2581					b^1G
a^1S	a^2S	37227	2415					
a^1D	c^2D	38164	2360					b^1D
a^1F	c^2F	44915	2018					
	(b^2F)	(31999)	(−10898)					

as shown in table 4, where only the leading components of the terms are given. The designations of the related terms in Fe III and Fe I are added for reference. For the doublets derived from singlets the differences II–III are closely in line with the means, weighted in accordance with the multiplicities, in the other cases. These are found in the last column but one of the table. This comparison shows also that d^2D of Fe II does not come from d^6s, as Dobbie tentatively suggests, but from d^7, where there is a place for it (table 3).

The nature of c^3F is settled by the Zeeman data. It may be assigned with confidence as the lowest term from d^8. For elements of neighboring atomic number, in which the lowest terms from d^n are well-determined, the differences between these and the lowest terms of $d^{n-1}s$ are as follows for the leading components:

Cr I	$d^6\ {}^5D - d^5s\ {}^7S$	35399
Co I	$d^9\ {}^2D - d^8s\ {}^4F$	24016
Ni I	$d^{10}\ {}^1S - d^9s\ {}^3D$	14525

The corresponding term in Fe I should be from 25000 to 30000 above a^5F_5. For c^3F_4 (32873) this difference is 25955. The over-all separations of the terms are 243 and 973 for Cr I and Co I. The value 892 for c^3F also falls into line.

The next lowest term of the d^8 configuration should be ^1D, followed by ^3P. The separations of $^1D_2$ and $^3P_2$ from $^3F_4$ are 6868, 8215 for d^2s^2 in Ti I, and 13521 (raised by perturbations) and 15609 for d^8s^2 in Ni I. It is probable that these terms in Fe I lie about 10000 and 13000 cm^{-1} above c^3F.

The list of low even terms of Fe I is complete up to the level where combinations would be very difficult to find—except for two singlet terms from d^6s^2.

(9) RELATED HIGHER TERMS. IONIZATION POTENTIAL

The strongest lines of Fe I arise from terms based on the three lowest terms of Fe II and form a distinctive group, as follows (table 5):

TABLE 5

Limit Fe II		Added Electron			
		4s	4p		
d^6s	a^6D		z^7P°	z^7D°	z^7F°	e^7D
		a^5D	z^5P°	z^5D°	z^5F°	e^5D
			y^5P°	x^5D°	x^5F°	g^5D
d^6s	a^4D		z^3P°	z^3D°	z^3F°	e^3D
d^7	a^4F	a^5F	y^5D°	y^5F°	z^5G°	e^5F
		a^3F	y^3D°	y^3F°	z^3G°	e^3F

These terms and their mutual relations have long been recognized. They include all the lower terms belonging to each group (low even, odd, high even) and are almost completely separated from the higher terms of each group except for g^5D, e^3D, which are intermingled with others. Because of this isolation, they should be little perturbed, and comparison of the 4s and 5s terms should give a good approximation to the ionization potential. In making this, it is possible to improve considerably on a simple Rydberg formula by assuming that the difference in the denominators n* for 4s and 5s is not exactly unity, but has a value interpolated between the results for homologous terms in the spectra of elements of neighboring atomic numbers, in which the limits and values of n* are accurately determined by longer series.

For the configuration $d^{n-1}4s$ and $d^{n-1}5s$, each low term has a high one uniquely in series with it, but for $d^{n-2}4s^2$ and $d^{n-2}4s5s$, both the higher terms such as e^7D, e^5D are, in a sense, in series with a^5D. The differences Δn^* are smaller for the change involving an increase in multiplicity, but both run smoothly with atomic numbers, and may be used as is illustrated in table 6.

TABLE 6

VALUES OF Δn^* (5s−4s)

Low Configuration		$d^{n-1}s$	$d^{n-2}s^2$	
Change in Multiplicity		0	+2	0
At. No.	Element			
24	Cr	1.075		
25	Mn		0.960	1.080
26	Fe			
27	Co		0.999	1.108
28	Ni	1.077	1.007	1.120
29	Cu	1.066	1.013	1.142
30	Zn		1.025	

The value of Δn^* for $d^{n-1}s$ is for the higher multiplicity. Ill-determined or perturbed terms are omitted, and some which do not exist are left blank.

The values of Δn^* in Fe I may now be estimated as 1.072 for a^5F, e^5F, 0.982 for a^5D, e^7D, and 1.094 for a^5D, e^5D. The corresponding limits are easily found with the aid of a detailed Rydberg Table,[25] changing approximate values until the desired Δn^* is obtained. The results for the component of greatest J-value in each term are as follows (table 7):

TABLE 7

Terms	n*	n*	Δn*	Limit	Ionization
a^5D_4, e^7D_5	1.3170	2.2950	0.9820	63650	63650
a^5D_4, e^5D_4	1.3135	2.4075	1.0940	63630	63630
a^5F_5, e^5F_5	1.3699	2.4419	1.0723	65408	63535

The limit given in the last column but one is $a^6D_{4\frac{1}{2}}$ in Fe II (the ground-level) for the first two lines, but

[25] *Rydberg Interpolation Table.* Princeton Univ. Observ., 1934.

$a^4F_{4\frac{1}{2}}$ (which is higher by 1873) for the last line. Subtracting this gives the value of $a^6D_{4\frac{1}{2}} - a^5D_4$, which measures the ionization potential. The three determinations are remarkably concordant with mean 63605.

An independent and probably a still better determination can be obtained from the higher ^7D terms which were found during the revision of the spectrum. The term g^7D evidently arises from $(a^6D)6s$. Search for the 7s term found the level here called h^7D_5. The remaining components were not located, but this is determined by three good combinations with the 4p triad, and is trustworthy. Fitting a Ritz formula to e^7D_5, g^7D_5, h^7D_5 (by adjustment of the limit) gives:

$$n^7D_5 = 63732 - T; \quad T = \frac{R}{n - 1.6454 - 3.07 \times 10^{-6}T}$$

This corresponds to an ionization potential of 7.862 volts, 0.016 volts higher than the first determination. The value 63700, corresponding to 7.858 volts, may be adopted. This is 0.02 volt higher than the previously accepted value and is probably reliable to 0.01 volt.

V. ELECTRON CONFIGURATIONS

(10) THE HIGH EVEN TERMS

The terms arising from configurations involving 5s electrons have been discussed. A 4d electron added to the three principal limit-terms should give pentads, of which the following terms have been identified:

a^6D; e^7G, e^7F, f^7D, e^7P, e^7S; e^5G, f^5F, f^5D, e^5P, e^5S
a^4F; e^5H, f^5G, g^5F, h^5D, f^5P; e^3H, e^3G, f^3F, f^3D, e^3P
a^4D; g^5G, . . . i^5D . . .

All but one of the 42 levels arising from a^6D and of the 38 from a^4F have been found in each case. They form closely packed groups—the first between 50342 and 52067 and the second from 53061 to 55726. A large majority of the levels can be conclusively identified from the multiplet intensities and g-values, but the assignments are uncertain for the levels of small J. These complicated groups were first unravelled by Catalán. His conclusion that e^5S_2 and e^7F_3 are practically coincident is confirmed by the writers, but the present arrangement of the $(a^4F)4d$ group differs from his. In Co I, Ni I, and Cu I, where components of the limit-terms are widely separated, there is definite evidence that in the configura-

TABLE 8

TERM VALUES AND DENOMINATORS

Limits	$a^6D_{4\frac{1}{2}}$ 63700			$a^4F_{4\frac{1}{2}}$ 65573, $a^4F_{3\frac{1}{2}}$ 66130			$a^4D_{3\frac{1}{2}}$ 71655		
Electron	Desig	Term	n*	Desig	Term	n*	Desig	Term	n*
4s	a^5D_4	63700	1.312	a^5F_5	58645	1.368	a^5D_1	71655	1.238
				a^3F_4	53597	1.431			
5s	e^7D_5	20884	2.292	e^5F_5	18567	2.431	g^5D_4	20305	2.325
	e^5D_4	19023	2.402	e^3F_4	17612	2.496	e^3D_3	20361	2.321
6s	g^7D_5	9899	3.330						
7s	h^7D_5	5803	4.349						
4d	e^7S_3	12130	3.008	f^5P_3	12413	2.973			
	e^7P_4	13225	2.881	h^5D_4	12418	2.973			
	f^7D_5	13322	2.870	g^5F_5	12512	2.961	i^5D_4	13957	2.804
	e^7F_6	13358	2.866	f^5G_6	12404	2.974			
	e^7G_7	13048	2.905	e^5H_7	12298	2.987	g^5G_6	13653	2.835
	e^5S_2	12551	2.957	e^3P_2	11250	3.123			
	e^5P_3	11863	3.041	f^3D_3	12382	2.977			
	f^5D_4	13277	2.875	f^3F_4	11447	3.096			
	f^5F_5	12597	2.961	e^3G_5	12391	2.976			
	e^5G_6	13177	2.886	e^3H_6	12289	2.988			
4p	$z^7P_4^\circ$	39989	1.657	$y^5D_4^\circ$	32477	1.838	$y^5P_3^\circ$	34888	1.774
	$z^7D_5^\circ$	44349	1.573	$y^5F_5^\circ$	31878	1.855	$x^5D_4^\circ$	32029	1.851
	$z^7F_6^\circ$	41050	1.635	$z^5G_6^\circ$	30729	1.890	$x^5F_5^\circ$	31398	1.870
	$z^5P_3^\circ$	34644	1.780	$y^3D_3^\circ$	27398	2.001	$z^3P_2^\circ$	37708	1.706
	$z^5D_4^\circ$	37800	1.704	$y^3F_4^\circ$	28887	1.949	$z^3D_3^\circ$	40333	1.650
	$z^5F_5^\circ$	36825	1.726	$z^3G_5^\circ$	30194	1.906	$z^3F_4^\circ$	40348	1.649
5p	$u^5P_3^\circ$	12008	3.023						
	$t^5D_4^\circ$	12623	2.948						
	$u^5F_5^\circ$	12683	2.941						

tion $d^{n-2}s \cdot d$ the leading components of all the ten terms belonging to a pentad are derived from the component of highest J-value of the limiting terms and all lie at almost the same level—while in $d^{n-1} \cdot d$ the two leading components of the terms of higher multiplicity both go to the limit with greatest J and lie close together, while the leading components of the terms of lower multiplicity go to the next component of the limit and are near the level of the third components of the terms of higher multiplicity. Catalán has assumed that the former rule applies to both sets of pentads in Fe I. For those derived from $d^6s(^6D)$ the Zeeman data confirm this, but for those from $d^7(^4F)$ the other arrangement gives a better representation of the intensities and also of the g-values. For the smaller values of J the mutual perturbations of the levels are great. The g's differ widely from the theoretical values (§ 29), and large perturbations of the intensities and levels are to be anticipated. There can be no doubt that these levels belong to the pentad as a whole, but the assignment of individual term-designations has little significance.

Of the third pentad, with limit a^4D, only the terms g^5G (discovered by Miss Weeks from the Zeeman effect) and part of i^5D have been identified. There are unclassified lines in the appropriate region of the spectrum which may come from other levels belonging to this pentad, but not enough to locate them.

The term-values for the leading components of the terms based on the three lowest terms of Fe II, referred to these as limits, and the corresponding values of n* are given in table 8. The triplets of $(a^4F)4d$ are referred to $a^4F_{3\frac{1}{2}}$ as limit, for reasons described above. The limits are referred to the ground-level of Fe I as origin. Table 8 shows that few more high even terms should be observable even though more complete line lists were available. The group $(a^4D)4d$ could probably be filled up. The $^5D_4$ term from $(a^6D)6s$ should have n* about 3.45, and be at 63700–9200 or 54500, and its combinations might be found. The ^5F, ^3F of $(a^4F_4)6s$ may be expected near 56000 but should give faint lines and those from $(a^4D)6s$ still fainter ones.

A 5d electron should give n* about 3.9 and levels about 7000 below the limit. The unidentified levels 1, 2, and 3, which combine like septets, may belong to the pentad group with limit a^6D.

Terms having limits in Fe II higher than a^4D are hardly to be expected. The next lowest limits are $a^4P_{2\frac{1}{2}}$ at 13477 and $a^2P_{1\frac{1}{2}}$ at 18361 above $a^6D_{4\frac{1}{2}}$. The addition of any even electron but 5s to them would produce states lying above the principal ionization level, subject to auto-ionization and giving faint and diffuse lines, if any. The terms arising from $(a^4P)5s$ have been searched for unsuccessfully.

The energy of binding of a 4s electron to the various states of Fe II to produce the known low even terms and the corresponding values of n* are given in table 9.

TABLE 9

VALUES OF n* FOR 4s ELECTRON

Fe II		Fe I			Fe II		Fe I		
d^7	Level	d^7s	Term	n*	d^6s	Level	d^6s^2	Term	n*
a^4F	65573	a^5F	58645	1.368	a^6D	63700	a^5D	63700	1.313
		a^3F	53597	1.431	a^4D	71655		71655	1.238
a^4P	77174	a^5P	59624	1.357	b^4P	84531	a^3P	66153	1.288
		b^3P	54336	1.421	b^2P	89487		71109	1.242
a^2G	79545	a^3G	57829	1.378	a^4H	84952	a^3H	65562	1.294
		a^1G	54970	1.413	b^2H	89870		70480	1.248
a^2P	82061	c^3P	57725	1.379	b^4F	86337	b^3F	65696	1.292
		a^1P	54518	1.419	a^2F	91015		70374	1.249
a^2H	84040	b^3H	57934	1.376	a^4G	89129	b^3G	65345	1.296
		a^1H	55220	1.410	b^2G	94089		70305	1.249
a^2D	84217	a^3D	57992	1.376	b^4D	95183	b^3D	65811	1.291
		a^1D	55612	1.405	b^2D	99953		70581	1.247
b^2F	95699	d^3F	58653	1.368	a^2I	96576	a^1I	67263	1.277
		a^1F	55165	1.410	c^2G	97167	b^1G	67368	1.276
d^6s^2					a^2S	100927			
a^6S	87017				c^2D	101864	b^1D	67227	1.278
					c^2F	108614			

TABLE 10

PREDICTED ODD TERMS ARISING FROM 4p

Limits in Fe II		Septets			Quintets							Triplets								Singlets							
Config	Desig	P	D	F	S	P	D	F	G	H	I	S	P	D	F	G	H	I	K	S	P	D	F	G	H	I	K
d^6s	a^6D	x	x	x		x	x	x																			
d^7	a^4F						x	x	x					x	x	x											
d^6s	a^4D					x	x	x					x	x	x												
d^7	a^4P				x	x	x					x	x	x													
d^7	a^2G														x	x	x						x	x	x		
d^7	a^2P											x	x	x						x	x	x					
d^7	a^2H															x	x	x						x	x	x	
d^7	a^2D												x	x	x						x	x	x				
d^6s	b^4P				x	x	x					x	x	x													
d^6s	a^4H								x	x	x					x	x	x									
d^6s	b^4F						x	x	x					x	x	x											
d^5s^2	a^6S	x				x																					
d^6s	a^4G							x	x	x					x	x	x										
d^6s	b^2P											x	x	x						x	x	x					
d^6s	b^2H															x	x	x						x	x	x	
d^6s	a^2F													x	x	x						x	x	x			
d^6s	b^2G														x	x	x						x	x	x		
d^6s	b^4D					x	x	x					x	x	x												
d^7s	b^2F													x	x	x						x	x	x			
d^6s	a^2I																x	x	x						x	x	x
d^6s	c^2G														x	x	x						x	x	x		
d^6s	b^2D												x	x	x						x	x	x				
d^6s	a^2S												x								x						
d^6s	c^2D												x	x	x						x	x	x				
d^6s	c^2F													x	x	x						x	x	x			
Totals		2	1	1	2	6	7	6	4	2	1	4	10	14	14	12	8	4	1	2	6	8	9	8	6	3	1
Observed		2	1	1	2	7	7	6	4	1		2	5	8	7	10	7	3			1	4	4	5	3	1	
Exc. 7 Highest Limits		2	1	1	2	6	7	6	4	2	1	4	7	10	9	9	6	3		2	3	4	4	5	4	2	

Each d^7 limit has two associated d^7s terms and the first six d^6s^2 terms have two associated d^6s limits. The means of n* for the pairs, weighted according to multiplicity, range from 1.378 to 1.392 in the first group and 1.273 to 1.284 in the second. The values of n* for the d^6s^2 terms with only doublet limits lie in the latter range. The general means are: for $d^7 \rightarrow d^7s$ 1.385 with average deviation ± 0.004 and for $d^6s \rightarrow d^6s^2$ 1.277, A.D. ± 0.003. The greater binding energy for d^6s^2 corresponds to the completion of a pair of 4s electrons.

(11) THE HIGHER ODD TERMS

The addition of a 4p electron to the various limits gives rise to numerous triads of terms. The observed odd terms should include these (except for some of the highest limits) and a few terms involving a 5p electron. For the known 5p triad from a^6D the mean value of n* is 2.971—greater by 1.25 than for the corresponding 4p triad. For the triads from a^4F, n* should be about 3.11 for the quintets and 3.20 for the triplets, leading to levels near 54000 and 55000. These have not been found. Apart from these, the observed odd terms are almost certainly from 4p.

The predicted terms are listed in table 10.

The numbers of predicted and observed terms of each type are given at the bottom of the table. The deficiency of the latter indicates that many terms with high-lying limits have been missed. Omitting the seven highest limits gives the predictions in the lowest line, which are not far from the observed numbers.

The mean levels at which the various triads may be expected can probably best be estimated by assuming that the difference Δn^* between the mean $\overline{n^*}$ for the terms of a 4p triad and those for the related 4s term are the same as in the cases already discussed. This gives the results in table 11.

TABLE 11

DIFFERENCES OF n*

Limit	High Multiplicity		Low Multiplicity	
	$\overline{n^*}$	Δn^*	$\overline{n^*}$	Δn^*
d^7 a^4F	1.861	+0.493	1.952	+0.521
d^6s a^6D	1.622	+0.310	1.737	+0.425
d^6s a^4D	1.832	+0.594	1.668	+0.430

For the remaining d^7 limits the values of n* (4s) are almost identical with their means—1.372 for the terms of higher multiplicity in a related pair and 1.413 for those of lower multiplicity. Adding the appropriate Δn^* gives 1.865 and 1.934, corresponding to term-values of 31516 and 29338. Proceeding similarly with d^6s, the differences in level between the limit and the 4p triad come out as follows (table 12):

TABLE 12

APPROXIMATE TERM VALUES—4p TRIADS

Limit	High Mult.	Low Mult.
d^7	31500	29300
d^6s High Mult.	42500	38000
d^6s Low Mult.	32400	39000

The term-values for d^6s apply to the groups of four triads related to the low triplet terms of d^6s^2. The singlet d^6s^2 terms have but two related triads. For these the values 42500 and 39000 have been adopted, as the other pair of term-values are affected by the mutual repulsion of terms of the same multiplicity.

The assignment of the terms to specific triads is complicated by considerable perturbations shown by the g-values (Part II, § 29) which indicate a good deal of sharing of identity among the levels, especially those with small J-values. The guiding principles of the present arrangement have been as follows:

(*a*) *Level.* The terms of a given triad are likely to be within 2000 cm^{-1} of the levels determined as above (which it is useless to predict within 1000 cm^{-1}).

(*b*) *Separation.* In general, limit-terms and low related terms of a wide internal separation are associated with odd terms of wide separation.

(*c*) ***Intensity.*** The terms of a triad will in general combine most strongly with the related low even term having the same limit in Fe II. Each d^7s term in Fe I has one related d^7p triad; most d^6s^2 terms have four—two of the same multiplicity and one each of multiplicity higher and lower by 2. The latter give intersystem combinations with the low term, which may be relatively faint. Combinations of d^7p terms with d^7s terms having different limits are usually stronger than those with d^6sp terms, and vice versa. On account of the Boltzmann factor, the population of high states is less than that of lower states, and their combinations with the same low term tend to be fainter. The combinations of a d^7s term with its related d^7p triad are always strong. Those of a d^6s^2 term with the lower of the related d^6sp triads of the same multiplicity are much weaker; with the higher triad they are stronger, but hardly comparable with d^7s–d^7p. This is clearly shown by King's determinations of f-values for Fe I and Ti I [26] and is a general property of the spectra of the iron group, as is shown by the incidence of ultimate lines.[27] Further evidence of this has appeared in the present work. The combinations of the singlet d^7s terms are much stronger than those of the d^6s^2 singlets, so that the latter were among the last terms to be detected in the analysis.

[26] King, R. B., and A. S. King, *Mount Wilson Contr.* No. 581, 1938; *Astrophys. Jour.* **87**: 24, 1938.

(12) ODD SEPTETS AND QUINTETS

The predicted septet terms have been identified with certainty. The term y^7P° is fully confirmed by the Zeeman effect. Its components are arranged, as was to be expected, in the normal order, with the smallest J lowest, while almost all terms of Fe I are inverted. The term-value 46596 ($n^* = 1.535$) is unusually great for 4p. Very few examples of the configuration $d^{n-3}\,s^2p$ are known in other spectra.

The quintet terms may be assigned with considerable confidence. By applying the approximate term-values of table 12 to the limits in table 9, the estimated mean levels of the triads in table 13 are obtained. For the limit a^6S the separation for d^6sp has been taken as a rough guide. The d^7p triad with limit a^4P should combine more strongly with a^5P, and the d^6sp with b^4P. This puts v^5P°, u^5D° in the former, and leaves x^5P°, w^5D° for the latter—all near the predicted levels. The relative levels then place y^5S°, z^5S°.

The $^5P^\circ$ term from a^6S should be related to y^7P°, much like the $^5P^\circ$, $^7P^\circ$ terms from $3d^54p$ in Mn II. This $^7P^\circ$ term has normal separations (−264, −176) while the $^5P^\circ$ term is inverted (+114, +72). In Fe I y^7P° has separations −215, −155 and w^5P° +176, +97—much smaller than the other $^5P^\circ$ terms. The difference $^5P_3^\circ - {}^7P_4^\circ$ is 4564 in Mn II and 5715 in Fe I. This identification is conclusive.

Their levels suffice to assign y^5G°, z^5H° to a^4H. The related $^5I^\circ$ could be detected only by combinations with a^5F, which should be very faint. The triad with limit b^4F is reasonably filled by v^5D°, w^5F°, x^5G°. The higher one with limit a^4G evidently contains v^5F° and w^5G°. The $^5H^\circ$ term has not been found. Two lines at 38628.18 and 38713.63 cm^{-1} of intensities 8 III and 6 III by their behavior in the furnace come from a^5F or a level a little higher. If they are transi-

TABLE 13

ODD QUINTET TERMS

Limit	Est. Level	Adopted Terms					
a^4P	46000	y^5S°	44512	v^5P°	47967	u^5D°	46721
b^4P	42000	z^5S°	40895	x^5P°	42532	w^5D°	43499
a^4H	42000	y^5G°	42784	z^5H°	43321	$^5I^\circ$	
b^4F	44000	v^5D°	44415	w^5F°	44243	x^5G°	45608
a^6S	44000:			w^5P°	46137		
a^4G	46000	v^5F°	47606	w^5G°	47363	$^5H^\circ$	
b^4D	53000	t^5P°	53388	$^5D^\circ$	(53891)	$^5F^\circ$	(54013)

[27] Meggers, W. F., *Jour. Optical Soc. America* **31**: 39, 1941.

tions to the $^5H^\circ$ term, it must be near 46000. The high terms u^5P°, t^5D°, u^5F° have already been assigned to $(a^6D)5p$. The corresponding septet triad should be near 50000, but has not been identified. This leaves only t^5P°, which may be assigned tentatively to the triad with b^4D as limit. The unclassified levels 10_3° and 12_5° have positions and g-values agreeing well with $^5D_3^\circ$, and $^5F_5^\circ$ and are given in parentheses; but the other levels of these terms have not been identified. The quintets are now well accounted for.

TABLE 14

ODD TRIPLET TERMS FROM 4p

Limit	Est. Level	Adopted Terms			Related Low Term
a^4F	36000	y^3D° 38175	y^3F° 36686	z^3G° 35379	a^3F
a^4D	33000	z^3P° 33947	z^3D° 31322	z^3F° 31307	[a^5D]
a^4P	48000	y^3S° 47556	y^3P° 46727	w^3D° 47017	b^3P
a^2G	48000	x^3F° 46889	v^3G° 49460	y^3H° 49434	a^3G
a^2P	51000	8_1° 52857?	v^3P° 52916:	u^3D° 51969	c^3P
a^2H	53000	u^3G° 51373	w^3H° 52431	y^3I° 52655	b^3H
a^2D	53000	w^3P° 50186:	t^3D° 52213:	u^3F° 56592:	a^3D
b^4P	47000	z^3S° 46601	x^3P° 48304	x^3D° 45220	a^3P
b^2P	57000	$^3S^\circ$	$^3P^\circ$	$^3D^\circ$	
a^4H	47000	y^3G° 45294	z^3H° 46982	z^3I° 45978	a^3H
b^2H	57000	$^3G^\circ$	u^3H° 56334:	x^3I° 57027	
b^4F	48000	v^3D° 49135	w^3F° 49108	x^3G° 47834	b^3F
a^2F	58000	$^3D^\circ$	$^3F^\circ$	$^3G^\circ$	
a^4G	51000	v^3F° 51304:	t^3G° 53983:	x^3H° 51023:	b^3G
b^2G	62000	$^3F^\circ$	$^3G^\circ$	$^3H^\circ$	
b^4D	57000	$^3P^\circ$	$^3D^\circ$	$^3F^\circ$	b^3D
b^2D	68000	$^3P^\circ$	$^3D^\circ$	$^3F^\circ$	
b^2F	64000	$^3D^\circ$	$^3F^\circ$	$^3G^\circ$	d^3F
a^2I	54000:	$^3H^\circ$	$^3I^\circ$	$^3K^\circ$	a^1I
c^2G	54000:	$^3F^\circ$	$^3G^\circ$	$^3H^\circ$	b^1G
a^2S	58000:		$^3P^\circ$		
c^2D	59000:	$^3P^\circ$	$^3D^\circ$	$^3F^\circ$	
c^2F	64000:	$^3D^\circ$	$^3F^\circ$	$^3G^\circ$	

(13) TRIPLETS

Table 14 is similar to the last, except that the low even terms in Fe I which are related to the triads are given. The triads already located are included. The numbers of odd triplet terms to be anticipated below 54000 and of those observed are as follows:

Term	$^3S^\circ$	$^3P^\circ$	$^3D^\circ$	$^3F^\circ$	$^3G^\circ$	$^3H^\circ$	$^3I^\circ$
Predicted	3	5	7	6	6	4	2
Observed	2	5	8	5	7	4	2

The agreement is surprisingly good in view of the roughness of the prediction. From 54000 to 62000 the comparison gives:

Term	$^3S^\circ$	$^3P^\circ$	$^3D^\circ$	$^3F^\circ$	$^3G^\circ$	$^3H^\circ$	$^3I^\circ$	$^3K^\circ$
Predicted	1	4	4	5	4	4	2	1
Observed				2	3	3	1	
Omitting faint combinations	1	2	3	3	3	2	1	

Many of the predicted terms have evidently been missed.

In assigning the observed terms, so far as practicable, to their limits, the intensities are important, since each triad has an associated low level of the same multiplicity with which its combinations should be strong. Table 15 gives that of the strongest line in each multiplet, omitting predicted lines present in the sun (§ 23). Values not in parentheses are on King's scale and fairly comparable. Those in parentheses are on a great variety of scales. Most of these lines are faint. For the combinations with a^5D, which lie far in the ultra-violet, the intensities are on a more open scale. The symbol "†" denotes that the intensities in the multiplet are seriously abnormal. The maximum separation of each term is given. When only two components are known, the separation is followed by +. Inverted separations are listed as positive, the few normal ones as negative. The separations of the odd terms average much smaller than for the even terms, doubtless because the 4p electron by itself would produce a considerable separation in the normal direction. Terms already assigned are omitted. Of the three $^3I^\circ$ terms, y^3I° combines strongly with b^3H, and belongs to its triad; and the others to a^3H. All three are near the predicted levels. Intensities and levels connect u^3G°, w^3H°, with b^3H and place y^3G°, z^3H° in the lower a^3H triad. The one related to a^3G consists of y^3H°, v^3G° and either x^3F° or w^3F°; strong combinations with a^5P connect y^3P°, w^3D° with a^4P; x^3P°, x^3D° combine well with a^3P; u^3D° is related to c^3P, and either w^3P° or v^3P° would fit, though the wide separation of the latter is favorable. The two known $^3S^\circ$ terms should belong to the lower two of these triads. The uncertain choice between them depends on the strong combination of z^3S° with a^5D. Intensities and levels suggest assigning x^3G°, w^3F°, v^3D° to b^3F (leaving x^3F° for a^3G). Beyond this, assignments are difficult; x^3H°, t^3G°, and v^3F° may be provisionally connected with b^3G and w^3P°, t^3D°, u^3F° with a^3D.

It is unprofitable to attempt specific configuration assignments for the higher terms, and some of those already made and marked with colons in the table are doubtful. The lowest missing term is the $^3S^\circ$ related to c^3P, which should be near 51000. The only unassigned odd level with $J = 1$, 8_1°, is in the expected place, and combines mainly though irregularly with the low ^3P term. The well-determined value of g for this level is 1.246—greatly perturbed from a theoretical 2.000—but the neighboring levels z^1P°, $u^3D_1^\circ$,

TABLE 15

INTENSITIES OF TRIPLET COMBINATIONS

d^6sp, d^7p		d^6s^2						d^7s									d^8
Desig		a^5D	a^3P	b^3D	b^3F	b^3G	a^3H	a^5P	a^5F	b^3P	c^3P	a^3D	a^3F	d^3F	a^3G	b^3H	c^3F
	Sep	978	1660	−52	398	555	398	377	1226	213	756	398	992	−105	534	522	892
z^3S°		(10)	1					2		6	(2)						
y^3S°		(1n)	3					2		10	3	(1)					
y^3P°	229	(3)	4	(1)				10	(2)	12	(2)						
x^3P°	211	(25)	6	(2)				4	6	6	5	(2)	(1)				1
w^3P°	−236		5	(1)	(1)			(1)		4	7	5					(1)
v^3P°	892+		10	3				(3)		7	2†	(1)†					
x^3D°	331	(15)	8		4			2	4	12	5†	(1)	12		(1)*		10
w^3D°	255	(3)	1	(1)†	1	(1)		15	(7)	12	(1)	(1)	5†	(1)†	4		40
v^3D°	163		(2)*	(1)†	3	(1)*		3	(4)	4†	(1)	(2)†		3*	3		(1)
u^3D°	543		5	(1)	4		(1)		(1)	(1)	12	2	(2)†	2	(1)		(3)
t^3D°	502		5†	(1)	6†		(1)			(1)†	(1)	8		2			(2)
s^3D°	322+	(1)			(2)				(2)		(1)	3					(1)†
x^3F°	308	(2)	(1)	(1)	4	(3)	(1)	20	(6)	5	(1)		8	(2)	10		8
w^3F°	324		(1)	(2)	8†	2	(1)	2*	(4)*		(1)		(2)	3*	12		4
v^3F°	−103			(2)	4	3	2	(1)?		(1)	2	5		3	6	3	(1)
u^3F°	266				1	(2)	(2)	(3)				8	(4)		(3)†	(3)†	
t^3F°	159				4	(2)				(1)?			(2)		(3)		
y^3G°	268	(2)	(2)		3	6	10		6				10		3	(1)	
x^3G°	−22		(1)		6	(1)	(1)		(20)				5	(1)†	3†	(1)	
w^3G°	244				2	(2)	(1)	8	(2)		(1)		3†		3	(1)	(1)
v^3G°	390				4	7	4		(1)			(1)	3		20†	(1)	4
u^3G°	452				4	(2)	6			(1)	1	(1)	(5)	4	10	10	(2)
t^3G°	617			(1)		4	(3)		(20)†			(2)	(2)	(1)	6	10	(2)
s^3G°	192				(3)	3	(2)						(4)		7		
r^3G°	438				6	(2)											
q^3G°	129				(4)	(1)					(3)†	(2)†	(2)?†			(2)†	
z^3H°	124	(1)			2	5	8	3	(6)				2		15	(4)	5
y^3H°	293				(1)	5†	(2)						2		20	3	4*
x^3H°	45+					3	3	(1)					(1)		(2)	(1)	(1)
w^3H°	337				(2)	5	15								(2)	12	(−)†
v^3H°	−60				(2)	(2)†	(4)						(1)		10	5	
u^3H°	89					(1)†	4					(3)			5	5	
t^3H°	392				(4)		(25)						(1)		(1)	10	
z^3I°	158				(1)		8		1							5	
y^3I°	244					(1)*	3									20	
x^3I°	77					(−)†	5								(2n)	10	

* Blend. † Multiplet intensities abnormal.

and $u^5P_1^\circ$ have g-values which are too great by 0.266, 0.200, and 0.133, enough to make up four-fifths of the difference. This possibility is noted in table 14.

(14) SINGLETS

The odd singlet terms to be expected from limits below 95000 are given in table 16. The estimated differences from the limits are 29000 for d^7 and 39000 for d^6s. The predicted and observed numbers of terms are given at the bottom. Some terms with large and small J have been missed. The predicted levels are closely packed, and assignment to configurations must depend mainly on the intensities. The first four triads in table 16 have related low even singlet terms; the others from d^6sp are related to low triplets and likely to give fainter combinations. The observed intensities of combinations between the known odd singlets and the relevant low terms are given in table 17. It is clear that z^1H°, z^1F° belong with a^1G; z^1I°, y^1H° with a^1H; y^1D°, z^1P° with a^1P; and w^1D° probably with a^1D. The first two triads may be filled out with z^1G° and y^1G° (though the latter is very low). These nine levels account for all the strongest singlet combinations. It is hardly practicable to assign the others. Many of them must come from d^6sp.

TABLE 16

ODD SINGLET TERMS

Limit	Est. Level	Adopted Terms						Rel. Low Term
a^2G	50000	$z^1F°$	50587	$z^1G°$	47453	$z^1H°$	48383	a^1G
a^2P	53000	$^1S°$		$z^1P°$	53230	$y^1D°$	51708	a^1P
a^2H	55000	$y^1G°$	48703:	$y^1H°$	53722	$z^1I°$	53094	a^1H
a^2D	55000	$^1P°$		$w^1D°$	55754	$^1F°$		a^1D
b^2P	50000	$^1S°$		$^1P°$		$^1D°$		a^3P
b^2H	51000	$^1G°$		$^1H°$		$^1I°$		a^3H
a^2F	52000	$^1D°$		$^1F°$		$^1G°$		b^3F
b^2G	55000:	$^1F°$		$^1G°$		$^1H°$		b^3G
Type		S	P	D	F	G	H	I
Numbers predicted		2	3	4	4	5	4	2
Numbers found			1	4	4	5	3	1

(15) UNCLASSIFIED LEVELS

The even levels called 1, 2, and 3 combine with septet terms and are at the right level for the pentad $(a^6D)5d$, while 4 may fit into the incomplete quintet pentad $(a^4D)4d$. The high odd levels fall in a region containing many anticipated terms; but $2_2°$, which is undoubtedly real, lies low enough to be puzzling.

Two even levels at 40871 ($J = 3$) and 41178 ($J = 2$) present a perplexing problem. Both are confirmed by several combinations, as is shown in table 18, which gives the intensities and the residuals in 0.01 cm^{-1}. These are given in the sense o–c for transitions from lower terms, but c–o for transitions to higher terms, so that a change in these levels affects all residuals in the same sense. Predicted solar lines are here included, since the majority of them are probably real (§ 22). The intensities strongly suggest $^3D_3$, $^3D_2$; but the levels are far too low for any configuration including a 5s or 4d electron. The only ^3D term from the lower configurations which has not already been unequivocally located comes from d^7s and has d^2D for limit. The regularities shown in table 3 indicate that this term should be near 53000. It appears very improbable that so enormous a displacement can be due to perturbations, especially since those of the lower terms of the overlapping configurations d^6s^2 and d^7 are small.

There are predicted ^3P and ^3F terms of d^6s^2, related to b^3F and b^3P of Fe III, but by table 2 these should be near 48000.

The singlet $^1F_3$ from d^6s^2 should be close to 41200, and $^1D_2$ from d^8 roughly at 43000; but the combination-intensities of the observed levels are irreconcilable with their being singlets, though they do indicate that the two are related.

The writers can suggest no solution of the difficulty. In recognition of it, they have deliberately departed from the usual nomenclature to the extent of calling these two levels X_3 and X_2 rather than assigning numbers as usual.

The general result of this analysis is that the iron arc spectrum, despite its complexity, is highly regular. All the low even terms predicted by theory have been found close to their anticipated position except two which should give few and faint lines and others whose combinations should be in the infra-red. The high even terms are also well identified, though not so completely. The lower odd terms (below about 55000) are also satisfactorily accounted for except for a couple which could be detected only through highly improbable combinations, and a few of low J-value.

TABLE 17

SINGLET COMBINATIONS

	a^1P	a^1D	a^1G	a^1H	c^3P	a^3D	a^3G	b^3H	a^3P	b^3D	b^3F	b^3G	a^3H	b^1D	b^1G	a^1I
$z^1P°$	3	3			4	1			(2)					(2)		
$z^1D°$		7			(1)	(1)			(1)	(1)		(1)*				
$y^1D°$	7	3			2	9	(1)*			(2)*	(1)			(2)		
$x^1D°$	3n*	(1)			(1)	(1)	2			(1)		1*		(1)		
$w^1D°$	(4)	20			(1)*	3			(1)	(1)				(1)		
$z^1F°$		4	8		(1)	(2)				(1)	(1)	(1)		(1)	(1)	
$y^1F°$		2	3			(1)			(2)					m	(2)	
$x^1F°$		3	4		(1)	(1)*				(1)	(–)	(1)			(2)	
$w^1F°$		2	2			(2)*	5			(1)		2		(1)		
$z^1G°$			7	3			10	(1)		(2)	1	(1)			(1)	
$y^1G°$			15	4			10	(2)*			(1)	(1)	(1)		(2)	
$x^1G°$			7	(2)		(1)*	(2)	(2)			(1)		(1)		(3)*	
$w^1G°$			m	2			(–)	(2)*				(1)			2	
$v^1G°$				6				(1)							(1)	
$z^1H°$			20	m			3	(2)				5	(1)		(2)	4
$y^1H°$				10			10					2*				
$x^1H°$			(3)	6			(2)						(2)			2
$z^1I°$				15				10					(1)			(1)

* Blend. m Masked.

TABLE 18

COMBINATIONS OF X-LEVELS

	X_3	X_2		X_3	X_2
$w^1F_3^\circ$	*+04 (−3⊙)	+01 (−3⊙)	$z^5P_3^\circ$	−09 (−3⊙)	
			$z^5P_2^\circ$	−03 (−3N⊙)	
$v^1G_4^\circ$	+05 (1)				
			$z^5D_4^\circ$	−01 (−3⊙)	
$v^3P_2^\circ$	−04 (−1⊙)		$z^5D_3^\circ$	−11 (−3⊙)	+13 (−3⊙)
			$z^5D_2^\circ$	+08 (−3⊙)	+05 (−3⊙)
$z^3D_3^\circ$	+03 (20V)	+12 (−3⊙)	$z^5D_1^\circ$		+26 (−3⊙)?
$z^3D_2^\circ$	−05 (3)	+11 (10)			
$z^3D_1^\circ$		+01 (3)	$z^5F_2^\circ$		−07 (−3⊙)
			$z^5F_1^\circ$		+06 (−3N⊙)
$u^3D_3^\circ$	−06 (2)				
$t^3D_3^\circ$	0 (2)				
$t^3D_2^\circ$	+01 (−3⊙)	*−04 (−1⊙)			
$s^3D_3^\circ$	−11 (6IV?)				
$s^3D_2^\circ$		−05 (3V)			
$z^3F_4^\circ$	+10 (5)				
$u^3F_3^\circ$		−07 (1)?			

* Blend. ⊙ Predicted line present in solar spectrum.

Not many of the numerous predicted odd terms above this level have been identified. Others may be found by future intensive observations of faint lines.

(16) UNCLASSIFIED LINES

The comprehensive list of lines upon which the present analysis was based includes great numbers of faint lines which have not been classified. It is probable that many of these are not really due to Fe I. In preparing a list for this purpose, much more trouble will be caused by excluding one real iron line as doubtful than by including several impurity lines. As the analysis advances, the latter will accumulate more and more in the unclassified residuum—while, in a spectrum as complex as this, a few of them will coincide by chance with wave-lengths predicted from term values and creep into the "classified" list. In the region between 6600 and 2975A the more obviously dubious "observed" lines were rejected, and a statistical comparison of the rest was made with the solar spectrum (§ 24), with the conclusion that the majority are accounted for in the sun and are probably really due to iron.

The strongest unclassified lines of wave-length greater than 2000A are listed in table 19. All those for which King has given a temperature classification are included with the exception of four which are clearly due to Fe II. For other observers the limit was set at intensity 3 (on the narrow scale) except in the far ultra-violet, where the scale was much more open. Of these 100 lines, 20 lie between λ9700 and λ6600, 26 between λ6600 and λ2975, and 54 short of this. Of the 46 which are accessible in the sun, 42 are present or accounted for by blending or masking. It is probable, therefore, that almost all the lines in this list are really due to iron. Almost, if not quite, all the lines of wave-length less than 3000A must arise from transitions from known low even terms to unknown high odd ones (of which there are still many). The temperature class shows that the lines of longer wave-length arise from higher levels.

VI. THE SUN AS A SOURCE FOR THE IRON SPECTRUM

(17) PREDICTED IRON LINES IN THE SOLAR SPECTRUM

It has long been recognized that the arc spectrum of iron is more fully exhibited among the Fraunhofer lines of the solar spectrum than by any laboratory measures so far published. Many lines, predicted from the term values, agree so closely with unidentified solar lines that the coincidences cannot reasonably be attributed to mere chance.[28] Many accidental coincidences must, however, occur in a general comparison of predicted with solar wave-lengths, and an investigation of their probable number is in order. The present work offers ample material for statistical study.

(18) ACCEPTANCE OF PREDICTED SOLAR LINES

In the course of the analysis of the spectrum, wave numbers were computed for all combinations from low even to odd levels and for the more probable transitions from odd to high even levels, not pro-

[28] Russell and Moore, *Mount Wilson Contr.* No. 365, 1928; *Astrophys. Jour.* **68**: 151, 1928.

hibited by the inexorable inner-quantum and parity rules. This gave about 6500 predicted positions of possible lines—with the certainty that a large majority, though not theoretically forbidden, must be far too faint to be observable. The resulting wave-lengths were compared with the "Revised Rowland Table" [29] (R. R.). This line list is known to be practically complete, except near the limit in the ultraviolet set by atmospheric ozone, and in the red beyond λ6600, where Rowland's plates were greatly inferior to modern ones.

Much progress has, however, been made in the past fifteen years in the identification of solar lines then unassigned, and in the improvement of earlier identifications. A systematic study of these has been made by one of the writers (C. E. M.) in connection with the preparation of a revised and extended edition of the Multiplet Table.[30] This investigation has added about 2850 identifications to those in the R. R. Though not yet finished, it is complete enough to provide a reliable basis for this statistical investigation. Of the predicted iron lines, 1928 agreed with observed solar lines closely enough to warrant calling them coincident, blended, or masked, in the same sense in which these terms have been applied to other lines in the general study.

The examination was made multiplet by multiplet, noting first the absences and apparent coincidences. The limits of tolerance were necessarily a matter of judgment, depending on the wave-length, intensity, and diffuseness of the solar line; but they represent the product of years of experience, and are believed to be fairly consistent.

The coincidences were divided into three classes:

(*a*) *Unblended*, when the solar line is not otherwise identified and the agreement in wave-length and estimated intensity is such that a line reliably observed in the laboratory would be described as present in the sun.

(*b*) *Blended*, when the solar line (especially if diffuse) disagrees slightly in wave-length, or is somewhat too strong, and is probably a blend of the line under consideration with some other line of known or unknown origin.

(*c*) *Masked*, when some neighboring strong line would conceal the presence of a line of the expected position and intensity. The decisive test was, however, based on the position of the predicted line in its multiplet. If the intensity anticipated on this basis was very small, the line was rejected, especially in the case of blends.

The predicted lines which passed all these tests, and were therefore accepted as "present," or "present blended," were graded as good, fair, or poor, with the position in the multiplet, the agreement in wave-

[29] Revision of Rowland's Preliminary Table of Solar Spectrum Wave-lengths. *Publ. Carnegie Inst. Washington*, No. 396, 1928.

[30] *A Multiplet Table of Astrophysical Interest.* Princeton, 1933.

TABLE 19

THE LEADING UNCLASSIFIED LINES OF Fe I

Ref†	I A	Int	T C	Ref	I A	Int	T C
F	9666.59	2	V	W	2786.81	(3)	
F	9637.55	2	V	G	2778.842	3	III
F	9529.31	2n	V	V	2778.075	3	III
E	9430.08	3	IV*	G	2773.232	2	III
O	8145.47	4	V	G	2757.856	(3)	
O	8024.50	3n	V	V	2737.833	(3)	
E	7994.473	20	IV*	G	2698.162	(4)	
O	7808.04	6n	V	V	2695.542	(3w)	
O	7573.53	2n	V	V	2664.042	(3w)	
L	7546.177	4	IV*	G	2615.420	(3)	
R	§7376.434	3n	V	G	2608.576	(3)	
V	7254.649	2	IV*	G	2606.644	(4)	
V	6975.46	3n	V	G	2604.864	(3)	
V	6902.80	3n	V	G	2604.751	(3)	
M	6881.46	1	V	G	2603.553	(4)	
V	6838.86	3n	V	G	2600.202	(3)	
M	6793.62	1	V	V	2594.046	1	III
V	6755.609	3	IV*	G	2592.285	(3)	
D	6726.78	(3)		G	2591.252	(3)	
V	6609.56	1	V	V	§2588.010	8	III
V	6528.53	2	V	G	2582.297	6	III
V	6501.681	4	IV*	G	2578.825	(3)	
J	6042.092	2	V	C	2575.744	(4)	
R	5036.294	6		G	2553.193	(7)	
J	4552.544	(3)		C	2551.094	(8)	
J	4237.162	2	V	G	2546.864	(4)	
W	4100.17	(3)		G	2533.802	4	IV
W	3851.58	(4)		W	2527.16	(5)	
J	3739.527	3	IV	G	2525.021	(7)	
V	3681.774	1	IV	G	2523.658	(6)	
V	3680.801	1	IV	W	2523.11	(5)	
V	3656.227	3n	IV	G	2520.968	(4)	
G	3634.698	4n	IV	G	2513.328	(3)	
J	3617.317	2	IV	G	2505.627	(4)	
J	3616.572	3n	IV	G	2505.485	(5)	
G	3614.550	2n	IV	W	2460.31	(4)	
J	3587.752	3	IV	G	2436.344	(10)	
W	3506.40	(3)		G	2435.865	(3)	
V	3438.306	(3w)		C	2431.025	(20)	
V	3262.284	4	IV	G	2301.171	(6)	
V	3179.538	3	IV	C	2165.861	(20)	
V	3139.908	4n	V	C	2163.368	(10)	
W	3136.17	(3)		X	2158.49	(6)	
G	3126.175	8n	IV	N	2111.274	(20)	
W	3102.71	(4)		N	2109.861	(25)	
G	2991.632	5n	IV	N	2077.507	(20)	
G	2945.050	3	IV	N	2041.204	(25)	
W	2927.55	(3)		N	2017.090	(15)	
V	2865.191	(3)		N	2007.215	(15)	
G	2799.149	1	III	N	2006.260	(15)	

† The reference numbers in this column are the same as those used in table B—see bibliography, page 169.

§ Blend with Fe II.

length, and the intensity as criteria. This grading was decidedly severe; for example, no members of multiplets containing only predicted lines and no lines involving improbable combinations, were called "good." This whole process is substantially the same that was applied to lines of less abundant elements observed in the laboratory, except that laboratory intensities were not available as guides.

(19) Selection of Group for Statistical Study

The essential condition in this case is that the group shall be selected *impartially*, by some method which neither favors nor discriminates against the characteristics under investigation. A desirable, though not necessary, condition is that the grouping should exclude irrelevant material in which these characteristics are not present.

The complete list of predicted possible lines satisfies the first condition, but not the second, for it is overloaded with transitions so improbable that there is no hope of their giving solar lines. To escape this the statistical study has been confined to multiplets in which at least one line (observed in the laboratory or predicted) satisfies the criteria of acceptance just described. This meets the second condition efficiently; but care is required about the first. When one or more lines of the multiplet have been observed in the laboratory, there is no assignable reason why the remaining lines should be more or less likely than the average to coincide accidentally with solar lines—though the chance of real coincidences is good. When no observed line is present, *one* coincidence with a predicted line in each multiplet is forced by the method of selection, but any other coincidences should be statistically free, just as in the previous case. The exclusion of one predicted line from each multiplet of the second type leaves statistically unbiassed material. All components of each multiplet, except observed (laboratory) lines, or the one predicted line, were included in the discussion, irrespective of their anticipated intensity.

A large group of predicted lines was thus obtained, free from statistical bias, and yet likely to contain a

TABLE 20

Predicted Lines of Fe I in the Solar Spectrum

λ		2975–4000		4000–5000		5000–6000		6000–6600		All	
Accepted Unblended	Good	82		82		94		10		268	
	Fair	127		211		161		39		538	
	Poor	46		59		43		10		158	
	Total		255		352		298		59		964
Blended with unidentified lines	Good	5		5		3		1		14	
	Fair	6		10		2		4		22	
	Poor			3		2				5	
	Total		11		18		7		5		41
Blended with identified lines	Good	22		11		7				40	
	Fair	40		26		13				79	
	Poor	15		7		7				29	
	Total		77		44		27		0		148
Rejected	Unblended		42		28		9		4		83
	Blended		94		74		63		19		250
Masked			250		122		51		19		442
Absent			167		240		278		142		827
Total			896		878		733		248		2755
Lines excluded (one per multiplet	Unblended		55		77		60		9		201
	Blended unidentified				1						1
	Blended identified		9				1				10
Available accepted lines	Unblended		200		275		238		50		763
	Blended unidentified		11		17		7		5		40
	Blended identified		68		44		26		0		138
Total available lines			832		800		672		239		2543

fair proportion of lines present in the sun. It includes members of 614 multiplets, of which 402 contained observed lines, and 212 predicted lines only. Counts of these lines were made separately for four spectral regions, as summarized in table 20. When two predicted lines coincided within the tolerance with one solar line, both were counted. The numbers of lines excluded (one per multiplet, as described above) and the numbers remaining as available for statistical discussion are given at the foot of the table.

(20) THEORETICAL PROBABILITY OF ACCIDENTAL COINCIDENCES

The elementary theory of accidental coincidences between spectral lines is well known.[31] If M lines (described as Group I) are distributed at random over an interval of X units, the probability that an arbitrarily chosen wave-length will fall within a distance x from one (or more) of them is $P(x) = 1 - e^{-2mx}$ where $m = M/X$; and the probability that the interval i between two successive lines of Group I lies between i and $i + di$ is $q(i) = me^{-mi}di$.

The observed solar lines are, however, not distributed at random, for pairs or groups too close to be resolved are measured as single lines. Fortunately, this simplifies the analysis. We may treat the lines of Group I approximately as if they were sharp-edged strips of width y which will be blended if the separation of centers is less than y. Suppose that we have another set of lines, Group II, of width z, which are distributed at random. Two lines, one of each group, will merge if their centers are separated by less than the coincidence-interval $c = \frac{1}{2}(y + z)$. If the lines of Group II are the fainter, we may assume $y > z$. We may now select from the whole range X a coincidence-range C, defined as follows. From the center of every unblended line of Group I lay off an interval c in both directions, and from the outer components of every blend lay off c outward. The whole range built up of their elements, disregarding overlapping, constitutes C. The probability that an arbitrary line of Group II will "coincide" with something belonging to Group I is then $\frac{C}{X}$, and the chance that it will not, $1 - \frac{C}{X}$. If there were no overlapping, C could be very easily found. Consider any individual element of the coincidence-interval of width w and the lines of Group II which, in a large number of trials, "coincide" with it. Their centers must fall in this interval and are distributed at random. Hence if $\bar{d}$ is the average distance, regardless of sign, of one of them from the center of the element, $\bar{d} = \frac{1}{4}w$. Summing for all the separate elements, $C = \Sigma w = 4\Sigma\bar{d}$.

For a single trial, there will usually be only one value for d in each element, and only a part of the elements of C will be used, but the general mean value D of $\bar{d}$ will approximate that for many trials, within the usual uncertainty arising from finite sampling. Let M' be the number of elements (which will be less than M on account of blending in Group I).

$$\text{Then} \quad D = \Sigma\bar{d}/M' \quad \text{and} \quad C = 4M'D. \tag{1}$$

If overlapping occurs, equation (1) still holds good provided that each of the confluent portions is divided into as many sections as there were originally discrete elements, and if the residual distances d are measured from the midpoint of each section. In practice, d is the difference between the wave-length of some predicted line (of Group II) and that of some component (presumably the nearest) of Group I as blended—that is, of some solar line. For the most frequent case where each element is a single line let the separation of their centers be $2s$. The range of d will be from 0 to c on one side, and 0 to s on the other, and the numerical mean $\bar{d} = \frac{c^2 + s^2}{2(c + s)} = \frac{c + s}{4} + \frac{v^2}{4(2c - v)}$ where $v = c - s$. The first term gives the approximation just discussed, the second, a small correction. The minimum value of s is $\frac{1}{2}y$, so that v ranges from 0 to $\frac{1}{2}z$. Assuming the distribution to be uniform over this range (which is good enough for this small term), the correction is found to be $\frac{z^2}{96c} + \cdots$. The ratio of this to the leading term is $\frac{z^2}{24c(c + s)}$ which at maximum, when $z = y$, is less than 0.03. As the number of confluent elements is a rather small part of the whole, the net correction to equation (1) will be less than 1 per cent. The statistical uncertainty of $\bar{D}$ is much greater than this unless the number of lines in Group II exceeds 1000, so that equation (1) is adequate in practice and it is needless to investigate the less probable case of confluence between a wide and a narrow element.

If a second group of lines (Ib) appears in the part of the spectrum left free by the original group (Ia), the coincidence-range for these may be found in exactly the same way. If M_b' is the observed number of elements in the group and D_b the mean residual, $C_b = 4M_b'D_b$. This automatically eliminates the complications arising from masking of lines of Group Ib by those of Ia. The probabilities of coincidence of a line of Group II with elements of Group Ia, Ib, $\cdots$ will be:

$$P_a = 4M_a'D_a/X; \quad P_b = 4M_b'D_b/X \cdots. \tag{2}$$

If there are N lines in Group II, the probable numbers of accidental coincidences will be NP_a, NP_b.

[31] *Cf.* Russell and Bowen, *Mount Wilson Contr.* No. 375, 1929; *Astrophys. Jour.* 69: 196, 1929.

(21) PREDICTED NUMBERS OF ACCIDENTAL COINCIDENCES

The numbers M' of solar lines have been counted in a copy of the Revised Rowland in which new identifications were entered to date. As this study is not quite completed, the results are not definitive—though adequate statistically—and only a single count was made. The solar lines were divided into two groups.

Group I*a*. *Identified Lines*: All solar lines which agree, whether as unblended or blended, under the rules of acceptance already described, with lines, of whatever origin, which have been observed in the laboratory. Observed lines of Fe I, predicted lines of other elements which pass the conditions of acceptance, and all lines shown by observation to be produced in the earth's atmosphere, are included. Blends are counted only once.

Group I*b*. *Unidentified Lines*: All solar lines for which no identification as just defined is available. Predicted lines of Fe I were completely disregarded in both groupings. The numbers of these lines for the four ranges of wave-length are given at the top of table 21. These sums should equal the numbers of lines between the given limits in the R. R. which may be found by inspection. The small differences represent the errors of the single approximate count and are statistically negligible.

Next follow the mean residuals $\bar{d}$ (in Angströms) between the predicted and solar wave-lengths for the groups of lines described in table 20. They increase moderately from the "good" to the "poor" cases, and considerably more with wave-length—mainly because the widths of the solar lines increase. They are greater for the rejected lines, since poor agreement was one factor in rejection, and greater still for the masked lines, since those which mask them are strong. Finally come the probabilities [eq. (2)] of accidental coincidences for a wave-length selected at random.

(22) PHYSICAL REALITY OF PREDICTED LINES

The numbers of predicted and observed coincidences are given in table 22. The sums for the separate regions are statistically preferable to the values calculated for the whole range together, which are given in the last line. The discordances are small, and indicate that the present subdivision is adequate.

The evidence of this table is decisive. The numbers of observed coincidences with unidentified solar lines are more than three times greater, in each of the four parts of the spectrum, than can be accounted for by chance, while those with otherwise identified lines show only a small excess, and the "absences" are much fewer than for a chance distribution. This puts it beyond doubt that a large majority of the unidentified solar lines which passed the conditions of acceptance are really due to iron.

TABLE 21

PROBABILITIES OF CHANCE COINCIDENCES

λ			2975–4000	4000–5000	5000–6000	6000–6600	All
Numbers of solar lines	Identified	M_a'	5507	3131	2280	688	11606
	Unidentified	M_b'	2300	2602	1651	584	7137
	Sum		7807	5733	3931	1272	18743
	R. R.		7803	5737	3960	1280	18780
Mean residuals $\bar{d}$ between predicted lines of Fe I and identified solar lines	Blended (accepted)	G	0.011	0.014	0.017		0.0125
		F	0.008	0.013	0.010		0.0099
		P	0.011	0.006	0.016		0.0107
		All	0.0092	0.0118	0.0133		0.0108
	Blended (rejected)		0.0138	0.0158	0.0170	0.024	0.0156
	Masked		0.0225	0.0256	0.0373	0.0410	0.0260
Mean for all identified lines (D_a)			0.0180	0.0200	0.0236	0.0312	0.0202
Mean residuals for unidentified solar lines	Unblended (accepted)	G	0.007	0.009	0.012	0.006	0.0093
		F	0.008	0.010	0.013	0.016	0.0111
		P	0.009	0.013	0.015	0.029	0.0137
		All	0.0081	0.0105	0.0131	0.0162	0.0110
	Blended (accepted)		0.010	0.013	0.010	0.034	0.014
	Blended (rejected)		0.0224	0.0200	0.0267	0.0175	0.0218
Mean for all unidentified lines (D_b)			0.0101	0.0113	0.0137	0.0177	0.0120
Probability of accidental coincidence	Identified lines		0.39	0.25	0.22	0.14	0.258
	Unidentified lines		0.09	0.12	0.09	0.07	0.094
	No coincidence		0.52	0.63	0.69	0.79	0.648

TABLE 22

PREDICTED AND OBSERVED COINCIDENCES

λ	Available Lines N	Identified Lines			Unidentified Lines			No Coincidence		
		Pred	Obs	$O-P$	Pred	Obs	$O-P$	Pred	Obs	$O-P$
2975–4000	832	324	412	+88	75	253	+178	433	167	−266
4000–5000	800	200	240	+40	96	320	+224	504	240	−264
5000–6000	672	148	140	−8	60	254	+194	464	278	−186
6000–6600	239	34	38	+4	17	59	+42	188	142	−46
Sums	2543	706	830	+124	248	886	+638	1589	827	−762
2975–6600	2543	656	830	+174	239	886	+647	1648	827	−821

TABLE 23

NUMBERS OF PHYSICALLY SIGNIFICANT COINCIDENCES

λ	With Unidentified Lines					With Identified Lines				
	N	N'	$G+F$	Astrophysical	Statistical	N	N'	$G+F$	Astrophysical	Statistical
2975–4000	266	211	220	177	178	77	68	62	55	88
4000–5000	370	292	308	243	224	44	44	37	37	40
5000–6000	305	245	260	209	194	27	26	20	19	−8
6000–6600	64	55	54	47	42	19	19	0	0	4
Sums				676	638				111	124

The probable numbers of such real lines are given in the columns headed $O-P$. It is of much interest to compare these with the numbers of lines which had already been classified as good or fair, upon astrophysical grounds quite independent of the statistical study. In doing this, the recorded numbers of these lines should be diminished in the ratio which the number N' of coincidences in each group available for the statistical study bears to the whole number N of such coincidences (since it is unfair to impose any special characteristic upon the $N-N'$ lines which were used to pick out the multiplets). The results are shown in table 23.

The numbers of physically real coincidences derived astrophysically are $(G + F)N'/N$, while the statistical estimates are $O - P$ of table 22. The agreement of the two columns is remarkable.

It is clear that the identifications of predicted iron lines in the sun which have been classified as good or fair are physically significant. The list doubtless includes some accidental coincidences and omits about as many real ones, but should be generally trustworthy. The excess of observed coincidences with otherwise identified lines, shown in table 22, is substantially accounted for by those which showed recognizable evidence of blending.

The whole number of physically real predictions (by the statistical test), according to this table, is 762. The actual number of predicted iron lines in the solar spectrum must be considerably greater. To begin with, the 212 predicted lines excluded from the statistical study should include about the same proportion which are physically real as the 960 which were available, that is, about 170. Also, the proportion of physically real coincidences should be the same for the 706 predicted lines which coincide by accident with lines of other elements as for the other 1837. The latter give 762 coincidences; there must be about 290 among the former. If there were no masking, there should be some 1200 lines of iron which, though not yet produced in the laboratory, should be observable in the solar spectrum between λ2975 and λ6600, while nearly 1600 lines in the same multiplets are too faint to appear.

The proportion of lines present decreases toward the red, where the multiplets arise from higher energy levels, and average fainter. Despite this, many more predicted iron lines should be found in the sun when the study of its spectrum from λ6600 to λ12000 has been completed.

(23) TABLE OF PREDICTED SOLAR LINES OF FE I

Table C (p. 170) contains those predicted lines of Fe I for which the evidence of presence in the solar spectrum was adjudged to be good or fair, 1254 in all. Those in the range from 6600–2975A have just been discussed. Those of longer wave-length are taken from the "Monograph on the Red and Infra-Red Solar Spectrum," which is under preparation at Mount Wilson by H. D. Babcock and others. The identifications have been made in the manner described above, by one of the writers. The writers greatly appreciate the use of this material. The wave-lengths

TABLE 24

PRESENCE IN THE SUN OF IRON LINES OBSERVED IN THE LABORATORY

λ	2975–4000	4000–5000	5000–6000	6000–6600	All
Classified Lines					
Unblended	928	593	357	107	1985
Blended	243	95	39	15	392
Masked	25	11	10	3	49
Absent	10	6	20	13	49
Total	1206	705	426	138	2475
Unclassified Lines					
Unblended	138	69	34	11	252
Blended	67	12	10	3	92
Masked	2	1	2	2	7
Absent	39	30	104	57	230
Total	246	112	150	73	581

and intensities are from this monograph or the Revised Rowland but include some changes in Rowland's intensities.[32] The third column gives the grade g or f assigned as above. A "b" added denotes that the line is blended in the solar spectrum.

The limitations of this table should be borne in mind. First, many lines (probably almost 300 between 6600–2975A) have perforce been omitted, owing to masking in the sun. Second, some accidental coincidences are doubtless present, and some real iron lines omitted, among those graded poor and rather drastically excluded. The great majority of the tabular lines must, however, be real.

(24) OBSERVED LINES OF IRON IN THE SUN

For comparison, the solar behavior of those iron lines which have been observed in the laboratory is listed in table 24. The lines which have and have not been classified in the present analysis are listed separately.

Only 2 per cent of the former fail to appear in the sun, but 40 per cent of the latter. The unclassified lines are faint, and more of them might be absent, but not so many more. Most of the "absent" lines among these are probably due to impurities (§ 16). Most of the classified lines of Fe I which do not appear in the sun are recorded as very faint in the laboratory. Four of them (at λλ2980.532, 2981.446, 2994.50 and 3020.643) are strong lines. A letter from H. D. Babcock states that all are present in the sun with about the anticipated strength.

The remaining 47 lines of wave-length less than 6600A (two of which are blended pairs) were graded on the same system as the predicted lines. Three were graded "good," 11 "fair," and 33 "poor." For many of the last, the discrepancies in wave-length are large, and it is doubtful whether they really belong in the classified places, nor is it certain whether they are due to iron at all. Indeed, if a line attributed to iron in the laboratory is absent from the sun, this is strong presumptive evidence that it is not really due to the metal.

[32] C. E. Moore, *Atomic Lines in the Sun-Spot Spectrum.* Princeton, 1933.

(25) DESIRABILITY OF FUTURE OBSERVATIONS

It is evident that the spectrum of the iron arc is very far from being fully observed. If a laboratory source could be discovered which was as efficient in producing it as is the absorption in the sun's atmosphere, at least 1500 lines would probably be added to the list. This estimate takes account only of transitions among terms already known, and would be increased by allowance for unknown high-lying levels. No such progress is likely to be made by repeating observations with familiar sources such as the arc in air. The advantage of the solar atmosphere probably consists in the great depth of highly rarefied gas. This can only be feebly imitated in the laboratory, but experiments would be attractive.

Once the source was found, observations with pure material in the usual region (say from λ2100 to λ11000) would be easy, and measurement by known apparatus rapid. The "vacuum region" has only been reconnoitered, and must contain much of interest. The infra-red, beyond the limit of photography, appears never to have been so much as examined with modern equipment of good resolving power. Observations might lead to the identification of many new levels, and lines predicted from known levels should provide abundant standards. Much could be done here with existing sources.

All told, it is evident that a great deal of work will still be required before this "familiar" spectrum is really thoroughly known, and that it still offers attractive and remunerative problems to the observer.

(26) THE TABLES OF CLASSIFIED LINES

The long tables of classified lines of Fe I already mentioned in the text (pp. 115 and 131) conclude Part I of this paper. Table B contains all classified lines of this spectrum that have been observed in the laboratory, 3606 in all. In some cases, where the laboratory line is measured only to 0.1A, the solar wave-length is entered in the table. For such lines there is good agreement between the solar and predicted wave-lengths, and it seems reasonable to assume that the line observed in the laboratory is the one to which the assigned designation applies. Consequently, the laboratory intensity is entered instead of the solar intensity. When these poorly measured lines are absent or masked in the sun, the wave-lengths are given to 0.1A and referred to the proper source.

The letters in column 1 refer to the various sources from which the laboratory wave-length (column 2) is taken. At the end of the table a complete bibliography is given with the various letters.

The selection of the best source to use for each wave-length has not been an easy task. For the many accurately measured lines, it is difficult to select the most accurate value. Those adopted for international standards have been the first choice throughout. For less accurate measures, the sources which have appeared to be consistently satisfactory for use in many multiplets have dominated the selection. The experience gained from studying the literature for the selection of wave-lengths for all elements included in the "Revised Multiplet Table" has been the guide throughout this work. All wave-lengths have been selected by one of the writers (C. E. M.). While there is no doubt that a more homogeneous list is highly desirable, yet it is hoped that the present table will suffice for those interested in using it (compare § 3).

The intensity and temperature class are in the next two columns. These have already been described in §§ 4 and 5. Then follow the wave-number, which corresponds to the observed wave-length in column 2, and the difference between the observed wave-number and that calculated from the term values of table A. The unit of o−c is 0.01 cm^{-1}, but only one digit is given if the line has been measured only to 0.1A, as described above. Finally, the last column contains the multiplet designation, which is self-explanatory.

The symbols used in table B are described at the end of the table. An asterisk denotes that the line is a blend, i. e. that the lines designated probably all contribute sensibly to the observed intensity. For a line which is probably of observable intensity but is masked by a stronger neighbor, the designation is given in parentheses. Lines blended with Fe II are marked "§." Other special notes, made by A. S. King in the course of his work on temperature classification, are indicated by a double asterisk.

Table C contains 1254 predicted lines of Fe I that have been accepted as present in the solar spectrum and graded "good" or "fair" (see § 23). It is arranged similarly to table B, except that the solar spectrum is the source used for all wave-lengths and intensities. If the solar line is a blend, a "b" follows the grade "g" or "f" in the third column.

TABLE A

TERMS OF FE I

Config	Desig	Term	Diff	Source	g
$3d^7(a^2P)4s$	a^1P_1	27543.00		10	0.817
$3d^7(a^2D)4s$	a^1D_2	28604.61		10	1.028
$3d^64s^2$	b^1D_2	34636.82		11	
$3d^7(b^2F)4s$	a^1F_3	40534.18:		11	
$3d^7(a^2G)4s$	a^1G_4	24574.690		7	1.001
$3d^64s^2$	b^1G_4	29798.96		11	0.979
$3d^7(a^2H)4s$	a^1H_5	28819.98		10	1.000
$3d^64s^2$	a^1I_6	29313.04		11	1.014
$3d^64s^2$	a^3P_2	18378.215	1174.278	5	1.506
	a^3P_1	19552.493	485.37	5	1.500
	a^3P_0	20037.86		5	
$3d^7(a^4P)4s$	b^3P_2	22838.360	108.500	5	1.498
	b^3P_1	22946.860	104.930	5	1.489
	b^3P_0	23051.790		5	
$3d^7(a^2P)4s$	c^3P_2	24335.804	436.256	7	1.484
	c^3P_1	24772.060	319.56	7	1.466
	c^3P_0	25091.62		10	
$3d^7(a^4F)4d$	e^3P_2	54879.720	496.397	7	1.459
	e^3P_1	55376.117	350.42	7	1.459
	e^3P_0	55726.54:		10	
$3d^7(a^2D)4s$	a^3D_3	26225.03	398.70	10	1.335
	a^3D_2	26623.73	−217.24	10	1.178
	a^3D_1	26406.49		10	0.731
$3d^64s^2$	b^3D_3	29371.86	− 15 08	10	1.326
	b^3D_2	29356.78	− 36.73	10	
	b^3D_1	29320.05		11	
$3d^64s(a^4D)5s$	e^3D_3	51294.262	445.702	5	1.345
	e^3D_2	51739.964	299.975	5	1.125
	e^3D_1	52039.939		5	0.801
$3d^7(a^4F)4d$	f^3D_3	53747.547	319.274	4	1.258
	f^3D_2	54066.821	382.51	4	
	f^3D_1	54449.33		4	
$3d^7(a^4F)4s$	a^3F_4	**11976.260**	584.693	1	1.254
	a^3F_3	**12560.953**	407.620	1	1.086
	a^3F_2	**12968.573**		1	0.670
$3d^64s^2$	b^3F_4	20641.144	233.377	5	1.235
	b^3F_3	20874.521	164.500	5	1.073
	b^3F_2	21039.021		5	0.663
$3d^8$	c^3F_4	32873.68	539.10	11	1.264
	c^3F_3	33412.78	352.55	11	1.066
	c^3F_2	33765.33		11	0.677
$3d^7(b^2F)4s$	d^3F_4	37046.00	− 70.36	11	
	d^3F_3	36975.64	− 35.04	11	
	d^3F_2	36940.60		11	
$3d^7(a^4F)5s$	e^3F_4	47960.973	570.923	4	1.288
	e^3F_3	48531.896	396.527	4	1.107
	e^3F_2	48928.423		4	0.622

Config	Desig	Term	Diff	Source	g
$3d^7(a^4F)4d$	f^3F_4	54683.39	441.58	7	1.141
	f^3F_3	55124.974	253.868	7	1.071
	f^3F_2	55378.842		7	0.676
$3d^7(a^2G)4s$	a^3G_5	21715.770	283.397	6	1.197
	a^3G_4	21999.167	250.294	6	1.051
	a^3G_3	22249.461		6	0.756
$3d^64s^2$	b^3G_5	23783.654	335.200	5	1.200
	b^3G_4	24118.854	219.951	5	1.048
	b^3G_3	24338.805		5	0.761
$3d^7(a^4F)4d$	e^3G_5	53739.488	327.08	5	1.248
	e^3G_4	54066.57	312.87	5	1.096
	e^3G_3	54379.44		5	0.842
$3d^64s^2$	a^3H_6	19390.197	230.839	5	1.163
	a^3H_5	19621.036	167.244	5	1.038
	a^3H_4	19788.280		5	0.811
$3d^7(a^2H)4s$	b^3H_6	26105.95	245.14	10	1.165
	b^3H_5	26351.09	276.55	10	1.032
	b^3H_4	26627.64		10	0.811
$3d^7(a^4F)4d$	e^3H_6	53840.68:	426.08	10	1.225
	e^3H_5	54266.76:	288.69	10	1.109
	e^3H_4	54555.45:		8	0.871
$3d^64s(a^6D)4d$	e^5S_2	51148.892		7	1.952
$3d^7(a^4P)4s$	a^5P_3	**17550.210**	176.797	1	1.666
	a^5P_2	**17727.017**	200.394	1	1.820
	a^5P_1	**17927.411**		1	2.499
$3d^64s(a^6D)4d$	e^5P_3	51837.279	230.17	7	1.664
	e^5P_2	52067.45	− 47.74	11	
	e^5P_1	52019.706		7	2.432
$3d^7(a^4F)4d$	f^5P_3	53160.53	408.19	8	
	f^5P_2	53568.72	356.54	7	
	f^5P_1	53925.26		8	
$3d^64s^2$	a^5D_4	**0.000**	415.933	1	1.490
	a^5D_3	**415.933**	288.070	1	1.497
	a^5D_2	**704.003**	184.129	1	1.494
	a^5D_1	**888.132**	89.942	1	1.498
	a^5D_0	**978.074**		1	
$3d^64s(a^6D)5s$	e^5D_4	**44677.010**	384.324	1	1.502
	e^5D_3	**45061.334**	272.546	1	1.508
	e^5D_2	**45333.880**	175.275	1	1.503
	e^5D_1	**45509.155**	85.929	1	1.518
	e^5D_0	**45595.084**		1	
$3d^64s(a^6D)4d$	f^5D_4	50423.185	111.250	7	1.514
	f^5D_3	50534.435	164.231	7	1.615
	f^5D_2	50698.666	181.486	7	1.614
	f^5D_1	50880.152	100.87	7	1.662
	f^5D_0	50981.02		8	
$3d^64s(a^4D)5s$	g^5D_4	**51350.505**	420.072	3	1.487
	g^5D_3	**51770.577**	279.24	3	1.492
	g^5D_2	52049.82	164.51	3	1.57:
	g^5D_1	52214.33	43.00	3	
	g^5D_0	52257.33		3	

TABLE A—(*Continued*)

Config	Desig	Term	Diff	Source	g
$3d^7(a^4F)4d$	h^5D_4	53155.13	390.75	7	1.435
	h^5D_3	53545.882	420.838	7	
	h^5D_2	53966.720	165.76	7	
	h^5D_1	54132.48		10	
	h^5D_0				
$3d^64s(a^4D)4d$	i^5D_4	57697.59	116.38	11	1.384
	i^5D_3	57813.97	160.19	11	1.415
	i^5D_2	57974.16		11	
	i^5D_1				
	i^5D_0				
$3d^7(a^4F)4s$	a^5F_5	**6928.280**	448.495	1	1.404
	a^5F_4	**7376.775**	351.296	1	1.349
	a^5F_3	**7728.071**	257.724	1	1.248
	a^5F_2	**7985.795**	168.930	1	0.995
	a^5F_1	**8154.725**		1	−0.014
$3d^7(a^4F)5s$	e^5F_5	**47005.510**	372.457	2	1.421
	e^5F_4	**47377.967**	377.571	1	1.331
	e^5F_3	**47755.538**	281.129	1	1.236
	e^5F_2	**48036.667**	184.656	1	0.991
	e^5F_1	**48221.323**		1	0.007
$3d^64s(a^6D)4d$	f^5F_5	51103.237	358.470	7	1.384
	f^5F_4	51461.707	142.439	7	1.355:
	f^5F_3	51604.146	100.906	7	
	f^5F_2	51705.052	49.482	7	0.967
	f^5F_1	51754.534		7	
$3d^7(a^4F)4d$	g^5F_5	53061.28	332.44	7	
	g^5F_4	53393.715	437.24	7	
	g^5F_3	53830.96	426.56	7	
	g^5F_2	54257.52	128.64	7	
	g^5F_1	54386.16		8	
$3d^64s(a^6D)4d$	e^5G_6	50522.94	180.97	7	1.351
	e^5G_5	50703.912	275.715	7	1.360
	e^5G_4	50979.627	239.432	7	1.238
	e^5G_3	51219.059	151.125	7	1.294
	e^5G_2	51370.184		7	0.953
$3d^7(a^4F)4d$	f^5G_6	53169.21	112.53	7	1.323
	f^5G_5	53281.735	487.285	7	1.221
	f^5G_4	53769.020	392.162	7	
	f^5G_3	54161.182	214.537	7	1.142
	f^5G_2	54375.719		7	
$3d^64s(a^4D)4d$	g^5G_6	58001.88	269.62	11	1.40:
	g^5G_5	58271.50:	248.68	11	
	g^5G_4	58520.18:	189.91	11	
	g^5G_3	58710.09:	114.72	11	
	g^5G_2	58824.81		11	0.343
$3d^7(a^4F)4d$	e^5H_7	53275.20:	77.82	8	1.30:
	e^5H_6	53353.02:	521.28	10	1.191
	e^5H_5	53874.30:	362.90	10	1.102
	e^5H_4	54237.20	253.88	10	0.90:
	e^5H_3	54491.08		10	0.484
$3d^64s(a^6D)4d$	e^7S_3	51570.16		7	1.92:
$3d^64s(a^6D)4d$	e^7P_4	50475.32	135.98	7	1.585
	e^7P_3	50611.303	250.02	7	1.687
	e^7P_2	50861.32		7	
$3d^64s(a^6D)5s$	e^7D_5	**42815.855**	347.472	2	1.585
	e^7D_4	**43163.327**	271.306	2	1.655
	e^7D_3	**43434.633**	198.902	2	1.755
	e^7D_2	**43633.535**	130.447	2	2.009
	e^7D_1	**43763.982**		2	3.002
$3d^64s(a^6D)4d$	f^7D_5	50377.92	430.13	7	1.510
	f^7D_4	50808.053	53.80	7	1.574
	f^7D_3	50861.85	136.84	8	
	f^7D_2	50998.686	49.41	7	1.844
	f^7D_1	51048.10		7	
$3d^64s(a^6D)6s$	g^7D_5	53800.90	323.72	11	1.586
	g^7D_4	54124.62	289.12	11	1.65:
	g^7D_3	54413.74:	197.98	11	
	g^7D_2	54611.72	136.02	11	
	g^7D_1	54747.74:		11	
$3d^64s(a^6D)7s$	h^7D_5	57897.17		11	
$3d^64s(a^6D)4d$	e^7F_6	50342.180	491.305	7	1.490
	e^7F_5	50833.485	358.835	7	1.505
	e^7F_4	51192.320	− 43.45	7	1.617
	e^7F_3	51148.87	182.22	7	1.499
	e^7F_2	51331.090	−123.05	7	
	e^7F_1	51208.04		7	2.490
	e^7F_0				
$3d^64s(a^6D)4d$	e^7G_7	50651.76:	316.11	8	
	e^7G_6	50967.873	260.722	7	1.415
	e^7G_5	51228.595	106.34	7	1.379
	e^7G_4	51334.94	125.59	7	1.338
	e^7G_3	51460.53	79.24	7	1.244
	e^7G_2	51539.77	27.09	7	
	e^7G_1	51566.86		10	−0.374
	X_3	40871.46	306.90	11	
	X_2	41178.36		11	
	1_6	56428.06		11	
	$2_{4,5}$	56452.04		11	
	3_4	56842.70		11	
	4_2	58213.17		11	
$3d^7(a^2P)4p$	$z^1P_1^\circ$	53229.94		10	1.266
	$z^1D_2^\circ$	49477.10		10	0.92:
$3d^7(a^2P)4p$	$y^1D_2^\circ$	51708.33		10	1.025
	$x^1D_2^\circ$	51762.12		11	0.883
$3d^7(a^2D)4p$:	$w^1D_2^\circ$	55754.29		11	0.990
$3d^7(a^2G)4p$:	$z^1F_3^\circ$	50586.89		8	1.018
	$y^1F_3^\circ$	53661.13		11	1.21:
	$x^1F_3^\circ$	53763.28		10	1.079
	$w^1F_3^\circ$	55790.72		11	0.908
$3d^7(a^2G)4p$:	$z^1G_4^\circ$	47452.770		8	1.025
$3d^7(a^2H)4p$:	$y^1G_4^\circ$	48702.57		7	1.063
	$x^1G_4^\circ$	50614.02		10	0.978
	$w^1G_4^\circ$	54810.82		11	1.001
	$v^1G_4^\circ$	56951.27		11	1.053

TABLE A—(*Continued*)

Config	Desig	Term	Diff	Source	g
$3d^7(a^2G)4p$	$z^1H_5^{\circ}$	48382.63		10	1.018
$3d^7(a^2H)4p$	$y^1H_5^{\circ}$	53722.44		10	1.03:
	$x^1H_5^{\circ}$	55525.58		11	1.018
$3d^7(a^2H)4p$	$z^1I_6^{\circ}$	53093.60		10	1.010
$3d^64s(b^4P)4p$	$z^3S_1^{\circ}$	46600.884		7	1.888
$3d^7(a^4P)4p$	$y^3S_1^{\circ}$	47555.63		5	1.884
$3d^64s(a^4D)4p$	$z^3P_2^{\circ}$	33946.965	415.925	1	1.493
	$z^3P_1^{\circ}$	34362.890	192.75	2	1.496
	$z^3P_0^{\circ}$	34555.64		2	
$3d^7(a^4P)4p$	$y^3P_2^{\circ}$	46727.137	174.755	5	1.444
	$y^3P_1^{\circ}$	46901.892	−229.32	5	1.600
	$y^3P_0^{\circ}$	46672.57		10	
$3d^64s(b^4P)4p$	$x^3P_2^{\circ}$	48304.707	211.44	7	1.263
	$x^3P_1^{\circ}$	48516.15	− 56.03	7	1.547
	$x^3P_0^{\circ}$	48460.12		10	
$3d^7(a^2D)4p$:	$w^3P_2^{\circ}$	50186.87	−143.62	7	1.469
	$w^3P_1^{\circ}$	50043.25	− 91.89	10	1.389
	$w^3P_0^{\circ}$	49951.36		10	
$3d^7(a^2P)4p$:	$v^3P_2^{\circ}$	52916.33	892.04	10	1.495
	$v^3P_1^{\circ}$	53808.37		11	1.418
	$v^3P_0^{\circ}$				
$3d^64s(a^4D)4p$	$z^3D_3^{\circ}$	**31322.639**	363.738	2	1.321
	$z^3D_2^{\circ}$	**31686.377**	250.973	2	1.168
	$z^3D_1^{\circ}$	**31937.350**		2	0.513
$3d^7(a^4F)4p$	$y^3D_3^{\circ}$	**38175.382**	502.685	1	1.324
	$y^3D_2^{\circ}$	**38678.067**	317.697	1	1.151
	$y^3D_1^{\circ}$	**38995.764**		1	0.493
$3d^64s(b^4P)4p$	$x^3D_3^{\circ}$	45220.738	61.151	7	1.352
	$x^3D_2^{\circ}$	45281.889	269.944	7	1.200
	$x^3D_1^{\circ}$	45551.833		8	0.556
$3d^7(a^4P)4p$	$w^3D_3^{\circ}$	47017.239	118.903	5	1.346
	$w^3D_2^{\circ}$	47136.142	135.953	5	1.216
	$w^3D_1^{\circ}$	47272.095		2	0.767
$3d^64s(b^4F)4p$	$v^3D_3^{\circ}$	49135.08	107.60	7	1.211
	$v^3D_2^{\circ}$	49242.68	54.98	8	0.954
	$v^3D_1^{\circ}$	49297.66		7	0.562
$3d^7(a^2P)4p$	$u^3D_3^{\circ}$	51969.14	327.82	10	1.306
	$u^3D_2^{\circ}$	52296.96	215.50	10	1.156
	$u^3D_1^{\circ}$	52512.46		7	0.700
$3d^7(a^2D)4p$:	$t^3D_3^{\circ}$	52213.29	469.64	10	1.317
	$t^3D_2^{\circ}$	52682.93	−502.11	10	1.145
	$t^3D_1^{\circ}$?	52180.82		10	0.801
	$s^3D_3^{\circ}$	52953.68:	321.59	11	1.231
	$s^3D_2^{\circ}$	53275.27		11	
	$s^3D_1^{\circ}$				
$3d^64s(a^4D)4p$	$z^3F_4^{\circ}$	**31307.272**	497.825	1	1.250
	$z^3F_3^{\circ}$	**31805.097**	328.917	1	1.086
	$z^3F_2^{\circ}$	**32134.014**		1	0.682
$3d^7(a^4F)4p$	$y^3F_4^{\circ}$	**36686.204**	476.566	1	1.246
	$y^3F_3^{\circ}$	**37162.770**	358.416	1	1.086
	$y^3F_2^{\circ}$	**37521.186**		1	0.688
$3d^7(a^2G)4p$	$x^3F_4^{\circ}$	46889.207	203.569	7	1.344
	$x^3F_3^{\circ}$	47092.776	104.298	2	1.159
	$x^3F_2^{\circ}$	47197.074		2	0.743
$3d^64s(b^4F)4p$	$w^3F_4^{\circ}$	49108.94	134.01	7	1.181
	$w^3F_3^{\circ}$	49242.950	190.23	7	1.165
	$w^3F_2^{\circ}$	49433.18		10	0.677
$3d^64s(a^4G)4p$:	$v^3F_4^{\circ}$	51304.65	60.65	10	1.122
	$v^3F_3^{\circ}$	51365.30	−163.97	10	1.096
	$v^3F_2^{\circ}$	51201.33		11	0.803
$3d^7(a^2D)4p$:	$u^3F_4^{\circ}$	56592.76	190.57	10	1.148
	$u^3F_3^{\circ}$	56783.33	75.32	10	1.077
	$u^3F_2^{\circ}$	56858.65		10	0.687
	$t^3F_4^{\circ}$	57550.09	90.97	11	1.235
	$t^3F_3^{\circ}$	57641.06	67.70	11	
	$t^3F_2^{\circ}$	57708.76		11	0.698
$3d^7(a^4F)4p$	$z^3G_5^{\circ}$	**35379.237**	388.354	1	1.248
	$z^3G_4^{\circ}$	**35767.591**	311.804	1	1.100
	$z^3G_3^{\circ}$	**36079.395**		1	0.791
$3d^64s(a^4H)4p$	$y^3G_5^{\circ}$	45294.86	133.60	5	1.207
	$y^3G_4^{\circ}$	45428.456	134.570	5	1.053
	$y^3G_3^{\circ}$	45563.026		5	0.765
$3d^64s(b^4F)4p$	$x^3G_5^{\circ}$	47834.622	− 22.44	7	1.203
	$x^3G_4^{\circ}$	47812.18	22.08	7	1.061
	$x^3G_3^{\circ}$	47834.26		10	0.668
	$w^3G_5^{\circ}$	48231.33	130.59	7	1.27:
	$w^3G_4^{\circ}$	48361.92	113.82	7	0.934
	$w^3G_3^{\circ}$	48475.74		7	0.584
$3d^7(a^2G)4p$	$v^3G_5^{\circ}$	49460.92	167.00	7	1.163
	$v^3G_4^{\circ}$	49627.92	222.69	7	0.914
	$v^3G_3^{\circ}$	49850.61		7	0.763
$3d^7(a^2H)4p$	$u^3G_5^{\circ}$	51373.96	294.26	10	1.140
	$u^3G_4^{\circ}$	51668.22	157.58	10	1.067
	$u^3G_3^{\circ}$	51825.80		10	0.801
$3d^64s(a^4G)4p$:	$t^3G_5^{\circ}$	53983.30	254.16	8	1.234
	$t^3G_4^{\circ}$	54237.46	362.89	8	1.183
	$t^3G_3^{\circ}$	54600.35		10	0.922
	$s^3G_5^{\circ}$	55907.22	− 1.66	8	1.145
	$s^3G_4^{\circ}$	55905.56	192.29	10	
	$s^3G_3^{\circ}$	56097.85		10	0.857
	$r^3G_5^{\circ}$	59926.62:	245.44	10	1.190
	$r^3G_4^{\circ}$	60172.06	192.70	10	1.030
	$r^3G_3^{\circ}$	60364.76:		10	0.780
	$q^3G_5^{\circ}$	60677.23:	77.48	10	
	$q^3G_4^{\circ}$	60754.71:	52.01	10	
	$q^3G_3^{\circ}$	60806.72		10	

TABLE A—(*Continued*)

Config	Desig	Term	Diff	Source	g
$3d^64s(a^4H)4p$	$z^3H_6^\circ$	46982.383	26.045	7	1.200
	$z^3H_5^\circ$	47008.428	98.116	7	1.060
	$z^3H_4^\circ$	47106.544		7	0.880
$3d^7(a^2G)4p$	$y^3H_6^\circ$	49434.20	170.25	7	1.17:
	$y^3H_5^\circ$	49604.45	122.61	7	1.075
	$y^3H_4^\circ$	49727.058		7	0.929
$3d^64s(a^4G)4p$:	$x^3H_6^\circ$	51023.19	45.58	10	1.161
	$x^3H_5^\circ$	51068.77		8	1.038
	$x^3H_4^\circ$				
$3d^7(a^2H)4p$	$w^3H_6^\circ$	52431.47	181.61	10	1.177
	$w^3H_5^\circ$	52613.08	155.70	10	1.033
	$w^3H_4^\circ$	52768.78		10	0.810
	$v^3H_6^\circ$	55489.81	− 59.92	10	1.169
	$v^3H_5^\circ$	55429.89	16.17	10	1.057
	$v^3H_4^\circ$	55446.06		10	0.804
$3d^64s(b^2H)4p$:	$u^3H_6^\circ$	56334.01	48.68	10	1.166
	$u^3H_5^\circ$	56382.69	40.64	10	1.029
	$u^3H_4^\circ$	56423.33		10	0.859
	$t^3H_6^\circ$	60365.70:	183.48	10	1.163
	$t^3H_5^\circ$	60549.18	208.50	10	1.040
	$t^3H_4^\circ$	60757.68		10	0.805
$3d^64s(a^4H)4p$	$z^3I_7^\circ$	45978.04:	48.94	10	1.149
	$z^3I_6^\circ$	46026.98	108.94	10	1.040
	$z^3I_5^\circ$	46135.92		10	0.833
$3d^7(a^2H)4p$	$y^3I_7^\circ$	52655.04:	−141.45	10	1.147
	$y^3I_6^\circ$	52513.59	385.47	10	1.019
	$y^3I_5^\circ$	52899.06		10	0.830
$3d^64s(b^2H)4p$	$x^3I_7^\circ$	57027.56:	42.69	10	1.145
	$x^3I_6^\circ$	57070.25	34.01	10	1.028
	$x^3I_5^\circ$	57104.26		10	0.832
$3d^64s(b^4P)4p$	$z^5S_2^\circ$	**40895.022**		2	1.985
$3d^7(a^4P)4p$	$y^5S_2^\circ$	44511.86		7	1.888
$3d^64s(a^6D)4p$	$z^5P_3^\circ$	**29056.341**	412.692	1	1.657
	$z^5P_2^\circ$	**29469.033**	263.716	1	1.835
	$z^5P_1^\circ$	**29732.749**		1	2.487
$3d^64s(a^4D)4p$	$y^5P_3^\circ$	**36766.998**	390.596	1	1.661
	$y^5P_2^\circ$	**37157.594**	251.981	1	1.836
	$y^5P_1^\circ$	**37409.575**		1	2.502
$3d^64s(b^4P)4p$	$x^5P_3^\circ$	42532.76	327.07	2	1.650
	$x^5P_2^\circ$	42859.829	219.22	2	1.822
	$x^5P_1^\circ$	43079.05		2	2.464
$3d^54s^2(a^6S)4p$	$w^5P_3^\circ$	46137.14	176.47	5	1.658
	$w^5P_2^\circ$	46313.61	96.83	5	1.822
	$w^5P_1^\circ$	46410.44		5	2.436
$3d^7(a^4P)4p$	$v^5P_3^\circ$	47966.63	196.86	2	1.646
	$v^5P_2^\circ$	48163.49	126.40	2	1.740
	$v^5P_1^\circ$	48289.89		2	2.213

Config	Desig	Term	Diff	Source	g
$3d^64s(a^6D)5p$	$u^5P_3^\circ$	51691.98:	253.33	9	
	$u^5P_2^\circ$	51945.31:	165.0	9	
	$u^5P_1^\circ$	52110.3:		9	2.633
$3d^64s(b^4D)4p$	$t^5P_3^\circ$	53388.68:	723.62	9	
	$t^5P_2^\circ$	54112.30	158.81	9	1.70:
	$t^5P_1^\circ$	54271.11		9	
$3d^64s(a^6D)4p$	$z^5D_4^\circ$	**25900.002**	240.191	1	1.502
	$z^5D_3^\circ$	**26140.193**	199.515	1	1.500
	$z^5D_2^\circ$	**26339.708**	139.685	1	1.503
	$z^5D_1^\circ$	**26479.393**	71.102	1	1.495
	$z^5D_0^\circ$	**26550.495**		1	
$3d^7(a^4F)4p$	$y^5D_4^\circ$	**33095.962**	411.182	1	1.496
	$y^5D_3^\circ$	**33507.144**	294.451	1	1.492
	$y^5D_2^\circ$	**33801.595**	215.532	1	1.495
	$y^5D_1^\circ$	**34017.127**	104.496	1	1.492
	$y^5D_0^\circ$	**34121.623**		1	
$3d^64s(a^4D)4p$	$x^5D_4^\circ$	**39625.829**	344.051	1	1.489
	$x^5D_3^\circ$	**39969.880**	261.485	1	1.504
	$x^5D_2^\circ$	**40231.365**	173.179	1	1.501
	$x^5D_1^\circ$	**40404.544**	86.768	1	1.498
	$x^5D_0^\circ$	**40491.312**		1	
$3d^64s(b^4P)4p$	$w^5D_4^\circ$	43499.54	423.16	1	1.492
	$w^5D_3^\circ$	43922.70	260.94	1	1.481
	$w^5D_2^\circ$	44183.64	227.54	1	1.533
	$w^5D_1^\circ$	44411.18	47.78	10	1.315
	$w^5D_0^\circ$	44458.96		10	
$3d^64s(b^4F)4p$	$v^5D_4^\circ$	44415.13	136.31	7	1.401
	$v^5D_3^\circ$	44551.44	112.69	7	1.386
	$v^5D_2^\circ$	44664.13	96.66	7	1.378
	$v^5D_1^\circ$	44760.79	66.13	7	1.389
	$v^5D_0^\circ$	44826.92		11	
$3d^7(a^4P)4p$	$u^5D_4^\circ$	46720.85	24.18	7	1.341
	$u^5D_3^\circ$	46745.03	143.55	7	1.397
	$u^5D_2^\circ$	46888.582	288.67	7	1.260
	$u^5D_1^\circ$	47177.25	− 5.73	10	1.410
	$u^5D_0^\circ$	47171.52:		7	
$3d^64s(a^6D)5p$	$t^5D_4^\circ$	51076.68	284.78	9	1.486
	$t^5D_3^\circ$	51361.46	268.61	9	
	$t^5D_2^\circ$	51630.07:	206.80	9	
	$t^5D_1^\circ$	51836.87:	104.89	9	
	$t^5D_0^\circ$	51941.76:		9	
$3d^64s(a^6D)4p$	$z^5F_5^\circ$	**26874.562**	292.275	1	1.399
	$z^5F_4^\circ$	**27166.837**	227.866	1	1.355
	$z^5F_3^\circ$	**27394.703**	164.895	1	1.250
	$z^5F_2^\circ$	**27559.598**	106.764	1	1.004
	$z^5F_1^\circ$	**27666.362**		1	−0.012
$3d^7(a^4F)4p$	$y^5F_5^\circ$	**33695.418**	344.122	1	1.417
	$y^5F_4^\circ$	**34039.540**	289.235	1	1.344
	$y^5F_3^\circ$	**34328.775**	218.460	1	1.244
	$y^5F_2^\circ$	**34547.235**	144.937	1	0.998
	$y^5F_1^\circ$	**34692.172**		1	−0.016
$3d^64s(a^4D)4p$	$x^5F_5^\circ$	40257.367	337.10	2	1.390
	$x^5F_4^\circ$	40594.47	247.66	1	1.328
	$x^5F_3^\circ$	40842.13	165.93	1	1.254
	$x^5F_2^\circ$	41018.06	112.56	1	0.998
	$x^5F_1^\circ$	41130.62		1	−0.006

TABLE A—(*Continued*)

Config	Desig	Term	Diff	Source	g
$3d^64s(b^4F)4p$	$w^5F_5^\circ$	44243.67		7	1.382
	$w^5F_4^\circ$	44022.55	−221.12	7	1.444
	$w^5F_3^\circ$	44166.24	143.69	7	1.351
	$w^5F_2^\circ$	44285.48	119.24	1	1.117
	$w^5F_1^\circ$	44378.42:	92.94	10	0.283
$3d^64s(a^4G)4p$	$v^5F_5^\circ$	47606.10		10	1.317
	$v^5F_4^\circ$	47930.04	323.94	8	1.264
	$v^5F_3^\circ$	48122.97	192.93	10	1.236
	$v^5F_2^\circ$	48238.903	115.93	8	1.267
	$v^5F_1^\circ$	48350.62	111.72	9	0.230
$3d^64s(a^6D)5p$	$u^5F_5^\circ$	51016.72:		9	
	$u^5F_4^\circ$	51381.48	364.76	9	
	$u^5F_3^\circ$	51619.14:	237.66	9	
	$u^5F_2^\circ$	51827.59:	208.45	9	
	$u^5F_1^\circ$	51945.86:	118.27	9	
$3d^7(a^4F)4p$	$z^5G_6^\circ$	**34843.980**		2	1.332
	$z^5G_5^\circ$	**34782.448**	− 61.532	1	1.218
	$z^5G_4^\circ$	**35257.345**	474.897	1	1.103
	$z^5G_3^\circ$	**35611.649**	354.304	1	0.887
	$z^5G_2^\circ$	**35856.424**	244.775	1	0.335
$3d^64s(a^4H)4p$	$y^5G_6^\circ$	42784.387		4	1.342
	$y^5G_5^\circ$	42911.918	127.531	1	1.203
	$y^5G_4^\circ$	43022.998	111.080	2	1.024
	$y^5G_3^\circ$	43137.511	114.513	2	0.905
	$y^5G_2^\circ$	43210.044	72.533	2	0.331
$3d^64s(b^4F)4p$	$x^5G_6^\circ$	45608.35:		5	1.336
	$x^5G_5^\circ$	45726.18	117.83	5	1.269
	$x^5G_4^\circ$	45833.24	107.06	5	1.158
	$x^5G_3^\circ$	45913.53	80.29	5	0.928
	$x^5G_2^\circ$	45964.98	51.45	5	0.323
$3d^64s(a^4G)4p$	$w^5G_6^\circ$	47363.39		10	1.306
	$w^5G_5^\circ$	47420.23	56.84	10	1.305
	$w^5G_4^\circ$	47590.07	169.84	10	1.145
	$w^5G_3^\circ$	47693.289	103.22	8	0.931
	$w^5G_2^\circ$	47831.20	137.91	11	0.472

Config	Desig	Term	Diff	Source	g
$3d^64s(a^4H)4p$	$z^5H_7^\circ$				
	$z^5H_6^\circ$	43321.12:		10	
	$z^5H_5^\circ$	42991.66	−329.46	8	1.054
	$z^5H_4^\circ$	43108.944	117.28	7	0.871
	$z^5H_3^\circ$	43325.98	217.04	10	0.509
$3d^64s(a^6D)4p$	$z^7P_4^\circ$	**23711.467**		2	1.747
	$z^7P_3^\circ$	**24180.876**	469.409	2	1.908
	$z^7P_2^\circ$	**24506.928**	326.052	2	2.333
$3d^54s^2(a^6S)4p$	$y^7P_4^\circ$	40421.89		11	1.75:
	$y^7P_3^\circ$	40207.12	−214.77	11	1.908
	$y^7P_2^\circ$	40052.08	−155.04	11	2.340
$3d^64s(a^6D)4p$	$z^7D_5^\circ$	**19350.894**		2	1.597
	$z^7D_4^\circ$	**19562.457**	211.563	2	1.642
	$z^7D_3^\circ$	**19757.040**	194.583	1	1.746
	$z^7D_2^\circ$	**19912.511**	155.471	1	2.008
	$z^7D_1^\circ$	**20019.648**	107.137	1	2.999
$3d^64s(a^6D)4p$	$z^7F_6^\circ$	**22650.427**		2	1.498
	$z^7F_5^\circ$	**22845.880**	195.453	2	1.498
	$z^7F_4^\circ$	**22996.686**	150.806	2	1.493
	$z^7F_3^\circ$	**23110.948**	114.262	2	1.513
	$z^7F_2^\circ$	**23192.508**	81.560	2	1.504
	$z^7F_1^\circ$	**23244.847**	52.339	2	1.549
	$z^7F_0^\circ$	**23270.392**	25.545	2	

Desig	Level	Source	g	
1_2°	47419.72	8	1.137	
2_2°	49052.93	11		
3_3°	49227.16	11		
4_4°	51409.18	10	0.953	
5_3°	51435.90:	11		
6_5°	51630.23	10	1.061	
7_2°	51756.16	11		
8_1°	52857.84	10	1.246	$^3S_1^\circ$
9_4°	53328.87	8		
10_3°	53891.54	11	1.476	$^5D_3^\circ$?
11_3°	54004.82	11		
12_5°	54013.78	10	1.356	$^5F_5^\circ$?
13_4°	54301.36	8		

NOTE.—Term values in heavy type are those recommended by the International Astronomical Union (see § 7).

REFERENCES

1. Walters, *Jour. Optical Soc. America* **8**: 245, 1924.
2. Laporte, *Zeit. f. Physik* **23**: 135, and **26**, 1, 1924.
3. Walters (quoted by Meggers, *Astrophys. Jour.* **60**: 60, 1924).
4. Laporte, *Proc. Nat. Acad. Sci.* **12**: 496, 1926.
5. Walters, unpublished (Moore and Russell, *Astrophys. Jour.* **68**: 151, 1928).
6. Russell, *Astrophys. Jour.* **68**: 151, 1928.
7. Burns and Walters, *Publ. Allegheny Observ.* **6**: 159, 1929.
8. Catalán, *Anales Soc. Española Fisica y Quimica* **28**: 1239, 1930.
9. Green, *Phys. Rev.* **55**: 1209, 1939.
10. Catalán, private communication, unpublished.
11. Present work, unpublished.

TABLE B

CLASSIFIED LINES OF FE I

Ref	λ I A	Int	T C	Wave Number Observed	o−c	Desig	Ref	λ I A	Int	T C	Wave Number Observed	o−c	Desig
D	11973.01	8		8349.84	+05	$a^5P_3 - z^5D_4^\circ$	E	9753.129	10	V	10250.31	−05	$y^3D_2^\circ - e^3F_2$
D	11884.12	3		8412.29	−01	$a^5P_1 - z^5D_2^\circ$	F	9747.24	2		10256.50	+03	$d^3F_2 - x^3F_2^\circ$
D	11882.80	7		8413.22	+04	$a^5P_2 - z^5D_3^\circ$	E	9738.624	200	V	10265.58	+01	$x^5F_5^\circ - e^5G_6$
D	11783.28	6		8484.28	00	$b^3P_2 - z^3D_3^\circ$	F	*9699.70	6n	V	10306.77	+01	$x^5F_3^\circ - e^5S_2$
D	11689.98	8		8551.99	+01	$a^5P_1 - z^5D_1^\circ$						+03	$x^5F_3^\circ - e^7F_3$
D	11638.25	7		8590.00	+02	$a^5P_3 - z^5D_3^\circ$	F	9693.69	1		10313.16	+13	$x^5F_2^\circ - e^7F_2$
D	11607.57	12		8612.70	+01	$a^5P_2 - z^5D_2^\circ$	F	9683.57	1		10323.94	−10	$z^5S_2^\circ - e^5G_3$
D	11593.55	5		8623.13	+05	$a^5P_1 - z^5D_0^\circ$	F	9676.42	1		10331.57	+15	$w^5D_4^\circ - g^5F_3$
D	11439.06	15		8739.59	+07	$b^3P_1 - z^3D_2^\circ$	F	9673.16	1n		10335.05	−06	$b^3H_5 - y^3F_4^\circ$
D	11422.30	6		8752.41	+03	$a^5P_2 - z^5D_1^\circ$	F	9658.94	3		10350.26	+07	$x^5F_3^\circ - e^7F_4$
D	11374.02	3		8789.56	+06	$a^5P_3 - z^5D_2^\circ$	F	9657.30	4	V	10352.03	−09	$x^5F_2^\circ - e^5G_2$
D	11355.97	1		8803.53	+01	$b^3D_3 - y^3D_3^\circ$	E	9653.143	20	V	10356.48	−03	$y^3D_3^\circ - e^3F_3$
D	11298.83	3		8848.05	+03	$b^3P_2 - z^3D_2^\circ$	Y	9636.69	(1)		10374.17	−06	$d^3F_4 - w^5G_5^\circ$
D	11251.09	3		8885.60	+04	$b^3P_0 - z^3D_1^\circ$	F	9634.22	5	V	10376.82	−11	$x^5F_3^\circ - e^5G_3$
D	11149.34	2		8966.69	−05	$b^3P_2 - z^3F_3^\circ$	E	9626.562	30n	V	10385.08	−08	$x^5F_4^\circ - e^5G_4$
D	11119.80	10		8990.51	+02	$b^3P_1 - z^3D_1^\circ$	F	9602.07	2		10411.57	−02	$y^7P_4^\circ - e^7F_5$
D	11013.27	1		9077.47	00	$y^3D_2^\circ - e^5F_3$	E	9569.960	40n	V	10446.50	−04	$x^5F_5^\circ - e^5G_5$
D	10925.80	1		9150.14	+09	$w^5F_5^\circ - g^5F_4$	F	9556.56	1		10461.14	−03	$a^3D_3 - y^3F_4^\circ$
D	10896.30	3		9174.92	+02	$c^3P_1 - z^3P_2^\circ$	F	9550.90	2		10467.35	+05	$x^5D_2^\circ - f^5D_2$
D	10884.30	3		9185.03	−05	$z^3D_2^\circ - X_3$	F	9527.7	1		10492.8	0	$x^5F_3^\circ - e^7G_4$
D	10881.65	1		9187.27	+12	$b^3P_1 - z^3F_2^\circ$	E	9513.24	10n	V	10508.78	+01	$x^5F_4^\circ - f^5F_5$
D	10863.60	5		9202.53	−05	$y^3D_3^\circ - e^5F_4$	F	9462.97	2	V	10564.61	+05	$x^5D_3^\circ - f^5D_3$
D	10849.68	2		9214.34	+05	$e^7G_4 - t^3H_5^\circ$?	F	9454.24	4n	V	10574.37	−06	$x^5F_1^\circ - f^5F_2$
D	10818.36	3		9241.02	+01	$z^3D_1^\circ - X_2$	F	*9452.45	2	V	10576.37	+25	$x^5F_5^\circ - e^7F_5$
D	10783.09	3		9271.24	−03	$c^3P_0 - z^3P_1^\circ$						−11	$x^5D_1^\circ - f^5D_0$
D	10752.99	3		9297.20	+05	$e^7G_3 - t^3H_4^\circ$?	F	9443.98	10n	V	10585.85	−22	$x^5F_2^\circ - f^5F_3$
D	10532.21	10		9492.09	+11	$z^3D_2^\circ - X_2$	F	9437.91	2		10592.66	−11	$y^3F_3^\circ - e^5F_3$
D	10469.59	20	V	9548.85	+03	$z^3D_3^\circ - X_3$	F	9414.14	20n	V	10619.41	−17	$x^5F_3^\circ - f^5F_4$
D	10452.70	5		9564.29	+10	$z^3F_4^\circ - X_3$	F	9410.1	1n		10624.0	+ 1	$x^5F_1^\circ - f^5F_1$
D	10435.36	0N		9580.19	+03	$y^3D_3^\circ - e^5F_3$	F	9401.09	10n	V	10634.15	+02	$x^5F_4^\circ - e^7G_5$
D	10423.65	3		9590.95	+12	$c^3P_1 - z^3P_1^\circ$	F	9394.71	3n	V	10641.37	−05	$x^5D_3^\circ - e^7P_3$
D	10422.99	0		9591.55	+05	$a^3G_5 - z^3F_4^\circ$	F	9388.28	3n		10648.66	−13	$x^5D_2^\circ - f^5D_1$
F	10395.75	8		9616.68	+05	$a^5P_3 - z^5F_4^\circ$	F	9382.83	3n		10654.84	+11	$y^7P_3^\circ - f^7D_3$
SS	10353.82	(2n)		9655.63	+04	$w^5D_4^\circ - h^5D_4$	E	9372.900	6	IV*	10666.13	00	$b^3F_4 - z^3F_4^\circ$
F	10348.16	4n		9660.91	−08	$w^5D_4^\circ - f^5P_3$	E	9362.370	4	IV*	10678.13	00	$a^3P_2 - z^5P_3^\circ$
F	10340.77	4		9667.81	+12	$a^5P_2 - z^5F_3^\circ$	E	9359.420	3	IV*	10681.50	00	$b^3F_4 - z^3D_3^\circ$
F	10218.36	3		9783.63	+05	$c^3P_1 - z^3P_0^\circ$	E	9350.46	10	V	10691.73	−03	$y^3F_4^\circ - e^5F_4$
E	10216.351	100	V	9785.55	−04	$y^3D_3^\circ - e^3F_4$	F	9343.40	3	V	10699.80	+01	$x^5F_4^\circ - e^3D_3$
F	10195.11	2		9805.94	+01	$a^3G_4 - z^3F_3^\circ$	F	9333.94	2		10710.65	+14	$x^5F_5^\circ - e^7G_6$
F	10167.4	1		9832.64	+06	$a^5P_2 - z^5F_2^\circ$	O	9324.07	(1)		10721.99	+09	$x^5F_2^\circ - e^3D_2$
E	10145.601	80	V	9853.79	−04	$y^3D_2^\circ - e^3F_3$	F	9318.13	3	V	10728.82	+03	$x^5D_3^\circ - f^5D_2$
F	10142.82	2		9856.49	−05	$x^5F_3^\circ - f^5D_2$	F	9307.94	2		10740.57	+10	$x^5F_4^\circ - e^7G_4$
F	10113.86	2		9884.71	+16	$a^3G_3 - z^3F_2^\circ$	F	9294.66	2	V	10755.92	−11	$x^5F_4^\circ - g^5D_4$
E	10065.080	60	V	9932.62	−04	$y^3D_1^\circ - e^3F_2$	F	9259.05	15	V	10797.29	−07	$x^5D_4^\circ - f^5D_4$
F	10057.64	3	V	9939.97	00	$x^5F_4^\circ - f^5D_3$	E	9258.30	20	V	10798.16	−04	$y^3F_3^\circ - e^3F_4$
F	9980.55	2n		10016.74	−09	$x^5F_4^\circ - e^7P_3$	F	9246.54	2	IV*	10811.89	+03	$b^3F_3 - z^3D_2^\circ$
F	9977.52	1		10019.78	+06	$x^5F_3^\circ - f^7D_3$	F	9242.32	2		10816.82	+08	$x^5D_2^\circ - f^7D_1$
F	9944.13	3n		10053.43	00	$y^7P_4^\circ - e^7P_4$	SS	9233.18	(1)		10827.54	−03	$y^5G_5^\circ - e^3G_5$
F	9917.93	2	IV*	10079.99	+15	$a^1F_3 - x^1G_4^\circ$	O	9225.55	(1)		10836.49	−05	$d^3F_3 - x^3G_4^\circ$
E	9889.082	40	V	10109.39	−05	$x^5F_4^\circ - e^5G_5$	F	9217.54	5n	V	10845.91	+04	$x^5F_5^\circ - f^5F_5$
SS	9881.54	(1)		10117.11	−03	$d^3F_3 - x^3F_3^\circ$	F	9214.45	6	V	10849.54	+05	$x^5D_4^\circ - e^7P_4$
F	*9868.09	3	V	10130.90	+07	$x^5F_2^\circ - e^5S_2$	E	9210.030	6	IV*	10854.75	+01	$b^3P_1 - y^5D_2^\circ$
					+09	$x^5F_2^\circ - e^7F_3$	F	9199.52	2n		10867.15	−09	$x^5F_4^\circ - f^5F_4$
E	9861.793	30	V	10137.37	−13	$x^5F_3^\circ - e^5G_4$	F	9178.57	1n		10891.96	−01	$x^5D_3^\circ - f^7D_3$
F	9839.38	1		10160.46	−04	$d^3F_3 - w^3D_2^\circ$	U	9173.83	(1)		10897.58	+12	$a^3D_2 - y^3F_2^\circ$
F	9834.04	3n	V	10165.97	+15	$x^5F_5^\circ - f^5D_4$	F	9173.46	4nd		10898.02	+19	$x^5F_3^\circ - e^3D_2$
F	9811.36	2		10189.47	+06	$y^7P_4^\circ - e^7P_3$	U	9164.51	(1)		10908.67	+06	$x^5D_4^\circ - f^5D_3$
E	9800.335	20	V	10200.94	−06	$x^5F_2^\circ - e^5G_3$	SS	*9156.94	(2)		10917.68	+15	$x^5D_2^\circ - e^5S_2$
F	9786.62	2		10215.24	+04	$y^3F_3^\circ - e^5F_4$						+17	$x^5D_2^\circ - e^7F_3$
F	9783.96	3	V	10218.01	+06	$x^5F_5^\circ - e^7P_4$	F	9155.9	1		10918.9	−3	$x^5F_1^\circ - g^5D_2$
E	9763.913	15	V	10238.99	−02	$x^5F_4^\circ - e^7F_5$	E	9147.800	5n	V	10928.59	+14	$x^5F_3^\circ - g^5D_3$
E	9763.450	15	V	10239.48	−08	$x^5F_1^\circ - e^5G_2$	F	9146.11	3	IV*	10930.61	+03	$b^3F_3 - z^3F_3^\circ$

TABLE B—(*Continued*)

Ref	λ I A	Int	T C	Wave Number		Desig
				Observed	o−c	
E	9118.888	20	IV*	10963.24	00	$b^3P_2 - y^5D_2^\circ$
F	9117.10	2		10965.39	+05	$b^3P_0 - y^5D_1^\circ$
F	9103.64	1		10981.60	+01	$y^5F_5^\circ - e^5D_4$
F	9100.50	5n	V	10985.40	−07	$x^5D_4^\circ - e^7P_3$
E	9089.413	30	IV*	10998.80	+01	$b^3G_5 - z^5G_5^\circ$
E	9088.326	50	IV*	11000.11	+01	$b^3P_1 - z^3P_2^\circ$
SS	9084.22	(1)		11005.09	−01	$y^5F_3^\circ - e^5D_2$
F	*9080.48	3n		11009.62	−13 −06	$x^5D_3^\circ - e^5G_4$ $x^5F_4^\circ - f^5F_3$
E	9079.599	8	V	11010.68	−03	$y^3F_2^\circ - e^3F_3$
F	*9070.42	2		11021.82	+03 −06	$y^5F_4^\circ - e^5D_3$ $x^5F_2^\circ - e^3D_1$
F	9062.24	2		11031.77	+01	$x^5F_2^\circ - g^5D_2$
F	9052.6	1		11043.5	−1	$y^5G_4^\circ - e^3G_4$
SS	9036.72	(1)		11062.93	+03	$x^5D_2^\circ - e^3D_3$
F	9030.67	1		11070.34	+07	$b^3P_1 - y^5D_1^\circ$
F	9024.47	15	V	11077.94	−14	$x^5D_4^\circ - e^5G_5$
F	9019.84	2		11083.63	−08	$x^5F_1^\circ - g^5D_1$
F	9013.90	1		11090.93	+11	$a^3P_2 - z^5P_2^\circ$
E	9012.098	30	V	11093.15	+01	$x^5F_5^\circ - g^5D_4$
F	9010.55	2	IV*	11095.06	+07	$b^3F_2 - z^3F_2^\circ$
F	9008.37	2		11097.74	+06	$X_3 - u^3D_3^\circ$
F	9006.72	1		11099.78	+06	$x^5D_2^\circ - e^7F_2$
E	8999.561	100	III	11108.61	+01	$b^3P_2 - z^3P_2^\circ$
F	8984.87	3	V	11126.77	+06	$x^5F_1^\circ - g^5D_0$
E	8975.408	10	IV*	11138.50	+01	$b^3G_4 - z^5G_4^\circ$
F	8946.25	1		11174.80	+04	$b^3P_1 - y^5D_0^\circ$
E	8945.204	20	V	11176.11	00	$x^5F_4^\circ - g^5D_3$
F	8943.00	3	IV*	11178.86	+09	$b^3P_2 - y^5D_1^\circ$
F	8929.04	5	V	11196.34	+07	$x^5F_2^\circ - g^5D_1$
F	8919.95	10	V	11207.75	+06 +09	$x^5F_3^\circ - g^5D_2$ $(x^5D_4^\circ - e^7F_5)$
F	8916.26	1	IIA	11212.39	+09	$a^3F_2 - z^7P_3^\circ$
F	8876.13	2	IV*	11263.08	−14	$x^5D_0^\circ - f^5F_1$
F	8868.42	3	IV*	11272.87	+03	$b^3G_3 - z^5G_3^\circ$
E	8866.961	150	V	11274.73	−04	$y^3F_4^\circ - e^3F_4$
F	8863.64	1p?		11278.95	−06	$y^7P_2^\circ - e^7F_2$
F	8846.82	5	V	11300.40	−11	$x^5D_1^\circ - f^5F_2$
E	8838.433	30	IV*	11311.12	+02	$b^3P_0 - z^3P_1^\circ$
E	8824.227	200	II	11329.33	+01	$a^5P_2 - z^5P_3^\circ$
F	8814.5	2		11341.8	0	$X_3 - t^3D_3^\circ$
SS	8808.173	4n		11349.98	−01	$x^5D_1^\circ - f^5F_1$
E	8804.624	10	IV*	11354.55	+02	$a^3P_2 - z^5P_1^\circ$
F	8796.42	2		11365.14	+08	$x^5D_3^\circ - e^7G_4$
E	8793.376	120	V	11369.08	−05	$y^3F_3^\circ - e^3F_3$
F	8790.62	10n	V	11372.64	−14	$x^5D_2^\circ - f^5F_3$
F	8784.44	5	V	11380.64	+02	$x^5D_3^\circ - g^5D_4$
E	8764.000	100	V	11407.19	−05	$y^3F_2^\circ - e^3F_2$
E	8757.192	50	IV	11416.05	+02	$b^3P_1 - z^3P_1^\circ$
F	8747.32	2		11428.94	+15	$b^3G_3 - z^3G_4^\circ$
SS	8729.171	(2)		11452.70	−06	$a^1P_1 - y^3D_1^\circ$
F	*8713.19	10	V	11473.70	+01 +01	$b^3G_5 - z^5G_4^\circ$ $x^5D_2^\circ - f^5F_2$
F	8710.29	20n	V	11477.53	+12	$x^5D_4^\circ - f^5F_5$
O	8699.43	4n	V	11491.85	+02	$x^5D_3^\circ - f^5F_4$
E	8688.633	1500	II	11506.13	00	$a^5P_3 - z^5P_3^\circ$
E	8674.751	60	III	11524.55	+02	$b^3P_2 - z^3P_1^\circ$
E	8661.908	600	II	11541.63	+01	$a^5P_1 - z^5P_2^\circ$
E	8621.612	10	IV*	11595.58	00	$b^3G_5 - z^3G_5^\circ$
E	8611.807	40	III	11608.78	00	$b^3P_1 - z^3P_0^\circ$
O	8598.79	4	V	11626.35	+08	$z^3G_5^\circ - e^5F_5$
O	8592.97	2n	V	11634.23	−04	$x^5D_3^\circ - f^5F_3$
E	8582.267	15	IV*	11648.74	00	$b^3G_4 - z^3G_4^\circ$
W	8559.98	(1)		11679.06	−05	$a^1F_3 - t^3D_3^\circ$
E	8526.685	8	V	11724.67	−01	$x^5D_4^\circ - g^5D_4$

Ref	λ I A	Int	T C	Wave Number		Desig
				Observed	o−c	
O	8515.08	20	IV*	11740.65	+06	$b^3G_3 - z^3G_3^\circ$
E	8514.075	150	II	11742.04	+02	$a^5P_2 - z^5P_2^\circ$
O	8497.00	8	V	11765.63	−02	$y^3F_3^\circ - e^3F_2$
O	8471.75	2	V	11800.70	00	$x^5D_3^\circ - g^5D_3$
E	8468.413	300	II	11805.35	+01	$a^5P_1 - z^5P_1^\circ$
E	8439.603	20	V	11845.65	−04	$y^3F_4^\circ - e^3F_3$
O	8424.14	2n	V	11867.39	−01	$x^5D_3^\circ - e^5P_3$
O	8422.95	2	V	11869.07	−04	$c^3F_3 - x^3D_2^\circ$
O	8401.42	2	IV*	11899.49	00	$a^3P_0 - z^3D_1^\circ$
E	8387.781	1200	II	11918.83	+01	$a^5P_3 - z^5P_2^\circ$
E	8365.642	25	IV?	11950.38	+03	$a^3D_3 - y^3D_3^\circ$
E	8360.822	8	V	11957.26	−01	$z^3G_3^\circ - e^5F_2$
R	8342.95	(−)		11982.88	−09	$x^5D_2^\circ - g^5D_1$
E	8339.431	80	V	11987.94	−01	$z^3G_4^\circ - e^5F_3$
E	8331.941	200	V	11998.71	−02	$z^3G_5^\circ - e^5F_4$
E	8327.063	1200	II	12005.74	+01	$a^5P_2 - z^5P_1^\circ$
E	8293.527	20	V	12054.29	−05	$a^3D_2 - y^3D_2^\circ$
O	8275.91	4n	V	12079.95	+01	$x^5D_3^\circ - g^5D_2$
O	8274.28	6	IV?	12082.33	+11	$X_3 - s^3D_3^\circ$
M	8264.27	3	V	12096.96	+05	$X_2 - s^3D_2^\circ$
E	8248.151	30	V	12120.60	−02	$z^5G_4^\circ - e^5F_4$
E	8239.130	8	IV*	12133.87	−01	$a^3P_1 - z^3D_2^\circ$
E	8232.347	50	V	12143.87	−02	$z^5G_3^\circ - e^5F_3$
E	8220.406	1500	V	12161.51	−02	$z^5G_6^\circ - e^5F_5$
E	8207.767	40	V	12180.24	00	$z^5G_2^\circ - e^5F_2$
E	8198.951	80	V	12193.34	−04	$z^3G_4^\circ - e^3F_4$
O	8186.80	10nd?	V	12211.43	−02	$x^5D_4^\circ - e^5P_3$
O	8179.03	(1)	IV*?	12223.03	−03	$z^5G_5^\circ - e^5F_5$
O	*8149.59	3	V	12267.18	+14 −13	$d^3F_3 - v^3D_2^\circ$ $d^3F_3 - w^3F_3^\circ$
E	8096.874	10	IV*	12347.06	00	$c^3F_4 - x^3D_3^\circ$
E	8085.200	500	V	12364.88	−02	$z^5G_2^\circ - e^5F_1$
E	8080.668	10nd?	V	12371.82	−21	$a^3D_2 - y^3D_1^\circ$
O	8075.13	4	II	12380.30	+04	$a^5F_4 - z^7D_3^\circ$
O	8047.60	15	II	12422.65	+04	$a^5F_5 - z^7D_5^\circ$
E	8046.073	600	V	12425.01	−01	$z^5G_3^\circ - e^5F_2$
E	8028.341	50	V	12452.45	−05	$z^3G_3^\circ - e^3F_3$
E	7998.972	700	V	12498.17	−02	$z^5G_4^\circ - e^5F_3$
O	7959.21	(1)		12560.61	−05	$x^5F_4^\circ - h^5D_4$
O	7955.81	(1)		12565.98	−08	$x^5F_4^\circ - f^5P_3$
E	7945.878	600	V	12581.68	−06 +16	$z^3G_5^\circ - e^3F_4$ $(a^3P_1 - z^3F_2^\circ)$
O	7941.09	10	IV*	12589.27	00	$a^3D_1 - y^3D_1^\circ$
E	7937.166	700	V	12595.50	−02	$z^5G_5^\circ - e^5F_4$
E	7912.866	6	IIA	12634.18	00	$a^5F_5 - z^7D_4^\circ$
O	7879.84	1	V	12687.13	−14	$x^5F_4^\circ - f^5G_5$
O	7869.65	4	V	12703.55	−08	$z^5G_4^\circ - e^3F_4$
O	7855.48	4n	V	12726.47	−12	$x^5F_3^\circ - f^5P_2$
O	7844.66	2	V	12744.02	−18	$y^3D_1^\circ - e^3D_2$
E	7832.224	400	V	12764.26	−04	$z^3G_4^\circ - e^3F_3$
E	7780.586	300	V	12848.97	−06	$z^3G_3^\circ - e^3F_2$
O	7751.18	5n	V	12897.72	−04	$x^5F_5^\circ - h^5D_4$
E	7748.281	125	IV	12902.54	−01	$b^3G_5 - y^3F_4^\circ$
O	7742.71	4n	V	12911.83	−01	$x^5F_5^\circ - f^5G_6$
O	7723.20	4	IV*	12944.44	+02	$a^3P_2 - z^3D_3^\circ$
E	7710.390	25	V	12965.95	−02	$y^5F_4^\circ - e^5F_5$
E	7664.302	80	IV	13043.92	00	$b^3G_4 - y^3F_3^\circ$
E	7661.223	30	V	13049.16	−03	$y^5F_3^\circ - e^5F_4$
L	7653.783	6	V	13061.84	−06	$y^3D_2^\circ - e^3D_2$
E	7620.538	25	V	13118.83	−05	$y^3D_3^\circ - e^3D_3$
O	7605.32	2n	V	13145.08	+06	$x^5F_4^\circ - e^3G_5$
E	7586.044	150	V	13178.48	−04	$z^5G_5^\circ - e^3F_4$
E	7583.796	50	IV*	13182.38	00	$b^3G_3 - y^3F_2^\circ$
E	7568.925	30	V	13208.28	−02	$y^5F_2^\circ - e^5F_3$
O	7563.03	1n	V	13218.58	+01	$y^3D_1^\circ - g^5D_1$

TABLE B—(*Continued*)

Ref	λ I A	Int	T C	Wave Number Observed	Wave Number o−c	Desig
O	7559.68	1n	V	13224.44	00	$x^5F_3^\circ - e^3G_4$
U	7541.61	(1)		13256.14	−10	$z^3F_3^\circ - e^5D_3$?
E	7531.171	60	V	13274.50	−05	$z^5G_4^\circ - e^3F_3$
E	7511.045	800	V	13310.07	−02	$y^5F_5^\circ - e^5F_5$
L	7507.300	8	V	13316.69	−08	$z^5G_3^\circ - e^3F_2$
V	7498.56	1	IV*	13332.23	−02	$c^3F_3 - u^5D_3^\circ$
E	7495.088	400	V	13338.40	−03	$y^5F_4^\circ - e^5F_4$
L	7491.678	12	V	13344.48	−02	$y^5F_1^\circ - e^5F_2$
SS	7481.934	(1)		13361.85	−02	$y^3D_2^\circ - e^3D_1$
SS	7476.376	(1)		13371.79	+04	$y^3D_2^\circ - g^5D_2$
O	7473.56	(1)		13376.82	−02	$y^5P_2^\circ - f^5D_3$
V	7461.534	(1)		13398.40	00	$b^3F_4 - y^5F_4^\circ$
V	7454.02	(1)		13411.89	−03	$c^3F_2 - u^5D_1^\circ$
V	7447.43	1	V	13423.76	−08	$x^5D_3^\circ - g^5F_4$
E	7445.776	200	V	13426.74	−02	$y^5F_3^\circ - e^5F_3$
					−14	$(a^3P_2 - z^3F_3^\circ)$
L	7443.031	2	IV*	13431.69	−05	$c^3F_2 - x^3F_2^\circ$
V	7440.98	2n	V	13435.40	−05	$x^5D_4^\circ - g^5F_5$
M	7430.90	1−	IV*	13453.62	−09	$y^5P_2^\circ - e^7P_3$
O	7430.73	(1)		13453.93	+01	$w^5F_5^\circ - i^5D_4$
M	7430.58	1−	IV*	13454.20	−05	$b^3F_3 - y^5F_3^\circ$
O	7421.60	1−	V	13470.48	−10	$y^5P_1^\circ - f^5D_1$
E	7418.674	5	IV*	13475.79	−01	$c^3F_3 - u^5D_2^\circ$
E	7411.178	100	V	13489.42	−01	$y^5F_2^\circ - e^5F_2$
E	7401.689	4	IV*	13506.72	−05	$c^3F_2 - w^3D_1^\circ$
E	7389.425	80	V	13529.13	−02	$y^5F_1^\circ - e^5F_1$
L	7386.394	8n	V	13534.68	−02	$x^5D_4^\circ - f^5P_3$
V	7382.99	1n	V	13540.92	−15	$y^5P_2^\circ - f^5D_2$
O	7370.16	1	V	13564.50	−08	$y^3D_3^\circ - e^3D_2$
O	7366.37	1	V	13571.48	+04	$y^5P_1^\circ - f^5D_0$
O	7363.96	1n	V	13575.92	−08	$x^5D_3^\circ - h^5D_3$
V	7353.528	1	V	13595.17	−03	$y^3D_3^\circ - g^5D_3$
V	7351.56	4	V	13598.81	−03	$x^5D_3^\circ - f^5P_2$
V	7351.160	2n	V	13599.55	−05	$x^5D_2^\circ - g^5F_3$
V	7333.62	1n	V	13632.08	−12	$y^5F_3^\circ - e^3F_4$
L	§*7320.694	5n	V	13656.15	−04	$y^5P_3^\circ - f^5D_4$
					+24	$x^5D_4^\circ - f^5G_5$
I	7311.101	12	V	13674.07	−02	$y^5F_2^\circ - e^5F_1$
L	§7307.938	8	IV*	13679.99	−01	$c^3F_3 - x^3F_3^\circ$
V	7306.61	3	V	13682.47	−08	$y^5F_5^\circ - e^5F_4$
O	7300.47	1n	V	13693.98	+08	$x^5D_2^\circ - f^5P_1$
V	7295.00	1−	V	13704.25	−01	$y^5P_2^\circ - f^7D_3$
I	7293.068	15	V	13707.88	−01	$y^5F_3^\circ - e^5F_2$
V	7292.856	3n	V	13708.28	−04	$y^5P_3^\circ - e^7P_4$
I	7288.760	10	V	13716.02	+02	$y^5F_4^\circ - e^5F_3$
V	7285.286	1	V	13722.52	−04	$y^5P_2^\circ - f^5D_1$
L	7284.843	4	IV*	13723.35	−01	$c^3F_3 - w^3D_2^\circ$
V	7282.39	1n	V	13727.98	+04	$x^5D_1^\circ - h^5D_1$
V	7261.54	3n	V	13767.40	−04	$y^5P_3^\circ - f^5D_3$
SS	7256.142	1−	V	13777.64	−03	$x^5D_3^\circ - f^3D_3$
V	7244.86	2n	V	13799.09	−05	$x^5D_3^\circ - f^5G_4$
I	7239.885	6	V	13808.58	+01	$z^3P_2^\circ - e^5F_3$
O	7228.69	1	IV*	13829.96	+03	$a^3G_3 - z^3G_3^\circ$
I	7223.668	12	IV*	13839.58	00	$c^3P_2 - y^3D_3^\circ$
V	*7222.88	(1)		13841.08	−18	$x^5F_3^\circ - f^3F_4$
					−01	$y^5P_2^\circ - f^7D_2$
V	7221.22	2n	V	13844.27	−03	$y^5P_3^\circ - e^7P_3$
I	7219.686	5	IV*	13847.21	+04	$c^3F_4 - u^5D_4^\circ$
V	7212.47	1n	V	13861.06	−02	$x^5D_3^\circ - g^5F_3$
E	7207.406	500	V	13870.80	−02	$y^5D_3^\circ - e^5F_4$
V	7207.123	6	IV*	13871.34	−01	$c^3F_4 - u^5D_3^\circ$
O	7194.92	1	V	13894.87	+02	$x^5D_0^\circ - g^5F_1$
O	7191.66	(1)		13901.17	+05	$x^5D_2^\circ - h^5D_1$
V	7189.17	3	IV*	13905.98	−03	$c^3P_1 - y^3D_2^\circ$
E	7187.341	800	V	13909.52	−03	$y^5D_4^\circ - e^5F_5$
V	7181.93	1n	V	13920.00	−05	$x^5D_4^\circ - h^5D_3$
L	7181.222	10	V	13921.38	−05	$y^5F_4^\circ - e^3F_4$
V	7180.020	1	IV*	13923.71	−03	$a^3F_4 - z^5D_4^\circ$
V	7176.886	2n	V	13929.78	−04	$x^5D_2^\circ - f^5G_3$
V	7175.937	3	V	13931.64	−03	$y^5P_3^\circ - f^5D_2$
E	7164.469	250	V	13953.93	−01	$y^5D_2^\circ - e^5F_3$
V	7158.502	1	V	13965.56	−04	$z^5P_2^\circ - e^7D_3$
V	7155.64	3n	V	13971.14	−04	$x^5D_1^\circ - f^5G_2$
V	7151.495	1	IV*	13979.24	−03	$a^3P_0 - y^5D_1^\circ$
R	*7148.69	(−)		13984.73	+07	$y^5F_2^\circ - e^3F_3$
					+03	$z^5S_2^\circ - e^3P_2$
V	*7145.317	5	V	13991.33	+05	$y^5P_2^\circ - e^7F_3$
					+03	$y^5P_2^\circ - e^5S_2$
V	7142.522	4n	V	13996.80	−04	$x^5D_3^\circ - h^5D_2$
I	7132.989	8	IV*	14015.51	−02	$c^3F_4 - x^3F_4^\circ$
I	7130.942	150	V	14019.52	−02	$y^5D_1^\circ - e^5F_2$
I	7112.176	3	IV*	14056.52	−01	$b^3G_4 - y^3D_3^\circ$
I	7107.461	4	IV*	14065.84	−03	$c^3F_2 - w^5G_2^\circ$
I	7095.425	3	V	14089.73	+03	$z^3P_2^\circ - e^5F_2$
SS	7091.942	(1)		14096.68	−01	$x^5D_3^\circ - e^3G_4$
O	7091.83	(1)		14096.86	−08	$x^5D_3^\circ - f^3D_2$
I	7090.404	40	V	14099.69	−01	$y^5D_0^\circ - e^5F_1$
V	*7086.76	2	V	14106.94	+03	$x^5F_2^\circ - f^3F_3$
					−05	$z^5P_3^\circ - e^7D_4$
V	7083.396	1n	V	14113.64	−02	$x^5D_4^\circ - e^3G_5$
V	7071.88	1	V	14136.62	−05	$y^5P_2^\circ - e^3D_3$
SS	7069.54	1−	IV*	14141.30	00	$b^3F_4 - z^5G_5^\circ$
I	7068.415	40	IV*	14143.54	−02	$c^3F_4 - w^3D_3^\circ$
O	7044.60	(1)		14191.36	+06	$x^5D_3^\circ - f^5G_3$
V	7038.818	2	V	14203.02	−10	$y^5F_3^\circ - e^3F_3$
I	7038.251	40	V	14204.16	−04	$y^5D_1^\circ - e^5F_1$
V	7027.60	(1)		14225.69	00	$d^3F_3 - v^3F_2^\circ$
V	7024.649	10n	V	14231.66	−03	$y^5P_3^\circ - f^7D_2$
V	7024.084	5	IV*	14232.81	−05	$c^3F_4 - z^3H_4^\circ$
L	7022.976	50	V	14235.06	−01	$y^5D_2^\circ - e^5F_2$
V	7016.436	60	V	14248.41	+02	$y^5D_3^\circ - e^5F_3$
V	7016.075	20	IV*	14249.06	−04	$a^3P_1 - y^5D_2^\circ$
O	7014.99	(1)		14251.26	00	$a^3H_4 - y^5F_4^\circ$
V	7011.364	3	IV*	14258.64	−01	$d^3F_4 - v^3F_4^\circ$
V	7010.362	2	IV*	14260.67	−06	$d^3F_2 - v^3F_2^\circ$
V	7008.014	5	V	14265.45	−11	$y^5F_5^\circ - e^3F_4$
V	7000.633	3	IV*	14280.49	−02	$c^3F_3 - w^5G_3^\circ$
I	6999.902	30	V	14281.98	−02	$y^5D_4^\circ - e^5F_4$
I	6988.530	5	IV*	14305.23	+01	$a^3H_6 - y^5F_5^\circ$
I	6978.855	100	III	14325.06	+03	$a^3P_0 - z^3P_1^\circ$
V	6977.445	4	IV*	14327.95	−01	$d^3F_4 - u^3G_5^\circ$
V	6976.934	3	IV*	14329.00	−01	$d^3F_3 - v^3F_4^\circ$
V	6976.306	1	V	14330.29	−10	$y^5P_1^\circ - e^3D_2$
V	6971.95	1	IV*	14339.24	−02	$b^3G_3 - y^3D_2^\circ$
U	6960.343	2	V	14363.15	−03	$d^3F_4 - 4_4^\circ$
SS	6951.656	1−	V	14381.10	−09	$y^5F_2^\circ - e^3F_2$
I	*6951.261	25	V	14381.92	+03	$y^5P_3^\circ - e^5S_2$
					+05	$y^5P_3^\circ - e^7F_3$
V	*6947.501	3	V	14389.70	+04	$d^3F_3 - v^3F_3^\circ$
					−20	$d^3F_4 - 5^\circ$
I	6945.208	150	III	14394.45	−02	$a^3P_1 - z^3P_2^\circ$
L	*6933.628	6	IV*	14418.49	−01	$a^3H_5 - y^5F_4^\circ$
					+07	$c^3F_3 - w^5G_2^\circ$
U	6933.04	1	V	14419.72	−01	$y^5D_2^\circ - e^5F_1$
V	6930.64	1	V	14424.71	+01	$d^3F_2 - v^3F_3^\circ$
I	6916.702	60	V	14453.78	−05	$y^5D_3^\circ - e^3F_4$
V	6911.52	1	IV*	14464.61	−02	$a^3P_1 - y^5D_1^\circ$
V	6898.31	3	V	14492.31	−05	$y^5F_4^\circ - e^3F_3$
L	6885.772	20	V	14518.70	−05	$y^3F_2^\circ - e^3D_1$
M	6881.74	1	V	14527.21	−05	$y^5P_3^\circ - e^3D_3$

TABLE B—(*Continued*)

Ref	λ I A	Int	T C	Wave Number Observed	o−c	Desig
V	6880.65	2	V	14529.51	−01	$y^5D_3^\circ - e^5F_2$
V	6875.98	1	IV*	14539.38	00	$c^3F_2 - x^3P_2^\circ$
V	6875.45	1	IV*	14540.50	00	$a^3H_4 - y^5F_3^\circ$
V	6862.481	4n	V	14567.98	+04	$y^5P_3^\circ - e^7G_4$
V	6861.93	2	IV*	14569.15	+02	$a^3P_1 - y^5D_0^\circ$
SS	6860.953	1−	IV*	14571.22	00	$b^3P_2 - y^5P_1^\circ$
V	6860.29	1	IV*	14572.63	00	$b^3F_2 - z^5G_3^\circ$
K	6858.164	40	V	14577.15	−04	$y^3F_3^\circ - e^3D_2$
V	6857.25	4	IV*	14579.09	00	$c^3F_4 - z^1G_4^\circ$
V	6855.74	2	V	14582.30	−07	$y^5P_2^\circ - e^3D_2$
I	6855.176	150	V	14583.50	−01	$y^5P_3^\circ - g^5D_4$
V	6854.82	2	IV*	14584.26	+03	$d^3F_4 - 6_5^\circ$
SS	6851.652	1−	IV*	14591.00	−02	$a^3F_2 - z^5F_2^\circ$
SS	6847.603	1−	V	14599.63	−02	$y^5F_3^\circ - e^3F_2$
SS	6844.683	1−	IIIA?	14605.86	−02 +02	$a^3F_3 - z^5F_4^\circ$ $(b^1D_2 - v^3D_2^\circ)$
I	6843.671	60	V	14608.02	−04	$y^3F_4^\circ - e^3D_3$
V	6842.668	6n	V	14610.16	+03	$y^5P_1^\circ - e^5P_1$
I	6841.349	80	V	14612.97	−01	$y^5P_2^\circ - g^5D_3$
V	6839.828	4	IV*	14616.22	+02	$b^3F_4 - z^5G_4^\circ$
O	6837.00	3	IV*	14622.27	+05	$d^3F_4 - u^3G_4^\circ$
V	6833.24	1	V	14630.32	−04	$y^5P_1^\circ - e^3D_1$
I	6828.610	50	V	14640.24	00	$y^5P_1^\circ - g^5D_2$
SS	*6822.042	1−	V	14654.33	+02 −10	$a^3P_0 - y^5F_1^\circ$ $d^3F_3 - t^5D_2^\circ$
O	6820.43	8n	V	14657.80	−07	$y^5P_1^\circ - e^5P_2$
SS	6819.595	(1)	V	14659.59	+01	$y^5D_4^\circ - e^5F_3$
V	6810.28	20n	V	14679.64	−04	$y^5P_2^\circ - e^5P_3$
L	6806.851	10	IV	14687.03	−01	$a^3G_4 - y^3F_4^\circ$
V	6804.27	3	IV*	14692.61	+03	$d^3F_3 - u^3G_4^\circ$
V	6804.020	5	V	14693.15	+01	$y^3F_2^\circ - g^5D_1$
V	6796.11	2	V	14710.25	+06	$c^3F_3 - v^5F_3^\circ$
V	6793.26	2	V	14716.42	+03	$c^3F_4 - w^5G_4^\circ$
V	6786.88	5	V	14730.25	−05	$y^5D_2^\circ - e^3F_3$
V	6783.71	2	IV*	14737.14	+01	$b^3F_3 - z^5G_3^\circ$
V	*6777.44	1	V	14750.77	−05 +06	$c^3F_2 - x^3P_1^\circ$ $c^3F_3 - v^5P_2^\circ$
L	6752.724	10	V	14804.76	+01	$y^5P_1^\circ - g^5D_1$
I	6750.152	100	III	14810.41	+01	$a^3P_1 - z^3P_1^\circ$
V	6745.11	1	IV*	14821.47	−05	$d^3F_2 - x^1D_2^\circ$
V	6739.54	1	IIIA	14833.72	−03	$a^3F_3 - z^5F_3^\circ$
U	6738.02	4nl	V	14837.07	−08	$y^5P_3^\circ - f^5F_3$
L	6733.164	6	V	14847.77	+02	$y^5P_1^\circ - g^5D_0$
V	6732.06	1	IV*	14850.20	+04	$d^3F_3 - u^3G_3^\circ$
V	6726.668	20n	V	14862.10	−01	$y^5P_2^\circ - e^5P_1$
V	6725.39	2	V	14864.93	−08	$y^5D_4^\circ - e^3F_4$
V	6717.556	3	V	14882.27	−07	$y^5P_2^\circ - e^3D_1$
V	6716.24	3	IV*	14885.18	−02	$d^3F_2 - u^3G_3^\circ$
V	6715.410	5	V	14887.02	−03	$y^3F_3^\circ - g^5D_2$
V	6713.76	3n	V	14890.68	+03	$y^3D_2^\circ - f^5P_2$
V	*6713.14	6d	V	14892.06	+13 −17	$c^3F_3 - x^3P_2^\circ$ $y^5P_2^\circ - g^5D_2$
V	6710.31	2	III?	14898.34	+04	$a^3F_4 - z^5F_5^\circ$
I	6705.117	15n	V	14909.88	+02	$y^5P_2^\circ - e^5P_2$
SS	6704.500	1	V	14911.25	−05	$y^5D_1^\circ - e^3F_2$
L	6703.573	10	IV	14913.31	00	$a^3G_3 - y^3F_3^\circ$
V	6699.14	2	V	14923.18	+04	$d^3F_4 - u^3D_3^\circ$
R	6692.5	(1)		14937.98	−07	$y^5P_3^\circ - f^5F_2$
B	6677.993	600	III	14970.43	00	$a^3G_5 - y^3F_4^\circ$
W	6671.36	(2)		14985.32	+07	$y^5G_5^\circ - h^7D_5$
V	6667.73	(1)		14993.48	−02	$d^3F_3 - u^3D_3^\circ$
SS	6667.455	1−	IV*	14994.09	−08	$a^3H_4 - z^5G_5^\circ$
U	6665.48	(−)		14998.54	−10	$a^3F_3 - z^5F_2^\circ$
B	6663.446	80	III	15003.12	−03	$a^3P_1 - z^3P_0^\circ$
V	6663.26	(1)		15003.53	−05	$y^5P_3^\circ - g^5D_3$

Ref	λ I A	Int	T C	Wave Number Observed	o−c	Desig
V	6653.88	(1)		15024.68	−07	$y^5D_3^\circ - e^3F_3$
SS	6648.121	1−	IIA	15037.70	−08	$a^5F_1 - z^7F_2^\circ$
V	6646.98	(1)		15040.28	−09 −18	$b^3F_2 - z^3G_3^\circ$ $(z^3G_4^\circ - f^7D_4)$
SS	6639.897	2	V	15056.32	−04	$c^3F_4 - v^5F_4^\circ$
SS	6639.717	4	V	15056.73	−01	$y^5P_2^\circ - g^5D_1$
SS	6634.123	4n	V	15069.41	−07	$y^3D_2^\circ - f^3D_3$
K	6633.764	50	V	15070.24	−04	$y^5P_3^\circ - e^5P_3$
M	6633.44	4n	V	15070.98	−08	$y^3D_1^\circ - f^3D_2$
L	6627.558	5	V	15084.36	−01	$y^3F_4^\circ - g^5D_3$
V	6625.04	1	IIA	15090.09	−03	$a^5F_1 - z^7F_1^\circ$
SS	6613.808	1−	IIIA	15115.72	+05	$a^5F_1 - z^7F_0^\circ$
I	6609.116	30	III	15126.45	00	$b^3F_4 - z^3G_4^\circ$
V	6608.03	2	IV*	15128.93	00	$a^3P_2 - y^5D_3^\circ$
V	6604.67	(1)		15136.63	−09	$y^3D_1^\circ - h^5D_1$
V	6597.607	15n	V	15152.83	−06	$y^3D_2^\circ - g^5F_3$
I	6593.878	60	III	15161.40	−01	$a^3H_5 - z^5G_5^\circ$
B	6592.919	300	III	15163.61	+01	$a^3G_4 - y^3F_3^\circ$
V	6591.32	2	V	15167.29	00	$d^3F_4 - t^3D_3^\circ$
V	6581.22	2	III?	15190.56	−02	$a^3F_4 - z^5F_4^\circ$
I	6575.022	30	IV	15204.88	+01	$b^3F_3 - z^3G_3^\circ$
V	6574.238	3	IIA	15206.70	−01	$a^5F_2 - z^7F_2^\circ$
U	6571.22	(1)		15213.68	−11	$b^1D_2 - v^3G_3^\circ$
I	6569.231	50n	V	15218.29	−04	$y^3D_3^\circ - g^5F_4$
U	6556.79	(1)		15247.16	−03	$y^3D_2^\circ - f^5P_1$
W	6552.77	(2)		15256.51	−03	$a^1F_3 - w^1F_3^\circ$
B	6546.245	200	III	15271.72	00	$a^3G_3 - y^3F_2^\circ$
U	6543.98	(1)		15277.01	−08	$z^5G_4^\circ - f^5D_3$
W	6539.72	(2)		15286.96	−06	$b^3G_3 - x^5D_4^\circ$?
V	6533.97	8n	V	15300.41	−04	$y^5P_3^\circ - e^5P_2$
I	6518.376	20	IV	15337.02	00	$b^3P_2 - y^3D_3^\circ$
V	6509.56	(1)		15357.79	+14	$c^3F_4 - w^3G_5^\circ$
V	6498.950	5	IIA	15382.86	−02	$a^5F_3 - z^7F_3^\circ$
K	6496.456	20n	V	15388.77	+02	$y^3D_2^\circ - f^3D_2$
U	6495.779	3	V	15390.37	−03	$y^3D_1^\circ - g^5F_1$
B	6494.985	1000	II	15392.25	00	$a^3H_6 - z^5G_5^\circ$
I	6481.878	20	III	15423.37	−01	$a^3P_2 - y^5D_2^\circ$
I	6475.632	12	IV	15438.25	00	$b^3F_4 - z^3G_3^\circ$
V	6474.61	(1)		15440.69	−05	$b^3D_1 - v^5D_1^\circ$
I	6469.214	15n	V	15453.57	00 −21	$y^3D_1^\circ - f^3D_1$ $(a^3H_6 - z^5G_6^\circ)$
I	6462.731	30	II	15469.07	+01 −03	$a^3H_4 - z^5G_4^\circ$ $(a^5F_4 - z^7F_5^\circ)$
U	6451.587	(2)		15495.79	−11	$b^1G_4 - y^3G_5^\circ$
V	6450.99	(1)		15497.22	+04	$y^5G_4^\circ - g^5G_4$
U	6438.775	(1)		15526.62	−05	$z^3G_4^\circ - e^3D_3$
V	6436.43	(1)		15532.28	−05	$c^3F_2 - v^3D_1^\circ$
B	6430.851	300	II	15545.75	00	$a^5P_3 - y^5D_4^\circ$
U	6428.793	(1)		15550.73	+02	$z^5G_4^\circ - f^7D_4$
B	6421.355	200	II	15568.74	−01	$a^3P_2 - z^3P_2^\circ$
K	6419.982	30n	V	15572.07	−09	$y^3D_3^\circ - f^3D_3$
I	6411.658	400	IV	15592.29	−01	$z^5P_2^\circ - e^5D_3$
SS	6411.125	1n	V	15593.59	−05	$y^3D_3^\circ - f^5G_4$
I	6408.031	60	V	15601.11	−02	$z^5P_1^\circ - e^5D_2$
W	6406.42	(1)		15605.04	+07	$X_2 - u^3F_3^\circ$?
W	6402.4	(1)		15614.8	0	$y^5G_2^\circ - g^5G_2$
U	6400.318	(50)	IA	15619.91	00	$a^5F_4 - z^7F_4^\circ$
I	6400.010	800	IV	15620.67	00	$z^5P_3^\circ - e^5D_4$
B	6393.605	400	II	15636.31	00	$a^3H_5 - z^5G_4^\circ$
U	6392.547	(1)		15638.90	−01	$a^3P_2 - y^5D_1^\circ$
I	6380.748	3	V	15667.82	−03	$c^3F_2 - w^3F_2^\circ$
V	6364.717	(1)		15707.28	−01	$d^3F_3 - t^3D_2^\circ$
V	6364.384	(1)		15708.11	+02	$y^3D_2^\circ - g^5F_1$
V	6362.889	(2)		15711.80	+03	$c^3F_2 - z^1D_2^\circ$
I	6358.692	3	IA	15722.17	+02	$a^5F_5 - z^7F_6^\circ$

TABLE B—(*Continued*)

Ref	λ I A	Int	T C	Wave Number		Desig
				Observed	o−c	
U	6356.293	(1)		15728.10	+12	$b^3F_2 - y^5P_3^\circ$
I	6355.038	4	III	15731.21	00	$b^3P_1 - y^3D_2^\circ$
I	6344.154	2	III	15758.19	−01	$a^3H_5 - z^3G_5^\circ$
V	6338.896	(1n)		15771.27	+01	$y^3D_2^\circ - f^3D_1$
K	6336.835	12	V	15776.40	−01	$z^5P_1^\circ - e^5D_1$
B	6335.335	10	III	15780.13	00	$a^5P_2 - y^5D_3^\circ$
U	6330.856	(1n)		15791.29	−05	$y^3D_3^\circ - h^5D_2$
I	6322.693	5	III	15811.74	+06	$b^3F_3 - y^3F_4^\circ$
B	6318.022	10	III	15823.37	00	$a^3H_4 - z^5G_3^\circ$
V	6315.814	(2)		15828.90	+01	$c^3F_4 - y^1G_4^\circ$
J	6315.316	(3)		15830.15	−02	$c^3F_3 - w^3F_3^\circ$
V	6311.506	(1)		15839.71	00	$b^3P_2 - y^3D_2^\circ$
U	6310.543	(1)		15842.13	−05	$b^3G_5 - x^5D_4^\circ$
V	6303.46	(1n)		15859.93	00	$z^5G_6^\circ - e^5G_5$
K	6302.507	6	V	15862.32	−02	$z^5P_1^\circ - e^5D_0$
K	6301.515	15	IV	15864.82	−03	$z^5P_2^\circ - e^5D_2$
I	6297.800	5	III	15874.18	00	$a^5P_1 - y^5D_2^\circ$
I	6290.968	3n	V	15891.42	−02	$y^3D_3^\circ - f^3D_2$
I	6280.625	2	IA	15917.59	−01	$a^5F_5 - z^7F_5^\circ$
U	6271.289	(1)		15941.29	00	$z^5F_5^\circ - e^7D_5$
J	6270.238	(2)		15943.96	−01	$b^3P_0 - y^3D_1^\circ$
U	6267.845	(1)		15950.05	−02	$b^1D_2 - z^1F_3^\circ$
B	6265.140	6	III	15956.93	00	$a^5P_3 - y^5D_3^\circ$
I	6256.370	4	III	15979.30	−01	$a^3H_4 - z^3G_4^\circ$
I	6254.262	6	III	15984.69	+01	$a^3P_2 - z^3P_1^\circ$
B	6252.561	20	III	15989.03	−01	$a^3H_6 - z^3G_5^\circ$
K	6246.334	15	V	16004.97	−02	$z^5P_3^\circ - e^5D_3$
V	6245.84	(1)		16006.24	+07	$y^7P_4^\circ - 1_5$
I	6240.656	(2)		16019.53	−02	$a^5P_1 - z^3P_2^\circ$
U	6240.266	(1)		16020.54	+14	$c^3F_3 - w^3F_2^\circ$
Q	6232.735	(−)		16039.89	−04	$z^5F_3^\circ - e^7D_3$
K	6232.661	5	V	16040.08	−04	$z^5P_2^\circ - e^5D_1$
B	6230.728	25	III	16045.06	00	$b^3F_4 - y^3F_4^\circ$
V	6229.234	(1)		16048.91	+01	$b^3P_1 - y^3D_1^\circ$
U	6226.756	(1)		16055.29	−04	$z^3D_3^\circ - e^5F_4$
U	6221.661	(−)		16068.44	+03	$a^5F_5 - z^7F_4^\circ$
U	6221.405	(1)		16069.10	−06	$z^3D_2^\circ - e^5F_3$
U	6220.774	(1)		16070.73	+03	$z^3F_4^\circ - e^5F_4$
I	6219.290	6	III	16074.57	−01	$a^5P_2 - y^5D_2^\circ$
U	6217.283	(1)		16079.76	−05	$X_3 - v^1G_4^\circ$
J	6215.152	(2)		16085.27	−01	$c^3F_2 - v^3G_3^\circ$
I	6213.438	5	III	16089.71	−01	$a^5P_1 - y^5D_1^\circ$
U	6212.045	(1)		16093.32	+16	$z^5G_4^\circ - g^5D_4$
J	6200.323	4	IV	16123.74	−01	$b^3F_2 - y^3F_3^\circ$
U	6199.475	(1)		16125.95	+10	$b^3F_4 - y^5P_3^\circ$
B	6191.562	20	II	16146.56	00	$a^3H_5 - z^3G_4^\circ$
V	6188.037	(2ld)		16155.75	−13	$z^3F_3^\circ - e^3F_4$
U	6180.212	(2)		16176.21	−01	$a^3G_4 - y^3D_3^\circ$
J	6173.343	3	III	16194.21	00	$a^5P_1 - y^5D_0^\circ$
K	6170.492	4n	V	16201.69	+04	$y^3D_2^\circ - e^3P_2$
J	6165.366	(2)		16215.16	+02	$c^3F_3 - v^3G_4^\circ$
U	6163.544	(1)		16219.95	00	$a^5P_2 - z^3P_2^\circ$
U	6159.409	(1n)		16230.84	−10	$y^3F_3^\circ - g^5F_4$
J	6157.734	4	V	16235.26	00	$c^3F_4 - w^3F_4^\circ$
L	6151.624	(2)		16251.38	00	$a^5P_3 - y^5D_2^\circ$
V	6147.85	(−)		16261.36	−04	$c^3F_4 - v^3D_3^\circ$
K	6141.734	4	V	16277.54	00	$z^5P_3^\circ - e^5D_2$
B	6137.696	18	III	16288.26	+01	$b^3F_3 - y^3F_3^\circ$
J	6136.999	(2)		16290.11	00	$a^5P_2 - y^5D_1^\circ$
B	6136.620	20	III	16291.12	00	$a^3H_4 - z^3G_3^\circ$
SS	6130.358	(1)		16307.75	+02	$a^3D_3 - x^5P_3^\circ$
J	*6127.913	(2)		16314.26	−02 +18	$c^3F_3 - y^3H_4^\circ$ $y^5F_2^\circ - e^7P_2$
U	6109.308	(1)		16363.95	−07	$b^3H_4 - z^5H_5^\circ$

Ref	λ I A	Int	T C	Wave Number		Desig
				Observed	o−c	
SS	6107.104	(1)		16369.85	−04	$y^5F_3^\circ - f^5D_2$
K	6103.190	3	V	16380.35	−01	$y^3D_1^\circ - e^3P_1$
K	6102.178	5	V	16383.07	−01	$y^3D_1^\circ - f^3F_2$
SS	*6100.284	(1)		16388.15	+02 −14	$y^5P_3^\circ - h^5D_4$ $y^5P_2^\circ - h^5D_3$
V	6096.689	(1)		16397.82	−06	$z^3F_2^\circ - e^3F_3$
U	6094.419	(1)		16403.92	−15	$y^3F_2^\circ - f^5P_1$
V	6093.66	(1)		16405.97	+02	$y^3F_3^\circ - f^5P_2$
L	6089.566	(1)		16416.99	−10	$a^1F_3 - v^1G_4^\circ$
V	6085.267	(1)		16428.59	−02	$a^3G_3 - y^3D_2^\circ$
U	6082.709	(1)		16435.50	+02	$a^5P_1 - z^3P_1^\circ$
V	6079.02	(1)		16445.48	−05	$y^3F_2^\circ - h^5D_2$
K	6078.496	4n	V	16446.90	−01	$y^3D_2^\circ - f^3F_3$
B	6065.487	15	III	16482.17	+01	$b^3F_2 - y^3F_2^\circ$
V	6062.89	(1)		16489.23	−10	$a^5P_3 - y^5F_4^\circ$
K	6055.987	4	V	16508.02	+01	$y^3D_3^\circ - f^3F_4$
U	6054.100	(2)		16513.17	−06	$z^5G_4^\circ - g^5D_3$
U	6043.738	(1)		16541.48	−19	$b^3D_3 - x^5G_3^\circ$?
SS	6034.057	(2)		16568.02	−04	$z^5G_5^\circ - g^5D_4$
V	6032.67	(1)		16571.83	+07	$y^5F_4^\circ - e^7P_3$
B	6027.057	4	V	16587.26	+02	$c^3F_4 - v^3G_5^\circ$
K	6024.066	15	V	16595.50	−03	$y^3F_4^\circ - f^5G_5$
W	*6021.82	(2n)		16601.69	+06 +03 −07	$y^5F_2^\circ - e^7F_3$ $y^5F_2^\circ - e^5S_2$ $a^5P_2 - y^5F_3^\circ$
K	6020.173	10n	V	16606.23	−02	$y^3F_3^\circ - f^5G_4$
W	6016.66	(2)		16615.93	−20	$a^1D_2 - x^3D_3^\circ$
K	6008.577	9	V	16638.28	−05	$z^3D_3^\circ - e^3F_4$
K	6007.961	(3n)		16639.98	−02	$y^3F_2^\circ - f^5G_3$
V	*6005.53	(1)		16646.72	+06 −04	$b^3F_3 - y^3F_2^\circ$ $y^5F_5^\circ - e^7F_6$
K	6003.033	8	V	16653.64	−06	$z^3F_4^\circ - e^3F_4$
U	5997.805	(1)		16668.16	−03	$y^3F_3^\circ - g^5F_3$
K	5987.057	6	V	16698.08	+03	$y^3D_2^\circ - e^3P_1$
K	5984.805	8	IV	16704.37	+03	$y^3D_3^\circ - e^3P_2$
K	5983.704	6	V	16707.44	−07	$y^3F_4^\circ - g^5F_4$
K	5976.799	5	V	16726.74	−06	$z^3F_3^\circ - e^3F_3$
J	*5975.355	4	V	16730.78	00 +01	$y^3D_1^\circ - e^3P_0$ $c^3F_4 - y^3H_5^\circ$
U	5969.554	(2)		16747.04	+01	$y^5F_2^\circ - e^3D_3$
V	5963.25	(1)		16764.74	−02	$a^5P_1 - y^5F_1^\circ$
U	5959.878	(1)		16774.23	+14	$c^3F_3 - w^3P_2^\circ$
SS	*5958.246	(2)		16778.82	−04 −06	$a^5F_3 - z^7P_2^\circ$ $y^5P_3^\circ - h^5D_3$
J	5956.702	(3)		16783.17	−02	$a^5F_5 - z^7P_4^\circ$
U	5955.682	(1)		16786.05	+05	$z^3P_1^\circ - e^5S_2$
V	5952.749	3	V	16794.32	−09	$z^3F_2^\circ - e^3F_2$
V	*5949.35	(2)		16803.91	−19 −04	$a^5F_4 - z^7P_3^\circ$ $y^3F_3^\circ - h^5D_2$
SS	5947.517	(1)		16809.09	−04	$y^5P_2^\circ - h^5D_2$
V	5940.972	(2)		16827.61	+09	$y^5F_5^\circ - e^5G_6$
K	5934.658	5	V	16845.51	−01	$z^3D_2^\circ - e^3F_3$
K	5930.173	8	V	16858.25	00	$y^3F_2^\circ - e^3G_3$
U	5929.700	(1)		16859.60	−08	$y^3F_4^\circ - h^5D_3$
V	5927.798	(2wd)		16865.01	+04	$y^3F_2^\circ - g^5F_1$
V	5920.520	(2)		16885.74	+03	$b^3H_6 - z^5H_5^\circ$
U	5919.024	(1)		16890.01	−27	$y^5F_3^\circ - e^5G_3$?
V	5916.250	(3)		16897.93	+01	$a^3H_4 - y^3F_4^\circ$
V	*5914.16	8	V	16903.90	+10 −15	$y^3F_3^\circ - e^3G_4$ $y^3F_3^\circ - f^3D_2$
U	5909.986	(3)		16915.84	−01	$z^5D_4^\circ - e^7D_5$
U	5908.252	(2)		16920.80	−15	$z^7D_1^\circ - d^3F_2$
K	5905.673	3n	V	16928.19	+05	$y^3F_2^\circ - f^3D_1$
U	5902.527	(1)		16937.21	−09	$d^3F_4 - t^3G_5^\circ$
U	5898.212	(1)		16949.60	+01	$y^3D_3^\circ - f^3F_3$

TABLE B—(*Continued*)

Ref	λ I A	Int	T C	Wave Number Observed	o−c	Desig
U	5895.007	(1)		16958.82	00	$d^3F_4 - 11_3^\circ$
W	5892.71	(2)		16965.43	−06	$y^5F_3^\circ - e^3D_3$
SS	5891.896	(1)		16967.77	−01	$d^3F_4 - 12_5^\circ$
U	5891.12	(1)		16970.01	−02	$b^3H_5 - z^5H_6^\circ$
K	5883.838	4	V	16991.01	−06	$z^3D_1^\circ - e^3F_2$
V	5880.00	(2wd)		17002.10	+08	$y^5P_3^\circ - f^5G_4$
U	5877.770	(1)		17008.55	+06	$y^5F_5^\circ - e^5G_5$
U	5873.219	(2)		17021.73	00	$y^5F_3^\circ - g^5D_4$
U	5871.289	(1)		17027.33	+04	$y^5D_3^\circ - f^5D_3$
V	5871.04	(1)		17028.05	−04	$z^7D_2^\circ - d^3F_2$
SS	5864.252	(1)		17047.76	−03	$y^5F_1^\circ - e^3D_2$
K	5862.357	8	V	17053.27	−01	$y^3F_4^\circ - e^3G_5$
K	5859.608	5	V	17061.27	−07	$y^3F_4^\circ - f^3D_3$
U	5859.197	(1)		17062.47	+11	$y^5F_1^\circ - f^5F_1$
U	5856.081	(2)		17071.55	+04	$b^1D_2 - y^1D_2^\circ$
U	5855.130	(1)		17074.32	−11	$y^3F_3^\circ - e^5H_4$
U	5853.195	(1)		17079.96	−12	$a^3F_4 - z^5P_3^\circ$
W	5852.19	(2n)		17082.90	+08	$y^3F_4^\circ - f^5G_4$
W	*5848.09	(2n)		17094.87	−05	$z^5D_2^\circ - e^7D_3$
					+12	$y^3F_3^\circ - g^5F_2$
U	5844.879	(1)		17104.26	+10	$y^5D_3^\circ - e^7P_3$
U	5838.418	(1)		17123.19	−14	$z^3F_3^\circ - e^3F_2$
U	5837.703	(1)		17125.29	−01	$b^1D_2 - x^1D_2^\circ$
V	5816.36	(3d)		17188.13	+03	$y^3F_4^\circ - e^5H_5$
V	5815.16	(1)		17191.68	+16	$y^5D_3^\circ - f^5D_2$?
U	5814.816	(1)		17192.69	−03	$y^5F_2^\circ - e^3D_2$
U	5811.936	(1)		17201.21	−03	$c^3F_3 - x^1G_4^\circ$
U	5809.245	(2)		17209.18	−08	$z^3D_3^\circ - e^3F_3$
V	5806.727	(2)		17216.64	−03	$y^3F_3^\circ - e^3G_3$
SS	5805.774	(1)		17219.47	−03	$x^5F_4^\circ - i^5D_3$
U	5804.478	(1)		17223.31	−03	$y^5F_2^\circ - g^5D_3$
U	5804.072	(1)		17224.52	−10	$z^3F_4^\circ - e^3F_3$
V	5798.194	(2)		17241.98	−07	$z^3D_2^\circ - e^3F_2$
V	5793.932	(2)		17254.66	−06	$y^5F_4^\circ - e^3D_3$
V	5791.044	(2)		17263.27	−05	$z^5D_4^\circ - e^7D_4$
V	5784.69	(1)		17282.23	−08	$z^5F_3^\circ - e^5D_4$
V	*5780.83	(1)		17293.77	−05	$z^5D_2^\circ - e^7D_2$
					+08	$z^3G_4^\circ - g^5F_5$
					−05	$b^1G_4 - x^3F_3^\circ$
V	5780.621	(2)		17294.40	−04	$z^5D_3^\circ - e^7D_3$
V	5778.47	(1)		17300.83	−03	$b^3F_3 - y^3D_3^\circ$
J	5775.090	(5)		17310.96	00	$y^5F_4^\circ - g^5D_4$
SS	5769.336	(1)		17328.22	−09	$y^3F_3^\circ - e^5H_3$
K	5762.992	10	V	17347.30	00	$z^3P_2^\circ - e^3D_3$
V	5762.434	(1)		17348.98	−01	$b^3D_3 - u^5D_4^\circ$
U	5761.246	(1)		17352.56	+04	$b^3D_1 - y^3P_0^\circ$
V	5760.351	(1)		17355.25	−03	$b^3D_3 - y^3P_2^\circ$
SS	*5759.550	(2)		17357.67	+02	$y^5F_1^\circ - g^5D_2$
					+05	$y^5P_3^\circ - g^7D_4$
U	5759.270	(1)		17358.51	−02	$y^3F_2^\circ - e^3P_2$
V	5754.41	(1)		17373.17	00	$b^3D_3 - u^5D_3^\circ$
J	5753.136	5	V	17377.02	−05	$z^3P_1^\circ - e^3D_2$
J	5752.043	(2)		17380.32	−05	$y^3F_4^\circ - e^3G_4$
U	5747.959	(1)		17392.67	−01	$y^3F_3^\circ - e^3H_4$
SS	5742.972	(1)		17407.77	−05	$y^5F_5^\circ - f^5F_5$
V	5741.861	(2)		17411.14	−05	$y^5F_3^\circ - e^3D_2$
J	5731.771	(3)		17441.79	−01	$y^5F_3^\circ - g^5D_3$
U	5727.75	(1)		17454.03	−10	$y^5P_2^\circ - g^7D_2$
U	5724.445	(1)		17464.11	+04	$z^3P_0^\circ - e^5P_1$
SS	5723.673	(1)		17466.47	−02	$z^3G_3^\circ - h^5D_3$
W	5720.8	(1n)		17475.2	−1	$y^7P_4^\circ - h^7D_5$
L	5717.845	(3)		17484.27	−03	$z^3P_0^\circ - e^3D_1$
V	*5715.107	(1)		17492.64	−06	$y^5F_2^\circ - e^3D_1$
					−03	$y^5D_2^\circ - e^3D_3$
U	5712.145	(2)		17501.71	−03	$z^5F_2^\circ - e^5D_3$
V	5711.867	(2)		17502.57	−01	$y^5F_2^\circ - g^5D_2$
K	5709.378	10	IV	17510.20	+03	$z^5F_4^\circ - e^5D_4$
V	5708.109	(1)		17514.09	−05	$z^3G_4^\circ - f^5G_5$
U	5707.055	(1)		17517.33	−02	$b^3D_3 - x^3F_4^\circ$
U	5705.992	(2)		17520.59	−03	$y^3F_3^\circ - f^3F_4$
U	5705.475	(1)		17522.18	+02	$y^5F_1^\circ - g^5D_1$
U	5702.434	(1)		17531.52	−28	$b^3D_2 - u^5D_2^\circ$?
J	5701.553	7	III?	17534.23	−01	$b^3F_4 - y^3D_3^\circ$
W	5698.37	(2)		17544.02	+02	$b^1D_2 - t^3D_1^\circ$
W	5698.05	(1)		17545.01	−10	$b^3D_2 - y^3P_1^\circ$
V	5691.509	(1)		17565.17	+01	$y^5F_1^\circ - g^5D_0$
V	5686.532	(3)		17580.54	−02	$y^3F_4^\circ - e^3H_5$
W	5680.26	(1)		17599.96	−01	$c^3F_2 - v^3F_3^\circ$
V	5679.023	(2)		17603.79	00	$y^3F_2^\circ - f^3F_3$
SS	5672.273	(1)		17624.74	+03	$d^3F_3 - t^3G_3^\circ$
U	*5666.837	(1)		17641.64	−11	$y^5D_3^\circ - e^5S_2$
					−09	$y^5D_3^\circ - e^7F_3$
U	5662.938	(1)		17653.79	−02	$b^1G_4 - z^1G_4^\circ$
B	5662.525	6	V	17655.08	−01	$y^5F_5^\circ - g^5D_4$
W	5661.36	(1)		17658.71	+02	$z^3P_0^\circ - g^5D_1$
W	5660.79	(1)		17660.49	+03	$b^3D_2 - w^3D_3^\circ$
B	5658.826	10	IV	17666.62	−01	$z^5F_3^\circ - e^5D_3$
U	5658.537	(1)		17667.52	00	$z^5F_1^\circ - e^5D_2$
V	*5655.506	4	V	17676.99	−06	$z^3P_1^\circ - e^3D_1$
					−04	$x^5F_4^\circ - g^5G_5$
V	5655.179	(2)		17678.01	−04	$x^5F_3^\circ - g^5G_4$
U	5653.889	(1w)		17682.05	+01	$z^3G_5^\circ - g^5F_5$
U	5652.317	(1)		17686.96	+03	$z^3P_1^\circ - g^5D_2$
U	5650.721	(1)		17691.96	−07	$x^5F_2^\circ - g^5G_3$
V	5650.01	(1)		17694.18	−01	$x^5F_1^\circ - g^5G_2$
V	5649.66	(1)		17695.28	−11	$a^1I_6 - z^3H_5^\circ$
U	5641.453	(2)		17721.02	−02	$y^5F_3^\circ - g^5D_2$
W	5640.46	(1n)		17724.14	+06	$y^5P_3^\circ - e^5H_3$
I	5638.266	3	V	17731.04	00	$y^5F_4^\circ - g^5D_3$
U	5636.693	(1)		17735.99	−01	$b^3D_2 - x^3F_3^\circ$
U	5635.845	(1)		17738.66	−01	$y^5F_3^\circ - e^5P_2$
V	5633.970	(2)		17744.56	+05	$x^5F_5^\circ - g^5G_6$
U	5631.72	(2)		17751.65	+09	$z^3G_3^\circ - g^5F_3$
B	5624.549	10	IV	17774.28	00	$z^5F_2^\circ - e^5D_2$
V	5624.056	(1)		17775.84	−05	$z^3G_5^\circ - h^5D_4$
V	5620.527	(1)		17787.00	−12	$y^5D_3^\circ - e^3D_3$
W	*5620.04	(1)		17788.54	+09	$y^5P_3^\circ - e^3H_4$
					−01	$c^3F_3 - v^3F_2^\circ$
V	5619.60	(1)		17789.94	−03	$z^3G_5^\circ - f^5G_6$
U	5618.633	(1)		17793.00	00	$z^3P_2^\circ - e^3D_2$
W	5617.22	(1)		17797.47	−05	$a^3D_3 - w^5F_4^\circ$
B	5615.652	50	IV	17802.44	−01	$z^5F_5^\circ - e^5D_4$
U	5615.301	(2)		17803.55	00	$b^3F_3 - y^3D_2^\circ$
J	5602.955	10	IV	17842.78	−01	$z^5F_1^\circ - e^5D_1$
U	5602.770	(2)		17843.37	+01	$y^5D_3^\circ - g^5D_4$
V	*5600.242	(1)		17851.43	−01	$z^3P_1^\circ - g^5D_1$
					−04	$b^3D_1 - u^5D_0^\circ$
J	5598.303	4	IV?	17857.61	−05	$y^3F_2^\circ - f^3F_2$
U	5594.670	(2)		17869.21	−04	$y^3F_4^\circ - e^3H_4$
U	5587.576	(1)		17891.89	+02	$c^3F_3 - v^3F_4^\circ$
B	5586.763	40	IV	17894.50	00	$z^5F_4^\circ - e^5D_3$
U	5584.766	(1)		17900.90	+03	$a^1H_5 - u^5D_4^\circ$
J	5576.097	10	IV	17928.73	+01	$z^5F_1^\circ - e^5D_0$
U	5573.105	(1)		17938.35	−02	$y^5D_2^\circ - e^3D_2$
B	5572.849	30	IV	17939.17	−01	$z^5F_3^\circ - e^5D_2$
B	5569.625	20	IV	17949.56	00	$z^5F_2^\circ - e^5D_1$
U	5568.81	(1)		17952.18	+13	$b^3D_1 - w^3D_1^\circ$?
U	5567.403	(2)		17956.72	−02	$b^3F_2 - y^3D_1^\circ$
I	5565.708	4	V	17962.19	−01	$y^3F_3^\circ - f^3F_3$
I	5563.604	3	V	17968.98	00	$y^5D_2^\circ - g^5D_3$

TABLE B—(*Continued*)

Ref	λ I A	Int	T C	Wave Number Observed	o−c	Desig
V	*5562.712	(2)		17971.86	−04	$z^3G^\circ_4 - e^3G_5$
					−07	$a^3D_1 - w^5F^\circ_1$
V	5560.230	(1)		17979.89	−07	$z^3G^\circ_4 - f^3D_3$
U	5557.962	(1)		17987.22	+05	$z^3G^\circ_3 - e^3G_4$
I	5554.895	4	V	17997.16	−03	$y^3F^\circ_4 - f^3F_4$
V	5553.586	(1)		18001.40	−03	$z^3G^\circ_4 - f^5G_4$
U	5549.94	(2)		18013.22	00	$b^1G_4 - x^3G^\circ_4$
W	5547.00	(2)		18022.77	−04	$y^5D^\circ_1 - e^3D_1$
U	5546.486	(1)		18024.44	+05	$z^5G^\circ_4 - f^5G_5$
J	5543.930	(2)		18032.75	+06	$y^5D^\circ_1 - g^5D_2$
V	5543.184	(2)		18035.18	−12	$b^1G_4 - x^3G^\circ_3$
U	5539.831	(1)		18046.09	−02	$b^1D_2 - t^3D^\circ_2$
U	5539.27	(1)		18047.92	+06	$b^3D_3 - 1^\circ_2$
V	*5538.54	(1)		18050.30	−02	$y^5D^\circ_1 - e^5P_2$
					−05	$a^1I_6 - w^5G^\circ_6$
W	5737.71?	(1)		18053.00	+09	$y^5D^\circ_4 - e^7F_3$
J	*5535.419	(2)		18060.48	+03	$a^3D_3 - w^5F^\circ_2$
					+01	$c^3F_2 - u^3G^\circ_3$
U	*5534.64	(1)		18063.02	00	$y^5D^\circ_3 - e^7S_3$
					+08	$b^3D_2 - 1^\circ_2$
U	5532.742	(1)		18069.21	−02	$a^1H_5 - x^3F^\circ_4$
U	5531.949	(1)		18071.80	+04	$x^5D^\circ_4 - i^5D_4$
U	5529.13	(2)		18081.02	+11	$b^3D_3 - z^1G^\circ_4$
V	5525.552	(3)		18092.73	+02	$y^5D^\circ_0 - g^5D_1$
					−24	$(z^3P^\circ_2 - e^3D_1)$
SS	5524.273	(1)		18096.92	−08	$y^5D^\circ_3 - f^5F_3$
V	5522.46	(2)		18102.86	00	$z^3P^\circ_2 - g^5D_2$
SS	5521.141	(1)		18107.18	−01	$a^1I_6 - w^5G^\circ_5$
W	5517.08	(1n)		18120.51	+02	$z^3P^\circ_2 - e^5P_2$
V	5512.277	(1)		18136.30	−07	$z^5G^\circ_4 - g^5F_4$
B	5506.782	18	IB	18154.40	00	$a^5F_2 - z^5D^\circ_3$
T	5505.893	(−)		18157.33	−04	$z^5G^\circ_3 - f^5G_4$
B	5501.469	12	IB	18171.93	00	$a^5F_3 - z^5D^\circ_4$
B	5497.519	15	IB	18184.98	00	$a^5F_1 - z^5D^\circ_2$
U	5494.462	(1)		18195.10	+01	$c^3F_4 - x^3H^\circ_5$
T	*5493.850	(0)		18197.13	−07	$y^5D^\circ_1 - g^5D_1$
					+17	$c^3P_2 - x^5P^\circ_3$
U	5493.511	(1)		18198.25	−06	$y^5D^\circ_4 - e^3D_3$
W	5491.84	(2)		18203.79	−02	$c^3F_2 - u^3D^\circ_3$
K	5487.747	(8)		18217.37	+08	$c^3F_3 - t^5D^\circ_2$
U	5487.144	(1)		18219.37	+06	$z^5G^\circ_3 - g^5F_3$
U	5483.116	(1)		18232.75	−07	$y^5D^\circ_3 - e^3D_2$
T	5481.451	(3)		18238.29	−05	$y^5D^\circ_2 - e^3D_1$
U	5481.256	(2)		18238.94	−04	$y^5D^\circ_4 - e^7G_4$
U	5480.873	(2)		18240.21	+01	$y^5D^\circ_1 - g^5D_0$
U	5478.463	(1)		18248.24	+02	$y^5D^\circ_2 - g^5D_2$
J	5476.571	10	IV	18254.54	00	$y^5D^\circ_4 - g^5D_4$
J	5476.298	(2)		18255.45	+01	$c^3F_3 - u^3G^\circ_4$
J	5473.908	(3)		18263.42	−01	$y^5D^\circ_3 - g^5D_3$
U	5472.729	(1)		18267.36	−01	$z^3P^\circ_2 - g^5D_1$
W	5470.17	(1)		18275.90	−16	$z^5G^\circ_2 - h^5D_1$
V	*5466.993	(1)		18286.52	+02	$z^5P^\circ_2 - e^5F_3$
					−04	$a^1H_5 - z^3H^\circ_4$
J	5466.404	(3)		18288.49	−05	$z^5G^\circ_4 - h^5D_3$
W	5465.1	(1)		18292.9	−2	$a^1I_6 - v^5F^\circ_5$
V	5464.286	(1)		18295.58	+03	$c^3F_3 - y^1D^\circ_2$
J	5463.282	10n	V	18298.94	−04	$z^3G^\circ_4 - e^3G_4$
J	5462.970	(2)		18299.99	−05	$z^3G^\circ_3 - e^3G_3$
U	5461.553	(1n)		18304.73	−03	$z^5G^\circ_2 - f^5G_3$
U	5460.909	(1)		18306.90	−09	$c^3P_1 - x^5P^\circ_1$
U	5456.468	(1)		18321.80	+17	$z^5P^\circ_3 - e^5F_4$
B	5455.613	40	IB	18324.67	00	$a^5F_1 - z^5D^\circ_1$
K	5455.433	(5)		18325.27	+04	$z^5G^\circ_6 - f^5G_6$
U	5452.119	(1)		18336.41	−10	$b^3D_2 - w^5G^\circ_3$

Ref	λ I A	Int	T C	Wave Number Observed	o−c	Desig
B	5446.920	40	IB	18353.91	00	$a^5F_2 - z^5D^\circ_2$
					−16	$(a^3F_2 - z^3D^\circ_3)$
J	5445.045	15n	V	18360.23	−02	$z^3G^\circ_5 - e^3G_5$
U	5441.321	(1)		18372.80	+12	$z^5G^\circ_5 - h^5D_4$
V	5436.594	(2)		18388.77	−01	$a^3P_2 - y^5P^\circ_3$
U	5436.299	(1)		18389.77	−01	$z^3G^\circ_5 - f^5G_4$
B	5434.527	30	IB	18395.77	00	$a^5F_1 - z^5D^\circ_0$
U	5432.950	(2n)		18401.11	+01	$z^5G^\circ_2 - g^5F_2$
B	5429.699	40	IB	18412.12	00	$a^5F_3 - z^5D^\circ_3$
I	5424.072	45n	V	18431.22	00	$z^5G^\circ_6 - e^5H_7$
U	5417.045	(1)		18455.13	−04	$z^5G^\circ_3 - f^3D_2$
I	5415.201	35n	V	18461.42	−02	$z^3G^\circ_5 - e^3H_6$
I	5410.913	15n	V	18476.05	00	$z^3G^\circ_3 - e^3H_4$
V	5409.125	(1)		18482.15	+01	$z^5G^\circ_4 - e^3G_5$
B	5405.778	40	IB	18493.60	00	$a^5F_2 - z^5D^\circ_1$
I	5404.144	30n	V	18499.19	+02	$z^3G^\circ_4 - e^3H_5$
					−10	$(z^5G^\circ_5 - f^5G_5)$
U	5403.819	(1)		18500.30	+02	$c^3F_4 - u^3G^\circ_5$
J	5400.509	(5)		18511.64	−04	$z^5G^\circ_4 - f^5G_4$
U	5398.280	(1)		18519.28	−02	$z^5G^\circ_2 - f^5G_2$
U	5397.616	(1)		18521.56	−02	$a^1I_6 - x^3G^\circ_5$
B	5397.131	40	IB	18523.23	00	$a^5F_4 - z^5D^\circ_4$
W	5395.25	(1n)		18529.69	−05	$z^5G^\circ_2 - g^5F_1$
T	5394.682	(−)		18531.63	00	$c^3F_2 - u^3D^\circ_2$
I	5393.174	10	IV	18536.82	00	$z^5D^\circ_3 - e^5D_4$
U	5391.470	(1)		18542.67	−01	$y^5D^\circ_3 - g^5D_2$
K	5389.461	(5)		18549.59	+06	$z^5G^\circ_3 - f^5G_3$
R	5387.51	3		18556.30	−06	$c^3F_3 - u^3D^\circ_3$
T	5386.958	(1)		18558.20	+02	$b^3D_3 - v^5F^\circ_4$
V	5386.341	(1)		18560.33	+02	$y^5D^\circ_3 - e^5P_2$
I	5383.374	35n	V	18570.56	−01	$z^5G^\circ_5 - e^5H_6$
T	5382.750	(−)		18572.72	+08	$a^1D_2 - u^5D^\circ_1$
J	5379.580	(2)		18583.66	−01	$b^1G_4 - z^1H^\circ_5$
U	5376.849	(2)		18593.10	−02	$b^1D_2 - z^1P^\circ_1$
V	5373.704	(1)		18603.98	−01	$z^3G^\circ_3 - f^3F_4$
B	5371.493	50	IB	18611.64	00	$a^5F_3 - z^5D^\circ_2$
					−21	$(z^3G^\circ_4 - e^3G_3)$
I	5369.965	25n	V	18616.93	−03	$z^5G^\circ_4 - e^5H_5$
I	5367.470	20n	V	18625.59	+04	$z^5G^\circ_3 - e^5H_4$
J	5365.403	3	V	18632.76	−03	$a^1H_5 - z^1G^\circ_4$
I	5364.874	15n	V	18634.60	−06	$z^5G^\circ_2 - e^5H_3$
U	5361.637	(1)		18645.85	−02	$z^5G^\circ_3 - g^5F_2$
U	5353.389	(2)		18674.58	−04	$y^5D^\circ_4 - g^5D_3$
T	5349.742	(3)		18687.31	−02	$z^3G^\circ_5 - e^3G_4$
B	5341.026	20	II	18717.80	00	$a^3F_2 - z^3D^\circ_2$
I	5339.935	12	V	18721.63	00	$z^5D^\circ_2 - e^5D_3$
J	5332.903	4	IB?	18746.31	−01	$a^3F_3 - z^3F^\circ_4$
U	5332.681	(1)		18747.09	−04	$c^3F_2 - u^3D^\circ_1$
J	5329.994	(2)		18756,54	−01	$c^3F_4 - 6^\circ_5$
B	5328.534	15	II	18761.68	−01	$a^3F_3 - z^3D^\circ_3$
I	5328.042	50	IB	18763.42	00	$a^5F_4 - z^5D^\circ_3$
T	5326.793	(−)		18767.82	+03	$z^5G^\circ_3 - e^3G_3$
V	*5326.154	(1)		18770.07	−02	$a^1H_5 - w^5G^\circ_4$
					−07	$b^3G_3 - z^5H^\circ_4$?
I	5324.185	30	IV	18777.01	00	$z^5D^\circ_4 - e^5D_4$
SS	5323.510	(1)		18779.39	+01	$a^3P_2 - y^5P^\circ_2$
U	5322.054	(2)		18784.53	−03	$a^3P_2 - y^3F^\circ_3$
V	5321.106	(1)		18787.87	+01	$z^3G^\circ_4 - e^3H_4$
U	5320.046	(1)		18791.62	−01	$b^3D_3 - v^5P^\circ_2$
V	5317.394	(1)		18800.99	+17	$b^3H_4 - y^3G^\circ_4$
U	5315.080	(1)		18809.17	−06	$z^5G^\circ_4 - e^3G_4$
T	5313.839	(−)		18813.57	−12	$d^3F_2 - w^1D^\circ_2$
B	5307.365	2	III?	18836.52	00	$a^3F_2 - z^3F^\circ_3$
W	5304.1	(1)		18848.1	0	$z^3D^\circ_2 - f^5D_3$
I	5302.307	10	V	18854.48	−01	$z^5D^\circ_1 - e^5D_2$

TABLE B—(*Continued*)

Ref	λ I A	Int	T C	Wave Number Observed	o−c	Desig	Ref	λ I A	Int	T C	Wave Number Observed	o−c	Desig
U	5298.779	(1)		18867.04	00	$b^3D_3 - v^5F_2^\circ$	B	5171.599	20	II	19331.01	00	$a^3F_4 - z^3F_4^\circ$
U	5295.316	(1)		18879.38	+05	$z^5G_3^\circ - e^5H_3$	B	5168.901	4	IA	19341.10	−01	$a^5D_3 - z^7D_3^\circ$
T	5294.555	(−)		18882.09	−03	$b^3D_2 - v^5F_2^\circ$	B	5167.491	40	II	19346.38	00	$a^3F_4 - z^3D_3^\circ$
U	5293.965	(1)		18884.19	+01	$c^3F_3 - u^3D_2^\circ$	J	5166.286	4	IA	19350.89	00	$a^5D_4 - z^7D_5^\circ$
U	5288.537	(2)		18903.58	−03	$b^1G_4 - y^1G_4^\circ$	J	5165.422	(4)		19354.13	−05	$y^5F_4^\circ - g^5F_4$
W	5285.6	(1)		18914.1	0	$z^3F_2^\circ - f^7D_1$	S	5164.922	(−)		19356.00	00	$c^3F_3 - w^3H_4^\circ$
T	5284.416	(−)		18918.32	+03	$a^1I_6 - w^3G_5^\circ$	W	5164.56	(1)		19357.36	00	$z^3G_4^\circ - f^3F_3$
I	5283.628	18	IV	18921.14	00	$z^5D_3^\circ - e^5D_3$	J	5162.288	10n	IV?	19365.88	+02	$y^5F_5^\circ - g^5F_5$
I	5281.796	10	IV	18927.70	00	$z^7P_2^\circ - e^7D_3$						+36	$(b^3F_2 - x^5D_1^\circ)$
V	5280.364	(1)		18932.83	−02	$b^3D_3 - x^3P_2^\circ$	V	5159.066	(2w)		19377.97	−05	$y^5F_2^\circ - f^5P_1$
W	5277.6	(1)		18942.7	−1	$z^3D_1^\circ - f^5D_1$	J	5151.915	4	IB	19404.87	00	$a^5F_1 - z^5F_2^\circ$
U	5275.021	(1n)		18952.01	−11	$c^3F_4 - u^3G_3^\circ$	B	5150.843	6	IB	19408.91	00	$a^5F_2 - z^5F_3^\circ$
J	5273.379	4	IV	18957.91	+01	$a^3P_0 - y^3D_1^\circ$	U	5148.234	(3)		19418.74	−03	$y^5F_3^\circ - f^3D_3$
K	5273.176	(5)		18958.64	−02	$z^5D_0^\circ - e^5D_1$	V	5148.061	(3)		19419.40	−08	$y^5F_2^\circ - h^5D_2$
B	5270.360	30	II	18968.77	−01	$a^3F_2 - z^3D_1^\circ$	T	5145.105	(−)		19430.55	−03	$a^5P_2 - y^5P_2^\circ$
I	5269.541	60	IB	18971.72	00	$a^5F_5 - z^5D_4^\circ$	J	5142.932	6	IB	19438.76	−01	$a^5F_3 - z^5F_4^\circ$
I	5266.562	30	IV	18982.45	00	$z^7P_3^\circ - e^7D_4$	J	*5142.541	(3w)		19440.24	00	$y^5F_3^\circ - f^5G_4$
U	5263.870	(1)		18992.16	−04	$a^1H_5 - x^3G_4^\circ$						−07	$y^5F_1^\circ - h^5D_1$
J	5263.314	8	V	18994.16	−01	$z^5D_2^\circ - e^5D_2$	U	5141.747	(2)		19443.24	−03	$a^3P_1 - y^3D_1^\circ$
V	5254.956	1	IA	19024.37	−01	$a^5D_1 - z^7D_2^\circ$	J	5139.468	20	IV	19451.86	00	$z^7P_4^\circ - e^7D_4$
					+04	$(b^1D_2 - y^1F_3^\circ)$	J	5139.260	10	IV	19452.65	−01	$z^7P_3^\circ - e^7D_2$
V	5253.479	(2)		19029.72	−04	$z^5D_1^\circ - e^5D_1$	J	5137.388	6n	V	19459.74	+03	$y^5F_5^\circ - h^5D_4$
B	5250.650	6	IV	19039.98	00	$a^5P_2 - y^5P_3^\circ$	W	5136.09	(1)		19464.66	+05	$c^3F_2 - z^1P_1^\circ$
U	5250.211	1	IA	19041.57	−01	$a^5D_0 - z^7D_1^\circ$	J	5133.692	20n	V	19473.75	−04	$y^5F_5^\circ - f^5G_6$
U	5249.099	(1n)		19045.60	+02	$z^3G_3^\circ - f^3F_3$	J	5131.475	(2)		19482.16	00	$a^5P_1 - y^5P_1^\circ$
U	5247.065	1	IA	19052.98	−06	$a^5D_2 - z^7D_3^\circ$	T	5129.658	(1)		19489.06	−10	$z^3F_3^\circ - e^3D_3$
U	5243.789	(1)		19064.89	−05	$y^5F_3^\circ - g^5F_4$	B	5127.363	5	IB	19497.79	00	$a^5F_4 - z^5F_5^\circ$
B	5242.495	4	IV	19069.59	00	$a^1I_6 - z^1H_5^\circ$	U	5126.598	(1)		19500.70	−08	$z^3F_4^\circ - f^7D_4$
SS	5241.931	(1)		19071.65	−09	$z^5G_3^\circ - f^3F_4$	T	5126.218	(1)		19502.14	−04	$y^5F_3^\circ - g^5F_3$
U	5236.204	(1)		19092.50	−01	$c^3F_2 - 8_1^\circ$	J	5125.130	6n	V	19506.28	−06	$y^5F_4^\circ - h^5D_3$
V	*5235.392	(2)		19095.47	+11	$b^3F_3 - x^5D_3^\circ$	W	5124.1	(1)		19510.2	+3	$c^3F_2 - s^3D_2^\circ$
					+01	$c^3F_4 - u^3D_3^\circ$	B	5123.723	6	IB	19511.64	00	$a^5F_1 - z^5F_1^\circ$
I	5232.946	40	III	19104.39	00	$z^7P_4^\circ - e^7D_5$	U	5121.636	(2n)		19519.59	00	$y^5F_2^\circ - f^3D_2$
U	5231.41	(1)		19110.00	−06	$a^1H_5 - v^5F_4^\circ$	T	5115.788	(1)		19541.90	−04	$a^1H_5 - w^3G_4^\circ$
J	*5229.857	5n	V	19115.67	−02	$z^5D_1^\circ - e^5D_0$	B	5110.414	10	IB	19562.45	−01	$a^5D_4 - z^7D_4^\circ$
					+08	$y^5F_4^\circ - h^5D_4$						−20	$(a^1H_5 - z^1H_5^\circ)$
U	5228.391	(1n)		19121.03	+04	$y^5F_4^\circ - f^5P_3$	U	5109.646	(2)		19565.39	+04	$y^5F_1^\circ - g^5F_2$
B	5227.192	40	II	19125.42	00	$a^3F_3 - z^3D_2^\circ$	J	5107.645	8	II	19573.06	00	$a^3F_3 - z^3F_2^\circ$
					−15	$(a^3P_1 - y^3D_2^\circ)$	J	5107.452	6	IB	19573.79	−01	$a^5F_2 - z^5F_2^\circ$
J	5226.868	15	IV	19126.61	00	$z^7P_2^\circ - e^7D_2$	SS	5104.441	(1)		19585.34	+10	$y^5F_2^\circ - h^5D_1$
SS	5226.063	(1)		19129.55	−02	$a^1P_1 - y^3P_0^\circ$	U	5104.21	(1)		19586.23	−09	$y^5F_5^\circ - f^5G_5$
U	5225.531	1	IA	19131.50	−02	$a^5D_1 - z^7D_1^\circ$	T	5104.038	(−)		19586.89	−01	$c^3P_2 - w^5D_3^\circ$
U	5223.193	(1)		19140.06	−01	$b^3D_1 - x^3P_0^\circ$	T	5099.091	(1)		19605.89	−06	$z^3F_2^\circ - e^3D_2$
W	5221.8	(1)		19145.2	−1	$a^3D_1 - x^3D_1^\circ$?	J	5098.703	8	IV	19607.38	00	$a^5P_3 - y^5P_2^\circ$
T	5217.927	(2)		19159.38	+01	$b^3D_2 - x^3P_1^\circ$	K	5098.594	(3)		19607.80	−08	$z^3D_2^\circ - e^3D_3$
J	5217.395	5	V	19161.33	00	$z^5D_4^\circ - e^5D_3$	J	5096.998	(6)		19613.94	−01	$y^5F_2^\circ - f^5G_3$
B	5216.278	10	II	19165.44	00	$a^3F_2 - z^3F_2^\circ$	K	5090.787	(6n)		19637.87	−07	$y^5F_3^\circ - h^5D_2$
J	5215.185	6	IV	19169.45	00	$z^5D_2^\circ - e^5D_1$	SS	5088.159	(1)		19648.02	+03	$y^5D_3^\circ - h^5D_4$
J	5208.601	7	IV	19193.68	−01	$z^5D_3^\circ - e^5D_2$	B	5083.342	7	IB	19666.63	00	$a^5F_3 - z^5F_3^\circ$
SS	5207.937	(1)		19196.13	+03	$b^3D_1 - x^3P_1^\circ$	J	5079.742	4	IB	19680.57	00	$a^5F_2 - z^5F_1^\circ$
J	**5204.582	2	IA	19208.51	00	$a^5D_2 - z^7D_2^\circ$	J	5079.226	6	IV	19682.57	+01	$a^5P_2 - y^5P_1^\circ$
B	5202.339	8	IV	19216.79	00	$a^5P_3 - y^5P_3^\circ$	U	5078.983	(1n)		19683.51	−04	$y^5F_1^\circ - f^5G_2$
W	5202.27?	(1)		19217.04	−07	$y^5F_3^\circ - h^5D_3$	T	5076.288	(2)		19693.96	−03	$y^5F_1^\circ - g^5F_1$
V	5198.843	(1)		19229.71	+06	$a^1D_2 - x^3G_3^\circ$	J	5074.757	10n	V	19699.90	−05	$y^5F_4^\circ - e^3G_5$
B	5198.714	4	IV	19230.19	+01	$a^5P_1 - y^5P_2^\circ$	T	5072.690	(1)		19707.93	−08	$y^5F_4^\circ - f^3D_3$
V	5196.100	(2w)		19239.86	−08	$y^5F_3^\circ - f^5P_2$	K	5072.077	(1)		19710.31	+03	$y^5F_2^\circ - g^5F_2$
K	5195.471	(8)		19242.19	−01	$y^5F_4^\circ - f^5G_5$	J	5068.774	10	V	19723.15	−02	$z^7P_4^\circ - e^7D_3$
I	5194.943	10	IB	19244.15	+01	$a^3F_3 - z^3F_3^\circ$	V	5067.162	(1)		19729.43	−05	$y^5F_4^\circ - f^5G_4$
I	5192.350	30	IV	19253.76	00	$z^7P_3^\circ - e^7D_3$	U	5065.213	(2)		19737.02	−06	$b^3D_3 - w^3F_4^\circ$
J	5191.460	20	IV	19257.06	+01	$z^7P_2^\circ - e^7D_1$	J	5065.020	6n	V	19737.77	−02	$y^5F_3^\circ - e^3G_4$
U	5187.924	(2)		19270.18	+03	$c^3F_3 - t^3D_2^\circ$	T	5063.296	(−)		19744.49	+20	$y^5D_2^\circ - h^5D_3$
U	5184.292	(3n)		19283.68	−04	$y^5F_2^\circ - g^5F_3$	T	*5060.079	(1)		19757.05	+01	$a^5D_4 - z^7D_3^\circ$
T	5180.065	(−)		19299.42	−03	$z^3G_3^\circ - f^3F_2$						−11	$y^5F_1^\circ - f^3D_1$
U	5178.798	(1n)		19304.14	−01	$z^3G_5^\circ - f^3F_4$	U	5058.507	(1)		19763.18	−04	$b^3D_3 - v^3D_3^\circ$
T	5177.230	(−)		19309.99	+01	$b^1G_4 - w^3F_4^\circ$	W	5058.00	(1)		19765.16	+10	$z^3F_3^\circ - e^7S_3$

TABLE B—(*Continued*)

Ref	λ I A	Int	T C	Wave Number Observed	o−c	Desig
W	*5057.49	(1)		19767.16	+04	$y^5D_2^\circ - f^5P_2$
					−03	$z^5G_3^\circ - f^3F_2$
U	5056.856	(1)		19769.64	+05	$z^3P_1^\circ - h^5D_1$
U	5056.023	(1)		19772.89	−11	$z^5G_5^\circ - e^3H_4$
T	5054.647	1		19778.28	−02	$b^3D_2 - v^3D_3^\circ$
B	5051.636	10	IB	19790.07	+01	$a^5F_4 - z^5F_4^\circ$
B	5049.825	15	III	19797.16	−01	$a^3P_2 - y^3D_3^\circ$
U	5048.457	(2)		19802.53	−08	$z^3D_1^\circ - e^3D_2$
T	5044.221	(2)		19819.16	−01	$z^7F_4^\circ - e^7D_5$
B	5041.759	10	III	19828.83	−01	$a^3F_4 - z^3F_3^\circ$
J	5041.074	7	IB	19831.53	00	$a^5F_3 - z^5F_2^\circ$
V	*5040.902	(2)		19832.20	00	$y^5F_2^\circ - e^3G_3$
					−21	$y^5F_3^\circ - f^5G_3$
U	5039.261	(2)		19838.66	−01	$z^5F_4^\circ - e^5F_5$
R	§5036.931	2		19847.84	00	$c^3P_2 - w^5D_2^\circ$
R	5035.025	3		19855.35	+05	$b^3D_3 - 3_3^\circ$
R	5031.901	8		19867.68	+05	$z^5G_4^\circ - f^3F_3$
R	*5031.030	2		19871.12	−01	$a^1D_2 - w^3G_3^\circ$
					+03	$b^3D_3 - w^3F_3^\circ$
R	§5030.784	5		19872.09	00	$b^3H_6 - z^3I_7^\circ$
V	5029.623	(1)		19876.68	−04	$a^1P_1 - 1_2^\circ$
J	5028.129	4	V	19882.58	−01	$a^1H_5 - y^1G_4^\circ$
T	5027.785	(−)		19883.95	−05	$z^3P_2^\circ - g^5F_3$
V	5027.212	(1)		19886.21	+04	$b^3D_2 - w^3F_3^\circ$
J	5027.136	5n	V	19886.51	−06	$y^5D_3^\circ - g^5F_4$
T	5023.476	(−)		19901.00	+06	$z^5G_5^\circ - f^3F_4$
T	5023.226	(−)		19901.99	−10	$y^5F_2^\circ - f^3D_1$
J	5022.244	6	V	19905.88	−04	$z^3F_2^\circ - e^3D_1$
V	5021.894	(1)		19907.27	+15	$a^3D_1 - w^5P_2^\circ$
U	5020.819	(1)		19911.53	−01	$a^1D_2 - x^3P_1^\circ$
J	5014.950	10	V	19934.83	−04	$z^3F_3^\circ - e^3D_2$
B	5012.071	12	IB	19946.28	00	$a^5F_5 - z^5F_5^\circ$
J	*5007.289	(3n)		19965.33	−15	$z^3F_3^\circ - g^5D_3$
					+01	$y^5D_4^\circ - g^5F_5$
I	5006.126	20	III	19969.97	−01	$z^7F_5^\circ - e^7D_5$
J	5005.720	10	V	19971.59	−03	$z^3D_3^\circ - e^3D_3$
T	5004.034	(1)		19978.32	+02	$z^3P_2^\circ - f^5P_1$
J	5002.800	(6)		19983.25	−01	$z^5F_3^\circ - e^5F_4$
B	5001.871	12	V	19986.96	−03	$z^3F_4^\circ - e^3D_3$
T	4999.114	(1)		19997.98	+03	$c^3F_2 - x^1F_3^\circ$
B	4994.133	8	IB	20017.93	00	$a^5F_4 - z^5F_3^\circ$
U	4993.687	(1)		20019.72	−04	$z^3P_2^\circ - h^5D_2$
J	4991.277	(3)		20029.38	+02	$y^5D_2^\circ - g^5F_3$
J	4988.963	(6)		20038.67	−07	$y^5D_3^\circ - h^5D_3$
U	4986.223	(1)		20049.68	−01	$y^5D_1^\circ - f^3D_2$
J	4985.553	7	V	20052.38	00	$z^7F_3^\circ - e^7D_4$
J	4985.261	7	V	20053.55	−04	$z^3D_2^\circ - e^3D_2$
J	4983.855	6n	V	20059.21	+04	$y^5D_4^\circ - h^5D_4$
J	4983.258	5n	V	20061.61	+03	$y^5D_3^\circ - f^5P_2$
J	4982.507	8n	V	20064.64	+07	$y^5D_4^\circ - f^5P_3$
U	4979.586	(1)		20076.41	+01	$b^3D_2 - w^3F_2^\circ$
J	4978.606	2	V	20080.36	+04	$z^3F_2^\circ - g^5D_1$
U	4977.653	(1)		20084.20	00	$z^3D_2^\circ - g^5D_3$
U	4975.415	(1)		20093.24	+03	$b^3H_4 - u^5D_4^\circ$
J	4973.108	3	V	20102.56	−03	$z^3D_1^\circ - e^3D_1$
SS	4972.398	(1)		20105.43	−05	$y^5F_5^\circ - g^7D_5$
U	4970.493	(2)		20113.13	00	$b^3D_1 - w^3F_2^\circ$
J	4969.927	(3)		20115.42	+07	$y^5D_1^\circ - h^5D_1$
U	4968.702	(1)		20120.38	+06	$b^3D_2 - z^1D_2^\circ$
J	4967.899	(3)		20123.64	−02	$y^5D_2^\circ - f^5P_1$
B	4966.096	8	V	20130.94	−01	$z^5F_5^\circ - e^5F_5$
U	4962.564	(1)		20145.27	+01	$y^5F_5^\circ - e^3H_6$
U	4961.908	(1)		20147.93	+05	$a^1I_6 - v^3G_5^\circ$
I	4957.603	60	III	20165.43	00	$z^7F_6^\circ - e^7D_5$
J	4957.302	20	III	20166.65	+01	$z^7F_4^\circ - e^7D_4$

Ref	λ I A	Int	T C	Wave Number Observed	o−c	Desig
V	*4952.646	(1n)		20185.61	−16	$y^5D_4^\circ - f^5G_5$
					+09	$z^3P_2^\circ - h^5D_1$
J	4950.112	(2)		20195.94	00	$z^5F_2^\circ - e^5F_3$
J	4946.394	4	IV	20211.12	−01	$z^5F_4^\circ - e^5F_4$
U	4945.63	(1)		20214.25	+03	$z^3P_2^\circ - f^5G_3$
B	4939.690	4	IB	20238.55	−01	$a^5F_5 - z^5F_4^\circ$
J	*4939.244	(2)		20240.38	−02	$y^5D_3^\circ - f^3D_3$
					−01	$y^5D_1^\circ - g^5F_2$
J	4938.820	10	IV	20242.12	00	$z^7F_2^\circ - e^7D_3$
K	4938.183	(2)		20244.73	+01	$z^3F_3^\circ - g^5D_2$
K	4934.023	(2n)		20261.80	−08	$y^5D_3^\circ - f^5G_4$
Q	4933.878	(1)		20262.39	+04	$z^3F_3^\circ - e^5P_2$
K	4933.348	(2n)		20264.57	+03	$y^5D_0^\circ - g^5F_1$
K	4930.331	(2)		20276.97	−01	$z^3D_1^\circ - g^5D_1$
U	4927.447	(1)		20288.84	−12	$a^1H_5 - w^3F_4^\circ$
U	4925.293	(1)		20297.71	−04	$y^5D_4^\circ - g^5F_4$
B	4924.776	3	V	20299.84	−01	$a^3P_2 - y^3D_2^\circ$
I	4920.509	60	III	20317.45	00	$z^7F_5^\circ - e^7D_4$
B	4918.999	30	III	20323.68	00	$z^7F_3^\circ - e^7D_3$
U	4918.023	(1)		20327.72	+01	$y^5D_0^\circ - f^3D_1$
U	4917.242	(1)		20330.95	+07	$y^5D_2^\circ - h^5D_1$
U	4911.786	(1)		20353.53	−03	$z^3D_2^\circ - e^3D_1$
J	4910.570	(1w)		20358.57	−02	$y^5D_1^\circ - f^5G_2$
J	4910.328	(1w)		20359.57	−02	$y^5D_2^\circ - f^5G_3$
J	4910.027	(2)		20360.82	−02	$z^5F_3^\circ - e^5F_3$
J	4909.387	(1)		20363.47	+03	$z^3D_2^\circ - g^5D_2$
K	4907.743	(1)		20370.29	−01	$z^5F_1^\circ - e^5F_2$
W	4905.15	(1)		20381.06	−01	$z^3D_2^\circ - e^5P_2$
B	4903.317	12	III	20388.68	−01	$z^7F_1^\circ - e^7D_2$
U	4896.437	(1)		20417.33	+01	$z^3D_3^\circ - e^3D_2$
U	4892.866	(1)		20432.23	+03	$y^5D_1^\circ - f^3D_1$
I	4891.496	50	III	20437.95	00	$z^7F_4^\circ - e^7D_3$
J	4890.762	25	III	20441.02	−01	$z^7F_2^\circ - e^7D_2$
U	4889.113	(2)		20447.91	−03	$z^3D_3^\circ - g^5D_3$
U	*4889.009	(1)		20448.35	−01	$a^5P_2 - y^3D_3^\circ$
					+03	$a^1D_2 - 2_2^\circ$
V	4888.651	(1)		20449.85	−08	$y^5D_4^\circ - h^5D_3$
K	4887.189	(−)		20455.96	+04	$y^5D_2^\circ - g^5F_2$
J	4886.335	(1)		20459.54	−04	$y^5D_3^\circ - h^5D_2$
J	4885.435	2	V	20463.31	+01	$z^3F_4^\circ - g^5D_3$
J	4882.151	(2)		20477.07	00	$z^5F_2^\circ - e^5F_2$
J	*4881.726	(2)		20478.85	−05	$b^3H_4 - z^3H_4^\circ$
					+09	$c^3F_3 - 10_3^\circ$
B	4878.218	12	III	20493.58	−01	$z^7F_0^\circ - e^7D_1$
U	4875.897	(1)		20503.34	−06	$z^5F_5^\circ - e^5F_4$
I	4872.144	20	III	20519.13	−01	$z^7F_1^\circ - e^7D_1$
I	4871.323	25	III	20522.59	00	$z^7F_3^\circ - e^7D_2$
J	4863.653	(2)		20554.95	−01	$z^5F_1^\circ - e^5F_1$
SS	4860.994	(1)		20566.20	−07	$z^5F_3^\circ - e^3F_4$
B	4859.748	15	III	20571.47	00	$z^7F_2^\circ - e^7D_1$
U	4859.142	(1)		20574.04	−08	$y^5D_2^\circ - f^5G_2$
J	4855.683	(3)		20588.69	−01	$z^5F_4^\circ - e^5F_3$
U	4854.888	(1n)		20592.06	+02	$c^3F_3 - 11_3^\circ$
U	4848.885	(1)		20617.55	00	$a^3P_2 - y^3D_1^\circ$
V	*4845.656	(2)		20631.29	00	$b^3H_5 - z^3H_6^\circ$
					−02	$b^3D_1 - w^3P_0^\circ$
U	4844.004	(2)		20638.33	−01	$a^1D_2 - w^3F_3^\circ$
J	4843.155	(3)		20641.95	−01	$z^5F_3^\circ - e^5F_2$
U	4842.788	(1)		20643.51	−02	$y^5D_4^\circ - e^3G_5$
W	4841.80	(1)		20647.72	−01	$y^5D_2^\circ - f^3D_1$
U	4840.319	(1n)		20654.04	00	$y^5D_3^\circ - f^5G_3$
J	4839.549	(3)		20657.34	00	$b^3H_5 - z^3H_5^\circ$
J	4838.519	(2n)		20661.72	00	$z^5F_2^\circ - e^5F_1$
K	4835.862	(3)		20673.08	+02	$y^5D_4^\circ - f^5G_4$
V	4834.511	(1)		20678.86	−01	$a^3P_1 - x^5D_2^\circ$

TABLE B—(*Continued*)

Ref	λ I A	Int	T C	Wave Number Observed	o−c	Desig
J	*4832.734	(2)		20686.46	−01	$b^3D_2 - w^3P_1^\circ$
					−21	$y^5F_1^\circ - f^3F_2$
U	4824.165	(1)		20723.20	00	$b^3D_1 - w^3P_1^\circ$
U	4817.773	(1)		20750.70	+04	$a^5P_1 - y^3D_2^\circ$
U	4813.115	(1)		20770.78	+02	$a^3D_1 - u^5D_1^\circ$
V	4811.04	(1)		20779.74	−03	$c^3P_1 - x^3D_1^\circ$
U	4809.950	(1)		20784.45	−02	$a^1H_5 - y^3H_5^\circ$
W	4809.3	(1)		20787.3	−1	$c^3F_4 - y^1F_3^\circ$
U	4809.154	(1)		20787.89	−04	$b^1G_4 - z^1F_3^\circ$
U	4808.159	(1)		20792.19	−02	$a^3D_3 - w^3D_3^\circ$
K	4807.725	(2)		20794.06	−08	$z^5F_4^\circ - e^3F_4$
S	*4807.243	(−)		20796.15	−05	$y^5F_3^\circ - f^3F_3$
					+16	$a^3D_2 - 1_2^\circ$
W	4804.6	(1)		20807.6	0	$a^1P_1 - v^5F_1^\circ$
U	4804.531	(1)		20807.89	−05	$a^1H_5 - v^3G_4^\circ$
J	*4802.883	(3)		20815.03	+02	$b^3D_3 - w^3P_2^\circ$
					−03	$b^1G_4 - x^1G_4^\circ$
J	4800.652	(2)		20824.70	+02	$c^3F_3 - t^3G_4^\circ$
U	4800.137	(1)		20826.94	−01	$z^7P_2^\circ - e^5D_2$
U	4799.414	(1)		20830.07	−02	$b^3D_2 - w^3P_2^\circ$
U	4798.735	(1)		20833.02	00	$a^3F_2 - y^5D_2^\circ$
U	4798.271	(1)		20835.03	+01	$c^3F_2 - t^3G_3^\circ$
W	4794.0	(1)		20853.6	−2	$a^1G_4 - y^3G_4^\circ$
U	4791.248	(1)		20865.57	−04	$a^3D_1 - w^3D_1^\circ$
B	4789.654	7	V	20872.52	+03	$a^1D_2 - z^1D_2^\circ$
J	4788.757	(4)		20876.43	00	$b^3H_6 - z^3H_6^\circ$
U	4787.839	(1)		20880.43	−03	$z^7P_3^\circ - e^5D_3$
B	4786.810	5	IV?	20884.92	−01	$c^3P_2 - x^3D_3^\circ$
U	4785.959	(1)		20888.63	+05	$c^3F_3 - 13_4^\circ$
J	4779.444	(1)		20917.10	−02	$a^1P_1 - x^3P_0^\circ$
V	4776.34	(1n)		20930.70	+11	$y^5P_3^\circ - i^5D_4$
U	4776.074	(1)		20931.86	−04	$a^3D_2 - y^3S_1^\circ$
B	*4772.817	3	III	20946.15	+07	$c^3P_2 - x^3D_2^\circ$
					−04	$a^3F_3 - y^5D_3^\circ$
J	4771.702	(1)		20951.04	−01	$a^5P_2 - y^3D_2^\circ$
V	4768.397	3n	V	20965.56	+02	$z^7P_4^\circ - e^5D_4$
V	4768.334	(1)		20965.84	−08	$z^5P_1^\circ - f^5D_2$
U	4765.482	(1)		20978.39	00	$a^3F_2 - z^3P_2^\circ$
J	*4757.582	(2)		21013.22	−01	$z^3P_1^\circ - e^3P_1$
					−01	$a^3D_1 - 1_2^\circ$
V	4749.93	(1)		21047.07	+10	$y^5P_3^\circ - i^5D_3$
B	*4745.806	3n	V	21065.36	−04	$z^5P_2^\circ - f^5D_3$
					+14	$y^5D_4^\circ - f^5G_3$
U	4745.129	(1)		21068.37	+02	$a^5P_1 - y^3D_1^\circ$
B	4741.533	3	V	21084.35	+01	$b^3P_2 - w^5D_3^\circ$
J	4741.081	(1)		21086.36	−05	$z^5F_5^\circ - e^3F_4$
J	4740.343	(1)		21089.64	−01	$b^3G_3 - y^3G_4^\circ$
J	4737.633	(1)		21101.70	+02	$b^3H_5 - z^1G_4^\circ$
I	4736.780	12	II?	21105.50	−01	$z^5D_4^\circ - e^5F_5$
J	4735.846	(2)		21109.67	+05	$c^3F_4 - t^3G_5^\circ$
J	4734.100	(1)		21117.45	−02	$b^1D_2 - w^1D_2^\circ$
B	4733.596	4	IB?	21119.70	00	$a^3F_4 - y^5D_4^\circ$
V	4729.699	(1)		21137.10	−09	$z^5F_3^\circ - e^3F_3$
J	4729.028	(1)		21140.10	00	$c^3F_4 - 12_5^\circ$
J	4728.555	3n	IV	21142.21	−06	$z^5P_2^\circ - e^7P_3$
J	4727.405	3n	IV	21147.36	−04	$z^5P_1^\circ - f^5D_1$
U	4726.160	(1)		21152.93	−07	$z^7P_3^\circ - e^5D_2$
U	4725.945	(1n)		21153.89	−01	$b^1D_2 - w^1F_3^\circ$
J	*4720.997	(1)		21176.06	+05	$b^3G_4 - y^3G_5^\circ$
					−19	$y^5D_3^\circ - f^3F_4$
V	4714.182	(1n)		21206.67	+05	$b^3H_4 - x^3G_3^\circ$
V	4714.074	(1n)		21207.16	00	$y^5P_3^\circ - i^5D_2$
V	4712.104	(1)		21216.03	00	$c^3P_2 - x^3D_1^\circ$
B	4710.286	5	IV	21224.21	−01	$b^3G_3 - y^3G_3^\circ$
J	4709.092	(3)		21229.59	−04	$z^5P_2^\circ - f^5D_2$

Ref	λ I A	Int	T C	Wave Number Observed	o−c	Desig
V	4708.972	(1)		21230.14	+03	$b^3D_2 - z^1F_3^\circ$
J	4707.487	(2)		21236.83	+05	$b^3P_1 - w^5D_2^\circ$
B	4707.281	8	IV	21237.76	−01	$z^5D_3^\circ - e^5F_4$
J	4705.464	(1)		21245.96	−04	$a^1D_2 - v^3G_3^\circ$
J	4704.958	(5)		21248.25	−02	$z^5P_1^\circ - f^5D_0$
J	4701.052	(1)		21265.90	−04	$z^5P_1^\circ - f^7D_2$
J	4700.171	(2n)		21269.89	+08	$b^1G_4 - x^3H_5^\circ$
B	4691.414	6	IV	21309.59	−01	$b^3G_4 - y^3G_4^\circ$
J	4690.146	(3)		21315.35	00	$z^5P_1^\circ - f^7D_1$
J	4687.387	(1)		21327.90	+02	$b^3P_2 - w^5F_3^\circ$
J	4683.565	(2)		21345.30	+02	$b^3P_2 - w^5D_2^\circ$
U	4682.583	(1)		21349.78	−09	$z^7P_4^\circ - e^5D_3$
V	4680.475	(1)		21359.39	00	$b^3P_0 - w^5D_1^\circ$
J	4680.297	(2)		21360.21	+01	$a^3F_2 - y^5F_3^\circ$
V	4679.229	(1)		21365.08	+02	$z^5F_4^\circ - e^3F_3$
B	4678.852	7	V	21366.80	−04	$z^5P_3^\circ - f^5D_4$
J	4673.169	(4)		21392.78	−04	$z^5P_2^\circ - f^7D_3$
J	4669.174	(4)		21411.09	−03	$z^5P_2^\circ - f^5D_1$
J	4668.142	6	IV	21415.82	−01	$z^5D_2^\circ - e^5F_3$
B	4667.459	6	V	21418.96	−02	$z^5P_3^\circ - e^7P_4$
J	4663.183	(1)		21438.60	−04	$a^1D_2 - w^3P_1^\circ$
J	4661.975	(2)		21444.15	−02	$b^3G_4 - y^3G_3^\circ$
J	4661.538	(2n)		21446.16	−01	$y^5P_3^\circ - 4_2$
U	4658.29	(1)		21461.12	+03	$b^3H_5 - x^3G_4^\circ$
U	4657.596	(1)		21464.31	−01	$b^3P_1 - w^5D_1^\circ$
J	*4654.628	5	V	21478.00	+04	$z^5D_4^\circ - e^5F_4$
					−09	$z^5P_3^\circ - f^5D_3$
J	4654.501	5	II?	21478.59	00	$a^3F_3 - y^5F_4^\circ$
U	4649.828	(1)		21500.17	+02	$b^3H_6 - v^5F_5^\circ$
B	4647.437	6	IV	21511.23	+02	$b^3G_5 - y^3G_5^\circ$
J	4643.468	(2)		21529.62	−03	$z^5P_2^\circ - f^7D_2$
J	4638.016	3	IV	21554.92	−04	$z^5P_3^\circ - e^7P_3$
J	4637.512	3	IV	21557.27	00	$z^5D_1^\circ - e^5F_2$
J	4635.846	(1)		21565.01	+01	$b^3P_1 - y^5S_2^\circ$
U	4633.764	(1)		21574.71	−02	$b^3G_3 - x^5G_3^\circ$
J	4632.915	2	III?	21578.65	−01	$a^3F_2 - y^5F_2^\circ$
U	4631.501	(1)		21585.25	−11	$z^5G_4^\circ - 3$
U	4630.785	(1)		21588.58	−04	$z^3F_3^\circ - g^5F_4$
J	4630.125	(2)		21591.66	00	$a^3P_2 - x^5D_3^\circ$
S	4626.758	(−)		21607.37	+04	$b^3G_4 - x^5G_5^\circ$
I	4625.052	3	IV	21615.35	+01	$z^5D_3^\circ - e^5F_3$
J	4619.294	3n	IV	21642.29	−03	$z^5P_3^\circ - f^5D_2$
J	4618.765	(2)		21644.77	−03	$b^3G_5 - y^3G_4^\circ$
V	4618.568	(2w)		21645.69	+08	$z^5G_5^\circ - 1$
J	4614.216	(1)		21666.10	−06	$a^3D_2 - v^5P_1^\circ$
J	4613.210	2n	V	21670.83	00	$z^5D_0^\circ - e^5F_1$
I	4611.285	5n	III	21679.88	+02	$z^5P_2^\circ - e^5S_2$
					+31	$(a^5F_4 - z^5P_3^\circ)$
					+04	$(z^5P_2^\circ - e^7F_3)$
J	*4607.655	3n	V	21696.95	−01	$z^5D_2^\circ - e^5F_2$
					00	$z^3F_2^\circ - g^5F_3$
V	4603.956	(1)		21714.39	00	$b^3G_4 - x^5G_4^\circ$
B	4602.944	9	IB?	21719.16	00	$a^3F_4 - y^5F_5^\circ$
J	4602.005	(2)		21723.59	−01	$a^3F_2 - y^5F_1^\circ$
J	4600.937	(1)		21728.64	−03	$b^3H_6 - x^3G_5^\circ$
J	4598.122	(2n)		21741.94	+01	$z^5D_1^\circ - e^5F_1$
U	4596.433	(1)		21749.93	+10	$z^5P_2^\circ - e^5G_3$
K	4596.059	(2n)		21751.70	−01	$z^5P_3^\circ - f^7D_4$
J	4595.363	(2)		21754.99	00	$b^3H_4 - z^1H_5^\circ$
U	4594.957	(2)		21756.91	−09	$a^3D_1 - v^5P_2^\circ$
U	4593.544	(1)		21763.61	−01	$z^3F_3^\circ - f^5P_2$
B	4592.655	5	IB	21767.82	00	$a^3F_3 - y^5F_3^\circ$
J	4587.132	(2)		21794.03	−01	$a^1H_5 - x^1G_4^\circ$
K	4584.824	(2)		21805.00	+02	$z^5P_3^\circ - e^7P_2$
U	4584.723	(1)		21805.48	−03	$z^5P_3^\circ - f^7D_3$

TABLE B—(*Continued*)

Ref	λ I A	Int	T C	Wave Number Observed	o−c	Desig
U	4582.941	(1)		21813.96	+03	$b^3P_1 - v^5D_1^\circ$
J	4581.517	(2)		21820.74	−04	$z^5D_3^\circ - e^3F_4$
K	4580.600	(2)		21825.10	−13	$z^5P_2^\circ - e^3D_3$
V	4579.825	(1)		21828.80	−02	$c^3P_1 - z^3S_1^\circ$
V	*4579.344	(1)		21831.09	−04	$z^7F_5^\circ - e^5D_4$
					−18	$b^1G_4 - 6_5^\circ$
U	*4575.80	(1)		21848.00	−10	$b^3H_4 - w^3G_3^\circ$
					+14	$z^3F_4^\circ - h^5D_4$
J	4574.724	(2)		21853.14	−01	$a^3P_2 - x^5D_2^\circ$
V	4574.240	(1)		21855.45	−09	$z^5D_4^\circ - e^5F_3$
W	4572.9	(1)		21861.9	−2	$z^5P_2^\circ - e^7F_2$
U	4568.840	(1)		21881.28	00	$b^3D_1 - v^3F_2^\circ$
U	4568.787	(1)		21881.53	−09	$z^5D_2^\circ - e^5F_1$
U	4566.988	(1)		21890.15	−03	$a^1P_1 - w^3F_2^\circ$
J	4566.520	(2)		21892.40	−02	$a^3D_2 - x^3P_1^\circ$
U	4565.667	(2)		21896.49	+02	$z^5D_3^\circ - e^5F_2$
V	4565.324	(2n)		21898.13	−09	$a^3D_1 - x^3P_2^\circ$
V	4564.832	(1)		21900.49	−02	$c^3P_1 - y^3P_0^\circ$
U	4564.713	(1)		21901.06	−09	$z^5P_2^\circ - e^5G_2$
J	4560.096	(2)		21923.24	−05	$z^5P_3^\circ - e^5G_4$
J	*4558.108	(1)		21932.80	+01	$b^3D_3 - v^3F_4^\circ$
					−01	$z^3F_2^\circ - f^3D_2$
J	4556.939	(1)		21938.43	−03	$a^3D_3 - v^5P_2^\circ$
J	*4556.129	4n	V	21942.32	−02	$z^5P_3^\circ - f^7D_2$
					−13	$z^3F_3^\circ - f^3D_3$
					−21	$b^3G_5 - x^5G_5^\circ$
U	4554.465	(1)		21950.34	−05	$z^7F_3^\circ - e^5D_3$
U	4551.667	(1)		21963.83	−09	$z^3F_3^\circ - f^5G_4$
B	4547.851	4	V	21982.26	−02	$a^1D_2 - z^1F_3^\circ$
J	4547.022	(2)		21986.27	−01	$a^3F_3 - y^5F_2^\circ$
V	4542.720	(1)		22007.09	−13	$z^5P_1^\circ - e^3D_2$
U	4542.420	(2)		22008.55	+03	$b^3D_2 - v^3F_3^\circ$
U	4541.953	(1)		22010.81	−02	$b^3H_5 - w^3G_4^\circ$?
W	4538.84	(2)		22025.91	+05	$z^3F_3^\circ - g^5F_3$
V	4538.764	(1)		22026.28	−05	$a^3P_2 - x^5D_1^\circ$
J	4537.677	(1n)		22031.55	+01	$b^3H_5 - z^1H_5^\circ$
U	4536.509	(1)		22037.22	−10	$b^3D_3 - 4_4^\circ$
V	4533.143	(1n)		22053.59	−04	$a^3D_1 - x^3P_0^\circ$
J	*4531.633	(2)		22060.93	+01	$a^1I_6 - u^3G_5^\circ$
					−04	$z^5D_4^\circ - e^3F_4$
					−24	$z^3D_2^\circ - f^3D_3$
B	4531.152	8	II	22063.28	00	$a^3F_4 - y^5F_4^\circ$
V	4529.562	(1)		22071.02	−06	$z^3D_3^\circ - g^5F_4$
B	4528.619	18	II	22075.62	00	$a^5P_3 - x^5D_4^\circ$
W	4527.9	(1)		22079:1	0	$b^3D_2 - 5^\circ$
U	4527.784	(1)		22079.69	+01	$a^3D_3 - x^3P_2^\circ$
J	4526.563	(2)		22085.64	+01	$c^3P_0 - u^5D_1^\circ$
U	4525.868	(1)		22089.04	+01	$z^7F_1^\circ - e^5D_2$
I	4525.142	5n	IV	22092.58	+06	$z^5P_3^\circ - e^5S_2$
					+05	$(z^5P_3^\circ - e^7F_3)$
J	4523.403	(2)		22101.07	−06	$z^5P_2^\circ - e^7S_3$
W	§4520.3	(1)		22116.2	−3	$c^3P_1 - u^5D_2^\circ$
U	4518.45	(1)		22125.30	−08	$b^3H_6 - w^3G_5^\circ$
B	4517.530	(2)		22129.80	−03	$c^3P_1 - y^3P_1^\circ$
SS	4515.178	(1)		22141.33	−04	$z^7F_2^\circ - e^5D_2$
J	4514.189	(2)		22146.18	+02	$a^1G_4 - u^5D_4^\circ$
U	*4509.306	(1)		22170.16	−02	$b^1G_4 - u^3D_3^\circ$
					−18	$a^1G_4 - u^5D_3^\circ$
J	4504.838	(2)		22192.15	−04	$z^5D_2^\circ - e^3F_3$
U	4502.590	(1)		22203.23	+02	$a^1H_5 - x^3H_6^\circ$
J	4495.966	(1)		22235.95	−07	$z^5P_2^\circ - f^5F_2$
J	4495.566	(1)		22237.92	00	$z^5P_3^\circ - e^3D_3$
V	*4495.386	(1)		22238.81	+05	$z^7F_0^\circ - e^5D_1$
					+20	$z^3F_4^\circ - h^5D_3$
B	4494.568	12	III	22242.86	00	$a^5P_2 - x^5D_3^\circ$

Ref	λ I A	Int	T C	Wave Number Observed	o−c	Desig
W	4493.3	(1)		22249.1	+3	$a^1H_5 - x^3H_5^\circ$?
V	4492.693	(1n)		22252.14	−01	$z^3F_2^\circ - g^5F_1$
J	*4490.773	(2n)		22261.66	+19	$z^3F_3^\circ - e^3G_4$
					−06	$z^3F_3^\circ - f^3D_2$
J	4490.084	(2)	IV	22265.07	−01	$c^3P_2 - z^3S_1^\circ$
B	4489.741	3	IA	22266.77	00	$a^5D_0 - z^7F_1^\circ$
J	*4488.917	(2)	IV	22270.86	+09	$b^3F_4 - y^5G_5^\circ$
					−07	$z^5P_2^\circ - e^3D_2$
J	4488.140	(2n)		22274.71	−04	$z^5P_3^\circ - e^7F_2$
J	4485.679	(2)	IV	22286.93	−03	$z^5P_1^\circ - e^5P_1$
I	4484.227	4	IV	22294.15	−01	$z^5P_3^\circ - g^5D_4$
V	4482.750	(2)		22301.50	−04	$z^5P_2^\circ - g^5D_3$
J	4482.257	6	I	22303.95	00	$a^5P_1 - x^5D_2^\circ$
J	4482.171	4		22304.38	00	$a^5D_1 - z^7F_2^\circ$
J	4481.621	(2)		22307.12	−07	$z^5P_1^\circ - e^3D_1$
J	4480.142	(3)	IV	22314.48	−04	$a^1G_4 - x^3F_4^\circ$
J	*4479.612	(3)	IV	22317.12	+05	$z^5P_1^\circ - g^5D_2$
					−07	$a^1I_6 - 6_5^\circ$
U	4478.040	(1)		22324.95	−11	$a^5P_2 - y^7P_2^\circ$
Q	4476.082	(4)		22334.72	+02	$z^5P_1^\circ - e^5P_2$
I	4476.021	10	III	22335.03	00	$b^3P_1 - x^3D_2^\circ$
J	*4472.721	(2)		22351.50	+02	$b^3H_5 - y^1G_4^\circ$
					−05	$b^3D_2 - y^1D_2^\circ$
SS	4471.810	(1)		22356.06	−02	$z^3F_3^\circ - f^5G_3$
SS	4471.685	(1)		22356.68	−04	$a^5D_1 - z^7F_1^\circ$
I	4469.381	5n	IV	22368.21	−04	$z^5P_2^\circ - e^5P_3$
U	4467.446	(1)		22377.89	−05	$c^3F_3 - w^1F_3^\circ$
V	4466.939	(2)		22380.43	−01	$z^3D_2^\circ - f^3D_2$
B	4466.554	12	II	22382.36	−02	$b^3P_2 - x^3D_3^\circ$
					+10	$(a^5D_1 - z^7F_0^\circ)$
U	4466.181	(1)		22384.23	−07	$b^3D_3 - 7_2^\circ$
W	4465.3	(1)		22388.6	+1	$y^5F_4^\circ - 1_5$
U	4464.766	(2)	IV	22391.33	00	$c^3P_2 - y^3P_2^\circ$
J	*4461.989	(4)	IV	22405.26	−11	$z^5P_3^\circ - f^5F_4$
					+07	$c^3P_1 - u^5D_1^\circ$
					−08	$b^3D_2 - x^1D_2^\circ$
B	4461.654	8	I	22406.94	00	$a^5D_2 - z^7F_3^\circ$
U	4461.373	(1)		22408.36	00	$a^1P_1 - w^3P_0^\circ$
V	4461.205	(2)		22409.20	−03	$c^3P_2 - u^5D_3^\circ$
B	4459.121	10	III	22419.67	00	$a^5P_3 - x^5D_3^\circ$
V	4458.101	(3)		22424.80	−11	$z^3D_3^\circ - f^3D_3$
J	4456.331	(1)		22433.71	−03	$a^1G_4 - z^3H_5^\circ$
J	4455.032	(2)		22440.25	−03	$z^3F_4^\circ - f^3D_3$
J	4454.655	(1)		22442.15	+08	$b^3D_1 - x^1D_2^\circ$
B′	4454.383	5	III	22443.52	−01	$b^3P_2 - x^3D_2^\circ$
J	4450.320	(3)		22464.01	00	$c^3P_0 - y^3S_1^\circ$
B	4447.722	9	III	22477.13	00	$a^5P_1 - x^5D_1^\circ$
J	4447.134	(2)Mn?	IV	22480.10	00	$a^5P_2 - y^7P_3^\circ$
J	4446.842	(2)		22481.58	00	$z^5P_1^\circ - g^5D_1$
U	4445.48	(1)	IA	22488.47	−03	$a^5D_2 - z^7F_2^\circ$
B	4443.197	7	III	22500.02	−02	$b^3P_0 - x^3D_1^\circ$
J	4442.835	(2)	IV	22501.85	−02	$a^5P_3 - y^7P_2^\circ$
B	4442.343	12	III	22504.35	00	$a^5P_2 - x^5D_2^\circ$
V	4440.972	(2)		22511.29	−06	$a^3D_2 - v^3D_3^\circ$
U	4440.838	(1)		22511.97	−01	$z^3D_1^\circ - f^3D_1$
V	4440.479	(1)		22513.79	−03	$z^5P_3^\circ - e^7S_3$
J	4439.883	(2)	IV	22516.81	00	$a^3P_2 - z^5S_2^\circ$
V	4439.643	(1)		22518.03	−06	$a^1G_4 - x^3F_3^\circ$
K	4438.353	(2)		22524.58	00	$z^5P_1^\circ - g^5D_0$
V	4436.931	(2)		22531.80	−05	$a^1G_4 - z^3H_4^\circ$
J	4435.151	2	IIA	22540.84	00	$a^5D_2 - z^7F_1^\circ$
J	4433.793	(3n)		22547.74	−06	$z^5P_3^\circ - f^5F_3$
J	4433.223	3n	IV	22550.64	−03	$z^5P_2^\circ - e^5P_1$
J	4432.572	(3)		22553.95	−03	$a^1H_5 - u^3G_5^\circ$
B	4430.618	6	III	22563.90	00	$a^5P_1 - x^5D_0^\circ$

TABLE B—(*Continued*)

Ref	λ I A	Int	T C	Wave Number Observed	o−c	Desig
V	4430.197	(2)	IV	22566.04	−05	$c^3P_2 - y^3P_1^\circ$
U	4429.32	(1)		22570.51	−11	$z^3F_3^\circ - f^5G_2$
B	4427.312	10	I	22580.75	00	$a^5D_3 - z^7F_4^\circ$
					−04	$(z^5P_2^\circ - g^5D_2)$
U	4425.660	(1)		22589.18	−02	$a^1H_5 - 4_4^\circ$
U	4424.192	(1)		22596.67	−05	$a^1D_2 - v^3F_2^\circ$
V	4423.858	(2l)		22598.38	−04	$z^5P_2^\circ - e^5P_2$
U	4423.142	(1)		22602.04	+04	$b^3G_4 - u^5D_4^\circ$
U	4422.884	(1n)		22603.36	−07	$a^3D_2 - 3_3^\circ$
B	4422.570	6	III	22604.96	−01	$b^3P_1 - x^3D_1^\circ$
U	4418.429	(1)		22626.15	−03	$b^3G_4 - u^5D_3^\circ$
B	4415.125	20	II	22643.08	00	$a^3F_2 - z^5G_3^\circ$
SS	4414.464	(1)		22646.47	+03	$a^3D_1 - 2_2^\circ$
J	4409.123	(2l)		22673.90	−03	$a^3D_2 - v^3D_1^\circ$
B	4408.419	6	III?	22677.52	−01	$a^5P_2 - x^5D_1^\circ$
J	4407.714	5	III?	22681.15	−01	$a^5P_3 - x^5D_2^\circ$
B	4404.752	30	II	22696.40	+01	$a^3F_3 - z^5G_4^\circ$
U	4401.450	(2)		22713.43	−04	$b^3P_2 - x^3D_1^\circ$
J	4401.293	(5)		22714.24	00	$z^5P_3^\circ - g^5D_3$
V	*4395.514	(1w)		22744.10	+17	$z^3D_3^\circ - e^3G_4$
					−08	$z^3D_3^\circ - f^3D_2$
U	4395.286	(2)		22745.28	−02	$z^5P_2^\circ - g^5D_1$
U	4392.58	(1)		22759.29	−01	$z^3F_4^\circ - e^3G_4$
B	4390.954	4	IV	22767.72	−02	$b^3G_3 - z^3H_4^\circ$
U	4390.458	(1)		22770.29	−06	$b^3G_4 - x^3F_4^\circ$
J	4389.244	2	IIA	22776.59	+01	$a^5D_3 - z^7F_2^\circ$
J	4388.412	4n	IV	22780.91	−03	$z^5P_3^\circ - e^5P_3$
J	4387.897	3	IV	22783.58	+01	$c^3P_1 - y^3S_1^\circ$
W	4386.6	(1n)		22790.3	0	$b^3D_1 - u^5P_1^\circ$
U	4385.258	(1)		22797.29	−05	$b^3G_3 - w^3D_2^\circ$
V	4384.682	(1)		22800.29	−05	$c^3P_2 - w^3D_2^\circ$
B	4383.547	45r	II	22806.19	00	$a^3F_4 - z^5G_5^\circ$
U	4382.773	(2)		22810.22	−03	$a^1H_5 - 6_5^\circ$
U	4377.793	(1)		22836.16	−03	$a^3D_1 - v^3D_2^\circ$
U	4377.330	(1)		22838.58	+04	$z^3D_3^\circ - f^5G_3$
V	*4376.782	(1)		22841.44	−01	$c^3P_2 - u^5D_1^\circ$
					+01	$b^3D_3 - t^3D_3^\circ$
B	4375.932	9	I	22845.88	00	$a^5D_4 - z^7F_5^\circ$
U	4374.491	(1)		22853.40	+03	$a^3D_2 - z^1D_2^\circ$
J	*4373.563	(2)		22858.25	−15	$b^3F_4 - w^5D_4^\circ$
					−02	$b^3G_3 - x^3F_2^\circ$
U	4372.991	(1)		22861.24	−03	$c^3P_2 - x^3F_2^\circ$
B	4369.774	7	III	22878.07	−01	$a^1G_4 - z^1G_4^\circ$
J	4367.906	2	IIIA	22887.86	+01	$a^3F_2 - z^5G_2^\circ$
J	4367.581	5	IV	22889.56	−01	$b^3G_4 - z^3H_5^\circ$
U	4365.899	(1)		22898.38	00	$b^3G_4 - w^3D_3^\circ$
U	4360.810	(1)		22925.10	00	$b^3D_3 - u^3D_2^\circ$
B	4358.505	3	IV	22937.22	+02	$b^3G_5 - u^5D_4^\circ$
B	4352.737	9	III	22967.62	+01	$a^5P_1 - z^5S_2^\circ$
J	4351.549	3	IV	22973.88	−04	$b^3G_4 - x^3F_3^\circ$
J	4348.939	(1)		22987.67	−02	$b^3G_4 - z^3H_4^\circ$
U	4347.851	(1)		22993.43	−05	$z^5P_3^\circ - g^5D_2$
V	4347.239	(1)		22996.67	−02	$a^5D_4 - z^7F_4^\circ$
J	4346.558	(2)		23000.27	−01	$b^3H_4 - v^3G_4^\circ$
J	4343.699	(2)		23015.40	+02	$a^1G_4 - w^5G_4^\circ$
J	4343.257	(2)		23017.75	+10	$a^3D_3 - v^3D_2^\circ$
W	*4340.5	(1)		23032.4	0	$a^3G_3 - x^3D_2^\circ$
					+1	$z^5F_1^\circ - f^5D_2$
J	4338.260	(2)		23044.26	00	$a^5P_3 - x^5F_4^\circ$
B	4337.049	10	II	23050.69	−01	$a^3F_3 - z^5G_3^\circ$
U	4335.89	(1)		23056.86	+06	$z^3D_3^\circ - e^3G_3$
U	4330.959	(1)		23083.10	−01	$b^3H_5 - y^3H_6^\circ$
W	4327.92	(2)		23099.31	−11	$b^3H_4 - y^3H_4^\circ$
J	4327.100	3	V	23103.69	−03	$a^1D_2 - y^1D_2^\circ$
U	4326.760	(2)		23105.51	−04	$b^3G_5 - x^3F_4^\circ$
B	4325.765	35	II	23110.82	00	$a^3F_2 - z^3G_3^\circ$
					−13	$(a^5D_4 - z^7F_3^\circ)$
U	4324.966	(1)		23115.09	−02	$a^5P_2 - x^5F_3^\circ$
U	4320.52	(1)		23138.88	−19	$z^5F_2^\circ - f^5D_2$
SS	4320.376	(1)		23139.65	−08	$z^5F_3^\circ - f^5D_3$
SS	4317.067	(1)		23157.39	−12	$a^1D_2 - x^1D_2^\circ$
B	4315.087	10	III	23168.01	+01	$a^5P_2 - z^5S_2^\circ$
U	4309.380	4	IV	23198.69	−04	$b^3G_5 - z^3H_6^\circ$
J	4309.036	(2)		23200.54	−01	$a^1I_6 - y^3I_6^\circ$
B	4307.906	35	II	23206.63	−01	$a^3F_3 - z^3G_4^\circ$
SS	4306.601	(1)		23213.66	−13	$z^5F_1^\circ - f^5D_1$
B	4305.455	3	IV	23219.84	+01	$c^3P_2 - y^3S_1^\circ$
U	4305.20	(1)		23221.22	+03	$a^1D_2 - u^3G_3^\circ$
J	4304.552	(1)		23224.71	−06	$b^3G_5 - z^3H_5^\circ$
J	4302.191	(2)		23237.46	−03	$a^1G_4 - x^3G_4^\circ$
U	4300.825	(1)		23244.84	+01	$z^3F_2^\circ - f^3F_2$
U	4299.635	(1)		23251.27	00	$b^3G_3 - w^5G_4^\circ$
I	4299.242	18	III	23253.39	−01	$z^7D_4^\circ - e^7D_5$
					+03	$(b^3H_5 - y^3H_5^\circ)$
B	4298.040	(2)	IV	23259.90	−03	$a^1G_4 - x^3G_5^\circ$
V	4294.939	(1w)		23276.69	−14	$b^3H_5 - v^3G_4^\circ$?
B	4294.128	15	II	23281.09	+01	$a^3F_4 - z^5G_4^\circ$
U	4292.290	(1)		23291.06	+02	$a^5P_2 - x^5F_2^\circ$
I	*4291.466	4	IA	23295.53	+06	$a^3F_3 - z^5G_2^\circ$
					00	$a^5D_3 - z^7P_4^\circ$
J	4290.870	(1)		23298.76	−02	$b^3P_2 - w^5P_3^\circ$
J	4290.382	(2)		23301.41	+03	$b^3G_4 - w^5G_5^\circ$
U	4288.965	(1)		23309.11	−01	$b^3F_3 - w^5D_2^\circ$
J	4288.148	(2)		23313.55	−01	$a^3G_3 - y^3G_3^\circ$
U	4286.992	(1)		23319.84	−04	$z^3F_3^\circ - f^3F_3$
U	4286.437	(1)		23322.86	−03	$b^3G_5 - z^3H_4^\circ$
U	4285.829	(1)		23326.17	+02	$b^3D_2 - t^3D_2^\circ$
B	4285.445	3	IV	23328.26	+01	$b^3H_6 - y^3H_6^\circ$
U	4284.415	(1)		23333.87	−05	$b^3G_4 - z^1G_4^\circ$
B	4282.406	12	III	23344.81	00	$a^5P_3 - z^5S_2^\circ$
U	4280.53	(1)		23355.04	+07	$b^3H_6 - v^3G_5^\circ$
J	4279.864	(1)		23358.68	+03	$b^3P_0 - w^5P_1^\circ$
V	4279.480	(1)		23360.77	+02	$z^3D_3^\circ - f^3F_4$
J	4278.234	(1)		23367.58	−02	$z^5F_4^\circ - f^5D_3$
U	4277.68	(1)		23370.60	−02	$a^3H_5 - z^5H_5^\circ$
J	4276.684	(1)		23376.05	−07	$z^3F_4^\circ - f^3F_4$
U	4275.72	(1)		23381.32	−09	$b^3F_4 - w^5F_4^\circ$
W	4273.87	(1)		23391.44	+01	$c^3P_1 - v^5P_2^\circ$
B	4271.764	35	II	23402.97	−01	$a^3F_4 - z^3G_5^\circ$
J	4271.159	20	III	23406.28	−01	$z^7D_3^\circ - e^7D_4$
J	4268.744	2	IV	23419.53	+01	$a^3D_2 - w^3P_1^\circ$
B	4267.830	5	IV	23424.54	+01	$c^3P_0 - x^3P_1^\circ$
J	4266.968	3	IV	23429.27	−02	$a^3G_4 - y^3G_4^\circ$
J	*4265.260	(2)		23438.66	+06	$z^3D_2^\circ - f^3F_3$
					−11	$z^3D_1^\circ - e^3P_1$
U	4264.743	(1)		23441.50	+01	$z^3D_1^\circ - f^3F_2$
J	4264.209	(2)		23444.43	−04	$z^5F_4^\circ - e^7P_3$
H	4260.479	35	III	23464.96	00	$z^7D_5^\circ - e^7D_5$
V	4260.135	(1)		23466.85	+01	$c^3P_1 - v^5F_2^\circ$
U	4260.003	(2)		23467.58	−04	$z^5F_5^\circ - e^7F_6$
					−27	$(a^5P_3 - x^5F_2^\circ)$
J	4258.956	(1)		23473.34	−04	$b^3G_3 - x^3G_4^\circ$
J	4258.619	(1)		23475.20	−05	$b^3P_2 - w^5P_2^\circ$
J	4258.320	2	IA	23476.85	−02	$a^5D_2 - z^7P_3^\circ$
U	4256.79	(1)		23485.29	+10	$y^5F_3^\circ - i^5D_3$
V	4256.212	(3)		23488.48	−02	$z^5F_2^\circ - f^7D_1$
U	4255.496	(1)		23492.43	+03	$b^3G_3 - w^5G_2^\circ$
V	*4254.938	(1)		23495.51	+05	$b^3G_3 - x^3G_3^\circ$
					+11	$c^3P_2 - w^5G_2^\circ$
B	4250.790	25	II	23518.44	00	$a^3F_3 - z^3G_3^\circ$

TABLE B—(*Continued*)

Ref	λ I A	Int	T C	Wave Number Observed	o−c	Desig
J	4250.125	25	III	23522.12	00	$z^7D_2^\circ - e^7D_3$
J	4248.228	4	IV	23532.62	−03	$c^3P_1 - x^3P_2^\circ$
I	4247.432	12	III	23537.03	−05	$z^5F_4^\circ - e^5G_5$
SS	4246.572	(1)		23541.80	+12	$z^5F_1^\circ - e^7F_1$
J	4246.090	3	V	23544.47	00	$b^3D_3 - v^3P_2^\circ$
M	4245.358	tr?	IV?	23548.53	−09	$z^5F_5^\circ - f^5D_4$
I	4245.258	6	III	23549.09	00	$b^3P_0 - z^3S_1^\circ$
V	4243.786	(1w)		23557.25	+17	$z^3D_3^\circ - e^3P_2$
U	4243.370	(2)		23559.57	+02	$b^3D_2 - v^3P_2^\circ$
J	4242.730	(2)		23563.12	−02	$a^3D_2 - w^3P_2^\circ$
U	4242.592	(1)		23563.89	+03	$a^3G_4 - y^3G_3^\circ$
V	4241.112	(1)		23572.11	+03	$b^3P_2 - w^5P_1^\circ$
J	4240.372	(2)		23576.22	+01	$a^1D_2 - t^3D_1^\circ$?
J	*4239.847	2	III	23579.14	+05	$a^3G_5 - y^3G_5^\circ$
					−06	$a^5F_3 - z^3F_4^\circ$
U	4239.735	3	IV	23579.76	+02	$b^3G_5 - w^5G_6^\circ$
I	4238.816	10n	IV	23584.88	−04	$z^5F_3^\circ - e^5G_4$
J	*4238.027	4	IV	23589.27	−02	$z^5F_2^\circ - e^5S_2$
					00	$z^5F_2^\circ - e^7F_3$
M	4237.085	(2)	IIIA	23594.52	−05	$a^5F_3 - z^3D_3^\circ$
U	4236.76	(1)		23596.32	+04	$b^3D_1 - v^3P_2^\circ$
I	4235.942	25	III	23600.88	+01	$z^7D_4^\circ - e^7D_4$
I	4233.608	18	III	23613.89	00	$z^7D_1^\circ - e^7D_2$
U	4232.732	1	IA	23618.78	−02	$a^5D_1 - z^7P_2^\circ$
V	4231.525	(1)		23625.51	−07	$a^3D_3 - v^3G_3^\circ$
U	4230.584	(1)		23630.77	−06	$c^3P_2 - v^5P_3^\circ$
J	4229.760	(1)	III	23635.37	−02	$a^3F_4 - z^5G_3^\circ$
J	*4229.516	(1gn)		23636.73	+15	$b^3G_5 - w^5G_5^\circ$
					−03	$a^3D_1 - w^3P_1^\circ$
SS	4228.722	(1)		23641.17	−05	$z^5F_4^\circ - f^7D_4$
J	4227.434	30	III	23648.37	−01	$z^5F_5^\circ - e^5G_6$
					−07	$(z^5F_2^\circ - e^7F_1)$
J	4226.426	3	IV	23654.01	−01	$b^3P_1 - z^3S_1^\circ$
J	4225.956	3	IV	23656.65	+01	$a^1G_4 - w^3G_5^\circ$
J	4225.460	6n	IV	23659.42	−04	$z^5F_2^\circ - e^5G_3$
U	4224.517	3n	IV	23664.70	−03	$z^5F_1^\circ - e^7F_2$
J	4224.176	6n	IV	23666.61	−04	$z^5F_4^\circ - e^7F_5$
J	4222.219	12	III	23677.58	−01	$z^7D_3^\circ - e^7D_3$
J	4220.347	4	IV	23688.09	+03	$c^3P_1 - x^3P_0^\circ$
B	4219.364	12	IV	23693.60	−01	$a^1H_5 - y^3I_6^\circ$
J	4217.551	7n	IV	23703.79	−03	$z^5F_1^\circ - e^5G_2$
B	4216.186	8	I	23711.46	−01	$a^5D_4 - z^7P_4^\circ$
U	4215.970	(1)		23712.68	−01	$a^3G_5 - y^3G_4^\circ$
J	*4215.430	2	IV	23715.72	+20	$a^3G_3 - x^5G_2^\circ$
					−05	$b^3G_4 - x^3G_5^\circ$
B	4213.650	5	IV	23725.73	+02	$b^3P_1 - y^3P_0^\circ$
J	4210.352	15	III	23744.32	−01	$z^7D_1^\circ - e^7D_1$
J	*4208.610	3n	V	23754.15	−02	$z^5F_3^\circ - e^7F_3$
					−04	$z^5F_3^\circ - e^5S_2$
J	4207.130	4	IV	23762.50	−02	$b^3P_2 - z^3S_1^\circ$
J	4206.702	3	IA	23764.92	−02	$a^5D_3 - z^7P_3^\circ$
J	4205.546	(2)		23771.45	−04	$z^5F_2^\circ - e^7F_2$
B	4203.987	10	III	23780.27	−01	$b^3P_1 - y^3P_2^\circ$
V	4203.953	(1)		23780.46	−10	$a^1I_6 - z^1I_6^\circ$
V	4203.570	(1)		23782.63	+01	$a^5F_1 - z^3D_1^\circ$
U	4203.30	(1)		23784.15	−02	$b^3G_3 - v^5F_3^\circ$
V	*4202.755	(1)		23787.24	+07	$c^3P_2 - v^5F_3^\circ$
					+01	$a^1G_4 - w^3G_4^\circ$
B	4202.031	30	I	23791.33	00	$a^3F_4 - z^3G_4^\circ$
W	4201.73	(1)		23793.04	−06	$a^1H_5 - w^3H_5^\circ$
J	4200.930	3n	V	23797.57	−05	$z^5F_3^\circ - e^7F_4$
W	4199.97	1	IA	23803.01	+09	$a^5D_2 - z^7P_2^\circ$
J	4199.098	20	III	23807.95	+01	$a^1G_4 - z^1H_5^\circ$
J	4198.645	4n	V	23810.52	−07	$z^5F_2^\circ - e^5G_2$
J	4198.310	20	III	23812.42	−01	$z^7D_5^\circ - e^7D_4$

Ref	λ I A	Int	T C	Wave Number Observed	o−c	Desig
V	4198.268	(1)		23812.66	−13	$z^5F_4^\circ - e^5G_4$
V	4196.533	(1)		23822.51	+06	$b^3G_5 - v^5F_5^\circ$
J	4196.218	4	IV	23824.29	−07	$z^5F_3^\circ - e^5G_3$
J	4195.615	(3)		23827.72	+03	$c^3P_2 - v^5P_2^\circ$
J	4195.337	5	IV	23829.30	−05	$z^5F_5^\circ - e^5G_5$
J	4191.685	(2)		23850.05	−05	$b^3P_0 - y^3P_1^\circ$
J	4191.436	15	III	23851.47	00	$z^7D_2^\circ - e^7D_1$
U	4189.564	(2)		23862.13	−04	$b^1G_4 - y^1F_3^\circ$
J	4187.802	20	III	23872.17	−01	$z^7D_4^\circ - e^7D_3$
U	4187.589	(1)		23873.38	−03	$z^5F_1^\circ - e^7G_2$
J	4187.044	20	III	23876.49	−01	$z^7D_3^\circ - e^7D_2$
B	4184.895	10	III	23888.75	−03	$b^3P_2 - y^3P_2^\circ$
U	4184.22	(1)		23892.61	+03	$a^3G_5 - x^5G_6^\circ$
V	4183.025	(1)		23899.43	−13	$z^5F_3^\circ - e^3D_3$
U	4182.770	(2b)		23900.89	−04	$z^5F_2^\circ - e^7G_3$
J	4182.384	4	IV	23903.09	−01	$c^3P_2 - v^5F_2^\circ$
J	4181.758	15	III	23906.67	00	$b^3P_2 - u^5D_3^\circ$
J	4177.597	4	IIA	23930.48	−02	$a^5F_4 - z^3F_4^\circ$
SS	4177.084	(1)		23933.42	−07	$z^5F_5^\circ - f^7D_4$
J	4176.571	7n	IV	23936.36	−04	$z^5F_4^\circ - f^5F_5$
					−01	$(z^5F_3^\circ - e^7F_2)$
B	4175.640	10	III	23941.70	−02	$b^3P_1 - u^5D_2^\circ$
J	4174.917	5	IIA	23945.85	−01	$a^5F_4 - z^3D_3^\circ$
U	4174.419	(1)		23948.70	−10	$a^1H_5 - w^3H_4^\circ$
J	4173.926	2	IIA	23951.53	−03	$a^5F_2 - z^3D_1^\circ$
J	4173.322	2	IV	23955.00	−03	$b^3P_1 - y^3P_1^\circ$
J	4172.749	4	IIA	23958.29	−02	$a^5F_3 - z^3D_2^\circ$
V	4172.641	(1)		23958.91	−01	$z^5F_5^\circ - e^7F_5$
J	4172.126	5	IV	23961.86	+02	$a^3D_3 - w^3P_2^\circ$
V	4171.904	(2)		23963.14	−02	$a^3D_2 - z^1F_3^\circ$
J	4171.696	(2)		23964.34	+02	$b^1G_4 - x^1F_3^\circ$
B	4170.906	5	IV	23968.87	−03	$c^3P_2 - x^3P_2^\circ$
U	4169.766	(1)		23975.43	−05	$z^5F_3^\circ - e^5G_2$
U	4168.946	(1w)		23980.14	−03	$z^5F_2^\circ - e^7G_2$
V	4168.625	(1w)		23981.99	−04	$z^5F_4^\circ - e^7F_3$
V	4167.862	(2)		23986.38	00	$b^3H_4 - x^1G_4^\circ$
U	4164.80	(1)		24004.01	−11	$b^3G_4 - v^5F_3^\circ$
V	*4163.676	(1)		24010.49	−07	$z^5F_2^\circ - e^7S_3$
					+08	$a^3G_5 - x^5G_5^\circ$
V	4161.488	(1)		24023.12	00	$b^3G_3 - w^3G_4^\circ$
V	4161.080	(1)		24025.47	−01	$z^5F_4^\circ - e^7F_4$
V	4160.561	(1)		24028.47	−06	$b^3G_6 - x^3G_4^\circ$
J	4158.798	5n	V	24038.66	−03	$z^5F_1^\circ - f^5F_2$
J	4157.788	8n	IV	24044.49	−06	$z^5F_2^\circ - f^5F_3$
B	4156.803	12	III	24050.19	−03	$b^3P_2 - u^5D_2^\circ$
V	4156.670	(1)		24050.96	−01	$b^3G_5 - x^3G_5^\circ$
V	4156.460	(1)		24052.18	−04	$z^5F_4^\circ - e^5G_3$
J	4154.812	9n	IV	24061.72	−04	$z^5F_4^\circ - e^7G_5$
J	4154.502	12	III	24063.51	−02	$b^3P_2 - y^3P_1^\circ$
J	4154.109	(1)		24065.79	−04	$z^5F_3^\circ - e^7G_3$
J	4153.906	10n	IV	24066.97	−03	$z^5F_3^\circ - f^5F_4$
J	4152.172	4	IIA	24077.02	−01	$a^5F_3 - z^3F_3^\circ$
V	4151.957	(1)		24078.26	−06	$a^1D_2 - t^3D_2^\circ$
J	4150.258	(4)		24088.12	−05	$z^5F_1^\circ - f^5F_1$
SS	4149.767	(1)		24090.97	−03	$a^5D_3 - z^7P_2^\circ$
J	4149.372	5n	V	24093.26	−05	$z^5F_5^\circ - e^7G_6$
B	4147.673	10	III	24103.13	−01	$a^3F_4 - z^3G_3^\circ$
U	4146.070	(2)		24112.45	−03	$b^3G_4 - w^3G_5^\circ$
U	4145.206	(1)		24117.48	+01	$a^3G_5 - x^5G_4^\circ$
B	4143.871	30	I	24125.24	−01	$a^3F_3 - y^3F_4^\circ$
J	4143.418	15	III	24127.88	00	$a^1G_4 - y^1G_4^\circ$
U	4142.625	(1N)		24132.50	−14	$y^5F_1^\circ - g^5G_2$
V	4141.862	(1)		24136.95	+01	$b^3G_3 - w^3G_3^\circ$
U	4141.352	(1)		24139.92	−02	$c^3P_2 - w^3G_3^\circ$

TABLE B—(*Continued*)

Ref	λ I A	Int	T C	Wave Number		Desig
				Observed	o−c	
V	*4140.441	(1)		24145.23	−22	$z^5F_2^\circ - f^5F_2$
					+16	$z^5F_3^\circ - e^7G_2$
J	4139.933	2	IIA	24148.20	−02	$a^5F_2 - z^3F_2^\circ$
U	4138.84	0		24154.57	+03	$a^3P_2 - x^5P_3^\circ$
J	4137.002	7	IV	24165.30	−03	$a^1P_1 - y^1D_2^\circ$
J	4136.512	(1)		24168.16	+06	$z^5F_4^\circ - e^7G_4$
U	4135.77	(1)		24172.50	−06	$y^5D_2^\circ - i^5D_2$
B	4134.681	12	IV	24178.87	−01	$b^3P_2 - w^3D_3^\circ$
V	*4134.433	(1)		24180.32	−05	$z^5F_2^\circ - e^3D_2$
					−03	$c^3P_2 - x^3P_1^\circ$
U	4134.340	(1)		24180.86	−02	$a^5D_4 - z^7P_3^\circ$
J	4133.869	(2)		24183.62	−05	$z^5F_4^\circ - g^5D_4$
J	4132.903	8	III	24189.27	−01	$b^3P_1 - w^3D_2^\circ$
					+25	$(a^3F_2 - y^5P_2^\circ)$
B	4132.060	25	II	24194.20	00	$a^3F_2 - y^3F_3^\circ$
U	*4130.035	(1)		24206.06	+02	$a^3F_3 - y^5P_3^\circ$
					+02	$c^3P_0 - v^3D_1^\circ$
U	4129.22	(1)		24210.84	−14	$z^5F_2^\circ - g^5D_3$
J	*4127.807	3n	V	24219.13	−14	$z^5D_1^\circ - f^5D_2$
					+01	$a^1P_1 - x^1D_2^\circ$
B	4127.612	7	IV	24220.27	−03	$b^3P_0 - w^3D_1^\circ$
U	4126.88	(1)		24224.57	−09	$b^3P_1 - u^5D_0^\circ$
J	4126.192	3n	IV	24228.61	−07	$z^5F_5^\circ - f^5F_5$
J	4125.884	(2)		24230.42	+03	$b^3P_1 - u^5D_1^\circ$
J	4125.622	(1)		24231.96	00	$y^5F_4^\circ - g^5G_5$
J	*4123.748	(1)		24242.97	+10	$b^3F_2 - x^3D_2^\circ$
					−10	$b^3G_4 - w^3G_4^\circ$
J	4122.522	4	IV	24250.18	−03	$b^3P_1 - x^3F_2^\circ$
B	4121.806	5	IV	24254.39	−03	$b^3P_2 - x^3F_3^\circ$
J	4120.211	5	IV	24263.78	00	$b^3G_4 - z^1H_5^\circ$
V	4118.904	(1)		24271.48	−12	$z^5D_2^\circ - e^7P_3$
B	4118.549	15	IV	24273.57	−05	$a^1H_5 - z^1I_6^\circ$
V	*4117.872	(1)		24277.56	−12	$z^5F_2^\circ - e^5P_3$
					−01	$y^5F_2^\circ - g^5G_2$
U	4117.71	(1)		24278.52	+01	$z^5P_2^\circ - f^3D_3$
U	4117.32	(1)		24280.82	−05	$c^3P_1 - 2_2^\circ$
U	4116.97	(1)		24282.88	−11	$z^5D_3^\circ - f^5D_4$
J	4114.957	(1w)		24294.76	−11	$z^5F_4^\circ - f^5F_4$
B	4114.449	5	IV	24297.76	−02	$b^3P_2 - w^3D_2^\circ$
J	4112.972	3n	V	24306.48	+02	$y^5F_5^\circ - g^5G_6$
V	4112.35	(1)		24310.16	−19	$z^5F_3^\circ - f^5F_2$
W	4111.1?	(1)		24317.6	−2	$z^5F_5^\circ - e^7F_4$
J	4109.808	9	IV?	24325.20	−04	$b^3P_1 - w^3D_1^\circ$
J	4109.070	(1)		24329.57	−09	$z^5D_0^\circ - f^5D_1$
B	4107.492	12	III	24338.91	+02	$b^3P_2 - u^5D_1^\circ$
V	4106.437	(1)		24345.17	−09	$z^5F_3^\circ - e^3D_2$
U	4106.265	(1)		24346.19	−03	$b^3F_3 - x^3D_3^\circ$
U	4104.97	(1)		24353.86	−17	$z^5F_5^\circ - e^7G_5$
K	*4104.132	3	V	24358.84	−12	$z^5D_2^\circ - f^5D_2$
					+13	$b^3P_2 - x^3F_2^\circ$
U	4101.681	(1)		24373.39	+07	$a^3P_0 - w^5D_1^\circ$
J	4101.272	(1)		24375.82	−05	$z^5F_3^\circ - g^5D_3$
J	4100.745	3	IIA	24378.96	−03	$a^5F_5 - z^3F_4^\circ$
U	*4099.08	(1)		24388.86	−22	$b^3H_4 - u^5F_5^\circ$?
					−13	$a^3D_3 - x^1G_4^\circ$
J	4098.183	4n	IV	24394.20	−04	$z^5D_3^\circ - f^5D_3$
J	4097.099	(1)		24400.65	−11	$z^5D_1^\circ - f^5D_1$
U	4096.114	(1)		24406.52	+02	$b^3D_2 - x^1F_3^\circ$
I	4095.975	4	IV	24407.35	−02	$b^3F_3 - x^3D_2^\circ$
V	4092.512	(1)		24428.00	−32	$a^5F_4 - z^3F_3^\circ$
J	4091.561	(1)		24433.68	−06	$b^3P_2 - w^3D_1^\circ$
V	4090.984	(1w)		24437.12	00	$z^5F_4^\circ - f^5F_3$
U	4090.077	(1)		24442.54	−04	$z^5F_3^\circ - e^5P_3$
J	4089.225	(1)		24447.64	−04	$b^3G_5 - w^3G_5^\circ$
V	4088.567	(1)		24451.57	−02	$b^3D_2 - v^3P_1^\circ$

Ref	λ I A	Int	T C	Wave Number		Desig
				Observed	o−c	
J	4087.099	(1)		24460.35	−03	$z^5F_5^\circ - e^7G_4$
W	4085.98	(1)		24467.05	+03	$y^5D_3^\circ - i^5D_2$
J	4085.312	4	IV	24471.05	−05	$z^5D_3^\circ - e^7P_3$
J	4085.011	4	IV	24472.85	−01	$b^3P_1 - 1_2^\circ$
J	4084.498	6	IV	24475.93	−01	$z^5F_5^\circ - g^5D_4$
J	4083.780	(1)	IV	24480.23	−11	$z^5F_2^\circ - e^3D_1$
V	4083.554	(1)		24481.59	−02	$a^3P_2 - x^5P_2^\circ$
U	4082.432	(2)		24488.31	−01	$b^3D_1 - v^3P_1^\circ$
J	4082.125	(1)		24490.15	−07	$z^5F_2^\circ - g^5D_2$
V	4080.886	(1w)		24497.59	−01	$z^5D_0^\circ - f^7D_1$
J	4080.226	2n	IV	24501.55	−08	$z^5D_1^\circ - f^5D_0$
J	4079.848	4	IV	24503.82	−02	$b^3P_0 - y^3S_1^\circ$
J	4078.365	4	IV	24512.73	−08	$b^3F_2 - x^3D_1^\circ$
SS	4076.884	(1)		24521.64	+03	$z^5D_2^\circ - e^7P_2$
J	4076.810	(1w)		24522.08	−06	$z^5D_2^\circ - f^7D_3$
J	4076.636	8n	IV	24523.13	−05	$z^5D_4^\circ - f^5D_4$
V	4076.498	(1)		24523.96	−04	$b^3F_2 - y^3G_3^\circ$
J	4076.232	(1)		24525.56	−04	$c^3P_1 - v^3D_1^\circ$
J	4074.794	5	IV	24534.22	−03	$a^1G_4 - w^3F_4^\circ$
K	4073.760	4n	IV	24540.44	00	$z^5D_2^\circ - f^5D_1$
V	4072.518	(2)		24547.93	−04	$z^5F_1^\circ - g^5D_1$
H	4071.740	40	II	24552.62	+01	$a^3F_2 - y^3F_2^\circ$
U	4071.52	(1)		24553.94	00	$b^3F_3 - y^3G_4^\circ$
J	4070.766	5n	III	24558.49	+02	$z^5D_3^\circ - f^5D_2$
U	4069.08	(1)		24568.67	−04	$z^5D_1^\circ - f^7D_1$
J	4067.984	8n	III	24575.29	−03	$z^5D_4^\circ - e^7P_4$
B	4067.275	4	III	24579.57	−02	$b^3F_4 - x^3D_3^\circ$
B	4066.979	6	III	24581.36	00	$b^3P_2 - 1_2^\circ$
U	4066.590	(1)		24583.71	−01	$b^3G_4 - y^1G_4^\circ$
U	4065.392	(2)		24590.95	−02	$z^5F_1^\circ - g^5D_0$
U	4064.45	(2)		24596.66	+02	$a^3F_3 - y^5P_2^\circ$
H	4063.597	45	II	24601.82	00	$a^3F_3 - y^3F_3^\circ$
J	4063.286	(3)		24603.70	−04	$z^5F_4^\circ - g^5D_3$
J	4062.446	10	III	24608.78	+01	$b^3P_1 - y^3S_1^\circ$
V	4059.726	3	V	24625.27	−06	$a^1D_2 - z^1P_1^\circ$
V	4058.766	3	IV	24631.10	−05	$a^3P_1 - w^5D_2^\circ$
K	**4058.227	4n	IV	24634.37	−06	$z^5D_4^\circ - f^5D_3$
U	4057.346	2	V	24639.72	−03	$a^3G_3 - x^3F_4^\circ$
U	4056.53	(1)		24644.67	+08	$z^7F_3^\circ - e^5F_3$
W	4055.98	(1)		24648.02	−02	$b^3D_2 - 11_3^\circ$
U	4055.039	3	V	24653.74	+02	$b^3F_4 - y^3G_5^\circ$
V	4054.883	3	V	24654.68	−05	$z^5F_2^\circ - g^5D_1$
V	4054.833	(1)		24654.99	−13	$z^5F_3^\circ - g^5D_2$
W	4054.18	(1)		24658.96	−02	$z^5D_2^\circ - f^7D_2$
W	4053.82	(1)		24661.15	+03	$c^3P_1 - w^3F_2^\circ$
SS	4052.724	(1)		24667.82	−04	$z^5D_3^\circ - f^7D_4$
V	4052.664	(1)		24668.18	−08	$a^1G_4 - w^3F_3^\circ$
V	4052.466	(1)		24669.39	−11	$z^5D_1^\circ - e^5S_2$
J	*4052.312	(1)		24670.33	−11	$z^5F_4^\circ - e^5P_3$
					+07	$a^1I_6 - t^3G_5^\circ$
J	4051.923	(2)		24672.70	−05	$z^5F_3^\circ - e^5P_2$
U	4049.331	(1)		24688.49	−01	$b^3F_3 - y^3G_3^\circ$
V	*4047.315	(1)		24700.79	−04	$a^3P_2 - x^5P_1^\circ$
					+05	$a^1I_6 - 12_5^\circ$
V	4046.629	(1)		24704.97	−07	$c^3P_1 - z^1D_2^\circ$
B	4045.815	60r	II	24709.94	00	$a^3F_4 - y^3F_4^\circ$
V	4045.139	(1)		24714.07	−06	$b^3G_3 - 2_2^\circ$
J	4044.614	6	IV	24717.28	+01	$b^3P_2 - y^3S_1^\circ$
V	*4043.901	5n	IV	24721.64	−04	$a^3G_4 - u^5D_4^\circ$
					−02	$z^5D_3^\circ - f^7D_3$
S	4041.911	(−)		24733.81	−01	$b^3H_4 - t^5D_3^\circ$
V	*4041.288	(1)		24737.62	−04	$b^3H_4 - v^3F_3^\circ$
					−11	$a^3D_2 - t^5D_3^\circ$
J	4040.650	4	V	24741.53	−04	$a^3D_2 - v^3F_3^\circ$
W	4039.94	(1)		24745.88	+02	$a^3G_4 - u^5D_3^\circ$

TABLE B—(*Continued*)

Ref	λ I A	Int	T C	Wave Number Observed	Wave Number o−c	Desig
Q	*4038.622	(−)		24753.95	+11	$b^3H_4 - u^5F_4^{\circ}$
					−01	$a^1P_1 - u^3D_2^{\circ}$
V	4037.725	(1)		24759.45	+15	$a^3P_2 - y^5G_3^{\circ}$?
SS	4033.190	(1)		24787.29	−02	$b^3F_4 - y^3G_4^{\circ}$
U	4032.630	4	III	24790.73	−01	$a^3F_4 - y^5P_3^{\circ}$
U	4032.469	(1)		24791.72	−10	$z^7F_1^{\circ} - e^5F_2$
U	4031.965	4	V	24794.82	−02	$a^3D_1 - v^3F_2^{\circ}$
V	4031.243	(2)		24799.26	−02	$c^3P_2 - v^3D_3^{\circ}$
J	4030.499	(6)	IV	24803.84	−07	$z^5D_4^{\circ} - e^5G_5$
V	4030.194	(3)		24805.72	−02	$a^5P_2 - x^5P_3^{\circ}$
V	*4029.640	3n	V	24809.13	−05	$z^5D_2^{\circ} - e^5S_2$
					−03	$z^5D_2^{\circ} - e^7F_3$
J	4024.735	6n	V	24839.36	−07	$z^5D_3^{\circ} - e^5G_4$
J	4024.109	(1)		24843.22	−10	$a^3G_3 - x^3F_3^{\circ}$
U	*4022.744	(1)		24851.65	−05	$z^5D_1^{\circ} - e^7F_2$
					00	$a^3D_3 - t^5D_4^{\circ}$
W	4022.45	(1)		24853.47	00	$a^3H_6 - w^5F_5^{\circ}$
I	4021.869	12	III	24857.06	−02	$a^3G_3 - z^3H_4^{\circ}$
V	*4021.622	(1)		24858.59	+10	$z^5D_3^{\circ} - f^7D_2$
					−10	$a^3P_1 - w^5D_1^{\circ}$
V	4020.490	(1)		24865.59	−01	$b^3D_3 - t^3G_4^{\circ}$
U	4019.05	(1)		24874.50	−01	$b^3F_2 - x^5G_3^{\circ}$
J	4018.282	(4)		24879.25	−10	$z^5D_2^{\circ} - e^5G_3$
J	4017.156	6	III	24886.22	−01	$a^1G_4 - v^3G_5^{\circ}$
U	4017.093	(1)		24886.61	−07	$a^3G_3 - w^3D_2^{\circ}$
W	4016.54	(1)		24890.04	00	$a^3G_4 - x^3F_4^{\circ}$
U	4016.429	(2)		24890.73	−06	$z^5D_1^{\circ} - e^5G_2$
B	4014.534	10	III	24902.48	+02	$a^1H_5 - y^1H_5^{\circ}$
W	*4014.28	(1)		24904.05	+17	$b^3G_3 - v^3D_2^{\circ}$
					−09	$b^3G_3 - w^3F_3^{\circ}$
J	4013.822	2	V	24906.89	+01	$c^3P_2 - v^3D_2^{\circ}$
V	4013.798	(1)		24907.04	−11	$c^3P_2 - w^3F_3^{\circ}$
J	4013.641	(2)		24908.02	−03	$z^5D_4^{\circ} - f^7D_4$
W	4012.16	(1)		24917.21	−03	$b^3H_6 - x^3H_6^{\circ}$
W	4011.71	(1)		24920.01	+04	$z^7D_3^{\circ} - e^5D_4$
U	4011.412	(1)		24921.86	−02	$b^3F_4 - y^3G_3^{\circ}$
W	*4010.77	(1)		24925.85	+13	$z^7F_3^{\circ} - e^5F_2$
					−11	$b^3F_2 - x^5G_2^{\circ}$
W	4010.18	(1)		24929.51	+01	$b^3D_3 - 13_4^{\circ}$
I	4009.714	10	III	24932.41	−01	$a^5P_1 - x^5P_2^{\circ}$
J	4007.277	6	IV	24947.57	−04	$a^3G_3 - x^3F_2^{\circ}$
V	4007.233	(1gn)		24947.85	+09	$a^3P_2 - z^5H_3^{\circ}$
V	4006.768	(1w)		24950.74	−19	$z^7F_0^{\circ} - e^5F_1$
J	4006.631	2	IV	24951.60	−03	$c^3P_0 - w^3P_1^{\circ}$
J	4006.314	3	IV	24953.57	+01	$b^3H_5 - v^3F_4^{\circ}$
B	4005.246	25	II	24960.23	00	$a^3F_3 - y^3F_2^{\circ}$
J	*4004.976	(1)		24961.91	+05	$c^3P_2 - v^3D_1^{\circ}$
					+06	$z^5D_4^{\circ} - f^7D_3$
J	4004.832	(1)		24962.81	−01	$b^3H_6 - x^3H_5^{\circ}$
J	4003.764	2	V	24969.46	00	$a^1P_1 - u^3D_1^{\circ}$
V	*4002.665	(1)		24976.32	−16	$z^7F_1^{\circ} - e^5F_1$
					+02	$a^3D_3 - v^3F_2^{\circ}$
J	4001.666	5	III	24982.55	00	$a^5P_3 - x^5P_3^{\circ}$
J	4000.466	2	V	24990.05	−04	$b^3G_4 - w^3F_4^{\circ}$
J	4000.266	(1)		24991.30	−08	$z^5D_2^{\circ} - e^7F_2$
W	4000.02	(1)		24992.83	−01	$b^3P_2 - w^5G_2^{\circ}$
I	**3998.054	10	III	25005.12	+04	$a^3G_5 - u^5D_4$
I	3997.394	15	III	25009.25	−01	$a^3G_4 - z^3H_5^{\circ}$
J	3996.968	2	V	25011.92	+06	$b^1G_4 - w^1G_4^{\circ}$
J	3995.996	4	IV	25018.00	−07	$a^3G_4 - w^3D_3^{\circ}$
J	3995.199	(1w)		25022.99	+12	$b^3H_5 - u^3G_5^{\circ}$
J	3994.117	2	V	25029.77	+01	$a^1G_4 - y^3H_5^{\circ}$
U	3992.395	(1)		25040.57	−01	$b^3H_4 - u^3G_4^{\circ}$
J	3990.379	2	V	25053.22	−01	$a^1G_4 - v^3G_4^{\circ}$
J	3989.859	(2d)	V	25056.48	−04	$a^1D_2 - y^1F_3^{\circ}$

Ref	λ I A	Int	T C	Wave Number Observed	Wave Number o−c	Desig
J	3986.176	5	IV	25079.63	+01	$a^3D_3 - v^3F_4^{\circ}$
					+01	$(z^5D_4^{\circ} - e^5G_4)$
J	3985.393	3	IV	25084.56	−04	$a^3D_2 - y^1D_2^{\circ}$
I	3983.960	10	III	25093.58	−03	$a^3G_4 - x^3F_3^{\circ}$
U	3983.35	(1)		25097.42	+04	$c^3P_2 - w^3F_2^{\circ}$
J	3981.775	7	III	25107.35	−03	$a^3G_4 - z^3H_4^{\circ}$
U	3981.104	(1)		25111.58	−06	$a^3P_1 - v^5D_2^{\circ}$
W	3980.65	(1)		25114.45	−10	$z^7D_4^{\circ} - e^5D_4$
U	3979.630	(1)		25120.88	+06	$z^5D_2^{\circ} - e^7G_3$
U	3978.464	(1)		25128.25	−02	$b^3P_2 - v^5P_3^{\circ}$
I	3977.743	12	III	25132.80	−01	$a^5P_2 - x^5P_2^{\circ}$
J	*3976.865	(1)		25138.35	+05	$b^3G_3 - z^1D_2^{\circ}$
					−04	$a^3D_2 - x^1D_2^{\circ}$
J	3976.615	4	IV	25139.93	00	$a^1P_1 - t^3D_2^{\circ}$
U	3976.562	(1)		25140.26	−01	$a^3D_3 - v^3F_3^{\circ}$
U	3976.390	(1)		25141.35	+05	$c^3P_2 - z^1D_2^{\circ}$
U	3975.842	(1)		25144.82	+05	$z^3F_4^{\circ} - 2$
W	3975.21	(1)		25148.82	00	$z^7D_2^{\circ} - e^5D_3$
U	3974.764	(1)		25151.64	00	$a^5P_1 - x^5P_1^{\circ}$
J	3974.397	(1)		25153.96	−11	$z^5D_3^{\circ} - e^3D_3$
J	3973.655	3	V	25158.66	−01	$a^1D_2 - x^1F_3^{\circ}$
U	3972.918	(1)		25163.32	00	$a^1H_5 - t^3G_5^{\circ}$
U	3971.82	(1)		25170.28	+02	$a^3G_3 - 1_2^{\circ}$
I	3971.325	9	III	25173.42	−02	$a^3G_5 - x^3F_4^{\circ}$
J	3970.391	4	IV	25179.34	+04	$c^3P_1 - w^3P_0^{\circ}$
J	3969.628	(1)		25184.18	+03	$a^3D_3 - 4_4^{\circ}$
B	3969.261	30	II	25186.51	00	$a^3F_4 - y^3F_3^{\circ}$
J	3967.964	4n	IV	25194.74	−01	$z^5D_3^{\circ} - e^7G_4$
B	3967.423	8	IV	25198.17	+01	$b^3H_4 - u^3G_3^{\circ}$
J	3966.824	(1)		25201.98	−09	$a^3D_2 - u^3G_3^{\circ}$
J	*3966.630	10n	IV	25203.21	−03	$z^5D_4^{\circ} - f^5F_5$
					−10	$a^3G_3 - z^1G_4^{\circ}$
V	*3966.532	(1n)		25203.83	−03	$a^3D_2 - u^5F_2^{\circ}$
					+07	$a^1D_2 - v^3P_2^{\circ}$
					−21	$z^5D_0^{\circ} - f^5F_1$
B	3966.066	10	III	25206.80	−01	$a^3F_2 - y^3D_3^{\circ}$
J	3965.511	(1)		25210.32	+01	$z^5D_3^{\circ} - g^5D_4$
U	3965.431	(1)		25210.83	−04	$a^3D_3 - 5^{\circ}$
J	3964.522	3	V	25216.61	−02	$b^3P_1 - v^5P_2^{\circ}$
J	3963.108	6n	V	25225.61	−05	$z^5D_1^{\circ} - f^5F_2$
V	3962.353	(2)		25230.42	−03	$z^5D_2^{\circ} - e^7S_3$
J	3961.147	(2)		25238.10	00	$b^3P_0 - v^5P_1^{\circ}$
J	3960.284	(1)		25243.60	+03	$b^3D_2 - t^3G_3^{\circ}$
W	3957.62	(1)		25260.59	+02	$z^5D_1^{\circ} - e^3D_2$
J	3957.027	4n	IV	25264.37	−07	$z^5D_2^{\circ} - f^5F_3$
B	3956.681	12	III	25266.58	−03	$a^3G_5 - z^3H_6^{\circ}$
J	3956.459	9	IV	25268.00	−01	$b^3H_6 - u^3G_5^{\circ}$
J	3955.956	2	V	25271.21	+02	$c^3P_1 - w^3P_1^{\circ}$
J	3955.352	(3)	IV	25275.07	−07	$z^5D_1^{\circ} - f^5F_1$
U	3954.715	(1)		25279.14	00	$b^3H_5 - 6_5^{\circ}$
U	3953.861	(1)		25284.60	−01	$b^3P_2 - v^5F_3^{\circ}$
J	3953.156	4	IV	25289.11	−01	$b^3G_3 - v^3G_4^{\circ}$
U	3952.702	(1)		25292.02	−02	$b^3P_1 - v^5F_2^{\circ}$
I	3952.606	8	IV	25292.63	−03	$a^3G_5 - z^3H_5^{\circ}$
I	3951.164	9	IV	25301.86	+02	$a^3D_1 - y^1D_2^{\circ}$
I	3949.954	10	III	25309.61	−01	$a^5P_3 - x^5P_2^{\circ}$
U	3949.156	(1)		25314.73	−11	$a^1P_1 - 8_1^{\circ}$
B	3948.779	10	IV	25317.14	+01	$b^3H_5 - u^3G_4^{\circ}$
J	3948.105	6n	IV	25321.47	−04	$z^5D_3^{\circ} - f^5F_4$
J	*3947.533	5	IV	25325.14	+01	$b^3P_2 - v^5P_2^{\circ}$
					−15	$b^3G_5 - w^3F_4^{\circ}$
U	3947.391	(1)		25326.05	−07	$z^7D_5^{\circ} - e^5D_4$
J	3947.002	4n	IV	25328.54	−05	$z^5D_4^{\circ} - e^7G_5$
J	3945.119	4	IV	25340.63	+02	$a^3G_3 - w^5G_4^{\circ}$
J	3944.890	3	IV	25342.10	+03	$b^3G_4 - v^3G_5^{\circ}$

TABLE B—(*Continued*)

Ref	λ I A	Int	T C	Wave Number Observed	o−c	Desig
J	3944.748	(2)		25343.02	−01	$b^3P_1 - v^5P_1^\circ$
J	3943.339	2	IV	25352.07	+04	$a^5P_2 - x^5P_1^\circ$
B	3942.443	6	IV	25357.83	−02	$b^3P_1 - x^3P_2^\circ$
J	3941.283	(3)		25365.29	−05	$z^5D_2^\circ - f^5F_2$
B	**3940.882	5	II	25367.87	−02	$a^5F_3 - y^5D_4^\circ$
V	3940.044	(1)		25373.27	−06	$a^1P_1 - v^3P_2^\circ$
J	3937.329	3	IV	25390.77	00	$a^3G_5 - z^3H_4^\circ$
B	3935.815	8	III	25400.53	−01	$b^3P_2 - v^5F_2^\circ$
					+27	$(z^5D_2^\circ - e^3D_2)$
U	3935.306	(2)		25403.82	+06	$b^3P_1 - v^5F_1^\circ$
J	*3933.606	(2)	IV	25414.80	−01	$c^3P_1 - w^3P_2^\circ$
					−03	$z^5D_2^\circ - f^5F_1$
J	*3932.629	4	IV	25421.11	+01	$a^3D_1 - u^5F_2^\circ$
					+05	$a^5G_4 - w^5G_5^\circ$
					−26	$(z^7D_2^\circ - e^5D_2)$
J	3931.122	(3)		25430.86	−01	$z^5D_2^\circ - g^5D_3$
B	3930.299	25R	I	25436.18	−01	$a^5D_2 - z^5D_3^\circ$
J	3929.208	(1)		25443.24	+05	$a^3D_3 - u^3G_4^\circ$
J	3929.114	(1)		25443.85	+02	$a^3G_3 - w^5G_3^\circ$
J	3928.085	(1)		25450.52	+02	$z^5D_4^\circ - g^5D_4$
B	3927.922	30R	I	25451.57	−01	$a^5D_1 - z^5D_2^\circ$
					+04	$(b^3P_2 - v^5P_1^\circ)$
V	3926.001	(1)		25464.03	+08	$z^5D_3^\circ - f^5F_3$
J	3925.946	6	IV	25464.38	+02	$b^3P_0 - x^3P_1^\circ$
J	3925.646	4	IV	25466.33	−02	$b^3P_2 - x^3P_2^\circ$
V	3925.201	(1)		25469.22	+01	$z^5D_0^\circ - e^5P_1$
B	3922.914	25R	I	25484.06	−01	$a^5D_3 - z^5D_4^\circ$
U	3921.27	(1)		25494.75	−03	$b^3F_4 - z^3I_5^\circ$
J	3920.839	(1)		25497.55	−02	$z^5D_2^\circ - e^5P_3$
U	3920.645	(1)		25498.81	−07	$z^7D_4^\circ - e^5D_3$
B	3920.260	20r	I	25501.32	00	$a^5D_0 - z^5D_1^\circ$
J	3919.069	3	IV	25509.07	00	$b^3G_4 - v^3G_4^\circ$
J	3918.644	6	IV	25511.83	+02	$b^3G_3 - v^3G_3^\circ$
J	3918.418	4	IV	25513.30	+04	$b^3P_1 - x^3P_0^\circ$
J	3918.319	3		25513.95	−02	$a^3P_0 - x^3D_1^\circ$
B	3917.185	8	II	25521.33	−02	$a^5F_2 - y^5D_3^\circ$
I	3916.733	6	IV	25524.28	00	$b^3H_6 - 6_5^\circ$
W	3914.73	(1)		25537.34	+25	$a^3D_3 - x^1D_2^\circ$
V	3914.273	(1)		25540.32	+01	$z^5D_1^\circ - e^5P_1$
J	3913.635	4	III	25544.48	00	$a^3P_2 - w^5D_3^\circ$
U	3911.699	(1)		25557.13	+04	$a^3D_2 - t^3D_1^\circ$?
SS	3911.005	(1)		25561.66	−04	$z^5D_4^\circ - f^5F_4$
J	3910.845	(3)	IV	25562.71	−01	$a^3G_3 - x^3G_4^\circ$
J	3909.830	3	III	25569.34	+05	$b^3P_1 - x^3P_1^\circ$
J	3909.664	(1)	V	25570.43	00	$z^5D_1^\circ - g^5D_2$
B	3907.937	4	IV	25581.73	−01	$a^3G_3 - w^5G_2^\circ$
J	3907.464	(1)	III	25584.83	+03	$a^3G_3 - x^3G_3^\circ$
J	3906.748	2	V	25589.52	−04	$a^3D_2 - t^3D_3^\circ$
B	3906.482	8	I	25591.26	00	$a^5D_1 - z^5D_1^\circ$
J	3903.902	5	IV	25608.17	−03	$b^3G_4 - y^3H_4^\circ$
B	**3902.948	20	II	25614.43	00	$a^3F_3 - y^3D_3^\circ$
J	3900.519	2	V	25630.38	00	$z^5D_3^\circ - g^5D_3$
B	3899.709	30R	I	25635.70	00	$a^5D_2 - z^5D_2^\circ$
J	3899.037	2	IV	25640.12	−06	$a^3H_4 - y^3G_4^\circ$
K	3898.012	10	II	25646.86	−01	$a^5F_1 - y^5D_2^\circ$
J	3897.896	8	IV	25647.62	00	$a^3G_5 - w^5G_6^\circ$
J	3897.449	(2)	IV	25650.57	+02	$b^3G_5 - y^3H_6^\circ$
B	3895.658	25r	I	25662.36	00	$a^5D_1 - z^5D_0^\circ$
SS	3895.450	(1)		25663.73	−10	$z^5D_0^\circ - g^5D_1$
U	3894.49	(1)		25670.06	−10	$z^5D_4^\circ - e^7S_3$
J	3894.005	(2)	III	25673.25	+02	$a^3D_2 - u^3D_2^\circ$
J	3893.924	(2)	IV	25673.79	−03	$a^3H_5 - y^3G_5^\circ$
I	3893.391	7	IV	25677.30	+03	$b^3G_5 - v^3G_5^\circ$
V	3893.316	(1)		25677.80	+01	$b^3P_2 - x^3P_1^\circ$
U	3892.98	(1)		25680.01	+01	$z^5D_2^\circ - e^5P_1$

Ref	λ I A	Int	T C	Wave Number Observed	o−c	Desig
U	3892.894	(1)		25680.58	00	$a^3G_3 - v^5F_4^\circ$
J	3891.928	3	V	25686.95	+01	$a^1P_1 - z^1P_1^\circ$
J	3890.844	2	IV	25694.11	−01	$a^3G_4 - w^5G_3^\circ$
U	3890.39	(1)		25697.11	+02	$z^5D_3^\circ - e^5P_3$
SSb	3889.931	(1)		25700.14	−09	$z^5D_2^\circ - e^3D_1$
V	3888.825	3	IV	25707.45	00	$c^3P_2 - w^3P_1^\circ$
B	3888.517	20	II	25709.49	00	$a^3F_2 - y^3D_2^\circ$
SS	3888.424	(1)		25710.10	−01	$z^5D_2^\circ \quad g^5D_2$
B	3887.051	15	I	25719.18	−01	$a^5F_4 - y^5D_4^\circ$
B	3886.284	40R	I	25724.26	00	$a^5D_3 - z^5D_3^\circ$
I	3885.512	5	III	25729.37	−03	$a^3P_1 - x^3D_2^\circ$
V	3885.154	(1)		25731.74	−02	$b^3G_4 - v^3G_3^\circ$
W	3884.66	(1)		25735.01	+07	$z^5D_1^\circ - g^5D_1$
J	3884.359	3	IV	25737.00	00	$a^3G_5 - z^1G_4^\circ$
J	3883.282	(4)		25744.14	+03	$a^3D_3 - u^3D_3^\circ$
U	3878.740	(2)		25774.29	−04	$a^3D_1 - t^3D_1^\circ$?
U	3878.676	(8)		25774.71	−04	$a^3H_4 - y^3G_3^\circ$
B	3878.575	100r	II	25775.39	00	$a^5D_2 - z^5D_1^\circ$
B	3878.021	60	II	25779.07	00	$a^5F_3 - y^5D_3^\circ$
W	3876.67	(1)		25788.05	+03	$a^3P_2 - w^5F_3^\circ$
J	3876.043	4	III	25792.22	−02	$a^5F_1 - z^3P_2^\circ$
V	3874.053	(1)		25805.47	+05	$a^3P_2 - w^5D_2^\circ$
B	3873.763	8	IV	25807.40	−02	$a^3H_5 - y^3G_4^\circ$
V	3872.923	1	IV	25813.00	−01	$a^3G_4 - x^3G_4^\circ$
B	3872.504	60	II	25815.79	−01	$a^5F_2 - y^5D_2^\circ$
J	3871.750	4	IV	25820.82	+02	$b^3G_5 - y^3H_5^\circ$
U	3869.609	(1)	IV	25835.11	+02	$a^3G_4 - x^3G_3^\circ$
K	3869.562	(2)	IV	25835.42	−04	$a^3G_4 - x^3G_5^\circ$
V	3868.243	(1)		25844.23	−04	$b^3G_5 - v^3G_4^\circ$
V	3867.925	1	IV	25846.36	+03	$b^3F_3 - u^5D_4^\circ$
B	3867.219	7	IV	25851.07	00	$c^3P_2 - w^3P_2^\circ$
B	3865.526	30	II	25862.40	00	$a^5F_1 - y^5D_1^\circ$
V	3863.745	2	IV	25874.32	+02	$a^3G_5 - w^5G_4^\circ$
U	3861.60	(1)		25888.69	−04	$a^3D_2 - u^3D_1^\circ$
J	*3861.341	2	IV	25890.42	+09	$a^3G_5 - v^5F_5^\circ$
					−05	$a^3D_1 - u^3D_2^\circ$
B	3859.913	300R	I	25900.01	+01	$a^5D_4 - z^5D_4^\circ$
I	3859.214	10	III	25904.69	+03	$a^3H_6 - y^3G_5^\circ$
B	3856.373	50r	IA	25923.78	00	$a^5D_3 - z^5D_2^\circ$
J	3855.846	(1w)		25927.32	+06	$z^5D_3^\circ - e^5P_2$
V	3855.329	(1w)		25930.80	−07	$a^3G_4 - v^5F_4^\circ$
U	3854.375	(1)		25937.22	−06	$z^5D_4^\circ - e^5P_3$
V	3853.462	(1)		25943.36	−04	$b^3G_5 - y^3H_4^\circ$
I	3852.574	6	IV	25949.34	+01	$a^5P_3 - w^5D_4^\circ$
B	3850.820	12	II	25961.16	−01	$a^5F_2 - z^3P_2^\circ$
B	3849.969	40	II	25966.90	00	$a^5F_1 - y^5D_0^\circ$
V	3846.949	(1)		25987.28	−03	$a^3H_5 - x^5G_6^\circ$
B	3846.803	8	IV	25988.27	+01	$a^3D_3 - t^3D_3^\circ$
J	3846.412	2	IV	25990.91	+07	$a^1H_5 - w^1G_4^\circ$
V	3846.001	(1w)		25993.69	00	$z^5F_4^\circ - f^5P_3$
J	3845.692	(1)		25995.78	+04	$a^1D_2 - t^3G_3^\circ$
K	3845.170	(5)		25999.30	−04	$a^3P_1 - x^3D_1^\circ$
B	3843.259	8	IV	26012.23	+03	$a^1G_4 - z^1F_3^\circ$
U	3842.975	(1)		26014.16	+10	$b^3F_3 - u^5D_2^\circ$
B	3841.051	80r	II	26027.19	00	$a^3F_2 - y^3D_1^\circ$
B	3840.439	80r	II	26031.33	00	$a^5F_2 - y^5D_1^\circ$
U	3839.630	(2w)		26036.82	+01	$z^3D_1^\circ - i^5D_2$
B	3839.259	7	IV	26039.33	00	$a^1G_4 - x^1G_4^\circ$
J	3837.132	1	IV	26053.77	+01	$b^3F_2 - x^3F_3^\circ$
I	3836.332	4	IV	26059.20	00	$a^3D_2 - t^3D_2^\circ$
B	3834.225	100r	II	26073.52	00	$a^5F_3 - y^5D_2^\circ$
J	3833.311	5	IV	26079.74	+03	$b^3F_4 - u^5D_4^\circ$
J	3830.850	1	IV	26096.49	+08	$a^3G_5 - x^3G_4^\circ$
J	3830.757	1	IV	26097.12	00	$b^3F_2 - w^3D_2^\circ$
U	3829.771	(2)		26103.84	−05	$b^3F_4 - u^5D_3^\circ$

TABLE B—(*Continued*)

Ref	λ I A	Int	T C	Wave Number Observed	o−c	Desig
V	*3829.458	1	IV	26105.98	+01 −09	$a^3D_1 - u^3D_1^\circ$ $b^3P_1 - 2_2^\circ$
J	3829.125	(1)		26108.25	−01	$b^1G_4 - s^3G_5^\circ$
V	3828.510	(1n)		26112.44	−02	$a^3G_3 - w^3G_4^\circ$
B	3827.825	75r	II	26117.11	00	$a^3F_3 - y^3D_2^\circ$
J	3820.572	1	IV	26118.84	−01	$a^3G_5 - x^3G_5^\circ$
J	3826.836	1	IV	26123.86	+06	$a^3G_4 - v^5F_3^\circ$
B	3825.884	200R	II	26130.36	−01	$a^5F_4 - y^5D_3^\circ$
J	3825.404	(1)		26133.64	00	$a^3P_2 - y^5S_2^\circ$
B	3824.444	50r	IA	26140.20	+01	$a^5D_4 - z^5D_3^\circ$
V	3824.306	2	IV?	26141.14	00	$b^3H_4 - w^3H_4^\circ$
J	3824.074	1	IV	26142.73	+01	$b^3F_3 - w^3D_3^\circ$
J	3821.834	3	IV	26158.05	00	$b^3F_2 - x^3F_2^\circ$
I	3821.181	10	IV	26162.52	+02	$b^3H_5 - y^3I_6^\circ$
I	3820.428	250R	II	26167.68	00	$a^5F_5 - y^5D_4^\circ$
U	3817.650	3n	IV	26186.72	00	$z^5F_5^\circ - g^5F_5$
J	3816.340	4	IV	26195.71	+03	$a^5P_2 - w^5D_3^\circ$
B	3815.842	100r	II	26199.13	+01	$a^3F_4 - y^3D_3^\circ$
J	3814.526	5	IIIA	26208.17	+01	$a^5F_1 - z^3P_1^\circ$
V	3813.891	2	V	26212.53	−01	$a^1I_6 - x^1H_5^\circ$
J	3813.638	2	IV	26214.27	00	$a^3G_5 - v^5F_4^\circ$
J	3813.059	5?	IV	26218.25	−01 +10	$b^3F_3 - x^3F_3^\circ$ $(a^3H_6 - x^5G_6^\circ)$
G	3812.964	40	II	26218.92	+03	$a^5F_3 - z^3P_2^\circ$
J	3811.892	2	IV	26226.27	−01	$a^3G_3 - w^3G_3^\circ$
U	*3811.05	(1)		26232.07	+05 −09	$b^3F_3 - z^3H_4^\circ$ $a^3G_4 - w^3G_5^\circ$
J	3810.759	2	IV	26234.07	−04	$a^3D_2 - 8_1^\circ$
V	3809.043	(1)		26245.89	+02	$b^3P_0 - v^3D_1^\circ$
J	3808.731	4	IV	26248.04	−02	$b^3F_4 - x^3F_4^\circ$
J	3808.286	(1)		26251.11	+02	$c^3P_2 - z^1F_3^\circ$
G	3807.534	7	III	26256.31	+08	$a^5P_1 - w^5D_2^\circ$
G	3806.697	10	III	26262.08	+09 +46	$b^3H_5 - w^3H_5^\circ$ $(b^3F_3 - w^3D_2^\circ)$
J	3806.203	2	IV	26265.47	+10	$a^1P_1 - v^3P_1^\circ$
B	3805.345	12	IV	26271.40	−02	$b^3H_4 - y^3I_5^\circ$
J	3804.013	(2)		26280.59	+02	$z^5F_5^\circ - h^5D_4$
J	3802.283	(1)		26292.55	−05	$a^3D_2 - v^3P_2^\circ$
J	3801.975	(2w)		26294.68	+03	$z^5F_5^\circ - f^5G_6$
J	3801.804	1	IV	26295.86	+04	$b^3P_1 - v^3D_2^\circ$
J	3801.681	3	IV	26296.71	−01	$b^3P_2 - v^3D_3^\circ$
B	3799.549	50	II	26311.47	00	$a^5F_3 - y^5F_4^\circ$
B	3798.513	40	II	26318.65	+01	$a^5F_4 - y^5F_5^\circ$
J	3797.948	(1)		26322.56	+01	$b^3F_3 - x^3F_2^\circ$
B	3797.517	12	III	26325.55	+03	$b^3H_6 - w^3H_6^\circ$
U	3796.90	(1)		26329.83	−12	$a^3D_2 - s^3D_3^\circ$
U	3796.00	(1)		26336.07	+09	$a^3H_6 - x^5G_5^\circ$
B	3795.004	60	II	26342.98	00	$a^5F_2 - y^5F_3^\circ$
J	3794.340	8	III	26347.59	−05	$a^3H_4 - z^3I_5^\circ$
J	3793.872	1	IV	26350.84	+04	$b^3P_1 - v^3D_1^\circ$
J	3793.478	(1)		26353.58	+02	$z^7P_3^\circ - f^5D_3$
U	3793.360	1	IV	26354.40	+01	$z^7P_2^\circ - e^7P_2$
U	3792.833	1	IV	26358.06	−01	$a^5P_1 - w^5F_2^\circ$
J	3792.156	2	IV	26362.77	+02	$a^3G_4 - w^3G_4^\circ$
U	3791.73	(1)		26365.73	+07	$z^5F_2^\circ - f^5P_1$
J	3791.504	(1)		26367.30	+02	$b^3F_4 - z^3H_5^\circ$
J	*3790.756	1	IVA	26372.50	+01 −08	$a^5P_3 - w^5D_3^\circ$ $a^3P_0 - w^5P_1^\circ$
J	3790.656	(1)		26373.20	−02	$z^7P_2^\circ - f^5D_1$
B	3790.095	12	II	26377.10	00 +100	$a^5F_2 - z^3P_1^\circ$ $(b^3F_4 - w^3D_3^\circ)$
J	3789.570	(1)		26380.75	+05	$b^3F_2 - 1_2^\circ$
J	3789.178	3	IV	26383.48	+02	$a^3G_4 - z^1H_5^\circ$
B	3787.883	50	II	26392.50	−01	$a^5F_1 - y^5F_2^\circ$
J	3787.164	(1)		26397.51	00	$b^3D_2 - w^1D_2^\circ$

Ref	λ I A	Int	T C	Wave Number Observed	o−c	Desig
J	3786.678	8	III	26400.90	−02	$a^5F_1 - z^3P_0^\circ$
J	**3786.176	4	IV	26404.40	+08	$b^3P_2 - v^3D_2^\circ$
G	3785.950	6	IV	26405.98	+04	$a^3H_5 - z^3I_6^\circ$
J	3785.706	(1)		26407.68	+04	$b^3H_6 - y^3I_6^\circ$
J	3782.608	(1)		26429.31	+04	$c^3P_1 - v^3F_2^\circ$
J	3782.450	1	IV	26430.41	−02	$z^7P_3^\circ - e^7P_3$
J	3781.938	(1)		26433.99	+05	$b^3D_2 - w^1F_3^\circ$
J	3781.188	2	IV	26439.23	+01	$a^5P_2 - w^5F_3^\circ$
J	3779.486	2	IV	26451.14	+13	$a^5P_1 - w^5F_1^\circ$
V	3779.444	(2n)	IV	26451.43	+08	$a^3D_1 - 8_1^\circ$
V	3779.424	(1)		26451.57	−06	$b^3F_4 - x^3F_3^\circ$
U	3779.213	(1)		26453.05	−06	$a^3G_3 - y^1G_4^\circ$
J	3778.697	(1)		26456.66	+04	$a^5P_2 - w^5D_2^\circ$
J	3778.509	4	IV	26457.98	+08	$a^3D_3 - t^3D_2^\circ$
U	3778.320	(1)		26459.30	00	$b^3P_2 - v^3D_1^\circ$
J	3777.448	2	IV	26465.41	+01	$b^3F_4 - z^3H_4^\circ$
J	3777.061	(1)		26468.12	+08	$b^3G_4 - z^1F_3^\circ$
G	3776.454	6	IV	26472.38	+04	$a^5P_3 - w^5F_4^\circ$
J	3775.860	(1)		26476.54	−03	$a^3G_4 - w^3G_3^\circ$
G	3774.823	5	IV	26483.81	+04	$a^5P_1 - w^5D_1^\circ$
J	3773.699	1	IV	26491.70	−06	$z^7P_2^\circ - f^7D_2$
V	3773.364	(1)		26494.05	−03	$a^1G_4 - x^3H_5^\circ$
V	3770.405	3	IV	26514.84	−04	$a^3H_5 - z^3I_5^\circ$
V	3770.305		IV	26515.55	−01	$a^3G_5 - w^3G_5^\circ$
V	3769.995	4	IV	26517.73	−06	$z^7P_3^\circ - f^5D_2$
W	3768.23	(1)		26530.15	−09	$b^3P_1 - z^1D_2^\circ$
J	3768.030	3	IV	26531.56	+01	$a^5P_1 - w^5D_0^\circ$
B	3767.194	80r	II	26537.44	−01	$a^5F_1 - y^5F_1^\circ$
V	3766.665	1	IV	26541.17	00	$z^7P_2^\circ - f^7D_1$
V	3766.092	(1)		26545.21	+01	$b^3F_3 - 1_2^\circ$
W	3765.70	(1)		26547.97	00	$b^3H_5 - y^3I_5^\circ$
B	3765.542	20	IV	26549.09	00	$b^3H_6 - y^3I_7^\circ$
B	3763.790	100r	II	26561.44	00	$a^5F_2 - y^5F_2^\circ$
V	3762.205	(1)		26572.63	−02	$z^5F_4^\circ - e^3G_5$
J	3761.416	1	IV	26578.21	−04	$b^3F_3 - z^1G_4^\circ$
G	3760.534	6	III	26584.44	−01	$a^5P_1 - y^5S_2^\circ$
B	3760.052	8	III	26587.85	+01	$a^3H_6 - z^3I_7^\circ$
V	3759.155	(1)		26594.20	+02	$a^1I_6 - s^3G_5^\circ$
B	3758.235	150R	II	26600.70	00	$a^5F_3 - y^5F_3^\circ$
J	3757.459	1	IV	26606.20	−01	$a^3D_2 - z^1P_1^\circ$
J	3756.939	4	IV	26609.88	−03	$a^1H_5 - v^3H_5^\circ$
J	3756.069	1	IVA	26616.04	+01	$a^5P_3 - w^5F_3^\circ$
J	3754.506	1	IV	26627.12	−06	$z^7P_3^\circ - f^7D_4$
G	**3753.610	8	III	26633.48	+05	$a^5P_3 - w^5D_2^\circ$
J	3753.154	(1gn)		26636.71	−07	$a^3H_6 - z^3I_6^\circ$
J	*3752.420	(1w)		26641.92	−02 −04	$z^7P_2^\circ - e^7F_3$ $z^7P_2^\circ - e^5S_2$
J	3751.820	(1)		26646.19	+04	$a^3G_5 - w^3G_4^\circ$
J	3751.059	(1)		26651.59	+05	$a^3D_2 - s^3D_2^\circ$
V	3750.677	(1)		26654.31	+04	$b^3F_2 - w^5G_3^\circ$
B	3749.487	200R	II	26662.76	−01	$a^5F_4 - y^5F_4^\circ$
J	3748.969	5	IV	26666.45	00	$z^7P_4^\circ - f^7D_5$
V	§3748.492	7	IV?	26669.84	+01	$a^1H_5 - v^3H_6^\circ$
B	3748.264	60R	IA	26671.46	−01	$a^5D_1 - z^5F_2^\circ$
J	3746.931	6	IV	26680.95	−02	$z^7P_3^\circ - f^7D_3$
J	3745.486	1	IV	26684.12	−04	$a^5P_2 - w^5D_1^\circ$
J	3745.901	40r	IA	26688.29	00	$a^5D_0 - z^5F_1^\circ$
J	3745.561	100R	I	26690.71	+01	$a^5D_2 - z^5F_3^\circ$
J	3744.105	4	IV	26701.09	−02	$z^7P_2^\circ - e^7F_1$
SS	3743.781	(0)		26703.40	00	$a^3G_4 - y^1G_4^\circ$
J	3743.468	6	IV?	26705.63	+03	$a^1H_5 - x^1H_5^\circ$
G	3743.364	20	IIA	26706.38	00	$a^5F_2 - y^5F_1^\circ$
V	3742.937	(0)		26709.42	+06	$z^5F_1^\circ - f^5G_2$
J	3742.621	4	IV	26711.68	−04	$z^7P_4^\circ - f^5D_4$
W	3742.07	(1)		26715.61	+06	$b^3F_3 - w^5G_4^\circ$

TABLE B—(*Continued*)

Ref	λ I A	Int	T C	Wave Number Observed	o−c	Desig
J	3740.247	3	IV	26728.63	−02	$a^3D_3 - s^3D_3^\circ$
V	*3740.061	(1)		26729.96	+04	$z^5F_3^\circ - g^7D_4$
					00	$a^1G_4 - v^3F_4^\circ$
V	3739.317	1	IV	26735.28	+01	$a^5P_3 - w^5F_2^\circ$
J	3739.120	1	IV	26736.69	−03	$a^5P_1 - v^5D_2^\circ$
B	3738.308	10	IV	26742.49	−02	$b^3H_6 - z^1I_6^\circ$
B	3737.133	150R	I	26750.90	00	$a^5D_3 - z^5F_4^\circ$
J	3735.325	6	IV	26763.85	00	$z^7P_4^\circ - e^7P_4$
B	3734.867	300r	II	26767.13	−01	$a^5F_5 - y^5F_5^\circ$
B	3733.319	40r	IA	26778.23	00	$a^5D_1 - z^5F_1^\circ$
B	3732.399	10	III	26784.83	−01	$a^5P_2 - y^5S_2^\circ$
J	3731.374	2	IV	26792.19	+01	$b^3F_2 - w^5G_2^\circ$
J	3730.945	2	IV	26795.27	+03	$b^3F_2 - x^3G_3^\circ$
G	3730.386	3	IV	26799.28	+01	$a^1G_4 - u^3G_5^\circ$
J	3728.668	1	IV	26811.63	00	$b^3F_4 - z^1G_4^\circ$
J	3727.809	3	IV	26817.81	00	$z^7P_3^\circ - f^7D_2$
B	3727.621	50r	II	26819.16	00	$a^5F_3 - y^5F_2^\circ$
J	3727.096	4	IV	26822.94	−03	$z^7P_4^\circ - f^5D_3$
J	3726.927	6	IV	26824.16	00	$z^7P_2^\circ - e^7F_2$
					−26	$(a^5P_2 - v^5D_3^\circ)$
J	3725.498	(1)		26834.45	−04	$a^1G_4 - 4_4^\circ$
B	3724.380	8	III	26842.50	−02	$a^3P_2 - x^3D_3^\circ$
B	3722.564	50r	IA	26855.59	−01	$a^5D_2 - z^5F_2^\circ$
J	3722.028	(1)		26859.46	−02	$a^3G_3 - w^3F_4^\circ$
V	3721.606	(1)		26862.51	−02	$b^3G_3 - v^3F_2^\circ$
U	3721.510	2	IV	26863.20	−06	$z^7P_2^\circ - e^5G_2$
V	3721.396	1	IV	26864.02	−01	$a^3P_0 - y^3P_1^\circ$
V	*3721.278	2	IV	26864.87	−06	$z^5F_5^\circ - e^3G_5$
					−05	$a^5P_3 - v^5D_4^\circ$
V	3721.189	(1)		26865.50	−01	$c^3P_2 - v^3F_2^\circ$
B	3719.935	250R	I	26874.57	+01	$a^5D_4 - z^5F_5^\circ$
J	3718.407	3	IV	26885.62	00	$a^3G_3 - v^3D_3^\circ$
G	*3716.442	12	IV	26899.83	−01	$z^7P_4^\circ - e^7P_3$
					+10	$z^5F_4^\circ - e^3G_4$
G	3715.911	4	IV	26903.68	+01	$a^3P_2 - x^3D_2^\circ$
J	3711.411	2	IV	26936.29	+02	$c^3P_1 - y^1D_2^\circ$
J	3711.225	3	IV	26937.65	−01	$b^3F_3 - x^3G_4^\circ$
U	3709.665	(1)		26948.97	+04	$b^3F_4 - w^5G_4^\circ$
J	3709.535	(1)		26949.92	00	$b^3G_4 - x^3H_5^\circ$
G	3709.246	75r	II	26952.02	+02	$a^5F_4 - y^5F_3^\circ$
V	*3708.602	(1)		26956.70	−05	$a^3H_4 - u^5D_3^\circ$
					+02	$b^3F_3 - w^5G_2^\circ$
G	3707.918	8	III	26961.67	+02	$a^5P_3 - y^5S_2^\circ$
U	3707.824	20	I	26962.35	−01	$a^5D_2 - z^5F_1^\circ$
I	*3707.048	8	IV	26968.00	+01	$z^7P_3^\circ - e^7F_3$
					−02	$z^7P_3^\circ - e^5S_2$
B	3705.567	100r	I	26978.77	00	$a^5D_3 - z^5F_3^\circ$
B	3704.463	10	IV	26986.81	+01	$a^3G_5 - y^1G_4^\circ$
V	3704.336	(1)		26987.74	+09	$b^3H_6 - z^1I_6^\circ$
U	3704.021	(1)		26990.04	−02	$c^3P_1 - x^1D_2^\circ$
J	3703.824	3	IV	26991.47	+01	$b^3P_0 - w^3P_1^\circ$
J	3703.697	3	IV	26992.40	−04	$z^7P_4^\circ - e^5G_5$
J	***3703.556	5	IV?	26993.42	−07	$a^3G_3 - w^3F_3^\circ$
					+20	$a^3G_3 - v^3D_2^\circ$
J	*3702.500	1	IVA	27001.12	−19	$a^3F_2 - x^5D_3^\circ$
					−11	$a^5P_3 - v^5D_3^\circ$
J	3702.033	3	IV	27004.53	+03	$b^3P_1 - w^3P_0^\circ$
G	3701.086	20	IV	27011.44	00	$z^7P_3^\circ - e^7F_4$
J	3699.147	1	IV	27025.60	−06	$c^3P_2 - t^5D_3^\circ$
J	3698.611	2	IV	27029.51	+01	$c^3P_2 - v^3F_3^\circ$
U	3697.536	(1w)		27037.37	−03	$a^3D_2 - y^1F_3^\circ$
G	3697.426	6n	IV	27038.18	00	$z^7P_3^\circ - e^5G_3$
U	3696.03	(1)		27048.39	00	$a^3P_1 - z^3S_1^\circ$
V	*3695.507	(1)		27052.21	+07	$b^3F_4 - w^5G_3^\circ$
					+09	$z^5F_2^\circ - g^7D_2$
B	*3695.054	8	IV	27055.53	+01	$b^3F_3 - v^5F_4^\circ$
					−01	$a^1G_4 - 6_5^\circ$
G	3694.005	20	IV	27063.21	−02	$z^7P_2 - e^7S_3$
J	3693.008	1	IV	27070.52	+14	$b^3G_3 - 4_4^\circ$
J	3690.730	4	IV	27087.23	−01	$a^1H_5 - s^3G_5^\circ$
V	*3690.450	(1)		27089.29	+09	$c^3P_0 - t^3D_1^\circ$?
					−04	$z^5D_1^\circ - f^5P_2$
S	3690.095	(−)		27091.89	−22	$b^3F_3 - v^5P_3^\circ$
V	3689.897	(1w)		27093.34	−19	$a^1G_4 - u^3G_4^\circ$
G	*3689.457	12	IV	27096.58	−01	$z^7P_4^\circ - f^7D_4$
					+19	$b^3P_1 - w^3P_1^\circ$
V	3688.877	(1)		27100.84	−09	$a^3H_4 - x^3F_4^\circ$
V	3688.476	(1w)		27103.78	−06	$a^3D_3 - 9_4^\circ$
J	3687.656	4	III	27109.81	+04	$a^3G_4 - w^3F_4^\circ$
B	3687.458	40r	I	27111.27	+01	$a^5F_5 - y^5F_4^\circ$
J	3687.100	2	IV	27113.90	−02	$a^5P_3 - v^5D_2^\circ$
J	3686.260	2	IV	27120.07	−01	$a^3P_1 - y^3P_0^\circ$
G	3685.998	15n	IV	27122.01	−01	$z^7P_4^\circ - e^7F_5$
G	3684.108	15	IV	27135.91	00	$a^3G_4 - v^3D_3^\circ$
V	*3683.616	(1)		27139.54	+15	$a^3P_0 - u^5D_1^\circ$
					−01	$a^3D_2 - x^1F_3^\circ$
G	3683.054	10	IA	27143.68	+02	$a^5D_3 - z^5F_2^\circ$
J	3682.226	20	IV	27149.78	+10	$a^1D_2 - w^1D_2^\circ$
U	3681.88	(1)		27152.34	+03	$b^1G_4 - v^1G_4^\circ$
U	3681.651	(1)		27154.03	−03	$z^7P_3^\circ - e^7G_4$
U	3680.675	2n	IV	27161.23	−05	$z^5D_4^\circ - g^5F_5$
B	3679.915	40r	IA	27166.83	−01	$a^5D_4 - z^5F_4^\circ$
W	*3679.53	(1)		27169.68	+05	$z^7P_3^\circ - g^5D_4$
					−02	$c^3P_1 - t^5D_0^\circ$
W	3679.33	(1)		27171.15	+11	$b^3F_4 - x^3G_4^\circ$
W	3678.98	(1)		27173.74	+12	$a^3P_2 - x^3D_1^\circ$
J	3678.863	3	IV	27174.60	−04	$a^3P_1 - y^3P_2^\circ$
B	3677.630	12	IV	27183.71	−01	$a^3G_3 - w^3F_2^\circ$
V	3677.477	(2)		27184.85	+04	$a^3P_2 - y^3G_3^\circ$
J	3677.309	2	IV	27186.09	−02	$a^1D_2 - w^1F_3^\circ$
V	3676.879	(1)		27189.27	−04	$z^7P_3^\circ - e^5G_2$
B	3676.314	6	IV	27193.44	−04	$b^3F_4 - x^3G_5^\circ$
J	3674.766	2	IV	27204.90	+01	$b^3P_2 - w^3P_1^\circ$
U	3672.722	1	IV	27220.04	−11	$a^3H_4 - z^3H_5^\circ$
W	3671.51	(1)		27229.02	+01	$z^5D_2^\circ - f^5P_2$
V	3670.810	1	IV	27234.22	−02	$a^3P_0 - w^3D_1^\circ$
J	3670.071	3	IV	27239.70	+16	$b^3G_5 - x^3H_6^\circ$
U	3670.028	3	IV	27240.02	+01	$b^3P_1 - w^3P_2^\circ$
B	3669.523	10	IV	27243.77	−01	$a^3G_4 - w^3F_3^\circ$
J	3669.151	3	IV	27246.53	+08	$b^3G_4 - v^3F_3^\circ$
U	3668.893	(1)		27248.45	00	$b^3F_3 - v^3F_3^\circ$
W	3668.6	(1)		27250.6	−3	$b^3F_2 - v^5P_1^\circ$?
U	3668.214	(1)		27253.49	−03	$z^5D_3^\circ - g^5F_4$
V	*3667.999	1	IV	27255.09	−04	$z^5D_4^\circ - h^5D_4$
					−02	$b^3G_4 - u^3G_5^\circ$
G	3667.252	3n	IV	27260.64	+11	$z^5D_4^\circ - f^5P_3$
U	3666.944	(1)		27262.93	+14	$a^3F_2 - x^5D_2^\circ$
W	*3666.24	1	IV	27268.17	00	$a^3H_5 - x^3F_4^\circ$
					+01	$z^7P_4^\circ - e^5G_4$
U	3664.694	(1)		27279.67	+02	$z^7P_3^\circ - e^7G_3$
G	3664.537	2	IV	27280.83	00	$z^7P_3^\circ - f^5F_4$
W	3663.95	(1)		27285.21	+09	$b^3G_5 - x^3H_5^\circ$
V	*3663.458	1	IV	27288.87	−05	$b^3F_3 - v^5P_2^\circ$
					−03	$b^3F_4 - v^5F_4^\circ$
W	3663.25	(1)		27290.42	+09	$b^3G_4 - 4_4^\circ$
W	3661.36	(1)		27304.50	00	$a^3H_4 - x^3F_3^\circ$
W	3660.33	(1)		27312.19	−05	$z^7F_3^\circ - f^5D_4$
G	3659.516	8	IV	27318.27	+01	$a^3H_4 - z^3H_4^\circ$
U	3658.55	(1)		27325.48	−01	$b^3F_4 - v^5P_3^\circ$
W	3657.89	1	IV	27330.41	+06	$z^7P_2^\circ - e^5P_3$

TABLE B—(*Continued*)

Ref	λ I A	Int	T C	Wave Number		Desig
				Observed	o−c	
U	3657.139	1	IV	27336.02	−07	$a^3P_1 - u^5D_2^\circ$
J	3655.465	4	IV	27348.54	+03	$b^3P_2 - w^3P_2^\circ$
W	3654.66	(1)		27354.56	+08	$a^5P_1 - x^3D_2^\circ$
V	3653.763	1	IV	27361.28	−07	$a^3H_5 - z^3H_6^\circ$
B	3651.469	20	IV	27378.47	+01	$a^3G_3 - v^3G_4^\circ$
W	*3651.10	(1)		27381.23	00	$z^7F_4^\circ - f^7D_5$
					+14	$a^3D_2 - 11_3^\circ$
J	3650.280	5	IV	27387.38	−01	$a^3H_6 - z^3H_6^\circ$
J	3650.031	4	IV	27389.25	−03	$z^7P_3^\circ - e^7S_3$
B	3649.508	12	IV	27393.18	+01	$a^3G_5 - w^3F_4^\circ$
J	3649.304	5	IA	27394.71	+01	$a^5D_4 - z^5F_3^\circ$
B	3647.844	100R	I	27405.67	00	$a^5F_4 - z^5G_5^\circ$
J	3647.427	3	IV	27408.81	−12	$a^3F_3 - x^5D_3^\circ$
J	3645.822	6	IV	27420.87	+03	$c^3P_0 - u^3D_1^\circ$
V	*3645.494	1	IV	27423.34	−15	$z^7F_3^\circ - f^5D_3$
					+02	$b^3G_3 - x^1D_2^\circ$
					+07	$z^7P_3^\circ - f^5F_3$
V	*3645.090	2	IV	27426.38	−12	$z^7F_4^\circ - f^5D_4$
					+06	$c^3P_2 - x^1D_2^\circ$
U	3644.798	(1)		27428.58	+05	$z^5D_3^\circ - f^5P_2$
SS	*3643.812	(1)		27435.99	+02	$a^3F_2 - x^5D_1^\circ$
					−11	$a^3D_3 - y^1F_3^\circ$
V	3643.716	1	IV	27436.72	00	$b^3F_2 - w^3G_3^\circ$
J	3643.627	2	IV	27437.39	−01	$z^7P_4^\circ - e^7F_3$
G	3640.388	15	IV	27461.80	+05	$a^3G_4 - v^3G_5^\circ$
G	3638.296	12	IV	27477.59	−01	$a^3G_3 - y^3H_4^\circ$
J	3637.862	3n	IV	27480.87	+02	$z^7P_4^\circ - e^7F_4$
W	3637.73	(1)		27481.87	+04	$b^3F_4 - v^5F_3^\circ$
V	3637.251	1	IV	27485.49	−02	$a^3H_5 - z^3H_4^\circ$
J	3636.995	2	IV	27487.42	+02	$b^3F_3 - w^3G_4^\circ$
V	3636.650	1	IV	27490.03	+03	$c^3P_2 - u^3G_3^\circ$
U	3636.234	(1)		27493.17	−07	$a^1D_2 - s^3G_3^\circ$
V	*3636.186	2	IV	27493.54	−18	$a^5P_2 - x^3D_3^\circ$
					−17	$z^5D_4^\circ - g^5F_4$
W	3635.19	2	IVA	27501.07	00	$c^3P_2 - t^5D_1^\circ$
G	3634.326	6n	IV	27507.61	+02	$z^7P_4^\circ - e^5G_3$
U	3633.833	1	IV	27511.34	−04	$b^3G_4 - 6_5^\circ$
SS	3633.077	(1)		27517.06	−07	$z^7P_4^\circ - e^7G_5$
J	3632.979	3	IV	27517.80	+03	$a^3P_0 - y^3S_1^\circ$
J	3632.558	3	IV	27521.00	00	$b^3G_5 - v^3F_4^\circ$
J	3632.042	10	IV	27524.90	+10	$c^3P_1 - u^3D_2^\circ$
B	3631.464	125R	I	27529.28	+01	$a^5F_3 - z^5G_4^\circ$
J	3631.103	7	IV	27532.02	−02	$z^7F_5^\circ - f^7D_5$
J	3630.353	4	IV	27537.71	−04	$z^7F_4^\circ - f^5D_3$
V	3628.094	1	IV	27554.85	−02	$a^5P_2 - x^3D_2^\circ$
U	3627.06	(1)		27562.71	00	$a^1H_5 - u^3H_5^\circ$
G	3625.140	6	IV	27577.31	+01	$z^7F_5^\bullet - f^5D_4$
U	3624.31	(1)		27583.62	−03	$a^3P_1 - w^3D_2^\circ$
U	3623.772	2	IV	27587.72	00	$z^7F_3^\circ - f^5D_2$
G	*3623.440	1	IV	27590.24	+05	$b^3F_4 - w^3G_5^\circ$
					−07	$b^3G_5 - u^3G_5^\circ$
G	3623.187	8	IV	27592.17	−02	$a^3H_6 - z^3H_6^\circ$
G	3622.001	12	IV	27601.21	+06	$a^3G_3 - v^3G_3^\circ$
V	3621.718	(2)		27603.36	+01	$a^1H_5 - u^3H_4^\circ$
B	3621.463	15	IV	27605.30	+02	$a^3G_4 - y^3H_5^\circ$
U	3620.228	(1)		27614.72	+10	$z^7F_4^\circ - e^7P_3$
U	3619.772	(1)		27618.20	−03	$a^3H_6 - z^3H_5^\circ$
B	3618.769	125R	I	27625.86	+01	$a^5F_2 - z^5G_3^\circ$
J	*3618.392	2	IV	27628.74	−01	$a^3G_4 - v^3G_4^\circ$
					−09	$z^5D_3^\circ - f^5G_4$
B	3617.788	12	IV	27633.35	+01	$c^3P_2 - u^3D_3^\circ$
W	3617.09	(1)		27638.68	+08	$a^1G_4 - t^3D_3^\circ$
U	3616.326	(1)		27644.52	−06	$a^3P_1 - x^3F_2^\circ$
SS	3616.157	(1)		27645.81	−07	$z^5D_4^\circ - h^5D_3$
U	3615.665	(1)		27649.57	00	$a^3F_4 - x^5D_4^\circ$

Ref	λ I A	Int	T C	Wave Number		Desig
				Observed	o−c	
W	3615.19	(1)		27653.21	+12	$z^5D_1^\circ - h^5D_1$
W	3613.15	(1)		27668.82	+01	$z^7F_2^\circ - e^7P_2$
J	*3612.940	1	IVA	27670.43	+02	$a^3F_3 - x^5D_2^\circ$
					−10	$a^5P_3 - x^3D_3^\circ$
G	3612.068	8	IV	27677.11	+05	$z^7F_5^\circ - e^5G_6$
J	3610.703	2	IV	27687.57	−07	$z^7F_2^\circ - f^5D_1$
G	3610.159	20	III	27691.75	00	$z^7F_6^\circ - e^7F_6^\circ$
B	3608.861	100r	I	27701.70	00	$a^5F_1 - z^5G_2$
J	*3608.146	3	IV	27707.19	−04	$z^7F_4^\circ - e^5G_5$
					+24	$b^3G_4 - u^3G_3^\circ$
G	3606.679	20	III	27718.46	+03	$a^3G_5 - y^3H_6^\circ$
G	**3605.450	15	IV	27727.91	+02	$a^3G_4 - y^3H_4^\circ$
					+42	$(z^7F_6^\circ - f^7D_5)$
U	3604.383	(1)		27736.12	−05	$z^7F_1^\circ - f^5D_0$
J	3603.828	1	IV	27740.39	−01	$c^3P_1 - u^3D_1^\circ$
U	3603.572	(1)		27742.36	+01	$a^3H_5 - w^5G_6^\circ$
G	3603.205	10	IV	27745.18	+03	$a^3G_5 - v^3G_5^\circ$
G	*3602.534	3	IV	27750.36	−01	$z^7F_3^\circ - e^7P_2$
					+12	$z^7P_4^\circ - f^5F_4$
U	3602.46	2	IV	27750.92	+02	$z^7F_3^\circ - f^7D_3$
U	3602.08	1	IVA	27753.85	+01	$z^7F_1^\circ - f^7D_2$
G	3599.624	3	IV	27772.79	+01	$a^1H_5 - u^3F_4^\circ$
U	3598.98	(1)		27777.75	+04	$z^7F_0^\circ - f^7D_1$
W	3598.93?	(1)		27778.14	+01	$z^5D_1^\circ - g^5F_2$
U	3598.721	1	IVA	27779.75	−04	$a^3D_3 - 11_3^\circ$
W	3597.05	3n	IV	27792.66	−11	$z^5D_2^\circ - h^5D_1$
U	3596.198	1	IV	27799.24	+05	$a^3H_5 - w^5G_5^\circ$
U	3595.857	(1)		27801.88	+09	$a^3H_4 - w^5G_4^\circ$
U	3595.66	(1)		27803.40	+15	$z^7F_1^\circ - f^7D_1$
U	3595.308	2	IV	27806.12	−06	$z^7F_2^\circ - f^7D_2$
G	3594.632	8	IV	27811.35	−02	$z^7F_4^\circ - f^7D_4$
U	3593.329	(1)		27821.44	−03	$z^5D_2^\circ - f^5G_3$
U	3592.881	(1)		27824.91	+09	$a^5P_2 - x^3D_1^\circ$
W	3592.68	(1)		27826.46	−07	$z^5D_3^\circ - h^5D_2$
U	3592.486	(1)		27827.96	−09	$b^3F_3 - y^1G_4^\circ$
U	3591.485	(1)		27835.72	+06	$z^5D_0^\circ - g^5F_1$
U	3591.345	(1)		27836.80	00	$z^7F_4^\circ - e^7F_5$
W	3590.99	(1)		27839.56	+07	$z^5D_4^\circ - e^3G_5$
W	3590.66	(1)		27842.12	+02	$b^1G_4 - t^3F_3^\circ$
U	3590.086	(1)		27846.57	−01	$b^3G_5 - 6_5^\circ$
G	3589.456	3	IV	27851.46	+02	$a^3G_4 - v^3G_3^\circ$
B	3589.107	8	III	27854.16	−01	$a^5F_5 - z^5G_5^\circ$
J	3588.918	2	IV	27855.63	+04	$z^7F_2^\circ - f^7D_1$
G	3588.615	3	IV	27857.98	−05	$z^7F_5^\circ - e^5G_5$
J	3587.424	2	IV	27867.23	00	$a^3P_1 - 1_2^\circ$
U	3587.240	2	IV	27868.66	−02	$z^7F_3^\circ - e^5G_4$
G	3586.985	30	II	27870.64	+01	$a^5F_2 - z^5G_2^\circ$
SS	3586.751	(2)		27872.46	−05	$z^7F_6^\circ - e^5G_6$
B	3586.114	10	IV	27877.41	+06	$b^3H_6 - t^3G_5^\circ$
					−08	$(c^3P_2 - t^3D_3^\circ)$
J	3585.708	20	II	27880.57	00	$a^5F_4 - z^5G_4^\circ$
B	3585.320	30	II	27883.58	00	$a^5F_3 - z^5G_3^\circ$
V	3585.193	(2)		27884.57	00	$b^3G_5 - u^3G_4^\circ$
J	*3584.960	4	IV	27886.38	+01	$b^3H_5 - t^3G_4^\circ$
					−19	$z^7P_3^\circ - e^5P_2$
J	3584.790	1	IV	27887.71	−03	$z^7F_3^\circ - f^7D_2$
B	3584.663	8	IV	27888.70	+02	$a^3G_5 - y^3H_5^\circ$
J	3583.337	2	IV	27899.01	+18	$z^5D_0^\circ - f^3D_1$?
W	3582.69	(2)		27904.05	+01	$z^7F_1^\circ - e^5S_2$
W	3582.56	(1)		27905.06	+05	$a^3H_4 - w^5G_3^\circ$
J	3582.201	5	IV	27907.86	+03	$b^3H_6 - 12_6^\circ$
U	3581.816	(1)		27910.86	−01	$c^3P_1 - t^3D_2^\circ$
J	3581.645	1	IV	27912.19	+04	$a^3G_5 - v^3G_4^\circ$
B	3581.195	250R	I	27915.70	00	$a^5F_5 - z^5G_6^\circ$
U	3578.380	(1)		27937.66	+01	$z^7F_0^\circ - e^7F_1$

TABLE B—(*Continued*)

Ref	λ I A	Int	T C	Wave Number Observed	o−c	Desig
B	3576.760	2	IV	27950.31	+04	$b^3H_5 - 13_4^\circ$
G	*3575.976	2	IV	27956.45	+09 +07	$z^7F_2^\circ - e^7F_3$ $z^7F_2^\circ - e^5S_2$
J	3575.374	4	III	27961.15	−01	$c^3P_2 - u^3D_2^\circ$
J	3575.249	2	IV	27962.13	−04	$z^7F_5^\circ - f^7D_4$
U	3575.118	(1)		27963.15	−04	$z^7F_1^\circ - e^7F_1$
G	3573.896	4	IV	27972.71	00	$b^3H_4 - t^3G_3^\circ$
U	3573.836	3	IV	27973.18	−01	$a^3H_6 - w^5G_6^\circ$
U	3573.400	2	IV	27976.60	−02	$a^3D_2 - t^3G_3^\circ$
U	3572.60	(1)		27982.86	−08	$z^7F_4^\circ - e^5G_4$
G	3571.995	6	IV	27987.60	00	$z^7F_5^\circ - e^7F_5$
V	3571.228	2	IVA	27993.61	−01	$a^3F_4 - x^5D_3^\circ$
V	3570.243	20	III	28001.33	00	$z^7F_6^\circ - e^7G_7$
G	3570.100	100R	I	28002.46	00	$a^5F_4 - z^3G_5^\circ$
W	3569.99	(1)		28003.32	+18	$a^3P_1 - y^3S_1^\circ$
J	3568.977	4	IV	28011.26	−03	$a^3G_5 - y^3H_4^\circ$
V	3568.828	(2)		28012.43	00	$a^3D_3 - t^3G_4^\circ$
U	3568.423	(1)		28015.61	+08	$z^7F_2^\circ - e^7F_1$
W	3567.36	(1)		28023.96	+06	$a^3H_4 - x^3G_4^\circ$
U	3567.038	2	IV	28026.49	−06	$z^7F_2^\circ - e^5G_3$
W	3566.59	(1)		28030.01	−02	$a^3H_6 - w^5G_5^\circ$
J	*3565.583	3	IV	28037.93	+01 −01	$z^7F_3^\circ - e^7F_3$ $z^7F_3^\circ - e^5S_2$
B	3565.381	60r	II	28039.52	00	$a^5F_3 - z^3G_4^\circ$
W	3564.11	(1)		28049.51	+02	$a^3F_2 - x^5F_2^\circ$
V	3560.705	5	IV	28076.34	+01	$a^3D_3 - 13_4^\circ$
J	3559.506	2	IV	28085.79	+01	$c^3P_1 - 8_1^\circ$
B	3558.518	30	II	28093.59	−01	$a^5F_2 - z^3G_3^\circ$
G	3556.877	7	IV	28106.55	00	$z^7F_4^\circ - f^5F_5$
W	3556.68	(1)		28108.11	00	$z^7F_3^\circ - e^5G_3$
G	3554.922	40	III	28122.01	+02	$z^7F_5^\circ - e^7G_6$
W	3554.50	2	IV	28125.35	+01	$z^7F_1^\circ - e^5G_2$
J	3554.122	4	IIIA	28128.34	−01	$a^5F_3 - z^5G_2^\circ$
G	3553.741	6	IV	28131.36	+07	$a^1H_5 - v^1G_4^\circ$
G	3552.828	3	IV	28138.58	00	$z^7F_2^\circ - e^7F_2$
U	3552.42	(1)		28141.81	+05	$a^3H_4 - v^5F_4^\circ$
V	3552.112	1	IV	28144.25	−02	$c^3P_1 - v^3P_2^\circ$
J	3549.868	4	III	28162.04	−01	$a^3F_2 - x^5F_1^\circ$
U	3548.037	(2)		28176.58	−08	$c^3P_2 - u^3D_1^\circ$
J	*3547.203	(2)		28183.20	+14 +02	$z^7F_6^\circ - e^7F_5$ $b^3H_4 - w^1G_4^\circ$
U	3546.21	(1)		28191.09	−05	$a^3H_5 - x^3G_4^\circ$
U	3545.832	(1)		28194.10	+01	$a^1G_4 - w^3H_4^\circ$
G	3545.639	5	IV	28195.63	00	$z^7F_4^\circ - e^7F_4$
U	3544.631	(2)		28203.65	−01	$b^3F_2 - v^3D_2^\circ$
J	3543.669	(4)		28211.31	+02	$a^1P_1 - w^1D_2^\circ$
U	3543.392	(1)		28213.51	−08	$a^3H_5 - x^3G_5^\circ$
V	3542.243	1	IV	28222.66	−01	$a^3P_2 - z^3S_1^\circ$
G	3542.076	15	IV	28224.00	+01	$z^7F_3^\circ - e^7G_4$
G	3541.083	15	IV	28231.91	00	$z^7F_4^\circ - e^7G_5$
J	3540.709	3	IIIA	28234.89	+02	$a^5F_4 - z^5G_3^\circ$
G	3540.121	3	IV	28239.58	+02	$z^7F_3^\circ - g^5D_4$
U	3538.79	(1)		28250.20	−07	$a^1H_5 - x^3I_6^\circ$
W	3538.55	(1)		28252.12	+09	$a^3P_0 - v^5P_1^\circ$
W	3538.31	(1)		28254.04	00	$a^1D_2 - u^3F_2^\circ$
J	3537.896	4	IV	28257.34	−02	$z^7F_6^\circ - f^5F_5$
J	3537.729	3	IV	28258.67	+03	$b^3F_2 - v^3D_1^\circ$
J	§3537.491	1	IV	28260.58	+02	$b^3F_3 - v^3D_3^\circ$
G	3536.556	15	IV	28268.05	+03	$z^7F_2^\circ - e^7G_3$
U	*3534.914	(1)		28281.18	+07 00	$a^3F_4 - x^5F_5^\circ$ $a^3F_3 - x^5F_3^\circ$
U	3534.53	(1)		28284.25	−03	$a^1H_5 - x^3I_5^\circ$
G	3533.201	10	IV	28294.89	−03	$z^7F_1^\circ - e^7G_2$
J	3533.008	5	IV	28296.43	−04	$z^7F_0^\circ - e^7G_1$
U	3531.446	(1)		28308.95	−05	$a^3H_6 - v^5F_4^\circ$

Ref	λ I A	Int	T C	Wave Number Observed	o−c	Desig
G	3530.385	2	IV	28317.46	+01	$z^7F_6^\circ - e^7G_6$
G	3529.818	6	III	28322.01	00	$z^7F_1^\circ - e^7G_1$
U	3529.531	(1)		28324.31	−06	$a^1G_4 - y^3I_5^\circ$
G	3527.792	4	IV	28338.27	+02	$z^7F_4^\circ - e^7G_4$
J	3526.673	5	IV	28347.26	00	$z^7F_2^\circ - e^7G_2$
J	3526.465	4	IV	28348.93	+01	$a^3P_2 - y^3P_2^\circ$
J	3526.377	4	IV	28349.64	+06	$z^7F_3^\circ - e^7G_3$
W	3526.23	(3)		28350.82	+06	$z^7F_3^\circ - f^5F_4$
J	3526.167	15	II	28351.33	+01	$a^5F_3 - z^3G_3^\circ$
J	3526.039	20	I	28352.36	+02	$a^5D_2 - z^5P_3^\circ$
V	3526.016	1	IV	28352.54	−10	$b^3F_3 - 3_3^\circ$
U	3525.856	(1)		28353.83	+01	$z^7F_4^\circ - g^5D_4$
G	3524.236	4	IV	28366.87	+06	$a^3P_2 - u^5D_3^\circ$
J	3524.075	3	IV	28368.16	00	$b^3F_3 - v^3D_2^\circ$
W	3523.30	(1)		28374.40	+05	$z^7F_2^\circ - e^7G_1$
U	3522.896	(1)		28377.65	00	$z^7F_2^\circ - e^7S_3$
G	3522.268	(3)		28382.71	−01	$z^7F_5^\circ - e^7G_5$
J	3521.833	2	IVA	28386.22	+02	$a^5P_1 - w^5P_2^\circ$
B	3521.264	25	II	28390.81	−01	$a^5F_4 - z^3G_4^\circ$
U	3520.855	(1)		28394.10	−06	$b^3F_2 - w^3F_2^\circ$
W	3518.86	(2)		28410.20	+08	$a^5P_2 - w^5P_3^\circ$
W	3518.68	(1)		28411.65	+01	$z^7F_2^\circ - f^5F_3$
W	3516.55	(1)		28428.86	+04	$z^7F_3^\circ - e^7G_2$
G	3516.403	5	IV	28430.05	+07	$b^3G_3 - w^3H_4^\circ$
U	3514.626	(1)		28444.42	00	$a^3H_6 - x^3G_5^\circ$
B	3513.820	30	II	28450.95	−01	$a^5F_5 - z^3G_5^\circ$
U	3513.065	(1)		28457.06	−05	$a^3F_3 - x^5F_2^\circ$
U	3512.97	(1)		28457.83	−05	$c^3P_1 - z^1P_1^\circ$
U	3512.239	(1)		28463.76	−08	$z^7F_4^\circ - e^7G_3$
W	3512.08	(1)		28465.04	+02	$z^7F_4^\circ - f^5F_4$
U	3511.748	(1)		28467.74	−06	$b^3F_4 - w^3F_4^\circ$
U	3510.446	(2)		28478.29	00	$a^3P_0 - x^3P_1^\circ$
U	3509.870	(1)		28482.97	−06	$a^5P_1 - w^5P_1^\circ$
W	3509.12	(1)		28489.05	−01	$z^7F_5^\circ - e^7G_4$
W	3508.52	(1)		28493.93	−01	$b^3F_4 - v^3D_3^\circ$
J	3508.494	5	IV	28494.14	−09	$b^3G_4 - w^3H_5^\circ$
W	§3507.39	(1)		28503.11	−10	$c^3P_1 - s^3D_2^\circ$
G	3506.498	6	IV	28510.36	−01	$a^3P_2 - u^5D_2^\circ$
W	3506.23	(1)		28512.54	00	$z^7F_2^\circ - f^5F_2$
V	3505.065	2	IV	28522.01	−03	$c^3P_2 - 8_1^\circ$
U	3504.859	2	IV	28523.69	+01	$a^3P_2 - y^3P_1^\circ$
U	3504.455	(1)		28526.97	+03	$b^3P_2 - v^3F_3^\circ$
G	3500.564	2	IV	28558.69	+03	$b^3F_3 - w^3F_2^\circ$
A	3497.843	40	I	28580.90	00	$a^5D_1 - z^5P_2^\circ$
V	3497.137	(1)		28586.67	+08	$a^5P_2 - w^5P_2^\circ$
J	3497.110	10	III	28586.89	−04	$a^5P_3 - w^5P_3^\circ$
U	3496.19	(1)		28594.41	+06	$a^3H_4 - z^1H_5^\circ$
G	3495.285	8	IV	28601.81	00	$b^3F_4 - w^3F_3^\circ$
U	3494.170	(1)		28610.94	−06	$a^3P_1 - v^5P_2^\circ$
U	3493.698	(2)		28614.81	−04	$a^3G_4 - x^1G_4^\circ$
U	3493.290	(1)		28618.15	−03	$a^3F_4 - x^5F_4^\circ$
A	3490.575	100r	I	28640.41	00	$a^5D_3 - z^5P_3^\circ$
J	3489.670	4	IV	28647.84	+02	$b^3G_5 - w^3H_6^\circ$
U	3486.556	(1)		28673.42	−05	$a^5P_1 - z^3S_1^\circ$
A	3485.342	7	IV	28683.41	−01	$a^5P_2 - w^5P_1^\circ$
U	3484.972	(1)		28686.45	+04	$a^3P_1 - v^5F_2^\circ$
U	3484.858	(1)		28687.39	−07	$a^3H_4 - w^3G_3^\circ$
G	3483.006	3	IIIA	28702.64	+02	$a^5F_4 - z^3G_3^\circ$
V	3481.558	(1)		28714.58	+02	$a^3P_2 - x^3F_3^\circ$
V	*3479.683	(1)		28730.06	+12 −05	$b^3G_5 - y^3I_6^\circ$ $a^1H_5 - t^3F_4^\circ$
V	3478.788	(1)		28737.45	+05	$a^3P_1 - v^5P_1^\circ$
U	3478.374	(1gn)		28740.87	−01	$a^3H_5 - w^3G_4^\circ$
U	3477.856	(2)		28745.15	−01	$a^5P_1 - y^3P_0^\circ$
V	3477.007	(1)		28752.17	−04	$a^3P_1 - x^3P_2^\circ$

TABLE B—(*Continued*)

Ref	λ I A	Int	T C	Wave Number Observed	o−c	Desig
J	3476.853	(2)		28753.44	+04	$b^3F_3 - v^3G_4^\circ$
A	3476.704	40	I	28754.68	00	$a^5D_0 - z^5P_1^\circ$
V	*3476.336	(2w)		28757.71	−22 +08	$a^3P_2 - w^3D_2^\circ$ $z^5P_3^\circ - i^5D_3$
V	*3475.867	(1)		28761.60	+10 +13	$a^3H_5 - z^1H_5^\circ$ $b^3P_1 - y^1D_2^\circ$
J	3475.651	6	IV	28763.38	−02	$a^5P_3 - w^5P_2^\circ$
G	3475.450	70r	I	28765.04	+01	$a^5D_2 - z^5P_2^\circ$
V	3473.497	(1)		28781.22	+02	$a^5F_2 - y^5P_3^\circ$
U	3471.350	6	IV	28799.02	−01	$a^3P_2 - u^5D_1^\circ$
V	3471.27	5	IV	28799.68	−05	$a^5P_1 - y^3P_2^\circ$
V	3469.834	2	IV	28811.60	+01	$b^3F_2 - v^3G_3^\circ$
V	3469.390	(1)		28815.29	+03	$b^3P_1 - x^1D_2^\circ$
V	3469.012	(2)		28818.43	+01	$b^3H_4 - v^3H_4^\circ$
V	3468.849	4	IV	28819.78	00	$b^3F_4 - v^3G_5^\circ$
V	3466.501	3	IIIA	28839.30	−01	$a^5F_5 - z^3G_4^\circ$
V	3466.279	(1)		28841.15	+02	$a^3H_6 - w^3G_5^\circ$
A	3465.863	60r	I	28844.61	−01	$a^5D_1 - z^5P_1^\circ$
V	3464.914	(1)		28852.51	−03	$b^3F_3 - y^3H_4^\circ$
V	3463.305	2	IV	28865.91	+04	$a^3F_4 - x^5F_3^\circ$
V	3462.808	(1)		28870.06	+09	$b^3P_2 - y^1D_2^\circ$
J	3462.353	2	IV	28873.85	−02	$a^5P_2 - z^3S_1^\circ$
G	3459.911	4	IV	28894.23	+09 +35	$c^3P_2 - z^1P_1^\circ$ $(a^3P_2 - w^3D_1^\circ)$
V	3459.429	(2)		28898.26	+01	$a^3G_5 - x^1G_4^\circ$
J	3458.304	4	IV	28907.66	+03	$a^3P_1 - x^3P_0^\circ$
V	3457.512	(1)		28914.28	−01	$a^3H_4 - y^1G_4^\circ$
V	*3457.090	(3w)		28917.81	−01 +01	$z^5P_3^\circ - i^5D_2$ $b^3P_2 - 7_2^\circ$
J	3453.022	(2)		28951.87	00	$a^3G_3 - v^3F_2^\circ$
G	3452.273	10	III	28958.15	+02	$a^5F_3 - y^3F_4^\circ$
G	3451.915	10	IV	28961.16	−01	$a^5P_1 - u^5D_2^\circ$
J	3451.628	2	IV	28963.57	−09 +26	$a^3P_1 - x^3P_1^\circ$ $(b^3F_4 - y^3H_5^\circ)$
G	3450.328	10	IV	28974.48	00	$a^5P_1 - y^3P_1^\circ$
V	3448.869	(1)		28986.74	−04	$b^3F_4 - v^3G_4^\circ$
V	3448.786	(1)		28987.44	00	$b^3P_2 - u^3G_3^\circ$
U	3448.472	(1)		28990.07	00	$b^3G_3 - 9_4^\circ$
G	3447.278	8	IV	29000.12	00	$a^5P_2 - y^3P_2^\circ$
U	3446.947	(1)		29002.90	+03	$a^5F_1 - y^5P_2^\circ$
V	3446.791	(1)		29004.21	−02	$b^3F_2 - w^3P_1^\circ$
A	3445.151	20	III	29018.02	+01	$a^5P_2 - u^5D_3^\circ$
A	3443.878	50r	I	29028.74	−01	$a^5D_2 - z^5P_1^\circ$
V	*3442.979	(1)		29036.32	+01 −13	$c^3P_1 - v^3P_1^\circ$ $a^1D_2 - t^3F_3^\circ$
J	3442.672	3	IIIA	29038.91	−02	$a^5F_3 - y^5P_3^\circ$
V	3442.364	5	IV	29041.51	+01	$a^3P_2 - 1_2^\circ$
J	3440.989	75R	I	29053.12	+02	$a^5D_3 - z^5P_2^\circ$
J	3440.610	150R	I	29056.32	−02	$a^5D_4 - z^5P_3^\circ$
U	3439.039	(1)		29069.59	−01	$a^5G_4 - x^3H_5^\circ$
U	3437.952	(2)		29078.78	−02	$b^3H_5 - v^3H_5^\circ$
V	3437.631	(1)		29081.49	−04	$a^3H_5 - y^1G_4^\circ$
G	3437.046	3	IV	29086.44	00	$a^1G_4 - y^1F_3^\circ$
V	3436.045	(1)		29094.92	−05	$b^3H_5 - v^3H_4^\circ$
V	3434.029	(1w)		29112.00	00	$a^3G_3 - t^5D_3^\circ$
V	3432.023	(1)		29129.01	−02	$b^3P_0 - t^3D_1^\circ$?
J	*3431.815	3	IV	29130.78	00 +22	$b^3P_2 - u^3D_3^\circ$ $a^3D_2 - w^1D_2^\circ$
V	3428.746	(2)		29156.85	+02	$z^5P_3^\circ - 4_2$
G	3428.192	8	III	29161.56	00	$a^5P_2 - u^5D_2^\circ$
A	3427.121	20	III	29170.66	+02	$a^5P_3 - u^5D_4^\circ$
J	3427.002	2	IIIA	29171.69	−11	$a^5F_2 - y^5P_2^\circ$
J	3426.637	5	IV	29174.80	−08	$a^5P_2 - y^3P_1^\circ$
J	*3426.383	5d	IIIA	29176.96	+03 −02	$a^5P_3 - y^3P_2^\circ$ $a^5F_2 - y^3F_3^\circ$
U	3426.337	(2)		29177.35	−06	$a^3P_2 - y^3S_1^\circ$
G	3425.009	4	IV	29188.67	+08	$a^1G_4 - x^1F_3^\circ$
G	3424.284	10	III	29194.84	+02	$a^5P_3 - u^5D_3^\circ$
G	3422.656	7	IV	29208.73	00	$a^5P_1 - w^3D_2^\circ$
J	3422.499	3	IV	29210.07	+05	$b^3G_4 - 9_4^\circ$
V	3419.706	(1)		29233.93	−03	$b^3P_1 - t^3D_1^\circ$?
U	3419.154	(1)		29238.65	00	$z^5D_3^\circ - f^3F_2$
G	3418.507	10	III	29244.18	+07	$a^5P_1 - u^5D_0^\circ$
U	3418.176	(2w)		29247.01	−14	$z^5D_1^\circ - e^3P_0$?
G	3417.842	12	III	29249.87	+03	$a^5P_1 - u^5D_1^\circ$
J	3417.273	(1gn)		29254.74	−11	$a^5F_1 - y^5P_1^\circ$
U	3416.679	(1)		29259.83	+03	$a^3P_0 - v^3D_1^\circ$
G	3415.530	4	IV	29269.67	+01	$a^5P_1 - x^3F_2^\circ$
A	3413.135	15	III	29290.21	−01	$a^5P_2 - w^3D_3^\circ$
G	3411.353	3	IV	29305.51	+03	$a^3G_4 - v^3F_4^\circ$
V	3411.134	(1)		29307.39	−03	$a^3G_5 - x^3H_6^\circ$
V	3410.905	(1)		29309.36	−07	$a^5F_4 - y^3F_4^\circ$
G	3410.171	3	IV	29315.66	+01	$a^1P_1 - u^3F_2^\circ$
U	3410.031	(1)		29316.87	+02	$a^1G_4 - 10_3^\circ$
U	3409.218	(2)		29323.86	−08	$b^3H_6 - v^3H_5^\circ$
A	3407.461	20d	III	29338.98	+08	$a^5P_3 - x^3F_4^\circ$
J	3406.803	6	IV	29344.64	−04	$a^5P_1 - w^3D_1^\circ$
J	3406.442	3	IV	29347.75	−05	$a^3D_1 - w^1D_2^\circ$
W	3405.83	(2)		29353.02	+02	$a^3G_5 - x^3H_5^\circ$
U	3404.923	(1)		29360.85	−06	$a^3G_5 - t^5D_4^\circ$
V	3404.755	(1)		29362.29	00	$a^3G_4 - t^5D_3^\circ$
G	3404.357	6	IV	29365.73	−03	$a^5P_2 - x^3F_3^\circ$
V	*3404.301	3	IIIA	29366.21	+08 −25	$a^3G_4 - v^3F_3^\circ$ $a^5F_1 - y^3F_2^\circ$
U	3403.299	(2)		29374.86	+07	$a^3G_4 - u^3G_5^\circ$
G	3402.256	5	IV	29383.86	00	$b^3H_6 - v^3H_6^\circ$
A	3401.521	6	III	29390.21	−01	$a^5F_4 - y^5P_3^\circ$
A	3399.336	15	III	29409.10	−02	$a^5P_2 - w^3D_2^\circ$
V	3399.230	(1)		29410.02	+01	$a^3G_4 - 4_4^\circ$
U	3398.220	(1)		29418.76	00	$a^3G_3 - u^3G_4^\circ$
V	3397.642	2	IIIA	29423.76	−02	$a^5F_2 - y^5P_1^\circ$
V	3397.560	(1)		29424.47	−01	$b^3G_3 - x^1F_3^\circ$
V	3397.221	(1)		29427.41	−07	$c^3P_2 - x^1F_3^\circ$
A	3396.978	4	IIIA	29429.59	−01	$a^5F_3 - y^5P_2^\circ$
V	3396.386	(1)		29434.64	−06	$a^5F_3 - y^3F_3^\circ$
G	3394.583	5	IV	29450.28	+05	$a^5P_2 - u^5D_1^\circ$
V	3394.085	(1)		29454.60	−07	$a^3H_4 - w^3F_3^\circ$
V	3393.915	(1)		29456.07	+03	$a^3P_2 - x^3G_3^\circ$
V	*3393.609	(1w)		29458.73	+13 −14	$b^3P_2 - u^3D_2^\circ$ $a^3G_3 - y^1D_2^\circ$
V	3393.382	(1)		29460.70	+03	$b^3P_0 - u^3D_1^\circ$
G	3392.652	15	III	29467.04	+01	$a^5P_3 - w^3D_3^\circ$
G	3392.304	8	IV	29470.06	00	$a^5P_2 - x^3F_2^\circ$
U	3392.014	2	IV	29472.58	+01	$c^3P_2 - v^3P_1^\circ$
V	3389.748	2	IV	29492.28	−03	$a^5P_1 - 1_2^\circ$
V	3388.966	(1w)		29499.09	+04	$c^3P_1 - t^5P_1^\circ$
W	3388.8	(1)		29500.5	+1	$a^3P_1 - 2_2^\circ$
V	3387.410	2	IV	29512.64	−02	$a^3G_3 - x^1D_2^\circ$
C	3383.981	8	IV	29542.54	−03	$a^5P_3 - x^3F_3^\circ$
G	*3383.692	5	IV	29545.07	−01 −15	$a^5P_2 - w^3D_1^\circ$ $b^3G_5 - 9_4^\circ$
V	3383.387	(1)		29547.73	−14	$b^3F_2 - z^1F_3^\circ$?
G	3382.403	3	IV	29556.32	−01	$a^5P_3 - z^3H_4^\circ$
V	*3381.340	(2)		29565.61	+01 −08	$b^3P_1 - u^3D_1^\circ$ $a^3D_3 - w^1F_3^\circ$
C	3380.111	8	IV	29576.36	+02	$a^3G_3 - u^3G_3^\circ$
V	3380.004	(1)		29577.30	−18	$z^5F_5^\circ - 2$
G	3379.017	6	IV	29585.94	+01	$a^5P_3 - w^3D_2^\circ$
G	3378.676	6	IV	29588.92	+04	$a^3G_5 - v^3F_4^\circ$
V	3374.221	(1)		29627.99	−23	$a^5P_1 - y^3S_1^\circ$

TABLE B—(*Continued*)

Ref	λ I A	Int	T C	Wave Number Observed	o−c	Desig
V	3373.874	(1)		29631.04	−02	$a^3G_4 - 6_5^\circ$
U	3372.352	(1)		29644.41	−02	$b^3G_4 - x^1F_3^\circ$
G	3372.070	3	IV	29646.89	+03	$a^5P_3 - x^3F_2^\circ$
A	3370.786	10	IV	29658.18	−01	$a^3G_5 - u^3G_5^\circ$
G	**3369.549	8	III	29669.07	+02 +05	$a^3G_4 - u^3G_4^\circ$ $(c^3P_2 - 11_3^\circ)$
V	3368.983	(1)		29674.05	−05	$b^3P_2 - u^3D_1^\circ$
U	3367.159	(1)		29690.13	−06	$a^3P_1 - v^3D_2^\circ$
U	3366.867	5	IV	29692.70	00	$a^5P_2 - 1_2^\circ$
U	3366.789	. 5	IV	29693.39	−02	$a^3G_5 - 4_4^\circ$
V	3364.639	(1)		29712.37	00	$b^3F_3 - z^1F_3^\circ$
V	3363.815	(1)		29719.64	−04	$a^3G_3 - u^3D_3^\circ$
V	3361.959	(1)		29736.05	−02	$b^3P_1 - t^3D_2^\circ$
U	3360.922	(1)		29745.22	+05	$a^3P_1 - v^3D_1^\circ$
V	3359.814	(2)		29755.03	−02	$b^3H_4 - u^3H_5^\circ$
U	3359.491	2	IIIA	29757.89	−03	$a^5F_5 - y^3F_4^\circ$
H	3356.407	3	IV	29785.03	−24	$a^3P_2 - v^5P_2^\circ$
U	3356.323	1	IVA	29785.98	−02	$a^5F_4 - y^3F_3^\circ$
V	3355.517	(1)		29793.14	+02	$a^5F_3 - y^3F_2^\circ$
C	3355.228	6	IV	29795.70	+01	$b^3H_4 - u^3H_4^\circ$
U	3354.064	3	IV	29806.04	−01	$b^3P_0 - 8_1^\circ$
U	3353.267	(1)		29813.13	−03	$a^3H_5 - y^3H_6^\circ$
V	3352.929	(1)		29816.13	−04	$a^3H_4 - y^3H_5^\circ$
V	3351.750	3	IV	29826.62	−01	$a^3G_4 - u^3G_3^\circ$
V	3351.529	2	IV	29828.58	−03	$a^5P_2 - y^3S_1^\circ$
V	*3350.284	(3)		29839.67	+03 −21	$a^3H_4 - v^3G_4^\circ$ $a^3H_5 - v^3G_5^\circ$
V	3349.739	(1)		29844.52	−05	$b^3P_2 - t^3D_2^\circ$
A	3347.927	6	IV	29860.68	−01	$a^3P_2 - v^5F_2^\circ$
V	3347.507	(1)		29864.42	−03	$b^3G_4 - t^3G_5^\circ$
U	3346.936	1	IV	29869.52	+01	$a^5P_3 - 1_2^\circ$
V	3345.679	(1)		29880.74	+05	$a^3P_1 - w^3F_2^\circ$
V	3343.678	(1)		29898.62	−04	$b^3G_3 - t^3G_4^\circ$
U	3343.240	(1)		29902.54	−02	$a^5P_3 - z^1G_4^\circ$
V	3342.298	4	V	29910.96	−02	$b^3P_1 - 8_1^\circ$
U	3342.216	5	IV	29911.70	+03	$a^3P_2 - v^5P_1^\circ$
G	3341.906	5	IIIA	29914.48	+02	$a^3G_5 - 6_5^\circ$
A	3340.566	6	IV	29926.47	−02	$a^3P_2 - x^3P_2^\circ$
U	3339.582	(1w)		29935.29	−02	$c^3P_2 - t^5P_1^\circ$
U	*3339.195	2	IV	29938.76	−02 −03	$a^3H_4 - y^3H_4^\circ$ $b^3G_5 - y^1H_5^\circ$
U	3338.638	(3w)		29943.75	+01	$z^7P_3^\circ - g^7D_4$
C	3337.666	6	IV	29952.48	+03	$a^3G_5 - u^3G_4^\circ$
U	3336.254	(3)		29965.15	+03	$b^3H_4 - u^3F_4^\circ$
V	3335.776	4	IV	29969.44	−03	$b^3P_1 - v^3P_2^\circ$
V	3335.513	(1)		29971.81	00	$a^3F_3 - x^5P_3^\circ$
V	3335.403	(1)		29972.80	−08	$b^3F_4 - x^1G_4^\circ$
V	3334.278	(1)		29982.91	−01	$b^3H_5 - u^3H_6^\circ$
V	3334.223	(3)		29983.40	−01	$a^3H_5 - y^3H_5^\circ$
V	3331.778	(2)		30005.40	+01	$a^3P_0 - w^3P_1^\circ$
U	3331.612	3	IV	30006.90	+02	$a^3H_5 - v^3G_4^\circ$
V	3329.532	(2)		30025.64	−02	$a^1G_4 - t^3G_3^\circ$
C	3328.867	5	IV	30031.64	+04	$b^3H_5 - u^3H_5^\circ$
V	3327.961	(1)		30039.82	−04	$a^5P_3 - w^5G_4^\circ$
V	3327.498	(2)		30044.00	00	$a^3H_6 - y^3H_6^\circ$
V	3325.468	4	IV	30062.34	+01	$a^3H_4 - v^3G_3^\circ$
V	3324.541	4	IV	30070.72	00	$a^3H_6 - v^3G_5^\circ$
V	3324.372	(2)		30072.25	+01	$b^3H_5 - u^3H_4^\circ$
C	3323.737	7	IV	30077.99	+02	$b^3P_2 - v^3P_2^\circ$
G	3322.474	5n	IV	30089.42	−01	$z^7P_4^\circ - g^7D_5$
U	3320.779	(2n,gn)		30104.78	−01	$z^7P_2^\circ - g^7D_2$
V	3320.650	(2)		30105.95	−07	$a^3H_5 - y^3H_4^\circ$
V	3319.258	2	IV	30118.58	−03	$b^3G_4 - t^3G_4^\circ$
G	3317.121	3	IV	30137.98	+05	$a^3P_2 - x^3P_1^\circ$
C	3314.742	7	IV	30159.61	+01	$a^3D_2 - u^3F_3^\circ$
U	3314.441	(2)		30162.35	+04	$b^3F_2 - v^3F_2^\circ$
V	§3314.070	(1)		30165.73	−03	$a^1P_1 - t^3F_2^\circ$
V	3313.723	(1)		30168.89	−05	$a^3F_2 - y^5G_3^\circ$
U	3312.224	(1)		30182.54	+03	$b^3G_4 - 13_4^\circ$
V	3311.451	(1)		30189.58	−01	$a^5F_2 - y^3D_3^\circ$
V	3310.496	(3)		30198.29	−01	$a^3D_3 - u^3H_4^\circ$
V	3310.347	4	IV	30199.65	00	$b^3G_5 - t^3G_5^\circ$
G	3307.234	5	IV	30228.07	+01	$b^3H_6 - u^3H_6^\circ$
U	3307.008	(1)		30230.14	+01	$b^3G_5 - 12_5^\circ$
S	3306.703	(−)		30232.93	+07	$z^7P_3^\circ - g^7D_3$
U	3306.490	(1)		30234.88	−04	$a^3D_2 - u^3F_2^\circ$
C	3306.356	20	III	30236.10	+02 −03	$a^5P_1 - v^5P_2^\circ$ $(a^1G_4 - w^1G_4^\circ)$
C	3305.971	20	III	30239.62	+01	$a^5P_2 - v^5P_3^\circ$
V	3303.574	(2)		30261.56	+01	$b^3G_3 - t^3G_3^\circ$
U	3301.917	(1)		30276.75	+01	$b^3H_6 - u^3H_5^\circ$
V	3301.227	(2)		30283.08	00	$b^3P_1 - z^1P_1^\circ$
U	3299.509	(1)		30298.85	−03	$a^3F_3 - x^5P_2^\circ$
U	3299.077	(1w)		30302.81	−08	$z^5F_3^\circ - i^5D_4$
A	3298.133	6	IV	30311.49	00	$a^5P_1 - v^5F_2^\circ$
V	§3296.806	(1)		30323.69	+06	$b^3H_4 - v^1G_4^\circ$
U	3296.467	(1)		30326.80	−01	$b^3F_3 - v^3F_2^\circ$
U	3293.142	(1)		30357.43	+02	$a^3F_2 - z^5H_3^\circ$
G	3292.590	8	IV	30362.51	+03	$a^5P_1 - v^5P_1^\circ$
G	3292.022	8	IV	30367.76	+03	$a^3D_3 - u^3F_4^\circ$
G	3290.988	5	IV	30377.29	−01	$a^5P_1 - x^3P_2^\circ$
U	3290.714	(2)		30379.82	−01	$a^5P_3 - v^5F_4^\circ$
V	3289.442	(2)		30391.57	−01	$b^3P_2 - z^1P_1^\circ$
U	3288.967	2	IV	30395.96	+01	$a^5P_2 - v^5F_3^\circ$
U	3288.651	(2)		30398.88	+01	$a^3P_1 - w^3P_0^\circ$
V	3287.117	(1w)		30413.07	−08	$z^7P_4^\circ - g^7D_4$
A	3286.755	20	III	30416.42	00	$a^5P_3 - v^5P_3^\circ$
U	3286.444	(2w)		30419.30	+03	$z^5F_3^\circ - i^5D_3$
U	3286.022	2	IV	30423.20	−01	$a^5P_1 - v^5F_1^\circ$
U	3285.20	(1)		30430.81	−03	$z^7P_3^\circ - g^7D_2$
A	3284.588	5	IV	30436.48	+01	$a^5P_2 - v^5P_2^\circ$
V	3283.430	(1)		30447.21	−10	$a^5F_3 - y^3D_3^\circ$
G	3282.891	(2)		30452.21	+05	$a^3D_1 - u^3F_2^\circ$
U	3282.720	(1)		30453.80	−01	$b^3G_5 - t^3G_4^\circ$
V	3280.763	(1)		30471.96	−06	$b^3G_3 - w^1G_4^\circ$
C	3280.261	8	IV	30476.62	00	$b^3H_4 - x^3I_5^\circ$
U	3279.739	(2l)		30481.48	−02	$b^3G_4 - t^3G_3^\circ$
V	*3278.741	4	IV	30490.76	00 −02	$a^3P_1 - w^3P_1^\circ$ $b^3F_3 - v^3F_3^\circ$
U	3276.468	4	IV	30511.91	+02	$a^5P_2 - v^5F_2^\circ$
V	3275.848	(1)		30517.68	−03	$b^3G_5 - 13_4^\circ$
V	3275.685	(1)		30510.20	−12	$a^3G_3 - w^3H_4^\circ$
U	3274.453	(2w)		30530.68	−07	$z^5F_4^\circ - i^5D_4$
U	3272.71	(1)		30546.95	+14	$z^5F_1^\circ - 4_2$
U	3271.683	(2)		30556.53	+03	$a^3F_4 - x^5P_3^\circ$
U	3271.487	(2)		30558.36	+06	$a^3D_3 - u^3F_3^\circ$
A	3271.002	15	III	30562.89	+02	$a^5P_2 - v^5P_1^\circ$
V	3269.964	(1)		30572.59	−17	$a^5P_3 - v^5F_3^\circ$
U	3269.235	(1w)		30579.41	−05	$z^5F_3^\circ - i^5D_2$
G	3268.234	5	IV	30588.78	+04	$a^5P_1 - x^3P_1^\circ$
G	3265.616	15	III	30613.30	+02	$a^5P_3 - v^5P_2^\circ$
G	3265.046	8	IA	30618.64	00	$a^5D_2 - z^3D_3^\circ$
U	3264.710	(2)		30621.79	−14	$z^7D_2^\circ - f^5D_3$
U	3264.512	5	IV	30623.65	+05	$a^5P_2 - v^5F_1^\circ$
V	3263.378	(2)		30634.29	−09	$a^3P_1 - w^3P_2^\circ$
V	3262.009	(2)		30647.15	+02	$z^5F_4^\circ - i^5D_3$
V	3261.332	(2)		30653.51	−06	$z^5F_2^\circ - 4_2$
U	3260.261	4	IV	30663.58	+07	$b^3F_4 - v^3F_4^\circ$
G	3259.991	6	IV	30666.12	−02	$z^7D_3^\circ - f^5D_4$
A	3257.594	8	IV	30688.68	−01	$a^5P_3 - v^5F_2^\circ$

TABLE B—(*Continued*)

Ref	λ I A	Int	T C	Wave Number Observed	o−c	Desig
V	*3257.244	2	IV	30691.98	+01 −29	$b^3G_4 - w^1G_4^\circ$ $a^5F_2 - y^3D_2^\circ$
U	3254.734	(2)		30715.65	−05	$a^3G_5 - w^3H_6^\circ$
C	3254.363	10	IV	30719.15	−01	$b^3H_6 - x^3I_6^\circ$
V	3254.261	(1)		30720.11	−21	$b^3F_4 - t^5D_3^\circ$
U	3253.949	(2)		30723.06	−04	$b^3F_2 - x^1D_2^\circ$
U	3253.834	(1)		30724.15	−01	$b^3F_4 - v^3F_3^\circ$
V	3253.610	4	IV	30726.26	+02	$a^3D_3 - v^1G_4^\circ$
U	3252.926	4	IV	30732.72	−10	$b^3F_4 - u^3G_5^\circ$
G	3251.236	8	IV	30748.69	−03	$a^5P_2 - w^3G_3^\circ$
U	3250.623	4	IV	30754.49	−01	$a^5P_3 - x^3P_2^\circ$
U	*3250.394	(2)		30756.66	+08 −20	$b^3P_0 - v^3P_1^\circ$ $a^3P_2 - v^3D_2^\circ$
U	3249.191	3	IV	30768.05	+01	$b^3F_4 - 4_4^\circ$
V	3249.037	(1)		30769.50	−11	$a^3G_4 - w^3H_4^\circ$
G	3248.206	10	IV	30777.38	−02	$z^7D_3^\circ - f^5D_3$
U	3247.278	3	IV	30786.17	+01	$z^7D_2^\circ - f^5D_2$
U	3246.962	6	IV	30789.17	+04	$a^5P_2 - x^3P_1^\circ$
U	3246.482	3	IV	30793.72	+02	$b^3F_3 - u^3G_4^\circ$
G	3246.005	8	I	30798.24	00	$a^5D_1 - z^3D_2^\circ$
V	3245.984	(2)		30798.44	−17	$a^5F_4 - y^3D_3^\circ$
A	3244.190	15	IV	30815.47	+01	$z^7D_4^\circ - f^7D_5$
V	*3243.406	3	IV	30822.92	−11 +15	$z^5F_5^\circ - i^5D_4$ $b^3P_2 - y^1F_3^\circ$
U	3243.109	(1)		30825.75	+01	$a^3H_4 - x^1G_4^\circ$
V	3242.268	(1)		30833.74	−07	$b^3F_3 - y^1D_2^\circ$
U	3241.52	(1)		30840.86	−18	$a^5F_1 - y^3D_1^\circ$
U	3240.013	(1)		30855.20	00	$a^1G_4 - v^3H_5^\circ$
A	3239.436	15	IV	30860.70	−02 +20	$z^7D_4^\circ - f^5D_4$ $(z^7D_1^\circ - f^5D_1)$
V	*3239.029	(1)		30864.57	+11 −17	$a^3P_2 - v^3D_2^\circ$ $a^3P_2 - w^3F_3^\circ$
S	3238.535	(−)		30869.28	+09	$z^7P_2^\circ - e^3P_1$
V	3237.234	(1)		30881.69	+05	$b^3F_3 - 7_2^\circ$
A	3236.223	8	IA	30891.33	−01	$a^5D_3 - z^3F_4^\circ$
U	3235.592	(1)		30897.36	+05	$a^3G_5 - w^3H_5^\circ$
G	3234.614	7	IA	30906.70	−01	$a^5D_3 - z^3D_3^\circ$
G	3233.967	12	IV	30912.88	+02	$z^7D_4^\circ - e^7P_4$
V	3233.304	(1)		30919.22	−22	$a^3P_2 - v^3D_1^\circ$?
G	3233.053	8	IV	30921.62	+01	$b^3H_6 - x^3I_7^\circ$
U	3231.576	(1)		30935.75	+09	$a^3F_4 - y^5G_5^\circ$
G	3230.963	10	IV	30941.62	−01	$z^7D_3^\circ - f^5D_2$
V	3230.210	6	IV	30948.84	+03	$z^7D_2^\circ - e^7P_2$
U	3229.994	(3)		30950.91	+02	$a^1G_4 - x^1H_5^\circ$
W	3229.78?	(1)		30952.96	−11	$b^3F_3 - u^5F_2^\circ$
G	3229.123	4	IIA	30959.26	−02	$a^5D_0 - z^3D_1^\circ$
G	3228.900	3	IV	30961.39	+02	$z^7D_1^\circ - f^5D_0$
U	3228.254	5	IV	30967.59	−05	$z^7D_2^\circ - f^5D_1$
V	3228.003	(2)		30970.00	−01	$b^3P_2 - v^3P_1^\circ$
G	3227.798	15	IV	30971.96	−02	$z^7D_4^\circ - f^5D_3$
U	3227.063	3	IV	30979.02	−02	$z^7D_1^\circ - f^7D_2$
U	3226.720	2	IIIA	30982.31	−06	$a^5D_2 - z^3D_2^\circ$
A	3225.789	25	III	30991.25	−04	$z^7D_5^\circ - e^7F_6$
U	*3225.607	(1)		30993.00	+02 +10	$a^3H_5 - x^1G_4^\circ$ $b^3D_3 - r^3G_3^\circ$
U	3223.844	(1)		31009.95	−02	$a^5F_2 - y^3D_1^\circ$
V	3223.273	(1)		31015.44	+04	$a^3F_4 - z^5H_5^\circ$
A	3222.069	20	III	31027.03	00 −14	$z^7D_5^\circ - f^7D_5$ $(b^3G_5 - w^1G_4^\circ)$
U	3221.931	2	IV	31028.36	−09	$z^7D_1^\circ - f^7D_1$
G	3219.806	10	III	31048.84	00 −39	$z^7D_4^\circ - e^7P_3$ $(a^5D_1 - z^3D_1^\circ)$
G	3219.581	12	IV	31051.01	00	$z^7D_3^\circ - f^7D_4$
A	3217.380	10	IV	31072.24	−05	$z^7D_5^\circ - f^5D_4$
A	3215.940	12	IV	31086.16	−02	$z^7D_2^\circ - f^7D_2$
V	3215.637	(3)		31089.09	+03	$z^7F_6^\circ - e^3G_5$
V	3214.624	(1)		31098.88	00	$a^3P_2 - z^1D_2^\circ$
G	3214.396	8	IA	31101.09	00	$a^5D_2 - z^3F_3^\circ$
V	*3214.044	20	III	31104.50	−16 −31 +22	$z^5F_4^\circ - g^5G_5$ $z^7D_3^\circ - f^7D_3$ $(z^7D_3^\circ - e^7P_2)$
U	3213.754	(1)		31107.30	+04	$b^3G_3 - v^3H_4^\circ$
G	3211.989	10	IV	31124.40	−03	$z^7D_5^\circ - e^7P_4$
V	*3211.872	4	IV	31125.53	+01 +05	$a^5P_1 - 2_2^\circ$ $z^5F_3^\circ - g^5G_4$
U	3211.683	8	IV	31127.36	+04	$z^5F_5^\circ - g^5G_6$
U	3211.487	4	IV	31129.25	+01	$z^7D_1^\circ - e^5S_2$
G	3210.830	10	IV	31135.63	+04	$z^7D_2^\circ - f^7D_1$
G	3210.230	8	IV	31141.45	00	$z^7D_4^\circ - e^5G_5$
G	*3209.297	6	IV	31150.50	+01 +03	$z^5F_2^\circ - g^5G_3$ $z^7F_6^\circ - g^7D_5$
V	3209.115	(1)		31152.27	−09	$a^5P_3 - y^1G_4^\circ$
G	3208.470	4	IV	31158.53	+08	$z^5F_1^\circ - g^5G_2$
V	3207.649	(1w)		31166.51	+05	$b^3P_2 - 11_3^\circ$
U	3207.089	2	IV	31171.95	−10	$z^7D_5^\circ - e^5G_6$
A	3205.400	15	IV	31188.37	−02	$z^7D_1^\circ - e^7F_1$
V	3202.562	2	IV	31216.01	−02	$a^1G_4 - w^1F_3^\circ$
S	3201.891	(−)		31222.55	−04	$z^7D_3^\circ - e^5G_4$
U	3200.784	2	IIIA	31233.35	00	$a^5D_2 - z^3D_1^\circ$
A	*3200.475	15	IV	31236.37	+01 −01	$z^7D_2^\circ - e^7F_3$ $z^7D_2^\circ - e^5S_2$
G	3199.530	15	II	31245.59	00 −29	$z^7D_4^\circ - f^7D_4$ $(a^5D_1 - z^3F_2^\circ)$
U	3198.266	(1)		31257.94	00	$b^3F_2 - u^3D_2^\circ$
U	3197.521	(1)		31265.22	+01	$z^5F_2^\circ - g^5G_2$
U	3196.977	(2)		31270.54	+10	$a^5D_3 - z^3D_2^\circ$
A	**3196.930	20	II	31271.00	−02	$z^7D_4^\circ - e^7F_5$
V	3196.147	2	IV	31278.66	−08	$z^7F_5^\circ - g^7D_4$
V	3194.422	3	IV	31295.55	+02	$z^7D_2^\circ - e^7F_1$
U	3193.303	8	IV	31306.52	−03	$z^7D_2^\circ - e^5G_3$
U	3193.228	10	IA	31307.25	−02	$a^5D_4 - z^3F_4^\circ$
G	3192.799	8	IV	31311.46	+02 +42	$z^7D_1^\circ - e^7F_2$ $(b^3G_4 - v^3H_5^\circ)$
U	*3192.417	(1)		31315.20	−07 −19	$a^5P_1 - v^3D_2^\circ$ $z^5F_3^\circ - g^5G_3$
A	3191.659	7	IA	31322.64	00	$a^5D_4 - z^3D_3^\circ$
S	3191.180	(−)		31327.34	+13	$b^3G_4 - v^3H_4^\circ$
U	3191.116	(1)		31327.97	−03	$b^3F_4 - u^3D_3^\circ$
U	3190.816	(2)		31330.91	+04	$a^1G_4 - s^3G_4^\circ$
U	3190.651	(2)		31332.54	+01	$a^1G_4 - s^3G_5^\circ$
W	3190.02	(1)		31338.74	−03	$b^3F_3 - t^3D_3^\circ$
G	3188.819	7	IV	31350.54	+01	$z^7D_1^\circ - e^5G_2$
G	3188.567	4	IV	31353.02	00	$z^7D_5^\circ - e^5G_5$
A	3184.896	7	IA	31389.15	−01	$a^5D_3 - z^3F_3^\circ$
U	*3184.622	3	IV	31391.85	+02 00	$z^7D_3^\circ - e^7F_3$ $z^7D_3^\circ - e^5S_2$
G	3182.970	3	IV	31408.14	+08	$a^5P_2 - v^3D_3^\circ$
U	*3182.060	3	IV	31417.13	−04 +08	$z^7D_4^\circ - e^5G_4$ $z^7F_4^\circ - g^7D_3$
U	*3181.922	(2)		31418.49	00 −09	$c^3P_2 - w^1D_2^\circ$ $z^7D_2^\circ - e^7F_2$
U	3181.847	(3)		31419.22	+01	$z^7F_2^\circ - g^7D_2$
G	3181.522	4	IV	31422.44	00	$b^3F_3 - u^3D_2^\circ$
G	3180.756	5	IIA	31430.01	00	$a^5D_2 - z^3F_2^\circ$
G	3180.223	20	IV	31435.27	−01	$z^7D_3^\circ - e^7F_4$
U	§3179.479	(1)		31442.63	+02	$a^3F_2 - w^5D_1^\circ$
U	3178.967	3	IV	31447.70	−03	$a^3H_5 - x^3H_5^\circ$
V	3178.545	2	IV	31451.87	−05	$b^3G_3 - w^1F_3^\circ$
A	3178.015	10	IV	31457.11	−05 −56	$z^7D_5^\circ - f^7D_4$ $(z^7D_2^\circ - e^5G_2)$

TABLE B—(*Continued*)

Ref	λ I A	Int	T C	Wave Number Observed	o−c	Desig
Q	3177.54	(2)	Fe II	31461.82	−20	$(z^7D_3^\circ - e^5G_3)$
V	3176.366	2	IV	31473.44	00	$b^3F_2 - u^3D_1^\circ$
W	3175.97	(1)		31477.37	+02	$z^7F_0^\circ - g^7D_1$
A	3175.447	12	IV	31482.55	−04	$z^7D_5^\circ - e^7F_5$
V	3173.663	(3r)		31500.25	+11	$a^5P_2 - 3_3^\circ$
U	3173.608	(1)		31500.79	+02	$z^7F_3^\circ - g^7D_2$
W	3173.40	(1)		31502.86	−03	$z^7F_1^\circ - g^7D_1$
V	*3172.067	2	IV	31516.10	+17 −27	$a^5P_2 - w^3F_3^\circ$ $a^3H_4 - v^3F_4^\circ$
U	3171.663	2	IV	31520.11	−01	$z^7D_1^\circ - e^7G_2$
V	*3171.353	5	IV	31523.19	−09 +03	$a^3F_4 - w^5D_4^\circ$ $a^1G_4 - s^3G_3^\circ$
U	3168.858	2	IV	31548.01	−01	$z^7D_2^\circ - e^7G_3$
V	3167.907	(1)		31557.48	+08	$z^5D_3^\circ - i^5D_4$
G	3166.435	6	IV	31572.15	00	$b^3F_4 - t^3D_3^\circ$
G	3165.860	4	IV	31577.89	−01	$z^7D_3^\circ - e^7G_4$
V	3165.005	3	IV	31586.41	00	$z^7D_4^\circ - e^7F_3$
U	3164.308	(1)		31593.37	−09	$z^7D_3^\circ - g^5D_4$
U	*3162.335	2n	IV	31613.08	−06 −02	$z^7D_3^\circ - e^5G_2$ $a^3G_5 - 9_4^\circ$
G	3161.949	8	IV	31616.94	−04	$z^7D_5^\circ - e^7G_6$
V	3161.370	4	IV	31622.73	+04	$a^3F_3 - w^5D_2^\circ$
A	3160.658	10	IV	31629.86	00	$z^7D_4^\circ - e^7F_4$
U	3160.344	(2)		31633.00	+01	$a^3H_6 - x^3H_6^\circ$
V	3160.200	(2n)		31634.44	−01	$z^5D_2^\circ - i^5D_2$
W	3158.99	(2)		31646.56	+32	$b^3G_5 - v^3H_5^\circ$?
U	3157.992	(2)		31656.56	−04	$z^7D_4^\circ - e^5G_3$
K	3157.88	6	IV	31657.64	−01	$z^7D_2^\circ - e^7S_3$
A	3157.040	8	IV	31666.10	−04	$z^7D_4^\circ - e^7G_5$
U	3156.464	(1)		31671.88	+01	$b^3G_4 - w^1F_3^\circ$
G	3156.275	5n	IV	31673.77	−01	$z^5D_3^\circ - i^5D_3$
V	3155.293	2	IV	31683.63	+02	$a^3H_5 - v^3F_4^\circ$
SS	3155.131	(1)		31685.26	−14	$z^7D_1^\circ - f^5F_2$
U	3154.505	2	IV	31691.55	−09	$z^7D_2^\circ - f^5F_3$
SS	3154.421	(1)		31692.39	−08	$a^5P_3 - v^3D_2^\circ$
U	3153.322	(1)		31703.44	−05	$z^7D_3^\circ - e^7G_3$
G	3153.200	5	IV	31704.66	−01	$z^7D_3^\circ - f^5F_4$
S	*3153.064	(−)		31706.03	−03 −13	$b^3G_5 - v^3H_6^\circ$ $a^5P_2 - w^3F_2^\circ$
V	3151.867	(1)		31718.08	00	$a^5D_3 - z^3F_2^\circ$
G	3151.353	10	IV	31723.25	−02	$a^3G_4 - y^1H_5^\circ$
U	3150.304	(2n)		31733.81	+03	$z^5D_1^\circ - 4_2$
U	3148.420	(2)		31752.80	−12	$a^3H_5 - u^3G_5^\circ$
U	3147.793	(1)		31759.12	+07	$b^3G_3 - s^3G_3^\circ$
U	3146.475	(1)		31772.42	−06	$z^7D_4^\circ - e^7G_4$
V	3145.057	(2)		31786.74	+03	$b^3G_4 - s^3G_4^\circ$
V	3144.488	6n	IV	31792.50	−04	$z^7D_2^\circ - f^5F_2$
C	3143.990	8	IV	31797.54	−05	$z^5D_4^\circ - i^5D_4$
V	3143.242	2	IIIA	31805.10	00	$a^5D_4 - z^3F_3^\circ$
V	3142.888	5	IV	31808.69	+04	$a^3P_2 - w^3P_2^\circ$
U	3142.453	6	IV	31813.08	−04	$z^7D_3^\circ - e^7S_3$
U	3140.391	5n	V	31833.98	+01	$z^5D_3^\circ - i^5D_2$
U	3139.661	(1)		31841.38	−05	$z^7D_5^\circ - e^7F_4$
U	3135.863	(1)		31879.94	00	$a^3H_4 - u^3G_4^\circ$
A	**3134.111	10	III	31897.76	00	$a^5F_3 - x^5D_4^\circ$
V	3132.514	4n	V	31914.02	+05	$z^5D_4^\circ - i^5D_3$
V	3129.334	5	IV	31946.45	+01	$a^3F_4 - w^5D_3^\circ$
SS	3129.178	(1)		31948.04	+03	$z^7D_3^\circ - f^5F_2$
U	3128.901	1	IV	31950.87	−04	$a^3F_3 - y^5S_2^\circ$
C	*3125.653	15	III	31984.07	−01 +02	$a^5F_2 - x^5D_3^\circ$ $z^7D_5^\circ - e^7G_4$
U	3124.099	(1)		31999.98	−08	$z^7D_1^\circ - e^5P_1$
U	3123.353	(1)		32007.62	−08	$z^7D_4^\circ - e^7S_3$
R	3122.665	(−)		32014.68	+07	$a^3G_4 - 12_5^\circ$
W	3121.76	(1)		32023.96	+01	$a^5P_1 - w^3P_0^\circ$

Ref	λ I A	Int	T C	Wave Number Observed	o−c	Desig
G	3120.435	6	IV	32037.56	+04	$a^3H_4 - u^3G_3^\circ$
G	3119.495	6	IV	32047.21	+03	$a^3H_5 - u^3G_4^\circ$
U	3117.640	1	IIIA	32066.27	−01	$a^5F_2 - y^7P_2^\circ$
A	§3116.633	12	III	32076.64	00	$a^5F_1 - x^5D_2^\circ$
U	3116.250	(1)		32080.58	+34	$z^7D_3^\circ - e^5P_3$?
V	3112.079	3	IV	32123.57	00	$b^3G_5 - s^3G_5^\circ$
U	3111.686	(2)		32127.63	−01	$b^3F_4 - w^3H_4^\circ$
U	3109.05	(1)		32154.87	−07	$z^7D_2^\circ - e^5P_2$
U	§3106.542	(1)		32180.83	−03	$a^3H_4 - u^3D_3^\circ$
U	3101.004	(2)		32238.30	+01	$a^3G_4 - t^3G_4^\circ$
V	3100.838	(2)		32240.02	−01	$a^3H_6 - 6_6^\circ$
G	3100.666	20	II	32241.81	00	$a^5F_3 - x^5D_3^\circ$
G	3100.304	20	II	32245.57	00	$a^5F_2 - x^5D_2^\circ$
U	3099.971	15	II	32249.04	−01	$a^5F_4 - x^5D_4^\circ$
U	3099.897	20	II	32249.81	−01	$a^5F_1 - x^5D_1^\circ$
U	3098.192	6	IV	32267.55	+02	$a^3G_5 - t^3G_5^\circ$
V	3095.270	(2)		32298.01	00	$a^3G_5 - 12_5^\circ$
U	3094.870	(1)		32302.19	00	$a^3G_4 - 13_4^\circ$
U	3093.883	(2ld)		32312.49	−05	$b^3F_4 - s^3D_3^\circ$
V	3093.806	3	IVA	32313.30	−02	$a^3F_2 - x^3D_2^\circ$
U	**3092.778	2	III?	32324.04	+03	$a^5F_3 - y^7P_2^\circ$
A	3091.578	20	II	32336.58	−01	$a^5F_1 - x^5D_0^\circ$
V	3090.209	(1)		32350.91	+02	$a^3G_3 - t^3G_3^\circ$
A	3083.742	20	II	32418.75	00	$a^5F_2 - x^5D_1^\circ$
U	3083.152	(1)		32424.95	−06	$a^3H_4 - t^3D_3^\circ$
V	3078.436	3	IV	32474.62	+02	$a^3P_0 - u^3D_1^\circ$
G	3078.014	4	IVA	32479.08	+03	$a^5F_3 - y^7P_3^\circ$
A	3075.721	25r	II	32503.29	00	$a^5F_3 - x^5D_2^\circ$
V	3074.157	(2)		32519.82	−03	$b^3G_3 - u^3F_2^\circ$
V	3073.982	(1)		32521.68	−01	$a^3G_5 - t^3G_4^\circ$
S	3073.244	(−)		32529.48	−09	$a^1G_4 - x^3I_5^\circ$
S	3068.927	(−)		32575.24	+06	$a^3F_4 - v^5D_3^\circ$
G	3068.175	8	IV	32583.22	−04	$a^3F_2 - x^3D_1^\circ$
V	3067.952	(1)		32585.59	00	$a^3G_5 - 13_4^\circ$
A	3067.244	30r	II	32593.12	+02	$a^5F_4 - x^5D_3^\circ$
U	3067.120	8	IV	32594.43	−02	$a^3F_2 - y^3G_3^\circ$
U	3066.483	3	IV	32601.20	+02	$a^3G_4 - t^3G_3^\circ$
U	3063.933	(2)		32628.34	+01	$a^3P_1 - t^3D_1^\circ$?
S	3063.149	(1)		32636.69	+03	$a^5P_3 - w^3P_2^\circ$
S	3062.872	(1)		32639.64	−04	$b^3G_5 - u^3H_4^\circ$
G	3060.984	4	IV	32659.73	−05	$a^3F_3 - x^3D_3^\circ$
V	3060.545	(1)		32664.45	−03	$b^3G_4 - u^3F_3^\circ$
A	3059.086	100R	I	32680.03	00	$a^5D_3 - y^5D_4^\circ$
A	3057.446	40R	II	32697.56	+01	$a^5F_5 - x^5D_4^\circ$
C	3055.263	12	III	32720.92	−02	$a^3F_3 - x^3D_2^\circ$
S	3054.949	(−)		32724.28	+02	$b^3F_2 - x^1F_3$
W	*3053.44	(2)	–	32740.46	+16 −11	$a^5F_1 - z^5S_2^\circ$ $z^7P_4^\circ - 2$
G	3053.065	5	IV	32744.48	+01	$a^3P_1 - u^3D_2^\circ$
A	3047.605	100R	I	32803.14	00	$a^5D_2 - y^5D_3^\circ$
S	3047.201	(−)		32807.49	+06	$b^3P_1 - w^1D_2^\circ$
U	3047.050	(1)		32809.11	00	$b^3G_5 - u^3F_4^\circ$
U	3046.930	(1)		32810.41	−02	$a^3H_5 - w^3H_6^\circ$
S	3046.819	(−)		32811.60	−05	$a^3G_4 - w^1G_4^\circ$
V	3045.594	(1)		32824.80	00	$a^3H_4 - w^3H_5^\circ$
G	3045.077	5	III	32830.37	+03	$a^5F_4 - y^7P_3^\circ$
G	3042.666	15	III	32856.39	+06	$a^5F_2 - x^5F_3^\circ$
G	3042.020	15	III	32863.36	+02	$a^5F_1 - x^5F_2^\circ$
V	3041.745	15	III	32866.33	−07	$a^5F_3 - x^5F_4^\circ$
V	3041.639	10	IV	32867.48	−02	$a^3F_3 - y^3G_4^\circ$
C	3040.428	15	III	32880.57	−02	$a^5F_4 - x^5F_5^\circ$
V	3039.322	(2)		32892.53	−02	$a^3H_5 - y^3I_6^\circ$
V	3037.782	2	IVA	32909.21	−02	$a^5F_2 - z^5S_2^\circ$
A	3037.388	80R	I	32913.47	+01	$a^5D_1 - y^5D_2^\circ$
W	3034.51	(2n)		32944.69	−27	$a^3F_2 - x^5G_3^\circ$?

TABLE B—(*Continued*)

Ref	λ I A	Int	T C	Wave Number Observed	o−c	Desig
U	3033.101	(1)		32959.99	+02	$a^3P_1 - u^3D_1^\circ$
G	3031.638	15	III	32975.90	00	$a^5F_1 - x^5F_1^\circ$
G	3031.213	12	IV	32980.52	+02	$a^3H_4 - w^3H_4^\circ$
S	3030.757	(−)		32985.48	+07	$b^3G_4 - x^3I_5^\circ$
C	3030.149	15	IV	32992.10	+06	$a^3H_5 - w^3H_5^\circ$
V	3029.237	3	IV	33002.08	+01	$a^3F_3 - y^3G_3^\circ$
G	3026.462	15	III	33032.29	+03	$a^5F_2 - x^5F_2^\circ$
K	3025.843	50R	I	33039.05	00	$a^5D_0 - y^5D_1^\circ$
U	3025.638	15	IV	33041.29	+02	$a^3H_6 - w^3H_6^\circ$
V	3025.283	3	III	33045.16	+05	$a^5F_4 - y^7P_4^\circ$
C	3024.033	15r	IA	33058.82	−01	$a^5D_1 - z^3P_2^\circ$
S	3023.583	(−)		33063.74	−07	$a^5P_3 - x^1G_4^\circ$
G	3021.074	150R	I	33091.20	−01	$a^5D_3 - y^5D_3^\circ$
U	3020.640	200R	I	33095.96	00	$a^5D_4 - y^5D_4^\circ$
U	3020.487	100R	II	33097.63	+04	$a^5D_2 - y^5D_2^\circ$
U	3019.290	(1)		33110.75	−03	$a^3H_4 - y^3I_5^\circ$
G	3018.983	15r	III	33114.12	+06	$a^5F_3 - x^5F_3^\circ$
U	3018.134	(1)		33123.44	+05	$a^3H_6 - y^3I_6^\circ$
G	3017.628	15r	IA	33129.00	00	$a^5D_1 - y^5D_1^\circ$
G	3016.186	12	III	33144.83	+01	$a^5F_2 - x^5F_1^\circ$
C	3015.913	4	IV	33147.83	+09	$a^3H_5 - w^3H_4^\circ$
U	3014.175	3	IV	33166.92	−03	$a^5F_3 - z^5S_2^\circ$
S	3014.120	(−)		33167.54	−08	$b^3G_5 - v^1G_4^\circ$
G	3011.482	7	IV	33196.60	00	$a^3G_3 - v^3H_4^\circ$
C	3009.570	25r	II	33217.69	00	$a^5F_4 - x^5F_4^\circ$
V	3009.098	3	IV	33222.90	+02	$a^3H_6 - w^3H_5^\circ$
G	3008.139	60R	I	33233.50	+01	$a^5D_1 - y^5D_0^\circ$
U	3007.281	12r	I	33242.97	+01	$a^5D_2 - z^3P_2^\circ$
U	3007.145	8	III	33244.47	−01	$a^3F_4 - x^3D_3^\circ$
G	3005.302	3	IV	33264.85	+01	$a^3H_6 - y^3I_7^\circ$
U	3004.620	(1)		33272.41	+12	$a^3F_3 - x^5G_4^\circ$
V	3004.119	(2)		33277.96	−06	$a^3H_5 - y^3I_5^\circ$
C	3003.031	10	III	33290.01	+02	$a^5F_3 - x^5F_2^\circ$
SS	3001.663	(1)		33305.18	−08	$c^3P_2 - t^3F_3^\circ$
G	3000.950	100R	I	33313.10	−02	$a^5D_2 - y^5D_1^\circ$
G	3000.452	8	III	33318.63	+03	$a^3F_4 - y^3G_5^\circ$
A	2999.512	30R	II	33329.07	−02	$a^5F_5 - x^5F_5^\circ$
G	2996.386	5	IV	33363.84	00	$a^3P_1 - v^3P_2^\circ$
U	2995.838	(1)		33369.91	−05	$b^3G_3 - t^3F_2^\circ$
U	2994.507	100R	I	33384.77	−05	$a^5D_0 - z^3P_1^\circ$
G	2994.427			33385.66	00	$a^5D_3 - y^5D_2^\circ$
C	2990.392	6	IV	33430.70	−02	$a^3G_4 - v^3H_5^\circ$
W	2989.4	(1)		33441.8	−1	$a^3F_2 - w^5P_1^\circ$?
S	2988.942	(−)		33446.93	+04	$a^3G_4 - v^3H_4^\circ$
G	2988.468	2	IV	33452.23	+03	$a^3F_4 - y^3G_4^\circ$
A	2987.292	10	III	33465.40	+05	$a^5F_4 - x^5F_3^\circ$
U	§2986.653	(1)		33472.56	00	$a^3H_5 - z^1I_6^\circ$
G	2986.456	3	III	33474.77	+01	$a^5D_1 - z^3P_1^\circ$
V	§2984.785	10	IV	33493.51	−10	$a^5F_5 - y^7P_4^\circ$
G	2983.574	125R	I	33507.10	−04	$a^5D_4 - y^5D_3^\circ$
U	§2982.234	(1)		33522.16	+15	$b^3G_4 - t^3F_3^\circ$
G	2981.852	6	IV	33526.46	−01	$a^5P_3 - t^5D_4^\circ$
A	2981.446	20r	I	33531.02	−01	$a^5D_3 - z^3P_2^\circ$
G	2980.532	5	IV	33541.29	+03	$a^3G_3 - w^1F_3^\circ$
U	2976.922	(1)		33581.97	−21	$z^7F_5^\circ - 1$
W	2976.5	(1)		33586.7	−1	$a^3F_4 - y^3G_3^\circ$
G	2976.126	5	IV	33590.95	+03	$a^3P_2 - u^3D_3^\circ$
W	2974.78	(1)		33606.15	−01	$z^7F_5^\circ - 2$
U	2973.237	60R	I	33623.59	−02	$a^5D_3 - y^5F_4^\circ$
U	2973.134	60R	I	33624.75	−02	$a^5D_2 - y^5F_3^\circ$
G	2972.277	3	IV	33634.45	+01	$a^5P_2 - t^5D_3^\circ$
G	*2970.106	40R	I	33659.00	+11	$a^5D_2 - z^3P_1^\circ$
					−10	$a^5D_1 - y^5F_1^\circ$
G	2969.474	10	I	33666.19	00	$a^5F_5 - x^5F_4^\circ$
U	2969.362	5	II	33667.46	−05	$a^5D_1 - z^3P_0^\circ$
U	2968.481	(2)		33677.46	+01	$a^3P_1 - z^1P_1^\circ$
G	2966.901	125R	II	33695.39	−03	$a^5D_4 - y^5F_5^\circ$
U	2966.26	(2)		33702.68	+02	$a^5P_1 - t^5D_2^\circ$
U	2965.811	2	IV	33707.77	−06	$a^3H_5 - 9_4^\circ$
A	2965.255	20	II	33714.09	−01	$a^5D_0 - y^5F_1^\circ$
					−03	$(a^3G_5 - v^3H_6^\circ)$
U	2963.71	(1n)		33731.67	−08	$z^7F_3^\circ - 3$
W	2962.11	(2)		33749.89	−03	$a^3F_4 - x^5G_5^\circ$
U	2961.70	(1)		33754.56	+12	$a^5P_3 - v^3F_4^\circ$
U	2960.666	(2)		33766.35	−09	$b^3G_5 - t^3F_4^\circ$
U	2960.299	1	IV	33770.53	+02	$a^3P_0 - v^3P_1^\circ$
C	2959.992	10	IV	33774.03	−01	$a^3G_5 - v^3H_6^\circ$
G	2959.682	5		33777.57	−06	$z^7F_6^\circ - 1$
U	2957.491	(2)		33802.60	00	$a^3P_2 - t^3D_1^\circ$?
A	2957.365	30R	II	33804.04	00	$a^5D_1 - y^5F_1^\circ$
U	2956.86	(2n)		33809.81	00	$a^3G_5 - x^1H_5^\circ$
U	2956.71	(1)		33811.52	+27	$a^5P_3 - t^5D_3^\circ$
G	2954.651	5	IV	33835.08	+01	$a^3P_2 - t^3D_3^\circ$
A	2953.940	50R	II	33843.23	00	$a^5D_2 - y^5F_2^\circ$
U	2953.486	5	IV	33848.43	+04	$a^3G_3 - s^3G_3^\circ$
G	2950.240	20n	IV	33885.67	−02	$a^5P_3 - 5^\circ$
U	2948.733	(2)		33902.99	−06	$a^5P_2 - t^5D_2^\circ$
U	2948.433	4	IV	33906.44	+05	$a^3G_4 - s^3G_4^\circ$
G	2947.877	60R	I	33912.83	−01	$a^5D_3 - y^5F_3^\circ$
U	2947.363	(2)		33918.74	00	$a^3P_2 - u^3D_2^\circ$
U	2941.77	(1)		33983.24	−18	$z^7D_4^\circ - h^5D_3$
A	2941.343	15r	I	33988.16	−01	$a^5D_2 - y^5F_1^\circ$
G	2940.586	(3)		33996.92	+10	$z^7F_5^\circ - 3$
G	2939.072	(1)		34014.43	+08	$a^5P_1 - t^5D_0^\circ$
G	2937.806	10n	IV	34029.08	−06	$a^5P_2 - 7_2^\circ$
G	2936.904	60R	I	34039.53	−01	$a^5D_4 - y^5F_4^\circ$
W	2936.1	(1)		34048.9	+2	$a^3F_2 - w^3D_3^\circ$
U	2934.370	(1)		34068.93	00	$a^5P_3 - u^5F_3^\circ$
W	2931.8	(1)		34098.8	+1	$a^5G_4 - s^3G_3^\circ$
U	2931.420	(2)		34103.21	−05	$a^3H_4 - 10_3^\circ$
W	2930.6	(1)		34112.8	0	$z^7D_1^\circ - h^5D_1$?
V	2929.618	2	IV	34124.18	−02	$a^3F_2 - x^3F_3^\circ$
V	2929.118	6	IV	34130.01	−03	$b^3H_4 - t^3H_4^\circ$
A	2929.008	25r	I	34131.29	−01	$a^5D_3 - y^5F_2$
U	2928.753	(3)		34134.26	+02	$a^3P_2 - u^3D_1^\circ$
U	2928.105	(2)		34141.82	+05	$a^5P_3 - u^5P_3^\circ$
V	2925.899	4	IV	34167.56	−01	$a^3F_2 - w^3D_2^\circ$
G	2925.359	4	V	34173.86	−01	$a^3G_3 - u^3H_4^\circ$
W	2924.6	(1n)		34182.7	−2	$a^5P_1 - u^5P_1^\circ$
G	2923.851	7	IV	34191.49	+04	$a^3G_5 - s^3G_5^\circ$
G	2923.288	7	IV	34198.08	−01	$b^3H_5 - t^3H_5^\circ$
V	2922.62	(1n)		34205.89	−06	$a^5P_3 - 7_2^\circ$
U	2922.383	(1)		34208.66	−02	$a^3F_2 - u^5D_1^\circ$
C	2920.691	5	IV	34228.48	−02	$a^3F_2 - x^3F_2^\circ$
U	2920.29	(1)		34233.18	−07	$a^3P_0 - t^5P_1^\circ$
G	2919.838	(2)		34238.47	+03	$z^7D_4^\circ - g^7D_5$
V	2918.354	3	IV	34255.89	+01	$a^3P_1 - v^3P_1^\circ$
G	2918.023	10	IV	34259.77	+02	$b^3H_6 - t^3H_6^\circ$
G	2914.305	3	IV	34303.48	−04	$a^3F_2 - w^3D_1^\circ$
V	2912.257	3	IV	34327.60	−03	$a^3F_3 - u^5D_2^\circ$
A	2912.158	20r	I	34328.77	+01	$a^5D_4 - y^5F_3^\circ$
U	2910.930	(3)		34343.25	−05	$a^3G_3 - u^3F_4^\circ$
U	2909.313	(1)		34362.34	+08	$a^3H_5 - t^3G_5^\circ$
V	2908.864	(2)		34367.64	+06	$z^7D_3^\circ - g^7D_4$
G	*2907.518	5	V	34383.56	+3	$a^5P_2 - u^5P_1^\circ$
					+04	$a^3G_4 - u^3H_5^\circ$
U	2905.57	(1)		34406.60	+01	$b^3H_5 - t^3H_4^\circ$
G	2901.910	5	IV	34450.00	−01	$z^7D_5^\circ - g^7D_5$
G	2901.381	5	IV	34456.27	−02	$a^3F_3 - w^3D_3^\circ$
C	2899.416	8	IV	34479.63	+01	$a^3P_2 - 8_1^\circ$

TABLE B—(*Continued*)

Ref	λ I A	Int	T C	Wave Number Observed	o−c	Desig	Ref	λ I A	Int	T C	Wave Number Observed	o−c	Desig
W	2897.6	(1)		34501.2	0	$z^7D_2^\circ - g^7D_3$	A	2832.436	25r	II	35294.94	+01	$a^5F_3 - y^5G_4^\circ$
C	2895.035	8	III	34531.80	−02	$a^3F_3 - x^3F_3^\circ$	G	2828.808	7	III	35340.21	+03	$a^5F_2 - z^5H_3^\circ$
C	2894.505	10	III	34538.12	+01	$a^3P_2 - v^3P_2^\circ$	G	2827.892	5	III	35351.65	−01	$a^5D_3 - z^3G_4^\circ$
V	2893.882	2	IV	34545.56	−03	$a^3F_3 - z^3H_4^\circ$	W	2827.67	(2n)		35354.43	−05	$a^3G_5 - x^3I_6^\circ$
V	2893.763	1	IVA	34546.98	+02	$a^5F_2 - x^5P_3^\circ$	U	2826.50	(3)		35369.06	−03	$a^3F_3 - v^5F_4^\circ$
G	2892.479	(1)		34562.32	+16	$z^7D_4^\circ - g^7D_4$	U	§2825.995	(2)		35375.38	−01	$a^5D_2 - z^3G_3^\circ$
W	2891.73	(2)		34571.26	−02	$b^3H_6 - q^3G_5^\circ$	V	2825.687	6	II	35379.23	−01	$a^5D_4 - z^3G_5^\circ$
U	2891.410	(1)		34575.09	−10	$a^3F_3 - w^3D_2^\circ$	G	2825.557	20	II	35380.87	00	$a^5F_3 - z^5H_4^\circ$
U	2890.868	(2)		34581.57	−12	$a^3D_3 - q^3G_3^\circ$	U	2824.70	(2)		35391.60	00	$a^3G_3 - t^3F_3^\circ$
V	2889.991	(2)		34592.07	00	$z^7D_1^\circ - g^7D_2$	A	2823.276	20	II	35409.45	+01	$a^5F_3 - y^5G_3^\circ$
W	2889.89	(3)		34593.28	+18	$a^3H_6 - t^3G_5^\circ$	U	2821.63	(1)		35430.10	−05	$a^3P_2 - v^3P_1^\circ$
U	2887.961	(1)		34616.38	−04	$a^3H_5 - t^3G_4^\circ$	G	2820.801	2	IV	35440.51	+02	$a^5D_3 - z^5G_2^\circ$
G	2887.806	5	V	34618.24	00	$a^3G_5 - u^3H_6^\circ$	W	2819.5	(2)		35456.9	0	$b^3F_4 - s^3G_3^\circ$
W	2887.36	(1)		34623.59	+01	$a^3H_6 - 12_5^\circ$	G	§2819.286	(1)		35459.56	+26	$a^3G_3 - t^3F_2^\circ$
G	2886.316	3	IV	34636.11	−01	$a^3F_3 - x^3F_2^\circ$	G	2817.505	6	III	35481.98	+01	$a^5F_3 - y^5G_2^\circ$
V	2883.748	4	V	34666.95	+03	$a^3G_5 - u^3H_5^\circ$	G	2815.506	3	IV	35507.17	00	$a^3F_2 - w^3G_3^\circ$
G	2880.575	2	IV	34705.14	+04	$a^5F_1 - x^5P_2^\circ$	G	2815.017	(1)		35513.34	−05	$a^3P_2 - 10_3^\circ$
U	2879.461	(1)		34718.56	−06	$a^3P_1 - t^5P_1^\circ$	A	2813.288	30R	II	35535.15	+01	$a^5F_4 - y^5G_5^\circ$
C	2877.300	8	III	34744.64	+05	$a^3F_4 - u^5D_4^\circ$	U	2812.31	(1)		35547.51	−07	$a^3F_2 - x^3P_1^\circ$
G	§2875.302	5	IV	34768.78	+01	$a^3F_4 - u^5D_3^\circ$	G	2812.042	(1)		35550.90	−02	$a^3G_4 - t^3F_4^\circ$
W	2874.89	(3)		34773.76	+03	$z^7D_5^\circ - g^7D_4$	U	2811.160	(1n)		35562.06	+04	$a^3F_3 - v^5F_3^\circ$
C	2874.172	10	I	34782.44	−01	$a^5D_4 - z^5G_5^\circ$	G	2808.328	2	III	35597.91	00	$a^5F_3 - z^5H_3^\circ$
U	2873.655	(2)		34788.70	−05	$b^3F_4 - v^3H_5^\circ$	U	2807.96	(1)		35602.58	+04	$a^3F_3 - v^5P_2^\circ$
W	2872.5	(1)		34802.7	0	$b^3P_2 - t^3F_3^\circ$	U	2807.245	2	III	35611.65	00	$a^5D_4 - z^5G_3^\circ$
G	2872.333	7	III	34804.71	+02	$a^5F_3 - x^5P_3^\circ$	C	2806.984	20	II	35614.96	+08	$a^5F_4 - z^5H_5^\circ$
U	2871.73	(1)		34812.02	−05	$a^3H_4 - t^3G_3^\circ$	W	2806.5	(1n)		35621.1	0	$z^7F_6^\circ - g^5G_5$
U	2871.31	(1)		34817.11	−17	$z^7F_4^\circ - i^5D_3$	G	2806.072	(1)		35626.54	−06	$a^3P_2 - 11_3^\circ$
U	2869.833	(2)		34835.03	−20	$z^7D_2^\circ - g^7D_1$	G	2805.808	(1)		35629.88	+04	$a^3F_4 - v^5F_5^\circ$
A	2869.308	10	I	34841.41	00	$a^5D_3 - z^5G_4^\circ$	V	2804.865	(2)		35641.86	−03	$a^3G_4 - t^3F_3^\circ$
G	§*2868.454	3	IV	34851.78	{ +06	$a^3P_2 - z^1P_1^\circ$	A	2804.521	20	II	35646.23	+01	$a^5F_4 - y^5G_4^\circ$
					+07	$z^7F_5^\circ - i^5D_4$	G	2803.613	(2)		35657.78	00	$a^3H_4 - v^3H_4^\circ$
G	2868.213	(1)		34854.71	+03	$z^7D_3^\circ - g^7D_2$	G	2803.169	(1)		35663.43	−03	$a^5D_3 - z^3G_3^\circ$
U	2867.880	(1)		34858.76	−01	$a^3F_3 - 1_2^\circ$	C	2797.775	15	III	35732.18	+01	$a^5F_4 - z^5H_4^\circ$
G	2867.560	3	IV	34862.64	+01	$a^3F_2 - w^5G_2^\circ$	G	2796.871	(1)		35743.72	−03	$a^3F_3 - x^3P_2^\circ$
G	2867.311	3	IV	34865.68	−01	$a^3F_2 - x^3G_3^\circ$	G	2795.540	(2)		35760.75	+01	$a^5F_4 - y^5G_3^\circ$
G	2866.624	7	II	34874.02	−01	$a^5F_2 - x^5P_2^\circ$	G	2795.006	3	III	35767.58	−01	$a^5D_4 - z^3G_4^\circ$
C	2863.864	8	I	34907.63	−02	$a^5D_2 - z^5G_3^\circ$	G	2794.700	(1)		35771.49	+02	$a^5F_3 - w^5D_4^\circ$
G	2863.429	8	III	34912.94	−01	$a^3F_4 - x^3F_4^\circ$	U	2794.157	(1)		35778.44	−22	$a^5P_3 - 9_4^\circ$
G	2862.496	4	IV	34924.32	−01	$a^5F_1 - x^5P_1^\circ$	G	2792.397	1	III	35800.99	+02	$a^3F_3 - w^3G_4^\circ$
G	2858.896	5	II	34968.29	00	$a^5D_1 - z^5G_2^\circ$	G	2791.786	(2)		35808.82	+07	$a^3H_5 - v^3H_5^\circ$
W	§2857.20	(1)		34989.05	+13	$a^5P_1 - v^3P_2^\circ$	G	2789.803	(3)		35834.28	−04	$a^3G_5 - t^3F_4^\circ$
U	2853.774	(3)		35031.05	+01	$b^3F_3 - s^3G_4^\circ$	G	2789.477	(2)		35838.47	00	$a^5P_3 - t^5P_3^\circ$
V	2853.685	(2)		35032.14	−03	$a^3F_4 - z^3H_5^\circ$	G	2788.106	30	II	35856.09	−02	$a^5F_5 - y^5G_6^\circ$
G	2852.952	(1)		35041.14	+16	$a^3F_4 - w^3D_3^\circ$	U	2787.935	5	II	35858.29	−07	$a^3F_4 - x^3G_5^\circ$
A	2851.798	15r	II	35055.32	00	$a^5F_1 - y^5G_2^\circ$	U	2787.12	(1)		35868.77	00	$a^3H_5 - v^3H_6^\circ$
W	2851.52	(2)		35058.74	−09	$b^3F_2 - s^3G_3^\circ$	U	2786.18	(1)		35880.88	−08	$a^5P_1 - v^3P_1^\circ$
G	2848.713	5	III	35093.28	+30	$a^5F_2 - x^5P_1^\circ$	U	2784.346	(2)		35904.51	−03	$a^3H_5 - x^1H_5^\circ$
G	2846.830	3	IV	35116.50	−02	$a^3F_4 - x^3F_3^\circ$	U	2784.017	(2)		35908.75	−06	$b^3F_3 - u^3F_3^\circ$
U	2845.714	(2)		35130.26	−02	$a^3F_4 - z^3H_4^\circ$	U	2782.055	(1)		35934.07	−04	$a^5P_2 - y^1F_3^\circ$
C	2845.595	8	III	35131.74	−02	$a^5F_3 - x^5P_2^\circ$	C	2781.835	4	III	35936.92	+02	$a^5F_2 - w^5D_3^\circ$
U	2845.544	(1)		35132.37	+03	$a^3F_3 - w^5G_3^\circ$	U	2780.700	1	III	35951.58	−04	$b^3F_4 - u^3F_4^\circ$
G	2843.977	20r	II	35151.72	00	$a^5F_2 - y^5G_3^\circ$	A	2778.221	20	III	35983.66	+02	$a^5F_5 - y^5G_5^\circ$
U	2843.923	(3)		35152.39	−03	$a^5D_2 - z^5G_2^\circ$	G	§2774.730	3	III	36028.93	+01	$a^5F_1 - w^5D_2^\circ$
G	2843.631	10	III	35155.99	+01	$a^5F_4 - x^5P_3^\circ$	U	2774.15	(1)		36036.46	+20	$a^5P_2 - x^1F_3^\circ$
U	2840.932	(3)		35189.39	+08	$a^5P_2 - v^3P_2^\circ$	U	2773.907	(1)		36039.62	−07	$a^3H_6 - v^3H_5^\circ$
G	2840.422	6	II	35195.72	00	$a^5D_3 - z^5G_3^\circ$	W	2772.86	(2)		36053.22	+01	$b^3G_4 - r^3G_4^\circ$
A	2838.120	10	III	35224.26	+01	$a^5F_2 - y^5G_2^\circ$	G	2772.113	1	III	36062.94	+06	$a^5D_2 - y^5P_3^\circ$
G	2836.315	(1)		35246.67	−07	$z^7F_6^\circ - h^7D_5$	V	2772.083	20	II	36063.33	−05	$a^5F_5 - z^5H_5^\circ$
G	2835.948	(1)		35251.24	+01	$a^3F_3 - x^3G_4^\circ$	G	2770.695	(1)		36081.39	+04	$a^5P_2 - v^3P_1^\circ$
G	2835.457	6	I	35257.33	−01	$a^5D_4 - z^5G_4^\circ$	G	2769.670	1	III	36094.74	+02	$a^5F_5 - y^5G_4^\circ$
G	2834.755	(2)		35266.07	−01	$b^3F_4 - s^3G_5^\circ$	G	2769.297	(6)		36099.60	−01	$a^3H_6 - v^3H_6^\circ$
U	*2834.414	(1)		35270.31	{ −02	$a^3F_2 - v^5F_2^\circ$	G	2768.432	(2)		36110.88	−04	$a^5P_3 - y^1F_3^\circ$
					+06	$a^3F_3 - w^5G_2^\circ$	A	§2767.523	20	III	36122.75	−01	$a^5F_4 - w^5D_4^\circ$
U	2834.177	(1)		35273.26	−05	$a^5F_3 - x^3G_3^\circ$	G	2766.909	2	III	36130.76	00	$a^5F_1 - w^5F_2^\circ$
U	2833.401	(2)		35282.92	+01	$a^3P_2 - y^1F_3^\circ$	U	2766.03	(1)		36142.24	+10	$b^3F_4 - u^3F_3^\circ$

TABLE B—(*Continued*)

Ref	λ I A	Int	T C	Wave Number Observed	o−c	Desig
U	2765.70	(1)		36146.56	−15	$a^3F_4 - v^5F_3^\circ$
G	2764.323	3	III	36164.56	+04	$a^5P_2 - 10_3^\circ$
C	2763.108	4	III	36180.46	+02	$a^5F_2 - w^5F_3^\circ$
					−20	$(a^5F_5 - z^5H_4^\circ)$
G	2762.770	(3)		36184.89	00	$a^5P_1 - t^5P_2^\circ$
G	2762.027	15	III	36194.62	−01	$a^5F_3 - w^5D_3^\circ$
G	§2761.780	18	III	36197.86	+02	$a^5F_2 - w^5D_2^\circ$
W	2761.5	(1)		36201.5	−3	$a^3P_1 - w^1D_2^\circ$
G	2759.814	4	III	36223.65	−05	$a^5F_1 - w^5F_1^\circ$
G	2757.315	10	III	36256.47	+01	$a^5F_1 - w^5D_1^\circ$
G	2756.329	20	I	36269.45	−01	$a^5D_1 - y^5P_2^\circ$
U	2756.264	(3)		36270.30	+03	$a^5D_3 - y^3F_4^\circ$
U	2755.184	(3)		36284.51	−01	$a^3H_5 - s^3G_4^\circ$
G	2754.427	2	III	36294.48	00	$a^5F_3 - w^5F_4^\circ$
G	2754.030	3	III	36299.71	+03	$a^5F_2 - w^5F_2^\circ$
G	2753.687	3	III	36304.23	−01	$a^5F_1 - w^5D_0^\circ$
G	2750.872	5		36341.38	+05	$a^5P_3 - 10_3^\circ$
U	2750.708	(1)		36343.55	−15	$a^5P_1 - t^5P_1^\circ$
G	2750.140	25r	II	36351.06	00	$a^5D_3 - y^5P_3^\circ$
G	2747.553	(3)		36385.29	+01	$a^5P_2 - t^5P_2^\circ$
C	§2746.982	20	III	36392.84	00	$a^5F_5 - z^5H_6^\circ$
					+22	$(a^5F_2 - w^5F_1^\circ)$
G	2744.526	8	III	36425.42	+04	$a^5F_2 - w^5D_1^\circ$
G	2744.068	10	II	36431.50	00	$a^5D_0 - y^5P_1^\circ$
G	2743.564	3	III	36438.18	+01	$a^5F_3 - w^5F_3^\circ$
G	2742.406	30r	II	36453.57	−02	$a^5D_2 - y^5P_2^\circ$
U	2742.256	20	III	36455.56	−01	$a^5F_3 - w^5D_2^\circ$
U	2742.017	2	III	36458.75	−02	$a^5D_2 - y^3F_3^\circ$
U	2741.578	(2)		36464.57	−04	$a^3F_2 - w^3F_2^\circ$
W	2741.10	(3)		36470.94	+02	$c^3P_2 - q^3G_3^\circ$
G	2738.210	(2)		36509.42	+01	$a^5F_1 - v^5D_2^\circ$
V	§2737.643	(2)		36516.99	−03	$a^3H_6 - s^3G_5^\circ$
G	2737.310	20r	II	36521.43	−01	$a^5D_1 - y^5P_1^\circ$
G	§2736.960	(3)		36526.10	+04	$a^5F_2 - y^5S_2^\circ$
U	2735.614	8	III	36544.07	−02	$a^5P_2 - t^5P_1^\circ$
A	2735.475	8	III	36545.92	00	$a^5F_4 - w^5D_3^\circ$
G	2734.613	(1)		36557.44	+03	$a^5F_3 - w^5F_2^\circ$
G	2734.266	2	III	36562.08	−01	$a^5P_3 - t^5P_2^\circ$
G	2734.002	2	III	36565.62	−02	$a^5F_2 - v^5D_3^\circ$
G	2733.581	15	II	36571.25	−01	$a^5F_5 - w^5D_4^\circ$
U	2731.281	(2)		36602.04	00	$b^3F_2 - t^3F_3^\circ$
G	2730.981	2	III	36606.06	−01	$a^5F_1 - v^5D_1^\circ$
U	2728.973	(2)		36632.99	−06	$a^5D_1 - y^3F_2^\circ$
G	2728.819	2	III	36635.07	+02	$a^3H_4 - u^3H_4^\circ$
G	2728.020	3	III	36645.79	+02	$a^5F_4 - w^5F_4^\circ$
G	2726.237	(2)		36669.75	+01	$b^3F_2 - t^3F_2^\circ$
G	2726.054	6	III	36672.21	−01	$a^5F_1 - v^5D_0^\circ$
U	2725.805	(1)		36675.57	00	$b^3F_3 - t^3F_4^\circ$
U	2725.606	(2)		36678.24	−09	$a^5F_2 - v^5D_2^\circ$
G	§2724.951	10	III	36687.06	00	$a^5F_3 - v^5D_4^\circ$
A	2723.577	15	II	36705.57	00	$a^5D_2 - y^5P_1^\circ$
U	2722.032	(2)		36726.40	+09	$a^3F_4 - y^1G_4$
G	2720.902	40r	II	36741.65	−01	$a^5D_3 - y^5P_2^\circ$
U	2720.516	(1)		36746.86	+02	$a^5D_3 - y^3F_3^\circ$
G	2720.194	(3)		36751.21	+06	$a^5P_3 - 13_4^\circ$
G	2719.418	3	III?	36761.71	+06	$a^3H_5 - u^3H_5^\circ$
G	2719.027	60R	II	36766.98	−02	$a^5D_4 - y^5P_3^\circ$
					+44	$(b^3F_3 - t^3F_3^\circ)$
C	2718.435	6	III?	36774.99	00	$a^5F_2 - v^5D_1^\circ$
G	2717.786	2	III	36783.78	−01	$a^5F_3 - y^5S_2^\circ$
G	2717.368	(1)		36789.43	−03	$a^5F_4 - w^5F_3^\circ$
U	§2716.41	(1)		36802.41	+12	$a^3H_5 - u^3H_4^\circ$
V	2716.259	(2)		36804.45	−03	$a^3H_4 - u^3F_4^\circ$
U	2715.323	1	III	36817.14	−04	$a^5D_2 - y^3F_2^\circ$
G	2714.868	1	III	36823.32	−05	$a^5F_3 - v^5D_3^\circ$

Ref	λ I A	Int	T C	Wave Number Observed	o−c	Desig
U	2714.062	(2)		36834.25	+01	$b^3F_3 - t^3F_2^\circ$
C	2711.655	4	III	36866.94	+05	$a^5F_4 - w^5F_5^\circ$
G	2710.543	2	III	36882.07	+03	$a^3F_2 - v^3G_3^\circ$
G	2709.989	(2)		36889.60	+02	$z^7D_4^\circ - 2$
W	2709.7	(1)		36893.5	−1	$b^3G_5 - q^3G_5^\circ$
G	2708.570	4	IV	36908.93	−02	$b^3F_4 - t^3F_4^\circ$
C	2706.581	8	III	36936.04	−02	$a^5F_3 - v^5D_2^\circ$
G	2706.012	4	IV	36943.81	00	$a^3H_6 - u^3H_6^\circ$
U	2702.453	(2)		36992.47	−02	$a^3H_6 - u^3H_5^\circ$
G	2701.908	(2)		36999.92	00	$b^3F_4 - t^3F_3^\circ$
A	2699.107	6	III	37038.32	−03	$a^5F_4 - v^5D_4^\circ$
G	2697.019	2	III	37066.99	+02	$a^3F_3 - v^3G_4^\circ$
G	2696.284	(5)		37077.10	−07	$z^7D_5^\circ - 1$
U	2695.662	(2gn)		37085.65	−01	$z^7D_3^\circ - 3$
G	2695.032	1	III	37094.32	+05	$a^5F_5 - w^5F_4^\circ$
G	2694.536	(5)		37101.15	00	$z^7D_5^\circ - 2$
G	2694.222	(1)		37105.47	+22	$a^5D_3 - y^3F_2^\circ$
U	2692.658	(3)		37127.02	−14	$a^5F_1 - x^3D_2^\circ$
G	2692.247	(2)		37132.69	+01	$a^3F_4 - w^3F_4^\circ$
G	2690.067	2	III	37162.78	−01	$a^5D_4 - y^3F_3^\circ$
G	2689.827	2	III	37166.10	00	$a^3F_3 - y^3H_4^\circ$
A	2689.212	8	III	37174.60	+06	$a^5F_4 - v^5D_3^\circ$
U	2684.857	(2)		37234.89	−05	$a^5F_2 - x^3D_3^\circ$
G	2681.586	(2)		37280.31	+07	$z^7D_4^\circ - 3$
U	2680.91	(1)		37289.71	+05	$a^3F_3 - v^3G_3^\circ$
G	2680.452	2	III	37296.08	−01	$a^5F_2 - x^3D_2^\circ$
A	2679.062	10	III	37315.43	+04	$a^5F_5 - w^5F_5^\circ$
U	2674.71	(1)		37376.14	+07	$a^3P_2 - w^1D_2^\circ$
C	2673.213	1	III	37397.07	−04	$a^5F_1 - x^3D_1^\circ$
G	2669.492	2	IV	37449.20	−01	$a^3H_5 - x^3I_6^\circ$
G	2667.912	1	IIIA	37471.37	−01	$a^5D_2 - y^3D_3^\circ$
U	2666.970	3	III	37484.60	−06	$a^3F_4 - v^3G_5^\circ$
G	2666.811	8	III	37486.84	−01	$a^5F_5 - v^5D_4^\circ$
G	2666.398	2	III	37492.65	−02	$a^5F_3 - x^3D_3^\circ$
C	2662.056	3	III	37553.79	−03	$a^5F_3 - x^3D_2^\circ$
U	2661.196	(2)		37565.93	−11	$a^5F_2 - x^3D_1^\circ$
G	2660.396	1	III	37577.22	−01	$a^5F_2 - y^3G_3^\circ$
G	2656.792	(2)		37628.19	00	$a^3F_4 - y^3H_5^\circ$
G	2656.145	3	III	37637.36	00	$a^3H_6 - x^3I_7^\circ$
U	2655.14	(1)		37651.61	−05	$a^3F_4 - v^3G_4^\circ$
C	2651.706	2	III	37700.36	−02	$a^5F_3 - y^3G_4^\circ$
C	2647.558	3	III	37759.43	−02	$a^5D_3 - y^3D_3^\circ$
G	2645.422	1	IIIA	37789.92	−02	$a^5D_1 - y^3D_2^\circ$
C	2643.997	8	III	37810.27	+01	$a^5F_1 - x^5G_2^\circ$
G	2641.645	4	III	37843.94	−02	$a^5F_4 - x^3D_3^\circ$
G	2636.477	1	III	37918.11	+03	$a^5F_4 - y^3G_5^\circ$
A	2635.808	8	III	37927.73	00	$a^5F_2 - x^5G_3^\circ$
G	2632.593	2	IIIA	37974.05	−01	$a^5D_2 - y^3D_2^\circ$
G	2632.238	4	III	37979.18	00	$a^5F_2 - x^5G_2^\circ$
G	§2629.579	2	III	38017.58	−11	$a^5D_0 - y^3D_1^\circ$
G	2623.532	5	III	38105.19	+02	$a^5F_3 - x^5G_4^\circ$
U	2623.366	2	III	38107.62	−01	$a^5D_1 - y^3D_1^\circ$
G	2618.708	2	III	38175.40	+02	$a^5D_4 - y^3D_3^\circ$
G	2618.018	5	III	38185.46	00	$a^5F_3 - x^5G_3^\circ$
G	2614.494	1	III	38236.91	00	$a^5F_3 - x^5G_2^\circ$
G	2612.771	2	III	38262.13	00	$a^5D_3 - y^3D_2^\circ$
G	2610.750	1	III	38291.74	−02	$a^5D_2 - y^3D_1^\circ$
G	2606.826	6	III	38349.38	−02	$a^5F_4 - x^5G_5^\circ$
G	2605.656	6	III	38366.60	+02	$a^5F_5 - y^3G_5^\circ$
U	2599.565	6	III	38456.49	+03	$a^5F_4 - x^5G_4^\circ$
U	2598.855	(1)		38466.99	−34	$a^5F_2 - 5^\circ$?
G	2595.422	(2)		38517.86	+01	$a^5F_1 - y^3P_0^\circ$
G	2594.150	1	III	38536.76	+01	$a^5F_4 - x^5G_3^\circ$
G	2593.510	(3)		38546.27	−01	$z^7D_5^\circ - h^7D_5$
U	2586.557	(1)		38649.88	−05	$a^3G_5 - t^3H_6^\circ$

TABLE B—(*Continued*)

Ref	λ I A	Int	T C	Wave Number Observed	o−c	Desig
A	2584.536	8	III	38680.10	+03	$a^5F_5 - x^5G_6^\circ$
G	2580.450	(2)		38741.34	00	$a^5F_2 - y^3P_2^\circ$
G	2580.062	(2)		38747.17	00	$a^5F_1 - y^3P_1^\circ$
G	*2579.266	(4)		38759.12	−11 −02	$a^5F_2 - u^5D_3^\circ$ $a^5F_4 - z^3I_5^\circ$
G	2576.688	4	III	38797.91	+01	$a^5F_5 - x^5G_5^\circ$
G	2572.752	(4)		38857.26	+03	$a^3F_2 - u^3G_3^\circ$
W	§2571.57	(3)		38875.11	+16	$a^3F_3 - 5^\circ$
G	§2569.742	(4)		38902.77	−02	$a^5F_2 - u^5D_2^\circ$
G	2569.595	(6)		38904.99	+03	$a^5F_5 - x^5G_4^\circ$
G	§2568.862	(5)		38916.09	−01	$a^5F_2 - y^3P_1^\circ$
W	2567.86	(3)		38931.27	+03	$a^5P_1 - u^3F_2^\circ$
G	2564.555	(4)		38981.44	+02	$a^5F_1 - w^3D_2^\circ$
V	2563.820	(2)•		38992.62	−16	$a^5F_3 - u^5D_4^\circ$
G	*2562.224	(5)		39016.91	+11 −05	$a^5F_1 - u^5D_0^\circ$ $a^5F_3 - u^5D_3^\circ$
G	2561.852	(3)		39022.57	+04	$a^5F_1 - u^5D_1^\circ$
U	2561.262	(2)		39031.56	+12	$a^5F_2 - w^3D_3^\circ$
G	2560.556	(4)		39042.32	−03	$a^5F_1 - x^3F_2^\circ$
G	2557.268	(1)		39092.51	00	$a^3F_4 - x^3H_5^\circ$
G	2556.862	1	III	39098.72	+02	$a^5F_5 - z^3I_6^\circ$
U	2556.298	(4)		39107.35	+08	$a^3F_3 - u^3G_4^\circ$
G	2555.648	(1)		39117.29	−08	$a^5F_1 - w^3D_1^\circ$
G	2552.827	(4)		39160.51	00	$a^5F_3 - u^5D_2^\circ$
G	2552.604	2	III	39163.94	−01	$a^5D_1 - y^7P_2^\circ$
G	2549.612	10r	III	39209.89	−01	$a^5D_3 - x^5D_4^\circ$
G	2545.977	10r	III	39265.87	−01	$a^5D_2 - x^5D_3^\circ$
G	2544.706	6	IV	39285.48	00	$b^3F_4 - r^3G_5^\circ$
G	2543.920	6	IV	39297.61	+07	$b^3F_3 - r^3G_4^\circ$
C	2542.101	6	IV	39325.74	00	$b^3F_2 - r^3G_3^\circ$
G	2540.971	10R	III	39343.22	−01	$a^5D_1 - x^5D_2^\circ$
U	2539.575	(1)		39364.85	+15	$a^5F_3 - x^3F_3^\circ$
G	2539.355	(7)		39368.26	+01	$a^5F_4 - u^5D_3^\circ$
G	2537.454	(5)		39397.75	+05	$a^3F_4 - u^3G_5^\circ$
U	§2536.738	(5)		39408.87	−20	$a^5F_3 - w^3D_2^\circ$
G	2535.604	8r	III	39426.48	+01	$a^5D_0 - x^5D_1^\circ$
G	2535.128	(5)		39433.90	−02	$a^5F_2 - 1_2^\circ$
G	2532.874	(2)		39468.98	−02	$a^5F_3 - x^3F_2^\circ$
W	2531.5	(1)		39490.4	+2	$b^3F_3 - r^3G_3^\circ$
C	2530.694	3	III	39502.98	−14	$a^5D_2 - y^7P_3^\circ$
G	2529.833	3	III	39516.42	+01	$a^5D_1 - x^5D_1^\circ$
G	2529.134	10r	III	39527.35	−01	$a^5D_2 - x^5D_2^\circ$
W	2528.91	(3)		39530.85	−07	$b^3F_4 - r^3G_4^\circ$
G	2527.433	15r	II	39553.94	−01	$a^5D_3 - x^5D_3^\circ$
G	2524.290	8r	II	39603.20	+02	$a^5D_1 - x^5D_0^\circ$
G	2522.848	40R	II	39625.83	00	$a^5D_4 - x^5D_4^\circ$
G	2522.488	(6)		39631.48	−17	$a^5F_4 - z^3H_5^\circ$
G	2521.917	(7)		39640.46	00	$a^5F_4 - w^3D_3^\circ$
C	2519.628	(10)		39676.46	−02	$a^5F_1 - w^5G_2^\circ$
G	2518.100	12r	II	39700.53	−01	$a^5D_2 - x^5D_1^\circ$
G	2517.658	(8)		39707.50	+01	$a^5F_2 - w^5G_3^\circ$
G	2516.569	(5)		39724.68	−02	$a^5F_3 - z^1G_4^\circ$
G	2516.249	(2)		39729.74	−03	$a^5F_4 - z^3H_4^\circ$
G	2515.848	(2)		39736.07	+06	$a^3F_3 - u^3D_2^\circ$
G	2513.847	(3)		39767.70	00	$b^3F_2 - q^3G_3^\circ$
G	2512.361	5r	III	39791.22	+03	$a^5D_3 - y^7P_3^\circ$
G	2510.833	15R	II	39815.43	00	$a^5D_3 - x^5D_2^\circ$
G	2508.751	(5)		39848.47	+01	$a^5F_2 - x^3G_3^\circ$
C	2507.899	6	III	39862.01	+01	$a^5F_3 - w^5G_4^\circ$
G	2506.569	(4)		39883.16	00	$b^3F_3 - t^3H_4^\circ$
G	2505.004	(3)		39908.08	+04	$b^3F_4 - t^3H_5^\circ$
G	2503.491	(3)		39932.19	−01	$b^3F_3 - q^3G_3^\circ$
G	2501.692	(6)		39960.91	−02	$a^5F_5 - x^3F_4^\circ$
G	2501.130	20R	II	39969.88	00	$a^5D_4 - x^5D_3^\circ$
G	§2498.895	10	IV	40005.62	−34	$a^5D_3 - y^7P_4^\circ$

Ref	λ I A	Int	T C	Wave Number Observed	o−c	Desig
G	2496.992	(4)		40036.12	−07	$b^3F_4 - q^3G_5^\circ$
C	2496.532	6	III	40043.48	+03	$a^5F_4 - w^5G_5^\circ$
G	§2495.869	(5)		40054.12	+02	$a^5F_5 - z^3H_6^\circ$
G	2494.250	(5)		40080.13	−02	$a^5F_5 - z^3H_5^\circ$
G	*2493.998	(6)		40084.17	+06 −01	$a^5F_3 - x^3G_4^\circ$ $a^5F_1 - v^5F_2^\circ$
W	2492.64	(2)		40105.95	−24	$a^5F_3 - x^3G_3^\circ$
W	2492.17	(2)		40113.57	00	$b^3F_4 - q^3G_4^\circ$
G	2491.983	(8)		40116.58	+04	$b^3F_4 - t^3H_4^\circ$
G	2491.155	20R	II	40129.92	−01	$a^5D_1 - x^5F_2^\circ$
G	2490.642	30R	II	40138.18	+05	$a^5D_2 - x^5F_3^\circ$
G	2489.751	15r	II	40152.55	00	$a^5D_0 - x^5F_1^\circ$
G	2488.942	(6)		40165.58	00	$b^3F_4 - q^3G_3^\circ$
G	2488.143	40R	II	40178.49	−05	$a^5D_3 - x^5F_4^\circ$
G	2487.368	(4)		40191.01	−01	$a^5D_2 - z^5S_2^\circ$
C	2487.064	(12)		40195.92	+02	$a^5F_1 - v^5F_1^\circ$
G	2486.690	(10)		40201.97	00	$a^5F_3 - v^5F_4^\circ$
G	§2486.372	(10)		40207.11	−01	$a^5D_4 - y^7P_3^\circ$
G	2485.989	(10)		40213.31	+02	$a^5F_4 - w^5G_4^\circ$
G	§2484.186	15R	II	40242.49	00	$a^5D_1 - x^5F_1^\circ$
G	2483.531	10	II?	40253.10	−01	$a^5F_2 - v^5F_2^\circ$
G	2483.270	60R	II	40257.32	−05	$a^5D_4 - x^5F_5^\circ$
G	2479.775	20R	II	40314.06	00	$a^5D_2 - x^5F_2^\circ$
G	2479.478	6	III	40318.90	−01	$a^5F_2 - x^3P_2^\circ$
G	2476.861	(2)		40361.49	+06	$a^5F_1 - x^3P_1^\circ$
G	2476.654	3	III	40364.86	+04	$a^5F_2 - v^5F_1^\circ$
C	2474.813	(8)		40394.88	−02	$a^5F_3 - v^5F_3^\circ$
G	2473.156	(3)	II	40421.96	+07	$a^5D_4 - y^7P_4^\circ$
V	2472.910	12R	II	40425.97	−23	$a^5D_3 - x^5F_3^\circ$
V	2472.875	(5)		40426.54	−08	$a^5D_2 - x^5F_1^\circ$
G	*2472.343	5	III	40435.24	−16 +13	$a^5F_4 - x^3G_4^\circ$ $a^5F_5 - w^5G_6^\circ$
G	*2470.961	(4)		40457.86	+38 +01	$a^5F_4 - x^3G_3^\circ$ $a^5F_4 - x^3G_5^\circ$
C	2468.878	4	III	40491.98	+03	$a^5F_5 - w^5G_5^\circ$
G	2467.730	(5)		40510.83	00	$a^5F_3 - v^5F_2^\circ$
C	2465.148	6	III	40553.25	−01	$a^5F_4 - v^5F_4^\circ$
G	§2463.728	(6)		40576.62	−02	$a^5F_3 - x^3P_2^\circ$
G	2462.645	10r	II	40594.46	−01	$a^5D_4 - x^5F_4^\circ$
G	2462.178	4	III	40602.15	+02	$a^5D_3 - x^5F_2^\circ$
G	2458.564	(4)		40661.83	+04	$a^5F_5 - w^5G_4^\circ$
C	2457.596	6	II	40677.86	+04	$a^5F_5 - v^5F_5^\circ$
C	2453.475	5	III	40746.18	−01	$a^5F_4 - v^5F_3^\circ$
G	2452.590	(2)		40760.88	−02	$a^3H_4 - t^3H_5^\circ$
A	§2447.708	4	II?	40842.17	+04	$a^5D_4 - x^5F_3^\circ$
G	2445.210	(6)		40883.89	−01	$a^5F_5 - x^3G_4^\circ$
C	2443.871	(20)		40906.29	−05	$a^5F_5 - x^3G_5^\circ$
C	2442.567	(20)		40928.12	−02	$a^3H_5 - t^3H_5^\circ$
G	2440.106	(15)		40969.40	00	$a^3H_4 - t^3H_4^\circ$
G	2439.743	(25)		40975.49	−01	$a^3H_6 - t^3H_6^\circ$
C	2438.181	2	III	41001.74	−02	$a^5F_5 - v^5F_4^\circ$
G	2429.810	(4)		41142.98	+04	$a^5F_1 - v^3D_1^\circ$
G	*2423.094	(4)		41257.01	+13 −15	$a^5F_2 - v^3D_2^\circ$ $a^5F_2 - w^3F_3^\circ$
G	2420.390	(2)		41303.10	+05	$a^5F_5 - w^3G_5^\circ$
G	2419.879	(2)		41311.82	−04	$a^5F_2 - v^3D_1^\circ$
G	2419.058	(2)		41325.84	+05	$a^5F_4 - y^1G_4^\circ$
G	2417.490	(2)		41352.64	+03	$a^3F_4 - 9_4^\circ$
G	*2408.045	(3)		41514.83	+22 −05	$a^5F_3 - v^3D_2^\circ$ $a^5F_3 - w^3F_3^\circ$
U	2398.215	(1)		41684.98	+11	$a^3F_4 - y^1F_3^\circ$
C	2389.971	(25)		41828.75	−01	$a^5D_2 - x^5P_3^\circ$
W	2385.9	(1)		41900.1	+3	$a^5F_3 - v^3G_4^\circ$
U	2381.831	(1)		41971.69	−01	$a^5D_1 - x^5P_2^\circ$
U	2377.991	(2)		42039.46	+06	$a^3F_3 - t^3G_3^\circ$

TABLE B—(*Continued*)

Ref	λ I A	Int	T C	Wave Number Observed	o−c	Desig
C	2374.517	(10)		42100.96	−02	$a^5D_0 - x^5P_1^\circ$
G	2373.618	(20)		42116.90	+07	$a^5D_3 - x^5P_3^\circ$
C	2371.428	(15)		42155.79	−04	$a^5D_2 - x^5P_2^\circ$
G	2369.454	(8)		42190.91	−01	$a^5D_1 - x^5P_1^\circ$
U	2365.509	(1n)		42261.27	+07	$a^3F_4 - t^3G_4^\circ$
U	2355.915	(1)		42433.35	−16	$a^5D_2 - y^5G_3^\circ$
G	§2355.327	(2)		42443.95	+05	$a^5D_3 - x^5P_2^\circ$
G	*2350.408	(5)		42532.77	+13	$a^5F_5 - v^3G_5^\circ$
					+01	$a^5D_4 - x^5P_3^\circ$
U	2341.575	(1n)		42693.20	+19	$a^5D_3 - z^5H_4^\circ$
G	2329.637	(2)		42911.95	+03	$a^5D_4 - y^5G_5^\circ$
C	2320.356	(40)	III	43083.58	−03	$a^5D_3 - w^5D_4^\circ$
G	2317.892	(2)		43129.37	+09	$a^3F_2 - s^3G_3^\circ$
C	2313.102	(40)	III	43218.67	−03	$a^5D_2 - w^5D_3^\circ$
C	2308.997	(30)	III	43295.50	−01	$a^5D_P - w^5D_2^\circ$
G	2306.378	(4)		43344.66	+05	$a^3F_3 - s^3G_4^\circ$
G	2306.164	(2)		43348.68	+07	$a^5F_3 - t^5D_4^\circ$
G	2304.727	(5)		43375.71	+05	$a^5F_2 - t^5D_3^\circ$
C	2303.579	(20)	II	43397.32	−03	$a^5D_1 - w^5F_2^\circ$
C	2303.422	(15)		43400.27	−08	$a^5D_0 - w^5F_1^\circ$
C	2301.682	(20)		43433.09	−02	$a^5D_0 - w^5D_1^\circ$
U	2300.599	(1)		43453.53	−10	$a^3F_4 - v^3H_5^\circ$
C	2300.140	(30)		43462.21	−03	$a^5D_2 - w^5F_3^\circ$
U	2299.453	(1)		43475.19	−16	$a^5F_1 - t^5D_2^\circ$?
C	2299.218	(25)	III	43479.63	−01	$a^5D_2 - w^5D_2^\circ$
G	2298.657	(6)		43490.24	−05	$a^5D_1 - w^5F_1^\circ$
U	2298.175	10r	II	43499.37	−17	$a^5D_4 - w^5D_4^\circ$
C	2297.785	(35d)		43506.74	−03	$a^5D_3 - w^5D_3^\circ$
C	2296.925	(15d)		43523.04	−01	$a^5D_1 - w^5D_1^\circ$
U	2295.535	(1n)		43549.37	+05	$a^3F_4 - x^1H_5^\circ$
C	2294.406	(25)		43570.81	−02	$a^5D_1 - w^5D_0^\circ$
C	2293.845	(25)		43581.46	−02	$a^5D_2 - w^5F_2^\circ$
X	2292.79	(1)		43601.52	+08	$a^5F_1 - 7_2^\circ$
C	2292.523	(30)		43606.60	−02	$a^5D_3 - w^5F_4^\circ$
G	2291.624	(4)		43623.70	−03	$a^5D_1 - y^5S_2^\circ$
C	*2291.122	(15)		43633.26	−13	$a^5F_3 - t^5D_3^\circ$
					−08	$a^5F_2 - u^5F_3^\circ$
G	2290.771	(3)		43639.93	−01	$a^5F_4 - u^5F_5^\circ$
G	2290.546	(9)		43644.23	−04	$a^5F_2 - t^5D_2^\circ$
G	2290.064	(3)Ni?		43653.41	00	$a^5F_3 - u^5F_4^\circ$
G	2289.032	(10)		43673.09	+22	$a^5F_1 - u^5F_2^\circ$
C	2287.632	(15)		43699.82	−08	$a^5F_4 - t^5D_4^\circ$
C	2287.248	(30)		43707.16	−02	$a^5D_2 - w^5D_1^\circ$
C	2284.087	(40)		43767.63	−08	$a^5D_3 - w^5D_2^\circ$
C	2283.653	(12)		43775.95	−05	$a^5D_1 - v^5D_2^\circ$
G	2283.299	(9)		43782.74	+02	$a^5D_0 - v^5D_1^\circ$
G	2283.079	(9)		43786.96	−08	$a^5F_1 - t^5D_0^\circ$
G	2282.861	(4)		43791.14	00	$a^5F_1 - u^5F_1^\circ$
U	2281.986	(1)		43807.93	+07	$a^5D_2 - y^5S_2^\circ$
X	2281.66	(1)		43814.2	−3	$a^3F_4 - w^1F_3^\circ$
G	2280.222	(8)		43841.82	+03	$a^5F_2 - u^5F_2^\circ$
C	§2279.922	(10)		43847.58	+14	$a^5D_2 - v^5D_3^\circ$
U	2278.614	(2)		43872.75	+09	$a^5D_1 - v^5D_1^\circ$
G	2277.663	(12)		43891.07	00	$a^5F_3 - u^5F_3^\circ$
C	2277.098	(9)		43901.96	−04	$a^5F_3 - t^5D_2^\circ$
C	2276.025	(12)		43922.65	−05	$a^5D_4 - w^5D_3^\circ$
G	2275.593	(2)		43930.99	+03	$a^3F_4 - s^3G_5^\circ$
G	2275.189	(6)		43938.79	00	$a^5D_1 - v^5D_0^\circ$
C	*2274.088	(9)		43960.05	−01	$a^5F_2 - u^5F_1^\circ$
					−08	$a^5D_2 - v^5D_2^\circ$
G	2272.816	(8)		43984.66	−02	$a^5F_4 - t^5D_3^\circ$
C	2272.067	(15)		43999.16	−04	$a^5D_3 - v^5D_4^\circ$
C	2271.781	(40)		44004.69	−01	$a^5F_4 - u^5F_4^\circ$
C	2270.860	(18)		44022.54	−01	$a^5D_4 - w^5F_4^\circ$
G	2269.093	(18)		44056.82	+03	$a^5D_2 - v^5D_1^\circ$

Ref	λ I A	Int	T C	Wave Number Observed	o−c	Desig
G	2267.465	(15)		44088.45	+01	$a^5F_5 - u^5F_5^\circ$
G	2267.080	(9)		44095.93	00	$a^5D_3 - y^5S_2^\circ$
G	2266.903	(10)		44099.38	−14	$a^5F_3 - u^5F_2^\circ$
X	2265.61	(1)		44124.5	0	$a^5F_2 - u^5P_1^\circ$
C	2265.053	(20)		44135.39	−12	$a^5D_3 - v^5D_3^\circ$
C	2264.389	(45)		44148.32	−08	$a^5F_5 - t^5D_4^\circ$
U	2263.476	(6)		44166.14	−10	$a^5D_4 - w^5F_3^\circ$
G	2260.860	12	Fe II	44217.24	00	($a^5F_3 - u^5P_2^\circ$)
U	2260.594	(2)		44222.44	+06	$a^3F_3 - u^3F_3^\circ$
C	2259.511	15		44243.63	−04	$a^5D_4 - w^5F_5^\circ$
U	2259.279	(1)		44248.17	−03	$a^5D_3 - v^5D_2^\circ$
U	2256.750	(1)		44297.76	+06	$a^3F_3 - u^3F_2^\circ$
C	2255.861	(45)		44315.21	+01	$a^5F_4 - u^5P_3^\circ$
G	2251.865	(12)		44393.84	+08	$a^5D_1 - x^3D_2^\circ$
G	2250.784	(10)		44415.16	+03	$a^5D_4 - v^5D_4^\circ$
C	2248.858	(25)		44453.20	00	$a^5F_5 - u^5F_4^\circ$
U	2247.461	(1)		44480.82	−08	$a^5F_5 - 4_4^\circ$
C	2245.651	(15)		44516.67	−07	$a^5D_2 - x^3D_3^\circ$
X	2245.14	(1)		44526.8	+1	$a^5F_2 - u^3D_1^\circ$
U	2243.911	(1)		44551.19	−25	$a^5D_4 - v^5D_3^\circ$
U	2242.579	(15)		44577.65	−24	$a^5D_2 - x^3D_2^\circ$
X	2241.85	(1)		44592.1	+3	$a^5F_4 - u^3D_3^\circ$
C	2240.627	(4)		44616.48	−02	$a^3F_4 - u^3F_4^\circ$
U	2238.259	(2)		44663.68	−02	$a^5D_1 - x^3D_1^\circ$
U	2237.814	(2n)		44672.56	+07	$a^3F_2 - t^3F_3^\circ$
U	2234.432	(2)		44740.16	−03	$a^3F_2 - t^3F_2^\circ$
C	2231.211	(15)		44804.75	−05	$a^5D_3 - x^3D_3^\circ$
U	2229.066	(5)		44847.85	+02	$a^5D_2 - x^3D_1^\circ$
U	2228.489	(1)		44859.46	+44	$a^5D_2 - y^3G_3^\circ$
C	2228.170	(10)		44865.88	−08	$a^5D_3 - x^3D_2^\circ$
G	2222.75	(7)		44975.1	+1	$a^3F_4 - v^1G_4^\circ$
U	2220.912	(2)		45012.50	−02	$a^5D_3 - y^3G_4^\circ$
U	2217.744	(1)		45076.78	−07	$a^5D_1 - x^5G_2^\circ$
U	2217.578	(1n)		45080.17	+06	$a^3F_3 - t^3F_3^\circ$
C	2211.234	(7)		45209.48	−05	$a^5D_2 - x^5G_3^\circ$
C	2210.686	(9)		45220.69	−05	$a^5D_4 - x^3D_3^\circ$
C	2207.068	(6)		45294.81	−05	$a^5D_4 - y^3G_5^\circ$
C	2201.117	(4)		45417.26	−05	$a^5D_3 - x^5G_4^\circ$
C	2200.723	(15)		45425.39	−09	$a^5D_1 - w^5P_2^\circ$
U	2200.370	(10r) (5)		45432.67	+30	$a^5D_0 - w^5P_1^\circ$
U	2197.230	(1)		45497.59	−01	$a^5D_3 - x^5G_3^\circ$
C	2196.040	(50)		45522.24	−07	$a^5D_1 - w^5P_1^\circ$
U	2193.564	(2)		45573.62	−21	$a^3F_4 - t^3F_4^\circ$
U	2193.411	(2)		45576.80	−10	$a^5F_4 - s^3D_3^\circ$
G	2191.836	(60)		45609.55	−06	$a^5D_2 - w^5P_2^\circ$
C	2191.202	(10)		45622.74	−07	$a^5D_0 - z^3S_1^\circ$
U	2189.393	(1n)		45660.44	−17	$a^5F_3 - t^5P_3^\circ$
U	2189.183	(1)		45664.82	+02	$a^3F_4 - t^3F_3^\circ$
C	2187.192	(40)		45706.38	−06	$a^5D_2 - w^5P_1^\circ$
C	2186.890	(5)		45712.69	−06	$a^5D_1 - z^3S_1^\circ$
G	2186.483	(40)		45721.19	−02	$a^5D_3 - w^5P_3^\circ$
U	2186.241	(3)		45726.26	+08	$a^5D_4 - x^5G_5^\circ$
U	2183.465	(1)		45784.39	−05	$a^5D_1 - y^3P_0^\circ$
U	2181.133	(1n)		45833.33	+09	$a^5D_4 - x^5G_4^\circ$
C	2180.866	(4)		45838.94	−06	$a^5D_1 - y^3P_2^\circ$
U	*2178.073	(35)		45897.72	+04	$a^5D_3 - w^5P_2^\circ$
					−16	$a^5D_2 - z^3S_1^\circ$
C	2176.837	(6)		45923.77	−05	$a^5D_0 - y^3P_1^\circ$
U	2176.396	(1)		45933.08	+02	$a^5F_3 - y^1F_3^\circ$
C	2173.212	(8)		46000.37	−08	$a^5D_1 - u^5D_2^\circ$
C	2172.581	(6)		46013.72	−04	$a^5D_1 - y^3P_1^\circ$
U	2172.137	(2)		46023.12	−01	$a^5D_2 - y^3P_2^\circ$
G	2171.292	(40)		46041.04	+01	$a^5D_2 - u^5D_3^\circ$
G	2166.769	(100)		46137.13	−01	$a^5D_4 - w^5P_3^\circ$

TABLE B—(*Continued*)

Ref	λ I A	Int	T C	Wave Number Observed	Wave Number o−c	Desig
U	2165.537	(1n)		46163.38	−09	$a^5F_3 - 10_3^\circ$
C	2164.547	(7)		46184.49	−09	$a^5D_2 - u^5D_2^\circ$
C	2163.860	{(6), (1)}		46199.15	−03	$a^5D_0 - u^5D_1^\circ$
C	2161.577	(5)		46247.94	−07	$a^5D_1 - w^3D_2^\circ$
U	2160.236	(1)		46276.65	−10	$a^5F_3 - 11_3^\circ$
X	2159.92	(3)		46283.4	0	$a^5D_1 - u^5D_0^\circ$
U	2159.645	(4)		46289.30	+18	$a^5D_1 - u^5D_1^\circ$
U	2159.425	(2)		46294.02	00	$a^5D_0 - w^3D_1^\circ$
U	2158.922	(4)		46304.81	−11	$a^5D_3 - u^5D_4^\circ$
U	2158.622	{(1), (1)}		46311.25	+05	$a^5D_3 - y^3P_2^\circ$
C	2157.792	(5)		46329.06	−04	$a^5D_3 - u^5D_3^\circ$
U	*2155.238	(2)		46383.95	−28	$a^5F_3 - t^5P_2^\circ$
					−01	$a^5D_1 - w^3D_1^\circ$
U	2155.012	(3)		46388.82	+05	$a^5D_2 - x^3F_3^\circ$
C	2154.458	(2)		46400.74	+15	$a^5F_5 - 9_4^\circ$
C	2153.004	(5)		46432.07	−07	$a^5D_2 - w^3D_2^\circ$
C	*2151.099	{(3), (2)}		46473.19	−06	$a^5D_2 - u^5D_1^\circ$
					−08	$a^5D_3 - x^3F_4^\circ$
C	2150.182	(3)		46493.01	−06	$a^5D_2 - x^3F_2^\circ$
U	2149.416	(1)		46509.57	+18	$a^5F_3 - t^3G_4^\circ$
U	2149.170	(1)		46514.89	+13	$a^5F_4 - 10_3^\circ$
U	2148.394	(1n)		46531.69	+10	$a^5D_1 - 1_2^\circ$
U	2146.710	(2n)		46568.19	+10	$a^5D_2 - w^3D_1^\circ$
C	2145.188	(3)		46601.23	−08	$a^5D_3 - w^3D_3^\circ$
U	2144.576	(1)		46614.52	−03	$a^5F_2 - t^3G_3^\circ$
U	2142.141	(1n)		46667.51	+01	$a^5D_1 - y^3S_1^\circ$
C	2141.715	(1)		46676.79	−05	$a^5D_3 - x^3F_3^\circ$
U	2141.083	(1)		46690.56	−05	$a^5D_3 - z^3H_4^\circ$
U	2139.929	(2)		46715.74	+02	$a^5D_2 - 1_2^\circ$
C	*2139.695	{(3), (2)}		46720.85	00	$a^5D_4 - u^5D_4^\circ$
					+64	$a^5D_3 - w^3D_2^\circ$
C	2138.589	(3)		46745.01	−02	$a^5D_4 - u^5D_3^\circ$
U	2133.311	(1)		46860.64	−04	$a^5F_4 - t^3G_4^\circ$
C	2132.015	(4)		46889.13	−08	$a^5D_4 - x^3F_4^\circ$
U	2130.417	(1)		46924.29	−29	$a^5F_4 - 13_4^\circ$
U	2126.212	(1)		47017.08	−16	$a^5D_4 - w^3D_3^\circ$
N	2122.188	1		47106.2	−3	$a^5D_4 - z^3H_4^\circ$
N	2119.125	5		47174.3	+2	$a^5D_3 - w^5G_4^\circ$
C	2115.168	20		47262.54	−09	$a^5D_2 - v^5P_3^\circ$
N	2114.588	25		47275.5	+1	$a^5D_1 - v^5P_2^\circ$
N	2113.08	20		47309.1	−1	$a^5F_5 - t^3G_4^\circ$
C	2112.966	25		47311.79	−03	$a^5D_0 - v^5P_1^\circ$
C	2110.233	30		47373.06	+51	$a^5D_0 - v^5F_1^\circ$
					−02	$(a^5F_5 - 13_4^\circ)$
C	2108.955	30		47401.76	00	$a^5D_1 - v^5P_1^\circ$
N	2108.302	12		47416.4	−2	$a^5D_1 - x^3P_2^\circ$
N	2108.188	1p		47419.0	+7	$a^5D_3 - x^3G_3^\circ$
N	2108.139	12		47420.1	−1	$a^5D_4 - w^5G_5^\circ$
N	2106.380	25		47459.7	+2	$a^5D_2 - v^5P_2^\circ$
N	2106.260	20		47462.4	−1	$a^5D_1 - v^5F_1^\circ$
N	2103.964	1		47514.2	+1	$a^5D_3 - v^5F_4^\circ$
N	2103.048	25		47534.9	0	$a^5D_2 - v^5F_2^\circ$
N	2102.910	20		47538.0	+1	$a^5D_0 - x^3P_1^\circ$
C	2102.349	30		47550.69	−01	$a^5D_3 - v^5P_3^\circ$
C	2100.795	30		47585.86	−03	$a^5D_2 - v^5P_1^\circ$
N	2100.144	10		47600.6	−1	$a^5D_2 - x^3P_2^\circ$

Ref	λ I A	Int	T C	Wave Number Observed	Wave Number o−c	Desig
N	2098.953	25		47627.6	−4	$a^5D_1 - x^3P_1^\circ$
N	2098.081	15p		47647.4	+8	$a^5D_2 - v^5F_1^\circ$
N	2095.451	1		47707.2	+2	$a^5D_3 - v^5F_3^\circ$
N	2093.660	40		47748.0	+4	$a^5D_3 - v^5P_2^\circ$
N	2090.862	20		47811.9	−3	$a^5D_2 - x^3P_1^\circ$
N	2090.380	30		47822.9	−1	$a^5D_3 - v^5F_2^\circ$
N	2087.525	25		47888.3	−5	$a^5D_3 - x^3P_2^\circ$
N	2084.117	50		47966.6	0	$a^5D_4 - v^5P_3^\circ$
N	2058.100	1		48572.9	0	$a^3F_4 - t^3H_5^\circ$
N	2047.241	2		48830.5	0	$a^3F_4 - q^3G_3^\circ$?
N	2016.512	5		49574.5	0	$a^5F_4 - v^1G_4^\circ$
	λ *Vacuum*					
N	1974.059	1		50657.0	−5	$a^5D_2 - t^5D_3^\circ$
N	1973.911	1		50660.8	0	$a^5D_3 - t^5D_4^\circ$
N	1970.771	0		50741.6	−3	$a^5D_1 - t^5D_2^\circ$
N	1964.043	20		50915.4	+3	$a^5D_2 - u^5F_3^\circ$
N	1963.629	15		50926.1	0	$a^5D_2 - t^5D_2^\circ$
N	1963.110	25		50939.6	+1	$a^5D_1 - u^5F_2^\circ$
N	1962.871	20		50945.8	+3	$a^5D_3 - t^5D_3^\circ$
N	1962.746	15		50949.0	+3	$a^5D_1 - t^5D_1^\circ$
N	1962.100	30		50965.8	−8	$a^5D_3 - u^5F_4^\circ$
N	1962.031	25		50967.6	−2	$a^5D_0 - u^5F_1^\circ$
N	1961.236	20		50988.2	+2	$a^5D_2 - u^5P_3^\circ$
N	1960.129	25		51017.0	+3	$a^5D_4 - u^5F_5^\circ$
N	1958.739	15		51053.2	−4	$a^5D_1 - t^5D_0^\circ$
N	*1958.598	30		51056.9	−8	$a^5D_1 - u^5F_1^\circ$
					−3	$a^5D_1 - u^5P_2^\circ$
N	1957.831	25		51076.9	+2	$a^5D_4 - t^5D_4^\circ$
N	1956.026	30		51124.1	+5	$a^5D_2 - u^5F_2^\circ$
N	*1955.690	20		51132.8	−1	$a^5D_2 - t^5D_1^\circ$
					+6	$a^5D_0 - u^5P_1^\circ$
N	1952.997	20		51203.4	+2	$a^5D_3 - u^5F_3^\circ$
N	1952.596	30		51213.9	−2	$a^5D_3 - t^5D_2^\circ$
N	1952.262	20		51222.6	+4	$a^5D_1 - u^5P_1^\circ$
N	*1951.556	25		51241.2	−7	$a^5D_2 - u^5F_1^\circ$
					−1	$a^5D_2 - u^5P_2^\circ$
N	1950.223	20		51276.2	+2	$a^5D_3 - u^5P_3^\circ$
N	1946.978	25		51361.6	+1	$a^5D_4 - t^5D_3^\circ$
N	1946.219	10		51381.7	+2	$a^5D_4 - u^5F_4^\circ$
N	1945.294	25		51406.1	−2	$a^5D_2 - u^5P_1^\circ$
N	1945.070	20		51412.0	+3	$a^5D_3 - u^5F_2^\circ$
N	1940.649	25		51529.2	−2	$a^5D_3 - u^5P_2^\circ$
N	1937.274	35		51618.9	−2	$a^5D_4 - u^5F_3^\circ$
N	1934.528	25		51692.2	+2	$a^5D_4 - u^5P_3^\circ$
N	1903.37	1		52538.4	+6	$a^5D_3 - s^3D_3^\circ$
N	1888.32	12n		52957.2	+1	$a^5D_2 - y^1F_3^\circ$
N	1887.761	14		52972.8	0	$a^5D_3 - t^5P_3^\circ$
G	1880.14	5		53187.6	+1	$a^5D_2 - 10_3^\circ$
N	1878.849	2		53224.1	−1	$a^5D_1 - t^5P_2^\circ$
N	1876.421	10		53292.9	−1	$a^5D_0 - t^5P_1^\circ$
N	1873.259	15		53382.9	−1	$a^5D_1 - t^5P_1^\circ$
N	1873.052	12		53388.8	+1	$a^5D_4 - t^5P_3^\circ$
N	1872.359	15		53408.6	+3	$a^5D_2 - t^5P_2^\circ$
N	1866.815	10		53567.2	+1	$a^5D_2 - t^5P_1^\circ$
N	1866.07	12		53588.6	−3	$a^5D_3 - 11_3^\circ$
N	1863.54	0p		53661.2?	+1	$a^5D_4 - y^1F_3^\circ$
N	1862.318	15		53696.5	+1	$a^5D_3 - t^5P_2^\circ$
N	1855.58	15		53891.5	0	$a^5D_4 - 10_3^\circ$

NOTES TO TABLE B

* Blend.
§ Blend with *Fe* II.
( ) Masked.
** Notes by A. S. King as follows:

5204.582	Blend with *Cr*
4058.227	Blend with *Co*
3998.054	May be partly *Co*
3940.882	May be partly *Sr* and *Co*
3902.948	Blend with *Cr*
3786.176	Probably double
3753.610	Blend with *Ti*
3703.556	Blend with *V*
3605.450	Blend with *Cr* to violet
3369.549	Blend with *Ni*
3196.930	Blend with *Ni* to red
3134.111	Blend with *Ni*, but chiefly *Fe*
3092.778	Blend with *Al*

References for Wave-length and Intensity:

A International Standards, *Trans. Internat. Astron. Union* **6**: 80, 1938.

B International Standards, *Trans. Internat. Astron. Union* **3**: 86, 1928.

C Meggers and Humphreys, *Bur. Standards Jour. Research* **18**: 543 (RP 992), 1937.

D Kiess, *Bur. Standards Jour. Research* **20**: 33 (RP 1062), 1938.

E Meggers, *Bur. Standards Jour. Research* **14**: 33 (RP 755), 1935.

F Meggers and Kiess, *Bur. Standards Jour. Research* **9**: 309 (RP 473), 1932.

G Burns and Walters, *Publ. Allegheny Observ.* **8**: 39 (No. 4), 1931.

H Jackson, *Proc. Royal Soc.* **A 133**: 553, 1931.

I Babcock, *Mount Wilson Contr. No. 343*, 1927; *Astrophys. Jour.* **66**: 256, 1927.

J St. John and Babcock, *Mount Wilson Contr. No. 202*; *Astrophys. Jour.* **53**: 260, 1921. (Corrected by values in Table IV, Ref. I: mean of grating and interf. meas. used.)

K Burns and Walters, *Publ. Allegheny Observ.* **6**: 159 (No. 11), 1929.

L Meggers and Kiess, *Sci. Papers Bur. Standards* **19**: 273 (No. 479), 1924.

M King, *Mount Wilson Contr. No. 496*, 1934; *Astrophys. Jour.* **80**: 124, 1934.

N Green, *Phys. Rev.* **55**: 1209, 1939; and unpublished material, 1937.

O Meggers and Kiess, *Sci. Papers Bur. Standards* **14**: 642 (No. 324), 1918.

Q Babcock, unpublished material.

R King, A. S., unpublished material.

S Dobbie, unpublished material.

SS Solar Spectrum Wave-length (SSb denotes that the line is blended in the solar spectrum).

T Smith, Sinclair, unpublished material.

U Harrison, unpublished material, June 1942.

V Burns, *Lick Observ. Bull.* **8**: 27 (No. 247), 1913.

W Kayser, *Handbuch der Spectroscopie* **6**: 896, 1912.

X Schumacher, *Zeitschr. f. Wissen. Photographie* **19**: 149, 1919. (Corrected to agree with meas. in Ref. G and K.)

Y Dingle, *Monthly Notices Royal Astron. Soc.* **94**: 866, 1934.

For Intensity see also:

King, A. S., *Mount Wilson Contr. No. 247*, 1922; *Astrophys. Jour.* **56**: 318, 1922.

King, A. S., *Mount Wilson Contr. No. 66*, 1913; *Astrophys. Jour.* **37**: 239, 1913.

TABLE C

PREDICTED LINES OF FE I PRESENT IN THE SOLAR SPECTRUM

Solar λ	Solar Int	Grade	Wave Number		Desig	Solar λ	Solar Int	Grade	Wave Number		Desig
			Solar	o−c					Solar	o−c	
10987.02	1	fb	9099.16	+17	$b^3P_2 - z^3D_1^\circ$	8950.217	−1	g	11169.85	−02	$y^5D_3^\circ - e^5D_4$
10780.69	−2N	g	9273.30	+01	$b^3H_6 - z^3G_5^\circ$	8931.76	−2N	g	11192.93	+03	$a^1G_4 - z^3G_4^\circ$
10725.20	0	g	9321.28	−01	$b^3D_2 - y^3D_2^\circ$	8922.643	−2	g	11204.37	+03	$x^5F_5^\circ - f^5F_4$
10616.73	1	g	9416.52	+02	$b^3H_5 - z^3G_4^\circ$	8905.989	0	f	11225.32	00	$x^5F_3^\circ - e^5P_2$
10577.15	1	g	9451.75	−01	$b^3H_4 - z^3G_3^\circ$	8902.926	−3	g	11229.18	+01	$x^5D_2^\circ - e^7G_3$
10555.70	0	f	9470.95	−07	$w^5D_3^\circ - g^5F_4$	8895.98	−3	g	11237.95	+03	$z^3G_4^\circ - e^5F_5$
10388.77	−2	g	9623.14	−04	$w^5D_3^\circ - h^5D_3$	8892.11	−3	f	11242.84	+03	$x^5F_4^\circ - e^5P_3$
10379.04	−1	g	9632.16	−03	$a^5P_1 - z^5F_2^\circ$	8887.07	−3N	f	11249.22	+04	$x^5D_3^\circ - e^5G_3$
10364.05	0	f	9646.10	+08	$w^5D_3^\circ - f^5P_2$	8878.775	−2	g	11259.73	−01	$y^5D_2^\circ - e^5D_3$
10362.72	−1	f	9647.33	+01	$w^5D_2^\circ - g^5F_3$	8878.271	−1	g	11260.36	−02	$b^3G_4 - z^3G_5^\circ$
10333.21	−1	f	9674.88	+03	$d^3F_4 - u^5D_4^\circ$	8834.025	−2	g	11316.76	+01	$y^5D_1^\circ - e^5D_2$
10332.36	−1	g	9675.69	−02	$b^3D_1 - y^3D_1^\circ$	8828.103	−3	g	11324.36	−02	$x^5D_3^\circ - e^3D_3$
10311.96	−1	g	9694.82	−07	$a^3P_0 - z^5P_1^\circ$	*8819.51	−3	g	11335.38	−04	$x^5D_1^\circ - e^3D_2$
10307.46	−3N	f	9699.05	+02	$d^3F_4 - u^5D_3^\circ$	8816.876	−2	f	11338.78	−02	$x^5D_2^\circ - e^7S_3$
10283.87	−2	g	9721.30	00	$w^5D_1^\circ - h^5D_1$	8805.19	−2	f	11353.82	+02	$x^5D_4^\circ - e^5G_4$
10265.22	−2	g	9738.96	+01	$a^5P_1 - z^5F_1^\circ$	8798.07	−3	g	11363.01	−03	$y^7P_3^\circ - e^7S_3$
10156.56	0	g	9843.16	−05	$d^3F_4 - x^3F_4^\circ$	8779.08	−2	g	11387.59	+06	$y^5D_0^\circ - e^5D_1$
10155.19	1	g	9844.49	00	$a^5P_3 - z^5F_3^\circ$	8767.68	−3Nd?	f	11402.40	−03	$z^5P_2^\circ - X_3$
10149.13	−3	g	9850.37	−03	$x^5F_1^\circ - f^5D_0$	8727.19	−3	g	11455.30	−11	$c^3F_2 - x^3D_3^\circ$
10143.48	−3	f	9855.84	+12	$z^3D_3^\circ - X_2$	8700.314	−2	g	11490.69	+04	$x^5D_3^\circ - e^7G_3$
10137.14	−1	g	9862.01	−08	$x^5F_2^\circ - f^5D_1$	8698.717	0	g	11492.80	00	$b^3G_4 - z^5G_3^\circ$
10084.41	0	g	9913.58	+01	$d^3F_3 - x^3F_4^\circ$	*8689.788	−1	g	11504.61	−09	$a^1G_4 - z^3G_3^\circ$
10081.43	0	g	9916.51	−03	$a^3P_1 - z^5P_2^\circ$	*8686.75	3	g	11508.63	+03	$x^5D_2^\circ - e^3D_2$
10080.43	−1	f	9917.49	+01	$x^5F_1^\circ - f^7D_1$	8680.82	−3	g	11516.49	−07	$c^3F_2 - x^3D_2^\circ$
10070.58	−2	f	9927.20	00	$w^5D_0^\circ - g^5F_1$	8679.646	2	gb	11518.05	−03	$y^7P_2^\circ - e^7S_3$
10058.36	−3	f	9939.26	−08	$a^5P_2 - z^5F_1^\circ$	8671.879	0	g	11528.36	−03	$x^5D_0^\circ - e^5P_1$
10032.89	0	f	9964.49	−05	$w^5D_1^\circ - f^5G_2$	8663.723	−1	g	11539.22	+01	$x^5D_0^\circ - g^5D_3$
10019.81	0	f	9977.49	−05	$w^5D_2^\circ - f^5G_3$	8656.672	−1	g	11548.61	−02	$x^5D_0^\circ - e^3D_1$
10016.76	2	f	9980.54	−09	$x^5F_2^\circ - f^7D_2$	8654.436	−1	g	11551.60	−05	$a^3D_2 - y^3D_3^\circ$
9970.23	−2	g	10027.11	+02	$c^3P_2 - z^3P_1^\circ$	8652.475	−1	g	11554.21	+02	$y^5D_3^\circ - e^5D_3$
9967.30	3	fb	10030.06	+02	$x^5F_2^\circ - f^7D_1$	8632.424	0	g	11581.06	+01	$y^5D_4^\circ - e^5D_4$
9953.51	−1	g	10043.95	−07	$w^5D_3^\circ - h^5D_2$	8616.284	3	g	11602.75	−02	$x^5D_4^\circ - e^7G_5$
9951.19	0	g	10046.30	−04	$w^5D_4^\circ - h^5D_3$	8613.946	2	g	11605.89	−02	$x^5D_2^\circ - e^5P_3$
9950.62	0	g	10046.87	+09	$d^3F_4 - x^3F_3^\circ$	8610.609	2	g	11610.40	+02	$z^3G_4^\circ - e^5F_4$
9924.41	0	f	10073.40	−06	$a^1D_2 - y^3D_2^\circ$	8607.075	1	g	11615.16	00	$x^5D_1^\circ - e^5P_1$
9920.54	−1	g	10077.33	−09	$x^5F_1^\circ - e^7F_1$	8592.119	−2	g	11635.38	−02	$x^5D_1^\circ - e^3D_1$
9913.19	2	g	10084.81	00	$x^5F_5^\circ - e^7F_6$	8584.791	−2	g	11645.31	+03	$x^5D_1^\circ - g^5D_2$
9878.200	1	g	10120.53	−02	$x^5F_5^\circ - f^7D_5$	8576.48	−3	f	11656.60	+03	$d^3F_4 - y^1G_4^\circ$
9800.80	−3	g	10200.45	−02	$x^5F_1^\circ - e^7F_2$	8571.807	2	g	11662.95	+04	$x^5D_1^\circ - e^5P_2$
9771.07	−3	f	10231.49	−01	$d^3F_2 - w^3D_1^\circ$	8567.776	−1	g	11668.44	+01	$x^5D_4^\circ - e^3D_3$
9764.37	−3	f	10238.51	+03	$w^5D_3^\circ - f^5G_3$	8562.109	0	g	11676.16	+02	$z^3G_3^\circ - e^5F_3$
9608.93	−3	fb	10404.14	−04	$y^7P_3^\circ - e^7P_3$	8538.021	2	g	11709.10	−01	$x^5D_4^\circ - e^7G_4$
9573.65	−3N	g	10442.47	00	$x^5F_2^\circ - e^7G_3$	8527.847	0	g	11723.07	+05	$x^5D_0^\circ - g^5D_1$
9531.226	2	fb	10488.95	−01	$x^5F_3^\circ - e^7F_2$	8525.008	−1	f	11726.98	+05	$d^3F_3 - y^1G_4^\circ$
9433.34	−3	g	10597.80	−05	$x^5F_4^\circ - e^7F_4$	8519.10	−3	g	11735.11	−06	$x^5D_3^\circ - f^5F_2$
9409.59	−3NN	f	10624.54	−05	$x^5F_4^\circ - e^5G_3$	8509.65	−1	g	11748.15	−01	$z^5G_4^\circ - e^5F_5$
9383.423	−2	f	10654.17	−03	$y^7P_3^\circ - e^7P_2$	8496.483	−1	g	11766.35	+03	$z^5G_3^\circ - e^5F_4$
9297.14	0	g	10753.05	+01	$y^3D_3^\circ - e^3F_2$	8493.796	1	gb	11770.07	−01	$x^5D_3^\circ - e^3D_2$
9289.44	−2	f	10761.97	−05	$x^5F_3^\circ - f^5F_3$	8481.986	1	g	11786.46	−04	$c^3F_2 - x^3D_1^\circ$
9248.76	−3	f	10809.30	+06	$y^7P_2^\circ - e^7P_2$	8480.636	0	g	11788.33	−01	$x^5D_2^\circ - e^5P_1$
9203.21	−3d?	f	10862.80	−12	$x^5F_3^\circ - f^5F_2$	8466.510	−3	g	11808.00	+04	$c^3F_3 - x^3D_3^\circ$
9173.12	−3	g	10898.43	+10	$b^3F_2 - z^3D_1^\circ$	8466.102	−3	g	11808.57	00	$x^5D_2^\circ - e^3D_1$
9156.26	0	g	10918.50	−04	$b^3G_3 - z^5G_4^\circ$	8465.173	−2	g	11809.86	+07	$x^5D_1^\circ - g^5D_1$
9140.12	−3	f	10937.78	+04	$a^3D_3 - y^3F_3^\circ$	8461.472	−3	f	11815.03	−09	$z^5P_3^\circ - X_3$
9116.940	2N	fb	10965.58	−06	$x^5D_1^\circ - e^5G_2$	8458.99	−3N	g	11818.50	+04	$x^5D_2^\circ - g^5D_2$
9112.19	−3	g	10971.30	+07	$x^5F_5^\circ - e^7G_5$	8447.678	−3	g	11834.32	−07	$a^5F_3 - z^7D_4^\circ$
9038.79	−3	g	11060.39	+06	$b^3G_5 - z^5G_6^\circ$	8447.34	−3	f	11834.80	+10	$x^5D_4^\circ - e^7G_3$
9024 70	−2	g	11077.66	+09	$x^5F_5^\circ - e^7G_4$	8434.509	0d?	g	11852.80	+01	$x^5D_1^\circ - g^5D_0$
8994.66	−3d	f	11114.67	−03	$a^3D_1 - y^3F_2^\circ$	8425.889	−3	g	11864.93	+01	$a^5F_1 - z^7D_1^\circ$
*8978.16	−3	g	11135.08	−15	$x^5D_1^\circ - e^7G_2$	8414.084	0	gb	11881.57	−01	$z^3G_3^\circ - e^3F_4$
		g		+01	$a^1P_1 - y^3D_2^\circ$	8401.695	−2	g	11899.08	−03	$z^5G_2^\circ - e^5F_3$
8967.59	−3	g	11148.21	−06	$y^7P_4^\circ - e^7S_3$	8382.217	−3	g	11926.74	+02	$a^5F_2 - z^7D_2^\circ$
8956.30	−3	g	11162.26	−06	$x^5D_1^\circ - e^7G_1$	8369.858	−1?	f	11944.35	+02	$x^5D_4^\circ - e^7S_3$

TABLE C—(*Continued*)

Solar λ	Solar Int	Grade	Wave Number Solar	Wave Number o−c	Desig
8358.504	2	g	11960.58	+04	$b^3G_4 - z^3G_3^\circ$
8356.02	−3	f	11964.14	+08	$b^1D_2 - z^3S_1^\circ$
8355.15	−3	g	11965.38	+01	$y^5D_4^\circ - e^5D_3$
8349.02	3	gb	11974.17	+05	$a^5F_4 - z^7D_5^\circ$
8345.19	−3	g	11979.66	+01	$a^3G_5 - y^5F_5^\circ$
8342.290	3	fb	11983.82	−12	$b^3G_5 - z^3G_4^\circ$
8307.603	0	fb	12033.87	+02	$a^5F_2 - z^7D_1^\circ$
8303.17	−3	g	12040.29	−08	$a^3G_4 - y^5F_4^\circ$
8299.985	−1	g	12044.91	+04	$X_3 - v^3P_2^\circ$
8269.644	0	g	12089.09	+01	$d^3F_4 - v^3D_3^\circ$
8263.850	0	gb	12097.58	+01	$x^5D_3^\circ - e^5P_2$
8254.32	−3	f	12111.54	+03	$a^1G_4 - y^3F_4^\circ$
8204.95	0N	g	12184.42	−02	$a^5F_3 - z^7D_2^\circ$
8204.09	−2	g	12185.69	+01	$a^5F_4 - z^7D_4^\circ$
8196.51	−2	f	12196.96	+01	$d^3F_4 - w^3F_3^\circ$
8171.239	0	f	12234.69	+09	$a^1F_3 - w^3H_4^\circ$
8146.67	−3	g	12271.58	00	$a^3D_1 - y^3D_2^\circ$
8129.35	−3	g	12297.72	−05	$a^3G_3 - y^5F_2^\circ$
8112.179	−2	g	12323.76	−01	$a^3G_5 - y^5F_4^\circ$
8108.312	−2	g	12329.64	+03	$a^3G_4 - y^5F_3^\circ$
8072.162	−1	g	12384.85	−01	$a^3P_1 - z^3D_1^\circ$
8027.93	−1	g	12453.09	+05	$a^3D_3 - y^3D_2^\circ$
8016.523	−1	fb	12470.81	−01	$y^3D_2^\circ - e^5S_2$
8002.56	−2	f	12492.57	−01	$d^3F_2 - w^3F_2^\circ$
7965.55	−1N	f	12550.61	−05	$x^5F_2^\circ - f^5P_2$
7964.970	3	gb	12551.53	−06	$x^5F_3^\circ - g^5F_4$
7954.97	−1N	g	12567.30	−05	$b^3G_4 - y^3F_4^\circ$
7941.79	1N	fb	12588.16	+08	$a^1G_4 - y^3F_3^\circ$
7924.169	−2	g	12616.15	−05	$y^3D_2^\circ - e^3D_3$
7820.81	−1	f	12782.89	−01	$b^1D_2 - 1_2^\circ$
7813.67	−3NN	f	12794.57	−07	$x^5F_1^\circ - f^5P_1$
7810.815	1	g	12799.24	−01	$x^5F_4^\circ - g^5F_4$
7807.916	6	g	12804.00	+09	$x^5F_5^\circ - g^5F_5$
7802.51	1	g	12812.86	−04	$x^5F_2^\circ - g^5F_3$
7766.62	−3	f	12872.10	+19	$z^3F_3^\circ - e^5D_4$
7746.605	1	g	12905.33	−09	$x^5F_3^\circ - f^3D_3$
7745.521	1	g	12907.14	−06	$x^5F_2^\circ - f^5P_1$
7737.65	−1NN	g	12920.26	+01	$z^5G_3^\circ - e^3F_3$
7733.738	−2	gb	12926.81	−08	$x^5F_3^\circ - f^5G_4$
7720.72	−1	g	12948.60	−06	$x^5F_2^\circ - h^5D_2$
7719.046	3	g	12951.41	00	$x^5F_4^\circ - h^5D_3$
7689.04	−3	f	13001.95	+09	$x^5F_1^\circ - h^5D_1$
7664.18	−3	g	13044.12	−06	$y^3D_1^\circ - e^3D_1$
7661.48	−3	g	13048.72	−04	$x^5F_2^\circ - f^3D_2$
7647.84	−2NN	g	13072.00	00	$z^5G_2^\circ - e^3F_2$
7617.985	−2	g	13123.22	−03	$c^3F_2 - u^5D_2^\circ$
7617.245	−3	g	13124.49	−10	$x^5F_3^\circ - h^5D_2$
7588.310	3	g	13174.54	−01	$x^5F_4^\circ - f^5G_4$
7582.120	0	g	13185.30	+05	$x^5D_3^\circ - h^5D_4$
7573.72	−3	f	13199.92	+05	$z^3F_2^\circ - e^5D_2$
7552.795	−2	g	13236.49	00	$x^5F_4^\circ - g^5F_3$
7551.108	0	g	13239.45	−01	$x^5F_2^\circ - g^5F_2$
7547.904	0	g	13245.07	−03	$x^5F_1^\circ - f^5G_2$
7540.444	0	g	13258.17	−01	$a^3G_4 - z^5G_4^\circ$
7537.96	−3N	f	13262.54	+01	$v^5D_3^\circ - i^5D_3$
7528.18	−2N	f	13279.77	−06	$x^5F_4^\circ - e^5H_5$
7526.67	−3N	f	13282.43	−03	$v^5D_4^\circ - i^5D_4$
*7512.166	−2	g	13308.08	−08	$a^3P_2 - z^3D_2^\circ$
		g		+01	$c^3F_3 - u^5D_4^\circ$
7508.60	−3N	g	13314.40	−12	$x^5D_2^\circ - h^5D_3$
7506.030	1	g	13318.96	−09	$x^5F_3^\circ - f^5G_3$
7501.280	−2	g	13327.39	−06	$c^3F_2 - x^3F_3^\circ$
7495.66	−3	f	13337.39	+03	$x^5D_2^\circ - f^5P_2$
7494.74	−3	f	13339.02	−03	$a^3F_3 - z^5D_4^\circ$
7486.118	−3	g	13354.39	+02	$z^3D_3^\circ - e^5D_4$
7484.308	−1	g	13357.62	−04	$x^5F_2^\circ - f^5G_2$
7482.213	−1	g	13361.36	−02	$x^5F_2^\circ - e^3G_3$
7481.736	−3	g	13362.21	+02	$a^3G_3 - z^5G_3^\circ$
7477.595	0N	f	13369.60	−14	$z^3F_4^\circ - e^5D_4$
7476.87	−3	g	13370.90	+09	$c^3F_2 - w^3D_2^\circ$
*7474.513	−3N	f	13375.12	−02	$z^3F_2^\circ - e^5D_1$
		f		+16	$z^3D_2^\circ - e^5D_3$
7471.757	−2	g	13380.05	−02	$a^3G_4 - z^3G_5^\circ$
7463.395	−1	f	13395.04	−03	$x^5F_3^\circ - e^5H_4$
7461.25	−3	f	13398.89	+05	$v^5D_4^\circ - i^5D_3$
7452.110	−1	g	13415.33	−06	$x^5F_3^\circ - g^5F_2$
7447.912	0	fb	13422.88	+16	$v^5D_3^\circ - i^5D_2$
7443.25	−2	f	13431.30	+03	$x^5F_2^\circ - f^3D_1$
7431.97	−3N	gb	13451.68	−06	$y^5P_1^\circ - e^7P_2$
7420.241	−3N	f	13472.94	−08	$x^5F_2^\circ - e^5H_3$
7418.330	−3	g	13476.42	−01	$c^3F_3 - x^3F_4^\circ$
7415.193	−1	g	13482.12	00	$x^5F_5^\circ - e^3G_5$
7400.851	−3	g	13508.24	+03	$b^3F_2 - y^5F_2^\circ$
7398.96	−3N	g	13511.70	+05	$x^5F_5^\circ - f^5G_4$
7396.526	−1	g	13516.14	−04	$x^5D_2^\circ - f^3D_3$
7385.51	−3N	f	13536.30	+04	$y^3D_2^\circ - g^5D_1$
7385.00	−3	f	13537.24	−07	$x^5F_3^\circ - e^3G_3$
7382.614	0	g	13541.62	+04	$a^3G_5 - z^5G_4^\circ$
7373.011	2	fb	13559.08	−06	$a^3P_2 - z^3D_1^\circ$
7359.983	−3	f	13583.25	−06	$x^5F_5^\circ - e^3H_6$
7356.76	−3	f	13589.20	+09	$y^5P_1^\circ - f^7D_2$
7348.51	−2N	g	13604.46	00	$c^3F_3 - w^3D_3^\circ$
7347.15	−3	g	13606.98	+02	$a^3G_3 - z^5G_2^\circ$
7344.200	0	gb	13612.44	−04	$a^3G_4 - z^5G_3^\circ$
7341.78	−3	f	13616.93	00	$x^5F_5^\circ - e^5H_5$
7330.150	0	f	13638.53	+01	$y^5P_1^\circ - f^7D_1$
7325.28	−3	f	13647.60	+10	$z^3D_2^\circ - e^5D_2$
7317.43	−1	g	13662.24	−04	$x^5D_1^\circ - f^3D_2$
7316.739	1	g	13663.53	+06	$a^3G_5 - z^3G_5^\circ$
7312.08	−2N	f	13672.24	−05	$x^5F_4^\circ - e^3H_5$
7311.265	2	g	13673.76	−02	$z^3P_1^\circ - e^5F_2$
7300.548	−1	f	13693.83	+07	$c^3F_3 - z^3H_4^\circ$
7295.28	−3	g	13703.73	00	$y^5P_2^\circ - e^7P_2$
7278.526	−3	f	13735.27	−09	$x^5D_2^\circ - h^5D_2$
7268.566	1	f	13754.09	+03	$z^3F_4^\circ - e^5D_3$
7266.96	−3	f	13757.13	+07	$a^5P_3 - z^3F_4^\circ$
7261.30	−3	g	13767.85	−04	$x^5D_4^\circ - g^5F_4$
7261.016	1	g	13768.39	−03	$a^3G_4 - z^3G_4^\circ$
7225.79	−3N	g	13835.51	+05	$x^5D_2^\circ - f^3D_2$
7216.63	−1	g	13853.07	+09	$x^5D_1^\circ - g^5F_2$
7213.847	0	g	13858.42	−01	$z^3P_1^\circ - e^5F_1$
7205.536	−2	fb	13874.40	−04	$y^3D_3^\circ - g^5D_2$
7190.128	1	g	13904.13	−01	$c^3P_0 - y^3D_1^\circ$
7162.34	0	g	13958.08	+06	$x^5D_0^\circ - f^3D_1$
7160.859	−2	fb	13960.96	−02	$x^5F_4^\circ - e^3H_4$
7127.573	3	g	14026.16	00	$x^5D_2^\circ - g^5F_2$
7125.33	−2N	g	14030.58	−10	$d^3F_4 - t^3D_4^\circ$
7120.58	−3	g	14039.94	−05	$c^3F_3 - z^1G_4^\circ$
7120.03	1	fb	14041.02	−04	$y^5P_3^\circ - f^7D_4$
7118.105	0	g	14044.82	+03	$x^5D_1^\circ - f^3D_1$
7114.574	0	g	14051.79	−03	$a^3G_5 - z^3G_4^\circ$
7109.70	−3	f	14061.42	−04	$y^5P_2^\circ - e^5G_3$
7107.25	−3	f	14066.26	+09	$a^1F_3 - t^3G_3^\circ$
7105.87	−3	f	14069.00	+07	$c^3F_2 - x^3G_3^\circ$
7103.150	0	gb	14074.39	+01	$a^3H_5 - y^5F_5^\circ$
7101.31	−3	f	14078.04	−04	$a^5P_2 - z^3F_3^\circ$
7093.09	0	g	14094.35	+03	$y^5P_3^\circ - e^7P_2$
7089.71	−2N	gb	14101.07	+03	$d^3F_3 - t^5D_4^\circ$
7079.27	−1N	g	14121.86	+14	$x^5D_4^\circ - f^3D_3$
7074.485	−3N	g	14131.41	−08	$y^3F_3^\circ - e^3D_3$

TABLE C—(*Continued*)

Solar λ	Solar Int	Grade	Wave Number Solar	Wave Number o−c	Desig
7072.80	−2	f	14134.78	+03	$c^3F_4 - z^3H_5^\circ$
7068.64	0	f	14143.10	−09	$x^5D_4^\circ - f^5G_4$
7068.07	−1	f	14144.24	−11	$x^5D_2^\circ - f^5G_2$
7062.79	−3	f	14154.80	00	$x^5D_2^\circ - g^5F_1$
7057.92	−2	f	14164.58	+08	$z^5P_2^\circ - e^7D_2$
*7034.090	−2	{f f}	14212.57	{−06 −02	$y^5P_3^\circ - e^5G_4$ $y^5P_2^\circ - e^5G_2$
7031.40	−3N	g	14218.01	+04	$x^5D_2^\circ - f^3D_1$
7031.09	−2	g	14218.63	−15	$y^3F_2^\circ - e^3D_2$
7028.59	−1	gb	14223.69	−01	$c^3P_1 - y^3D_1^\circ$
7022.385	−1	g	14236.27	+02	$y^5F_1^\circ - e^3F_2$
6997.080	−1	f	14287.74	+10	$x^5D_3^\circ - g^5F_2$
6983.52	−3N	g	14315.49	+03	$d^3F_4 - t^5D_3^\circ$
6979.156	−3	f	14324.44	+03	$b^3P_2 - y^3F_3^\circ$
6970.495	1	g	14342.23	−03	$c^3P_2 - y^3D_2^\circ$
6963.01	−3N	f	14357.65	+01	$c^3F_2 - v^5F_3^\circ$
6953.057	1N	fb	14378.20	−09	$z^5P_3^\circ - e^7D_3$
6942.84	−2NN	f	14399.36	−04	$c^3F_3 - x^3G_4^\circ$
6936.496	−1	f	14412.53	−04	$y^5P_2^\circ - e^7S_3$
6932.498	−3	g	14420.84	−02	$d^3F_2 - t^5D_3^\circ$
6930.384	−3NN	f	14425.24	−08	$y^5P_3^\circ - e^7F_4$
6926.385	−3	f	14433.57	+03	$d^3F_3 - 4_4^\circ$
6920.168	−2	f	14446.54	−01	$y^5P_2^\circ - f^5F_3$
6881.054	−3N	fb	14528.65	+02	$y^3F_2^\circ - g^5D_2$
*6879.55	−3	g	14531.83	+08	$z^5D_2^\circ - X_3$
6864.324	−2N	f	14564.07	−02	$y^5P_3^\circ - e^7F_2$
6860.099	−3	g	14573.04	+08	$y^3D_1^\circ - f^5P_2$
6859.493	−3	f	14574.32	−01	$b^3P_1 - y^3F_2^\circ$
6848.87	−3	f	14596.93	−01	$y^5P_2^\circ - f^5F_1$
6845.98	−3	fb	14603.09	−10	$y^5P_3^\circ - e^5G_2$
6841.642	−3	f	14612.35	−01	$X_2 - w^1F_3^\circ$
6824.857	−3	f	14648.28	−08	$x^5D_2^\circ - e^3P_2$
6819.49	−3	f	14659.81	−15	$c^3P_2 - y^3D_1^\circ$
6808.769	−3N?	f	14682.90	+07	$b^3P_2 - y^3F_2^\circ$
6805.752	−3	g	14689.42	−05	$d^3F_2 - t^5D_2^\circ$
6803.854	−3	f	14693.50	−03	$y^5P_3^\circ - e^7G_3$
6803.27	−3	f	14694.77	+06	$y^5P_3^\circ - f^5F_4$
6801.849	−3	g	14697.83	+04	$a^3F_2 - z^5F_1^\circ$
6801.202	−3	f	14699.23	+26	$z^5D_1^\circ - X_2?$
6794.623	−2	f	14713.47	−04	$x^5D_3^\circ - f^3F_4$
6786.460	−3	g	14731.16	−11	$z^5D_3^\circ - X_3$
6785.88	−3N	f	14732.42	00	$c^3F_4 - v^5F_5^\circ$
6785.764	−2	f	14732.68	−01	$d^3F_3 - y^1D_2^\circ$
6783.28	−3	g	14738.07	−02	$b^3F_4 - z^3G_5^\circ$
6769.682	−3	f	14767.68	−05	$d^3F_2 - y^1D_2^\circ$
6764.19	−3	g	14779.66	−14	$d^3F_4 - u^3G_3^\circ$
6756.568	−3	fb	14796.34	−02	$b^1D_2 - w^3F_2^\circ ?$
6753.470	−2	f	14803.12	−04	$y^5P_3^\circ - e^7S_3$
6746.975	−2	g	14817.37	−03	$b^3F_2 - z^5G_2^\circ$
6745.984	−1	g	14819.55	−06	$c^3F_4 - w^5G_3^\circ$
6737.28	−3	g	14838.70	+05	$z^5D_2^\circ - X_2$
6736.546	−3	g	14840.31	+03	$b^1D_2 - z^1D_2^\circ$
6712.467	−3	f	14893.55	−06	$x^5D_2^\circ - f^3F_3$
6711.282	−3	g	14896.18	−09	$d^3F_2 - t^5D_1^\circ$
*6700.919	−3	f	14919.22	−04	$X_3 - w^1F_3^\circ$
6696.322	0	g	14929.46	−04	$y^3D_1^\circ - f^5P_1$
6682.24	−3	f	14960.92	−02	$c^3F_4 - x^3G_5^\circ$
6681.30	−3	f	14963.02	+08	$z^3G_5^\circ - e^7F_6$
*6677.54	−3	{g f}	14971.45	{−01 −12	$z^5D_4^\circ - X_3$ $x^5D_1^\circ - e^3P_1$
6676.89	−3	g	14972.90	−07	$y^5P_3^\circ - e^3D_2$
6647.856	−3	g	15038.30	+13	$z^5D_3^\circ - X_2$
6635.702	−1	f	15065.84	−05	$z^3G_4^\circ - e^7F_5$
6615.01	−3	f	15112.97	+05	$z^3G_3^\circ - e^7F_4$
6609.693	−2	g	15125.13	−02	$a^5F_2 - z^7F_3^\circ$

Solar λ	Solar Int	Grade	Wave Number Solar	Wave Number o−c	Desig
6601.14	−3	f	15144.72	−03	$x^5D_2^\circ - e^3P_1$
6555.864	−3N	f	15249.31	+02	$c^3F_4 - v^5F_3^\circ$
6551.714	−2	fb	15258.98	−07	$a^5F_2 - z^7F_1^\circ$
6524.749	−3	f	15322.03	+03	$x^5D_1^\circ - e^3P_0$
6494.510	1	g	15393.38	+04	$y^3D_3^\circ - f^5P_2$
6483.954	−1N	g	15418.44	00	$a^3F_4 - z^5F_3^\circ$
6472.152	−3N	f	15446.55	−02	$z^5G_4^\circ - e^5G_5?$
6468.842	−3	f	15454.45	+04	$y^3D_2^\circ - h^5D_1$
6456.874	−1N	f	15483.10	−02	$y^3D_2^\circ - f^5G_3$
6419.676	−1	f	15572.81	−06	$z^3F_3^\circ - e^5F_4$
6396.395	−3N	f	15629.49	−01	$b^1G_4 - y^3G_4^\circ$
6388.424	−2	f	15649.00	−02	$z^5F_4^\circ - e^7D_5$
6385.744	0	f	15655.56	−02	$y^3D_3^\circ - g^5F_3$
6376.180	−3	f	15679.05	+09	$z^5G_6^\circ - e^5G_6$
6353.856	−3	f	15734.13	−04	$a^5F_4 - z^7F_3^\circ$
6351.305	−3	f	15740.45	−04	$z^5G_5^\circ - e^5G_6$
6339.982	−3	f	15768.56	−06	$z^5F_3^\circ - e^7D_4$
6315.425	−1	g	15829.88	−02	$c^3F_3 - v^3D_2^\circ$
6307.885	−2N	f	15848.80	−08	$b^3D_3 - x^3D_3^\circ$
6293.952	−1d?	g	15883.88	−08	$y^3D_1^\circ - e^3P_2$
6290.547	−2N	g	15892.48	00	$b^3F_3 - y^5P_3^\circ$
6284.007	−3	f	15909.02	−01	$a^3D_2 - x^5P_3^\circ$
6253.847	−1	f	15985.74	−06	$y^3D_3^\circ - f^5G_3$
6251.292	−3	f	15992.28	−08	$y^3F_3^\circ - h^5D_4$
6249.657	−2	f	15996.46	−03	$z^5F_4^\circ - e^7D_4$
6219.528	−3	f	16073.95	+01	$z^5F_2^\circ - e^7D_2$
6209.760	−3N	f	16099.24	−08	$z^3D_1^\circ - e^5F_2$
6187.413	−1	g	16157.39	−01	$b^3P_2 - y^3D_1^\circ$
6171.010	−3	f	16200.33	−01	$y^3D_3^\circ - f^5G_2$
6157.427	−2	f	16236.07	−03	$a^3D_2 - x^5P_2^\circ$
6148.668	−2	f	16259.19	−07	$z^5G_6^\circ - f^5F_5$
6145.415	−2N	f	16267.80	00	$z^5F_4^\circ - e^7D_3$
6139.663	−2N	g	16283.04	−03	$b^3F_3 - y^5P_2^\circ$
6137.509	−2	f	16288.76	00	$z^5F_5^\circ - e^7D_4$
6124.084	−3	f	16324.46	−01	$a^1F_3 - u^3F_2^\circ$
6120.257	−1	f	16334.67	−02	$a^5F_4 - z^7P_4^\circ$
6114.396	−3N	f	16350.33	+04	$z^3D_2^\circ - e^5F_2$
6106.865	−3	g	16370.49	−06	$b^3F_2 - y^5P_1^\circ$
6105.144	0	g	16375.11	+03	$y^3F_4^\circ - g^5F_5$
6098.259	0	f	16393.59	+06	$y^5P_3^\circ - f^5P_3$
6097.106	−3	f	16396.70	−06	$a^5P_3 - z^3P_2^\circ$
6091.738	−2	f	16411.14	+01	$y^5P_2^\circ - f^5P_2$
6083.708	−3N	fb	16432.80	−10	$z^3D_3^\circ - e^5F_3$
6081.843	−3	f	16437.84	+01	$c^3F_3 - v^3G_3^\circ$
6081.723	−3	g	16438.17	00	$z^5G_3^\circ - g^5D_2$
6065.813	−3	f	16481.28	−02	$b^3H_4 - z^5H_4^\circ$
6060.829	−3	f	16494.84	−06	$y^5F_4^\circ - f^5D_3$
6051.037	−3N	f	16521.52	−11	$b^3F_4 - y^3F_3^\circ$
6019.386	−2	f	16608.40	−08	$a^1H_5 - y^3G_4^\circ$
6018.314	0	fb	16611.36	+07	$y^3F_2^\circ - h^5D_1$
6016.930	−3	f	16615.18	+05	$d^3F_4 - y^1F_3^\circ$
6015.264	−2	g	16619.78	−04	$a^5P_1 - y^5F_2^\circ$
6012.784	−2N	fb	16626.64	−08	$y^5P_3^\circ - g^5F_4$
6009.853	−3	fb	16634.74	−06	$a^3D_3 - x^5P_2^\circ$
6007.722	−2	f	16640.64	+07	$b^3H_5 - z^5H_5^\circ$
5996.510	−3	f	16671.76	−06	$y^5F_2^\circ - e^5G_3$
5996.230	−3	f	16672.54	−02	$a^3D_1 - x^5P_1^\circ$
5995.949	−2N	f	16673.32	−05	$y^5P_2^\circ - g^5F_3$
5991.575	−3	f	16685.49	00	$d^3F_3 - y^1F_3^\circ$
5981.398	−3	f	16713.88	−06	$a^1I_6 - z^3I_6^\circ$
5978.149	−2	g	16722.96	+06	$y^5P_1^\circ - h^5D_1$
5976.171	−3	f	16728.50	+02	$b^1D_2 - v^3F_3^\circ$
5974.596	−3N	f	16732.91	+07	$y^5D_2^\circ - f^5D_3$
5973.362	−3	g	16736.36	+03	$y^3F_2^\circ - g^5F_2$
5958.351	−1	g	16778.53	−03	$a^5P_3 - y^5F_3^\circ$

TABLE C—(*Continued*)

Solar λ	Solar Int	Grade	Wave Number		Desig
			Solar	o−c	
5955.117	−3	f	16787.64	00	$d^3F_3 - x^1F_3^\circ$
5952.192	−3	g	16795.89	−02	$x^5F_2^\circ - i^5D_3$
5950.142	−2	f	16801.68	−04	$y^5P_3^\circ - f^5P_2$
*5943.602	0	g	16820.16	−06	$a^5P_2 - y^5F_2^\circ$
5943.117	−3	f	16821.54	−02	$c^3F_2 - z^1F_3^\circ$
5942.737	−3	f	16822.61	−07	$d^3F_2 - x^1F_3^\circ$
5933.811	−2	f	16847.92	−02	$y^5P_1^\circ - g^5F_2$
5931.905	−3	f	16853.33	−05	$c^3F_4 - y^3H_4^\circ$
5928.527	−3	f	16862.93	−09	$y^5D_1^\circ - f^5D_1$
5901.533	−1	f	16940.07	−02	$y^5F_4^\circ - e^5G_4$
5893.243	−2	f	16963.89	00	$y^5D_1^\circ - f^5D_0$
5892.478	−2	f	16966.10	−04	$y^5P_1^\circ - f^5G_2$
5891.186	1	fb	16969.82	−07	$y^3F_2^\circ - e^5H_3$
5890.508	−1d?	gb	16971.77	−07	$x^5F_3^\circ - i^5D_3$
5887.478	−2	f	16980.51	−04	$y^5P_3^\circ - f^3D_3$
5881.728	−2	g	16997.11	+09	$a^5P_3 - y^5F_2^\circ$
5881.288	1	g	16998.38	−03	$y^3F_3^\circ - f^5G_3$
5879.503	0	f	17003.54	−05	$y^5P_2^\circ - f^5G_3$
5876.302	−1	g	17012.80	−08	$y^5F_1^\circ - f^5F_2$
5867.010	−3	f	17039.74	−01	$y^5P_1^\circ - f^3D_1$
5861.120	0	g	17056.87	−04	$y^5F_2^\circ - f^5F_3$
5859.965	−3	f	17060.24	−01	$y^5D_2^\circ - f^7D_3$
5858.790	0	g	17063.65	−05	$y^5F_4^\circ - f^5F_5$
5858.284	−2	f	17065.12	−05	$a^3H_5 - y^3F_4^\circ$
5849.698	0	f	17090.17	−08	$b^1G_4 - x^3F_4^\circ$
5845.298	0	g	17103.04	−08	$x^5F_4^\circ - i^5D_4$
5835.589	−1	f	17131.49	−03	$b^3P_2 - x^5D_3^\circ$
5835.434	−1	g	17131.95	−08	$x^5F_3^\circ - i^5D_2$
5835.114	0	g	17132.89	−04	$y^5F_3^\circ - f^5F_4$
5827.886	0	f	17154.14	00	$z^5D_1^\circ - e^7D_2$
5826.649	−2N	g	17157.78	−04	$y^5F_2^\circ - f^5F_2$
5815.229	(1)	g	17191.47	+01	$d^3F_4 - t^3G_4^\circ$
5813.341	−3N	f	17197.06	−03	$y^5D_2^\circ - f^7D_2$
5809.878	−2N	g	17207.31	+01	$y^5F_2^\circ - f^5F_1$
5807.995	−1	g	17212.88	−07	$y^3F_3^\circ - f^5G_2$
5807.804	−1	f	17213.45	−04	$z^5D_0^\circ - e^7D_1$
5807.253	−3	f	17215.08	−09	$b^3H_6 - z^5H_6^\circ$
5796.674	−2	f	17246.50	00	$y^5D_2^\circ - f^7D_1$
5791.539	0	g	17261.79	−03	$d^3F_3 - t^3G_4^\circ$
5787.277	−2	f	17274.50	−01	$a^3D_3 - w^5D_4^\circ$
5787.024	−1	gb	17275.26	−11	$y^5F_3^\circ - f^5F_3$
5778.814	−2	fb	17299.80	−02	$y^5P_3^\circ - f^3D_2$
5762.847	0	gb	17347.73	−04	$y^5F_1^\circ - e^3D_1$
5761.094	−2	f	17353.01	−05	$y^5D_1^\circ - e^5G_2$
5760.538	−2	f	17354.69	−02	$y^5D_3^\circ - f^7D_3$
5754.923	−2N	g	17371.62	−10	$a^3P_0 - y^5P_1^\circ$
5753.983	−2N	fb	17374.46	−03	$a^3H_4 - y^3F_3^\circ$
5753.400	−2N	g	17376.22	−06	$y^5F_3^\circ - f^5F_2$
5749.640	−2	f	17387.58	+04	$z^3G_4^\circ - h^5D_4$
5747.865	−2	f	17392.95	−05	$b^3P_2 - x^5D_2^\circ$
5739.810	−2	fb	17417.36	−10	$y^5D_2^\circ - e^5G_3$
5738.242	0	g	17422.12	−05	$y^5F_4^\circ - f^5F_4$
5732.886	−2	f	17438.40	−07	$y^5D_4^\circ - f^5D_3$
5732.311	0	g	17440.14	−08	$x^5F_5^\circ - i^5D_4$
*5721.717	−1	f	17472.44	−03	$y^5F_2^\circ - e^5P_1$
		f		−04	$y^5D_3^\circ - e^5G_4$
5720.902	0	g	17474.92	−06	$y^3F_4^\circ - f^5G_3$
5715.476	−2	f	17491.51	−03	$y^5D_3^\circ - f^7D_2$
5714.903	−3	f	17493.27	−07	$z^5D_3^\circ - e^7D_2$
5709.931	−1	f	17508.50	00	$y^5F_3^\circ - e^5P_3$
5707.249	−2	f	17516.73	+01	$b^3D_3 - u^5D_2^\circ$
5706.116	0	f	17520.21	00	$y^5F_2^\circ - e^5P_2$
5696.108	0	fb	17550.99	−01	$y^3F_4^\circ - e^5H_4$
5691.707	−1	g	17564.56	−05	$y^5F_4^\circ - f^5F_3$
5690.074	−3	f	17569.60	−02	$x^5D_1^\circ - i^5D_2$
5685.887	−2	f	17582.54	−07	$x^5D_2^\circ - i^5D_3$
5678.616	−2	g	17605.05	−05	$a^3P_1 - y^5P_2^\circ$
5678.407	−2	g	17605.70	−08	$z^3D_3^\circ - e^3F_2$
5677.705	−1	f	17607.87	−08	$y^5D_4^\circ - e^5G_5$
5661.988	−2	f	17656.75	−07	$z^3P_1^\circ - e^5P_1$
5661.028	−2	f	17659.75	00	$d^3F_2 - t^3G_3^\circ$
5658.672	0	g	17667.10	+01	$y^5F_2^\circ - g^5D_1$
5652.026	−3	g	17687.88	−04	$y^5D_1^\circ - f^5F_2$
5651.477	0	g	17689.59	−03	$z^3G_3^\circ - f^5G_4$
5650.287	−3	g	17693.32	+08	$y^3F_4^\circ - e^3G_3$
5648.917	−3	f	17697.61	−06	$a^3D_3 - w^5D_3^\circ$
5646.690	−1	f	17704.59	+03	$z^3P_1^\circ - e^5P_2$
5644.352	−2	f	17711.92	00	$y^5D_3^\circ - e^5G_3$
5643.945	−2	f	17713.20	−01	$c^3F_4 - z^1F_3^\circ$
5642.764	−1	f	17716.91	−04	$y^3F_3^\circ - e^3P_2$
5636.004	−3	f	17738.16	−01	$y^5D_2^\circ - e^7G_2$
5634.526	−3	fb	17742.81	+01	$x^5D_2^\circ - i^5D_2$
5627.100	−2	g	17766.22	−07	$y^5F_5^\circ - f^5F_4$
5623.644	−3	f	17777.14	−01	$a^3D_1 - w^5D_2^\circ$
5617.152	−1	f	17797.69	−05	$y^5F_4^\circ - e^5P_3$
5615.169	−3	g	17803.98	+04	$z^5G_4^\circ - g^5F_5$
5614.284	−1N	g	17806.78	+03	$x^5F_2^\circ - g^5G_2$
5613.716	−2N	f	17808.58	−05	$x^5D_1^\circ - 4_2$
5609.991	−2	f	17820.41	−06	$b^3D_2 - u^5D_1^\circ$
5608.982	−1	g	17823.61	00	$z^3P_2^\circ - g^5D_3$
5607.673	0	f	17827.77	−03	$y^5D_3^\circ - e^7G_4$
5602.569	−2	f	17844.01	−08	$x^5D_3^\circ - i^5D_3$
5595.069	−2	g	17867.93	−03	$x^5F_3^\circ - g^5G_3$
5583.992	−1N	g	17903.38	−08	$y^5D_2^\circ - f^5F_2$
5579.357	−1	f	17918.25	−07	$y^5D_0^\circ - e^3D_1$
5577.035	−1	g	17925.71	00	$x^5F_4^\circ - g^5G_4$
5570.070	−2	g	17948.13	−03	$b^3P_1 - z^5S_2^\circ$
5568.709	−3	f	17952.51	−01	$c^3F_3 - v^3F_3^\circ$
5568.466	−3	f	17953.30	−09	$y^5D_3^\circ - e^7G_3$
5568.081	−2	g	17954.53	−03	$y^5D_3^\circ - f^5F_4$
5566.819	−2	f	17958.61	00	$a^3D_3 - w^5D_2^\circ$
*5563.700	0	f	17968.67	−02	$a^3P_1 - y^3F_2^\circ$
		f		−03	$c^3F_3 - u^5F_4^\circ$
5562.128	−2	f	17973.75	−03	$z^3G_5^\circ - e^5H_6$
5559.652	−1	f	17981.76	−05	$x^5D_2^\circ - 4_2$
*5557.921	0	fb	17987.36	−07	$c^3P_0 - x^5P_1^\circ$
5555.180	−3	f	17996.23	−04	$a^1D_2 - z^3S_1^\circ$
5553.238	−2N	g	18002.53	−05	$y^5D_1^\circ - e^5P_1$
5552.854	−3	f	18003.77	00	$b^3P_2 - x^5F_3^\circ$
5552.702	−1	f	18004.26	−02	$x^5D_3^\circ - i^5D_2$
5551:781	−3N	g	18007.25	−03	$y^5D_4^\circ - f^5F_5$
5551.310	−3N	f	18008.78	−05	$a^1P_1 - x^3D_1^\circ$
5549.661	−1	g	18014.13	00	$x^5F_5^\circ - g^5G_5$
5549.534	−3	f	18014.54	+06	$z^3G_5^\circ - g^5F_4$
*5543.049	−2	g	18035.62	−04	$b^1G_4 - x^3G_5^\circ$
		g		−06	$y^5D_2^\circ - e^5P_3$
5541.592	−3	f	18040.36	−04	$a^3D_2 - v^5D_2^\circ$
5536.599	−1	g	18056.63	−03	$b^3P_2 - z^5S_2^\circ$
5529.791	−3	f	18078.86	+03	$b^3P_0 - x^5F_1^\circ$
5528.905	−1	g	18081.76	−03	$z^3G_3^\circ - f^5G_3$
5521.303	−1	f	18106.65	−06	$z^3G_4^\circ - e^5H_5$
5518.546	−3	gb	18115.70	+08	$x^5F_4^\circ - g^5G_3$
5516.306	−2	f	18123.05	−05	$y^5D_4^\circ - e^5G_3$
5512.414	−1	f	18135.85	−05	$z^5G_3^\circ - f^3D_3$
5510.243	−3	f	18142.99	−05	$c^3F_4 - u^5F_5^\circ$
5505.734	−3	f	18157.85	+05	$z^3G_2^\circ - e^5H_4$
5499.600	−1	f	18178.10	−02	$z^3G_3^\circ - g^5F_2$
5496.577	−1	f	18188.10	−04	$x^5D_4^\circ - i^5D_3$
5493.356	−3	f	18198.77	−08	$b^3D_2 - y^3S_1^\circ$
5489.872	−1	f	18210.31	−09	$z^5G_2^\circ - f^3D_2$

TABLE C—(*Continued*)

Solar λ	Solar Int	Grade	Wave Number		Desig	Solar λ	Solar Int	Grade	Wave Number		Desig
			Solar	o−c					Solar	o−c	
5488.173	−1d?	gb	18215.95	−12	$y^3F_3^\circ - f^3F_2$	5257.648	0	fb	19014.64	00	$a^1H_5 - x^3G_5^\circ$
*5487.524	−1N	g	18218.11	00	$y^5D_2^\circ - e^5P_1$	5255.747	−2	f	19021.51	+03	$y^5F_2^\circ - f^5P_2$
5482.268	−2N	f	18235.57	−01	$b^3D_1 - y^3S_1^\circ$	5255.666	−2	g	19021.80	+06	$y^5F_4^\circ - g^5F_5$
5479.984	−2	f	18243.17	−12	$x^5D_3^\circ - 4_2$	5253.256	−3	f	19030.53	−04	$b^3D_1 - v^5F_1^\circ$
5474.098	−3	g	18262.79	−02	$x^5F_5^\circ - g^5G_4$	5253.040	−1	f	19031.31	−05	$a^3P_2 - y^5P_1^\circ$
5473.172	0	gb	18265.88	+03	$y^5D_2^\circ - e^5P_2$	5246.007	−3	fb	19056.83	−03	$a^3D_3 - x^3D_2^\circ$
5469.283	−1N	gb	18278.87	+04	$z^5G_5^\circ - g^5F_5$	5245.738	−3	f	19057.80	−08	$a^1P_1 - z^3S_1^\circ$
5469.076	−3	f	18279.56	+05	$b^1D_2 - v^3P_2^\circ$	5245.638	−3	g	19058.17	−06	$z^5G_6^\circ - e^3H_6$
5461.824	−2N	f	18303.83	−09	$z^5P_1^\circ - e^5F_2$	5240.360	−2	g	19077.36	−01	$b^3H_5 - y^3G_4^\circ$
5455.094	−3	f	18326.41	00	$a^3D_3 - v^5D_3^\circ$	5238.253	−3N	f	19085.04	00	$z^3F_2^\circ - e^5G_3$
5453.996	−2	g	18330.10	−04	$y^5D_3^\circ - e^5P_3$	5236.386	−2	g	19091.84	−01	$z^5G_5^\circ - e^5H_5$
5443.425	−2	g	18365.70	−04	$y^5D_4^\circ - f^5F_4$	5226.388	−3	f	19128.36	+10	$b^3G_5 - y^5G_5^\circ$
5438.055	−2	f	18383.83	−06	$d^3F_4 - v^3H_5^\circ$	5218.516	−3	f	19157.22	−03	$d^3F_2 - s^3G_3^\circ$
5437.209	0	g	18386.69	−07	$z^5G_5^\circ - f^5G_6$	5217.677	−3N	g	19160.30	+05	$z^3F_2^\circ - e^3D_3$
5435.184	−1	f	18393.54	−05	$z^3G_4^\circ - f^5G_3$	5213.816	−2	f	19174.49	−10	$z^3F_3^\circ - e^5G_4$
5429.858	1	f	18411.58	−10	$z^3G_3^\circ - e^5H_3$	5213.353	−2	g	19176.19	−02	$z^3G_5^\circ - e^3H_4$
5429.513	1	g	18412.75	+02	$y^5D_2^\circ - g^5D_1$	5209.896	−2	g	19188.91	00	$b^3H_6 - y^3G_5^\circ$
5429.434	−2	g	18413.02	00	$c^3F_3 - u^3G_3^\circ$	5206.821	−2	g	19200.25	−06	$y^5F_2^\circ - f^3D_3$
5428.715	−2	f	18415.46	−03	$c^3F_2 - t^3D_1^\circ$?	5197.944	0	f	19233.04	−05	$y^5F_1^\circ - f^5P_1$
5423.760	−2	f	18432.29	−08	$b^1G_4 - w^3G_5^\circ$	5196.270	−3N	f	19239.23	−11	$b^3G_5 - y^5G_4^\circ$
5422.167	−1	g	18437.70	−06	$z^5G_6^\circ - f^5G_5$	5184.199	−2	f	19284.03	−09	$z^5G_5^\circ - e^3G_4$
5421.846	−1	g	18438.79	+02	$y^3F_4^\circ - f^3F_3$	5172.219	−1	f	19328.69	−05	$b^3F_4 - x^5D_3^\circ$
5412.795	−1	f	18469.62	+01	$z^3G_4^\circ - e^5H_4$	5169.302	−1	gb	19339.60	−01	$c^3F_4 - t^3D_3^\circ$
5412.577	−2	f	18470.37	−05	$d^3F_3 - v^3H_4^\circ$	*5168.193	−2d?	f	19343.75	−02	$z^3F_3^\circ - e^7F_3$
5406.781	1	fb	18490.16	−04	$z^5G_4^\circ - f^3D_3$	5167.718	−1	f	19345.53	−05	$a^1P_1 - u^5D_2^\circ$
5406.342	−1	f	18491.67	+05	$c^3F_4 - v^3F_3^\circ$	5164.687	−2	f	19356.88	+04	$b^3F_3 - x^5D_2^\circ$
5405.360	1	f	18495.03	−03	$z^3G_5^\circ - e^5H_5$	5159.971	−2	g	19374.57	−08	$y^5F_1^\circ - f^3D_2$
5401.272	0	g	18508.99	−05	$z^5G_6^\circ - e^5H_6$	5150.196	−1	g	19411.34	−01	$a^1H_5 - w^3G_5^\circ$
5396.908	−2	f	18523.99	−03	$c^3P_2 - x^5P_2^\circ$	5146.319	−2	g	19425.97	−07	$z^5G_4^\circ - f^3F_4$
5391.793	−3	f	18541.56	−04	$a^3G_5 - x^5F_5^\circ$	5145.740	−3N	f	19428.15	−05	$b^1G_4 - 3_3^\circ$
5385.591	−1	f	18562.92	−04	$b^1G_4 - w^3G_4^\circ$	5141.542	−3	f	19444.01	+02	$b^1G_4 - w^3F_3^\circ$
5384.204	−3	f	18567.70	+07	$z^5P_2^\circ - e^5F_2$	5130.936	−3	g	19484.21	−10	$z^5G_5^\circ - e^3H_5$
5374.771	−2	f	18600.28	+03	$a^1H_5 - w^5G_5^\circ$	5127.690	−1	g	19496.54	−04	$a^5D_3 - z^7D_2^\circ$
5358.120	−1	f	18658.09	−07	$a^3D_2 - x^3D_2^\circ$	5124.617	−1	g	19508.23	−05	$b^3H_4 - z^3I_5^\circ$
5346.341	−2	f	18699.19	−01	$z^5P_3^\circ - e^5F_3$	5123.290	−2N	g	19513.29	−03	$z^5G_3^\circ - f^3F_3$
5339.428	−1	f	18723.40	−09	$z^3G_4^\circ - e^5H_3$	5120.888	−3	g	19522.44	+02	$z^5G_2^\circ - f^3F_2$
5334.339	−2	g	18741.27	−05	$y^5D_4^\circ - e^5P_3$	5119.917	−3	f	19526.14	−07	$z^3F_4^\circ - e^7F_5$
5331.197	−3	f	18752.31	+02	$z^5P_2^\circ - e^5F_1$	5114.516	−2d?	fb	19546.76	00	$d^3F_4 - u^3F_4^\circ$
5327.895	−1	f	18763.93	−14	$z^5G_3^\circ - f^5G_2$	5096.187	−2	f	19617.06	−06	$d^3F_3 - u^3F_4^\circ$
5327.266	−2	f	18766.15	−04	$b^3D_2 - v^5F_3^\circ$	*5091.726	−2	f	19634.25	00	$a^1P_1 - u^5D_1^\circ$
5319.216	−2	g	18794.55	+01	$c^3F_4 - u^3G_4^\circ$	5086.776	−2	f	19653.35	−04	$y^5D_3^\circ - f^5P_3$
5318.040	−3	fb	18798.71	00	$b^3G_3 - y^5G_3^\circ$	5085.907	−3N	f	19656.71	+10	$z^3F_3^\circ - f^5F_4$
5317.549	−1d?	g	18800.44	−07	$c^3F_3 - t^3D_3^\circ$	5085.685	−2	fb	19657.58	−02	$y^5F_5^\circ - e^5H_6$
5315.785	−2N	f	18806.68	−03	$b^3D_2 - v^5P_2^\circ$	5084.563	−2	f	19661.91	−05	$b^1G_4 - v^3G_5^\circ$
5308.690	−1	f	18831.81	+06	$y^5F_3^\circ - f^5P_3$	5082.656	−3	f	19669.29	+12	$c^3P_0 - v^5D_1^\circ$
5301.314	−2	f	18858.02	+06	$z^3G_5^\circ - e^5H_4$	5081.845	−2	fb	19672.43	+07	$z^3F_4^\circ - e^5G_4$
5300.407	−2	f	18861.24	+02	$d^3F_4 - s^3G_5^\circ$	5080.937	−3	g	19675.94	+05	$b^3H_5 - z^3I_6^\circ$
5293.041	−1	g	18887.49	−03	$z^3G_5^\circ - e^3H_5$	5075.167	−3	gb	19698.31	+01	$y^5F_5^\circ - g^5F_4$
5288.379	−3	f	18904.14	00	$b^3G_4 - y^5G_4^\circ$	5069.627	−3	f	19719.82	−13	$b^3F_3 - x^5F_4^\circ$
5285.131	0	g	18915.76	−04	$z^3G_4^\circ - f^3F_4$	5064.975	1	g	19737.95	−10	$y^5F_3^\circ - f^3D_2$
5284.618	−1	g	18917.59	−01	$c^3F_2 - t^3D_2^\circ$	5052.993	−1	g	19784.75	−08	$b^3H_5 - z^3I_5^\circ$
5284.284	−3	f	18918.79	−06	$b^3D_1 - v^5F_2^\circ$	5051.311	−2N	g	19791.34	−08	$y^5F_4^\circ - g^5F_3$
5281.165	−2	f	18929.96	+04	$d^3F_3 - s^3G_4^\circ$	5050.139	−2	f	19795.93	−03	$z^3F_4^\circ - f^5F_5$
5279.675	−2	g	18935.31	−08	$b^3H_4 - y^3G_3^\circ$	5047.125	−2	f	19807.75	+06	$d^3F_3 - u^3F_3^\circ$
*5277.312	−1	g	18943.78	+01	$b^3H_5 - y^3G_5^\circ$	*5041.325	−1	f	19830.54	+02	$z^3P_0^\circ - g^5F_1$
		g		−02	$z^5G_3^\circ - e^3H_4$			f		−04	$a^1F_3 - r^3G_3^\circ$
5275.286	1	f	18951.06	+04	$a^1D_2 - y^3S_1^\circ$	5040.248	−2	f	19834.77	+01	$y^5F_4^\circ - e^5H_5$
5273.602	−3N	g	18957.11	+07	$z^5G_5^\circ - e^3G_5$	5027.531	−3	f	19884.95	−10	$z^3F_4^\circ - e^7F_4$
5270.067	−2	f	18969.83	−01	$b^3D_1 - v^5P_1^\circ$	5027.355	−1	f	19885.64	−05	$z^3F_2^\circ - e^5P_1$
5267.280	0	g	18979.86	00	$z^5G_4^\circ - e^5H_4$	5025.081	−1	f	19894.64	+01	$z^3P_1^\circ - g^5F_2$
5265.424	−3	g	18986.56	−01	$z^5G_5^\circ - f^5G_4$	5021.694	−2	f	19908.06	−07	$y^5D_1^\circ - f^5P_1$
5262.889	−1	f	18995.70	−01	$a^3D_3 - x^3D_3^\circ$	5021.603	0	fb	19908.42	00	$y^5F_3^\circ - e^5H_4$
5262.624	−2	g	18996.65	−05	$z^5G_6^\circ - e^3H_6$	5019.737	−1	f	19915.82	+01	$z^3F_2^\circ - g^5D_2$
5261.503	−3	f	19000.70	−03	$b^3G_5 - y^5G_6^\circ$	5019.189	−2	f	19918.00	−05	$d^3F_2 - u^3F_2^\circ$
5259.095	−3N	g	19009.40	−02	$z^5G_4^\circ - e^3H_5$	5016.484	0	g	19928.74	00	$y^5F_3^\circ - g^5F_2$

TABLE C—(*Continued*)

Solar λ	Solar Int	Grade	Wave Number		Desig	Solar λ	Solar Int	Grade	Wave Number		Desig
			Solar	o−c					Solar	o−c	
5015.301	−2	f	19933.44	00	$z^3F_2^\circ - e^5P_2$	4802.525	−1	g	20816.58	+01	$y^5P_2^\circ - i^5D_2$
5012.700	0	f	19943.78	−06	$y^5F_2^\circ - e^5H_3$	4801.622	−2	f	20820.49	+01	$z^3P_0^\circ - e^3P_1$
5012.160	1	g	19945.93	−02	$y^5D_2^\circ - f^3D_3$	4800.544	−3	g	20825.17	+04	$b^3H_4 - z^1G_4^\circ$
5011.209	−2	f	19949.72	+13	$y^5D_1^\circ - h^5D_2$	4799.071	−3	f	20831.56	−05	$y^5F_2^\circ - f^3F_2$
5010.331	−2	f	19953.21	−12	$b^3F_4 - x^5F_4^\circ$	4794.365	−1	g	20852.01	−04	$a^3P_1 - x^5D_1^\circ$
5006.695	−2	f	19967.70	+09	$b^3F_3 - x^5F_3^\circ$	4790.752	−1	f	20867.73	−02	$a^3D_3 - x^3F_3^\circ$
5003.881	−3	f	19978.93	−11	$b^3F_2 - x^5F_2^\circ$	4790.570	−1	f	20868.53	−05	$y^5D_3^\circ - f^5G_2$
4995.411	−1	f	20012.81	−02	$z^3P_1^\circ - f^5G_2$	4782.813	−2	g	20902.37	−11	$b^3H_6 - z^3H_5^\circ$
4992.787	−2	f	20023.32	+05	$z^3P_1^\circ - g^5F_1$	4780.822	−1	g	20911.08	−03	$a^3D_3 - w^3D_2^\circ$
4991.862	−1	fb	20027.03	00	$y^5F_4^\circ - e^3G_4$	4766.879	−1	g	20972.24	−06	$z^5F_2^\circ - e^3F_3$
4987.857	−3	fb	20043.11	−12	$z^3F_4^\circ - g^5D_4$	4760.076	−1	g	21002.21	−02	$z^7P_2^\circ - e^5D_1$
4987.654	−3	f	20043.95	−12	$y^5F_5^\circ - e^3G_5$	4749.260	−3N	f	21050.05	−02	$y^5F_3^\circ - f^3F_2$
4986.915	−1	g	20046.90	−04	$y^5F_3^\circ - f^5G_2$	4744.644	−2	f	21070.52	−03	$a^5F_2 - z^5P_3^\circ$
4985.992	−1	f	20050.61	−05	$y^5F_3^\circ - e^3G_3$	4742.939	−2	f	21078.10	−02	$y^5D_2^\circ - e^3P_2$
4979.840	−3	f	20075.38	00	$c^3P_2 - w^5D_1^\circ$	4727.003	−2	f	21149.16	+02	$a^3D_1 - y^3S_1^\circ$
4978.116	−2	f	20082.34	−02	$z^3D_1^\circ - e^5P_1$	4716.838	−1	f	21194.73	+04	$a^3D_3 - 1_2^\circ$
4972.914	−3	f	20103.34	−07	$a^3D_2 - y^3P_2^\circ$	4701.910	0N	g	21262.02	−04	$z^5F_1^\circ - e^3F_2$
4970.653	0	fb	20112.48	+01	$z^3D_1^\circ - g^5D_2$	4700.441	−1	f	21268.67	−08	$a^5P_2 - y^3D_1^\circ$
4966.286	−2	f	20130.17	+07	$z^3D_1^\circ - e^5P_2$	4690.382	0	f	21314.28	−03	$a^5F_1 - z^5P_2^\circ$
4957.705	−2	g	20165.01	−11	$y^5D_2^\circ - h^5D_2$	4688.382	0	fb	21323.37	−01	$y^5D_2^\circ - f^3F_3$
4954.298	−3	f	20178.88	00	$y^5F_5^\circ - e^5H_5$	4687.678	−1	f	21326.57	−06	$b^3P_0 - w^5F_1^\circ$
4945.284	−3	f	20215.66	+02	$c^3P_2 - v^5D_3^\circ$	4687.313	0	f	21328.23	−04	$a^5F_3 - z^5P_3^\circ$
4942.602	−2	f	20226.63	−04	$y^5F_3^\circ - e^3H_4$	4685.036	−1	fb	21338.60	−02	$b^3P_1 - w^5F_2^\circ$
4939.481	−3	f	20239.41	−08	$c^3F_2 - 11_3^\circ$	4679.985	−2	f	21361.63	−09	$y^5D_1^\circ - f^3F_2$
4933.193	0	g	20265.21	−02	$y^5D_2^\circ - f^3D_2$	4678.422	−1	g	21368.77	−05	$z^5F_2^\circ - e^3F_2$
4930.067	−2	f	20278.06	−10	$a^3D_2 - y^3P_1^\circ$	4677.604	0	f	21372.50	−08	$y^5D_3^\circ - e^3P_2$
4926.848	−2	f	20291.31	−10	$a^1I_6 - y^3H_5^\circ$	4674.658	0	f	21385.97	−04	$a^3F_3 - z^3P_2^\circ$
4922.162	−2	f	20310.62	+06	$z^3P_2^\circ - g^5F_2$	4673.280	1	g	21392.28	−01	$z^5P_2^\circ - e^7P_2$
4919.749	−2	f	20320.59	−06	$a^3D_1 - y^3P_2^\circ$	4672.839	1	f	21394.30	−02	$a^3F_2 - z^3P_1^\circ$
4916.678	−2	f	20333.28	−05	$z^3D_2^\circ - e^5P_1$	4672.038	−1	f	21397.96	−08	$c^3F_3 - w^1G_4^\circ$
4911.541	0	f	20354.54	−07	$y^5F_3^\circ - f^3F_4$	4668.072	1	g	21416.14	00	$z^5P_1^\circ - e^5S_2$
4908.608	−2	g	20366.70	+02	$a^3P_0 - x^5D_1^\circ$	4665.551	−1	f	21427.72	+04	$c^3F_4 - 13_4^\circ$
4906.775	−2	f	20374.31	+11	$y^5F_4^\circ - g^7D_3$	4665.259	−2	f	21429.05	−10	$z^3P_2^\circ - e^3P_1$
4893.707	−2	f	20428.72	−03	$z^3P_2^\circ - f^5G_2$	4661.336	−1N	f	21447.09	−03	$b^3P_2 - w^5F_2^\circ$
4893.572	−3	f	20429.29	+09	$y^5F_5^\circ - g^7D_4$	4653.505	−1	f	21483.18	−06	$a^5F_2 - z^5P_2^\circ$
4887.369	−1	f	20455.21	+02	$c^3F_4 - 9_4^\circ$	4643.217	−1	g	21530.78	−10	$a^3F_4 - y^5D_3^\circ$
4886.179	−3	g	20460.19	−02	$c^3P_0 - x^3D_1^\circ$	4642.593	−1N	g	21533.68	−04	$z^5F_3^\circ - e^3F_2$
4877.592	0	g	20496.21	+08	$z^7P_3^\circ - e^5D_4$	4641.218	0	f	21540.05	−01	$b^3P_2 - w^5F_1^\circ$
4876.204	−2	f	20502.05	−06	$a^3D_3 - y^3P_2^\circ$	4636.678	−1	f	21561.14	−09	$a^1G_4 - z^3I_5^\circ$
4875.741	−3	f	20503.99	−10	$d^3F_4 - t^3F_4^\circ$	4635.630	0	g	21566.02	−04	$z^7F_3^\circ - e^5D_4$
4874.363	0	g	20509.79	−04	$c^3P_1 - x^3D_2^\circ$	4632.818	1	g	21579.11	+04	$z^5P_2^\circ - f^7D_1$
4873.754	−1	g	20512.36	−05	$a^3D_2 - w^3D_2^\circ$	4632.147	0	fb	21582.24	−02	$a^1D_2 - w^3P_2^\circ$
4872.910	−2	f	20515.90	−01	$y^5F_4^\circ - e^3H_4$	4631.039	0N	f	21587.39	−04	$y^5D_4^\circ - f^3F_4$
4872.703	−3	f	20516.78	−05	$z^3P_1^\circ - e^3P_2$	4638.687	−1	f	21598.37	+03	$z^5P_1^\circ - e^7F_2$
4871.937	1	f	20520.00	00	$a^3D_3 - u^5D_3^\circ$	4625.440	−1N	g	21613.53	00	$z^3F_2^\circ - f^3D_3$
4870.049	−1	f	20527.96	+01	$z^3D_2^\circ - g^5D_1$	4621.622	−1	f	21631.39	+02	$z^3D_1^\circ - f^5P_2$
4869.471	0	f	20530.39	−08	$a^1D_2 - v^3D_3^\circ$	4620.140	−1	f	21638.32	−06	$c^3P_1 - w^5P_1^\circ$?
4868.38?	−1	g	20535.00	−01	$a^3F_3 - y^5D_4^\circ$	4612.620	−1	g	21673.60	+10	$b^3P_2 - y^5S_2^\circ$
4867.641	−3	f	20538.11	−01	$b^3H_5 - x^3F_4^\circ$	4611.194	0	g	21680.30	−02	$z^7F_4^\circ - e^5D_4$
4867.544	−1	g	20538.52	−05	$a^3F_2 - y^5D_3^\circ$	4611.075	−1	f	21680.86	−12	$a^3D_2 - x^3P_2^\circ$
4863.782	−3	g	20554.41	00	$z^7P_2^\circ - e^5D_3$	4607.100	−1N	f	21699.57	−11	$a^1P_1 - v^3D_2^\circ$
4862.553	−2	g	20559.60	−08	$y^5D_3^\circ - f^3D_2$	4606.015	−2N	f	21704.68	−14	$b^3D_3 - t^5D_4^\circ$
4859.306	−2	f	20573.34	00	$a^3D_2 - x^3F_2^\circ$	4605.105	−2	f	21708.97	−03	$b^3P_0 - v^5D_1^\circ$
*4858.264	−2	f	20577.75	+01	$y^5F_2^\circ - f^3F_3$	4604.852	−1	g	21710.17	+02	$a^1I_6 - x^3H_6^\circ$
4849.662	−1	g	20614.26	+04	$a^1H_5 - y^3H_6^\circ$	4604.247	−1N	f	21713.02	−06	$b^3P_2 - v^5D_3^\circ$
4843.370	−2	f	20641.03	+09	$a^1H_5 - v^3G_5^\circ$	4603.352	0	f	21717.24	−03	$b^3P_1 - v^5D_2^\circ$
4842.734	−3	f	20643.74	−11	$y^5F_4^\circ - f^3F_4$	4598.745	0	f	21738.99	−02	$z^5P_2^\circ - e^7F_1$
4841.675	−2	gb	20648.26	−11	$a^3D_2 - w^3D_1^\circ$	*4598.374	−1	f	21740.74	−22	$a^5F_3 - z^5P_2^\circ$
4839.790	−3	g	20656.30	−08	$y^5P_2^\circ - i^5D_3$			f		−04	$z^3F_3^\circ - h^5D_3$
4838.094	−1	f	20663.54	−01	$a^3D_3 - u^5D_2^\circ$	4597.038	−2	f	21747.06	+11	$a^5F_2 - z^5P_1^\circ$
4837.668	−3	fb	20665.36	−06	$d^3F_3 - t^3F_3^\circ$	4595.213	−1	g	21755.70	−03	$a^1I_6 - x^3H_5^\circ$
4822.676	−2	g	20729.60	−05	$a^3D_1 - w^3D_2^\circ$	4587.726	−1	f	21791.21	−04	$z^3F_2^\circ - f^5P_1$
4816.684	−3	g	20755.39	−06	$b^3H_5 - z^3H_4^\circ$	4585.601	−2	f	21801.30	−04	$c^3P_2 - w^5P_3^\circ$
4815.231	−1	f	20761.65	−06	$a^1P_1 - x^3P_2^\circ$	4583.721	−1	f	21810.24	−03	$c^3P_0 - y^3P_1^\circ$
4813.727	−3N	f	20768.14	−02	$d^3F_2 - t^3F_2^\circ$	4579.692	−1	f	21829.43	−04	$b^3D_3 - v^3F_2^\circ$

TABLE C—(*Continued*)

Solar λ	Solar Int	Grade	Wave Number Solar	Wave Number o−c	Desig
*4579.061	−1N	f f	21832.44	+03 −05	$a^3D_1 - v^5F_2^\circ$ $z^3D_3^\circ - h^5D_4$
4571.448	0	g	21868.80	−03	$z^7F_2^\circ - e^5D_3$
4569.073	−2	g	21880.17	−07	$b^3H_5 - w^3G_5^\circ$
4568.607	−1	f	21882.40	+06	$z^3D_2^\circ - f^5P_2$
4555.740	−2	f	21944.20	+07	$a^3D_1 - v^5F_1^\circ$
4546.682	−1	f	21987.92	+01	$z^3D_1^\circ - f^5P_1$
4546.479	−1	f	21988.90	−06	$c^3F_2 - w^1D_2^\circ$
4545.547	−2	g	21993.41	−03	$b^3D_3 - v^3F_3^\circ$
4544.490	−1N	f	21998.52	+05	$z^3F_2^\circ - h^5D_1$
4543.237	−1	f	22004.59	−09	$b^3D_2 - t^5D_3^\circ$
4538.958	−2	f	22025.34	−05	$c^3F_2 - w^1F_3^\circ$
4538.604	−1N	f	22027.05	−12	$z^3F_2^\circ - f^5G_3$
4538.185	−2N	f	22029.08	+07	$y^5D_4^\circ - f^3F_3$
4528.825	0	f	22074.61	−03	$c^3P_2 - w^5P_1^\circ$?
4528.764	0	g	22074.91	−02	$b^3H_4 - y^1G_4^\circ$
4526.414	1	f	22086.37	−07	$z^3F_4^\circ - g^5F_4$
4521.670	−2N	f	22109.54	−12	$a^3D_1 - x^3P_1^\circ$
4518.589	0	fb	22124.62	−05	$a^5P_1 - y^7P_2^\circ$
4517.600	−2	g	22129.46	−01	$z^3D_1^\circ - f^3D_2$
4516.464	−2	f	22135.03	−08	$z^5P_2^\circ - f^5F_3$
4516.273	1	f	22135.96	−02	$z^5P_3^\circ - e^7F_4$
4516.091	−2N	f	22136.86	−03	$a^3D_3 - w^3G_4^\circ$
4511.069	−1	f	22161.50	−12	$z^3F_3^\circ - h^5D_2$
4510.836	0	f	22162.64	−08	$z^5P_3^\circ - e^5G_3$
4504.208	−2N	f	22195.26	+13	$z^3D_1^\circ - h^5D_1$
4498.562	−1N	f	22223.11	−13	$z^3D_3^\circ - h^5D_3$
4494.064	0	f	22245.36	−07	$z^3F_2^\circ - e^3G_3$
4492.970	−2	f	22250.77	+06	$a^3D_3 - w^3G_3^\circ$
4490.236	−1	g	22264.32	+01	$z^7F_1^\circ - c^5D_1$
4487.754	0	g	22276.63	−05	$b^3H_6 - z^1H_5^\circ$
4487.006	−1	f	22280.35	+01	$z^3D_2^\circ - h^5D_2$
4485.978	0	f	22285.45	−05	$z^5P_2^\circ - f^5F_1$
4483.780	0	f	22296.37	+01	$b^3D_3 - u^3G_4^\circ$
4481.033	−1	fb	22310.04	+02	$b^3D_1 - t^5D_2^\circ$
4480.278	−1	fb	22313.80	−04	$z^5P_3^\circ - e^5G_2$
4479.971	1	g	22315.33	+01	$z^3F_2^\circ - f^3D_1$
*4479.001	−1N	f f	22320.16	+04 −01	$b^3D_3 - u^5P_3^\circ$ $z^3D_1^\circ - g^5F_2$
*4472.544	−2	g	22352.39	−13	$a^3F_4 - y^5F_3^\circ$
*4463.139	−1	f	22399.49	+03	$c^3P_1 - u^5D_0^\circ$
4461.822	−2N	f	22406.10	−13	$b^3G_3 - u^5D_3^\circ$
4452.618	0	f	22452.42	00	$z^3F_3^\circ - g^5F_2$
4450.764	0	fb	22461.77	+02	$z^3F_4^\circ - f^5G_4$
4441.557	−1	f	22508.33	+01	$z^3D_3^\circ - g^5F_3$
4438.526	−1N	f	22523.70	+01	$z^3F_4^\circ - g^5F_3$
4433.394	−1	f	22549.77	−01	$b^3G_3 - u^5D_2^\circ$
4429.207	−1	fb	22571.09	−05	$z^3D_2^\circ - g^5F_2$
4428.713	−1	fb	22573.61	+16	$b^3D_3 - u^5P_2^\circ$
4428.551	1	f	22574.43	+09	$z^3F_3^\circ - e^3G_3$
*4425.771	−1N	f f	22588.61	−11 +08	$z^5D_2^\circ - e^3F_2$ $b^3D_2 - u^5P_2^\circ$
4419.785	0	fb	22619.20	−02	$a^3D_2 - w^3F_3^\circ$
4418.576	−1d?	f	22625.39	+13	$b^3D_1 - u^5P_2^\circ$
4414.234	−1	fb	22647.65	−01	$c^3P_1 - 1_2^\circ$
4414.050	−2	f	22648.59	−12	$z^5P_3^\circ - f^5F_2$
4412.424	−1N	fb	22656.94	+03	$a^5P_3 - y^7P_3^\circ$
4405.420	−1	f	22692.96	−10	$z^3D_2^\circ - e^3G_3$
4405.035	1	gb	22694.94	−08	$a^5D_3 - z^7F_3^\circ$
4404.105	−1	f	22699.73	−05	$z^3D_2^\circ - g^5F_1$
4394.306	−2	f	22750.35	00	$z^3F_3^\circ - e^3H_4$
4393.701	0	f	22753.48	0	$b^3D_2 - u^5P_1^\circ$
4393.039	0	f	22756.91	−06	$c^3P_2 - x^3F_3^\circ$
4392.313	−1N	fb	22760.68	−01	$a^1D_2 - v^3F_3^\circ$
4391.877	0	gb	22762.94	−01	$z^3D_2^\circ - f^3D_1$

Solar λ	Solar Int	Grade	Wave Number Solar	Wave Number o−c	Desig
4382.003	−1	f	22814.22	+10	$b^1G_4 - w^3H_5^\circ$
4375.487	−1	g	22848.20	−04	$a^1H_5 - u^3G_4^\circ$
4373.899	0	f	22856.50	−01	$b^3D_2 - t^3D_3^\circ$
4369.716	0	gb	22878.38	+09	$z^3F_3^\circ - f^3F_4$
4368.644	0	fb	22883.98	+07	$a^3D_3 - w^3F_4^\circ$
4367.065	−2	f	22892.26	00	$z^3G_5^\circ - g^5G_5$
4358.926	−1	f	22935.01	+13	$z^3D_3^\circ - g^5F_2$
*4357.519	0Nd?	fb	22942.41	+04	$z^3D_1^\circ - e^3P_2$
4354.267	0Nd?	fb	22959.55	+06	$z^3F_4^\circ - e^3H_5$
4351.392	−1	f	22974.72	−12	$z^5F_2^\circ - f^5D_3$
4343.214	2	fb	23017.98	+06	$a^3D_3 - w^3F_3^\circ$
4341.565	−1	f	23026.72	+03	$a^3D_1 - w^3F_2^\circ$
4341.252	−1	fb	23028.38	−10	$z^5F_3^\circ - f^5D_4$
4338.835	0d	fb	23041.21	+02	$a^3P_0 - x^5P_1^\circ$
4333.053	−1	f	23071.95	+01	$b^1D_2 - t^3F_2^\circ$
4330.824	−1	f	23083.83	−09	$c^3P_2 - 1_2^\circ$
4329.552	−1	f	23090.61	−04	$a^5P_1 - x^5F_2^\circ$
4325.960	1	f	23109.78	−05	$b^3H_5 - v^3G_5^\circ$
4323.372	0	f	23123.61	−03	$a^3H_4 - y^5G_5^\circ$
4322.703	−1N	f	23127.19	−03	$b^3F_2 - w^5F_3^\circ$
4319.456	0	f	23144.58	−04	$b^3F_2 - w^5D_2^\circ$
4318.801	−1	f	23148.09	+06	$b^3F_3 - w^5F_1^\circ$
4315.956	−1	f	23163.35	00	$a^3H_5 - y^5G_5^\circ$
4313.037	1N	gb	23179.02	+02	$a^3G_3 - y^3G_4^\circ$
4310.381	1	g	23193.30	−04	$z^3D_2^\circ - e^3P_2$
4309.463	1	f	23198.24	−03	$c^3P_0 - v^5P_1^\circ$
4307.058	−2d?	f	23211.20	+12	$z^5F_4^\circ - f^7D_5$
4300.220	−1	f	23248.11	−07	$z^3F_4^\circ - e^3H_4$
4299.486	0	f	23252.08	+01	$a^3D_3 - z^1D_2^\circ$
4298.199	1	f	23259.04	+04	$c^3P_0 - v^5F_1^\circ$
4292.136	2	f	23291.89	−03	$a^5P_3 - x^5F_3^\circ$
4283.414	−1	f	23339.32	−08	$b^3F_2 - w^5F_1^\circ$
4281.601	−1	f	23349.20	−03	$a^3H_4 - y^5G_3^\circ$
4280.638	0	f	23354.45	−03	$b^3G_3 - w^5G_3^\circ$
4278.002	−3	f	23368.85	+04	$y^5F_3^\circ - i^5D_4$
4277.392	0	f	23372.18	+02	$b^3F_2 - w^5D_1^\circ$
4271.961	1N	fb	23401.89	−07	$a^3H_5 - y^5G_4^\circ$
4271.637	0Nd?	f	23403.66	+06	$a^5P_2 - x^5F_1^\circ$
4270.331	−1	f	23410.82	−14	$b^3F_3 - w^5F_2^\circ$
4269.862	2N	fb	23413.39	+04	$z^5F_3^\circ - f^7D_4$
4260.735	1	fb	23463.55	−03	$b^3P_1 - w^5P_1^\circ$
4259.309	1Nd?	fb	23471.40	+18	$b^3G_4 - w^5G_4^\circ$
4256.317	−1	f	23487.90	−01	$a^3H_5 - z^5H_4^\circ$
4253.914	1d	fb	23501.17	+11	$b^3D_2 - 8_1^\circ$
*4253.542	−1	f	23503.23	−13	$z^5F_5^\circ - f^7D_5$
4250.916	1	f	23517.74	−09	$c^3P_1 - v^5P_1^\circ$
4249.352	−1	f	23526.40	−16	$a^3P_1 - x^5P_1^\circ$
4248.416	1	f	23531.58	−07	$a^5F_1 - z^3D_2^\circ$
*4247.317	2	f f	23537.67	−03 −12	$a^3H_4 - z^5H_3^\circ$ $b^3D_1 - 8_1^\circ$
4246.023	0	g	23544.84	−03	$a^3D_1 - w^3P_0^\circ$
4239.958	1	fb	23578.52	−04	$c^3P_1 - v^5F_1^\circ$
4239.368	2	f	23581.81	−01	$b^3D_3 - s^3D_3^\circ$
4238.622	−1N	gb	23585.95	−07	$a^1I_6 - y^3I_5^\circ$
4237.680	1	f	23591.20	−04	$b^3G_3 - v^5F_4^\circ$
4236.645	−1	f	23596.97	+07	$b^3D_2 - s^3D_3^\circ$
4235.838	0	f	23601.45	−01	$a^3H_6 - z^5H_5^\circ$
4235.640	−1	f	23602.56	+03	$b^3F_4 - w^5F_5^\circ$
4225.717	1	fb	23657.98	−07	$y^5F_4^\circ - i^5D_4$
4224.634	−1	f	23664.05	−02	$a^3G_3 - x^5G_3^\circ$
4223.733	0	g	23669.10	−02	$b^3G_5 - z^1G_4^\circ$
4220.053	1	gb	23689.73	−01	$z^3D_2^\circ - e^3P_1$
4219.736	−1	f	23691.52	+02	$z^5P_2^\circ - f^5P_3$
4219.421	3	g	23693.28	−05	$b^3G_4 - x^3G_4^\circ$
4218.226	1	f	23700.00	−08	$a^3H_5 - z^5H_6^\circ$

TABLE C—(*Continued*)

Solar λ	Solar Int	Grade	Wave Number Solar	Wave Number o−c	Desig
4213.422	−1	f	23727.02	+01	$a^3G_4 - x^5G_5^\circ$
4212.043	−1N	f	23734.79	+13	$z^5F_2^\circ - c^3D_3$
4210.404	3	g	23744.02	−07	$c^3P_1 - x^3P_1^\circ$
4200.789	1	gb	23798.37	−05	$a^3F_2 - y^5P_3^\circ$
4200.104	−1	g	23802.25	−09	$z^3D_3^\circ - f^3F_3$
4199.379	−1	f	23806.36	−06	$b^3G_5 - w^5G_4^\circ$
4197.362	−1	g	23817.80	+10	$z^3F_4^\circ - f^3F_3$
4197.102	2	gb	23819.28	−02	$a^5F_2 - z^3F_3^\circ$
4194.491	0	f	23834.10	+03	$a^3G_4 - x^5G_4^\circ$
4181.549	1	f	23907.87	+02	$a^1D_2 - u^3D_1^\circ$
4181.194	0	f	23909.90	+01	$b^3D_1 - z^1P_1^\circ$
4180.404	1	f	23914.42	+06	$a^3G_4 - x^5G_3^\circ$
4175.914	1	g	23940.13	−11	$z^5F_3^\circ - e^7G_4$
4173.151	−1d?	g	23955.98	+18	$z^5F_3^\circ - g^5D_4$
*4172.978	1	{f	23956.97	{−04	$b^3D_3 - 9_4^\circ$
		f}		−06	$y^5D_1^\circ - i^5D_2$
4169.096	−1	g	23979.28	−01	$a^5F_1 - z^3F_2^\circ$
4164.265	0	fb	24007.10	−16	$z^5F_2^\circ - e^7G_1$
4163.358	−2	f	24012.32	−05	$y^5D_2^\circ - i^5D_3$
4162.910	−1	f	24014.91	+09	$c^3P_2 - v^5F_1^\circ$
4152.085	1	f	24077.52	−07	$c^3F_4 - v^1G_4^\circ$
4149.501	0	f	24092.51	−07	$b^1G_4 - 10_3^\circ$
4148.260	−1	f	24099.72	+03	$z^5P_2^\circ - f^5P_2$
4147.490	1	f	24104.20	+01	$z^5P_3^\circ - f^5P_3$
4147.347	2	g	24105.03	−03	$z^5F_5^\circ - e^5G_4$
4143.510	2	g	24127.35	−07	$z^5F_4^\circ - e^3D_3$
4137.984	0	f	24159.57	−06	$z^7F_5^\circ - e^5F_5$
4137.417	2	g	24162.88	+03	$y^5F_2^\circ - g^5G_3$
4134.196	0	g	24181.70	−02	$b^3F_2 - x^3D_3^\circ$
4132.540	3	g	24191.40	00	$y^5F_3^\circ - g^5G_4$
*4131.959	2	{g}b	24194.79	{+06	$z^5D_2^\circ - f^5D_3$
		g}		−15	$z^5F_2^\circ - f^5F_1$
4131.758	1	f	24195.97	−07	$y^5D_1^\circ - 4_2$
4129.466	2	g	24209.40	−04	$z^5F_3^\circ - f^5F_3$
4125.234	−2	f	24234.24	−03	$a^3H_4 - w^5F_4^\circ$
4121.991	−1	f	24253.30	+07	$a^1D_2 - 8_1^\circ$
4119.672	0	f	24266.95	−07	$z^7F_3^\circ - e^5F_4$
4115.896	−1	f	24289.22	−05	$b^3D_3 - y^1F_3^\circ$
4112.176	−2	f	24311.19	−02	$a^3G_5 - z^3I_6^\circ$
4112.083	−1	f	24311.74	+02	$a^1D_2 - v^3P_2^\circ$
4108.303	−2	f	24334.11	+04	$z^5P_1^\circ - f^3D_2$
4108.138	1	g	24335.09	−04	$z^5D_3^\circ - e^7P_4$
4105.065	−1	f	24353.30	−04	$z^5F_1^\circ - e^5P_1$
4104.472	0	g	24356.82	−07	$b^3G_4 - w^3G_3^\circ$
*4103.623	0N	g	24361.86	00	$a^3D_3 - z^1F_3^\circ$
4100.916	0	fb	24377.94	−02	$a^3H_4 - w^5F_3^\circ$
*4100.350	0	{f}b	24381.31	{+03	$z^7F_4^\circ - e^5F_4$
		g}		00	$y^5F_3^\circ - g^5G_3$
4099.996	0	g	24383.41	−05	$z^5F_1^\circ - g^5D_2$
4097.018	−1	f	24401.13	+04	$z^5F_1^\circ - e^5P_2$
4096.951	−1	f	24401.53	+02	$a^3H_5 - w^5F_4^\circ$
4096.217	1	g	24405.91	−03	$a^5F_3 - z^3F_2^\circ$
4095.646	−2	g	24409.30	−10	$a^1I_6 - y^1H_5^\circ$
4095.274	0	f	24411.53	−04	$y^5D_2^\circ - 4_2$
4090.773	−1	f	24438.38	−12	$b^1G_4 - t^3G_4^\circ$
4090.326	0N	gb	24441.05	+05	$a^3F_2 - y^5P_1^\circ$
4087.801	−1	f	24456.15	−08	$z^5P_2^\circ - f^5P_1$
4084.152	−2	gb	24478.00	+08	$z^5D_4^\circ - f^7D_5$
4079.186	2	f	24507.80	−05	$z^5F_2^\circ - e^5P_2$
4074.686	2	f	24534.86	+10	$b^3D_2 - 10_3^\circ$
4067.856	−1	g	24576.06	−02	$y^5F_5^\circ - g^5G_5$
4067.604	0	g	24577.59	−01	$a^3D_2 - v^3F_2^\circ$
4067.493	0	g	24578.25	−02	$b^3G_5 - w^3G_4^\circ$
4066.006	−1	g	24587.24	+10	$z^5F_5^\circ - f^5F_4$
4064.054	1	gb	24599.05	+07	$b^3G_5 - z^1H_5^\circ$
4058.469	−1	f	24632.90	−06	$b^3D_3 - 11_3^\circ$
4057.671	0	f	24637.75	−07	$a^1P_1 - t^3D_1^\circ$?
4046.465	1	f	24705.97	−06	$y^5D_3^\circ - 4_2$
4046.083	2	gb	24708.31	−08	$z^5D_2^\circ - f^7D_1$
4045.601	2	gb	24711.25	−05	$z^5D_4^\circ - e^7P_3$
4044.497	1	f	24717.99	−02	$y^5D_4^\circ - i^5D_3$
4043.993	2	gb	24721.08	−05	$z^5D_3^\circ - e^7P_2$
4043.692	0	fb	24722.92	−01	$a^3P_0 - v^5D_1^\circ$
4042.763	−1	f	24728.60	−05	$z^5D_1^\circ - e^7F_1$
4036.377	1	f	24767.72	−06	$a^3G_3 - w^3D_3^\circ$
4035.986	−1	fb	24770.12	−02	$b^3G_3 - w^3F_4^\circ$
4031.718	2	fb	24796.34	+06	$b^3G_3 - v^3D_3^\circ$
4030.901	2	fb	24801.37	−02	$b^1G_4 - t^3G_3^\circ$
4020.024	0	f	24868.47	+14	$z^5D_2^\circ - e^7F_1$
4011.901	−1	g	24918.82	−10	$b^3G_5 - y^1G_4^\circ$
4009.549	1	f	24933.44	−04	$z^5D_4^\circ - e^7F_5$
4006.159	1	g	24954.53	−02	$z^5D_2^\circ - e^3D_3$
4005.484	1	f	24958.74	+02	$b^3F_3 - x^5G_4^\circ$
4005.390	1	f	24959.33	−04	$a^3P_1 - y^5S_2^\circ$
3998.475	0	f	25002.49	−10	$b^3H_4 - 6_5^\circ$
*3997.493	2	{f	25008.63	{−05	$z^5D_3^\circ - e^7F_3$
		f}		−07	$z^5D_3^\circ - e^5S_2$
3996.790	1	f	25013.03	−01	$y^5D_3^\circ - g^5G_4$
*3996.265	1	{f	25016.32	{+09	$b^3G_4 - v^3D_3^\circ$
		g}		−04	$z^5D_0^\circ - e^7G_1$
3994.272	−2	f	25028.80	−02	$z^7F_2^\circ - e^5F_1$
3992.646	0	f	25038.99	−02	$b^3F_3 - x^5G_3^\circ$
3990.569	0d?	fb	25052.02	−11	$z^5D_3^\circ - e^7F_4$
3989.611	−1N	fb	25058.04	−05	$b^3H_5 - 4_4^\circ$
3989.262	−1N	gb	25060.23	−15	$z^5D_1^\circ - e^7G_2$
3986.298	0	g	25078.86	−01	$z^5D_3^\circ - e^5G_3$
3985.322	1	f	25085.01	−03	$b^3F_4 - x^5G_5^\circ$
3984.943	1	f	25087.39	−08	$z^5D_1^\circ - e^7G_1$
3984.451	−1	f	25090.49	+03	$b^3F_3 - x^5G_2^\circ$
3983.818	0	g	25094.48	+10	$b^3G_3 - w^3F_2^\circ$
3979.117	−1	g	25124.12	+02	$b^3G_4 - w^3F_3^\circ$
3974.636	2	fb	25152.45	+08	$a^1G_4 - y^3H_4^\circ$
3971.010	1N	f	25175.42	−12	$y^5D_4^\circ - g^5G_5$
3965.844	−1	f	25208.20	−10	$a^3P_1 - v^5D_1^\circ$
3962.651	−1	f	25228.52	+03	$b^3D_3 - t^3G_3^\circ$
3962.400	0	g	25230.12	+13	$z^5D_3^\circ - e^5G_2$
3959.454	−1	f	25248.89	+02	$z^5D_4^\circ - e^7F_3$
3955.768	0	f	25272.42	+03	$b^3F_4 - x^5G_3^\circ$
3955.221	1	g	25275.91	−01	$a^1G_4 - v^3G_3^\circ$
3950.801	−2	g	25304.19	−10	$z^7D_3^\circ - e^5D_3$
3949.235	1	g	25314.22	−01	$z^7D_1^\circ - e^5D_2$
3948.476	−1N	g	25319.09	+03	$z^5D_4^\circ - e^5G_3$
3948.284	1	g	25320.32	−02	$z^5D_3^\circ - e^7G_3$
3947.980	−2	f	25322.27	+14	$a^3D_2 - u^5F_1^\circ$
3936.772	0d?	gb	25394.36	+10	$z^5D_4^\circ - e^3D_3$
3927.610	0	f	25453.60	00	$a^3G_4 - z^1G_4^\circ$
3925.540	1	f	25467.02	+07	$a^3D_3 - u^5P_3^\circ$
3923.043	1	g	25483.23	−07	$a^3D_3 - y^1D_2^\circ$
*3922.086	1	{g	25489.44	{ 00	$z^5D_0^\circ - e^3D_1$
		g}		−07	$z^7D_1^\circ - e^5D_1$
3918.575	0	fb	25512.28	+02	$b^3P_2 - v^5F_1^\circ$
3914.428	1	f	25539.31	−06	$a^3D_1 - u^5F_1^\circ$
3910.536	2	g	25564.72	−14	$z^5D_3^\circ - f^5F_2$
3908.687	−1	g	25576.82	−02	$z^7D_3^\circ - e^5D_2$
3906.965	1	gb	25588.09	+03	$z^5D_1^\circ - e^5P_2$
3905.681	2N	gb	25596.50	−14	$z^7D_2^\circ - e^5D_1$
3905.191	1	g	25599.72	−05	$z^5D_3^\circ - e^3D_2$
3905.011	1	f	25600.90	−03	$z^5F_2^\circ - f^5P_3$
3889.360	1d?	fb	25703.91	+ 1	$a^3D_1 - u^5P_1^\circ$
3885.935	0	gb	25726.57	−05	$b^1G_4 - x^1H_5^\circ$

TABLE C—(*Continued*)

Solar λ	Solar Int	Grade	Wave Number Solar	Wave Number o−c	Desig
3885.758	1	g	25727.74	00	$z^5D_2^\circ - e^5P_2$
3878.196	1	g	25777.91	−03	$z^5D_1^\circ - g^5D_0$
3867.442	0	g	25849.58	+02	$b^3F_2 - u^5D_2^\circ$
*3864.307	3	{g g} b	25870.55	{+04 −03	$b^3F_3 - u^5D_3^\circ$ $z^5D_4^\circ - g^5D_3$
3863.703	1	g	25874.60	−02	$z^5D_2^\circ - g^5D_1$
3860.730	−2	f	25894.52	+08	$z^5F_4^\circ - g^5F_5$
3858.474	0	gb	25909.69	+06	$z^5D_3^\circ - g^5D_2$
3848.299	2	g	25978.17	−05	$b^3F_2 - w^3D_3^\circ$
3846.290	1	fb	25991.74	−02	$b^1G_4 - w^1F_3^\circ$
3845.224	1	fb	25998.94	−07	$z^5F_3^\circ - g^5F_4$
3843.717	2N	fb	26009.13	+01	$z^5F_2^\circ - f^5P_2$
3842.905	1	g	26014.63	−06	$b^3F_3 - x^3F_4^\circ$
3840.203	0	g	26032.93	−03	$a^3P_2 - w^5D_1^\circ$
3834.476	0	g	26071.81	−12	$a^3D_3 - u^3D_2^\circ$
3826.627	1N	fb	26125.29	+04	$a^3H_4 - x^5G_3^\circ$
3824.752	−1	f	26138.10	−13	$b^3F_2 - u^5D_1^\circ$
3819.496	2	f	26174.06	+04	$z^5F_3^\circ - f^5P_2$
3816.924	1	g	26191.73	−01	$z^7P_2^\circ - f^5D_2$
3813.930	1	f	26212.26	+06	$a^3H_5 - x^5G_4^\circ$
3811.809	2	fb	26226.85	−03	$z^5F_4^\circ - g^5F_4$
3810.902	0	gb	26233.09	+02	$b^3F_2 - w^3D_1^\circ$
3803.260	1	f	26285.80	−11	$a^3P_2 - v^5D_2^\circ$
3789.824	1	g	26378.98	−06	$z^5F_4^\circ - h^5D_3$
3785.792	1	f	26407.08	−09	$z^5F_5^\circ - f^5G_5$
3784.254	0	fb	26417.81	+12	$b^3H_5 - w^3H_4^\circ$
3771.499	1	g	26507.15	+02	$b^3H_6 - w^3H_5^\circ$
3764.223	1	g	26558.39	−07	$a^5P_2 - w^5F_2^\circ$
3763.573	1	g	26562.97	−05	$a^3P_0 - z^3S_1^\circ$
3761.069	1	f	26580.66	−05	$z^5F_4^\circ - f^3D_3$
3758.131	−1	f	26601.44	−14	$z^5F_2^\circ - f^5G_3$
3754.876	−1N	f	26624.50	+13	$b^1G_4 - u^3H_4^\circ$
3751.092	1	gb	26651.36	−04	$a^5P_2 - w^5F_1^\circ$
3748.907	0	fb	26666.89	+03	$a^3G_5 - z^1H_5^\circ$
3747.006	2	gb	26680.42	−02	$z^7P_3^\circ - e^7P_2$
3742.569	2	gb	26712.05	−08	$z^7P_2^\circ - e^5G_3$
3742.148	1	f	26715.06	−02	$z^3F_3^\circ - g^5G_4$
3735.702	0Nd?	fb	26761.15	+03	$a^3P_1 - w^5P_2^\circ$
3733.197	1	g	26779.10	+01	$b^3F_4 - w^5G_5^\circ$
3731.161	−1	f	26793.72	−08	$b^1G_4 - u^3F_4^\circ$
3729.341	0	f	26806.79	00	$a^1G_4 - u^5F_4^\circ$
3727.687	1	gb	26818.69	−08	$b^3F_3 - w^5G_3^\circ$
3727.533	1	f	26819.80	−04	$z^5F_2^\circ - e^3G_3$
3727.028	0	fb	26823.43	−02	$a^3D_1 - z^1P_1^\circ$
3726.067	−1	f	26830.35	−02	$b^3G_5 - x^1G_4^\circ$
3722.760	0	fb	26854.18	+04	$z^5F_2^\circ - g^7D_3$
*3722.238	1	{f f}	26857.95	{00 −06	$a^3P_1 - w^5P_1^\circ$ $c^3P_1 - t^5D_2^\circ$
3717.837	0	f	26889.74	+01	$z^5F_2^\circ - f^3D_1$
3717.188	−1	f	26894.43	−03	$z^5F_5^\circ - f^5G_4$
3709.032	1	g	26953.57	−03	$z^7P_2^\circ - e^7G_3$
3708.189	−2	g	26959.70	−04	$b^3F_3 - x^3G_3^\circ$
*3705.710	2	fb	26977.73	+03	$a^3G_3 - 3_3^\circ$
3705.264	−1N	f	26980.98	−04	$z^5F_3^\circ - f^5G_2$
3704.798	−2	f	26984.38	+01	$b^1G_4 - u^3F_3^\circ$
3703.449	−2	f	26994.20	−14	$z^5F_4^\circ - f^5G_3$
3700.601	1	g	27014.98	+04	$z^5D_3^\circ - h^5D_4$
3699.397	−2	f	27023.77	+06	$z^3D_2^\circ - g^5G_3$
3698.018	1N	gb	27033.85	+08	$a^5P_2 - v^5D_1^\circ$
*3693.784	0	{g f}	27064.83	{−05 +02	$a^3F_3 - x^5D_4^\circ$ $c^3P_1 - t^5D_1^\circ$
3691.536	−3	f	27081.32	−06	$z^5F_1^\circ - g^7D_1$
3691.177	−2N	g	27083.95	00	$b^3F_2 - v^5F_3^\circ$
3689.375	3	fb	27097.18	−04	$z^7P_2^\circ - f^5F_3$
3685.663	−3	f	27124.47	00	$b^3F_2 - v^5P_2^\circ$
3683.757	−3	f	27138.50	+07	$z^3D_2^\circ - g^5G_2$
*3682.174	2	{g g}	27150.17	{−04 −21	$z^7P_3^\circ - e^7F_2$ $z^7P_4^\circ - f^7D_3$
3675.767	0	f	27197.49	−05	$z^3D_3^\circ - g^5G_4$
3675.450	−1	g	27199.84	−04	$b^3F_2 - v^5F_2^\circ$
3673.684	−1	f	27212.91	00	$z^3F_4^\circ - g^5G_4$
3670.221	−3	f	27238.59	+04	$a^3F_2 - y^7P_3^\circ$
3666.850	−2	f	27263.63	−02	$z^7P_2^\circ - g^5D_3$
3666.285	2	f	27267.83	+02	$a^3D_2 - 10_3^\circ$
3662.738	−2	f	27294.24	−03	$c^3P_2 - t^5D_2^\circ$
3660.412	1	fb	27311.58	−02	$b^3F_2 - v^5F_1^\circ$
3658.025	1	g	27329.40	−02	$b^3G_3 - u^3G_4^\circ$
3656.358	1	g	27341.86	−07	$z^7F_2^\circ - f^5D_3$
3655.356	1	g	27349.32	−08	$a^3P_1 - y^3P_1^\circ$
*3653.353	1	{g f}	27364.35	{−03 −02	$b^3F_3 - v^5F_2^\circ$ $z^7F_3^\circ - e^7P_4$
3652.261	−1	g	27372.53	00	$c^3P_2 - y^1D_2^\circ$
3651.040	−1	f	27381.68	−05	$z^5D_1^\circ - f^5G_5$
3649.699	−1	f	27391.74	−03	$z^7P_4^\circ - f^5F_5$
3647.563	0	f	27407.78	−06	$z^5D_2^\circ - f^3D_3$
3646.098	−1	f	27418.80	00	$z^7F_2^\circ - e^7P_3$
3641.460	1	fb	27453.72	−10	$z^7F_1^\circ - f^5D_2$
3638.169	1	f	27478.55	−08	$z^7F_4^\circ - e^7P_4$
3637.059	1	g	27486.93	−07	$b^3G_3 - u^3G_3^\circ$
*3636.486	1	{f g}	27491.27	{+14 +02	$a^3F_3 - y^7P_2^\circ$ $z^5D_2^\circ - g^5F_3$
3635.829	−1	g	27496.23	−07	$z^7F_5^\circ - e^7F_6$
3634.536	−2	g	27506.02	−14	$z^7F_2^\circ - f^5D_2$
3633.653	−2	f	27512.70	−08	$z^7P_2^\circ - e^5P_1$
3628.829	2	gb	27549.27	−10	$b^3G_4 - u^3G_4^\circ$
3627.360	−2N	f	27560.43	−09	$z^7P_2^\circ - e^5P_2$
3624.065	2	g	27585.49	−06	$z^5D_2^\circ - f^5P_1$
3623.511	−1	f	27589.70	00	$z^7P_3^\circ - g^5D_3$
3621.202	2	fb	27607.30	−05	$z^5D_3^\circ - f^3D_3$
*3620.881	1	{f g}	27609.74	{−02 −08	$z^7F_0^\circ - f^5D_1$ $b^3H_1 - t^3G_4^\circ$
3619.671	−1	g	27618.97	−06	$a^3P_1 - u^5D_0^\circ$
3618.924	0	g	27624.68	−08	$a^3P_1 - u^5D_1^\circ$
3618.616	1N	g	27627.03	+02	$z^5D_2^\circ - h^5D_2$
3618.305	3	fb	27629.40	−04	$z^7F_5^\circ - e^7P_4$
3617.961	2N	fb	27632.03	+08	$a^3H_4 - w^5G_5^\circ$
3615.005	−3	g	27654.62	00	$z^7D_5^\circ - e^5F_5$
3613.953	−3	fb	27662.68	−01	$b^3H_5 - 12_5^\circ$
3613.450	2	f	27666.52	+01	$a^3D_3 - 10_3^\circ$
3613.110	2	fb	27669.13	−21	$z^7F_2^\circ - f^7D_3$
3612.520	−2	f	27673.64	−08	$b^3H_4 - 13_4^\circ$
3609.473	2	fb	27697.00	−10	$z^7F_3^\circ - f^7D_4$
3606.539	2	g	27719.54	−06	$a^3P_1 - w^3D_1^\circ$
3606.379	0	g	27720.76	−02	$b^3F_4 - w^3G_4^\circ$
3602.709	−2	f	27749.00	−06	$z^7P_4^\circ - e^7G_3$
3601.429	−2	f	27758.87	−05	$a^3P_2 - w^5P_3^\circ$
3597.253	−3d?	f	27791.09	−13	$a^1I_6 - x^3I_5^\circ$
3594.105	−2	g	27815.43	−08	$z^7D_4^\circ - e^5F_4$
3593.795	−2	f	27817.83	+01	$a^3H_4 - v^5F_5^\circ$
3588.535	3	g	27858.61	−08	$z^7P_4^\circ - e^7S_3$
3588.247	0	f	27860.84	−10	$a^3F_3 - y^7P_4^\circ$
3582.332	2	f	27906.84	+07	$z^5D_1^\circ - g^5F_1$
3579.836	2	f	27926.30	−08	$z^5D_3^\circ - e^3G_4$
3572.322	−3	f	27985.04	−02	$a^3H_5 - v^5F_5^\circ$
3570.598	−3	g	27998.55	+05	$z^7D_3^\circ - e^5F_3$
3566.316	0N	f	28032.16	−06	$a^3P_2 - w^5P_1^\circ$
3565.839	0	f	28035.91	−10	$z^5D_2^\circ - f^5G_2$
3564.566	2	g	28045.93	−05	$a^3H_4 - x^3G_3^\circ$
3564.525	3	g	28046.25	−09	$a^3H_4 - x^3G_5^\circ$
3563.612	−1	g	28053.43	−05	$z^7F_6^\circ - e^5G_6$

TABLE C—(*Continued*)

Solar λ	Solar Int	Grade	Wave Number Solar	Wave Number o−c	Desig
3562.608	−3	f	28061.34	−09	$b^3F_4 - y^1G_4^\circ$
3560.077	−1	gb	28081.29	−08	$z^7F_3^\circ - e^7F_4$
3559.465	1	g	28086.12	−12	$z^7F_1^\circ - e^7F_2$
3558.211	−2	g	28096.01	−05	$b^3F_2 - v^3D_3^\circ$
3554.649	1	g	28124.17	+01	$z^7D_2^\circ - e^5F_2$
3554.453	2	fb	28125.72	−09	$z^7P_4^\circ - e^5P_3$
3551.113	1	g	28152.17	−01	$z^7F_4^\circ - e^7F_3$
3544.860	−3	g	28201.83	+15	$z^7D_1^\circ - e^5F_1$
3543.102	−3	fb	28215.82	−08	$a^3H_6 - v^5F_5^\circ$
3542.572	−1	g	28220.04	−10	$z^7F_3^\circ - e^7F_2$
3541.243	−1N	fb	28230.63	−23	$a^3F_4 - y^7P_3^\circ$
3528.242	0	f	28334.66	−03	$a^3H_4 - v^5F_3^\circ$
3527.901	0	f	28337.40	−03	$a^3G_3 - z^1F_3^\circ$
3526.975	0	f	28344.83	−11	$z^5P_2^\circ - i^5D_3$
3523.185	0	f	28375.33	+01	$a^3D_3 - t^3G_3^\circ$
3522.738	−2	f	28378.92	−07	$a^1G_4 - s^3D_3^\circ$
3515.410	0	f	28438.08	00	$b^3F_2 - z^1D_2^\circ$
3514.469	0	f	28445.69	+06	$a^3F_4 - y^7P_4^\circ$
3513.605	−2	gb	28452.69	−12	$z^7F_6^\circ - f^5F_5$
3512.812	−1	f	28459.11	−10	$z^7F_3^\circ - e^7S_3$
3512.731	−1	f	28459.77	+04	$b^3H_5 - w^1G_4^\circ$
3510.193	1	fb	28480.35	−07	$z^5P_1^\circ - 4_2$
3509.732	1	g	28484.09	−05	$z^7F_0^\circ - f^5F_1$
3507.146	2	f	28505.09	−04	$z^5P_2^\circ - i^5D_2$
3506.595	1	f	28509.57	−12	$z^7F_1^\circ - f^5F_1$
3502.470	−1	f	28543.14	−06	$z^5D_3^\circ - f^3F_4$
3498.184	1N	gb	28578.11	−06	$z^7F_6^\circ - e^7G_6$
3494.263	−3	f	28610.19	−10	$a^3H_5 - w^3G_5^\circ$
3493.583	−1	g	28615.75	−08	$z^7F_5^\circ - f^5F_4$
3490.490	0	f	28641.10	−15	$z^5P_3^\circ - i^5D_4$
3477.986	−2	f	28744.07	−07	$z^5P_2^\circ - 4_2$
3473.228	−3	fb	28783.45	+06	$z^5D_4^\circ - f^3F_4$
3473.011	−2	f	28785.25	−02	$z^5D_2^\circ - f^3F_3$
3459.281	−2	fb	28899.49	+04	$z^5D_1^\circ - f^3F_2$
3450.141	−1	g	28976.05	−04	$b^3F_3 - v^3G_3^\circ$
3449.052	−3	f	28985.20	+07	$b^3G_5 - w^3H_4^\circ$
3448.206	0Nd?	fb	28992.31	−12	$a^3H_6 - z^1H_5^\circ$
3440.740	−1N	f	29055.22	+03	$a^3G_3 - v^3F_4^\circ$
3438.101	−2	f	29077.52	+01	$a^3G_4 - t^5D_4^\circ$
3434.967	−1	f	29104.04	−11	$a^1D_2 - t^3F_2^\circ$
3430.886	−2	gb	29138.67	−05	$b^3H_5 - v^3H_6^\circ$
*3429.818	−2	f	29147.74	−11	$b^3F_2 - w^3P_2^\circ$
		f		−01	$a^1G_4 - y^1H_5^\circ$
3428.021	0	f	29163.02	−06	$b^3H_4 - w^1F_3^\circ$
3426.674	2	f	29174.48	−01	$b^3H_5 - x^1H_5^\circ$
3426.093	−2	f	29179.43	−06	$c^3P_0 - t^5P_1^\circ$
3411.878	−2	f	29301.00	+05	$a^3G_5 - u^5F_5^\circ$
3410.565	−1	f	29312.28	−07	$b^3F_3 - w^3P_2^\circ$
3409.399	−2N	fb	29322.30	−03	$b^3G_3 - y^1F_3^\circ$
3407.562	3	gb	29338.11	−26	$a^5P_3 - u^5D_2^\circ$
3406.172	−1	g	29350.08	−02	$b^3P_1 - u^3D_2^\circ$
3398.111	−3N	f	29419.70	+07	$b^3H_6 - x^1H_5^\circ$
3391.842	0	f	29474.08	−04	$a^3D_2 - s^3G_3^\circ$
3369.152	0	gb	29672.56	−08	$a^3H_4 - v^3G_5^\circ$
3368.248	−2	f	29680.53	00	$a^3D_3 - s^3G_4^\circ$
3344.084	−2N	f	29894.99	+06	$b^3G_4 - 12_6^\circ$
3342.761	−1	g	29906.82	+01	$z^7P_2^\circ - g^7D_3$
3336.549	−1	f	29962.50	−06	$b^3G_3 - 13_4^\circ$
3315.176	0	f	30155.66	−03	$b^3H_4 - u^3F_3^\circ$
3308.761	2	gb	30214.13	−12	$a^3H_6 - y^3H_5^\circ$
3305.751	0	fb	30241.63	−04	$b^3H_5 - u^3F_4^\circ$
3304.366	0	g	30254.32	−05	$z^5F_2^\circ - i^5D_3$
3291.431	−1	f	30373.20	+10	$b^1G_4 - r^3G_4^\circ$
3274.227	1	g	30532.79	+08	$a^5P_1 - x^3P_0^\circ$
3272.607	1	g	30547.91	−08	$a^3F_3 - z^5H_4^\circ$
3270.672	0	fb	30565.98	+18	$b^1G_4 - r^3G_3^\circ$
3269.433	0	g	30577.56	−13	$a^5P_2 - x^3P_2^\circ$
3265.557	2	gb	30613.85	−06	$a^3G_4 - w^3H_5^\circ$
3263.467	−3	f	30633.46	−16	$a^3D_3 - u^3F_2^\circ$
3258.632	1N	g	30678.91	−11	$z^7D_1^\circ - f^5D_2$
*3256.497	1N	f	30699.02	+23	$z^7D_2^\circ - e^7P_3$
3254.471	2	g	30718.13	−15	$z^7D_3^\circ - e^7P_4$
3240.122	−1	g	30854.16	−10	$z^7D_3^\circ - e^7P_3$
3238.319	−1	g	30871.34	−03	$a^1G_4 - v^3H_4^\circ$
3235.328	0	f	30899.88	−01	$a^3G_4 - y^3I_5^\circ$
3230.098	0	g	30949.91	−09	$a^5F_3 - y^3D_2^\circ$
3227.177	−3	f	30977.92	−08	$b^3F_4 - u^5F_3^\circ$
3223.100	−2	f	31017.10	−23	$a^3D_2 - t^3F_3^\circ$
3219.370	−2N	g	31053.04	+03	$a^3G_5 - w^3H_4^\circ$
3216.050	−2	f	31085.10	+07	$a^3D_2 - t^3F_2^\circ$
3202.668	0	f	31214.98	−09	$a^3F_2 - w^5D_2^\circ$
3193.735	−1N	fb	31302.29	+02	$a^3D_1 - t^3F_2^\circ$
3191.415	0	fb	31325.03	−03	$a^3D_3 - t^3F_4^\circ$
3187.169	0	f	31366.76	−11	$z^7F_1^\circ - g^7D_2$
3183.582	0	gb	31402.11	−04	$a^3H_5 - x^3H_6^\circ$
3174.222	0	g	31494.70	−07	$z^5D_1^\circ - i^5D_2$
3172.298	0	f	31513.80	−02	$a^3G_3 - x^1F_3^\circ$
3169.076	−2	f	31545.84	+12	$a^1H_5 - t^3H_6^\circ$
3166.596	0	f	31570.54	−10	$a^5P_2 - v^3D_1^\circ$
3166.256	1	gb	31573.94	−11	$z^7D_3^\circ - e^7F_2$
3165.158	1	fb	31584.89	+02	$a^5P_3 - v^3D_3^\circ$
3165.085	−2	g	31585.62	−06	$a^3H_4 - u^3G_5^\circ$
3161.554	−1	f	31620.89	−01	$a^3H_4 - 4_4^\circ$
3160.924	0	g	31627.19	−07	$z^7D_2^\circ - e^7G_2$
3159.262	0	gb	31643.83	−08	$b^3F_2 - t^3D_2^\circ$
3157.144	1	gb	31665.06	+03	$a^3P_2 - w^3P_1^\circ$
3155.796	0	gb	31678.58	+01	$a^3H_6 - x^3H_5^\circ$
3154.121	−1	fb	31695.41	−15	$a^3F_2 - v^5D_2^\circ$
3149.498	−1	fb	31741.93	00	$b^3G_5 - x^1H_5^\circ$
3144.926	0	f	31788.08	−06	$a^3H_5 - 4_4^\circ$
3139.107	−2	f	31847.00	−11	$z^7D_3^\circ - f^5F_3$
3138.407	−2	f	31854.10	−08	$a^3F_3 - v^5D_4^\circ$
3136.086	−2	g	31877.67	−04	$z^7D_5^\circ - e^7G_6$
3133.967	−1d?	fb	31899.23	−02	$z^7D_4^\circ - f^5F_4$
3125.054	2	fb	31990.21	−28	$a^3F_3 - v^5D_3^\circ$
3119.036	−1	f	32051.92	+02	$a^3G_3 - 13_4^\circ$
3113.667	0	f	32107.19	−01	$z^7D_2^\circ - e^5P_1$
3102.644	0	g	32221.26	−06	$a^5F_3 - y^7P_3^\circ$
3097.491	−1	f	32274.86	+04	$z^7D_4^\circ - e^5P_3$
3094.069	−3	f	32310.55	+14	$z^7D_3^\circ - e^5P_2$
3081.841	0	f	32438.75	−12	$a^3F_4 - v^5D_4^\circ$
3079.826	−2	f	32459.96	+11	$a^5P_2 - w^3P_2^\circ$
3057.802	1N	g	32693.75	−07	$a^5F_3 - y^7P_4^\circ$
3035.238	0	f	32936.78	+08	$c^3P_1 - t^3F_2^\circ$
3018.258	−1	f	33122.08	−06	$b^3F_4 - x^1F_3^\circ$
*2990.336	0	f	33431.35	−02	$z^7F_4^\circ - 1$
		f		+11	$b^3G_4 - t^3F_4^\circ$
2980.586	−1	f	33540.69	+10	$a^3H_4 - 9_4^\circ$

* Fe I blend. Designations listed only if grade is "good" or "fair."
b (Column three) Solar line not entirely due to Fe I.

PART II

AN ANALYSIS OF THE ZEEMAN PATTERNS OF THE SPECTRUM OF Fe I

DOROTHY W. WEEKS

(27) OUTLINE OF THE WORK

This analysis of the Zeeman patterns of the spectrum of Fe I is based upon photographs taken at the Massachusetts Institute of Technology with the assistance of Works Progress Administration workers under the supervision of Dean George R. Harrison, who generously placed them at the disposal of the author. To obtain the Zeeman effect, the Bitter [33] electromagnet was used, with field strengths ranging from 83,000 to 87,000 oersteds for 5 sets of spectrograms containing approximately 12 plates each. The technique used in taking these spectrograms is described in papers dealing with other elements [34] that have been analyzed at the Massachusetts Institute of Technology. The plates were measured upon Harrison's automatic measuring machine,[35] which recorded on a film the wave-lengths and relative photographic intensities of the components. From the film-records thus produced the Zeeman patterns were studied according to the method developed by Russell.[36] These permanent film-records are noted for the high degree of accuracy of the measurements [35] and provide a convenient means for reexamination of the patterns.

The strong magnetic field resulted in a large number of resolved patterns with their components well separated. The magnification of these separations afforded by the automatic comparator made the identification of unblended patterns a comparatively easy task, but the spectrum is so rich that loss by overlapping of neighboring patterns was serious. Many lines originally measured were rejected for underexposure, and a few for overexposure, so that, of some 1250 originally recorded from the films, between λ6494 and λ2272, 1038 were adopted as a basis for the final study. Of this number, 519 were resolved completely enough to permit the determination of both the g-values involved, from the observed pattern alone; for 163 others, the pattern, though resolved, was so much blended that this was impracticable, though the difference $g_1 - g_2$ could be found from the separation of the components. There were also 345 lines with unresolved patterns, sufficiently clear of blends to permit the determination of one g-value when the other was known—as can also be done for the preceding group. The remaining 11 lines are of the "unaffected" type for which both g-values are very nearly zero. From these data, g-values were derived for 130 of the 184 known even levels, and 242 of the 280 odd levels. There are also 8 even and 12 odd levels for which $J = 0$, and g is effectively zero, though theoretically indeterminate—raising the whole number of available determinations to 392 levels out of 464. These values were computed by the usual process—starting with the values determined from fully resolved patterns, and using the mean g's so obtained for the even terms to obtain additional values for the odd terms from the unresolved or partially obscured patterns. A second approximation (as usual) was found to give stable values.

There were about 90 blended patterns for which only an unresolved maximum on one side of the mean position could be observed. These could be utilized by assuming that the correction to the wave-length recorded by the automatic machine was the same as for neighboring lines. These corrections are almost always small, but they vary from one well-observed line to another (perhaps because of Paschen-Back effect); enough to introduce quite sensible errors into work of this refinement. Such patterns were therefore rejected, except when they gave the only determination of g for the level involved—in which case the result is given to two instead of three decimals and marked with a colon. Almost all of these rejected observations were fully consistent with the pattern resulting from the g-values found from better data. Five lines, however, show patterns inconsistent with them and are evidently accidental blends.

(28) OBSERVED ZEEMAN PATTERNS OF FE I

Table D (pp. 184–202) contains the observed Zeeman patterns for the 1038 lines and the g-values deduced from each. Column 2 gives the number of different π-components observed (those equidistant from the middle being counted but once, no matter whether one or both were measured). Columns 3, 4, and 5 give the positions, in units of 0.001a, of the strongest of these (indicated by heavy type), and two adjacent ones. Others are omitted to save space. The letters B and D represent unresolved patterns, respectively, strongest at the outer edges and at the center.[37] The readings of the films correspond very

[33] G. R. Harrison and F. Bitter, *Phys. Rev.* **57**: 15, 1940.

[34] G. R. Harrison and J. Rand McNally, Jr., *Phys. Rev.* **58**: 703, 1940.

[35] G. R. Harrison, *Jour. Opt. Soc. Amer.* **25**: 169, 1935.

[36] H. N. Russell, W. Albertson, and D. N. Davis, *Phys. Rev.* **60**: 641, 1941.

[37] W. F. Meggers and H. N. Russell, *Bur. Standards Jour. Research* **17**: 131, 1936.

TABLE 25

AVERAGE ERRORS OF g-VALUES

N	Type of Level	Number of Levels	$\bar{N}$	$\bar{D}$	$\bar{D}'$
51 to 10	Low Even	30	21.5	5.4	5.5
10 to 5	Low Even	12	7.1	5.6	6.0
	Odd	86	6.3	6.4	7.0
	High Even	10	5.8	8.6	9.4
4 to 3	Odd	28	3.4	6.2	10.8
	High Even	14	3.3	9.0	7.4
2	Odd	84	2	5.6	7.9
	High Even	22	2	8.5	12.0

closely to the maximum of the actual pattern. Components masked by other lines or not capable of interpretation are indicated by "m". A dagger (†) denotes that the measures are insufficient to establish the position of the center, which has been estimated by comparison with the instrumental corrections for other lines (see above).

Similar data for the σ components are given in the next four columns. Unresolved patterns, strongest at the inner edge, the outer edge, or the center, are denoted [37] by A, B, or C. Sharp components are denoted by S, and "unaffected" lines (for which both g's are unresolvably small), by U. The values of J and g for the two terms involved in each line are given in the last four columns—g_1 corresponding to the level with larger J. These values are derived from the data for the individual line for completely resolved patterns; otherwise the mean value of g found from other lines for one term is used to find g for the other. The assumed values of g are given in parentheses, both for unresolved patterns and for those so badly blended that only Δg can be determined. A few values of lower accuracy are marked with colons. A double colon is used throughout the table to denote unusually doubtful entries.

(29) OBSERVED AND CORRECTED g-VALUES

A comparison of the observed and theoretical g-values for the principal groups of terms indicates that the observed values are systematically too large. Similar discrepancies have been found in other cases

TABLE 26

RESIDUALS OF g-VALUES FROM THEORY (UNIT 0.001a)

Term	J=6	5	4	3	2	1	All
d^6s^2							
a^5D			− 4	− 3	− 6	− 2	− 15
a^3P					+ 6	0	+ 6
b^3D				− 7	. . .	. . .	− 7
b^3F			−15	−10	− 4		− 29
b^3G		0	− 2	+11			+ 9
a^3H	− 4	+5	+11				+ 12
b^1G			−21				− 21
a^1I	+14						+ 14
Sum	+10	+5	−31	− 9	− 4	− 2	− 31
d^7s							
a^5P				− 1	−13	− 1	− 15
a^5F		+4	− 1	− 2	− 5	− 14	− 18
b^3P					− 2	− 11	− 13
c^3P					−16	− 34	− 50
a^3D				+ 2	+11	+231	+244
a^3F			+ 4	+ 3	+ 3		+ 10
a^3G		−3	+ 1	+ 6			+ 4
b^3H	− 2	−1	+11				+ 8
a^1P						−183	−183
a^1D					+28		+ 28
a^1G			+ 1				+ 1
a^1H		0					0
Sum	− 2	0	+16	+ 8	+ 6	− 12	+ 16
d^8							
c^3F			+14	−17	+10		+ 7
All low even terms	+ 8	+5	− 1	−18	+12	− 14	− 8

Term	J=6	5	4	3	2	1	All
z^7P°			− 3	− 9	0		− 12
z^7D°		− 3	− 8	− 4	+ 8	− 1	− 8
z^7F°	−2	− 2	− 7	+13	+ 4	+49	+ 55
z^5P°				−10	+ 2	−13	− 21
z^5D°			+ 2	0	+ 3	− 5	0
z^5F°		− 1	+ 5	0	+ 4	−12	− 4
Sum	−2	− 6	−11	−10	+21	+18	+ 10
y^5D°			− 4	− 8	− 5	− 8	− 25
y^5F°		+17	− 6	− 6	− 2	−16	− 13
z^5G°	−1	−49	−47	−30	+ 2		−125
y^3D°				− 9	−16	− 7	− 32
y^3F°			− 4	+ 3	+21		+ 20
z^3G°		+48	+50	+41			+139
y^5P°				− 6	+ 3	+ 2	− 1
x^5D°			−11	+ 4	+ 1	− 2	− 8
x^5F°		−10	−22	+ 4	− 2	− 6	− 36
z^3P°					− 7	− 4	− 11
z^3D°				−12	+ 1	+13	+ 2
z^3F°			0	+ 3	+15		+ 18
y^7P°			0	− 9	+ 7		− 2
z^5S°					−15		− 15
Sum	−1	+ 6	−44	−25	+ 3	−28	− 89
Odd terms	−3	0	− 55	−35	+24	−10	− 79
e^7D		−15	+ 5	+ 5	+ 9	+ 2	+ 6
e^5D			+ 2	+ 8	+ 3	+18	+ 31
e^5F		+21	−19	−14	− 9	+ 7	− 14
e^3F			+38	+24	−45		+ 17
Sum		+ 6	+26	+23	−42	+27	+ 40

and probably arise from an error in the calibration constant, which was determined from a few persistent lines of other elements which appeared on the plate. It is not certain that the mean field was the same for the regions where these lines and those of iron were emitted. A close agreement with theory can be obtained by multiplying the observed g-values by the factor 0.987.

Table E (pp. 203–206) gives the observed and corrected g-values for all levels of Fe I for which they have been determined. The first column gives the designation, the second, the mean g-value resulting from the second approximation described above (giving half weight to a few weak determinations). The third and fourth give the numbers R of resolved patterns, and U of unresolved patterns or patterns disturbed by blending, upon which this mean is based, and the fifth (A. D.), the average residual, without regard to sign, for the individual determinations, in units of 0.001a. The sixth column gives the g-value corrected as just described, and the last, the residual excess of this above the predictions of Lande's theory. The uncorrected g-values should be used for comparison with table D; the corrected values, for all other purposes.

The means, without regard to sign, of the individual discordances D, grouped according to the number of determinations N for the various levels, are given in table 25. The last column gives $\bar{D}' = \bar{D}\sqrt{\bar{N}/(\bar{N}-1)}$, which should be a fair, though not an exact, approximation to the average error of one determination of g. The weighted means of this (still in units of 0.001a) are 5.6 for the low even levels, 7.3 for the odd, and 11.1 for the high even levels. If two wildly discordant values are rejected, the last becomes 9.7. Multiplying by 0.85, the probable error of an average determination of g may be estimated as ±0.005a for the low even levels, ±0.006 for the odd, and ±0.009 for the high even levels.

The better-determined g-values in table E should have probable errors of 0.001 to 0.003a. The deviations from the theoretical values for most of the higher levels are very much greater, indicating the existence of large perturbations.

Perturbations may be anticipated between neighboring levels of the same parity, and with the same J —irrespective of multiplicity, L-value, and electron configuration. Within a group of isolated terms the g-sums for each J should be unaffected. The even levels below 34000 form such a group. There are two groups of odd terms, between 19000 and 30000, and 33000 and 42000, and one of high even terms, between 42000 and 49000. The residuals for these terms are collected in table 26. The theoretical g-sum for all the low even terms is 61.333; for the odd terms, 101.000; for the high even terms, 24.000. The observed sums are 61.225, 100.901, and 24.040. Taken separately, these indicate additional calibration factors of 0.9999, 0.9991 and 1.0017. Together they give 0.9995—a negligible correction.

Some interesting perturbations appear in this table. The terms involved are:

Desig	Level	Residual	Desig	Level	Residual
c^3P_1	24772	−37	$z^5G_5^\circ$	34782	−48
a^3D_1	26406	+228	$z^3G_5^\circ$	35379	+51
a^1P_1	27543	−181			
Sum		+10			+3
$z^5G_4^\circ$	35257	−43	$z^5G_3^\circ$	35611	−31
$z^3G_4^\circ$	35767	+53	$z^3G_3^\circ$	36079	+42
Sum		+10			+11

Perturbations of the g-values are accompanied by mutual repulsion of the levels involved and by strong intersystem combinations. The repulsion of a^1P_1 lowers a^3D_1 below a^3D_2, and that of $z^3G_5^\circ$ puts $z^5G_5^\circ$ below $z^5G_6^\circ$, interrupting the usual sequence of the components for both terms. The combination of z^5G° with a^3F produces the very strong line at λ 4383, and other intersystem lines are exceptionally strong; and a^1P and a^3D show a marked tendency to combine strongly with the same terms.

The higher levels, both odd and even, are closely packed, and the perturbations are great. For example, the configurations $(a^6D)4d$ and $(a^4D)5s$ give rise to 50 levels, lying between 50342 and 52257, all but one of which are known. For the higher J-values all the g's are known and the sum-rule holds—the algebraic sums and arithmetic means of the residuals being +18, ±13 for J = 6; +5, ±40 for J = 5; +8, ±64 for J = 4; for smaller J's some g's are missing. The identity of e^5G_2 appears to be conclusively settled by multiplet intensities, despite the enormous perturbation of +620, but in a case like this the levels of the same J share all their properties and an exact specification is illusory.

In conclusion; it is a pleasure to express gratitude to Dean George R. Harrison and to the Massachusetts Institute of Technology Works Progress Administration assistants for the photographs of the Zeeman patterns upon which the work of this paper has been based; to Professor Henry Norris Russell for his guidance in the preparation of the data and for preparing the final form of the manuscript for the press while the author has been engaged in war work; to Miss Helen P. Beard and Miss Margaret Aylesworth for assisting with the computations; to the American Philosophical Society for a grant from the Penrose Fund for an assistant to help with the computations; to the American Academy of Arts and Sciences for a grant from the Permanent Science Fund, also for computational assistance; and to Wilson College for co-operating in the support of the assistants.

TABLE D

OBSERVED ZEEMAN PATTERNS OF Fe I

λ	π Components				σ Components				J_1	Obs. g_1	J_2	Obs. g_2
	No. Meas.	π			No. Meas.	σ						
6494.985	1	0	D		1	975	A		6	(1178)	5	1219
6430.851	4	0	168	341	5	1011	1180	1348	4	1521	3	1692
6421.355	1	0			1	1522	C		2	(1526)	2	1518
6419.982	2		106	177	5	1282	1332	1475	3	1334	3	1275
6411.658	3	0	334	668	4	857	1193	1544	3	1528	2	1863
6408.031	2	0	1005		2	512†	1464†		2	1515	1	(2520)
6400.010	4	0	165	340	5	1021	1187	1321	4	1502	3	1665
6393.605	3	0	060	132	1	818	m	m	5	1058	4	1118
6380.748	1	0			1	687			2	(682)	2	692
6355.038	2	0	332		2	852	1178		2	1180	1	1509
6336.835	1	995			2	1532	2507		1	2517	1	1532
6335.335	3	023	332	661	3	791	1125	1517	3	1501	2	1848
6322.693	4	0	172	354	3	1493:	1609	1784	4	1257	3	1081
6318.022	3	0	068	157	2	622	691		4	826	3	894
6315.316	2			302	5	1292	1393	1425	3	1079	3	1180
6302.507	1	0	S		1	2507	S		1	2507	0	0
6301.515	2		340	664	4	1535	1897:	2179	2	1861	2	1528
6297.800	2	0	1009		2	509	1536		2	1518	1	2527
6270.238	1	0	S		1	509	S		1	509	0	0
6265.140	2		346	518	6	1517	1683	1850	3	1685	3	1515
6256.370	3	588	873	1166	5	825	1111	1421	4	1113	4	824
6254.262	1	0	D		1	1530	B		2	(1526)	1	1522
6252.561	6	0	083	167	1	758			6	(1178)	5	1261
6246.334	3	186	340	498	6	1518	1678	1841	3	1684	3	1528
6232.661	2	0	335		2		1876	2208	2	1875	1	1542
6230.728	1	047	B		1	1265	C		4	1271	4	1259
6219.290	2	332	665		4	1519	1842	2177	2	1847	2	1517
6215.152	3	0	103	179	1	940:	B		3	(768)	2	679
6213.438	1	1014			2	1517	2528		1	2528	1	1517
6200.323	3	0	430	848	3	1101	1521	1952	3	1100	2	675
6191.562	1	0	D		1	831	A		5	(1052)	4	1107
6180.212	2	0	283						4	(1065)	3	1348
6173.343	1	0	S		1	2531	S		1	2531	0	0
6170.492	1		621						2	1476	2	(1166)
6165.366	3	0	172	331	3	445	607	749	4	920	3	1080
6157.734	1			327	4	1022	m	1288	4	1202	4	1287
6141.734	2	0	155						3	(1677)	2	1522
6137.696	1	0			1	1102			3	(1087)	3	1117
6136.999	3	0	649	1355	1	2179:	B		2	(1844)	1	1509
6136.620	1	0	D		1	881	B		4	(822)	3	802
6102.178	2	0	166		1	879	B		2	682	1	(505)
6078.496	1	0			4	946	1022	m	3	1092	2	(1166)
6065.487	1	0			1	690			2	(672)	2	708
6055.987	4	0	185	396	2	615	m	m	4	1156	3	(1341)
6027.057	5	0	96	196	2	781†	942†		5	(1176)	4	1275
6024.066	1	0	D		1	1131	A		5	1237	4	(1263)
6008.577	1	0	D		2	1171	1226		4	1304	3	(1348)
6003.033	1	0			1	1288			4	(1267)	4	1309
5987.057	2	0	312						2	(1166)	1	1478
5984.805	3	0	126	268	3	1089	1250	1348	3	1349	2	1479

TABLE D—(*Continued*)

λ	π Components				σ Components				J_1	Obs. g_1	J_2	Obs. g_2
	No. Meas.	π			No. Meas.	σ						
5976.799	1	0			1	1087	1135	1211	3	1089	3	1130
5952.749	1	0			1	655	C		2	(690)	2	620
5934.658	1	0	D		1	1020	A		3	1129	2	(1183)
5930.173	2	0	208:		1	1166	B		3	853	2	(696)
5914.16	1	0	D		1	1176	B		4	1120	3	(1100)
5709.378	2		468	602					4	1520	4	(1370)
5701.553	1	0	D		1	998	A		4	(1251)	3	1335
5662.525	3	0	088	158	1	1203	A		5	(1442)	4	1512
5658.826	3	232	506	754					3	1519	3	(1267)
5624.549	2	523	1005		4	1028	1530	1960	2	1528	2	1023
5615.652	1	0	D		1	1035	A		5	(1418)	4	1514
5602.955	1	1509			1	60:	1545		1	1527	1	(−12)
5586.763	3	0	148	292	2	944	1058		4	1382	3	1528
5576.097	1	U			1	U			1	0	0	0
5572.849	3	0	250	498	4	780	1026	1273	3	1274	2	1522
5569.625	2	0	503		2	520	1042		2	1033	1	1546
5565.708	1	0			1	1089			3	(1100)	3	1078
5563.604	1	0			1	1519			3	1515	2	(1513)
5506.782	3	0	507	1008	2		2087	2521	3	1513	2	(1008)
5501.469	2	0	255						4	1519	3	(1264)
5497.519	2	0	1521		2		1559	3034	2	1514	1	−8
5476.571	1	0			1	1511			4	(1515)	4	1507
5455.613	1	1513			2	0	1522		1	1522	1	9
5446.920	2	508	1007		4	1021	1522	2028	2	1524	2	1020
5434.527	1	U			1	U			1	0:	0	0:
5429.699	3	252	506	753	6	1278	1525	1770	3	1523	3	1274
5424.072	1	0	D		2	1127	A		7	1315	6	(1346)
5415.201	1	0	D		1	1127	A		6	1247	5	(1264)
5410.913	1	0	D		1	1127	B		4	882	3	(801)
5405.778	2	0	504		3	514	1021	1522	2	1019	1	1523
5404.144	1	0	D		1	1151:	B		5	1124	4	(1117)
5397.131	4	309	456	600	7	1361	1519	1665	4	1520	4	1372
5393.174	1	0			1	1520			4	1518	3	(1517)
5383.374	1	0	D		1	1075	A		6	1207	5	(1233)
5371.493	4	0	249	499	5	773	1026	1271	3	1274	2	1524
5369.965	1	0	D		1	1115			5	1117	4	(1117)
5367.470	1	0	D		1	958†	B		4	913:	3	(898)
5365.403	1	0	D		1	955	A		5	(1013)	4	1027
5364.874	1	0	D		1	793	B		3	490	2	(339)
5341.026	2	494	983		4	671	1186	1692	2	1183	2	684
5328.534	3	205:	476:	720:	6	622†	861†	1147:†	3	1347	3	(1100)
5328.042	4	0	149	295	6	930	1072	1207	4	1365	3	1510
5324.185	1	0			1	1535			4	(1519)	4	1551
5302.307	1	0	D		1	1539	B		2	(1519)	1	1529
5283.628	1	0			1	1530			3	(1517)	3	1543
5281.796	4	0	574	1214	3	597	1188	1778	3	1779	2	2370
5273.379	1	0	S		1	511	S		1	511	0	0
5273.176	1	0	S		1	1542	S		1	1542	0	0
5270.360	2	0	165		3	533	680	848	2	687	1	527
5269.541	2	0	187		1	1064	A		5	(1422)	4	1512
5266.562	4	0	269	540	3	893	1159		4	1672	3	1936
5263.314	1	0			1	1525			2	(1521)	2	1529
5250.650	3	0	158	320	2	1366	1522		3	1681	2	1839
5242.495	1	0	D		1	991			6	1024	5	(1029)
5232.946	4	0	148	296	3	1045	1167	1275	5	1605	4	1751

TABLE D—(*Continued*)

λ	π Components				σ Components				J_1	Obs. g_1	J_2	Obs. g_2
	No. Meas.	π			No. Meas.	σ						
5227.192	1	**0**	D		1	991:	A		3	(1100)	2	1155::
5226.868	2	292	**664**		4	**2043**	**2368**	2717	2	2376	2	2033
5217.395	1	**0**	D		1	**1523**	A		4	(1519)	3	1518
5216.278	1	**0**			1	**688**	C		2	(679)	2	697
5208.601	1	**0**	D		1	**1528**	A		3	(1517)	2	1512
5202.339	1	**0**			2	**1693:**	m		3	(1688)	3	1698
5198.714	2	**0**	668		2	**1191**	1854		2	1855	1	2520
5194.943	1	**0**			1	**1102**			3	(1100)	3	1104
5192.350	2		364	**503**	2	**m**	2096	2232	3	1937	3	1785
5191.460	2	**0**	665		1	**1705**	A		2	2370	1	3035
5171.599	1	**0**			1	**1273**			4	(1271)	4	1275
5167.491	1	**0**	D		1	**1101**	A		4	(1271)	3	1328
5150.843	3	**0**	255	512	1	**1791†**	B		3	1268	2	(1008)
5142.932	3	**0**	111	219	2	**1682†**	B		4	1374	3	(1264)
5141.747	1	**1017**			2	**501**	**1513**		1	1513	1	501
5139.468	2		295	**380**	1	**1695:**	C		4	(1774)	4	1689
5137.388	1	**0**	D		2	**1395**	A		5	(1442)	4	1454
5133.692	1	**0**	D		5	**m**	1007	1110	6	1340	5	1452
5131.475	1	**0**			1	**2527**			1	(2526)	1	2528
5127.363	2	**0**	050		1	**m**	1566†	m	5	1416	**4**	1366
5123.723	1	U			1	U			1	0	1	0
5110.414	3	307	459	**607**	1	**m**	m	m	4	1669	4	(1516)
5107.645	3	**0**	407	820	3	1084:	1514	**1918**	3	1105	2	698
5107.452	1	**0**			1	**1015**			2	(1008)	2	1022
5098.703	3	**0**	164	332	4	**1357**	1517	1674	3	1678	2	1840
5096.998	4	**0**	142	306					3	1157	2	(1010)
5083.342	1	**0**			1	**1269**	C		3	(1264)	3	1274
5079.742	2	**0**	1010		2		1136	**2022**	2	998	1	−12
5079.226	2	**0**	686		2	**1171**	1849		2	1851	1	2533
5074.757	1	**0**	D		1	**878**	A		5	1264	4	(1361)
5068.774	1	**0**	D		1	**1782**			4	(1774)	3	1771
5065.020	3	**0**	169	321					4	1103:	3	(1263)
5051.636	1	**0**			1	**1370**	C		4	(1367)	4	1373
5049.825	3	**0**	181	361	4	**987**	1162	1341	3	1341	2	1520
5041.759	4	**0**	159	316	4	1431	1586	**1736**	4	1270	3	1113
5041.074	3	**0**	262	533	3	1238	1518	**1775**	3	1261	2	1004
5028.129	2	**0**	m	122	1	**m**	810		5	(1013)	4	1074
5022.244	1	**0**	D		1	**582**	B		2	(690)	1	798
5014.950	2	**0**		129:	1	**991**	A		3	(1101)	2	1156
5012.071	1	**0**			1	**1422**			5	(1422)	5	1422
5006.126	1	**452**	B		1	**m**	m	1779:	5	1608	5	(1518)
5001.871	2	0	**117**		2	**986**	1071		4	(1267)	3	1363
4994.133	3	0	102	195	3	1455	1572	**1678**	4	1364	3	1262
4985.261	1		B	**123**	1	**1255**	C		2	(1183)	2	1121
4973.108	1	**303**			2	**507**	**840**		1	825	1	522
4966.096	1	0			1	**1434**			5	(1418)	5	1450
4957.603					1	**1135**	A		6	(1522)	5	1600
4957.302	3	303	406	**674**					4	(1515)	4	1675
4891.496	4	0	262	501	4	**769**	1035	1297	4	1522	3	1773
4890.762	2	560	**964**		2	**1560**	**2084**	2561	2	2050	2	1538
4871.323	3	**0**	492	1009	4	**546**	1015	1533	3	1533	2	2031
4859.748	2	**0**	1511		3	**0**	1538	3000	2	1513	1	3028
4789.654	1	**0**			1	987	C		2	(1041)	2	933:
4786.810	3	**0**	127	248	2	**1135**	1226		3	1375	2	1500
4741.533	1	**0**	D		1	**1493**	A		3	1510.	2	(1518)

TABLE D—(*Continued*)

λ	π Components				σ Components				J_1	Obs. g_1	J_2	Obs. g_2
	No. Meas.	π			No. Meas.	σ						
4736.780	3	0	092	198	1	1096	A		5	1429	4	(1519)
4710.286	1	0			1	774	C		3	(774)	3	774
4707.487	1	0	D		1	1566	B		2	1551	1	(1509)
4707.281	4	0	170	314	3	841	991	1147	4	1348	3	1517
4691.414	1	0			1	1068			4	(1068)	4	1068
4678.852	3	0	127	260	3	1167	1264	1448	4	1549	3	1677
4668.142	3	0	266	547	3	742	978	1249	3	1253	2	1510
4654.501	3	0	265	529	3	1597:	1890	2159	4	1359	3	1093
4647.437	1	0			1	1222			5	1216	5	1228
4638.016					1	1697	C		3	1717	3	(1677)
4637.512	2	0	521		1	504	A		2	1004	1	(1519)
4630.125	1	0			1	1536			3	1529	2	(1526)
4625.052	2		521	788	1	1267†	m		3	(1517)	3	1250
4619.294	1	0	D		1	1754	B		3	(1677)	2	1638
4613.210	1	U			1	U			1	0	0	0
4611.285	1			230					2	(1860)	2	1975
4602.944	5	0	148	298	6	1712	1858	2006	5	1420	4	1274
4598.122	1	1515			2	0	1522		1	1522	1	7
4592.655	1	158	335	496	6	1105	1265	1429	3	1264	3	1102
4587.132	1	0	D		1	1090	B		5	(1013)	4	994
4556.129	2	0	162:		3	1312†	1498†	1702†	3	(1677)	2	1860
4547.851	1	0	D		1	1013	A		3	(1032)	2	(1041)
4531.152	2	m	298	389	4	1266†	1355†	1454†	4	1362	4	(1271)
4528.619	4	0	168	336	6	1024	1187	1357	4	1519	3	1684
4525.142	1	473	B		1	1201	m	m	3	(1677)	3	1519
4517.530	1	142			1	1573	C		1	1627	1	(1485)
4494.568	3	0	322	645	5	891	1215	1536	3	1537	2	1859
4490.084	2	0	413		3	1098	1508	1859:	2	1509	1	1920
4489.741	1	0	S		1	1573	S		1	1573	0	0
4484.227	4	0	178	369	5	967†	1146†	1314†	4	1503	3	1677
4482.257	2	0	1003		3	528	1526	2522	2	1525	1	2525
4482.171	1	0			1	1525			2	1522	1	(1518)
4480.142	3	731	1054	1416	5	1044†	1399†	1625:†	4	1360	4	(1013)
4479.612	2	0	892						2	1628	1	(2520)
4476.021	2	0	293		3	931	1225	1513	2	1223	1	1515
4469.381	3	0	173	351	3	1336	1465	1647	3	1686	2	1861
4466.554	3	0	145	288	4	1096	1226	1361	3	1369	2	1510
4461.654	1	0			1	1571	B		3	1533	2	(1514)
4459.121	3	174	332	492	6	1521	1664	1822	3	1679	3	1521
4454.383	2	306	607		4	1225	1520	1822	2	1523	2	1224
4447.722	1	1003			2	1520	2523		1	2523	1	1520
4443.197	1	0	S		1	572	S		1	572	0	0
4442.343	2	328	650		4	1526	1842	2172	2	1847	2	1526
4433.223	2	0	591		3	1260†	1865†	2549:†	2	(1860)	1	2460
4430.618	1	0	S		1	2528	S		1	2528	0	0
4430.197	2	0	113		1	1481:	A		2	(1504)	1	1617
4427.312§	1	0			1	1532			4	1522	3	(1518)
	2		305	597	2	1895†	2200†		2	(1860)	2	1560
4422.570	1	942			2	576	1508		1	1508	1	576
4415.125	3	0	213	446	5	912	1133	1374:	3	909	2	687
4408.419	2	0	332		3	1523	1855	2178	2	1852	1	1522
4407.714	3	0	160	326	4	1695	1873	2015	3	1693	2	1531
4404.752	1	0	D		1	1166	B		4	1116	3	(1100)
4401.293	2		371	521					3	(1677)	3	1510
4390.954	4	0	132	264	5	1038	1172	1301	4	905	3	773

§ Two superposed Zeeman Patterns.

TABLE D—(*Continued*)

λ	π Components				σ Components				J_1	Obs. g_1	J_2	Obs. g_2
	No. Meas.	π			No. Meas.	σ						
4388.412	1	0			1	1702	C		3	1727	3	(1677)
4387.897	1	430			2	1489	1933		1	1933	1	1489
4383.547	1	0	D		1	1161	A		5	1249	4	(1271)
4377.793	1	0	D		1	1193	B		2	967	1	(741)
4375.932	1	0			1	1524			5	1518	4	(1516)
4373.563	1	0			1	790			3	765	2	(752)
4369.774	1	99	B		1	1044	C		4	1031	4	1056
4367.581	1	0			1	1126	B		5	1075	4	(1062)
4358.505	5	0	141	282	3	652	794	935	5	1220	4	1362
4352.737	2	0	516		3	1496	2007	2518	2	2007	1	2520
4337.049	3	209	403	598	6	910	1108	1301	3	1103	3	908
4325.765	3	0	122	248	3	808	928	1049	3	806	2	684
4315.087	2	164	319		4	1842	2004	2148	2	2000	2	1847
4309.380	1	0			1	1213	A		6	1215	5	(1216)
4309.036	1	56	B		1	1034	C		6	1039	6	(1030)
4307.906	1	0	D		1	1133	B		4	(1117)	3	1112
4305.455	2	0	404		2	1110	1515		2	1514	1	1918
4302.191	1	0:	B		2	m	1132	1189	4	1070:	4	(1013)
4298.040	5	0	210	402	3	1631	1830	2035	5	1217	4	1013
4294.128	4	293	435	581	7	1132	1274	1406	4	1271	4	1129
4288.148	1	0			1	778	C		3	(766)	3	790
4285.445	1	0			1	1191			6	(1180)	6	1202
4282.406	3	0	320	639	5	1060	1379	1700	3	1695	2	2013
4271.764	1	0			1	1262			5	1269	4	(1271)
4271.159	3	0	097	203	3	1377	1478	1575	4	1675	3	1774
4267.830	1	0	S		1	1570	S		1	1570	0	0
4266.968	1	0			1	1063			4	(1065)	4	1061
4260.479	1	0			1	1617			5	(1621)	5	1613
4250.790	3	292†	595†	904†	6	812	1109	1405	3	1108	3	806
4250.125	3	0	285	545	4	1259	1533	1784	3	1783	2	2045
4247.432	1	0			1	1388	B		5	1374	4	(1370)
4246.090	3	0	160	324	1	1235:†	A		3	1343	2	(1504)
4245.258	1	0	S		1	1911	S		1	1911	0	0
4239.847	1	0			1	1222			5	(1213)	5	1231
4238.816	1	0	D		1	1215	A		4	1254	3	(1267)
4235 942	1	0			1	1680	C		4	(1661)	4	1699
4233.608	2	0	1005		3	1026	2032	3037	2	2032	1	3037
4227.434	1	0	D		1	1125	A		6	1369	5	(1418)
4225.460	3	0	307	559	2	1305†		1918†	3	1306	2	(1017)
4224.176	3	0	130	287					5	1517	4	(1370)
4222.219	1	0			1	1786	C		3	(1771)	3	1801
4220.347	1	0	S		1	1478	S		1	1478	0	0
4219.364					1	1074	B		6	1023	5	(1013)
4216.186	3	512	764	1016					4	(1774)	4	1522
4210.352	1	0			1	3050			1	(3033)	1	3067
4207.130	2	0	395		1	1138	m	m	2	1533	1	1928
4203.987	1	0	D		1	1445	A		2	1477	1	(1509)
4202.031	4	305	451	599	8	1123	1274	1419	4	1266	4	1121
4199.098	1	0	D		1	1075	B		5	1026	4	(1013)
4198.310	1	0	D		1	1498	A		5	(1621)	4	1652
4195.337	1	194	B		2	1382	m	1466	5	1418	5	1378
4191.436	2	0	1008		2	1032	2030		2	2032	1	3035
4187.802	2	0	96		2	1379	1464	m	4	(1661)	3	1756
4187.044	3	0	252	497	4	1283	1535	1774	3	1779	2	2027
4184.895	1	124	B		1	1552:†	C		2	1456	2	(1518)

TABLE D—(*Continued*)

λ	π Components				σ Comopnents				J_1	Obs. g_1	J_2	Obs. g_2
	No. Meas.	π			No. Meas.	σ						
4181.758	3	**0**	097	207	3	**1210**	1317	1436	3	1422	2	1526
4175.640	2	**0**	225		3	**1058**	1295	1471	2	1288	1	1511
4172.126	3	**0**	132	262	3	**1107**	1210	1316	3	1355	2	1486
4156.803	2	202	**476**		4	**1300**	**1559**	1778	2	1527	2	1289
4154.812	1	**0**	D		1	**1482**	B		5	1392	4	(1370)
4147.673	1	**0**	D		3	1732	2190	**2658**	4	1267	3	804
4143.871	4	**0**	164	332	6	1443	1594	**1751**	4	1268	3	1105
4137.002	2	**0**	209		2	1039	**1246**		2	1038	1	830
4134.681	3	**0**	156	300	2	**1066**	1202		3	1364	2	1517
4132.903	2	**0**	304		2	**956**	1226		2	1231	1	1511
4132.060	3	**0**	425	851	4	1109	1532	**1958**	3	1107	2	682
4127.612	1	**0**	S		1	**781**	S		1	781	0	0
4118.549	1	**0**	D		1	**1074**			6	1023	5	(1013)
4114.449	1	**590**	B		3	938	**1223**	**1516**	2	1518	2	1229
4109.808	1	**748**			2	**779**	**1514**		1	1514	1	779
4107.492	1	**m**	92		1	**1610**	B		2	1518	1	1426
4095.975					2	**828**	953		3	1078	2	1203
4085.011	2	**0**	354		2	**803**	1146		2	1151	1	1505
4084.498					1	**1106**	A		5	(1418)	4	1496
4079.848	1	**0**	S		1	1902	S		1	1902	0	0
4078.365	2	**0**	110		2	667	**784**		2	669	1	556
4076.636	1	**93**	B		1	**1514**	C		4	(1519)	4	1509
4074.794					5	**1016**	**1185**	1388	4	1197	4	1011
4071.740	1	**0**			1	**692**	C		2	(696)	2	688
4067.984	1	**313**	B		1	**1575**	C		4	(1519)	4	1597
4067.275	1	**0**	D		3	**869**	978	1097	4	(1251)	3	1378
4066.979	1	**734**	B		4	**1154**	**1507**	1868	2	1511	2	1148
4063.597	1	**0**			1	**1094**			3	(1100)	3	1088
4062.446	1	**406**			2	**1506**	**1922**		1	1915	1	1509
4055.039					1	**1123**	A		5	1226	4	(1251)
4045.815	1	**0**			1	**1263**			4	(1272)	4	1254
4021.869	4	**0**	135	270	6	1027	1160	**1286**	4	895	3	762
4017.156					5	1475	1636	**1791**	5	1167	4	1012
4014.534	1	**0**			1	**1014**	C		5	(1013)	5	1015
4009.714	2	**0**	841†		3	**1163†**	1854†	2520†	2	1848	1	(2526)
4007.277	1	**0**	D		1	**767**	B		3	(766)	2	765
4006.631	1	**0**	S		1	**1405**	S		1	1405	0	0
4005.246	3	**0**	421	842	5	1097	1506	**1921**	3	1095	2	678
4003.764	1	**122**			2	**716†**	**828†**		1	(830)	1	708
4001.666	1	**0**			1	**1684**	C		3	(1688)	3	1680
4000.466	4	298	399	**527**	8	**1061**	**1193**	1313	4	1190	4	1062
3998.054	4	**0**	145	282	7	**648**	789	927	5	1214	4	1356
3997.394	1	**0**	D		1	**1107**	B		5	1073	4	(1065)
3996.968					1	**1004**			4	(1006)	4	1002
3995.996	3	**0**	303	612	5	166	**464**	761	4	1063	3	1364
3994.117	4	**0**	77	148	1	**1379**	B		5	1079	4	1004
3990.379	1	**315**	B		3	861:	925:	**993:**	4	998	4	926
3986.176	4	**0**	218	440	7	**489**	708	921	4	1136	3	1353
3983.960	4	**0**	118	234	4	**690**	818	952	4	1072	3	1198
3981.775	3	325	492	**679**	6	**898**	**1031**	1227	4	1063	4	894
3977.743	1	**0**			1	**1846**			2	(1844)	2	1848
3976.615	2	**0**	351		2	1157	**1503**		2	1157	1	811
3973.655	1	**0**	D		1	**1197**	B		3	1093	2	(1041)
3971.325	4	**0**	154	298	7	**617**	761	912	5	1217	4	1370
3970.391	1	**0**	S		1	**1477**	S		1	1477	0	0

TABLE D—(*Continued*)

λ	π Components				σ Components				J₁	Obs. g₁	J₂	Obs. g₂
	No. Meas.	π			No. Meas.	σ						
3969.261	4	**0**	165	330	7	1425	1586	**1747**	4	1264	3	1101
3967.423	1	**0**	D		1	**842**	B		4	(822)	3	815
3966.066	2	**0**	958		3	1345†	2014†	**2683†**	3	(1341)	2	672
3965.511					1	**1397**	A		4	1487	3	(1517)
3964.522	2	**0**	266		3	1535†	1823†	**2060†**	2	1771	1	(1509)
3963.108					3	**483**	959	1382:	2	980	1	(1519)
3961.147	1	**0**	S		1	**2240**	S		1	2240	0	0
3956.681	1	**0**	D		1	**1220**	B		6	1214	5	(1213)
3956.459	1	**0**	D		1	**1325**	B		6	(1180)	5	1151
3955.956	1	061			1	**1436**			1	1467	1	1406
3953.861	2	**0**	280		3	698†	962†	1235†	3	1244	2	(1518)
3953.156	4	**0**	173	343	3	1096	1262	**1429**	4	932	3	764
3952.702	2	**0**	219		3	**1024**	1273	1503	2	1273	1	1503
3952.606	3	432	569	**706**	8	**1068**	**1209**	1310	5	1208	5	1069
3951.164	2	**0**	302		3	731	1036	**1335**	2	1034	1	732
3949.954	3	**0**	170	338	5	**1353**	1519	1683	3	1684	2	1852
3948.779	1	**0**	D		1	**949**	A		5	(1046)	4	1070
3948.105					2	**857**	1090		4	1373	3	(1520)
3947.533	1	**502**	B		4	1146	**1522**	**1898**	2	1769	2	(1518)
3945.119	4	**0**	407	806	7	1894	2273	**2638**	4	1145	3	755
3944.890	4	**0**	104	208	6	1397	1509	**1625**	5	1187	4	1076
3943.339	2	**0**	643		3	**1216**	1852	2465	2	1844	1	2481
3942.443	2	**0**	222		3	**1060**	1286	1558:	2	1285	1	1509
3940.882	2	**0**	257		3	**1772**	2014	**2266**	4	1517	3	1267
3937.329	4	**0**	324	650	7	1841	2158	**2479**	5	1203	4	883
3935.815	1	**461**	B		4	1059	**1285**	**1516**	2	1516	2	1287
3932.629	5	**0**	255	512	7	1819	2086	**2317**	5	1313	4	1059
3930.299	1	**0**			1	**1515**			2	(1514)	3	1514
3927.922	1	**0**			1	1514			2	1516	1	(1518)
3925.946	1	**0**	S		1	**1578**	S		1	1578	0	0
3925.646	1	**500**	B		4	**1283**	**1518**	1755	2	1519	2	1276
3922.914	1	**0**			1	**1517**			4	1516	3	(1516)
3920.260	1	**0**	S		1	**1518**	S		1	1518	0	0
3919.069	2		412	**543**	3	**921**	**1054**	1182	4	1052	4	921
3918.644	1	**0**			1	**771**			3	(771)	3	771
3917.185	3	**0**	513	1045	4	1517	2018	**2496:**	3	1514	2	1006
3916.733	2	**0**	111		3	1503	1618	1719	6	1175	5	1065
3914.273	1	**952†**			1	**1518**	C		1	2470::	1	(1518)
3913.635	1	**0**	D		1	**1477**	A		3	1510	2	(1526)
3909.830	1	**64**			1	**1541**	C		1	1573	1	1509
3907.937	2	**0**	302		3	794	1069	**1352**	3	780	2	489
3906.748	3	**0**	147	275	4	1342†	1493†	**1644†**	3	1341	2	(1194)
3903.902	3	256	368	**488**	7	**937**	**1055**	1176	4	1054	4	935
3902.948	3	254	501	**750**	6	**1096**	**1342**	1583	3	1341	3	1095
3899.709	1	**0**			1	**1518**	C		2	(1514)	2	1522
3898.012	2	**0**	1525		3	0	**1512**	**3022**	2	1518	1	−10
3897.896	5	**0**	110	228	4	1651		**1858**	6	1323	5	1213
3895.658	1	**0**	S		1	**1518**	S		1	1518	0	0
3893.924	3	502†	682†	**855†**	6	**m**	1378	1554	5	1218	5	1047
3893.391	1	**207**	B		1	**1205†**	C		5	1223	5	1182
3891.928	1	**458**			2	**1275**	**832**		1	1277	1	830
3890.844	4	**0**	124	252	6	1183	1306	**1424**	4	1065	3	943
3888.517	2	508	**1019**		4	**678**	**1177**	1666	2	1172	2	674
3887.051	4	311	461	**611**	8	**1367**	**1514**	1667	4	1517	4	1367
3886.284	1	**0**			1	**1519**			3	(1517)	3	1521

TABLE D—(*Continued*)

λ	π Components				σ Components				J_1	Obs. g_1	J_2	Obs. g_2
	No. Meas.	π			No. Meas.	σ						
3885.512	2	0	307		3	**899†**	1213†	1512†	2	1210	1	(1520)
3884.359	2	**0**	174		5	1563†	1726†	**1895†**	5	(1213)	4	1040
3883.282	1	**67**	B		1	**1335**	C		3	1346	3	1324
3878.575	2	**0**	74		1	**1522**	B		2	(1514)	1	1506
3878.021	3	258	509	**765**	6	**1262**	**1512**	1759	3	1511	3	1260
3876.043	2	**0**	1535		2	0	1516	**m**	2	1516	1	−19
3873.763	1	**0**	B		1	**1004**	A		5	(1052)	4	1064
3872.504	2	513	**1019**		4	**1009**	**1507**	2004	2	1508	2	1006
3871.750	2	m	520	**630**	7	**1080**	1208	1329	5	1208	5	1083
3869.562					7	1505	1664	**1792**	5	1217	4	1069
3865.526	1	**1536**			2	**0**	**1510**		1	1510	1	−26
3861.341	2	**0**	463		3	743	1168	**1600**	2	1170	1	740
3859.913	1	**0**			1	**1515**			4	(1516)	4	1514
3859.214	1	**0**	D		1	**1040**	A		6	(1178)	5	1206
3856.373	1	**0**			1	**1520**			3	(1517)	2	1515
3852.574	4	**0**	178	359	5	**979**	1153	1335	4	1511	3	1689
3850.820	2	523	**1024**		4	**1016**	**1505**	2021	2	1511	2	1014
3849.969	1	U			1	U			1	0	0	0
3846.803	1	**0**			1	**1340**			3	(1353)	3	1327
3846.412					1	**1017**			5	(1013)	4	1012
3845.170	1	**973**			2	**564**	**1514**		1	1520	1	558
3843.259	1	**0**	D		1	**992**	A		4	(1013)	3	1020
3841.051	2	**0**	172		3	**497†**	679†	**850†**	2	(679)	1	506
3840.439	2	**0**	514		3	**486**	1009	1514	2	1006	3	1516
3839.259	1	**81**	B		1	1044	C		4	993	4	(1013)
3837.132	3	**0**	506	1013	3	1164	1658	**2155**	3	1158	2	657
3836.332	1	0			1	**1178**			2	(1194)	2	1162
3834.225	3	**0**	254	506	5	**749**	1014	1261	3	1258	2	1511
3833.311	2	m	333	**432**	6	**1230**	**1339**	1461	4	1348	4	1246
3829.458	1	**0**			1	**732**	C		1	(741)	1	723
3827.825	1	**0**	D		1	989	A		3	(1100)	2	1155
3825.884	4	**0**	152	301	5	**938**		1363	4	1368	3	1512
3824.306	1	**0**			1	822			4	(822)	4	822
3821.834	1	**168**	B		2	**665†**	**743†**		2	753	2	(672)
3821.181	1	**0**	D		1	**951**	A		6	1030	5	(1046)
3816.340	1	**0**	D		4	**792†**	1133†	1495†	3	1493	2	(1844)
3812.964	3	**0**	255	507	5	**764**	1015	1265	3	1262	2	1513
3811.892	2		344	**542**	4	228†	414†	**603†**	3	(766)	3	586
3810.759	1	**0**	D		1	**1140**	A		2	(1194)	1	1248
3808.731	2	345	**467**		6	**1247**	**1358**	1462	4	1359	4	1244
3807.534	2	**0**	962		3	**585**	1552	2519	2	1552	1	2517
3806.697	1	**0**	S		1	**1046**	S		5	(1046)	5	1046
3805.345	1	0	D		1	**932**	B		5	844	4	(822)
3802.283	2	308	**638**		4	**1201**	**1514**	1801	2	1516	2	1197
3801.681	3	**0**	296	589	4	**633**	923	1223	3	1224	2	1520
3799.549	2	m	95	194	2	**1643**	m	1465	4	1358	3	(1264)
3797.948	1	**0**	D		2	1091†	1431†	**m**	3	(1087)	2	747
3797.517	1	**0**			1	**1184**			6	(1180)	6	1188
3795.004	3	**0**	254	507	5	1263	1515	**1762**	3	1264	2	1012
3794.340	1	**0**	D		1	**930**	B		5	**844**	4	(822)
3792.156	3	241	366	**471**	6	**939**	1059	1185	4	1063	4	948
3790.095	2	**0**	512		3	**507**	1012	1515	2	1011	1	1519
3789.178	1	**0**	D		1	**909**	A		5	1034	4	(1065)
3787.883	2	**0**	1030		3	0	1012	**2022**	2	1016	1	−14
3786.678	1	U			1	U			1	0	0	0

TABLE D—(*Continued*)

λ	π Components				σ Components				J_1	Obs. g_1	J_2	Obs. g_2
	No. Meas.	π			No. Meas.	σ						
3785.950	1	**0**			1	**1037**			6	1054	5	(1052)
3781.188	2	**0**	489		3	**415**	890	1364	3	1365	2	**1840**
3779.444	1	**534**			2	**743**	**1264**		1	1264	1	**743**
3778.509	3	**0**	202	411	2		1522	**1720**	3	(1353)	2	1161
3777.061	1	0	D		1	**1126**	B		4	1054	3	(1030)
3776.454	3	**0**	231	470	4	**777**	992	1232	4	1462	3	1693
3774.823	1	**1206**			2	**1334**	**2521**		1	2521	1	1334
3773.699	1		**1059†**		4	**1856**	**2348**	2843	2	2350	2	1862
3770.305	1			**364**	3	1115†	1203†	1314†	5	(1213)	5	1143
3769.995	3	**0**	385†	**698†**	4	1931	2237	**2521**	3	1934	2	1632
3768.030	1	**0**	S		1	**2521**	S		1	2521	0	0
3767.194	1	U			1	U			1	0	1	0
3765.542	1	**0**	D		1	**1054**	A		7	1162	6	(1180)
3763.790	1	**0**			1	**1009**			2	(1008)	2	1010
3760.534	2	**0**	624		3	**1304**	1910	2512	2	1907	1	2517
3760.052	1	**0**	D		1	**1077**	m	m	7	1164	6	(1178)
3758.235	1	**0**			1	**1263**	C		3	(1264)	3	1262
3756.069	3	311	651	**958**	5	**1355**	**1682**	1980	3	1680	3	1364
3753.610	3	0	130	253	5	1693	1818	**1940**	3	1691	2	1564
3749.487	1	**55**	B						4	(1367)	4	1354
3748.264	2	**0**	518		3	**512**	1080	1512	2	1012	1	1521
3745.901	1	U			1	U			1	0	0	0
3745.561	3	**0**	258	520	5	**768**	1015	1264	3	1265	2	1518
3743.468	1	**85**	B		1	**1022**	C		5	1014	5	1031
3743.364	2	**0**	1021		3	0	1010	**2024**	2	1008	1	−10
3737.133	2	m	307	460	5	**932**	m	1360	4	1362	3	1513
3734.867	1	**68**	B		1	**1435**	C		5	1428	5	1442
3733.319	1	**1542**			1	**0**	**1504†**		1	1518	1	−24
3732.399	1				1	**1871**	C		2	1898	2	(1844)
3731.374	2	365†	**562†**		4	**485**	**668**	890	2	675	2	472
3730.945	1	**0**	D		1	**708**	B		3	684	2	(672)
3730.386	5	**0**	139	278	8	1435	1556	**1692**	5	1144	4	1007
3728.668	3	450†	671†	**885†**	3	817†	**1026†**	**1258†**	4	(1251)	4	1033
3737.621	3	**0**	259	512	5	1264	1518	**1765**	3	1265	2	1011
3724.380	3	**0**	154	306	5	**1071**	1235	1393	3	1377	2	1530
3722.564	2	515	**1032**		4	**1013**	**1517**	2019	2	1516	2	1014
3719.935	5	**0**	104	206	4	**1004**	1105	1201	5	1407	4	1508
3718.407	1	**1169**	B		5	311†	**768†**	**1223†**	3	1226	3	(766)
3716.442	1	**0**	D		2	**1937**	B		4	(1774)	3	1720
3715.911	1	**638**	B		4	906	**1207**	**1522**	2	**1524**	2	1213
3711.411	2	**0**	446		2	**605**	1031		2	1041	1	1487
3711.225	1	**0**	D		1	**1046**	A		4	1077	3	(1087)
3709.246	4	m	97	208	2	m	m	**1682**	4	1367	3	1262
3707.918	3	**0**	234	462	5	**650**	776	904	3	1686	2	1919
3707.824	2	**0**	1523		3	0	1475:	**3047**	2	1514	1	− 9
3705.567	3	258	512	763	6	**1263**	**1511**	1762	3	1514	3	1264
3704.463	5	**0**	140	280	7	1480	1620	**1738**	5	1210	4	1076
3702.033	1	**0**	S		1	**1504**	S		1	1504	0	0
3701.086	4	**0**	310	620	7	**731**	1022	1315	4	1626	3	1932
3698.611	2	**0**	403		4	**313**	713†	1105	3	1110	2	1510
3697.426	3	671	1298	**1924**	3	**1308**	**1929**	2578	3	1935	3	1304
3695.054	1	**0**	D		1	**1290**	B		5	1068	4	(1013)
3694.005	3	**0**	444	878	5	**558**	788	1018	3	1920	2	2355
3690.730	1	**734**	B		4	**817†**	963†	m	5	(1013)	5	**1159**
3690.450	1	**0**	S		1	**820**	S		1	820	0	0

TABLE D—(*Continued*)

λ	π Components				σ Components				J₁	Obs. g₁	J₂	Obs. g₂
	No. Meas.	π			No. Meas.	σ						
3689.457	4	229	427	643	4	1592	1776	1961	4	1776	4	1581
3687.656	3	297†	416†	530†	3	802†	931†	m	4	1196	4	(1065)
3687.458	1	0	D		1	1594	B		5	(1422)	4	1379
3686.260	1	0	S		1	1512	S		1	1512	0	0
3685.998	5	0	261	518	6	492	751	m	5	1534	4	1794
3684.108	4	0	163	319	7	563	747	906	4	1060	3	1220
3683.054	3	0	513	988	4	1507	2019	2554	3	1521	2	1013
3682.226	1	66	B		1	1015	C		2	1032	2	999
3679.915	4	310	460	609	8	1366	1515	1662	4	1517	4	1368
3677.630	1	0	D		1	912	B		3	(766)	2	693
3677.309	3	0	129	243	2	672†	782†	m	3	921	2	(1041)
3676.314	1	0	D		1	1108	A		5	1222	4	(1251)
3672.722	4	0	254	526	6	1598	1834	2085	5	1078	4	823
3670.810	1	0	S		1	789	S		1	789	0	0
3670.071	1	0	D		1	1060	A		6	1190	5	(1216)
3670.028	1	0	D		1	1480	A		2	1495	1	(1509)
3663.458	1	96	B		1	1267	C		4	1279	4	1255
3659.516	1	285	B		1	820	890		4	892	4	(822)
3657.139	2	0	242		3	1032	1278	1543	2	1277	1	1521
3655.465	1	62	B		1	1501	C		2	1517	2	1486
3653.763	4	0	174	352	7	1759	1941	m	6	1223	5	1046
3651.469	4	0	170	338	7	1095	1263	1420	4	931	3	764
3649.508	1	0	D		1	1267	B		5	(1213)	4	1200
3649.304	3	0	264	540	2		2040†	2292†	4	(1576)	3	1252
3647.844	4	m	129	266	5	711	m	1226	5	1229	4	1360
3645.822	1	0	S		1	714	S		1	714	0	0
3640.388	4	0	103	204	4	1390	1484	1565	5	1172	4	1069
3638.296	4	0	183	364	7	1120	1299	1475	4	942	3	763
3636.995	4	0	146	291	3	509†	650†	810†	4	941	3	(1087)
3636.650	2	0	673		3	−577†	m	820†	3	821	2	1504
3636.234	1	0	D		1	523	A		3	868	2	(1041)
3636.186	3	0	474	940	3	412	900	1354	3	1365	2	1841
3632.042	2	0	312		3	885†	1175†	1479†	2	1182	1	(1485)
3631.464	2	m	m	294	5	692	m. . .	1113	4	1113	3	1256
3625.140	1	0	D		1	1440	A		5	(1518)	4	1538
3623.187	1	186	B		1	1198	C		6	1214	6	1183
3622.001	1	0	B		1	768	C		3	766	3	(770)
3621.463	1	0	D		1	1151	B		5	1082	4	(1065)
3618.769	3	0	116	230	2	677	788		3	902	2	1015
3618.392	2	m	394†	526†	3	801†	931†	1075†	4	(1065)	4	930
3617.788	3	0	179	358	4	981†	1153†	1327†	3	1328	2	(1504)
3610.159	1	0	B		1	1511	C		6	(1522)	6	1500
3608.861	2	0	351		3	0	341	680	2	338	1	− 7
3606.679	1	0	D		1	927::	A		6	1161::	5	(1213)
3605.450	3	257	388	490	7	949	1067	1185	4	1062	4	948
3603.828	1	787			2	717	1496		1	717	1	1496
3603.205	1	210	B		1	1201	C		5	1180	5	1222
3599.624	5	0	151	301	6	431	557	710	5	1006	4	1154
3594.632	1	295	B		1	1543	C		4	1580	4	1506
3592.486	1	0	D		1	1062	A		4	1081	3	(1087)
3590.086	3	439†	569†	701†					5	(1216)	5	1085
3589.456	3	0	283	593	6	1320	1620	1911	4	1059	3	773
3589.107	1	937	B						5	(1422)	5	1235
3586.985	2	681	1360		3	338	1006	1690	2	1011	2	333
3585.708	4	490	754	998	4	1123	1357	1847	4	1361	4	1115

TABLE D—(*Continued*)

λ	π Components				σ Components				J_1	Obs. g_1	J_2	Obs. g_2
	No. Meas.	π			No. Meas.	σ						
3585.320	3	360	774	**1122**	6	**887**	1265	1632	3	1262	3	890
3584.663	4	332†	464†	**585†**	8	**1107†**	**1233†**	1357†	5	(1213)	5	1088
3582.201	6	**0**	192	381	7	**168†**	359†	575†	6	(1180)	5	1368
3581.195	1	**0**	D		1	**987:**	A		6	1350	5	(1422)
3575.374	2	337	**667**		4	**1172**	**1504**	1828	2	1504	2	1172
3573.896	2	**0**	126		1	**508**	A		4	(822)	3	930
3571.995	1	**0**	B		1	**1523**	C		5	(1518)	5	1528
3570.100					1	**849**	A		5	1263	4	(1367)
3568.977	4	**0**	273	560	6	1806†	2070†	**2341†**	5	(1213)	4	940
3567.038					1	919			3	1322	2	(1524)
3565.381	4	**0**	151	302	7	**331**	406	474	4	1120	3	1272
3559.506	1	**217**			2	**1204†**	**1428†**		1	(1485)	1	1265
3558.518	3	**0**	217	431	5	**373**	584	793	3	793	2	1006
3556.877	2	**0**	132		2	**946**	**1056**		5	1402	4	(1515)
3554.922	1	**m**	88		1	**1049**	A		6	1440	5	(1518)
3554.122	1	**0**	D		3	1262†	2183†	**3094†**	3	(1264)	2	348
3552.112	1	**0**	D		1	**1518**	B		2	1502	1	(1485)
3549.868	2	**0**	682		3	0	677	**1356**	2	676	1	− 6
3547.203	3	385	588	**780**	4	**833**	**1017**	1197	4	826	4	1016
3545.639	1	**517**	B						4	1644	4	(1515)
3543.669	2	**0**	172		3	842	1011	**1167**	2	1007	1	840
3542.076	3	**0**	148	305	1	**1378†**	A		4	1383	3	(1534)
3541.083	1		102		1	**1005**	A		5	1414	4	1516
3540.709	1	**m**	474†		3	1834†	2309†	**m**	4	(1367)	3	889
3540.121	1	**0**	D		1	**1496**	A		4	1524	3	(1534)
3537.729	1	**0**	D		1	775	B		2	(672)	1	569
3536.556	5	**0**	238	502					3	1262	2	(1524)
3529.818	1	**1942**	B		1	**1188**			1	1565	1	−378
3527.792	1	**637**	B						4	(1515)	4	1355
3526.465	1	B			1	**1500**			2	(1526)	2	1474
3526.167	3	480	945	**1420**	6	**798**	**1264**	1760	3	1267	3	797
3526.039	3	**0**	171	336	4	1685	1860	**2008**	3	1685	2	1519
3524.236	2	**0**	114		2	**1201**	1295		3	1413	2	1523
3521.833					1	**1166**	A		2	1846	1	(2526)
3521.264	4	518	768	**1018**	8	1112	**1363**	1614	4	1364	4	1111
3518.86	3	**0**	171	332	3	**1358**	1514	1672	3	1675	2	1837
3516.403	1	**0**	D		1	**957**	B		4	817	3	(771)
3513.820	5	468	621	**766**	10	**1267**	**1416**	1568	5	1416	5	1263
3511.748	2			**207**					4	(1251)	4	1199
3510.446	1	**0**	S		1	**1571**	S		1	1571	0	0
3509.870	1	**47**			1	**2493**			1	2516	1	2470
3508.494	1	**0**			1	**1018**			5	1053	4	(1062)
3506.498	1		**501**		4	1286	**1526**	1764	2	1525	2	1278
3505.065	2	**0**	246		3	1316†	1531†	**1768†**	2	(1504)	1	1268
3504.859	1	**0**	D		1	1440	A		2	(1526)	1	1612
3500.564					3	1099	1490	**1900**	3	1086	2	676
3497.843	2	**0**	342		3	1523	1856	**2188**	2	1855	1	1518
3497.110	1	**0**			1	**1687**			3	(1688)	3	1686
3490.575	3	191	347	**513**	6	**1513**	**1681**	1846	3	1682	3	1512
3489.670	1	**0**	D		1	**1059**	A		6	1190	5	(1216)
3485.342	2	**0**	624		3	**1233**	1849	2460	2	1847	1	2466
3476.853					5	428	571	726	4	933	3	(1087)
3476.704	1	**0**	S		1	**2525**	S		1	2525	0	0
3475.450	2	345	**687**		4	**1516**	**1849**	2184	2	1852	2	1513
3471.350	1	**0**	D		1	**1617**	B		2	(1526)	1	1435

TABLE D—(*Continued*)

λ	π Components				σ Comopnents				J_1	Obs. g_1	J_2	Obs. g_2
	No. Meas.	π			No. Meas.	σ						
3471.27					3	**380**	1463	2530	2	1458	1	2530
3469.012	1	**0**			1	**824**			4	(822)	4	826
3468.849	1	0	D		1	**875**	A		5	1176	4	(1251)
3465.863	1	**1036**			2	**1516**	**2524**		1	2534	1	1512
3463.305	1	**0**			1	**1263**			4	(1271)	3	1274
3462.353	1	0	D		1	**1795**	A		2	(1844)	1	1893
3458.304	1	**0**	S		1	**1517**	S		1	1517	0	0
3453.022	1	**0**	D		1	**670**	A		3	(766)	2	814
3452.273	1	**0**	D		1	**1255**			4	1262	3	(1264)
3451.915	2	**0**	1265		3	0	1272	2527	2	1272	1	2527
3451.628	1	**44**			1	**1541**	C		1	1563	1	1519
3450.328	1	**926**			2	**1615**	**2522**		1	2527	1	1610
3447.278	2	407	**799**		4	**1448**	**1843**	2238	2	1845	2	1447
3445.151	3	**0**	442	882	5	**544**	978	1390	3	1408	2	1845
3443.878	2	**0**	1029		3	**508**	1511	2525	2	1515	1	2525
3442.364	1		**752**		2	1010	1395		2	(1526)	2	(1148)
3440.989	3	**0**	343†	680†	5	**844**	1180	1512	3	1513	2	1850
3440.610	4	0	165	350	7	1020	1172:	1344	4	1512	3	1679
3439.039	1	**0**	D		1	**1103**	B		5	1073	4	(1065)
3437.046	3	m		217	5	m	m	281	4	(1013)	3	1223
3431.815					3	**931**	1115	1323	3	1315	2	1507
3428.192	2	578	**1163**		4	**1266**	**1840**	2399:	2	1842	2	1265
3427.121	4	0	338	673	7	**355**	**686**	1019	4	1354	3	1689
3424.284	3	285	560	**834**	6	**1416**	**1689**	1963	3	1690	3	1416
3422.656	2	0	1298		3	−59	1237	2507	2	1238	1	2535
3419.706	1	**712**			2	**809**	**1509**		1	1512	1	806
3418.507	1	0	S		1	**2525**	S		1	2525	0	0
3417.842	1	**1115**			1	**1424**			1	2539	1	1424
3413.135	3	0	493	986	5	**389**	875	1361	3	1363	2	1853
3411.353	1	**274**	B		1	**1104**	C		4	1138	4	1070
3410.171	2	**0**	135		3	**554**	693	826	2	691	1	828
3407.461	4	**0**	328	661	7	**387**	707	1029	4	1357	3	1684
3404.357	3	**0**	689	1374	5	−186	486	1157	3	1160	2	1840
3402.256	1	**0**			1	**1182**			6	(1180)	6	1184
3401.521	3	**0**	327	649	5	**396**	723	1045	4	1364	3	1686
3399.336	2	627	**1249**		4	1229	**1843**	2458	2	1845	2	1227
3396.978	3	0	618	1212	5	**80**	677	1261	3	1270	2	1870
3392.652	3	323	665	**991**	6	**1358**	**1680**	2018	3	1684	3	1356
3392.304	1		**2176**		4	**759**	**1841**	2880	2	1845	2	762
3392.014	1	0			1	1551			2	(1504)	1	1457
3389.748	2	**0**	1408		3	**−210**	1161	2552	2	1161	1	2532
3387.410	3	**0**	144	282	5	**479**	611	751	3	754	2	895
3383.981	2		1060	**1555**	6	1195:	**1693**	2210	3	1690	3	1170
3383.692	2	**0**	1102		3	769	1855	**2913**	2	1846	1	763
3380.111	1	**113**	B		1	**781**	C		3	800	3	762
3379.017	2	**0**	469		2	2150	**2597**		3	1686	2	1228
3378.676	1	**0**	D		1	**1444**	B		5	(1213)	4	1155
3373.874	1	**0**	D		1	**1100**	B		5	1072	4	(1065)
3372.070	3	**0**	953	1877	4	1681	2619	**3554**	3	1681	2	744
3370.786	1	**315**	B		1	**1185**	C		5	1217	5	1154
3369.549	1	**45**	B		1	**1073**	C		4	1079	4	1068
3366.867	2	693	**1407**		4	**1167**	**1850**	2533	2	1852	2	1159
3366.789	5	**0**	247	515	8	1712	1934	**2184**	5	1212	4	966
3356.407	2	256	**503**		4	**1533**	**1762**	1993	2	1768	2	1526
3355.228	1	**183**	B		2	**809**	**854**		4	854	4	809

TABLE D—(*Continued*)

λ	π Components				σ Components				J₁	Obs. g₁	J₂	Obs. g₂
	No. Meas.	π			No. Meas.	σ						
3354.064	1	**0**	S		1	**1261**	S		1	1261	0	0
3351.750	4	**0**	270	531	6	1330	1597	**1828:**	4	1070	3	807
3351.529	1	**0**	D		1	**1808**	A		2	(1844)	1	1880
3347.927	1	**480**	B		4	**1290**	**1518**	1752	2	1520	2	1283
3346.936	2	**0**	547		3	1692	2219	**2743**	3	1682	2	1146
3342.298	1	**235**			2	**1280**	**1503**		1	1506	1	1277
3342.216	2	**0**	732		2	799	**1492:**	m	2	1531	1	2263
3341.906	2	m	617†	**728†**	5	933†	**1099:†**	1221:†	5	(1213)	5	1083
3340.566	1	**491**	B		3	1289	**1524**	1757	2	1523	2	1284
3337.666	4	**0**	128	264	6	1478	1608	**1728**	5	1219	4	1090
3336.254	3	708	1060	**1418**	6	798	1162	1527	4	1163	4	814
3335.776	1	**0**	D		1	**1503**			2	1506	1	(1509)
3334.223	1	**171**	B		1	**1088**	C		5	1105	5	1071
3331.778	1	**0**	S		1	**1409**	S		1	1409	0	0
3331.612	2	**0**	124		6	1274	1383	**1511**	5	1033	4	913
3328.867	1	**0**			1	**1045**			5	(1046)	5	1044
3327.498	1	**0**	D		1	**1180**	A		4	1445	3	(1534)
3325.468	1	**0**	D		1	**934**	B		4	(822)	3	785
3324.541	1	**0**	D		1	**1220**	B		6	(1180)	5	1172
3323.737	1	**0**			1	**1507**			2	(1518)	2	1496
3319.258	1	**504**	B		6	**1083**	**1196**	1303	4	1199	4	1076
3317.121	1	**0**	D		1	**1499**	A		2	(1526)	1	1553
3314.742	1	**0**			1	**911**			3	1100	2	(1194)
3310.496	3	**m**	476	981	5	**−587**	**−90**	354	4	866	3	1351
3310.347	1	**210**	B		1	**1220**	C		5	1241	5	1199
3306.356	2	**0**	762		3	**1004**	1763	2565:	2	1764	1	2525
3305.971	3	**0**	160	323	2	**1344**	1506		3	1667	2	1828
3301.227	1	**234**			2	**1294**	**1509**		1	1514	1	1289
3298.133	2	**0**	1259		2	**33**	1282		2	1284	1	2538
3292.590	1	**290**			2	**2232**	**2516**		1	2516	1	2232
3292.022	4	**0**	192	382	4	**600**	789	978	4	1171	3	1362
3290.988	2	**0**	1253		3	**0**	1246:	2530	2	1267	1	2520
3288.967	2	**0**	611		3	**0**	640	1243	3	1245	2	1852
3288.651	1	**0**	S		1	**1520**	S		1	1520	0	0
3286.755	1	**0**			1	**1678**			3	(1688)	3	1668
3284.588	1	**152**	B		1	**1791**	C		2	1829	2	1753
3282.891	1	**0**	D		1	**658**	A		2	700	1	(741)
3280.261	1	**0**	D		1	**926**	B		5	843	4	(822)
3278.741	1	**0**			1	1099	C		3	(1087)	3	1111
3276.468	2	574	**1148**		4	1288	**1846**	2402	2	1846	2	1288
3274.453					1	**1389**	C		4	(1370)	4	1408
3271.002	2	**0**	397		3	**1448**	1840	2230	2	1839	1	2233
3268.234	1	**970**			2	**1573**	**2521**		1	2526	1	1568
3265.616	1	**0**	D		1	**1559**	A		3	(1688)	2	1752
3265.046	2	**0**	162		3	**990**	1156	1342	3	1339	2	1515
3264.512	2	**0**	1602		2	1850	3450		2	1850	1	249
3263.378	1	**0**	D		1	**1465**	A		2	1492	1	(1520)
3260.261	1	**446**	B		1	**1119†**	**m**	m	4	(1251)	4	1139
3257.594	2	**0**	402		3	1698	2084	**2479**	3	1691	2	1295
3254.363	1	**0**	D		1	**1009**	A		6	1040	5	(1046)
3253.610					4	510	800	1080	4	1067	3	(1353)
3248.206	1	**425**	B		1	**1776**	C		3	1776	3	1634
3246.962	2	**0**	278		2		1834	**2129**	2	1839	1	1553
3244.190	2	**0**	142†		1	**1011†**	A		5	1519	4	(1661)
3243.109	2	m	514:†	**678:†**	1	776:†	m	**m**	4	(822)	4	987:

TABLE D—(*Continued*)

λ	π Components				σ Components				J_1	Obs. g_1	J_2	Obs. g_2
	No. Meas.	π			No. Meas.	σ						
3239.436	1	483	B		1	1563	C		4	(1661)	4	1540
3236.223	4	0	256	511	5	507	762	1015	4	1268	3	1522
3234.614	1	541			4	1224†	1390†	1572†	3	(1517)	3	1337
3233.967	1	255	B		1	1427:			4	1597	4	(1661)
3233.053	1	0	D		1	1066	A		7	1164	6	(1180)
3229.123	1	0	S		1	510	S		1	510	0	0
3228.254	2	0	353		3	1679†	2060†	2401†	2	(2041)	1	1684
3227.798	1	0	D		1	1712	B		4	(1661)	3	1644
3225.789	1	0	114		1	m	1045:		6	1519	5	(1621)
3222.069	1	315	B		1	1574	C		5	1606	5	1542
3219.806	1	0	D		1	1631	A		4	(1661)	3	1671
3219.581	2	0	174		2	1075:†	1254:†		4	1595	3	(1771)
3215.940	1	306	B		1	m	2058:†		2	(2041)	2	1888
3214.396	3	0	436	872					3	(1514)	2	1078
3211.989	1	0	D		1	1677	B		5	(1621)	4	1607
3211.683					1	1417†			6	1418::	5	(1418)
3208.470	2	0	343		2	m	360†	690†	2	348	1	(−12)
3205.400	1	526			2	2525	3029		1	3033	1	2521
3202.562	1	0	D		1	1268	B		4	(1013)	3	928
3200.475	1		127	B					2	(2041)	2	1978
3199.530	1	219	B		1	1574:	m	m	4	(1661)	4	1606
3196.930					1	1003	A		5	1529	4	(1661)
3193.228	3	512	780	1024	6	1272	1506	1771	4	1511	4	1256
3191.659	4	0	183	361	6	1703†	1875†	2048†	4	(1516)	3	1339
3188.819	2	0	2154			−1182	972	2837:	2	970	1	3022
3188.567	1	1190	B						5	(1621)	5	1383
3184.896	2		827	1245	6	1098	1519	1915	3	1515	3	1104
3182.970					4	56†	671†	1258†	3	1236	2	(1844)
3181.522	1	0	D		1	947	A		3	(1087)	2	1157
3180.756	1	835†	m		3	678	1522	2378	2	1526	2	678
3180.223		D			1	1292	A		4	1651	3	(1771)
3178.967	1	0			1	1052			5	(1052)	5	1052
3178.545	1	454	B						3	(771)	3	922
3178.015	1	0	D		1	1659	B		5	(1621)	4	1612
3176.366					1	665			2	(672)	1	679
3175.447	1	495	B		3	1489:†	1637†	1739†	5	(1621)	5	1521
3171.353	4	0	146	291	4	1160	1305	1433	4	1016	3	873
3167.907					1	926:	A		4	1369:	3	(1517)
3166.435					2	1025	m	m	4	(1251)	3	1326
3165.860	4	0	420	844	5	578	899	1349	4	1331	3	1754
3161.949	6	0	196	393					6	1428	5	(1621)
3160.658	1	90	B		1	1642	C		4	1653	4	1631
3160.344	1	0			1	1170			6	(1178)	6	1162
3157.88	1	0	D		1	1801	A		3	1967	2	(2041)
3157.040	5	0	319	550	6	325	568	839	5	1381	4	1648
3156.275	1	244	B		1	m	m	m	3	(1517)	3	1436
3155.293	1	0	D		1	794	A		5	(1052)	4	1116
3151.353	1	0	D		1	1116	B		5	1075	4	(1065)
3142.888	1	68	B		1	1500	C		2	1517	2	1483
3134.111	3	0	254	512	4	1772	2026	2272	4	1514	3	1260
3129.334	4	0	250	484	6	557	793	1015:	4	1266	3	1502
3125.653	2	m	516	1043	3	1534	2032	2543	3	1523	2	1010
3120.435	1	0	D		1	846	B		4	(822)	3	814
3119.495	1	0	D		1	965	A		5	(1052)	4	1074
3116.633	2	0	1538		3	21	1518	3036	2	1513	1	−15

TABLE D—(*Continued*)

λ	π Components				σ Components				J_1	Obs. g_1	J_2	Obs. g_2
	No. Meas.	π			No. Meas.	σ						
3112.079	1	**271**	B						5	(1216)	5	1162
3100.666	2		308:	**783**	6	**1264**	**1521**	1772	3	1519	3	1261
3100.304	2	463	**1036**		4	**1008**	**1520**	2027	2	1008	2	1522
3099.971	1	**605**	B		1	m	**1503†**		4	1518	4	(1367)
3099.897	1	**1542**			2	**51**	**1524**		1	1524	1	−18
3098.192	1	**0**			1	**1236**			5	(1213)	5	1259
3095.270	4	**451†**	610†	**742†**	6	**1073**	**1233**	1377	5	1226	5	1379
3093.883	1	**0**	D		1	**1263**	B		4	(1251)	3	1247
3093.806	2	537	**1087**		4	**675**	1211	1763	2	1218	2	678
3092.778	3	**0**	1122	2225	3	**196†**	1286†	2400†	3	(1264)	2	2371
3091.578	1	**0**	U		1	**0**	U		1	0	0	0
3090.209	1	**528**	B		5	**944**	1116	1291	3	939	3	763
3083.742	2	**0**	513		3	**551**	1048	1550	2	1010	1	1517
3078.436	1	**0**	S		1	**713**	S		1	713	0	0
3075.721	3	**0**	258	514	5	**756**	1007	1260	3	1262	2	1518
3073.982	1	**0**	D		1	**1279**	B		5	(1213)	4	1197
3067.244	5	**0**	157	304	1	**909**	A		4	1371	3	1525
3063.933	1	**727**	B		2	**814**	**1523**		1	1528	1	810
3060.984	3	286	571	**853**	6	**1099**	**1372**	1654	3	1378	3	1097
3060.545	1	**0**	D		1	**1003**	A		4	(1062)	3	1082
3059.086	1	**0**			1	**1514**			4	1516	3	(1517)
3057.446					1	1052			5	(1422)	4	1515
3055.263	1	**0**	D		1	**892**	A		3	(1100)	2	1204
3053.065	2	**0**	349		2	826	1175		2	1175	1	1524
3047.605	1	**0**			1	**1514**			3	1514	2	(1514)
3045.077	4	**0**	543	1165	5	**307:**	793	1358	4	1362	3	1927
3042.666	3	**0**	260	516	5	1265	1523	**1780**	3	1267	2	1011
3042.020	2	**0**	1018		2	951:	**2050**		2	1032	1	−14
3041.745					1	**1582**	B		4	1344	3	(1264)
3040.428	1	**0**	D		1	**1560**	B		5	1406	4	(1367)
3039.322	1	**0**	D		1	**950**	A		6	1035	5	(1052)
3037.388	1	**0**			1	**1516**			2	1517	1	(1518)
3033.101	1	**832**			2	**715**	**1519**		1	1526	1	708
3031.638	1	**0**	U		1	**0**	U		1	0	1	0
3031.213	1	**0**			1	**822**			4	(822)	4	822
3030.149	1	**0**			1	**1046**			5	(1052)	5	1040
3029.237	3	335	**698**	**1024**	6	766	**1107**	1436	3	1107	3	768
3025.843	1	**0**	**S**		1	**1510**	S		1	1510	0	0
3025.638	1	**0**			1	**1188**			6	(1178)	6	1198
3024.033	1	**0**			1	**1514**			2	1516	1	(1518)
3021.074	1	**0**			1	**1521**			3	(1517)	3	1525
3020.640	1	**0**			1	**1513**			4	(1516)	4	1510
3020.487	1	**0**			1	**1516**			2	(1514)	2	1518
3018.983	1	**0**			1	**1265**			3	(1264)	3	1266
3016.186	2	**0**	1018		3	0	1010	**2018**	2	1008	1	− 5
3015.913	3	m	238	m	6	1526	1729	**1954**	5	1049	4	822
3014.175	3	**0**	760	1513	3	**−224**	521	1274	3	1273	2	2022
3011.482	1	**0**	D		1	**974**	B		4	818	3	(766)
3009.570	1	**0**			1	**1364**			4	(1367)	4	1361
3009.098	3	m	243	377	3	m	1703:	**1802**	6	(1178)	5	1045
3008.139	1	**0**	S		1	**1518**	S		1	1518	0	0
3007.281	1	**0**			1	**1508**			2	(1514)	2	1502
3005.302	1	**0**	D		1	**1066**	A		7	1162	6	(1178)
3004.119	4	615	809	**1024**	10	**856**	**1053**	1235	5	1049	5	848
3003.031	3	**0**	253	534	4	1266	1518	**1757**	3	1267	2	1012

TABLE D—(*Continued*)

λ	π Components				σ Components				J_1	Obs. g_1	J_2	Obs. g_2
	No. Meas.	π			No. Meas.	σ						
3000.950	1	**0**			1	**1519**			2	(1514)	1	1509
2999.512	1	**0**			1	**1416**			5	1422	5	1410
2996.386	1	**0**			1	**1496**			2	(1526)	1	1511
2994.507	1	**0**	S		1	**1507**	S		1	1507	0	0
2994.427	1	**0**			1	**1530**			3	(1517)	2	1510
2990.392	1	**0**	D		1	**1095**	B		5	1071	4	(1065)
2988.468	3	402	**617**	**817**	8	**1072**	**1268**	1464:	4	1274	4	1071
2987.292	1	**0**	D		1	**1613**	B		4	(1367)	3	1285
2983.574	1	**0**	D		1	**1556**	B		4	(1516)	3	1503
2981.852	1	**0**	D		1	**958**	A		4	1506	3	(1688)
2981.446	1	**0**			1	**1520**			3	(1517)	2	1516
2980.532	2	m	322†	**459†**	6	**765†**	**910†**	**1055†**	3	910	3	(766)
2976.126	3	**0**	203	416	5	**928**	1122	1309	3	1325	2	1525
2973.237	2	**0**	266		4	**734**	992	1273	4	1255	3	1517
2970.106	2	**0**	514		3	**503**	1010	1516	2	1009	1	1519
2968.481	1	**245**			2	**1309†**	**1531†**		1	(1520)	1	1283
2966.901	1	**0**	D		1	1151	A		5	1443	4	(1516)
2965.255	1	U			1	U			1	0	0	0
2960.299					1	**1425**			1	1425	0	0
2959.992	1	0	D		1	**1069**	A		6	1189	5	(1213)
2957.365	1	**1542**			1	**0**	**1517**		1	1517	1	−25
2954.651	3	**0**	187	372	4	**979†**	1162†	1342†	3	1342	2	(1526)
2953.940	2	504	**1037**		4	**1004**	**1516**	2012:	2	1519	2	1009
2947.877	3	265	530	**782**	6	**1260**	**1515**	1773	3	1517	3	1258
2947.363	2	379	**756**		3	**1185**	**1547**	1902	2	1543	2	1172
2941.343	2	**0**	1531		3	24	1517	**3026**	2	1506	1	−14
2936.904	2		507	**625**	6	**1364**	**1514**	1662	4	1516	4	1361
2929.618	3	**0**	525†	993†	3	1167†	1658†	2129†	3	1179	2	(679)
2929.008	3	**0**	513	1028	4	1518	2025	**2528**	3	1518	2	1009
2925.899	2	562	**1144**		4	**668**	1225	1784	2	674	2	1228
2925.359	1	**0**	D		1	**1140**	B		4	860	3	(766)
2923.288	1	**0:**			1	**1046**			5	(1046)	5	1046
2920.691	1	**163**	B		1	**703**	C		2	744	2	662
2918.354	1	89	B		1	1474			1	1519	1	1430
2918.023	1	**0:**			1	**1178**			6	(1180)	6	1178
2914.305	1	0			1	**586**	A		2	(679)	1	772
2912.158	4	0	258	519	6	1775	2027	**2276**	4	1519	3	1267
2908.864					1	**1387†**	A		4	1675::	3	(1771)
2907.518	1	0	D		1	1020	A		2	(1844)	1	2668
2901.910	1	0			1	**1614**			5	(1621)	5	1607
2901.381	3	283	549	**810**	6	1098	**1361**	1624	3	1359	3	1095
2899.416	2	0	273		3	1252	1523	**1785**	2	1520	1	1250
2895.035	1	**206**	B		1	**1127**	C		3	1161	3	1093
2894.505	1	**0:**			1	**1514**			2	(1526)	2	1502
2893.882	2	0	217		3	298†	501†	713†	4	893	3	(1100)
2887.806	1	0	D		1	**1063**	A		6	1188	5	(1218)
2886.316	3	0	371	697	4	1091	1434	**1777**	3	1089	2	744
2877.300	1	**340**	B		1	**1334**	C		4	1376	4	1292
2874.172	5	0	288	578	6	93	379	667	5	1238	4	1525
2872.333	3	440	844	**1232**	5	**1264**	**1663**	2084	3	1674	3	1264
2869.308	3	0	402	805	6	69	316	729	4	1116	3	1508
2866.624	2	844	**1676**		3	168	**987**	**1813:**	2	1829	2	995
2858.896	2	0	1184		2	−958†	m	1635:	2	334	1	(1518)
2851.798	2	0	345		3	28	343	**679**	2	338	1	− 5
2851.52					3	880†	1073†	**1263†**	3	864	2	(672)

TABLE D—(*Continued*)

λ	π Components				σ Components				J_1	Obs. g_1	J_2	Obs. g_2
	No. Meas.	π			No. Meas.	σ						
2848.713	2	0	1510		3	−381	1018	2493	2	1009	1	2502
2846.830	1	0	D		1	1514	B		4	(1271)	3	1190
2845.595	3	0	599	1184	4	84	677	1264	3	1267	2	1859
2843.977	1	0	D		1	754	A		3	923	2	(1008)
2843.631	4	0	312	624	6	446	748	1060	4	1370	3	1680
2838.120	2	679	1363		3	334	1010	1681:	2	1011	2	333
2835.457	3	810†	1224†	1623†	6	1127†	1520†	1924†	4	(1516)	4	1115
2832.436	4	0	228	456	6	366	589	810	4	1039	3	1265
2828.808	3	0	486	981	5	−441:	0	441	3	511	2	1014
2827.892	3	0	409	819	5	m	m	721†	4	1115	3	(1517)
2825.557	4	0	387	777	6	−243	145	495	4	884	3	1261
2823.276	3	358	718	1077	6	911	1258	1608	3	1262	3	908
2820.801	3	0	1203	2329	5	1519	2687	3888	3	1520	2	342
2819.286	1	0	D		1	884	B		3	(766)	2	707
2817.505	3	0	953	1871	5	1266	2185	3104	3	1260	2	330
2815.506	1	0	D		1	433	A		3	597	2	(679)
2813.288	5	0	146	281	5	595	758	917	5	1219	4	1371
2808.328	3	764	1506	2234	6	535	1266	2011	3	1267	3	522
2806.984	5	0	294	588	8	−75	163:	497	5	1074	4	1361
2804.521	3	673	1002	1332	8	1039	1366	1694	4	1369	4	1039
2803.613	1	0:			1	812			4	(822)	4	802
2797.775	3	1385	1906	2439	5	880	1353	1834	4	1358	4	881
2796.871	2	0	178						3	(1100)	2	1278
2795.540	2	m	m	924	5	1818†	2313†	2694†	4	(1367)	3	920
2792.397	4	0	155	299	4	503	647	802	4	950	3	1100
2791.786	1	111	B		1	1060	C		5	1071	5	1049
2789.803					1	1078			5	(1213)	4	1247
2788.106					1	1048	A		6	1360	5	(1422)
2781.835	3	0	506	1003	5	1507	1992	2477	3	1501	2	1005
2778.221	5	617	820	995	10	1219	1414	1607	5	1416	5	1219
2774.730	2	0	1588		3	10	1562	3117	2	1554	1	−17
2772.113	2	m	166	348	5	1655	1839	2003	3	1666	2	1496
2772.083	4	1040	1399	1732	7	740	1089:	1471:	5	1417	5	1063
2769.670	4	m	399†	805†	6	2185†	2567†	2960†	5	1422	4	1024
2769.297	1	0:			1	1178			6	(1178)	6	1178
2766.909	2	0	1173		3	50	1146	2266	2	1133	1	−13
2764.323	3	0	342	684	4	782	1153	1466	2	1837	2	1495
2763.108	3	0	364	730	5	1376	1713	2064	3	1364	2	1006
2762.027	2	m	496	730	6	1262	1505	1734	3	1494	3	1253
2761.780	2	602	1124		4	997	1553	2116	2	1565	2	1002
2759.814	1	301			2	24	287		1	287	1	− 7
2754.427	4	0	202	400	3	1468	1670	1870	4	1469	3	1269
2754.030	1	244	B						2	(1008)	2	1130
2750.140	1	529	B		2	m	m	1977†	3	1686	3	(1517)
2744.526					1	692	A		2	(1008)	1	1324
2743.564	1	313	B		1	1300	C		3	1368	3	(1264)
2742.406	2	351	707		4	1520	m	2211	2	1869	2	1521
2742.256					3	691†	979†	1267†	3	(1264)	2	1552
2737.310	1	1061			2	1520	2535		1	2547	1	1509
2735.475	1	m	127		1	979	A		4	(1367)	3	1495
2734.002	2	0	400		4	1384†	1782†	2190†	3	1410	2	(1008)
2733.581	1	0	D		1	1097	A		5	(1422)	4	(1503)
2728.819	1	180	B		1	863	C		4	885	4	841
2728.020	1	348	B		1	m	1456†		4	1454	4	(1367)
2723.577	2	0	1017		3	505	1515	2528	2	1516	1	2530

TABLE D—(*Continued*)

λ	π Components				σ Comopnents				J_1	Obs. g_1	J_2	Obs. g_2
	No. Meas.	π			No. Meas.	σ						
2720.902	1	**m**	m	709	5	**798**	1220	1525	3	1520	2	1867
2719.418	1	**0**			1	**1046**			5	(1052)	5	1040
2718.435	2	0	399		2	690†	m	**1480†**	3	1407:	2	(1008)
2717.786	3	**0**	669	1389:	5	**0**	626	**1286**	3	(1264)	2	1929:
2714.868	1	**421**	B		1	**1299:†**			3	1404	3	(1264)
2711.655	1	**0**	D		1	**1514**	B		5	1396	4	(1367)
2710.543	1	**0**	D		1	**936**	B		3	765	2	(679)
2708.570	1	**0:**			1	**1253**			4	(1251)	4	1255
2706.581	2	**0**	158		1	**1000**	A		3	(1264)	2	1396
2706.012	1	**0:**			1	**1178**			6	(1178)	6	1178
2699.107	1	**148**	B		1	**1390**	C		4	1372	4	1408
2697.019	3	**0**	170	322	4	**435**	**577**	750	4	926	3	1093
2695.032	1	**0**	D		1	**1302**	A		5	(1422)	4	1452
2690.067	4	**0**	423	**868**	5	1943	2329	**2773**	4	1508	3	1086
2689.827	2	**0**	142	m	3	**445†**	**580†**	**761†**	4	942	3	(1100)
2689.212	1	**0**	D		1	**1294**	A		4	(1367)	3	1391
2680.452	2	201	**416**		4	**1016**	**1215**	1412	2	1214	2	1011
2679.062	1	**0:**			1	**1413**			5	(1422)	5	1404
2673.213	1	**580**	B		1	**0**	**564**		1	564	1	−16
2669.492	1	**0**	D		1	**1006**	A		6	1045	5	(1052)
2667.912	3	**0**	175	376	3	**1002**	1163	1322	3	1322	2	1482
2666.398	1	**331**	B		1	**1294**	C		3	1349	3	1239
2662.056	1	**0**	D		1	**1330**	B		3	(1264)	2	1231
2661.196	2	**0**	456		3	547	1007	**1465**	2	1006	1	548
2660.396	3	**0**	253	507	5	**267**	515	**767**	3	767	2	1019
2656.792	5	**0**	179	368	5	**388**	554	735	5	1099	4	1279
2656.145	1	**0**	D		1	**1069**	A		7	1162	6	(1178)
2651.706	2	**m**	409	597	3	**467**	668	m	4	1070	3	1271
2647.558	1			**534**	6	**1344**	**1518**	1684	3	1517	3	1348
2645.422	1	**0**	D		1	**820**	A		2	1169	1	(1518)
2643.997	2	**0**	353		3	0	350	690	2	343	1	−10
2641.645	1	**0:**			1	**1352**			4	(1368)	3	1372
2636.477	3	**0**	m	296	4	**641**	770	m	5	1227	4	1376
2635.808	1	**0**	D		1	**804**	A		3	940	2	(1008)
2632.593	2	331	**726**		3	m†	**1489†**	1861†	2	(1514)	2	1167
2632.238	2	698	**1373**		4	**330†**	**990†**	1694†	2	(1008)	2	325
2623.532	1	**0**	D		1	**901**	A		4	1173	3	(1264)
2618.708	3	**0**	180	338	1	1716†	m	m	4	(1516)	3	1346
2618.018	3		528	**856**	5	**924**	**1275**	1589	3	1269	3	937
2614.494	1	m	768†		3	1107†	2045†	3009†	3	(1264)	2	313
2612.771	3	**0**	351	722					3	(1517)	2	1156
2610.750	2	**0**	1027		3	431†	1522†	**2538†**	2	(1514)	1	492
2606.826					1	**985**	A		5	1291	4	(1367)
2594.150	4	**0**	444	**876**	3	1809†	2226†	**2632†**	4	(1367)	3	942
2584.536	1	**0**	D		1	**1013**	A		6	1354	5	(1422)
2580.062	1	**1653**			1	**0**	**1640**		1	1640	1	−13
2576.688	1	680	B		3	996†	**1144†**	m	5	(1422)	5	1280
2564.555	2	**0**	1270		3	0	1217†	**2454†**	2	1241	1	(−14)
2561.262	2	**0**	369						3	1377	2	(1008)
2560.556	2	**0**	773†		2		750	**1519**	2	752	1	−14
2556.298					2	1075			4	1094	3	(1100)
2552.827	1	**0**	D		1	**1266**			3	(1264)	2	1263
2549.612	1	**0**	D		1	**1378**	A		4	1482	3	(1517)
2545.977	1	**0**	D		1	**1527**	B		3	1518	2	(1514)
2544.706	1	**0**	D		1	**1028**	A		5	1206	4	(1251)

TABLE D—(*Continued*)

λ	π Components				σ Components				J_1	Obs. g_1	J_2	Obs. g_2
	No. Meas.	π			No. Meas.	σ						
2543.920	1	0	D		1	913	A		4	1043	3	(1087)
2542.101	1	0	D		1	1026	B		3	790	2	(672)
2540.971					1	1521			2	1520	1	(1518)
2535.604	1	0	S		1	1514	S		1	1514	0	0
2530.694	3	0	418	845	3	1926†	2340†	2775†	3	1936	2	(1514)
2527.433	1	0			1	1519			3	(1517)	3	1521
2524.290	1	0	S		1	1515	S		1	1515	0	0
2522.848	1	0			1	1512			4	(1516)	4	1508
2519.628	2	0	472		3	0	468	937	2	467	1	− 3
2518.100	1	0	D		1	1516			2	(1514)	1	1512
2516.569	4	0	236	471	5	268:	532	797	4	1025	3	(1264)
2510.833	1	0	D		1	1527†	B		3	(1517)	2	1512
2508.751	3	0	337	648	3	0	352	694	3	670	2	(1008)
2507.899	1	0	D		1	902	A		4	1173	3	(1264)
2501.130	1	0	D		1	1508			4	(1516)	3	1519
2496.532	1	0	D		1	1169	A		5	1327	4	(1367)
2493.998	2	0	1320		3	0	1297	2582	2	1287	1	−18
2490.642	1	m	m	512	4	608	896	1164	3	1258	2	(1514)
2487.368	2	508	1030						2	(1514)	2	2011
2486.690	1	0	D		1	1294	B		4	1272	3	(1264)
2486.372	1		421		1	1207			4	(1516)	3	1937
2485.989	3	383†	575†	816†	5	1141†	1320†	1533†	4	(1367)	4	1161
2479.775	2	534	975		4	1013	1507	2005	2	1508	2	998
2476.654	2	0	797		3	1007	1780	2533	2	1001	1	216
2473.156	2	m	694†	947†	1	1342:†			4	(1516)	4	1769
2468.878	1	473	B		1	1381†	C		5	(1422)	5	1327
2467.730	1	0	D		1	1247	A		3	(1264)	2	1273
2463.728	1	0	D		1	1234	A		3	(1264)	2	1279
2462.645	1	588	B		1	1405	C		4	1479	4	1331
2462.178	2	0	518		2	1499†	m	m	3	1506	2	1003
2457.596	1	375	B		1	1372	C		5	1410	5	1334
2453.475	1	0	D		1	1666	B		4	(1367)	3	1267
2443.871	5	659	849	1041	9	1208:	1409:	1647	5	1435	5	1221
2442.567	1	0			1	1057			5	(1052)	5	1062
2440.106	1	0			1	819			4	(822)	4	816
2439.743	1	0			1	1178			6	(1178)	6	1178
2438.181	1	0	D		1	1940	B		5	(1422)	4	1292
2389.971					1	1938			3	1655	2	(1514)
2371.428	2	373	689		4	1517	1851	2140	2	1843	2	1505
2369.454	1	988			1	1974::			1	(1518)	1	2506
2320.356	1	0	D		1	1499			4	1512	3	(1517)
2313.102	1	0			1	1491			3	1506	2	(1514)
2308.997	1	0	D		1	1579	B		2	1548	1	(1518)
2300.140	1	0	D		1	1117	A		3	1382	2	(1514)
2299.218	1	0			1	1531			2	(1514)	2	1548
2298.175	1	0			1	1518			4	(1516)	4	1520
2297.785	1	0			1	1504			3	(1517)	3	1491
2294.406	1	0	S		1	1523	S		1	1523	0	0
2292.523	1	0	D		1	1352	A		4	1476	3	(1517)
2287.248	1	0	D		1	1690	B		2	(1514)	1	1338
2284.087	1	0	D		1	1473	A		3	(1517)	2	1539
2276.025	1	0	D		1	1540	B		4	(1516)	3	1508
2272.067					1	1169			4	1430	3	(1517)

TABLE E

OBSERVED AND CORRECTED g-VALUES

Desig	Observed g	R	U	A.D.	Corrected g	Res
a^1P_1	0.828	5	1	±6	0.817	−.183
a^1D_2	1.041	1	2	7	1.028	+.028
a^1G_4	1.014	6	7	6	1.001	+.001
b^1G_4	0.992		1		0.979	−.021
a^1H_5	1.013	3	3	4	1.000	.000
a^1I_6	1.027	1	1	3	1.014	+.014
a^3P_2	1.526	12	9	6	1.506	+.006
a^3P_1	1.520	12	4	4	1.500	.000
b^3P_2	1.518	12	5	8	1.498	−.002
b^3P_1	1.509	16	4	4	1.489	−.011
c^3P_2	1.504	5	4	6	1.484	−.016
c^3P_1	1.485	6	3	8	1.466	−.034
e^3P_2	1.478	1	1	2	1.459	−.041
e^3P_1	1.478		1		1.459	−.041
a^3D_3	1.353	6	2	6	1.335	+.002
a^3D_2	1.194	2	3	4	1.178	+.011
a^3D_1	0.741	4	3	7	0.731	+.231
b^3D_3	1.343		1		1.326	−.007
e^3D_3	1.363		1		1.345	+.012
e^3D_2	1.140		2	16	1.125	−.042
e^3D_1	0.812	1	1	14	0.801	+.301
f^3D_3	1.275	1			1.258	−.075
a^3F_4	1.271	11	8	6	1.254	+.004
a^3F_3	1.100	16	11	6	1.086	+.003
a^3F_2	0.679	10	8	7	0.670	+.003
b^3F_4	1.251	4	11	4	1.235	−.015
b^3F_3	1.089	4	8	6	1.073	−.010
b^3F_2	0.672	4	6	6	0.663	−.004
c^3F_4	1.281	1	1	6	1.264	+.014
c^3F_3	1.080	2		1	1.066	−.017
c^3F_2	0.686		2	6	0.677	+.010
e^3F_4	1.305		2	0	1.288	+.038
e^3F_3	1.122		2	8	1.107	+.024
e^3F_2	0.630		2	10	0.622	−.045
f^3F_4	1.156		1		1.141	−.109
f^3F_3	1.085		2	7	1.071	−.012
f^3F_2	0.685		1		0.676	+.009
a^3G_5	1.213	11	15	6	1.197	−.003
a^3G_4	1.068	16	10	5	1.051	+.001
a^3G_3	0.766	7	11	6	0.756	+.006
b^3G_5	1.216	3	5	9	1.200	.000
b^3G_4	1.062	5	5	8	1.048	−.002
b^3G_3	0.771	2	4	4	0.761	+.011

Desig	Observed g	R	U	A.D.	Corrected g	Res
e^3G_5	1.264		1		1.248	+.048
e^3G_4	1.110		2	8	1.096	+.046
e^3G_3	0.853		1		0.842	+.092
a^3H_6	1.178	2	12	5	1.163	−.004
a^3H_5	1.052	8	12	8	1.038	+.005
a^3H_4	0.822	5	9	4	0.811	+.011
b^3H_6	1.180		7	3	1.165	−.002
b^3H_5	1.046		6	4	1.032	−.001
b^3H_4	0.822	1	7	4	0.811	+.011
e^3H_6	1.241		1		1.225	+.058
e^3H_5	1.124		1		1.109	+.076
e^3H_4	0.882		1		0.871	+.071
e^5S_2	1.978		1		1.952	−.048
a^5P_3	1.688	23	6	5	1.666	−.001
a^5P_2	1.844	32	7	5	1.820	−.013
a^5P_1	2.526	26	2	5	2.499	−.001
e^5P_3	1.686	1			1.664	−.003
e^5P_1	2.464		2	4	2.432	−.068
a^5D_4	1.516	3	18	5	1.496	−.004
a^5D_3	1.517	13	19	6	1.497	−.003
a^5D_2	1.514	17	19	7	1.494	−.006
a^5D_1	1.518	10	10	5	1.498	−.002
e^5D_4	1.522	2	3	13	1.502	+.002
e^5D_3	1.528	3	3	6	1.508	+.008
e^5D_2	1.523	4	4	5	1.503	+.003
e^5D_1	1.538	5		7	1.518	+.018
f^5D_4	1.534	1	3	12	1.514	+.014
f^5D_3	1.636		2	6	1.615	+.115
f^5D_2	1.635	1	1	3	1.614	+.114
f^5D_1	1.684		1		1.662	+.162
g^5D_4	1.507	1	5	11	1.487	−.013
g^5D_3	1.512		2	3	1.492	−.008
g^5D_2	1.59:		2	34	1.57:	+.07:
h^5D_4	1.454		1		1.435	−.065
i^5D_4	1.402		3	17	1.384	−.116
i^5D_3	1.433		2	3	1.415	−.085
a^5F_5	1.422	6	16	5	1.404	+.004
a^5F_4	1.367	18	20	5	1.349	−.001
a^5F_3	1.264	30	21	6	1.248	−.002
a^5F_2	1.008	33	8	5	0.995	−.005
a^5F_1	−0.014	21		6	−0.014	−.014
e^5F_5	1.440		2	10	1.421	+.021
e^5F_4	1.348	1			1.331	−.019
e^5F_3	1.252	1	1	2	1.236	−.014
e^5F_2	1.004		1		0.991	−.009
e^5F_1	0.007	1			0.007	+.007

TABLE E—(*Continued*)

Desig	Observed g	R	U	A.D.	Corrected g	Res
f^5F_5	1.402		1		1.384	−.016
f^5F_4	1.373		1		1.355	+.005
f^5F_2	0.980		1		0.967	−.033
e^5G_6	1.369		1		1.351	+.018
e^5G_5	1.378		3	3	1.360	+.093
e^5G_4	1.254		1		1.238	+.088
e^5G_3	1.311	1	2	8	1.294	+.377
e^5G_2	0.970	1			0.953	+.620
f^5G_6	1.340	1			1.323	−.010
f^5G_5	1.237		1		1.221	−.046
f^5G_3	1.157		1		1.142	+.225
g^5G_6	1.42:		1		1.40:	+.07:
g^5G_2	0.348		1		0.343	+.010
e^5H_7	1.32:		1		1.30:	+.01:
e^5H_6	1.207		1		1.191	−.023
e^5H_5	1.117		1		1.102	+.002
e^5H_4	0.91:		1		0.90:	.00:
e^5H_3	0.490		1		0.484	−.016
e^7S_3	1.94:	1	1	24	1.92:	−.08:
e^7P_4	1.606	1	2	6	1.585	−.165
e^7P_3	1.709		3	16	1.687	−.230
e^7D_5	1.606	1	3	4	1.585	−.015
e^7D_4	1.677	2	4	11	1.655	+.005
e^7D_3	1.778	3	4	10	1.755	+.005
e^7D_2	2.035	5		6	2.009	+.009
e^7D_1	3.041	3	1	13	3.002	+.002
f^7D_5	1.530	1	1	12	1.510	−.090
f^7D_4	1.595	2	3	12	1.574	−.076
f^7D_2	1.868	1	2	14	1.844	−.156
g^7D_5	1.607		1		1.586	−.014
g^7D_4	1.68:		1		1.65:	.00:
e^7F_6	1.510		3	7	1.490	−.010
e^7F_5	1.525	1	4	6	1.505	+.005
e^7F_4	1.638	2	2	10	1.617	+.117
e^7F_3	1.519		1		1.499	−.001
e^7F_1	2.521	1			2.490	+.990
e^7G_6	1.434		3	4	1.415	+.010
e^7G_5	1.397	1	2	14	1.379	+.012
e^7G_4	1.356	2	1	18	1.338	+.038
e^7G_3	1.262		1		1.244	+.077
e^7G_1	−0.378	1			−0.374	+.126
$z^1P_1^\circ$	1.283	2	1	4	1.266	+.266
$z^1D_2^\circ$	0.93:		1		0.92:	−.08:
$y^1D_2^\circ$	1.038	3		2	1.025	+.025
$x^1D_2^\circ$	0.895	1			0.883	−.117
$w^1D_2^\circ$	1.003	2		4	0.990	−.010
$z^1F_3^\circ$	1.031		3	7	1.018	+.018
$y^1F_3^\circ$	1.22:		1		1.21:	+.21:

Desig	Observed g	R	U	A.D.	Corrected g	Res
$x^1F_3^\circ$	1.093		1		1.079	+.079
$w^1F_3^\circ$	0.920		4	5	0.908	−.092
$z^1G_4^\circ$	1.038	1	4	8	1.025	+.025
$y^1G_4^\circ$	1.077	1	2	3	1.063	+.063
$x^1G_4^\circ$	0.991		3	3	0.978	−.022
$w^1G_4^\circ$	1.014	1	1	2	1.001	+.001
$v^1G_4^\circ$	1.067		1		1.053	+.053
$z^1H_5^\circ$	1.031		3	4	1.018	+.018
$y^1H_5^\circ$	1.04:		2	30	1.03:	+.03:
$x^1H_5^\circ$	1.031	1			1.018	+.018
$z^1I_6^\circ$	1.023		1		1.010	+.010
$z^3S_1^\circ$	1.913	3	1	11	1.888	−.112
$y^3S_1^\circ$	1.909	5	1	7	1.884	−.116
$z^3P_2^\circ$	1.513	3	4	4	1.493	−.007
$z^3P_1^\circ$	1.516	2	1	6	1.496	−.004
$y^3P_2^\circ$	1.463	2	3	11	1.444	−.056
$y^3P_1^\circ$	1.621	2	3	10	1.600	+.100
$x^3P_2^\circ$	1.280	5	2	6	1.263	−.237
$x^3P_1^\circ$	1.567	7	1	8	1.547	+.047
$w^3P_2^\circ$	1.488	5		4	1.469	−.031
$w^3P_1^\circ$	1.407	3		2	1.389	−.111
$v^3P_2^\circ$	1.505	1	6	5	1.495	−.005
$v^3P_1^\circ$	1.437	2	1	16	1.418	−.082
$z^3D_3^\circ$	1.338	1	3	5	1.321	−.012
$z^3D_2^\circ$	1.183	1			1.168	+.001
$z^3D_1^\circ$	0.520	3		6	0.513	+.013
$y^3D_3^\circ$	1.341	5	3	6	1.324	−.009
$y^3D_2^\circ$	1.166	2	4	8	1.151	−.016
$y^3D_1^\circ$	0.504	4	1	6	0.493	−.007
$x^3D_3^\circ$	1.370	6	2	6	1.352	+.019
$x^3D_2^\circ$	1.216	6	3	7	1.200	+.033
$x^3D_1^\circ$	0.563	6		8	0.556	+.056
$w^3D_3^\circ$	1.364	4	2	4	1.346	+.013
$w^3D_2^\circ$	1.232	6	1	5	1.216	+.049
$w^3D_1^\circ$	0.777	4	1	7	0.767	+.267
$v^3D_3^\circ$	1.227	2	2	5	1.211	−.122
$v^3D_2^\circ$	0.967		1		0.954	−.213
$v^3D_1^\circ$	0.569		1		0.562	+.062
$u^3D_3^\circ$	1.323	3	1	4	1.306	−.027
$u^3D_2^\circ$	1.171	4	2	5	1.156	−.011
$u^3D_1^\circ$	0.709	4	3	9	0.700	+.200

TABLE E—(*Continued*)

Desig	Observed g	R	U	A.D.	Corrected g	Res
$t^3D_3^\circ$	1.334		4	7	1.317	−.016
$t^3D_2^\circ$	1.160	1	2	2	1.145	−.022
$t^3D_1^\circ$	0.812	3		5	0.801	+.301
$s^3D_3^\circ$	1.247		1		1.231	−.102
$z^3F_4^\circ$	1.266	2	1	7	1.250	.000
$z^3F_3^\circ$	1.100	2	2	11	1.086	+.003
$z^3F_2^\circ$	0.691	2	1	9	0.682	+.015
$y^3F_4^\circ$	1.262	3	2	6	1.246	−.004
$y^3F_3^\circ$	1.100	4	2	8	1.086	+.003
$y^3F_2^\circ$	0.697	1	2	13	0.688	+.021
$x^3F_4^\circ$	1.362	3	1	4	1.344	+.094
$x^3F_3^\circ$	1.174	5	2	13	1.159	+.076
$x^3F_2^\circ$	0.753	5	3	7	0.743	+.076
$w^3F_4^\circ$	1.197	3	3	3	1.181	−.069
$w^3F_3^\circ$	1.180	1			1.165	+.082
$w^3F_2^\circ$	0.686	1	2	6	0.677	+.010
$v^3F_4^\circ$	1.137	2	3	9	1.122	−.128
$v^3F_3^\circ$	1.110		1		1.096	+.013
$v^3F_2^\circ$	0.814		1		0.803	+.136
$u^3F_4^\circ$	1.163	2	1	6	1.148	−.102
$u^3F_3^\circ$	1.091		2	9	1.077	−.006
$u^3F_2^\circ$	0.696	1	1	4	0.687	+.020
$t^3F_4^\circ$	1.251		2	4	1.235	−.015
$t^3F_2^\circ$	0.707		1		0.698	+.031
$z^3G_5^\circ$	1.264	1	3	3	1.248	+.048
$z^3G_4^\circ$	1.114	4	3	4	1.100	+.050
$z^3G_3^\circ$	0.801	5	1	4	0.791	+.041
$y^3G_5^\circ$	1.223	2	4	7	1.207	+.007
$y^3G_4^\circ$	1.067	2	3	3	1.053	+.003
$y^3G_3^\circ$	0.775	2	2	8	0.765	+.015
$x^3G_5^\circ$	1.219	2	2	2	1.203	+.003
$x^3G_4^\circ$	1.075		2	2	1.061	+.011
$x^3G_3^\circ$	0.677	1	1	7	0.668	−.082
$w^3G_5^\circ$	1.29:		1		1.27:	+.07:
$w^3G_4^\circ$	0.946	3		4	0.934	−.116
$w^3G_3^\circ$	0.592	1	1	6	0.584	−.166
$v^3G_5^\circ$	1.178	4	3	7	1.163	−.037
$v^3G_4^\circ$	0.926	7	2	5	0.914	−.136
$v^3G_3^\circ$	0.773	1	4	5	0.763	+.013
$u^3G_5^\circ$	1.155	2	2	8	1.140	−.060
$u^3G_4^\circ$	1.081	2	3	8	1.067	+.017
$u^3G_3^\circ$	0.811	2	3	6	0.801	+.051
$t^3G_5^\circ$	1.250	1	1	9	1.234	+.034
$t^3G_4^\circ$	1.198	1	1	1	1.183	+.133
$t^3G_3^\circ$	0.934	1	1	10	0.922	+.172
$s^3G_5^\circ$	1.160		2	2	1.145	−.055
$s^3G_3^\circ$	0.868	1	2	3	0.857	+.107
$r^3G_5^\circ$	1.206		1		1.190	−.010
$r^3G_4^\circ$	1.043		1		1.030	−.020
$r^3G_3^\circ$	0.790		1		0.780	+.030

Desig	Observed g	R	U	A.D.	Corrected g	Res
$z^3H_6^\circ$	1.216	2	2	3	1.200	+.033
$z^3H_5^\circ$	1.074	2	3	3	1.060	+.027
$z^3H_4^\circ$	0.892	4	2	5	0.880	+.080
$y^3H_6^\circ$	1.18:		2	20	1.17:	.00:
$y^3H_5^\circ$	1.089	4	2	8	1.075	+.042
$y^3H_4^\circ$	0.941	3	2	3	0.929	+.129
$x^3H_6^\circ$	1.176		2	14	1.161	−.006
$x^3H_5^\circ$	1.062		2	10	1.038	+.005
$w^3H_6^\circ$	1.192		3	4	1.177	+.010
$w^3H_5^\circ$	1.047		4	4	1.033	.000
$w^3H_4^\circ$	0.821	1	3	2	0.810	+.010
$v^3H_6^\circ$	1.184		3	4	1.169	+.002
$v^3H_5^\circ$	1.071		2	0	1.057	+.024
$v^3H_4^\circ$	0.815		3	9	0.804	+.004
$u^3H_6^\circ$	1.181		3	4	1.166	−.001
$u^3H_5^\circ$	1.043		2	3	1.029	−.004
$u^3H_4^\circ$	0.870		2	10	0.859	+.059
$t^3H_6^\circ$	1.178		2	0	1.163	−.004
$t^3H_5^\circ$	1.054		2	8	1.040	+.007
$t^3H_4^\circ$	0.816		1		0.805	+.005
$z^3I_7^\circ$	1.164		1		1.149	+.006
$z^3I_6^\circ$	1.054		1		1.040	+.016
$z^3I_5^\circ$	0.844		1		0.833	.000
$y^3I_7^\circ$	1.162		2	0	1.147	+.004
$y^3I_6^\circ$	1.032	1	3	5	1.019	−.005
$y^3I_5^\circ$	0.841	1	1	2	0.830	−.003
$x^3I_7^\circ$	1.160		2	2	1.145	+.002
$x^3I_6^\circ$	1.042		2	2	1.028	+.004
$x^3I_5^\circ$	0.843		1		0.832	−.001
$z^5S_2^\circ$	2.011	4	1	6	1.985	−.015
$y^5S_2^\circ$	1.913	2	2	11	1.888	−.112
$z^5P_3^\circ$	1.679	6	1	5	1.657	−.010
$z^5P_2^\circ$	1.859	6		7	1.835	+.002
$z^5P_1^\circ$	2.520	6		5	2.487	−.013
$y^5P_3^\circ$	1.683	3	2	8	1.661	−.006
$y^5P_2^\circ$	1.860	5		10	1.836	+.003
$y^5P_1^\circ$	2.535	3	1	6	2.502	+.002
$x^5P_3^\circ$	1.672	2	2	9	1.650	−.017
$x^5P_2^\circ$	1.846	4	2	7	1.822	−.011
$x^5P_1^\circ$	2.496	1	2	10	2.464	−.036
$w^5P_3^\circ$	1.680	1	1	6	1.658	−.009
$w^5P_2^\circ$	1.846		1		1.822	−.011
$w^5P_1^\circ$	2.468	2		2	2.436	−.064
$v^5P_3^\circ$	1.668	1	1	1	1.646	−.021
$v^5P_2^\circ$	1.763	3	3	7	1.740	−.093
$v^5P_1^\circ$	2.242	4		10	2.213	−.287
$u^5P_1^\circ$	2.668		1		2.633	+.133
$t^5P_2^\circ$	1.72:		1		1.70:	−.13:

TABLE E—(*Continued*)

Desig	Observed g	R	U	A.D.	Corrected g	Res
$z^5D_4^\circ$	1.522	1	6	9	1.502	+.002
$z^5D_3^\circ$	1.520	2	6	6	1.500	.000
$z^5D_2^\circ$	1.523	3	5	5	1.503	+.003
$z^5D_1^\circ$	1.515	3	2	7	1.495	−.005
$y^5D_4^\circ$	1.516	3	2	3	1.496	−.004
$y^5D_3^\circ$	1.512	5	3	5	1.492	−.008
$y^5D_2^\circ$	1.515	5	3	4	1.495	−.005
$y^5D_1^\circ$	1.512	4	2	3	1.492	−.008
$x^5D_4^\circ$	1.509	3	3	9	1.489	−.011
$x^5D_3^\circ$	1.524	5	4	5	1.504	+.004
$x^5D_2^\circ$	1.521	6	2	5	1.501	+.001
$x^5D_1^\circ$	1.518	4	2	4	1.498	−.002
$w^5D_4^\circ$	1.512	1	3	4	1.492	−.008
$w^5D_3^\circ$	1.501	3	7	6	1.481	−.019
$w^5D_2^\circ$	1.553	5	3	6	1.533	+.033
$w^5D_1^\circ$	1.332	2	1	5	1.315	−.185
$v^5D_4^\circ$	1.419	1	1	11	1.401	−.099
$v^5D_3^\circ$	1.404		3	4	1.386	−.114
$v^5D_2^\circ$	1.396	1			1.378	−.122
$v^5D_1^\circ$	1.407		1		1.389	−.111
$u^5D_4^\circ$	1.359	5		8	1.341	−.159
$u^5D_3^\circ$	1.415	4		4	1.397	−.103
$u^5D_2^\circ$	1.277	6	1	7	1.260	−.240
$u^5D_1^\circ$	1.428	2	1	4	1.410	−.090
$t^5D_4^\circ$	1.506		1		1.486	−.014
$z^5F_5^\circ$	1.417	2	2	6	1.399	−.001
$z^5F_4^\circ$	1.373	3	3	5	1.355	+.005
$z^5F_3^\circ$	1.266	4	4	6	1.250	.000
$z^5F_2^\circ$	1.017	6	1	7	1.004	+.004
$z^5F_1^\circ$	−0.012	2	2	6	−0.012	−.012
$y^5F_5^\circ$	1.435	2	1	10	1.417	+.017
$y^5F_4^\circ$	1.362	2	4	3	1.344	−.006
$y^5F_3^\circ$	1.260	5	2	4	1.244	−.006
$y^5F_2^\circ$	1.011	5	1	2	0.998	−.002
$y^5F_1^\circ$	−0.016	3		6	−0.016	−.016
$x^5F_5^\circ$	1.408		2	2	1.390	−.010
$x^5F_4^\circ$	1.345	1	2	10	1.328	−.022
$x^5F_3^\circ$	1.270	1	4	8	1.254	+.004
$x^5F_2^\circ$	1.011	4		11	0.998	−.002
$x^5F_1^\circ$	−0.006	2		1	−0.006	−.006
$w^5F_5^\circ$	1.400		2	4	1.382	−.018
$w^5F_4^\circ$	1.463	2	3	8	1.444	+.094
$w^5F_3^\circ$	1.369	3	2	6	1.351	+.101
$w^5F_2^\circ$	1.132	1	1	2	1.117	+.117
$w^5F_1^\circ$	0.287	1			0.283	+.283
$v^5F_5^\circ$	1.334	1			1.317	−.083
$v^5F_4^\circ$	1.281	1	2	7	1.264	−.086
$v^5F_3^\circ$	1.252	1	2	10	1.236	−.014
$v^5F_2^\circ$	1.284	7	1	6	1.267	+.267
$v^5F_1^\circ$	0.233	2		16	0.230	+.230

Desig	Observed g	R	U	A.D.	Corrected g	Res
$z^5G_6^\circ$	1.350		1		1.332	−.001
$z^5G_5^\circ$	1.234	2	3	8	1.218	−.049
$z^5G_4^\circ$	1.117	5	2	3	1.103	−.047
$z^5G_3^\circ$	0.899	5	1	8	0.887	−.030
$z^5G_2^\circ$	0.339	4	1	5	0.335	+.002
$y^5G_6^\circ$	1.360		1		1.342	+.009
$y^5G_5^\circ$	1.219	2		0	1.203	−.064
$y^5G_4^\circ$	1.037	2	1	7	1.024	−.126
$y^5G_3^\circ$	0.917	1	2	6	0.905	−.012
$y^5G_2^\circ$	0.335	3		4	0.331	−.002
$x^5G_6^\circ$	1.354		1		1.336	+.003
$x^5G_5^\circ$	1.286		2	6	1.269	+.002
$x^5G_4^\circ$	1.173		1		1.158	+.008
$x^5G_3^\circ$	0.940	1	2	2	0.928	+.011
$x^5G_2^\circ$	0.327	1	2	11	0.323	−.010
$w^5G_6^\circ$	1.323		1		1.306	−.027
$w^5G_5^\circ$	1.322	1	2	6	1.305	+.038
$w^5G_4^\circ$	1.160	1	2	10	1.145	−.005
$w^5G_3^\circ$	0.943	1			0.931	+.014
$w^5G_2^\circ$	0.478	2		11	0.472	+.139
$z^5H_5^\circ$	1.068	2		6	1.054	−.046
$z^5H_4^\circ$	0.882	2		2	0.871	−.029
$z^5H_3^\circ$	0.516	2		5	0.509	+.009
$z^7P_4^\circ$	1.770	3	2	8	1.747	−.003
$z^7P_3^\circ$	1.933	4	1	2	1.908	−.009
$z^7P_2^\circ$	2.364	5		9	2.333	.000
$y^7P_4^\circ$	1.77:		1		1.75:	.00:
$y^7P_3^\circ$	1.933	1	2	4	1.908	−.009
$y^7P_2^\circ$	2.371		1		2.340	+.007
$z^7D_5^\circ$	1.618	1	5	9	1.597	−.003
$z^7D_4^\circ$	1.664		9	10	1.642	−.008
$z^7D_3^\circ$	1.769	2	5	15	1.746	−.004
$z^7D_2^\circ$	2.034	1	3	7	2.008	+.008
$z^7D_1^\circ$	3.038	3	1	11	2.999	−.001
$z^7F_6^\circ$	1.518		2	8	1.498	−.002
$z^7F_5^\circ$	1.518		4	2	1.498	−.002
$z^7F_4^\circ$	1.513	2	4	7	1.493	−.007
$z^7F_3^\circ$	1.533	1	1	0	1.513	+.013
$z^7F_2^\circ$	1.524	2	1	9	1.504	+.004
$z^7F_1^\circ$	1.569	2		4	1.549	+.049
1_2°	1.152	5	1	6	1.137	
4_4°	0.966	1			0.953	
6_5°	1.075	1	4	8	1.061	
8_1°	1.262	4	3	8	1.246	
10_3°	1.495	1			1.476	
12_5°	1.374	1	1	6	1.356	

INDEX

TRANSACTIONS

OF THE

AMERICAN PHILOSOPHICAL SOCIETY

HELD AT PHILADELPHIA
FOR PROMOTING USEFUL KNOWLEDGE

NEW SERIES—VOLUME XXXIV, PART III

THE MAMMALIA OF THE DUCHESNE RIVER OLIGOCENE

WILLIAM B. SCOTT
Blair Professor Emeritus of Geology
Princeton University

THE AMERICAN PHILOSOPHICAL SOCIETY
INDEPENDENCE SQUARE
PHILADELPHIA 6

March, 1945

LANCASTER PRESS, INC., LANCASTER, PA.

THE MAMMALIA OF THE DUCHESNE RIVER OLIGOCENE

William Berryman Scott

STRATIGRAPHIC INTRODUCTION

The succession of sandstones and clays which Peterson and Kay named *Duchesne River* lies, in apparent conformity, upon the upper Uinta (Myton substage = Horizon C), though with a marked change in color and in the nature of the contained fossils. The northern border of the formation displays an unconformity by overlap, the Duchesne River beds extending over upon older rocks, Triassic and Jurassic, with very strongly marked angular unconformity. It had always been taken for granted, because of complete ignorance of their fossils, that these strata formed a part of the Uinta Eocene, for they are nearly barren and fossils are exceedingly rare. Almost all our knowledge of the Duchesne River mammals is due to the labors of the Carnegie Museum staff; with a few more or less doubtful exceptions, all the known material is preserved in that museum. Messrs. Peterson and Kay and Dr. John Clark labored strenuously, season after season, in the apparently thankless task of gathering the scanty remains of this fauna, which has proved to be of the utmost interest.

To the extreme kindness of Dr. A. Avinoff, Director, I owe the opportunity of studying this unique collection. Somewhat abbreviated quotations from the papers of Messrs. Peterson and Kay will form a suitable introduction to a study of the fauna.

The formation under discussion is a long and rather narrow belt along the southern base of the Uinta range. It has an extent of approximately eighty miles from east to west, reaching the foothills of the Wasatch range westward; while to the east it extends close to the Colorado-Utah state-line. In a north-south direction the formation seldom exceeds twelve to fifteen miles.

In looking at the eroded faces of the stratigraphic mass the panorama appears as of a pale brick-red. There are many other colors, but the red predominates. Along the southern borders of these upper red beds the strata have a northward and slightly westward dip of from two to four degrees. In proceeding northward through the Basin this dip constantly decreases, especially in the region where the section was made. At a point some three to four miles north of White River the strata become horizontal and very soon there is a noticeable southward dip, thus forming a syncline in this region of the Basin. The extreme southward dip observed along the northern contacts with the base of the Uinta range is seldom over twelve to fifteen degrees.

Lithologically the strata in the region, through which the geological section was taken, are made up, for the most part, of soft and hard sandstones, often enclosing lenses of clay. These sandstones alternate with soft and indurated layers of strongly arenaceous clays. Along the northern border of the Basin are numerous intercalations of masses of conglomerates of finer and coarser texture. These layers of conglomerates are sometimes twenty to thirty feet thick and of considerable geographic extent.

The most intensive search by Peterson (1893–95, 1912, 1929–30), Douglass (1907–29), Riggs (1910), and Kay (1928–30), revealed no clear stratigraphic break between the upper red beds of the Uinta [*i. e.* Duchesne River] and horizon C. . . . Furthermore the lithological gradation, color, and other conditions, along the course of the cross-section made through the Basin, are so gradual from the lower to the higher strata that it becomes extremely difficult to separate the two horizons. The exposures of the lower red beds (top of horizon C), to the west of where the section was made, or in the Duchesne Valley, are, however, capped by a decided hard banding of sandstone, which is quite persistent in the middle region and western end of the Basin. At Point Randlette about three miles above the junction of the Duchesne and the Uinta rivers, on the north bank of the Uinta, the upper part of horizon C is capped with a brown sandstone some ten to twenty feet thick. To the east of this point (Point Randlette) it is possible to trace this horizon to the north of "Pelican Lake" or the northern rim of "Leota Basin," thence crossing the region where the cross-section is taken . . . on Green River between the Baser and Leota bends of Green River . . .

From Point Randlette westward, the sandstones overlying horizon C are clearly distinct and more easily traced. They weather out to characteristic reddish brown cliffs along the streams and on the divides between Lake Fork, "Dry Gulch," Duchesne, and the course of other rivers. In places the sandstones are fine-grained; in other places coarse, with a tendency to become conglomerates. Cross-bedding is often shown, indicating stream action. In a lithological sense these Upper Uinta sandstones are distinctly different from the lower formations in the middle and western part of the Basin. There is more sandstone, less clay, and the sedimentary mass, as a whole, has less vertical thickness. There was, however, no distinct break found between horizon C and the Upper Uinta sediments along the watercourses and divides here mentioned. In the study and measurements of the strata from south to north further east in the Basin, where the section was taken, we are in fact forced to temporarily conclude that the entire tertiary sedimentation of the Uinta Basin including the uppermost beds, went on continuously.

In dealing with these uppermost beds from a faunistic standpoint we observe a sharp break. Well toward the top of the series, in the Titanothere quarry . . . worked by the Carnegie Museum during 1929–30, we have secured an abundance of material pertaining to a typical Oligocene form (*Teleodus*). In the same quarry were found the portion of a lower jaw of a Cameloid (Poëbrotherium) and remains of Hyaenodonts. With these was also discovered material representing a small rhinoceros provisionally referred to *Amynodon* and a fragment of a mandibular ramus, with M_2 in place, of a mesonychid comparable in size with *Mesonyx* [i. e. *Pachyaena*] *ossifragus*. Outside of the two latter genera the fauna, so far found in this new quarry, is certainly representative of a post-Eocene or basal Oligocene deposit.

. . . According to the faunistic evidence now in our possession we must regard the uppermost red series of the

Uinta sediments as pertaining to the basal Oligocene, which has been more or less anticipated, though not hitherto proven, by paleontologists who have visited and worked in the Tertiary sediments of the Uinta Basin.

The lower members of the Uinta sediments are apparently entirely absent along the northern border of the Uinta Basin. Horizons A, B, and C_1 ($=C_1$ of Osborn) so characteristic of the southern and middle part of the Uinta Basin, together with the rich fauna of horizons B and C, are entirely lacking along the northern borders. This apparent unconformity, observed by Peterson in the years 1893–1912, by Douglass in 1908–09, by Kay 1928–29, led to the fruitless search by these parties for a distinct break between the Upper and Lower Uinta sediments along the southern, or rather the middle exposures, in the Uinta Basin.[1]

In a later paper by Peterson,[2] the uppermost beds of the Uinta Basin were formally separated from the Uinta Eocene and erected into a distinct stage, a proposal which has been generally accepted.

It is quite clear that the name "Upper Uinta," which was used by Peterson and Kay in their publication, Ann. C. M., Vol. XX, pp. 293–305, might in the future be a source of confusion to students. The "Upper Uinta" is a term which was and will be commonly used by geologists and palaeontologists in referring to the upper series of the Uinta Eocene, especially "Horizon C." The name "Duchesne," . . . is therefore proposed for the Oligocene horizon, which rests upon the Upper Eocene (Horizon C) in the Uinta Basin. The Duchesne River in Duchesne County, Utah, has its source on the southern side of the Uinta Mountains. The stream traverses these upper beds, which are now determined to be Basal Oligocene, before its confluence with the Green River a half mile below Ouray, Utah. . . . The geographical area covered by these Oligocene strata has an east-west extent of approximately eighty miles, and seldom exceeds from twelve to fifteen miles in a north-south direction along the northern margin of the Uinta Basin. From Randlette westward, along the Duchesne river, these Oligocene beds are quite clearly defined from the underlying Uinta series (Horizon C). . . . The sandstones weather out to characteristic reddish brown cliffs, which rest on softer clays [Horizon C of the Uinta] along the streams and on the divides between Lake Fork, "Dry Gulch," Duchesne, and the course of other rivers. Although a tentative division was made by Peterson and Kay between the Duchesne beds and the underlying Uinta strata to the eastward from Randlette, the distinction between the two horizons is not so clear toward the eastern end of the basin.

The relationship of the fauna of the Duchesne Oligocene, as now known, is less sharply defined from that in the underlying Uinta sediments (Horizon C) than is usually the case in the superimposed horizons of other localities. There is, nevertheless, an advance corresponding to that of the lithological change noted. The Titanotheres of the Duchesne Oligocene so far known represent an advance upon those found in Horizon C of the Uinta series. *Teleodus uintensis* is a typical Oligocene form, with frontonasal horns well developed; with two upper and three lower incisors, bearing the characteristic button-shaped crowns found in all the known Titanotheres of the White River Oligocene. The horses, or Anchitheres, and Tapiroids have the lower premolars more like the molars. The Cameloids are very distinctly advanced in the direction of those found in the White River Oligocene. The genus *Hyænodon* of the eastern Oligocene, not heretofore found in the Eocene of America, is recognized. With this assembly of forms there are, so far found, remains of Anchitheres, Amynodonts, Homacodonts, Agrichœrids, and Mesonychids, as survivors from the underlying Eocene. The Amynodonts have the second lower premolar reduced to vestigial proportions. The Homacodonts and Mesonychids have advanced in their trend of evolution. It is quite safe to say, that, when more complete material is found representing the Agrichœrids and the Oreodonts, they will be found to have similarly advanced.

The Duchesne Oligocene of the Uinta Basin may thus be regarded as a horizon quite perfectly transitional between the Upper Eocene and the Chadron horizon of the White River series of South Dakota. We were gratified in finding in the Duchesne Oligocene the genus *Hyænodon*, hitherto not reported from so low a level in the American Oligocene. In this vast deposit of strata, over one thousand feet thick, we have not yet discovered any horizons which abruptly traverse the formation, such as the well known Metamynodon and Protoceras Sandstones; the two latter being stream deposits, which contain sudden breaks in the Oligocene fauna of the White River Badlands of South Dakota. The evidence of stream action is abundant in the Duchesne series, but the scarcity of fossil remains through the entire formation precludes at this time any clear comparison with what is known of the Oligocene of Dakota.

Of the fossils described in the following pages the fragment of an upper jaw as well as limb bones of *Hyænodon* were found in 1929 in the Titanothere quarry of the upper Duchesne Oligocene and the fragment of a lower jaw of the Amynodont described in the following pages was found in the same quarry in 1930. The rest of the material described in what follows was collected in different horizons of the Oligocene of the Uinta Basin.[3]

Much the most complete study of the stratigraphy of the Uinta Basin that has yet been published is a paper by Mr. J. Leroy Kay, of the Carnegie Museum, entitled "The Tertiary Formations of the Uinta Basin, Utah." The part of this paper which deals with the Duchesne River beds may be quoted, in somewhat abbreviated form, with advantage.

DUCHESNE RIVER (OLIGOCENE) FORMATION

The name "Duchesne" was proposed by Professor W. B. Scott for the Lower Oligocene Formation of the Uinta Basin, Utah. . . . Duchesne has been used in several publications by the Carnegie Museum. It appears that the name Duchesne is preoccupied, having been used by Keyes[4] for a limestone of Jurassic age. This duplication is rather unfortunate. A new name is required for the Oligocene formation in Utah, but the writer feels that there will be less confusion if the word "Duchesne" is retained in the new term. At Prof. Scott's suggestion the name **Duchesne River** is now proposed for the lower Oligocene formation of the Uinta Basin, Utah.

The Duchesne River Formation to the south lies conformably upon Horizon C of the Uinta Eocene, but with greater or less unconformity with the older rocks along its northern border. It is composed chiefly of alternating bands of sandstones and red, brown, and variegated clays, the sandstones often enclosing lenses of arenaceous clays.

[1] Peterson, O. A., and Kay, J. L., *Ann. Carnegie Mus.* **20**: 293–306, 1931.

[2] Peterson, O. A., *ibid.* **21**: 61–78, 1932.

[3] *Ibid.* **23**: 357–371, 1934.

[4] Keyes, C. R., Grand Staircase of Utah, *Pan. Amer. Geologist* **41**: 36, Feb. 1925.

Many of the clay members are deceptive as to their thickness. In weathering out they cover the sandstones below and appear to be several feet in thickness but if the surface is cleaned off, they are found to be only a few inches thick. The surfaces of the sandstones are more or less stained by these clays and appear to be reddish-brown in color, but are mostly grey or very light brown. In looking at the beds as a whole, they appear a brick red. The sandstones predominate and are often conglomeratic.

At Rock Creek, where the cross-section of the western part of the Uinta Basin was made, the Duchesne River Formation has a thickness of over fifteen hundred feet, which is probably the maximum thickness. The greatest north-south extent of this formation is east of Lake Fork.

In most places along the northern border, at or near the base of the Duchesne River sediments, heavy bedded, coarse conglomerates occur. These conglomerates probably attain their greatest thickness at Currant Creek but are well exposed at Lake Fork where they rest with a slight southern dip upon the Triassic and Jurassic. The latter formations at this point have a dip of about sixty degrees. . . .

In the Asphalt Ridge, about five miles south and west of Vernal, . . . specimens were found and identified by O. A. Peterson as lower Oligocene Agriochœrids. . . . Thus it appears that these conglomerates cannot always be considered the base of the Duchesne River Formation.

Along the northern border, the contact of the Duchesne River with the older rocks is more or less obscured by reworked glacial gravels which cap most of the benches and only in a few places are they consolidated. In several places, lying beneath the Duchesne River Formation, are exposed strata of undoubtedly Eocene age. Although these beds were prospected no identifiable fossils were found. The best exposures are between Rock Creek and the Duchesne River; Lupton[5] maps them as Wasatch. The writer is inclined to regard them as Uinta. This conclusion is based chiefly on their lithological character.

To facilitate correlation and reference of specimens now in the Carnegie Museum, the Duchesne River Formation is divided into three horizons. These are, from lowest to highest, the **Randlett**, the **Halfway**, and the **Lapoint**. The type locality for the Randlett Horizon . . . is north and east of the town of that name and consists of 478½ feet of brown, red, bluish grey and variegated clays alternating with brown and grey sandstones. These beds are capped by a ten foot stratum of conglomerate . . . which is taken as the base of the Halfway Horizon.

The strata of the Halfway Horizon are well exposed along Halfway Hollow, the basin which drains that district. . . . There are 557½ feet of sediments referred to this horizon, consisting chiefly of sandstones and shales with several conglomerate members. The sandstones in general are coarser than those of the Randlett. Overlying the Halfway is a layer, twenty-two feet thick, of bluish white clay . . . , which can be seen just north of the Vernal-Lapoint Road and is quite persistent throughout the area. This is considered the basal member of the Lapoint.

The Lapoint, the uppermost division of the Duchesne River Formation, is typically observed along the head of Halfway Hollow east and north of the town of Lapoint. This horizon includes 336 feet of sandstones, clays, and conglomerates. The conglomerates are coarse, the sandstones conglomeratic, and the shales quite sandy.

Much of the so-called "Uinta" of Beaver Divide, in central Wyoming, should, in my opinion, be included in the Duchesne River, at least provisionally. The exposures were first described by Granger, whose account may be here summarized.

The divide rises to a height above Beaver Creek of over 1,200 feet. . . . To the southward the land slopes gently down to the Sweetwater; to the northward, facing the Wind River Basin, it breaks off very abruptly [and precipitously] . . . along almost the whole southern border of the basin. . . . The beds of this bluff in the area examined, aggregating nearly 1,100 feet, are all Tertiary, lying nearly horizontal, and resting on the upturned edges of the Mesozoic strata. . . .

Lithologically, as well as faunally, the beds fall into three groups: (1) the red banded clays and coarse sands of the Lower Eocene, (2) the greenish, blue-gray and olive-colored shales . . . of the (?) Middle and Upper Eocene, and (3) the buff and light gray calcareous marls of the Lower Oligocene. The upper and lower groups are fairly constant in character throughout the area examined, but the middle one is variable, the various strata maintaining their character over a very limited area only. An exception to the uniformity of the Lower Oligocene is the presence, near Wagon-bed Spring, of a five-foot bed of volcanic material not observed at Green Cove, where the above section was taken [fig. 1, p. 238]. The bed is from fifty to one hundred feet above the unconformity and consists of gray ash, in which is [*sic*] imbedded in places numerous smooth, rounded masses of pumice an inch or two in diameter. . . .

No break in sedimentation was detected in the Eocene series, yet a careful examination by a competent stratigrapher might bring it to light. The Bridger formation in the Bridger Basin has a maximum thickness of 1,800 feet, and the only beds in this section which might be assigned to that formation are the 200 or 250 feet of unfossiliferous strata between the Lower Eocene banded beds and the layers lying immediately below the unconformity, which are unquestionably Upper Eocene.

It must be remembered that, when Granger wrote this description, the existence of the Duchesne River, as a separate unit, had not been suspected and the uppermost beds of the Uinta Basin were unhesitatingly referred to the Uinta stage. He continues:

Mammalian fossils were found on three levels, as noted in the section above. They were not at all abundant at any level nor in any locality, and it was only by painstaking and prolonged search that the number of forms listed below was obtained. . . .

On the Uinta level the fossils came, with one exception, from Wagon-bed Spring, and were found mostly in a pale yellowish, tough, sandy clay. . . .

It is doubtful, therefore, whether this horizon is to be correlated with the true Uinta (Horizon C) or with Uinta B and its equivalent (in part) of the Washakie Basin, Washakie B.[6]

Granger regarded the disconformity, which is clearly shown in his generalized section of the Tertiary at Beaver Divide, as the boundary between Eocene and Oligocene, but Dr. John Clark, who has examined the section and who referred the beds just below the break to the Uinta, not the Duchesne River, writes me (*in litt.*) as follows: The Uinta

[5] Lupton, Charles T., The Deep Creek District of the Vernal Coal Field, Uinta County, Utah, *U. S. Geol. Surv. Bull.* **471**: 579–594, 1912.

[6] Granger, W., *Bull. Amer. Mus. Nat. Hist.* **28**: 237–241, 1910.

is overlain by a series of tan and white agglomerates and tuffs, from the lower part of which [Granger] collected Chadron titanotheres. At this point there is some confusion, since the Carnegie Museum recently collected a skull of *Protoreodon* from one of the agglomerates. Naturally enough, Granger placed the boundary between the Uinta and the Chadron at the base of the agglomerates; occurrence of so definitely Eocene a form as *Protoreodon* within the agglomerates demonstrates that the contact lies higher, somewhere within the agglomerate series, and has not yet been accurately determined.

While not accepting Clark's dictum that the presence of *Protoreodon* demonstrates the Eocene age of the containing beds, we may admit that the exact boundary between the White River and the older formation remains to be determined.

Professor H. E. Wood, who has also examined the Beaver Divide section, reports that the *Amynodon* skull, the discovery of which was hailed with such interest, is definitely referable to the Uinta species, *A. advenus*, and that it lay below the levels at which the supposedly Duchesne River fossils were found, *Brachyhyops*, *Protoreodon*, and *Mesagriochoerus*. He agrees with me in thinking that in the Beaver Divide section are to be found at least three stages above the Wind River Eocene. In descending order these are: (1) the Chadron substage of the White River; (2) Duchesne River; (3) Uinta. Beneath the Uinta lie 185 feet of completely barren sandstones, in which no fossils have ever been found. These, Granger tentatively referred to the Bridger, or Middle Eocene, but they may equally well be assigned to the Uinta. As Granger noted, there is no perceptible stratigraphic break between the Wind River below and the topmost Uinta above, and there must be two, or more, hiatuses in time in these beds, which are only 540 feet thick and yet represent the entire Middle and Upper Eocene and part of the Lower. In the Bridger Basin of southwest Wyoming the Bridger formation alone measures 1800 feet in thickness, to say nothing of the thick masses of Wind River and Gray Bull in the Bighorn Basin.

DUCHESNE RIVER FAUNA

The subjoined list is taken, with some additions and corrections, from the posthumous paper by Peterson,[7] his last contribution to a subject which he had made so peculiarly his own.

REPTILIA

TESTUDINATA

Cymatholcus longus Clark

CROCODILIA

Crocodilus? acer Cope

[7] Peterson, O. A., *op. cit.* **23**: 373, 1934.

MAMMALIA

INSECTIVORA

LEPTICTIDAE

Protictops alticuspidens Peterson

CARNIVORA

CREODONTA

MESONYCHIDAE

Hessolestes ultimus Peterson

HYAENODONTIDAE

Hyaenodon sp.

FISSIPEDIA

UINTACYONIDAE

?Pleurocyon, or *?Uintacyon* sp.

?FELIDAE

Eosictis avinoffi gen. et sp. nov.

RODENTIA

PARAMYIDAE

Leptotomus kayi Burke

ADJIDAUMIDAE

Protadjidaumo typus Burke

LAGOMORPHA

LEPORIDAE

Mytonolagus sp. Peterson

ARTIODACTYLA

PALAEODONTA

HELOHYIDAE

?Helohyus sp.

DICHOBUNIDAE

Pentacemylus progressus Peterson

?CHOEROPOTAMIDAE

Brachyhyops wyomingensis Colbert

TYLOPODA

AGRIOCHOERIDAE

Mesagriochoerus primus Peterson
Diplobunops crassus sp. nov.

MERYCOIDODONTIDAE

Protoredon tardus sp. nov.

HYPERTRAGULIDAE

?Leptomeryx minutus Peterson

CAMELIDAE

Camelodon arapahovius Granger
Poabromylus kayi Peterson

PERISSODACTYLA

EQUIDAE

Epihippus (Duchesnehippus) intermedius Peterson

BRONTOTHERIIDAE

Teleodus uintensis Peterson
Protitanotherium sp.

HELALETIDAE

Heteraletes leotanus Peterson

HYRACODONTIDAE

Hyracodon primus Peterson

Mesamynodon medius Peterson

AMYNODONTIDAE

Megalamynodon regalis gen. et sp. nov. H. E. Wood

?RHINOCEROTIDAE

Epitriplopus medius Peterson

?CHALICOTHERIIDAE

?Chalicothere sp.

In spite of the extreme scarcity of its fossils, the Duchesne River fauna is decidedly a rich one, in the sense of the number of families and genera represented. There are two reptilian orders, and of mammals there are already known 6 orders, 22 families, and 26 genera. It will be observed that, with two exceptions, each family is represented by a single genus and each genus, again with one probable exception, has but one species. These numerical relations are strongly indicative of a fragmentary fauna, which is very incompletely known.

As to the Duchesne River reptiles, the fossils are even more scanty than those of mammals, but there is one point in connection with them, which is of some interest. The crocodile skull which Peterson provisionally referred to *C. acer* Cope, is one of the larger Eocene types, the specimen having an approximate length of 16–18 inches. After this time, crocodilians became very rare in the northern United States, and the large true crocodiles, which were abundant in the Eocene, were replaced by a dwarf species of alligator in the Chadron substage of the White River. Climatic change was almost certainly not the occasion of this revolution, which was more probably due to increased altitude of the Rocky Mountain region than to any general change of the North American climate.

Attention has been repeatedly called to the apparent conformity between the Duchesne River beds and the underlying Myton substage of the true Uinta; Peterson and Kay even expressed a provisional belief in the continuous deposition of all the Tertiary sediments, including those of the Duchesne River, of the Uinta Basin.[8] However, two considerations show that this apparent conformity is deceptive: (1) the overlapping of the Duchesne River beds upon Triassic and Jurassic along the northern border and (2) the sudden and almost complete break between the fossils of the topmost Uinta (Myton substage = Horizon C) and those of the Duchesne River implies a considerable hiatus in time. Very probably, this hiatus was greater than the faunal lists, *as at present constructed*, would lead one to believe, for, so fragmentary is much of the material, that specific or even generic determination of it is impracticable. Of the clearly determinable species, none is common to Horizon C of the Uinta and the Duchesne River, and only those of the genera *Epihippus*, *Diplobunops*, and *Protoreodon* are known in the Uinta. The lapse of time between the Myton and the Duchesne River may not be so great as that between the last-named and the Chadron, but the difference is not very marked. The faunal change is abrupt, as appears from the absence of a single definable Uinta species from the Duchesne River beds, and even the common genera are somewhat doubtful, except in the case of *Diplobunops*, of which a new and very distinct species is described in the following pages. Even the horselike genus *Epihippus*, Peterson believed, would, when more completely known, prove to be distinct from the Uinta genus, and, for this eventuality, he proposed the hypothetical term *Duchesnehippus;* the agriochoerid, *Diplobunops*, quite certainly survived from the upper Uinta. *Protoreodon* appears to be another survival, but is somewhat questionable and may, eventually, be assigned to a different genus; the species *P. tardus*, again, is very distinct and appreciably nearer to the White River *Merycoidodon*.

Because of the extreme scarcity of its fossils, but few palaeontologists have had an opportunity to study the Duchesne River fauna, but all of them have, "with one accord," emphasized its transitional character between the latest Eocene and the earliest Oligocene, as previously known, the Chadron of South Dakota and adjoining states. When the Duchesne River fauna is considered as a whole, this transitional character is very obvious. Several of the genera are clearly more or less modified survivals from the Uinta, others are new and akin to, or even congeneric with White River forms, such as *Teleodus*, *Hyracodon*, and *Hyaenodon*, seemingly their first appearance in North America. A very interesting possibility is suggested by the genus which I am proposing to name *Eosictis* and regard as the most ancient of American sabretooths. The type specimen is too incomplete to demonstrate that this animal was a fissiped and not a creodont; it might even prove, when more completely known, to be related to the remarkable sabretooth creodont, *Apataelurus*, of the Uinta. I believe, however, that it is more probably a machairodont. Several genera of this fauna, such as *Hessolestes*, the latest known member of the creodont family, Mesonychidae, and *Diplobunops*, of the Agriochoeridae, appear to have died out, leaving no descendants.

The individual genera illustrate, even better, the transitional nature of the fauna; *Teleodus*, a brontothere, is transitional between the Uinta members of the family, such as *Diplacodon* and *Protitanotherium*, and the White River forms, though such is the welter of variations in the Chadron brontotheres, that it is impossible to select the actual descendants of *Teleodus*, especially as migration from Eastern Asia probably played an important, though as yet an indeterminable, part in recruiting the White River brontotheres.

[8] Peterson, O. A., and Kay, J. L., *ibid.* 20: 295, 1931.

All of the three families of the Rhinocerotoidea, the Hyracodontidae, the Amynodontidae, and the Rhinocerotidae, appear to be represented in the Duchesne River fauna, though remains are insufficient for a generic determination in all of them. Peterson assigned certain fossils to the White River genus *Hyracodon*, and I have followed his example. The amynodonts, on the other hand, are admirably represented by the genus which H. E. Wood, in subsequent pages, proposes to name *Megalamynodon*. This is an ideally perfect transition between the ancestral *Amynodon*, of the Uinta, and the White River descendant, *Metamynodon*. There can be little doubt that the true rhinoceroses, of the family Rhinocerotidae, were represented in the Duchesne River fauna, as they almost certainly were in the antecedent Uinta and unquestionably were in the subsequent Chadron.

Of the horses, very few remains have been collected, the best of which is a right *ramus mandibuli*. This, Peterson felt assured, would eventually prove to be referable to a new genus, intermediate between the Uinta *Epiphippus* and *Mesohippus*, of the White River.

The aberrant, clawed Perissodactyla, which Cope named the Ancylopoda, are listed by Peterson, though only as a questionable chalicothere, but we may be sure of their presence in the fauna, for they occur in the preceding upper and middle Eocene and in the succeeding Oligocene and Miocene.

Artiodactyla are less common in the fauna than would have been expected, though six families, at least, are represented, most of them by a single genus each. A small bunodont was tentatively referred by Peterson to the Bridger genus *?Helohyus*, but the reference is not likely to be confirmed by future discovery, because of the long hiatus in time. *Helohyus* has never been found in the Uinta. *Pentacemylus* may well be referable to the European family of the Dichobunidae, in which Sinclair has included most of the Bridger artiodactyls, but the material is insufficient to make the reference certain.

Another questionable bunodont artiodactyl is the relatively large *Brachyhyops*, which was found at the Beaver Divide, Wyoming, and is, therefore, but tentatively included in the Duchesne River fauna. Colbert[9] has referred this strange creature to the European family, Choeropotamidae, of which it will be one of the very few American representatives, if the reference is confirmed. Stratigraphically, it is without assignable significance, for it stands alone.

The Camelidae are one of the rare families which are represented in this fauna by more than one genus; that is, on the assumption that the Beaver Divide fossils are properly referable to the Duchesne River. Both of these genera, *Camelodon* Granger, from the Beaver Divide, and *Poabromylus* Peterson, from the Uinta Basin, might well have been derived from the Uinta *Protylopus*, though only Peterson's genus can be regarded as transitional to the White River *Poëbrotherium*.

Of the two Oreodont families, only the Agriochoeridae have, as yet, been found in the Duchesne River of Utah, and this is another of the exceptional families with more than one genus in these beds. One of these genera, *Diplobunops*, is a survival from the Uinta and, so far as is known, the species, *D. crassus*, is peculiar and is the last of the series, which appears not to have extended into the White River. The true agriochoerids are represented by *Mesagriochoerus*, which is a most interesting and significant transition between the Uinta and the White River forms. *M. primus* Peterson is the only mammal which has, as yet, been found in both the Utah and the Wyoming areas.

The merycoidodonts have not been found in the Duchesne River beds of Utah, a lack which is, assuredly, merely an accident of collecting. In the Wyoming area of Beaver Divide a peculiar species of the Uinta genus *Protoreodon* has been obtained in strata above the disconformity which has been supposed to mark the lower boundary of the White River.

Attention has been called to the remarkable thickness of the Duchesne River beds, 1,500 feet, which is four or five times as great as the entire section of the White River, as exposed in the "Big Bad Lands" of South Dakota. In the latter there are three very distinct faunal zones, the *Brontotherium*, *Merycoidodon*, and *Protoceras-Leptauchenia* beds, corresponding to the stratigraphic divisions of the Chadron, Lower and Upper Brulé, respectively, and, within the limits of these substages, all of the families of mammals show marked evolutionary changes, so that hardly a single genus occurs in all three of the substages. There can be no reasonable doubt that, were the Duchesne River fossils at all comparable in numbers and variety with those of the White River, a similar development might be observed through consecutive strata. It is true that Kay was able to distinguish three substages, the Randlett, the Halfway, and the Leota, but he made these divisions on lithological grounds. In the excessive rarity of the fossils, it is quite impossible to distinguish faunal zones. The fossils collected near the bottom of the Randlett are entirely different from those found near the top of the Leota, but the meaning of this difference it is wellnigh impossible to interpret, for one cannot say how far it is accidental and due to the hazards of collecting and how far it is inherent in the stratigraphical succession. However, the list, incomplete as it is, does give a fair conception of the fauna and suffices to fix its position in the geological column. I am entirely in agreement with those members of the Carnegie Museum staff, Messrs. Peterson, Kay, and Clark, who have been most concerned in the collection and preparation of the Duchesne River fossils, in their determination of this fauna as the oldest known Oligocene. That it is intermediate in

[9] Colbert, E. H., *Ann. Carnegie Mus.* 27: 87–108, 1938.

time between the uppermost Uinta Eocene (C, or Myton) and the oldest White River (Chadron substage) is obvious. The only question that can arise in this connection is: Where should the line between Eocene and Oligocene be drawn? There is an almost complete continuum between Wind-River—Bridger—Washakie—Uinta—Duchesne-River—Chadron, and in this great thickness of beds there is hardly any stratigraphic break, except those made by geographical separation, e. g. between the Wind River and Bridger, and, perhaps, between the Bridger and Washakie, though Clark does regard the lower Washakie as equivalent to the topmost Bridger.

It is a generally accepted principle that, in dealing with stratigraphic problems, greater weight should be given to forms which are newcomers than to persistent older forms. The Duchesne River fauna includes several species which are referred to White River genera, such as *Hyaenodon*, *Hyracodon*, *Teleodus*, and several genera which, appearing for the first time, are clearly intermediate between Uinta and White River forms and are more advanced than the former, less so than the latter. Examples of this kind are *Mesagriochoenus*, *Poabromylus*, and *Megalamynodon*, which have an obviously intermediate position between the Uinta and the Chadron forms. The new genus, *Eosictis*, is, unfortunately, represented by a very incomplete type specimen, but is very probably a machairodont and, if so, is the first appearance of the group in North America and emphasizes the Oligocene reference.

Equally significant is the fact that the genera which survive from the upper Uinta are all represented by new species, and almost all persistent ancient families by new genera. It is probable that, if better preserved material were available, nearly all the genera which are now listed under Uinta names would prove to be so different as to require new terms.

To Peterson's list of Duchesne River mammals, I have added those found at the Beaver Divide in Fremont County, Wyoming. The list of these fossils is so very short, that correlation of the beds is uncertain, further than to say that they are referable either to the Duchesne River or to the uppermost Uinta. In the lower strata of the section occurs the typical Uinta species *Amynodon advenus;* above these strata and succeeding them conformably are beds that have yielded fossils which are distinctly more advanced than those of the Myton substage (Uinta C), such as a large species of *Protoreodon*, one of *Mesagriochoerus*, and the extraordinary skull which Colbert has named *Brachyhyops* and referred to the European family Choeropotamidae. The presumably Duchesne River beds are unconformably overlain by unmistakable Chadron.

The Sespé formation of southern California offers an instructive parallel to the succession of beds at Beaver Divide, Wyoming. The 7,000 feet of continental deposits, which make up the Sespé, appear to represent the whole of Oligocene time and to include some of the Upper Eocene. They rest unconformably upon marine Eocene and are unconformably overlain by marine Miocene. Most of the great thickness of the Sespé is barren of fossils, but a considerable fauna of mammals occurs at a level about 1,500 feet above the base of the deposits, and this fauna has much resemblance to that of the Duchesne River formation. Professor Chester Stock, of the California Institute of Technology, prefers to regard the fauna as latest Eocene, but, as already stated, I feel compelled to agree with the palaeontologists of the Carnegie Museum in assigning the Duchesne River and, with it, the lower Sespé, to the Oligocene. The fossils of the beds at Beaver Divide upon which the White River rests, though very few, probably suffice to show that these beds are intermediate in time between the Myton (Uinta C) and the Chadron, as is indicated by their stratigraphic position.

The continental Lower Eocene of North America has a mammalian fauna which contains a great many elements in common with that of Europe and Asia, a feature which is usually explained by postulating a land connection between Kamchatka and Alaska, across what is now the shallow Bering Sea. This connection, which is extremely probable, must have been severed or, at least, greatly restricted, in the Middle Eocene, for the Bridger and Washakie mammals differ more from their Old World contemporaries than was the case in any other part of Tertiary time. Again, in the Uinta, there was a great influx of selenodont artiodactyls, evidently immigrants from some other region, and this very probably points to a renewal of the land bridge to Asia, which, apparently, was not again interrupted until after the close of the Pleistocene.

The intermediate, transitional nature of the Duchesne River fauna has been repeatedly emphasized, and this characteristic is not likely to be changed by future discoveries, however much the list of genera may be increased.

ORDER **INSECTIVORA** LATREILLE

Small mammals are particularly rare in the Duchesne River beds; one insectivore, one lagomorph, and two rodents are all that the collectors have, so far, obtained.

FAMILY **LEPTICTIDAE** GILL

This characteristically North American family is most abundant and varied in the White River, beyond which it is not known to extend; it is also frequent in the Bridger and the Uinta.

PROTICTOPS PETERSON

Protictops Peterson, *Ann. Carnegie Mus.* **23**: 374, 1934.

Protictops alticuspidens Peterson

Protictops alticuspidens Peterson, *loc. cit.*

As I have nothing to add to Peterson's description (pp. 374–375), I reproduce it here.

Principal characters of the holotype. Trigonid relatively large and the heel small when compared with the molars of *Ictops acutidens* Douglass. The paraconid is larger and of greater functional value when compared with the latter species. In the present form, the heel of the molar is of a more nearly cross-crest structure but with the median region constricted and the median tubercle very indistinct or not at all indicated, while in *Ictops acutidens* the median tubercle is well formed. The specimen pertains to an animal approximately one-half the size of *Ictops acutidens.*

An upper premolar (?P^3) found in the same horizon and locality in which *Protictops* was found, is associated with this series of specimens and referred to the Insectivora. It is rather doubtful that the specimen pertains to the genus proposed above, because the posterior part of the prominent protocons is more nearly of a carnassial structure and has no true tritocone as in *Ictops.* There is, however, an antero-external basal tubercle as in P^3 of *Ictops.* The deuterocone is definitely formed, but basal in position and the crown is surrounded by prominent external, internal and posterior cingula.

Peterson gives no measurements, but his figure indicates that the antero-posterior diameter of m_1 and m_2, taken together, is 4.5 mm.

Order CARNIVORA Gray

Even in the scanty list of the Duchesne River, the rarity of predaceous mammals is noteworthy; both individually and specifically, beasts of prey are very uncommon, though two families of creodonts and two of fissipeds have been found. All the fossils of these groups, so far discovered, are in a very fragmentary state, and the generic identification of them is more or less uncertain.

Suborder CREODONTA Cope

Two families of creodonts are, almost certainly, represented in this fauna, one, the very ancient Mesonychidae, in what is believed to be its latest appearance; the other was identified by Peterson as the White River *Hyaenodon.*

Family MESONYCHIDAE Cope

This family is one of the oldest of all the creodonts and extends from the Paleocene through the entire Eocene into the lowest Oligocene. The Eocene genera are remarkably specialized for cursorial habits and have tetradactyl feet of paraxonic symmetry; the ungual phalanges are broad and almost hooflike, whence the name of the typical genus, the Middle Eocene *Mesonyx,* of the family. Each division of the Eocene had one or more characteristic genera. The dentition is a curious combination of primitive and specialized features; the upper molars are tritubercular of very primitive pattern, while the lower are reduced to two cusps of pseudo-sectorial shape. The Duchesne River genus *Hessolestes* Peterson is, so far as is yet known, the last representative of the family.

HESSOLESTES Peterson

Hessolestes Peterson, *Ann. Carnegie Mus.* **20**: 338, 1931.

The typical and only known specimen of this genus is a fragment of the right *ramus mandibuli* from the hinder end of the jaw, comprising the angle, condyle, coronoid, and part of the dentary portion, m_2 in place and the alveolus of m_3. This suffices to confirm Peterson's reference of the fossil to the Mesonychidae and also to demonstrate its generic distinction. The original definition is:

M_3 rudimentary; M_2 with posterior basal heel more symmetrically rounded, a mere trace of the anterior basal cusp, the tooth relatively larger, the angle shorter and deeper than in *Harpagolestes.*

The fragment upon which this genus is based furnishes characters which at once distinguish it from *Harpagolestes* found in the middle and upper Eocene. It seems quite evident that M_1 and M_2 have increased in size and otherwise changed, while M_3 has decreased to vestigial proportions being inserted in the jaw with probably only a single root.

Hessolestes ultimus Peterson

Hessolestes ultimus Peterson, *Ann. Carnegie Mus.* **20**: 338, 1931.

Peterson's figures do not altogether agree with his description. Figure 1, the side view of m_2, shows a complete tooth, with uninterrupted cingulum and anterior basal cusp reduced to "vestigial proportions," while figure 2, the crown view, shows that the front end of the tooth was broken away; the plane of fracture, as may be seen from the specimen itself, passing down through the anterior root. Evidently, the anterior basal tubercle was larger than Peterson supposed, and it may have been relatively as well developed as in *Harpagolestes;* the posterior cusp is an imperfectly shearing blade and is much smaller than the anterior cusp, or protoconid. In the preceding genera of the family, the lower molars are bicuspid, having lost the inner elements, while the two external cusps are of subequal size and are in the same fore-and-aft line, the two together forming a rather blunt shearing blade, but in *Hessolestes,* alone in the genera of the family, the reduction of the metaconid gives to m_2 a peculiar appearance. Peterson was assuredly right in inferring from the single, matrix-filled alveolus that m_3 was much more reduced than in the antecedent genera of the family and was a mere vestige.

The fossil is the posterior part of the right half of the lower jaw, of which the ascending ramus is preserved almost entire and the external side is covered with matrix. The lingual side is smooth, featureless, and nearly plane, except that the angle is slightly concave. This portion of the jaw displays many small differences from that of *Harpagolestes,* though the two are manifestly of the same family type. The following points of difference between the two genera should be noted.

In *Hessolestes:* (1) The anterior border of the coronoid is much less steeply inclined, and the whole ascending ramus is relatively wider; the posterior border is more nearly erect.

(2) The sigmoid notch is broader, shallower, and much less distinctly marked.

(3) The position of the condyle is remarkably low, far beneath the level of the molars, and relatively much lower than in any other known genus of the family. Compared with that of the machairodont subfamily of the Felidae, the mandibular condyle of *Hessolestes* has a more inferior position than in any but such extremely modified forms as *Eusmilus*. In the absence of the canines, it is impossible to make sure of the significance of this position of the condyle, but it certainly suggests that the upper tusks were laniary sabres. If so, the genus must be related to the extraordinary Uinta sabretooth creodont, *Apataelurus kayi*, which belongs either to the Oxyaenidae or to the Hyaenodontidae. As long as the dentition remains so incompletely known, the systematic position of *Hessolestes* must remain indeterminate and the genus may be left in the Mesonychidae, where Peterson placed it.

(4) The condyle projects much farther behind the coronoid, but is less upturned.

(5) The angle extends much farther below (ventrally) the condyles and, presumably, projected downward as a distinct process, set off from the horizontal ramus; there is not the least indication of an inflection of the angle.

MEASUREMENTS

Distance from m_3 to summit of coronoid	71 mm.
Ascending ramus, height from coronoid to angle	112
M_3, antero-posterior diameter (est.)	25
M_3, transverse diameter	11

Horizon: Lapoint substage of Duchesne River
Locality: 11 miles W. of Vernal, Utah

FAMILY **HYAENODONTIDAE** LEIDY

The American history of this family begins in the Lower Eocene with certain small, lightly built, long-tailed creatures, the classification of which is made according to the dentition; these genera continued through the Bridger, but have not been found in the Uinta, and, apparently at least, the family died out in North America with the end of the Bridger. In the White River large representatives of the family have long been known, and that they were immigrants from Asia has been generally agreed, especially as the type genus, *Hyaenodon*, is European.

The discovery of *Hyaenodon*, by the Carnegie Museum collectors, in the Duchesne River beds was of great interest, as marking the earliest known appearance in America of this typically European genus. Though the material is very scanty, it suffices for a probable identification. Another European genus of this family, *Pterodon*, occurs in the Sespé of California.

HYAENODON LAIZER ET PARIEU

Hyaenodon Peterson, *Ann. Carnegie Mus.* **21**: 64, 1931.

The original description is as follows:

> A portion of the maxillary of the right side, with the last two molars in place . . ., is referred to the well known genus *Hyænodon* of the Nebraska and Dakota Oligocene. The specimens were found in the Duchesne Oligocene together with the great mass of material representing *Teleodus uintensis* in the Titanothere Quarry eleven miles west of Vernal. The maxillary represents an animal smaller than *Hyænodon cruentus*, but larger than *Hyænodon crucians*. With the exception of a well-marked swelling on the external face of the shearing blade, midway between the anterior cusp and the posterior limit of the last tooth, the detailed structure of these most closely suggests those of *Hyænodon cruentus*. On the last tooth there is no indication of an antero-internal basal tubercle, while on the tooth in front there is a very slight indication of such a tubercle. The maxilla is broken off at the posterior alveolus of the larger tooth and in front of the first molar.

To this description, it may be added that each of the molars is implanted by three roots, two anterior and one posterior, and that the anterior half of the crown is more than twice as broad transversely as the posterior part. Both teeth are much more abraded on the lingual side, so that any antero-internal cusps which may have originally been present, have been worn away. On m^1 the lingual face has lost all its enamel, but on m^3 only the anterior cusp is so worn; on the hinder shearing blade the enamel is almost intact.

Hyaenodon *sp.* Peterson

Hyaenodon sp. (?) Peterson, *Ann. Carnegie Mus.* **21**: 64, 1932.

Despite its great interest and stratigraphic importance, Peterson did not propose a species name for this fossil, and his example is here followed.

In the subjoined table the dimensions given by Peterson have been verified and, where necessary, corrected, and one or two others have been added.

MEASUREMENTS

M^1, antero-posterior diameter, on base	12 mm.
M^1, transverse diameter, anterior face	7
M^1, transverse diameter, posterior face	4
M^2, antero-posterior diameter, on base	17
M^2, transverse diameter, anterior face	9
M^2, transverse diameter, posterior face	5

Peterson also describes and figures (pp. 64–65) an isolated humerus, which he refers, with much probability, to this genus. In any event, it almost certainly pertained to a creodont. Peterson's description reads:

> The right humerus . . . represents that of a smaller animal, about the size of *Hyænodon mustelinus*, though slenderer and relatively less expanded at the head and distal end. As in *Hyænodon* the articulation for the scapula faces almost as much backward as proximally. The tuberosities are wide apart, displaying a wide and shallow bicipital groove. The deltoid crest is not prominent. The supinator ridge is low and poorly developed and the external epicondyle is likewise poorly developed, while the inner condyle is large and perforated by the characteristic foramen. The supratrochlear and anconeal fossae com-

municate by a large foramen characteristic of the genus. The distal trochlea is distinctly typical of *Hyænodon* and quite perfectly answers the description of Professor Scott, who states that the trochlea "is divided into three facets, of which the inner one is both the widest and the highest, while the outer one is very narrow; the median facet is a broad and strongly convex [ridge]."

I can hardly agree with Mr. Peterson in saying that this humerus is of about the size of that of *H. mustelinus*, for, according to my measurements and Peterson's natural-size figure, this bone is but four-fifths as long as in the White River species. The shaft is slender, its diameter diminishing downward regularly from the proximal end to the distal expansion, and, for the lower third of its length, it is quite cylindrical. This shaft has a gentle longitudinal curvature, making the posterior face rather concave, though the anterior face is but slightly convex. The internal epicondyle is relatively very large, decidedly more so than in the White River hyaenodonts. On the whole, it seems that Peterson's reference of this humerus is probably correct, if not to the genus, at least to the family.

MEASUREMENTS

Humerus, length from head	80 mm.
Humerus, width of proximal end	13
Humerus, width of distal end, over epicondyle	18
Humerus, width of distal trochlea	13

Horizon: Lapoint, Duchesne River.

Locality: Jaw fragment and humerus, Teleodus Quarry, 11 miles N.W. of Vernal, Utah.

SUBORDER **FISSIPEDIA**, FISCHER DE WALDHEIM

SUPERFAMILY **MIACOIDEA** TEILHARD DE CHARDIN

FAMILY **MIACIDAE**

Having transferred this family from the Creodonta to the Fissipedia, I feel that it is necessary either to include it in one of Flower's superfamilies or to adopt Teilhard de Chardin's term, Miacoidea, as a group-name for it. This superfamily includes exceedingly ancient and primitive carnivores, which, there is much reason to believe, were, directly or indirectly, ancestral to all of the other fissiped superfamilies and, therefore, not to be included in any one of them. The Duchesne River beds have never yielded any determinable fossils of miacoids or cynoids, but scattered foot-bones, metapodials and phalanges, have been obtained, which indicate the presence of Fissipedia, and, very probably, these are either miacoids or cynoids.

SUPERFAMILY **AELUROIDEA** FLOWER

FAMILY **FELIDAE** GRAY

SUBFAMILY **MACHAIRODONTINAE** GILL

EOSICTIS gen. nov.

The type specimen of the species for which this genus is proposed is a most interesting and suggestive but exasperatingly incomplete fossil; it certainly seems to indicate a primitive sabre-tooth, though the upper canine is almost the only characteristic feature that is left, and that is of such exceptional form as to render the systematic position of the creature open to some question. The specimen is a fragment of the right upper jaw, with the canine, p^2 and ^{3} intact, the root of p^1 and the anterior half of p^4 the premaxillary and all the incisors have been lost. Associated with the jaw-fragment is a portion of the right squamosal, which retains the root of the zygomatic process and the complete glenoid cavity.

The upper *canine* is a large and formidable fang, or tusk, as long, proportionally, as in *Dinictis*, the sabre of which it resembles, though with certain striking differences. The external side of the tooth is simple and convex, but the lingual side is very different from that seen in any of the White River or later machairodonts, and decidedly peculiar. Prominent and conspicuous enamel ridges run down the anterior and posterior borders for the full height of the crown. The posterior ridge forms the border, but the anterior one is demarcated by a fine groove. Between the ridges the lingual face is convex, but only moderately so. The tusk is so much compressed laterally, that the antero-posterior diameter greatly exceeds the transverse, and the tooth, aside from the internal ridges, suggests the sabre form, although it is but slightly curved and is as straight as in *Dinictis*. A peculiar feature is the anterior face, which is not an edge, as it is in all other known machairodonts, but a narrow surface, and, seen from the front, resembles a rodent incisor. The canine, thus, has much about it that is peculiar; it suggests, nevertheless, a laniary sabre, at the first glance, chiefly because of its great length, projecting relatively far below the level of the premolars, and its thinness transversely, which is so different from the thick oval cross-section of the canine of true felines.

The *premolars* are four in number and are closely set, without diastemata. P^1 has lost its crown, but the root remains and is in close contact with that of the canine; this root is grooved, as though originally double, and, though curiously stout, it can have supported but a small and simple crown. P^2 is two-rooted and its crown is a simple, compressed, and trenchant cone, without accessory cusps or tubercles. P^3 is decidedly larger than p^2, but similar in form and equally simple; these three premolars are all small, while p^4 is very much larger, especially in dorso-ventral height of crown, and, though the posterior cusp has been broken off, the tooth was evidently a sectorial. The principal cusp is a high, sharp-pointed, and compressed cone, without anterior basal cusp, as is also true of the White River sabre-tooth *Dinictis*. The posterior part of the shearing blade, or tritocone, which was originally present, has been lost, and its size and shape can only be estimated.

This dentition, so far as it is preserved, is more simple and primitive than that in any of the later machairodonts, but, in spite of its incompleteness, the fossil is, most probably, one of the Fissipedia and not a creodont. One is tempted to identify it with the extraordinary *Apataelurus*, of the Uinta, the upper teeth of which are unknown, but that "sham sabre-tooth" is an undoubted creodont,[10] which *Eosictis*, almost certainly, is not; the very large size of p^4 is against any such reference. However, the loss of the upper molars and of all the lower teeth makes it impossible to be entirely confident of the reference.

On a preceding page it was mentioned that a small fragment of the squamosal bone was found in association with the jaw just described, and this is of considerable interest, for it includes an uninjured glenoid cavity, which is very catlike and might almost belong to the White River *Eusmilus*, which has the most advanced and highly specialized dentition and jaw-mechanism of all known sabre-tooths of whatever geological date.[11] In *Eosictis* the glenoid cavity is very deeply concave in the antero-posterior dimension, with preglenoid and postglenoid processes of almost equal height, thus giving to the cavity a remarkably catlike appearance, though the preglenoid process is higher than in *Felis* and more as in *Eusmilus*. Unless this skull fragment has a merely accidental association with the jaw, it lends strong support to the reference of this genus to the machairodonts. In view of the extreme rarity of fossils in the Duchesne River beds, such accidental association is very unlikely.

Dr. John Clark informs me that, when the type specimen was discovered, crumbling remains of a skull and impressions in the matrix suggested to him, even before the teeth were fully exposed, that the animal was a cat, in the broad sense of the term. Assuming the correctness of this reference, the fossil is of the greatest interest, as recording the first known appearance of the cat family and the sabre-tooth subfamily in the Western Hemisphere.

SYSTEMATIC POSITION OF *EOSICTIS*

Examination of this fossil immediately suggests the question of its possible relation to the "sham sabre-tooth" creodont, *Apataelurus*,[12] of the preceding Uinta formation. Comparison of the two supposedly distinct genera is made difficult by the fact that they have no structures in common, one having only the lower jaw and teeth, the other only an incomplete upper dentition. It is, however, possible to make certain inferences from each fossil as to the missing parts. *Apataelurus* is a creodont, which imitates a sabre-tooth cat in a really astonishing way, though not more surprisingly than the imitation of the same group displayed by the marsupial *Thylacosmilus*, of the Argentinian Pliocene. No doubt this creodont had two pairs of carnassial teeth, p^4 and m_1 and m^1 and m_2, of which the latter pair were much the larger and more specialized, while m_1 is notably small and is exceeded in size by p_4. From the small size of m_1 we may infer that its fellow in the shearing pair, p^4, was likewise small and that the enlarged and efficient pair of carnassials was made up of molars, m^1 and m^2. All this is typically creodont and different from the fissiped arrangement, in which there are either no carnassials, as in the bears, or else only a single pair, always made up of p^4 and m_1.

As to *Eosictis*, with the incomplete p^4 and in the absence of the mandible and all the lower teeth, it would be premature to make any positive statements. On the other hand, the very large relative size of p^4, which so conspicuously exceeds the other premolars, makes it extremely probable that this tooth was one of the single sectorial pair, typically the fissiped arrangement. If this is true, there can be no hesitation in referring *Eosictis* to the Fissipedia, and the character of the canine strongly suggests the machairodonts.

Eosictis avinoffi sp. nov.

The specific name is given in recognition of the extraordinary kindness of Director Avinoff to Mr. Horsfall and myself during our visit to the Carnegie Museum in October and November 1941.

As there is but one species, it is unnecessary to add a specific diagnosis, other than the dimensions: the catalogue number is C.M. No. 11,847.

MEASUREMENTS

Upper canine, antero-posterior diameter	16 mm.
Upper canine, height of crown (enamel cap)	34
Upper premolar series, length (est.)	40.7
P^2, antero-posterior diameter	10
P^2, height of crown	6.6
P^3, anterior-posterior diameter	10
P^3, height of crown	8
P^4, anterior-posterior diameter (est.)	18
P^4, height of protocone	13.6

Horizon: Duchesne River, Halfway substage.
Locality: West Fork of Halfway Hollow, Uinta County, Utah.

ORDER **RODENTIA** CUVIER

Like all other small mammals, rodents are disproportionally rare in the Duchesne River beds; Peterson lists only two species, which have been described by Burke.

FAMILY **ADJIDAUMIDAE** MILLER AND GIDLEY

As I have had no opportunity to examine the Duchesne River rodents, the only course open to me is to reproduce Burke's descriptions of the Carnegie Museum material.

[10] Scott, W. B., *Ann. Carnegie Mus.* 27: 113–120, 1938.
[11] Scott and Jepsen, *Trans. Amer. Philos. Soc.* 28: 141–146, 1936.
[12] Scott, W. B., *Ann. Carnegie Mus.* 27: 113, 1938.

PROTADJIDAUMO Burke

Protadjidaumo Burke, *Ann. Carnegie Mus.* **23**: 394, 1934.

Diagnosis: Pattern of cheek-teeth essentially as in *Adjidaumo*, but cusps more bunoid and connecting crests weaker, P_4 with low posterior cingulum connecting with hypolophid, anterior valley of M_1 opening freely externally.

If not directly ancestral to the genus *Adjidaumo* in later Oligocene horizons, *Protadjidaumo* is certainly closely related to the latter genus. By converting the posterior cingulum of P_4 into a cingulum crest (a process which appears to be under way already), developing a dam near the external exit of the anterior valley of M_1, raising and strengthening its connecting crests and attaining a more crescentic condition of its outer cusps, *Protadjidaumo typus* might easily give rise to a conservative species of *Adjidaumo* such as *Adjidaumo minutus* (Cope).[13]

Protadjidaumo typus Burke

Protadjidaumo typus Burke, *Ann. Carnegie Mus.* **23**: 394, 1934.

Except for the dimensions, Burke gives no diagnosis of this species:

Measurements

P_4 antero-posterior	1.1 mm.
P_4 transverse	1.0
M_1 antero-posterior	1.1
M_1 transverse	1.1
M_2 antero-posterior	1.2
M_2 transverse	1.1

Horizon: Duchesne River Oligocene Series, Lapoint Horizon.

Locality: Fourteen miles west of Vernal, Uinta County, Utah, one-half mile north of Vernal-Lapoint road.

Family PARAMYIDAE Miller and Gidley

LEPTOTOMUS Matthew

Leptotomus Matthew, *Bull. Amer. Mus. Nat. Hist.* **28**: 50, 1910. (Subgenus of *Paramys*.)

Leptotomus Burke, *Ann. Carnegie Mus.* **23**: 391, 1934.

Diagnosis: Cheek teeth brachyodont, the pattern recalling that of *Prosciurus*: low intermediate tubercles on protolophs and metalophs of P^4, M^1 and M^2, the metaconules more prominent than the protoconules; external intermediate cuspules on P^4 and the molars; molars with prominent anterior but much reduced posterior cingula and slight internal enamel invaginations. Low posterior transverse crest in M^3. P^4 molariform, its anterior cingulum more reduced than that of molars, the posterior cingulum prominent, the metacone and metaconule nearly equal in size. Species larger than *Leptotomus grangeri* Matthew.

Anterior to P^4 the maxillary is shattered and P^3 is not preserved. P^4 is almost as much enlarged as the molars, is not prolonged forward, and has a well-developed anterior cingulum. Its anterior cross-valley is not as wide as the corresponding cross-valleys in the molars, and the posterior cingulum is low. P^4 bears a very small cuspule between the external cusps.

In the molars the anterior cingula are well-developed. The posterior cingula were much reduced. In M^1 the external intermediate cuspule is prominent. A slight enamel invagination occurs in the internal face of this molar. M^2 is badly worn in the region of the protoloph, but the presence of a low protoconule is still indicated. The external intermediate cuspule is somewhat larger than the corresponding cuspule in M^1, and an enamel invagination in the postero-internal wall of the tooth is a little shallower than the invagination seen in M^1. M^3 bears a complete anterior transverse crest. The posterior basin has been badly worn, but a short crest runs internally from what appears to be a low external intermediate cuspule, and there are indications of a low posterior transverse crest.

The zygoma is heavy and sturdy, and begins to flare out laterally posterior to M^1. Anteriorly it extends obliquely upward and forward.[14]

Leptotomus kayi Burke

Leptotomus kayi Burke, *Ann. Carnegie Mus.* **23**: 391, 1934.

This species agrees best with a smaller form, probably a distinct species, which is found in the Myton Member of the Uinta Eocene Series, the lower jaw of which resembles that found in *Leptotomus grangeri* Matthew. P^4 was the only superior cheek-tooth found with the type of *Leptotomus grangeri* Matthew. From it, P^4 of both *Leptotomus kayi* m. and the Myton Member species differ in being more molariform, the trigonal outline seen in P^4 of *Leptotomus grangeri* Matthew is less evident, and the anterior cingulum is more reduced. In *Leptotomus kayi* m. the metacone and metaconule appear to be nearer the same size than in *Leptotomus grangeri* Matthew, and the Duchesne River Oligocene species exceeds both *Leptotomus grangeri* Matthew and the species from the Myton Member in size.

Measurements

P^4 antero-posterior	4.8 mm.
P^4 transverse	5.4
M^1 antero-posterior	4.8
M^1 transverse	5.7
M^2 antero-posterior	4.9
M^2 transverse	5.0
M^3 antero-posterior	5.1
M^3 transverse	5.0[15]

Horizon: Duchesne River near base of Randlett substage.

Locality: 2 miles N.E. of Randlett Point, Uinta County, Utah.

Order LAGOMORPHA Brandt

Family LEPORIDAE Gray

MYTONOLAGUS Burke

Mytonolagus Burke, *Ann. Carnegie Mus.* **23**: 400, 1934.

Mytonolagus sp. Peterson

Mytonolagus sp. Peterson, *Ann. Carnegie Mus.* **23**: 374, 1934.

Peterson lists, without description or specific name, this species of a Uinta genus, and, so far as I can discover, Burke merely mentions it in passing: "Both the Duchesne River and Wagonhound specimens are fragmentary, but are referred to the same genus as the Myton fossils."

Order ARTIODACTYLA Owen

Throughout the Lower and Middle Eocene of North America the preponderant ungulates were perisso-

[13] Burke, J. J., *Ann. Carnegie Mus.* **23**: 394, 1934.

[14] *Ibid.*: 392–393.

[15] *Ibid.*: 393.

dactyls and artiodactyls were rare. Between the Washakie and the Uinta there was a striking change in the fauna, though there is no distinct physical or stratigraphic break at this level. The outstanding feature of the change is the great increase in the number of artiodactyls, especially the selenodonts. This revolutionary change, which is apparently so sudden and unheralded, is obviously to be interpreted as an immigration, though whence that migration came is not altogether clear. Some of the families were undoubtedly derived from eastern Asia, but the two most abundant of the newly appearing families, the Merycoidodontidae and the Agriochoeridae, are not definitely known to have occurred in the Old World. These preëminently North American groups, which flourished so conspicuously throughout the Oligocene, Miocene, and early Pliocene, may have arrived in the western United States from some other part of North America. On the whole, however, it seems more probable that the region which is now Siberia, the early Tertiary mammals of which are still unknown, was the source of origin. Mongolia may have contributed some of the groups, as it almost certainly did to the White River immigration, but hardly the most abundant and characteristic families.

The list of artiodactyl families which form part of the Duchesne River fauna is quite considerable, especially when it is remembered how very scanty that fauna is. Other probably immigrant families which arrived in the Wasatch and continued through the Duchesne River, are the Dichobunidae and the Helohyidae. According to my views, the achaenodont subfamily of the Entelodontidae did not reach the White River, and whether it occurs in the Duchesne River, will depend upon the disposition made of Colbert's enigmatic *Brachyhyops*,[16] which he refers to the Choeropotamidae. If that reference is confirmed, *Brachyhyops* is assuredly an immigrant, but it may prove to be an achaenodont, in which case it is probably of North American origin. The uncertainty of determination arises from the abraded condition of the teeth, from which all trace of cuspidation has been worn away.

It is my belief that no forerunner of any pecoran family occurs in the Eocene or Oligocene of North America, for I refer all the selenodont groups of these epochs to the Tylopoda, in which I would include the Camelidae, Hypertragulidae, Merycoidodontidae, and Agriochoeridae.

Suborder **PALAEODONTA** Matthew

Matthew proposed this term to include the early and primitive artiodactyls, for which no appropriate place could be found in any of Flower's four suborders.

[16] Colbert, E. H., *Ann. Carnegie Mus.* **27**: 87–108, 1938.

Family **HELOHYIDAE** Marsh

HELOHYUS Marsh

?**Helohyus** sp. Peterson

Peterson refers,[17] though but tentatively, certain scattered teeth to this genus, without venturing to give any specific name. It seems unlikely that a typically Bridger genus should persist to so late a period as the Duchesne River, but it is not at all impossible.

According to Colbert's view, the European family of the Choeropotamidae is probably represented in this fauna by the extraordinary genus which he named *Brachyhyops*.

Family **?CHOEROPOTAMIDAE** Owen

The Carnegie Museum expedition of 1934, directed by Mr. J. LeRoy Kay, explored the Beaver Divide, Wyoming, and collected some new and extremely interesting fossils, which, as previously explained, are here provisionally included in the Duchesne River fauna. One of these, a well-preserved but much distorted skull, without mandible, was described by Dr. E. H. Colbert, of the American Museum of Natural History, who named it *Brachyhyops wyomingensis*.

BRACHYHYOPS Colbert

Brachyhyops Colbert, *Am. Jour. Sci.*, ser. 5, **33**: 473, 1937.
Brachyhyops Colbert, *Ann. Carnegie Mus.* **27**: 87, 1938.

Diagnosis: Of medium size, the skull being of a length comparable to that of the skull of a modern peccary. Dentition $^{3-(?)-1-4-3}$; lower dentition unknown. Cheek teeth closely comparable to those of *Chœropotamus*, *Helohyus*, *Achænodon* and *Parahyus*, being near to the latter in size. Cranium broad, having widely separated parietal crests. Orbit closed posteriorly; situated above the last two molars. Muzzle relatively short, so that the postorbital length exceeds somewhat the preorbital length of the skull. Zygomatic arch vertically expanded behind and below the orbit, somewhat in the manner of the entelodonts. Skull transversely broad, due to the lateral expansion of the arches. Glenoids shallow and broad and relatively low, being about on a level with the occlusal line of the upper cheek teeth. Paroccipital processes short and pterygoids weak. Posterior nares extending to the middle of the upper third molar. Basicranium primitive, the arrangement of the foramina being like that in the dichobunids.[18]

Colbert's excellent, detailed description leaves little room for additions and is, therefore, reproduced in somewhat shortened form, with the addition of merely a few minor details.

The holotype of the genus is a skull (Carnegie Museum, No. 12,048), which lacks the mandible and is much distorted by vertical, or dorso-ventral pressure. It is, however, not difficult to correct the distortion, and there is every reason to believe that the drawings made by the late Mr. Sidney Prentice for

[17] Peterson, O. A., *Ann. Carnegie Mus.* **23**: 375, 1934.
[18] Colbert, E. H., *op. cit.*: 87–88.

Dr. Colbert's description and by Mr. R. B. Horsfall for the present paper, give a sufficiently accurate conception of this specimen, which, so far, is entirely unique. It is a curious and unfortunate fact, that so many of the most interesting and significant fossil mammals should remain unique even after long periods of active and meticulous collecting. Marsh's skull of *Tillotherium*, from the Bridger, described in 1873, *Protoptychus* Scott, 1895, and *Eotrigonias* Wood, from the Uinta, and, most remarkable of all, Leidy's *Leptictis*, from the White River, which was named in 1868, are still represented by single specimens, despite the most careful and long continued search. So far, *Brachyhyops* belongs in the same category, and nothing else remotely resembling it has ever been found in any American formation.

DENTITION

The dental formula is: i?, c^1, p^4, m^3. Only upper teeth are known, and, of these, the *incisors*, together with both premaxillae, have been destroyed by weathering. The *canine*, missing on both sides, is shown, by the matrix-filled alveoli, to have been a comparatively large and formidable tusk. It was, evidently, much compressed laterally and seems to have been a laniary sabre. The grinding teeth are arranged in nearly straight lines, without diastemata, and are in such an advanced state of wear, that nothing can be made out concerning their cuspidation.

Of the *premolars*, p^1 and p^4 are in place, on one side or the other, while p^2 and p^3 are represented only by the alveoli. P^1 is small, evidently much the smallest of the cheek-teeth, two-rooted, and with a simple, compressed-conical crown. Likewise, p^2 and p^3 are two-rooted; their alveoli indicate that they are somewhat larger than p^1, which they probably resembled in form. Of p^4, the crown remains, but is so extremely abraded that the cuspidation is unrecognizable. Colbert says (1938: 94), however, that "it gives indications of having a large, single outer cone and a smaller inner cone." Differing from the other premolars, the crown of p^4 is wider transversely than elongate antero-posteriorly, and the alveolus of the inner root is conspicuously larger than the two external ones.

Cuspidation of the *molars* has been removed by abrasion, but it may be confidently assumed that the pattern was bunodont and, presumably, quadritubercular. The three teeth of the series are of different sizes, m^1 being the smallest and m^2 the largest. M^3 is asymmetrical, because the posterior half of the crown is much narrower than the anterior half.

SKULL

Partly on account of the advanced age of the animal before fossilization and, partly, because of its crushed and distorted condition, which has resulted in a multitude of matrix-filled cracks, sutures are seldom clearly visible, and most of those which are shown in the drawings are restorations made from faint suggestions. As the outlines of the various bones of the skull are so seldom distinctly marked, it will be advisable to describe the five different aspects of the skull.

Side View.—The disproportionate shortness of the preorbital and the elongation of the postorbital region are conspicuous features. Even after making all due allowance for the effects of downcrushing, this skull is very low dorso-ventrally. The orbit is relatively small, completely encircled in bone and placed high in the face, seemingly projecting above the level of the forehead. A comparison of the late Mr. Prentice's drawing of the side view of the skull (Colbert's fig. 3) with that which Mr. Horsfall has made for this paper (pl. VI, fig. 1) will show how the two artists have differed in several minor details in their manner of compensating the effects of distortion. Mr. Horsfall's drawing shows the roof of the orbit projecting above the level of the forehead, while Mr. Prentice has raised it less. Likewise, there are differences in the shape given to the anterior end of the zygomatic process of the squamosal, the squamoso-jugal suture, and that between the jugal and the postorbital process of the frontal. Mr. Horsfall has restored the sutures bounding the lachrymal, maxillary, nasals, and frontals in a way that Mr. Prentice has refrained from doing, but, in my judgment, these sutures are substantially correct, according to the faint indications. Among ungulates of so early a geological date, *Brachyhyops* is the only known instance of the closed orbit, and the flat-topped cranium is likewise unique. In shape, the orbit seems to be circular and to have suffered but slightly from distortion.

Another conspicuous feature of the side view of this skull is the relatively great dorso-ventral breadth of the jugal, which suggests an incipient suborbital process such as characterizes the White River genera, *Archaeotherium* and *Entelodon*.

The *maxillary* is long and low, extending beneath the orbit, of which it forms the floor; at the other end of the bone the canine alveolus forms a facial convexity. The infraorbital foramen is rather small and is placed low down and relatively far in front of the orbit, opening above the hinder part of p^3. The *lachrymal* is of somewhat uncertain size and shape, but appears to be large and well extended upon the face and to form a considerable part of the anterior boundary of the orbit; apparently the foramen is concealed within the orbit, and the spine is a vestigial tubercle.

The *jugal* is a comparatively large and massive bone and bounds the orbit below, in front and behind, making up a large part of the facial region; its ventral border is raised and thickened and converts the external surface into a shallow concavity. This border descends from the anterior end of the maxillary to the bottom of the suborbital process, whence it rises to the subzygomatic extension. Posteriorly, the jugal forms about half of the hinder orbital margin, has a relatively long suture with the postorbital process of the frontal, and is deeply notched by the zygomatic process of the squamosal, beneath which it extends almost to the glenoid cavity.

The zygomatic process of the *squamosal* is conspicuously short and stout, broad dorso-ventrally, and thick transversely. The external surface is concave, the concavity having, presumably, been deepened by the downcrushing; the postglenoid process, which is visible from the side, projects downward below the level of the zygoma. No distinct post-tympanic process is to be seen, but, if it was originally present, it must have coalesced with the paroccipital. The entrance to the auditory meatus is a relatively broad space between the postglenoid and paroccipital processes, but this region is somewhat damaged on both sides and is, therefore, difficult to interpret; the paroccipital is short and stout, ending distally in a bluntly rounded point.

The cranial portion of the squamosal is comparatively small and articulates with the parietal by one of the few distinctly visible sutures which this specimen displays. Except for the squamosal, the whole lateral wall of the brain-case appears to be formed by the *parietal*, for the limits of the supra- and exoccipitals are quite indistinguishable. The brain-case is of small capacity, and its external portion is but slightly convex on the outer side, becoming concave dorsally, where the temporal ridge bounds it above. Posteriorly, the parietals are drawn out into the winglike occipital processes, which, though less extended than in the White River *Merycoidodon*, are yet very conspicuous and overhang the condyles. The ventral portion of the exoccipitals extends strongly backward and is but partially concealed, in side view, by the paroccipital; the condyle projects obliquely posteriorly and ventrally and is a conspicuous feature in this aspect of the skull.

Top View.—This is, perhaps, the most peculiar and exceptional of all the different views and emphasizes the completely isolated position of *Brachyhyops* among the known North American artiodactyls. The most unusual feature of this skull is the broad, flattened cranial roof, the sagittal crest having been reduced to a very low ridge, the presence of which may be principally due to the great age of this individual. The crest appears to have been buried in the sinus-filled cranial roof. This flattening is all the more remarkable, because the brain capacity was so limited, and was brought about by the great expansion of the cranial sinuses and the diploetic structure of the skull roof, and was, no doubt, correlated with the growth of the temporal muscles. The perissodactyl family of the brontotheres offers an interesting parallel to this condition. As late as the Middle Eocene, the brontotheres all had a long, prominent sagittal crest; then, in the Upper Eocene and Lower Oligocene the crest disappeared and the top of the cranium became broad and flat, yet the brain did not keep pace with this development and remained absurdly small.

Returning to *Brachyhyops*, we may note that, in its present crushed state, the cranial roof is concave on each side of the sagittal ridge, but how far these concavities were produced by the distortion, it is difficult to estimate.

An even more striking feature in the dorsal aspect of this skull is its great breadth in proportion to the antero-posterior length; the ratio, making the necessary allowance for the missing premaxillaries, is, approximately, length 3 : 2 width. This relatively great breadth is due to the remarkable expansion of the zygomatic arches, which, in connection with the narrow brain-case, makes the temporal openings exaggeratedly large and of a somewhat irregularly circular shape. The postorbital part of the skull, whether seen from above or below, has a curiously rectangular outline (pl. VI, figs. 1*a*, 1*b*), the anterior border of which bisects the postorbital processes of the frontals and the hinder border passes through the postglenoids. The vertical aspect, dorsal or ventral, strongly suggests that of the White River entelodont, *Archaeotherium*, though the brain capacity is not quite so far reduced as in that genus. The small orbits seem to have a somewhat oblique position, presenting slightly upward, as well as laterally, although, here again, this may be a result of distortion. Assuming that the obliquity was original and not entirely due to crushing, and taking it together with the projection of the orbital roof above the plane of the forehead, the whole structure suggests that the animal was aquatic in habits. However, the evidence is insufficient to justify positive statements in this matter, especially since it is so difficult to estimate the amount of displacement.

No sutures are visible in the cranial roof and those which appear in the drawings of the facial region are largely conjectural. The two artists differ somewhat in their interpretation of the effects of distortion and shattering of the bones, but both agree in making the nasals narrow and transversely convex. In this view, the face narrows gradually forward from the orbits to a point above the second premolar, thence broadening to the muzzle, the breadth of which is increased by the canine sheaths. The comparative massiveness of the zygomatic arches is more conspicuous in this aspect of the skull than in the side view, as is also their great lateral extension, the cranial roof making up less than one-third of the total skull-width. The backward slope of the occiput is plainly shown and also the median notch in the dorsal margin of the foramen magnum.

Base View.—The ventral side of the skull shows decidedly less distortion than the dorsal side, suggesting that the downcrushing had pushed the comparatively thin and fragile roof down against an unyielding base. The outlines of the skull, zygomatic arches and temporal openings are, of course, the same as in the top view, but the massiveness of the arches and their wide extension from the sides of the skull, seem to be even more striking when seen from below. The bony palate is elongate, narrow and but slightly concave; it broadens anteriorly as the tooth-rows diverge. The sutures between the palatine bones and the palatine processes of the maxillaries remain distinct; the former are relatively small and extend forward only to the hinder edge of m^1. The posterior palatine foramina are comparatively large and open between the palatine bones and the maxillary plates. The posterior nares are relatively small and are placed far back, their front margin on a line with the middle of m^3; the narial canal is, consequently, short.

The basioccipital is broad and bears a keel; the condyles are small, transversely extended and widely separated below. No trace of a tympanic bulla remains and its original presence or absence is a matter of conjecture.

Of the *Cranial Foramina*, Colbert says:[19]

The ethmoid and optic foramina are not discernible, but the foramen lacerum anterius and foramen rotundum are plainly visible, enclosed in a common vestibule. The foramen ovale is situated just medially to the internal border of the glenoid, while the foramen lacerum medius [*sic*] and the opening for the eustachian canal are behind it and separated from it by a bony ridge. The foramen lacerum anterius is isolated and located opposite and median to the postglenoid process, while the stylomastoid foramen is postero-lateral to the foramen lacerum posterius. The condylar foramen would seem to be single.

The arrangement of the basicranial foramina described above is characteristic of a primitive mammal, in that there is little tendency toward fusion or concrescence of the various openings in the cranial floor.

This genus displays its relationships to the primitive bunodont artiodactyls by reason of the lack of any appreciable compression of the structures on the floor of the brain case. Thus it may be contrasted with the achaenodonts and entelodonts in which the basicranium is greatly compressed by the backward shift of the glenoids.

Rear View.—This aspect of the skull is quite as strange and exceptional as the top and side views and has suffered much less from the effects of compression, unintelligible as that fact may be. The downcrushing has affected principally the top of the cranial regions, especially along the median line and in front of the occiput. This view shows plainly how little of the great skull-width is made up by the brain-case; that breadth is principally due to the wide expansion of the zygomatic arches, each one of which is nearly as broad as the occiput. The sutures are so obliterated, that the limits of the various occipital elements are indistinguishable. In the rear view the skull is conspicuously low and wide and the effects of crushing appear especially in several matrix-filled cracks, but there is little dislocation or faulting. The dorsal half of the occiput is made quite deeply concave by the wing-like extensions of the parietals, which are accompanied by extensions of the supraoccipital. The crest of the inion would seem to have projected but little above the cranial roof, before the downcrushing had much increased its prominence, for it was little, or not at all involved in that movement. Distally, the occipital surface becomes moderately convex and slopes downward and backward to the foramen magnum, forming the prominence which is conspicuous in the lateral view. The dorsal border of the foramen has a median notch, on each side of which the exoccipital forms a projection, ending in a bluntly rounded tip; notch and projections are plainly visible in figures 1*a* and 1*b* of plate VI. On each side of the median convexity, the occiput is bounded by a deep groove, setting it off from the paroccipital process. These processes are short and stout and project but slightly below the condyles; though they are not quite complete, it is evident that the paroccipitals have lost but little of their length. The foramen magnum is unusually wide transversely and low dorso-ventrally, proportions which seem not to have been much affected by the downcrushing.

Front View.—It is the anterior portion of the skull which has suffered most from the crushing and distortion, yet, even so, the anterior aspect displays certain characteristics which are not unworthy of mention. Seen from before, this skull is remarkably broad and low, even after making all due allowance for distortion, and gives an altogether different picture from the rear view. In this position, the orbits are especially conspicuous and seem to be larger than when viewed from the side. Their obliquity of position is also peculiar, as they slope from behind forward and inward, as well as downward and outward; it is this double obliquity, which makes the orbits present upward and forward, as well as laterally, and to which their prominence in both top and front views is to be ascribed. The great expansion of the zygomatic arches is not conspicuous in this position, except on the left side, where the spread has been manifestly increased by the downward pressure; on the right side, where the bones are in almost their natural position and the outer face of the arch is nearly vertical, the zygoma is largely concealed from the front by the prominent orbit and is given great dorso-ventral extent by the suborbital process of the jugal. The flat cranial roof is conspicuous in anterior view, because of the upward slope from orbit to crest of the inion, and is highly characteristic, for no other known Eocene, or Oligocene artiodactyl has a skull of this shape, though the White River peccary, *Perchoerus*, approximates it. Otherwise, this skull, when seen from the front, has a distinct resemblance to that of one of the larger oreodonts, such as *Promerycochoerus*. To this effect, the closed orbits largely contribute.

[19] Colbert, E. H., *op. cit.*

SYSTEMATIC POSITION OF *BRACHYHYOPS*

Dr. Colbert had the very great kindness to permit me to reproduce his figures of the *Brachyhyops* skull in the second edition of my *Land Mammals*,[20] in advance of his own publication.[21] In that work I tentatively referred the genus to the achaenodont subfamily of the Entelodontidae, though feeling little confidence in the assignment. Colbert, after an elaborate analysis of resemblances and differences, preferred to include it temporarily in the European family of the Choeropotamidae, to which he also referred the Wasatch genus *Parahyus*. That the two opinions differ but slightly, is shown by the fact that *Parahyus* has never been satisfactorily distinguished from *Achaenodon*, and Colbert's diagram of the phylogeny of the bunodont artiodactyls shows the Achaenodontinae and the Choeropotamidae in close juxtaposition. As it is impossible to decide the issue and will remain so, until the unworn dentition shall have been discovered, I have no objection to following Colbert's example. Even that hoped-for discovery may not solve the problem of the flat-topped cranium. For an explanation of that most exceptional condition it will be necessary to find one or more intermediate steps of gradation between the cranium with long and prominent sagittal crest and that with flattened, disploë-filled roof. All that we can say, with assurance, is that *Brachyhyops* is one of the primitive bunodont artiodactyls which Matthew grouped together under the name of Palaeodonta.

Brachyhyops wyomingensis Colbert

Brachyhyops wyomingensis Colbert, *Amer. Jour. Sci.*, Ser. 5, **23**: 473, 1933.
Idem: Ann. Carnegie Mus. **27**: 87–108, 1938.

The author gives no separate specific diagnosis, merely saying that the description of the species is the same as that of the genus. In the subjoined table the measurements are mine and where those given by Colbert are different, his figures are added in parentheses.

MEASUREMENTS

Upper canine, antero-posterior diameter of alveolus (est.)	15 mm.
Upper canine, transverse diameter of alveolus (est.)	11
Upper cheek-tooth series, length	109
Upper premolar series, length	63
P^1, antero-posterior diameter	10
P^1, transverse diameter	5
P^2, antero-posterior diameter of socket	17
P^3, antero-posterior diameter of socket	?15
P^4, antero-posterior diameter	15
P^4, transverse diameter	15
Upper molar series, length	44
M^1, antero-posterior diameter	14
M^1, transverse diameter	17
M^2, antero-posterior diameter	17 mm.
M^2, transverse diameter	20
M^3, antero-posterior diameter	15
M^3, transverse diameter	18
Skull, length occ. cond. to hinder edge of canine	206
Face, length anterior margin of orbit to hinder edge of canine	76
Skull, width over suborbital processes	158(161)
Skull, width at postglenoid processes	150
Skull, width over paroccipitals	77
Occiput, width across crest of inion	62
Occiput, height fr. basiocc. between condyles	59
Cranium, least width over temporal ridges	47
Temporal opening, antero-posterior diameter (right)	56
Temporal opening, transverse diameter (right)	43
Jugal, suborbital depth	54 (45)
Palate, width at canines	36
Palate, width at m^3	20

Horizon: "Uppermost level of the Uinta formation, immediately beneath the basal Oligocene [White River] agglomerate," Colbert. Probably Duchesne River.
Locality: Beaver Divide, ½ mile north of Wagon Bed Spring, Wyoming.

FAMILY **DICHOBUNIDAE** GILL

This is a characteristic Eocene family and to it Sinclair[22] referred all the Wasatch and Bridger artiodactyls, except the swine-like achaenodonts. The European *Dichobune*, which Schlosser believed to be ancestral to all the Pecora and Tragulina (and which may nearly represent the ancestor of the Tylopoda, as well) differs from all the American genera which have quinque-tuberculate molars, in having the unpaired fifth cusp in the posterior half of the crown. It is, thus, a metaconule, not a protoconule, which it is in all of the American genera. The occurrence in the Duchesne River of representatives of the dichobunids is an interesting survival of Eocene mammals.

PENTACEMYLUS PETERSON

Pentacemylus Peterson, *Ann. Carnegie Mus.* **21**: 72, 1931.

Pentacemylus Type: Two upper molars of the left side, inner portion of P_4, M_1 and M_2 of right side.

Characters obtained from the type material. Upper molars five pointed with protoconule well developed; no evidence of the vestigial hypocone on M^2 which is present in *Bunomeryx*.

Cusps of molars distinctly more crescentic than in *Homacodon* Marsh from the Bridger Eocene, *Bunomeryx* Wortman, and other genera from the Uinta (Horizon C) described later by Peterson. Upper molars of more nearly equal size than those in earlier genera. P_4 with well-developed deuteroconid and paraconid. Lower molars with slightly more advanced selenodont and hypsodont condition than in genera from the Bridger and the Uinta.

With the advanced condition of the cheek-dentition from that of the *Homacodonts* found in the lower strata there still persists the antero-median tubercle, well developed and entirely separated from the protocone as in *Bunomeryx*. Furthermore, the para- and mesostyles and the cingulum are even better developed than in *Bunomeryx*.

[20] Scott, W. B., *A History of Land Mammals in the Western Hemisphere*, 2nd ed., Macmillan, 1937.
[21] Colbert, E. H., *op. cit.:* 105, fig. 5.
[22] Sinclair, W. J., *Bull. Amer. Mus. Nat. Hist.* **33**: 267–295, 1914.

The well developed meso- and parastyles closely suggest *Protoreodon*, but the latter does not have the heavy cingulum bounding the tooth anteriorly, internally, and posteriorly, as is the case in *Pentacemylus*.

The advanced condition of the lower dentition of *Pentacemylus* consists chiefly in the greater development of the para- and deuteroconids and of the greater height of the tubercles along the buccal side of the molars, when compared with those in *Bunomeryx* in the lower horizons.

The new genus here proposed evidently stands very close to *Bunomeryx* of the lower horizons of the Uinta, while *Hylomeryx*, *Sphenomeryx* and *Mesomeryx* stand perhaps closer to *Homacodon*.

The triangular shape of the upper molars in *Pentacemylus* should be mentioned and the resemblance of these teeth to those of *Bunomeryx* emphasized even more than Peterson has done. According to Wortman's conception of these teeth in the latter genus:

> The full six cusps are found in the first superior molar only, while in the second molar there is but a vestige of the postero-internal cusp or hypocone. The evidence appears to be conclusive, therefore, that the true homological hypocone is in process of retrogressive disappearance, and in proportion as this cusp is reduced, the posterior intermediate is pushed out to take its place.[23]

Stehlin has developed this view in a manner that will be more fully dealt with in the succeeding section of this paper, which describes the Uinta fauna. As a preliminary, it may be mentioned that Stehlin divides the artiodactyls into three groups; (1) Hypoconifera; (2) Cainotheridae; (3) Euartiodactyla, concluding that only in the families Entelodontidae and Dichobunidae is the real hypocone present, the metaconule taking its place in other artiodactyls.[24] The molars of *Pentacemylus* certainly seem to bear this interpretation, but the extension of this conception to the order, in general, is a very different matter, which need not be discussed here.

In the upper molars of *Pentacemylus* the outer crescents have ribbed faces and all three styles are present; they are most prominent on m^2, especially the parastyle. The lower teeth are comparatively simple; the single remaining premolar has lost the outer half of the crown, but the remnant suffices to show the cuspidation. The main cusp, or protoconid, is an acutely pointed, compressed cone, which was evidently trenchant; there is a very distinct anterior basal cusp and a relatively large deuteroconid, both of which are sharp-pointed, and an enamel pocket is enclosed between the latter and the external wall, but no posterior basal cusp is visible. In the lower molars, two of which are preserved, m_1 and m_2, the external cusps are crescentic and enclose the internal ones, which are conical.

Pentacemylus progressus Peterson

Pentacemylus progressus Peterson, *Ann. Carnegie Mus.* **21**: 72, 1931.

Peterson gave no separate description of this species, other than the dimensions, which are repeated below, after verification; where corrections are necessary, the original figures are added in parenthesis.

MEASUREMENTS

M^1, antero-posterior diameter	6.5 mm.
M^1, transverse diameter	8 (7.5)
M^2, antero-posterior diameter (approx.)	6.5
M^2, transverse diameter	8 (7.5)
P_4 to m_2, length	19
M_1, antero-posterior diameter	6 (6.5)
M_1, transverse diameter	4
M_2, antero-posterior diameter	6.5
M_2, transverse diameter	4.5 (4)

Horizon: Duchesne River, Randlett substage.
Locality: 3 miles north of Leota Ranch, Uinta County, Utah.

SUBORDER **TYLOPODA** FLOWER

As just stated, I use this suborder in a much more comprehensive sense than is common among palaeontologists. Briefly stated, it is my conclusion that, in North America, from the upper Eocene to the lower Miocene, the camel tribe took the place which, in the Old World, was held by the Pecora and Tragulina. Aside from dental and skeletal resemblances, there is a remarkable geographical unity among the families here referred to the Tylopoda, which appears in the fact that, with one exception, none of these families has ever been found outside of North America before the Pliocene migrations of the llamas to South America and the true camels to Asia. Most of these families and subfamilies are exclusively North American in known distribution, but I cannot doubt that the earliest origin of all these groups and the common ancestry of all the selenodonts will be found in Asia, perhaps in Mongolia, perhaps in Siberia.

So great is the diversity among the Tylopoda, as here interpreted, so extreme the aberrance of certain families and so great the evolutionary changes that took place within the recorded history of each family, that it is well-nigh impossible to frame a definition of the suborder without rewriting that history; such definition may be dispensed with here.

FAMILY **CAMELIDAE** GRAY

This is an entirely natural and clearly circumscribed group, which, from the early Eocene to the end of the Pleistocene, was conspicuous in North America and, until the late Pliocene, was confined to that continent. Indeed, throughout all that time, horses and camels were the most characteristic elements of the North American fauna. Within this family, there are several distinct tribes, or phyla, which apparently began their divergent careers in White River times. The giraffe-like camels, the gazelle-like camels and the giant llama-camels are the most important of these tribes, but in the Duchesne River, as in the Uinta, only the single stem has been found. The occurrence of the very primitive *Eotylopus* in the lowest White River, or

[23] Wortman, J. L., *Bull. Amer. Mus. Nat. Hist.* **10**: 101, 1898.
[24] Stehlin, H. G., *Abhand. d. schweiz. Paläont. Ges.* **27**: 1910.

Chadron substage, is an indication that other tribes remain to be discovered in the Duchesne River, if not in the Uinta, also. Two presumably Duchesne River genera have been named, one from the Uinta Basin in Utah, the other from Beaver Divide in Wyoming.

CAMELODON Granger

Camelodon Granger, *Bull. Amer. Mus. Nat. Hist.* **28**: 248, 1910.

The original description reads:

An extremely slender-jawed type of selenodont artiodactyl with the teeth, especially the premolars and the third molar, and *with a long diastema between the second and third premolars*, the latter character not being observed in other Uinta genera of this group. The second and third premolars are rather simple, compressed, and trenchant. Each has a long posterior crest and a shorter, more abrupt anterior one. On each tooth, but more marked on P_2, there is also a rudimentary internal posterior crest, running backward and somewhat inward from the tip of the protocone, and a long, narrow valley between it and the external posterior crest, a character well shown in the type of *Leptotragulus profectus* Matthew from the Titanotherium beds of Pipestone Springs, Montana; on P_3 the posterior and anterior basal cusps are barely indicated. In addition to strong fore and hind basal cusps there is [on P_4] a large cusp on the posterior inner face of the main cusp or protocone and a very small one on the inner side of the posterior basal cusp. The first and second molars are too much worn to show important characters. The last molar is very long and narrow, with well-rounded outer crescents and a strong fifth lobe with a median ridge and a small cusp on its inner side, as in *Protylopus*. There is a mental foramen below the second premolar. The symphysis extends backward to the posterior edge of the second premolar.

This genus shows strong resemblances to both *Leptotragulus* and *Protylopus*, especially in the premolar construction, and for that reason is placed in the Camelidae. It differs from *Leptotragulus* in the more simple P_2 and P_3 and the more complicated P_4. From *Protylopus* it is distinguished by the more selenodont character of the molars and the greater complication of P_4. *Leptomeryx* differs in having strong anterior and posterior basal cusps on P_2 and P_3, a broader P_4, and more highly specialized molars. *Hypertragulus* resembles the present genus in having a diastema back of P_2, but the premolars are much narrower antero-posteriorly, with high, sharp, pointed cusps, while in *Camelodon* these teeth are broad, with low, blunt cusps.

This description I have verified from the type-specimen and find it unnecessary to add anything, except to emphasize the difference from the Uinta camel, *Protylopus*, in the well-marked diastema between p_2 and p_3. Even in the White River *Poëbrotherium eximium* the diastemata are incipient. Granger gives no separate diagnosis of the species and, as there is but one species, *C. arapahovius*, the measurements of the type will suffice for that purpose. In the table below Granger's measurements are in parentheses.

Camelodon arapahovius Granger

Camelodon arapahovius Granger, *Bull. Amer. Mus. Nat. Hist.* **28**: 248, 1910.

This species is somewhat smaller than Peterson's *Poabromylus kayi*, which it resembles in many respects, yet manifestly belongs to a different genus.

In the following table, where Granger's measurements differ from mine, they are given in parentheses.

Measurements

Lower cheek-tooth series, length p_2–m_3 inc.	51 mm.
P_2, antero-posterior diameter	6
P_2, transverse diameter	2
Diastema between p_2 and p_3, length	5 (4.8)
P_3, antero-posterior diameter	7
P_3, transverse diameter	3
P_4, antero-posterior diameter	7
P_4, transverse diameter	4
Molar series, length	28 (27.5)
M_1, antero-posterior diameter	6
M_1, transverse diameter	5
M_2, antero-posterior diameter	9
M_2, transverse diameter	6
M_3, antero-posterior diameter	13
M_3, transverse diameter	5
Depth of jaw at p_2	5.7
Depth of jaw at m_2	10 (12) *

* This dimension probably makes allowance for the abraded ventral surface of the jaw.

Horizon: ? Uppermost Duchesne River: "From the Uinta beds (Diplacodon zone of the Beaver Divide)," Granger.

Locality: "Near Hailey, Wyoming."

POABROMYLUS Peterson

Poabromylus Peterson, *Ann. Carnegie Mus.* **21**: 75, 1931.

Peterson's description reads:

Generic Characters: Jaw slender; P_2, P_3, and P_4 of semi-equal length, as in *Poëbrotherium wilsoni*. P_3 and $_4$ relatively shorter and thicker; protoconid rounder and higher, and the basal accessories less distinct than in *Poëbrotherium*. There is a diastema in front of P_2, as in *Poëbrotherium wilsoni*. The molars are relatively shorter and less hypsodont than in *Poëbrotherium*. Animal nearly the size of *Poëbrotherium wilsoni*.

The diastema in front of P_2 is of considerable length (8 mm. from P_2 to where the jaw is broken). As already stated P_2 is close to P_3, the tooth was implanted by two roots, the alveoli of which are directly antero-posterior, indicating an antero-posterior diameter approximately equal to either of the posterior premolars. The protoconid of P_3 is rather high and terminates in a round pointed apex. The internal face of the main cusp (protoconid) has near its posterior base a cusp-like rib closely adhering to the main body of the protoconid. Together with the external body of the heel this rib on P_3 helps form a fossa which is open behind. Forward and inward from the protoconid there extends a sharp blade-like tubercle, which likewise helps to form a vertical shallow groove on the inner face of that portion of P_3, not unlike that in *Poëbrotherium*. P_4 is very similar to P_3, except that the posterior heel in P_4 is more pronounced, the plate-like tubercle on the inner face of the protoconid is slightly better indicated, and the anterior blade-like tubercle heavier. Altogether the premolars in *Poabromylus* are less trenchant than in *Poëbrotherium*.

The molars are narrow when compared with the known selenodonts of the Upper Eocene, with the possible exception of *Leptotragulus*. On the whole the molars as well as the two premolars just described, are most nearly like those of *Poëbrotherium wilsoni* of the White River Oligocene, though less hypsodont. The lingual face of M_2 and part of that of M_3 are destroyed, but enough remains to determine that, though the inner face of the molars have [*sic*] the vertical grooves and ridges more pronounced than in *Poëbrotherium*, they are on the whole most nearly comparable to those in that genus. A cingulum between the crescents on the external face of the molars is slightly indicated, especially on the first and second molars. The heel of M_3 is fully as large proportionally as that in *Poëbrotherium wilsoni*, but the pillar on the inner face at the junction with the postero-internal crescent in *Poëbrotherium* is in the present form represented only by a tubercle, as in *Leptotragulus* of the Uinta.

The proposed genus, just described, differs from "*Leptotragulus*" *profectus* Matthew (Bull. Amer. Mus. Nat. Hist., 1903, p. 224) by having the posterior internal crests of the premolars less developed and the anterior crests less pronounced. Judging from Matthew's illustration (*l. c.*) of the lower jaw, P_2 in "*Leptotragulus*" *profectus* is very probably a smaller tooth than that in *Poabromylus kayi* by having a diastema between P_2 and P_3, heavier heels, and the protoconid more nearly in the midbody of the premolars, and by a more complicated fifth lobe of M_3.

The proposed new genus *Poabromylus* is distinctly farther advanced than *Eotylopus reedi* from the Lower Oligocene of Wyoming and apparently cannot therefore be regarded as holding an intermediate position between *Protylopus* of the Uinta and *Poëbrotherium* of the White River Oligocene. The late Dr. Matthew has pointed out that *Eotylopus* stands closer to *Protylopus* of the Uinta Eocene than to *Poëbrotherium* of the Eastern Oligocene. From present evidence *Poabromylus* certainly stands closer to *Poëbrotherium* than to *Protylopus*.

No addition to Mr. Peterson's description is necessary, except to emphasize the distinction between *Poabromylus* Peterson and *Camelodon* Granger. The latter is the more specialized, as is shown in the conspicuous diastema between p_2 and p_3 and in the complication of the fifth lobe of m_3. It is unfortunate that in the type and only known specimen of *Camelodon* the molars are so worn as to make the cuspidation unrecognizable, but even in this unsatisfactory condition, the advanced specialization is plain and supports the reference, as here made, to a post-Uinta date.

There is considerable doubt and confusion concerning the phylogeny of the camels in Oligocene times, the difficulty being due to an *embarras de richesses* of the material at hand. The most ancient known member of the Camelidae is *Protylopus*, which is the only genus of the family yet found in the Uinta. In *Protylopus* the lower teeth form a continuous series, without diastemata; the upper teeth are more widely spaced, to permit the occlusion of the inferior premolars and canines. In the Duchesne River, we find *Poabromylus*, without diastemata, and *Camelodon*, in which there is a considerable space between p_2 and p_3. In the lower White River, the most ancient species of the commonest White River genus, *Poëbrotherium*, *P. eximium*, the spacing of the lower teeth is nearly the same as in *Protylopus*, but in the subsequent *P. wilsoni* and *P. labiatum* p^2 is isolated by considerable diastemata. As usually interpreted, the *law of orthogenesis would exclude Camelodon from the ancestry of Poëbrotherium*, for *P. eximium*, indubitably a member of the genus, retains the continuous arrangement of the lower teeth which characterizes the Eocene *Protylopus*. In a study recently made of the phylogeny of the horses and camels,[25] I found strong reasons to believe that in the anterior teeth, incisors, canines and premolars of both of these groups, there was much fluctuation and even reversal of development. If this conclusion is substantiated, the ancestral line of the camels may include *Camelodon* as well as *Poabromylus* and *Poëbrotherium eximium*.

Poabromylus kayi Peterson

Poabromylus kayi Peterson, *Ann. Carnegie Mus.* **21**: 75, 1931.

As but one species is known, the dimensions make a sufficient diagnosis. I have verified Peterson's figures, which, in the subjoined table, are put in parentheses, making a few changes, and have also added the measurements of a second specimen, of which only the lower molars are preserved.

MEASUREMENTS

	C. M. No. 11,856	*C. M. No. 11,753*
Total length of the jaw-fragment		75 (80) mm.
Depth of ramus at posterior part of symphysis		10
Depth of ramus beneath m_3 (est.)		18
P_3, antero-posterior diameter		8.5
P_3, greatest width		3.5 (3)
P_4, antero-posterior diameter		8
P_4, greatest width		5 (4)
Lower molar series, length	30 mm.	35
M_1, antero-posterior diameter	9	9 (9.5)
M_1, greatest width	8	6
M_2, antero-posterior diameter	8.5	11
M_2, greatest width	8	8 (8 approx.)
M_3, length	14	16 (21)
M_3, width of anterior crescent	8	9 (7)
M_3, width of posterior crescent	7	8
M_3, width of heel	4	5

Horizon: Duchesne River, Lapoint substage.
Locality: type, No. 11,753, 200 feet N. of *Teleodus* Quarry; No. 11,856, Uinta Co., Utah, near head of Halfway Hollow.

FAMILY ?**HYPERTRAGULIDAE** COPE

The materials which are supposedly referable to this family, so far found in the Duchesne River beds, are very rare and very fragmentary, even in this scanty fauna. Peterson listed a ?*Leptomeryx minutus*, of which the identification is extremely doubtful. The fossil here attributed to a hypertragulid *inc. sed.* is a jaw-fragment with three teeth which are selenodont, but so tiny, that analysis is very difficult. As these teeth are low-crowned, they cannot have belonged to *Hypisodus*, though they may indicate some ancestor of that genus. However, this is entirely conjectural and of small interest. Another individual, of a much

[25] Scott, W. B., *Trans. Amer. Philos. Soc.* **28**: 951, 1941.

larger animal, described below, is more significant, though of uncertain reference.

?Hypertragulid incertae sedis

(Pl. II, fig. 3)

The specimen in question is the fragment of a palate, with teeth on both sides; (C. M., No. 10,199) on the left are three teeth, p^3, p^4, and m^1, on the right the same two premolars. The third premolar is a sharp-pointed, compressed-conical tooth, with a prominent median rib on the buccal face and, on the lingual side, a strong cingulum, which thickens to a small cusp on the postero-internal angle; this cusp is carried on a separate, third root. P^4 is the usual bicrescentic tooth, so general among selenodonts; on the buccal face there are a median rib, a strong metastyle and much less distinct parastyle. The protocone is conical rather than crescentic, but the deuterocone is a very complete crescent, the horns of which reach the outer wall, embracing the protocone; the cingulum is prominent on the lingual face.

The first molar is very wide transversely, short antero-posteriorly and curiously asymmetrical, owing to the greater breadth of the anterior half of the crown; the external crescents have prominent median ribs and the para- and mesostyles are relatively large, but there is no metastyle. The antero-internal crescent is considerably larger than the postero-internal and the valley separating it from the outer cusp is notably wider than that separating the two posterior crescents, internal and external. The width of the anterior half of the crown is increased by the heavy internal cingulum, which surrounds the inner crescent and gives to the tooth an appearance of having three anterior cusps arranged in a transverse line, but none of these is a true conule. The hypocone is distinctly smaller than the protocone and its cingulum is much less prominent, except in the transverse valley, at the entrance of which it forms a pillar.

This tooth is distinctly quadritubercular in contradistinction from the quinque-tuberculate upper molars which characterize the upper molars of the oreodonts, agriochoerids and other Uinta selenodonts, except the Camelidae and Hypertragulidae. It is this feature which has occasioned the reference of the specimen to the hypertragulids.

MEASUREMENTS

P^3, antero-posterior diameter	6 mm.
P^3, transverse diameter	4
P^4, antero-posterior diameter	6
P^4, transverse diameter	8
M^1, antero-posterior diameter	7
M^1, transverse diameter (anterior half)	10
M^1, transverse diameter (posterior half)	9

Horizon: Duchesne River, Lapoint substage.
Locality: 2 miles N.E. of Randlett, Uinta Co., Utah.

FAMILY **?LEPTOMERYCIDAE** OSBORN

?LEPTOMERYX LEIDY

?Leptomeryx minutus Peterson

?*Leptomeryx minutus* Peterson, *Ann. Carnegie Mus.* **21**: 75, 386, 1931.

Under this term Peterson described a fragment of the right mandibular ramus of an extremely small selenodont artiodactyl, belonging, presumably, to this group, but entirely insufficient for any definitive reference. The typical specimen is a fragment of the lower jaw, containing three teeth, p_4, m_1 and m_2, in a well worn condition, though hardly to be called old. (C.M. No. 11,913) Peterson's description will suffice and it is, therefore, reproduced here.

Specific characters: The apices of proto- and deuteroconids of P_4 more completely separated and the postero-internal tubercle of less development than in the eastern forms (i. e. in South Dakota), while the antero-internal body of the tooth is nearly equally developed to that of other species. The molars appear to be equally hypsodont to those in the known species, the exit of the external cross-valley is also more solidly filled as in some forms of *Leptomeryx* [*L.* (?) *evansi*] and less styliform than in *Leptotragulus*, while the styloid-like cingulum on the antero-internal angle of M_2 is present as in *Leptotragulus*; absent in *Leptomeryx*. The specimen from the Duchesne River Oligocene is smaller than any hypertragulid heretofore discovered in the eastern Oligocene or in the Uinta Eocene.

MEASUREMENTS

Length P_4-M_2	13 mm.
Length P_4	4
Transverse diameter M_1	3
Length of M_1	4
Length of M_2	4.5
Transverse diameter of M_2	3

Horizon: Duchesne River Oligocene.
Locality: "North side of 'Red Narrows,' east of Tridell, Uinta County, Utah."

SUPERFAMILY **OREODONTOIDEA** GILL, 1872

Agriochoeroidae Hay, 1929
Merycoidodontoidea Thorpe, 1937

Whatever value the evidence may have, the most striking difference, on the face of that evidence, between the Uinta and the Duchesne River faunas is that, in the former, there is a great abundance of the family Merycoidodontidae, most representatives of which belong to the genus *Protoreodon*, while in the latter, they are relatively very rare. Making all due allowance for the extreme scarcity of Duchesne River fossils of all sorts, it seems unlikely that this difference is without meaning and due entirely to the accidents of collecting. The comparative rarity of merycoidodonts persists throughout the Chadron substage of the White River, but again gives place to an extraordinary abundance in the succeeding Brulé substage. The Agriochoeridae, on the other hand, are much better represented and are relatively more important than in any subsequent geological stage. In such an impoverished fauna, however, statistics are of little importance.

Of the Duchesne River agriochoerids, Peterson says:

From the mutilated skulls and jaws, as well as limb-bones, few important characters are obtained, but the detailed structure of the dentition affords nearly everything desired in the way of forms in the line to the Agriochoerids of the White River, John Day and other Oligocene deposits, while *Diplobunops* from the Uinta (Horizon C) . . . apparently represents a subfamily. The robustness of the premolar teeth, especially those anterior, in the latter genus is well marked when compared with those teeth in the new material noted in the following discussion. The step toward the molarization of P 4/4 in the new material is clearly an advance beyond *Protagriochoerus* and it is certainly a great advance upon that of *Diplobunops* when the similarities in the appendicular structure in the latter genus and *Agriochoerus* are considered.[26]

FAMILY **AGRIOCHOERIDAE** LEIDY

MESAGRIOCHOERUS PETERSON

Mesagriochoerus Peterson, *Ann. Carnegie Mus.* **23**: 377, 1934.

Principal Characters: $I\frac{3}{3}$, $C\frac{1}{1}$, $P\frac{4}{4}$, $M\frac{3}{3}$. P^1 two-rooted and crown laterally compressed. P^2 similar to P^1, but with posterior portion of greater transverse diameter. P^3 with a distinct deuterocone, placed well back, as well as a slight cingulum on the posterior face. P^4 with the apex of the protocone twinned and the initial step taken toward the postero-internal crescent. M^1 and M^2 with weak protoconule, while M^3 has little or no indication of such. P_1 with high, trenchant and caniniform crown characteristic of the group; P_2–P_3 with laterally compressed and simple crowns; P_4 with undeveloped deuteroconid, broad heel and the antero-internal body of the crown well indicated. Lower molars relatively long and narrow, limb bones slender when compared with those of *Diplobunops*. In size the animals range between *Protoreodon medius* and *P. parvus*.

The premaxillary is of fair proportionate size and in its alveolar border are inserted the roots of three incisors of nearly subequal size. The muzzle and facial region when compared with that of *Protoreodon* may be considered as medium long. There is a short diastema between canine and P^1 as in *Protagriochoerus*. The molar of the material under study perhaps suggests most nearly that of *Protagriochoerus;* it has not the abrupt lateral expansion seen in *Agriochoerus*, but the postorbital process is proportionally as robust, as is also the postorbital process of the frontal. The rather small and round orbit is, however, widely open posteriorly as in *Protagriochoerus* and later forms. The height of the alveolar border of the maxillary is relatively equal to that in *Agriochoerus* and the infraorbital foramen appears to be similarly located (above P^3).

As already stated, there are three incisors of approximately subequal size. The canine is well proportioned and of the characteristically D-shaped cross-section. $P\frac{1}{1}$ following the canine after a short diastema, has a simple crown, secant, laterally compressed, with an antero-posterior diameter somewhat less than in *Protagriochoerus*, but relatively greater than that in *Agriochoerus* in which P^1 is much reduced. The crown of P^2 is equally as sharp-pointed as that of P^1 but is of somewhat greater transverse diameter and has a prominent basal cingulum on the postero-internal angle, but not what can be regarded as a deuterocone. P^3 has a deuterocone fairly well defined, a weak antero-internal cingulum, and the indication of a vertical rib at the posterior part on the internal face of the protocone like the corresponding one in *Agriochoerus* which is, however, of greater prominence in the genus of the White River Oligocene. The latter genus also has the cingulum more strongly developed on the posterior face of P^3 and therefore the initiative step towards the development of a tetartocone or posterior inner crescent is more clearly before us. The most interesting and extraordinary feature in the genus here proposed is the stage in the development of p^4. The two distinct vertical ribs or styles on the internal face of the protocone of P^4 in *Protagriochoerus* heretofore overlooked by students, have in the present genus advanced farther. In the unworn tooth the apex of the protocone has reached the twinning stage, somewhat analogous to what we find in the molars of *Deltatherioides* and *Zalambdalestes* from the Mongolian Cretaceous described by Gregory and Simpson. . . . This structure of P^4 in our present specimen is so plainly evident that there can be no question as to the subsequent division of this tubercle into the proto- and tritocone, a step still further advanced in *Agriochoerus minimus* Douglass from the Thompson Creek Oligocene of Montana and completely effected in the agriochoerids of the White River Oligocene of Dakota. On the posterior horn of the deuterocone crescent of P^4 in our new material there is a distinct vertical style which cannot be regarded as anything but the initial rudiment of the postero-internal crescent again slightly advanced in *Agriochoerus minimus* and which, though small, is completely developed in the White River forms. The antero-external style of P^4 in the present genus is of greater prominence than in the White River representatives and suggests more closely that in *A, minimus*.[27]

The protoconules of M^1 and M^2 are reduced and that of M^3 slightly indicated or entirely absent. The para- and mesostyles in our specimen are reduced in size and the cingulum between them less firmly connected when compared with *Protagriochoerus* and *Diplobunops* from the upper Eocene, but the styles have not reached the sessile position and are not as gently rounded in shape as those in *Agriochoerus*, nor does the cross-valley between para- and metacone extend as far outward as in the latter genus, especially on M^1 and M^2. The lingual faces of the internal tubercles are also more sharply rounded than in *Agriochoerus.*

The ramus of the lower jaw is deep, thin, and the symphysis is long and robust. The second and third lower incisors, with slightly expanded crowns, are yet adhering to the front of the lower jaws and the median tooth was undoubtedly present. Though incisiform, the lower canine is slightly larger than the true incisors. P_1 has the characteristically high, sharply pointed and laterally compressed crown. There is a very slight diastema between the first and second premolars. P_2 is of relatively greater antero-posterior extent when compared with that of *Agriochoerus* and is also more compressed laterally. P_3 is much like P_2 except that it has a greater transverse diameter posteriorly. The primary tubercle (protoconid of P_4) is well developed and rises well above the rest of the crown; the metaconid though low, is distinctly formed, while the apex of the deuteroconid cannot be said to be entirely isolated from the internal face of the protoconid as is the case in *Agriochoerus.* The anterior crest of the protoconid appears to be almost as far along in its development as it is in *Agriochoerus*, while postero-internally there is yet no indication of a tubercle; in *Agriochoerus*, on the other hand, P_4 is completely molariform, plus the anterior crest of the trito-

[26] Peterson, O. A., *Ann. Carnegie Mus.* **23**: 377, 1934.

[27] *Ibid.:* 377 ff.

conid which was already developed before the quadritubercular structure of the tooth took place.

The lower molars are relatively long and narrow when compared with those of *Agriochoerus*. The basal styles on the lingual face of the molars, though already indicated are not as well developed as in the genus from the White River Oligocene. The heel of M_3 is of an equal proportionate length, but narrower than that in *Agriochoerus*.[28]

In comparing *Mesagriochoerus* with *Protagriochoerus*, it is at once observed that the latter genus is slightly over one-fifth larger, and P^4 not nearly as far along with regard to its indication of its molariform structure as that in *Mesagriochoerus*. While the general detailed structure of the molars is very similar, the protoconule in the species under description [i. e. *Mesagriochoerus primus*] is considerably less developed. The two genera here compared are quite similar and it seems quite probable that *Protagriochoerus* stands ancestral to *Mesagriochoerus*.[29]

In the Carnegie Museum there is a skull, with separate lower jaw (C.M. No. 12,080) collected after the publication of the paper just cited and (indeed, after Mr. Peterson's death) which is very much more complete than the holotype. It is, however, slightly distorted and largely covered by an extremely hard incrustation, which contains much hematite. This skull may be taken as the basis of a description more detailed than was possible from the original material.

DENTITION

The number of incisors cannot be determined from No. 12,080, but Peterson gave the formula as unreduced: $i\frac{3}{3}$, $c\frac{1}{1}$, $p\frac{4}{4}$, $m\frac{3}{3}$.

Upper Teeth.—The broken stumps of some *incisors* remain in the premaxillae and these are relatively larger than in *Agriochoerus* and the canine has the D-shaped cross-section that characterizes the oreodont superfamily, but is decidedly slender. This slenderness may, however, be sexual rather than specific. One of the most striking differences between the Duchesne and the White River genera is that, in the former, the tooth-rows are continuous and without diastemata, other than the narrow space into which the lower tusk (p_1) occluded. In this respect, *Protagriochoerus* and *Mesagriochoerus* agree with the merycoidodont family. In *Agriochoerus*, on the other hand, there are considerable diastemata between tusks and cheek-teeth, above and below, which is a departure from all the preceding members of the superfamily, save the aberrant *Diplobunops*.

The *premolars* are of especial interest, as being intermediate between those of the Uinta and those of the White River genus, a fact which Peterson emphasized by his choice of a name for the Duchesne River form. These teeth increase in size and complexity posteriorly, except that p^3 has a broader buccal face than have the teeth in front of and behind it. On the external side, the four premolars look much alike, compressed-conical and trenchant, with sharp edges and acute apices and having, in Leidy's phrase, cordate outlines; none of them has any basal cusps, or tubercles, anterior or posterior. Internally, however, matters are very different; p^1 is quite simple, with no cusps on the lingual face; p^2 is simple, except for two minute enamel pockets internally. P^1 and p^2 are each implanted by two roots, p^3 and p^4 by three, an inner root being added; p^3 has a distinct deuterocone, from which a ridge runs backward and outward to the external wall, enclosing a deep and relatively large enamel pocket; p^4 has the form usual among selenodonts, of two transversely placed crescents, but, as Peterson pointed out, the protocone has a twinned appearance, owing to the presence of an incipient deuterocone. Peterson also called attention to the existence of two "distinct vertical ribs, or styles on the internal face of the protocone," which occur in *Protagriochoerus* and are larger in *Mesagriochoerus*, and to the vertical style on the posterior horn of the inner crescent.

The *molars* are likewise of much interest, because of a character which is intermediate between the tooth pattern of *Agriochoerus* and that of *Protagriochoerus*, of the Uinta, the upper molars of the latter being much nearer to those of *Protoreodon*. Of the three molars, the first is decidedly the smallest and the third somewhat the largest; they retain, but in much reduced condition, the anterior intermediate cusps, or protoconules. The external crescents are distinctly more like those of *Agriochoerus*, especially those of m^3, in being more concave, in having more prominent external styles, especially on m^3, in which the mesostyle clearly approximates that of the latter genus in its rounded, bulbous form, the transverse valley extending into it and separating the adjacent horns of the outer crescents, while in m^1 and m^2 these horns are connected, as they are in *Protagriochoerus*, in which all three molars are alike in the junction of the external crescents. The internal crescents are very wide and low and have a decided resemblance to those of *Agriochoerus*.

Lower Teeth.—The *incisors* are small, with narrow roots and broadened crowns; the *canine* is like an incisor in form and function and appears to be one of that series, as in the Pecora. The *premolars* have the same transitional character as those of the upper jaw, between the Uinta and White River members of the family. P_1 is, of course, the functional canine, as is well-nigh universal throughout the superfamily, but is, like the upper canine, slender and compressed, again perhaps, a sexual feature; p_2 is a compressed and trenchant cone, without complications. P_3 is similar externally, but lower and more elongate antero-posteriorly; internally, this tooth differs slightly in the two specimens; in the type there is a minute deuteroconid and a very low ridge of enamel below it, but in No. 12,080 there is no distinct deuteroconid but a ridge parallel to the protoconid and enclosing with it a nar-

[28] *Ibid.*: 381.

[29] *Ibid.*: 384.

row enamel pocket. P_4 is evidently beginning to take on the molar pattern, which is fully attained in *Agriochoerus*, a feature that is very rare among artiodactyls. Again, there is a small difference between the two individuals, which is probably due to wear; in the unworn tooth there is a large deuteroconid, which rises almost as high as the principal cusp; there are also a distinct, basin-like heel and an anterior basal cusp. In the worn tooth the same elements may be detected, but they are somewhat obscured.

The lower *molars*, as Peterson noted, are narrow and elongated from before backward and are progressively larger posteriorly; they have a distinct resemblance to the corresponding teeth of *Agriochoerus*, but are relatively narrower and the longitudinal valley, between internal and external crescents, is also narrower; the inner cusps are thin plates rather than crescents. The heel of m_3 is large.

In the American Museum is a second individual from the Beaver Divide, Wyoming, which is presumably referable to this genus, though, for lack of m_3, the reference is not altogether certain. The fossil consists of the incomplete facial portion of a skull, with three teeth, dp^4, m^1 and m^2, on the left side, m^1 and m^2 and the unerupted anterior half of m^3, still in the alveolus, on the right. There is also part of the right ramus of the mandible containing dp_3, dp_4, m_1 and m_2. For the opportunity to describe this interesting fossil, I am indebted to the kindness of Dr. G. G. Simpson.

Milk Teeth.—In the upper jaw the only temporary tooth preserved is dp_4, which may be briefly dismissed, for it is completely molariform, though much smaller than any of the molars, and the protoconule is less reduced.

In the lower jaw, as mentioned above, dp_3 and dp_4 are preserved; dp_3 is of compressed-conical shape and somewhat peculiar. The free, cutting edge of the crown is considerably worn both in front and behind, though the apex remains acutely pointed, and the antero-internal valley is almost obliterated. There is a minute, but distinct deuteroconid, which has an enamel pocket on its hinder side. Seen from the buccal side, this tooth appears to be quite simple, as the incipient complications are confined to the lingual face.

The hindmost of the milk-premolars, dp_4, has the three pairs of cusps, universal among artiodactyls; the anterior pair of crescents are the smallest, the posterior pair the largest and the middle pair intermediate in size as well as position. Aside from the additional pair of cusps the tooth is smaller than m_1, but has the molar pattern. There is a prominent inner cingulum and a minute pillar between the first and second internal crescents.

SKULL

In both individuals the skull is badly damaged by crushing and largely hidden under ferruginous incrustation, but its outline may be discerned. Mr. Horsfall's drawings (pl. III) skillfully compensate these deficiencies and present the skull as it must have appeared before fossilization. It differs from the skull of *Agriochoerus* in the relatively shorter face and longer cranium and in the absence of diastemata from the dentition. The orbit is more widely open behind, because of the shorter postorbital processes, though these are already distinct, above and below. The zygomatic arches are much alike in the two genera, but the flat suborbital portion of the jugal is narrower in *Mesagriochoerus* and the whole arch relatively less robust. The bony palate is moderately concave transversely and broadens posteriorly because of the divergent tooth-rows, and the posterior nares extend forward between the first molars. The brain case is much crushed, but appears to be less capacious than in *Agriochoerus*.

The series *Protagriochoerus-Mesagriochoerus-Agriochoerus*, well-nigh demonstrates the common origin of the two families included in the Oreodontoidea, the Merycoidodontidae and the Agriochoeridae and, if this inference be accepted, it follows that the undoubted likeness of the molars of *Agriochoerus* to those of the bothriodonts is one of the many instances of convergent development, for which there is such strong evidence.

Those who have accepted Schlosser's suggestion that the agriochoerids were nearly related to the anthracotheres, have always ignored the merycoidodonts and yet their deer-like, or camel-like molars must be taken into account. There are several possible interpretations of the evidence, as it stands: (1) That the two families are entirely unrelated and that the many resemblances between them, including the most exceptional character of the caniniform first lower premolar, are all due to convergence; (2) that the two families are nearly related, but that one of them has convergently developed a molar pattern, the other changing little from the ancestral type of dentition; (3) that both families have acquired different molar-plans, convergently, one to the anthracotheres the other to the selenodonts.

The fossils of the Uinta and Duchesne River formations strongly support the second of these alternatives; the contemporary Uinta genera *Protoreodon* and *Protagriochoerus* are very closely related and there is little difference in molar teeth between them, yet in the Duchesne River genus, *Mesagriochoerus*, the separation is already distinct and the approximation to the White River *Agriochoerus* is unmistakable. Either one or the other of the two families must have developed a molar pattern convergently and, in view of the connecting series above described, it is far more probable that the convergently developed dentition is that of the agriochoerids. Unfortunately, the ungual phalanges are not known in the Uinta *Protagriochoerus* but in the allied, though aberrant *Diplobunops*, the unguals are claw-like. In *Protoreodon* of the Uinta the unguals are longer, narrower, more pointed and altogether more claw-like than in the succeeding merycoidodonts of the White River.

If these conclusions are well founded, it follows that the likeness of the upper molars of *Agriochoerus* to those of the anthracotheres—a likeness to which attention has repeatedly been directed—is not due to any close relationship between the groups, but has been independently acquired. The molars of *Protagriochoerus*, despite the presence of the fifth cusp (protoconule) are less like those of *Ancodus* [= *Bothriodon*] than are the molars of its White River successor.[30]

Mesagriochoerus primus Peterson

Mesagriochoerus primus Peterson, *Ann. Carnegie Mus.* **23**: 377, 1934.

The two available specimens of this species (Carn. Mus., Nos. 11,893, and 12,080; the former is the type specimen) were found in the same horizon of the Duchesne River beds, the Randlett, and the minute differences between their premolars are unimportant; in size, they are almost identical. A third individual, from the Beaver Divide, Wyoming, presumably referable to the same species, is in the collection of the American Museum of Natural History, No. 22,558.

MEASUREMENTS

	C.M. No. 11,893	*C.M. No. 12,080*	*A.M.N.H. No. 22,538*
Upper dentition, length, $c–m^3$	75 mm.	68 mm.	
Upper canine, antero-posterior diameter	6	6	
Upper cheek-tooth series, length	66	68	
Upper premolar series, length	32	32	
Upper molar series, length	35	32	
P^1, antero-posterior diameter	6	7	
P^1, transverse diameter	2	4	
P^2, antero-posterior diameter	9	8	
P^2, transverse diameter	3	4.5	
P^3, antero-posterior diameter	9	7	
P^3, transverse diameter	6	8	
P^4, antero-posterior diameter	9	7	
P^4, transverse diameter	9	9	
Dp^4 antero-posterior diameter			9 mm.
Dp^4, transverse diameter			9.5
M^1, antero-posterior diameter	10	10	10
M^1, transverse diameter	10	10	11.5
M^2, antero-posterior diameter	11	12	11
M^2, transverse diameter	12	13	13
M^3, antero-posterior diameter	12	12.5	
M^3, transverse diameter	13	13	
Lower dentition, length $c–m_3$		77	
Lower premolar series, length	33	34	
Lower molar series, length		38	
P_1, antero-posterior diameter	7	6	
P_1, transverse diameter	4		
P_2, antero-posterior diameter	7	7	
P_2, transverse diameter	3	4	
P_3, antero-posterior diameter	7.5	7.5	
P_3, transverse diameter	4	4.5	
P_4, antero-posterior diameter	9 mm.	9 mm.	
P_4, transverse diameter	5	5	
Dp_3 antero-posterior diameter			7 mm.
Dp_3, transverse diameter			4
Dp_4, antero-posterior diameter			10
Dp_4, transverse diameter			5
M_1, antero-posterior diameter	9	9.5	8
M_1, transverse diameter	6.5	7	7
M_2, antero-posterior diameter	10	11	10
M_2, transverse diameter	7	8	8
M_3, antero-posterior diameter		17	
M_3, transverse diameter		8	

Horizon: Duchesne River, Randlett substage.

Localities: Type, No. 11,893, Randlett Point; No. 12,080, 1 mile S. of Baser Bend, Uinta Co., Utah; No. 22,558, A.M.N.H., Beaver Divide, Lander, Wyo.

DIPLOBUNOPS PETERSON

Diplobunops Peterson, *Ann. Carnegie Mus.* **12**: 76, 1919.

The detailed description of this genus is deferred until publication of Part II of this monograph, which is to deal with the mammalian fauna of the Uinta age, because the material of the Uinta species is incomparably more complete than of those found in the Duchesne River beds. In the Carnegie Museum the type and paratype specimens of *D. uintensis* and *D. ultimus* afford material for the assembling of a nearly complete skeleton and even more perfect is a skeleton in the Princeton collection. The genus also occurs in the Duchesne River, where it is represented by a distinct species, *D. crassus* sp. nov., which is described below. The type of the genus was so very fragmentary that Peterson was misled into referring it to the European anoplotheres, comparing it especially with *Diplobune quercyi*. The name which he gave to his new genus, *Diplobunops*, shows his belief as to its relationships. The original generic description (1914) was drawn entirely from the limb-bones, for, in neither type nor paratype were there any complete upper teeth preserved, but, at a much later date (1931), after the discovery of the skeletons now in the Carnegie Museum, Peterson published an emended diagnosis, which is repeated below, when he had learned that the reference of *Diplobunops* to the anopolotheres was a mistake and had transferred it to the Agriochoeridae.

Generic characters: . . .
P^1 isolated by well defined diastemata; P^3 without distinct deuterocone, but with a heavy internal mass supported by a strong root. P^4 with well developed deuterocone. Molars with anterior intermediate tubercles. Short and stout limbs; short and broad feet with high and laterally compressed unguals.[31]

It should be added that the molar teeth have much resemblance to those of *Agriochoerus*, more so, in fact, than do those of *Protagriochoerus* of the Uinta, but are relatively narrower, and the external crescents are less extended across the crown than in the White River genus.[32]

[30] Scott, W. B., *Trans. Wagner Free Inst. Sci. Phila.* **6**: 110, 1899.

[31] Peterson, O. A., *Ann. Carnegie Mus.* **20**: 342, 1931.

[32] Scott, W. B., *Trans. Amer. Philos. Soc.* **28**: pl. 77, fig. 1, 1940.

So far as is known, *Diplobunops* is confined to the Duchesne River and the Myton substage of the Uinta (Horizon C) but, in all probability, some forerunner must have existed in the earlier Uinta.

Peterson named three species of *Diplobunops*, each at a different level in the Uinta beds: *D. matthewi* occurs near the base of the Myton; *D. uintensis* was found 250 feet above the base of the Myton and *D. ultimus* 150 feet above *D. uintensis*. *D. leotensis* [33] is very probably meant to apply to a fragmentary skeleton found "near the base of the Duchesne Oligocene, three miles north of Leota Ranch," though the author does not say so, and I can find no other mention of this species, of which no definition is given and which is plainly a *nomen nudum*. It may, however, be a synonym of *D. ultimus* Peterson q.v.

Diplobunops crassus sp. nov.

?*Diplobunops sp.* Peterson: *Ann. Carnegie Mus.* **21**: 74, 1931.
?*Diplobunops leotensis: nomen nudum, Ann. Carnegie Mus.* **21**: 74, 1931

The type specimen of this species is a skull, without mandible, in the Carnegie Museum (No. 2967) which is uncommonly complete and free from distortion for a Duchesne River fossil, though it is much cracked, covered with an incrustation and somewhat down-crushed; in consequence, few of the sutures are visible. Mr. Peterson died in November 1933 and appears never to have seen this skull, or the two referred specimens, Nos. 11,301 and 11,769. As the detailed description of the skull in this genus is expected to follow in Part II of this monograph, which is to deal with the Uinta fauna, it will be necessary, at this point, merely to enumerate the specific characters.

The only Duchesne River fossils which Peterson provisionally assigned to the present genus were an upper molar and the fragmentary skeleton above referred to. Of the isolated tooth he says:

> An upper molar tooth, C.M. No. 11,853, . . . is provisionally placed in the genus *Diplobunops*. My principal reason for doing this is the presence of the vestigial protoconule on the crown of the tooth, the poorly developed posterior horns of proto- and hypocones and the heavy and obliquely backward directed parastyle, as in the Eocene Oreodonts in general. That the tooth may pertain to a distinct new genus is entirely probable, but I prefer to wait until a more complete specimen is found before adding another genus from this new horizon.

Of the skeletal fragments, he writes:

> The fragments of the upper teeth indicate an animal the size of *Diplobunops uintensis*. The humerus, femur, and tibia appear to be proportionally lighter and the cnemial crest of the tibia not extending so low as in *Diplobunops*. The astragalus, on the other hand, is low and broad, as in the latter genus. The distal articulation of Mt. IV is not hemispherical on the dorsal face, as it is in *Diplobunops*, but more nearly like that in *Merycoidodon*. The two terminal phalanges present are not high, narrow and claw-like, as in *Diplobunops*, nor as much depressed, though fully as broad as in *Merycoidodon*. The lateral borders of the anterior half of the ungual is [*sic*] expanded near the plantar face, giving the bone a unique appearance. If this specimen pertains to one individual, the combination of the characters noted would certainly indicate a distinct species of the genus *Diplobunops*, if not a distinct genus, nearly allied to the latter. The material is, however, in my judgment, unsatisfactory to serve as a type.[34]

In the doubt concerning the provenance of this material and whether it was derived from more than one individual, any discussion of it would be unprofitable, further than to say that the metatarsal and ungual phalanges cannot have belonged to any species of *Diplobunops*.

The Carnegie Museum skull, which is the type specimen of *Diplobunops crassus*, differs from those of the Uinta species in several conspicuous respects and the enumeration of these differences may serve as a diagnosis of *D. crassus*.

(1) The skull, while nearly of the same length as that of *D. uintensis*, is considerably broader and more massive.

(2) The postorbital constriction is decidedly shallower, making the cranium much broader at this point.

(3) The nasal bones are relatively broader.

(4) The postorbital process of the jugal is much longer and more prominent.

(5) The diastema between the canine and p^2 is longer.

(6) On p^3 the deuterocone is smaller.

(7) On the upper molars the antero-intermediate cusp (protoconule) is much reduced, hardly more than vestigial, a fact which Peterson notes for the isolated upper molar from the Duchesne River beds, which he figured.

(8) The parastyle of m^3 is larger, making the anterior half of the crown relatively wider, while the postero-internal crescent (hypocone) is reduced.

(9) The general aspect of the upper tooth-row is distinctive, but in a way that is difficult to describe.

MEASUREMENTS

	C. M. No. 2,967	*C. M. No. 11,769*	*C. M. No. 11,301*
Upper dentition, length c–m^3 incl.	110 mm.		
Upper cheek-tooth series, length	92.5		
Upper premolar series, length*	41		
Upper molar series, length*	45	37 mm.	40 mm.
P^1, antero-posterior diameter	9.5		
P^1, transverse diameter	5		
P^2, antero-posterior diameter	10		
P^2, transverse diameter	5		
P^3, antero-posterior diameter	11		
P^3, transverse diameter	12		
P^4, antero-posterior diameter	11		
P^4, transverse diameter	14.5		
M^1, antero-posterior diameter	15	11	13

[33] Peterson, O. A., *op. cit.* **21**: 74, 1931.

[34] *Ibid.*: 74–75.

M^1, transverse diameter	17 mm.	15 mm.	14 mm.
M^2, antero-posterior diameter	17	12	13
M^2, transverse diameter	20	18	15
M^3, antero-posterior diameter	17.5	14	15
M^3, transverse diameter anterior half	22.5	21	18
M^3, transverse diameter posterior half	15		
Skull, length prmx. to occ. cond.	203		
Skull, width over zyg. arches	136		
Skull, max. width of braincase (approx.)	49		
Skull width of postorb. constr.	35		
Skull, width over orbits	76		
Skull, width of muzzle over canine alv.	51		

Horizon: Duchesne River, Randlett substage.
Locality: 1 mile S. of Baser Bend of Green River, Utah.
* Because of overlapping, the length of premolar and molar series is less than the sum of the measurements of individual teeth.

FAMILY MERYCOIDODONTIDAE THORPE

Oreodontidae Leidy, *Jour. Acad. Nat. Sci. Phila.* Ser. 2, **7**: 71, 1869.

In no respect is the contrast between the Uinta and Duchesne River faunas more striking than in the representation of this family. In the Uinta these fossils, especially the genus *Protoreodon*, are the commonest of mammals, while in the Duchesne River they are excessively rare. With a doubtful exception in the Upper Washakie, the family appears for the first time in the lower Uinta, or Wagonhound substage and must have been immigrants from some region, not as yet identified. In the upper Uinta (Horizon C, or Myton substage) there is a great increase of genera and species and, especially, of individuals. In the Duchesne River merycoidodonts are exceedingly rare and, indeed, it is not certain that they have been found in that formation at all. Peterson lists only one specimen, which he refers to ?*Ticholeptus* and says of it: "A fragment of a lower jaw with P^4 and M^1 . . . was perhaps not found in the Uinta Basin at all." [35] A fairly good skull, without mandible, collected by a Carnegie Museum party (C.M. No. 12,049) at the Beaver Divide, is referable to a very distinct species of *Protoreodon*, but, as previously pointed out, the stratigraphic position of the beds immediately below the lowest White River is not clearly established. In my opinion, these beds are, very probably, referable to the uppermost Duchesne River and I have, therefore, included in that fauna the fossils collected there. So very scanty are the fossils of the Duchesne River, both in the Uinta Basin and at the Beaver Divide, that statistical data are of very limited significance. It is not to be believed that the family died out, only to reappear in the lower White River, in which they continue to be rare, but in the succeeding Brulé substage they had a second culmination and became so abundant and characteristic that Wortman named the lower Brulé the "Oreodon Beds." The supposed absence of merycoidodonts from the Duchesne River of the Uinta Basin is assuredly merely an accident of collecting and has no particular significance.

[35] Peterson, O. A., *op. cit.* **23**: 374, 1934.

In the same formation at the Beaver Divide are found representatives of both oreodont families, *Protoreodon* and *Mesagriochoerus*, and, as pointed out before, the latter genus, which is characteristic of the Duchesne River, is intermediate between the White River *Agriochoerus* and the Uinta *Protagriochoerus* in the pattern of the upper molars and especially in the reduction of the protoconule, which is small on m^1 and m^2, vestigial or absent on m^3. The same is true of the species of *Protoreodon* from the Beaver Divide, which is here ascribed to the Duchesne River, though the genus has not yet been found in the Uinta Basin above the level of the Myton substage of the Uinta. In both families, therefore, there appears to be a complete transition between the Upper Eocene species, which have well-developed and conspicuous protoconules and the White River species, in which they have disappeared altogether.

The late Professor T. B. Loomis and, following him, Dr. Malcolm Thorpe, have maintained that the conules on the upper molars of *Protoreodon*, *Protagriochoerus*, etc., were additions to a primitively quadricrescentic molar, such as appears in the Mongolian *Archaeomeryx*. Loomis writes:

> In figure 3 I have shown a series of five types of upper molars. . . . I can see no reason for assuming that the selenodont illustrated in *E* [i. e. *Merycoidodon*] is derived from any of the others. It is the simplest of the whole series. . . . I can not but feel that . . . the four cusped molar [goes] back to an early Eocene stock as yet unknown. The earliest form which shows the characters of the modern artiodactyls is *Archaeomeryx* from Mongolia and gives no suggestion of there ever having been more than four cusps.[36]

Thorpe accepts this view and says:

> Loomis called attention to the fact that *Archaeomeryx* of the later Eocene of Mongolia is the earliest form which shows in its molar pattern the characters of the modern artiodactyls, and it affords no hint that the molars ever had more than four cusps. The lower first premolar is caniniform, and the true canine is incisiform, as they are in all of the merycoidodonts. *Archaeomeryx* is not ancestral to the family under consideration, but in my opinion it indicates that the true ancestors did not have the fifth lobe, or protoconule, and that *Protoreodon*, possessing the protoconule, is not in the direct line of ancestry. In other words, the true stem stock of the Merycoidodontidae has not yet been discovered.[37]

In his phylogenetic diagram [38] Thorpe shows *Protoreodon* as a side branch, but distantly related to the main stem of family descent.

[36] Loomis, F. B., *Bull. Geol. Soc. Amer.* **36**: 589, 1925.
[37] Thorpe, M. R., *Mem. Peabody Mus. Nat. Hist.* **3** (4): 24, 1937.
[38] *Ibid.*: 25, fig. 2.

In my judgment this view held by Loomis and Thorpe is quite untenable. Simplicity of molar pattern among mammals is rarely, if ever, a primitive character; but has resulted from a reduction in the number of cusps. Both Loomis and Thorpe emphasize the fact that the molars of *Archaeomeryx* show no sign of ever having had a fifth cusp, but why should such sign be expected to occur, in case the conule had been present in the ancestral form and subsequently lost? One can rarely be positively assured that any two genera of fossil mammals stand to each other in the relation of ancestor and descendant, but, subject to the necessary reservations implied in all phylogenetic discussions, we have in the Bothriodontinae, a subfamily of the anthracotheres, two extremely probable instances of the loss of the protoconule, leaving no sign of its former presence. All of the Oligocene bothriodonts, of which four genera have been described, have very conspicuous protoconules, but in the Miocene *Arretotherium*, which there is every reason to regard as a descendant of *Bothriodon*, there are only four cusps and no trace of the fifth. The same modification took place in India, where the genus *Merycopotamus*, with four cusps in the upper molars, appears to be descended from some bothriodont with five.

There is more direct and positive evidence, which was not known to either Loomis or Thorpe, that the White River *Merycoidodon* was, in literal fact, descended from the Uinta *Protoreodon*. The several species of the latter which occur in the Uinta all have the protoconule conspicuously developed, while in *P. tardus* sp. nov. (described in the following pages) from the Beaver Divide, the conules are much diminished in size and that of m^3 is no more than a vestige.

The same is true of the *Agriochoeridae;* in the Uinta genus, *Protagriochoerus*, the protoconules of the upper molars are as prominent and conspicuous as in *Protoreodon*, while in the succeeding *Mesagriochoerus*, the conules are greatly reduced, almost at the vanishing point. Incidentally, it is of interest to observe that the complete loss of the conules has taken place indepently in these three families and probably also in many others, for there is very strong evidence that in nearly all artiodactyls and perissodactyls the molar pattern has passed through the "quadritubercular" stage, in which there are four principal cusps symmetrically arranged in pairs and two much smaller cuspules, or conules, placed between the transverse pairs of principal cusps and, therefore, frequently called "the intermediates." The most ancient of known American artiodactyls, *Diacodexis*, seems to show that there was an antecedent tritubercular stage, but that need not be considered here. The point is, that there is every reason to believe that the absence of the conules, or intermediates, in the upper molars of any artiodactyl is not a primitive character, but has resulted from a process of reduction.

PROTOREODON Scott and Osborn

Protoreodon Scott and Osborn, *Proc. Amer. Philos. Soc.* **24**: 257, 1887.

?*Eomeryx* Marsh, *Amer. Jour. Sci.* Ser. 3, **14**: 364, 1877 (*nomen nudum*).

In the Carnegie and Princeton Museums are contained large numbers of specimens referable to several species of this genus, most of which are from the Myton substage of the Uinta. The material includes several skeletons in various degrees of completeness, so that the osteology is very fully known and will be described in Part II of this monograph, which is to deal with the Uinta mammals. Individually, *Protoreodon* includes the most abundant of Uinta fossils, but in the Duchesne River fauna there is a great change and the genus apparently disappears, so that Peterson did not include it in his list,[39] but where fossils are so exceedingly rare, that seeming absence has no great significance. At the Beaver Divide, in beds which are usually referred to the Chadron substage, was found a skull which I have referred to *Protoreodon tardus* sp. nov. and which is described formally below. I do not feel altogether confident, however, that it would not be better to erect a new genus for it. Under whatever name it be studied, it is most obviously intermediate between the Uinta species of *Protoreodon* and the White River *Oreonetes* and *Merycoidodon*. Perhaps, too, the fossil should be included in the Chadron, rather than in the Duchesne River fauna, but, if truly referable to the former, it must belong in the lowest and oldest portion of it.

For the present, it will suffice to say that all known species of this genus are small animals with unreduced dentition, in which, as in the Oreodontoidea generally, the lower canine appears to be one of the incisors and the first lower premolar is a caniniform tusk. The upper molars have an anterior intermediate conule and moderately concave external crescents, mesostyle not invaded by the transverse valley. The lower molars are much more primitive than those of the White River and succeeding genera of the family; the crescentic cusps are thicker and less complete than in the latter and plainly indicate their derivation from conical cusps.

The manus has vestigial, but complete pollex, ungual phalanges hoof-like.

Protoreodon tardus sp. nov.

(Pl. II, figs. 1, 1*a*)

The principal character which distinguishes this species from those of the Uinta stage is the reduced size of the intermediate conules of the upper molars, especially on m^3 when it is hardly more than vestigial; evidently, these intermediates were on the point of disappearance. If this line persisted into the middle and upper Chadron, it must have passed into *Oreonetes* and

[39] Peterson, O. A., *Ann. Carnegie Mus.* **23**: 374, 1934.

thence into *Merycoidodon* or, perhaps, *Eporeodon*, without a break. The summit of the protocone, or antero-internal cusp is bifid on all three molars, though this fact has no obvious morphological significance; on m^1 and m^2 the antero-internal crescent is incomplete, lacking the posterior horn; the ribs on the buccal faces of the outer crescents are less developed than in the Utah species of the genus, but the styles are more prominent, especially the parastyle of m^3, which is relatively very large. M^3 has an imperfect postero-internal crescent, without posterior horn, and has a sharp-pointed hypocone.

The animal was not quite mature; p^2 was in process of eruption, but p^3–m^3 were all in place and functional. The crown of p^1 has been broken away and lost and of p^2 only the sharp point of the protocone's apex is visible; p^3 was probably the largest of the premolars in antero-posterior diameter; it has a trenchant compressed-conical shape, with sharp edges and acute point; the deuterocone is larger than in the Uinta species of *Protoreodon*, but not so large as in *Protagriochoerus;* it is somewhat asymmetrical in position, placed behind the median transverse line of the crown and is part of the prominent cingulum. The cingulum is high behind and encloses a valley, but becomes very faint anteriorly. P^4 is a little shorter antero-posteriorly and wider transversely than p^3 and has the usual selenodont pattern of two crescents, placed transversely; the cingulum is prominent on the hinder face, but not on the other sides.

The only *milk-tooth* preserved is an upper canine, which resembles its permanent successor, except in being somewhat smaller and more slender.

It is particularly unfortunate that the *skull* was not found until it was so much damaged by the weather, for it is exceptionally little crushed or distorted. On the right side, the face is intact, except for the loss of the zygomatic arch, and most of the brain-case is preserved, but the left side and base of the cranium were destroyed. The short and complete postorbital process of the frontal shows that the orbit was incompletely closed behind, as is also the case in all the known Uinta species of the superfamily and is relatively longer than in the agriochoerids. In front of the orbit, the *maxillary* is conspicuously swollen, as though by an enlarged antrum; the infraorbital foramen is small, low in position and very far forward, opening above p^3. The maxillary is very broad dorsoventrally in the preorbital area, but has only a minute contact with the frontal, the lachrymal and nasal separating these bones except at one point.

The *premaxillary* is relatively very small; the horizontal ramus has a rather stout alveolar portion and the ascending ramus is extremely narrow, but forms the entire border of the anterior nares, rising to a contact with the nasal. The incisive foramina are very small and the premaxillary spines are long and narrow and are received into a deep, V-shaped notch of the palatine processes of the maxillaries; the notch extends back almost to the line of p^2.

The *nasals* are relatively large and are curved transversely and, more strongly, antero-posteriorly, especially in the anterior third of their length, which has a decided downward curvature. The nasals invade the frontals deeply and end behind in points; from these pointed posterior ends, they broaden gradually forward to their maximum width at the lachrymal suture. From that point, they narrow again for a short distance and, then, are of uniform width for the remainder of their course and have entire, bluntly rounded, decurved ends, which project freely for a quarter of an inch in front of the premaxillaries.

The *frontals* form a nearly plane forehead, with only a slight, transverse convexity; there is a deep, median notch between the frontals, into which the hinder ends of the nasals project and, on each side of this notch, the frontals send out long nasal processes; the anterior border of the frontal is made up almost entirely of the nasal and lachrymal sutures. The postorbital process of the frontal is much shorter than in *P. parvus* and the smooth forehead shows hardly a trace of the temporal ridges; this lack, however, may be due to the immaturity of the animal. The same explanation may account for the low sagittal crest, which does not reach any considerable height until it approaches the occiput.

The *lachrymal* is relatively large and conspicuous in the young animal and forms the antero-superior boundary of the orbit; it is of triangular shape, narrowing forward to a point; the base of the triangle is the orbital boundary, which is notched near the ventral angle and the longest side is the frontal suture. In *Merycoidodon*, of the White River, the lachrymal has a more quadrate shape and the antorbital fossa (or so-called "lachrymal pit") which covers nearly the whole lachrymal bone, is deep and conspicuous, but varies somewhat, apparently according to sex. As with so many other characters, the lachrymal of *Protoreodon tardus* is intermediate between that of the Uinta species of the genus and that of the White River *Merycoidodon;* the antorbital fossa is lacking in the former, large and deep in the latter, while in *P. tardus* it is present, though very shallow and hardly more than incipient.

MEASUREMENTS

	C.M. No. 12,049
Upper dentition length i^1–m^3 incl.	74 mm.
Upper canine, antero-posterior diameter	6
Upper canine, transverse diameter	4.5
Upper milk canine, antero-posterior diameter	5
Upper milk canine, transverse diameter	5
Upper cheek-tooth series, length	57
Upper premolar series, length	26
Upper molar series, length	31
P^3, antero-posterior diameter	8
P^3, transverse diameter	7

P^4, antero-posterior diameter	7 mm.
P^4, transverse diameter	9
M^1, antero-posterior diameter	10
M^1, transverse diameter	12
M^2, antero-posterior diameter	10
M^2, transverse diameter	12.5
M^3, antero-posterior diameter	11
M^3, transverse diameter	14

Horizon: ?Uppermost Duchesne River, or lowest Chadron.
Locality: Beaver Divide, Wyoming.

ORDER **PERISSODACTYLA** OWEN

In the Lower and Middle Eocene of North America, perissodactyls were the dominant ungulates, indeed, the dominant mammals, while artiodactyls were very rare. Presumably through immigration, artiodactyls increased immensely in the Uinta, especially in the Myton substage, but the perissodactyls showed no definite indication of decline other than the loss of relative importance, in spite of the arrival of several new artiodactyl families from the Old World. So far as one can judge from the very brief faunal list, the proportional numbers of the two orders persisted with no particular change through the Duchesne River, in which representatives of the following perissodactyl families have been found: (1) Equidae, (2) Brontotheriidae, (3) Helaletidae, (4) ?Tapiridae, (5) Amynodontidae, (6) Hyracodontidae, (7) Chalicotheriidae. The Rhinocerotidae were probably also present, but have not yet been certainly identified.

FAMILY **EQUIDAE** GRAY

For some unexplained reason, remains of horses are very rare in the Uinta, as compared with the White River; in the Duchesne River, they are as yet represented only by a couple of incomplete mandibles, which Peterson believed would prove to be different from the Uinta genus, *Epihippus*, when better material should have been obtained and, for that contingency, he suggested the term *Duchesnehippus*.

EPIHIPPUS MARSH

Epihippus Marsh, *Proc. Amer. Assoc. Adv. Sci.*, 26th Meeting, 236, footnote, 1877.

Intermediate, in regard to the molarization of the premolars, between *Mesohippus*, of the White River, and *Orohippus*, of the Bridger. Granger's account of the Uinta genus is as follows:

> The characters pointed out by Marsh do not serve to distinguish the Uinta from the Bridger forms. The distinctions are rather in the highly developed mesostyle, in the advanced condition of the second upper and lower premolars, in the crescentic external cusps of the upper molars and in the more perfect development of the cross-crests.[40]

Peterson named another species from the Duchesne River, referring it provisionally to the Uinta genus.

Epihippus intermedius Peterson

Epihippus intermedius Peterson, *Ann. Carnegie Mus.* **21**: 66, 1931.

Peterson described this Duchesne River species as follows:

> *Specific characters;* Antero-internal cusp on P_3 single; antero-internal cusp on P_4, on tooth not worn, very faintly twinned; on very slight wear twinning disappears. On M_1 and M_2 twinning of antero-internal tubercle practically the same as on P_4; P_1 single-rooted. No diastema between P_1 and P_2. Animal slightly larger than *Epihippus gracilis.*
>
> In comparing P_2 of the present specimen with fig. 5*b* on Pl. XVIII in Granger's paper on the "Revision of American Eocene Horses" it appears that the paraconid of *Epihippus gracilis* (= *uintensis*) is less developed and does not turn inward as in *E. intermedius.* Furthermore, when the lower premolars of the type of the new species are compared with those in a specimen, C.M., No. 3,397, referred to *Epihippus parvus*, it becomes quite apparent that the anterior cross-crests in the latter species are higher than the posterior cross-crests, while in *E. intermedius* the anterior cross-crests are no higher than the posterior, exactly the condition found in *Mesohippus bairdii.* P_2 in the latter species differs from that in *Epihippus intermedius.* These differences together with relatively smaller incisors, a larger canine, a longer diastema between P_1 and the canine in *E. intermedius* appear to be the chief differences between *Epihippus intermedius* and *Mesohippus bairdii.*
>
> When the upper dentition of *E. intermedius* is found in the Duchesne Oligocene, it is quite safe to predict that the antero-internal tubercle on P^2 will be much further advanced in its development of molariform structure than it is in *Epihippus parvus*, and that the fifth digit of the manus in *E. intermedius* will be found to be considerably more reduced than it is in the species from the lower levels in the Uinta Basin. That *Epihippus intermedius* may represent a distinct genus is entirely probable. When more satisfactorily determined this new genus may be called *Duchesnehippus.*

An interesting item in the differences between the Duchesne River and the White River horses is in the comparative size of the lower canine, which is distinctly larger in the more ancient genus, although it seems to belong to the incisor-series and follows i_3 without diastema. In a recent publication,[41] I have discussed the problem of the development of the canines in the horse-phylum and have shown that in the White River genera, *Mesohippus, Miohippus, Pediohippus*, the canines, upper and lower, are much reduced, as compared with those of the Eocene genera and the lower one is nearly incisiform. In the John Day, there began a re-enlargement of the canines, leading eventually to the Recent condition. The fossils, thus, seem to record a fluctuation in the phylogenetic development of the canines, which apparently contravenes Dollo's principle of the "irreversibility of evolution," which is so generally accepted and, yet, if such a degree of fluctuation in development be not admitted, it will be necessary to remove almost every one of the American Oligocene horses from the direct line of equine descent. Had Dollo's attention ever been

[40] Granger, W., *Bull. Amer. Mus. Nat. Hist.* **14**: 232, 1908.

[41] Scott W. B., G. L. Jepsen, and H. E. Wood, *Trans. Amer. Philos. Soc.* **28**: 912, 1941.

directed to this instance, he might have thought that such a fluctuation was too trivial a matter to constitute a real exception to his "law," according to the legal maxim: "*de minimis non curat lex.*"

Exactly the same situation is revealed in the developmental history of the true camels, in which all the anterior teeth, canines and incisors alike, are first much reduced in the Upper Eocene genera and then gradually re-enlarged in the Oligocene and Miocene forms.[42]

In the table of measurements which follows, Peterson's figures have been repeated and verified; in case of differences, his dimensions are given in parentheses.

MEASUREMENTS

	C.M. No. 11,845
Length of jaw-fragment, incl. incisors	86 (84) mm.
Length of diastema between canine and p_1	18 (13)
Cheek-teeth, length p_1–m_2	50 (49)
P_2, antero-posterior diameter	9
P_2, transverse diameter, posterior part of crown	5 (4.5)
P_3, antero-posterior diameter	8.5
P_3, transverse diameter	6
P_4, antero-posterior diameter	9
P_4, transverse diameter	7 (6)
M_1, antero-posterior diameter	9
M_1, transverse diameter	7 (6)
M_2, antero-posterior diameter	9
M_2, transverse diameter	6.5 (6)
Mandible, depth at p_2	15.5
Mandible, depth at m_2	19

Horizon: Duchesne River, Halfway stage.
Locality: Halfway Hollow, Uinta Co., Utah.

FAMILY BRONTOTHERIIDAE MARSH

The Bridger, Washakie, and Uinta stages of the Eocene are characterized by great numbers of brontotheres, which, in the first two stages named, are almost all of moderate or medium size, about equalling the American Tapir in stature. So marked is this resemblance, that the genera were long called by the noncommittal name of "tapiroid." In the Uinta, most of the medium-sized species seem to have died out and to have been replaced by animals that may fairly be called large, though far smaller than the huge creatures which were so abundant in the Chadron stage of the White River. Two genera of the comparatively large-sized brontotheres, of the Uinta have been named, *Diplacodon* and *Protitanotherium*. In the Duchesne River, only a single genus, the White River *Teleodus*, has so far been discovered, but there can be little doubt that this great reduction was more apparent than real and due rather to the rarity of fossilization than to such wholesale disappearance of the mammalian fauna. It is to be expected that the larger brontotheres of the Uinta *Protitanotherium* and *Diplacodon*, or their little modified descendants, will eventually be found in the Duchesne River beds.

In Bridger and Washakie times the brontotheres were nearly all hornless and such of the skulls as show signs of having had dermal, rhinoceros-like horns, had them in incipient, hardly recognizable form, but in the upper Uinta (Myton stage) conspicuous, bony knobs made their appearance, becoming, in the White River, immense and grotesque outgrowths of the skull. It is customary to call these protuberances "horns" and, very probably, such they were in function; evidences of fighting occur in the skeletons. In their earliest form, as in *Manteoceras* of the Washakie, the only evidence of horns is the paired roughening of the nasal bones, to which small dermal horns must have been attached. This may also have been true of the genera with small bony knobs, such as *Diplacodon*, of the Uinta and *Teleodus*, of the Duchesne River, but the monstrous "horns" of most White River genera must have been pedicles, greatly enlarged and with no indication of their having supported dermal horns. Their shape precludes the suggestion that they might have been cores with horny sheaths. The medium-sized, hornless animals, so characteristic of the Bridger and Washakie, have become few and rare in the Uinta and lacking, apparently at least, in the Duchesne River. The change from the Uinta to the last named formation may be largely deceptive because of the scantiness of the fossils, but it is probable that the comparatively small, hornless brontotheres had given way entirely to the larger, horned genera. Only two genera, *Teleodus* and ?*Protitanotherium*, have been obtained, as yet, and in surprisingly large numbers, but all the specimens have been obtained from a single "quarry," none whatever from any other locality. Under such circumstances, statistics are without significance, but the discovery of a White River genus in the Duchesne River beds is of great stratigraphic interest and importance.

TELEODUS MARSH

Teleodus Marsh, *Amer. Jour. Sci.*, ser. 3, **39**: 524, 1890.

Peterson's description of the Duchesne River species may be reproduced here with advantage.[43]

Generic Characters. $I\frac{2}{3}$, $C\frac{1}{1}$, $P\frac{4}{4}$–?$\frac{3}{3}$, $M\frac{3}{3}$.
Canines with short crowns, especially in the females. Skull brachycephalic. Bases of horns elongate oval in transverse section.

Teleodus avus Marsh (No. 10,321, Yale Museum) is based upon the greater part of the lower jaws and was found near the base of the Oligocene of Dakota. This specimen, the type of the genus, clearly pertains to a later form, the chief features of which are: (*a*) the crowded condition of the lower incisors, I_3 with an extremely short root and almost crowded out of its original position in the alveolar border; (*b*) the alveolar border of the incisors occupies a straight transverse line between the lower canines and does not extend in front of the canine teeth; (*c*) the cheek-teeth are relatively more developed than in other species, being broader, with more inwardly slanting external faces, especially in the case of the molars.

[42] *Ibid.*

[43] Marsh, *loc. cit.*

Teleodus primitivus (Lambe) from the Cypress Hills of Canada is an earlier or more retardant [*sic*] species. In this species the alveolar border of the incisors extends well in front of the canines with a more liberal space for the incisors. Whether or not the presence of P_1 in *T. primitivus* is due to the young, though adolescent, stage of the type specimen it is hard to determine; but the relatively long, narrow cheek-teeth, with vertical external faces, especially in the case of the molars, are characters of strong specific value. The longer diastema between the canines and P_1 and the relatively long and narrow mandible are additional distinctive features of *T. primitivus*. The delicately constructed canines in the type-specimen may well be due to sex.

Teleodus uintensis Peterson

Teleodus uintensis Peterson, *Ann. Carnegie Mus.* **20**: 308, 1931.

Type: Lower jaws, practically complete, but crushed in the symphysial region, No. 11,809, female, Carnegie Museum.

Paratypes: Skull complete, slightly depressed by crushing, No. 11,759, female, Carnegie Museum; lower jaws complete but crushed in the region of the angle, No. 11,761, Carnegie Museum. In addition there are over twenty individuals from the same quarry typical of the species, which will be consulted in connection with publications of the Carnegie Museum in the near future.

Specific Characters: Smallest known titanothere of the Oligocene. In anatomical structure the species is intermediate between *Teleodus avus* and *Teleodus primitivus*. Incisors as in *T. avus*, but less crowded and alveolar border slightly further advanced in front of the canines. P_1 present or absent. Diastema between canine and cheek-teeth present or absent. Molar-premolar dentition relatively broad and external faces of molars less vertical than in *T. primitivus* of the Canadian Oligocene.

The female skull, C.M. No. 11,759, one of the paratypes, is the most perfect specimen of the series.

Its contour is less brachycephalic than those of the males. The small, round-topped, button-like crowns of the two upper incisors of the left side are of nearly subequal size, separated from those on the right by a deep median invagination of the alveolar border. Incisor three is crowded close to the inner side of the canine. The latter is of small size (clearly a sexual character) with short and blunt crown and a prominent posterior cingulum. Following the canine, without diastema, P^1 appears nearly as broad as long and well worn, so that its detailed structure is practically obliterated. Other individuals show the structure of P^1 perfectly. P^2 has the tetartocone indistinctly separated from the deuterocone. P^3, in one or two cases, is seen to have the tetartocone poorly indicated, but in the great majority of the skulls and upper dentitions under study the deutero- and tetartocones are distinctly marked. The tetartocone on P^4 is small, though distinctly separated from the deuterocone by a well-marked constriction of the inner lobe. The inner cingula of the premolars are usually well indicated, while externally they vary in different individuals. In the paratype, No. 11,759, the hypocone of M^3 is distinct and of large size, but in most cases this element is small and has no diagnostic value, due to variation in the different specimens. The parietals, in all the skulls from the new quarry in the Upper Uinta, present a prominent convexity, which is especially well developed on male skulls. This convexity is observed in the crania of other Oligocene titanothere, *e. g. Brontops*, *Megacerops*, *Brontotherium*, but not so general or proportionately as large as in the species from the Upper Uinta.

In the type, No. 11,809, the mental foramen is located below P_4, while in the paratype No. 11,761, this foramen is below P_3. This discrepancy is, in part, due to crushing of both specimens. The anterior part of the symphysis in the paratype is much depressed by crushing. The incisor alveolar border of the type is imperfect, while in the paratype it is complete, and there is a space between the median incisors of about five millimeters. I_1 and I_3 are subequal in size. The canine is small, with a postero-internal basal cingulum projecting at the base of the crown. In the type the diastema between the canine and premolars is much shortened by crushing, while in the paratype it is equally clear that the diastema is lengthened and the symphysis in general has a procumbent and unnatural appearance due to crushing from above downward. This crushing is also observed in the region of the condyle and the coronoid process, *see* pl. XV. In the type P_1 is absent in both jaws, while in the paratype this tooth is represented by the root on the left and by an alveole on the right side. The rest of the cheek-teeth are much worn in both specimens. The cingula are rather weak, while the teeth are relatively broad, when compared with *T. primitivus* of Canada and relatively less broad than in *T. avus*.

Mr. Peterson's description of *Teleodus* was preliminary and he hoped to follow it with a more detailed account in the publications of the Carnegie Museum, but, unfortunately, his lamented death in November 1933 prevented the accomplishment of this intention. The kindness of Director Avinoff and Messrs. J. Leroy Kay and John Clark has given me the opportunity to carry out Mr. Peterson's plan and the following description was prepared during my visit to Pittsburgh in October and November 1941.

DENTITION

None of the twenty or more skulls found in this wonderful "quarry" in the Duchesne River beds, eleven miles west of Vernal, Utah, has its lower jaw attached to it, or even distinctly associated with it and the number of separate mandibles is surprisingly small, much less than that of the skulls; there are only three mandibles for twenty-three skulls and it is difficult to imagine how all of the lower jaws became detached. All three of the mandibles pertained to old animals with much worn teeth, while the skulls are of all ages except the juvenile.

The dental formula is $i\frac{2}{3}$, $c\frac{1}{1}$, $p\frac{4}{4}-\frac{3}{3}$, $m\frac{3}{3}$; the number of upper premolars is variable.

Upper Teeth.—The *incisors* are very small and, like those of the White River genera, have shoe-button-like spheroidal crowns, which can have been of little or no use to their possessors, for, even in old animals with much worn molars, they show no sign of abrasion.

The *canines* are relatively small, but were still functional tusks, projecting well below the level of the masticating surface of the grinding teeth; the anterior side of the crown is convex and the posterior concave, giving the tooth a recurved shape; at the base is a posterior cingulum, almost a heel. One of the most striking contrasts between the brontotheres of the

Bridger and those of the White River is in the relative size of the canines. In the Middle Eocene genera, such as *Palaeosyops* and *Telmatherium*, the canines are formidable tusks, proportionately as large as in the Grizzly Bear, while, in the White River, there is much variation in the size of these teeth, but in none of the genera are the tusks comparable to those of the Bridger forms and, in most White River species, the canines must have been useless as weapons, for their points do not project below, or above, the grinding surfaces of the premolars.

The *premolars* are very small in comparison with the molars and, except p^1, are what is usually called "molariform," but, in both the rhinoceroses and the brontotheres, the term is inexact, for, in both of these families, the external wall of the crown remains constantly different in molars and premolars; in the molars the two external lobes are deeply concave on the buccal side, while in the premolars, the outer lobes have convex and ribbed buccal sides. In the other perissodactyl families, horses, tapirs and palaeotheres, there is no such constant difference between molars and premolars; in the early Eocene and, in many instances, in the middle Eocene, none of the premolars has assumed the molar pattern, but, from the Bridger upward, the premolars, one by one, become molariform, until all of them, except the first, resemble the molars, except for the differences as noted above.

P^1 is much the smallest of the series in *Teleodus*, though it is carried on two roots, and its cuspidation is difficult to make out, for it is worn, even in those skulls in which the other teeth show little sign of abrasion. Very probably, this tooth was not changed. There appear to be four cusps, anterior and posterior, internal and external, with a small postero-internal valley, or pocket. The six following teeth increase in size posteriorly from p^2 to m^3, with sudden increment at m^1; each tooth is larger than the one in front of it and smaller than the one behind it. The premolars, p^2 to p^4, have all the elements of the molar crown, but in different degrees of approximation and, as above noted, in each of them the external wall is unlike that of the molars. In the latter the two external cusps are deeply and conspicuously concave, with very prominent mesostyle, while in the premolars these cusps are convex, or flat. The external cusps are ribbed, the anterior ones particularly so. Seen from the buccal side, the two categories of teeth are strikingly different, though their grinding surfaces are much alike.

P^2 has two external cusps, of which the posterior one (tritocone) is the larger; very short cross-crests are on the inner, or lingual, face of these cusps and an enamel pocket is enclosed by these crests; there are two internal cusps, not very distinctly separated, the deutero- and tetartocone, of which the former is much the larger; the abortive cross-crests do not reach these inner elements. The prominent cingulum passes around three sides of the crown and is especially heavy on the anterior and posterior faces.

P^3 is like p^2, but is larger and its cusps are more distinctly demarcated; the cross-crests, however, are even smaller and the larger cusps are the postero-external and the antero-internal and the enamel pocket has a more symmetrical position, near the middle of the fore-and-aft diameter of the tooth. The cingulum is complete, except externally and turns upward on the inner face of the deuterocone.

P^4 is the most molar-like of the series; the two external cusps are of nearly equal size and the anterior one has a median rib, while the posterior one is slightly concave and the three external styles are present in an incipient stage; the cingulum is as complete as on the other premolars. In one respect, however, p^4 is less completely molariform than p^3, namely the relatively smaller size of the postero-internal cusp, or tetartocone.

The *molars* are essentially of the same pattern as in the Lower Eocene *Lambdotherium* and in all other known members of the Brontheriidae; the two external cusps, para- and metacones of Osborn's nomenclature, have deeply concave buccal faces, without ribs, and very prominent mesostyles, formed by the junction of the adjacent horns of the outer crescents; para- and metastyles can hardly be said to exist, though the effect of styles is produced by the outward curvature of the crescent borders. The free, ventral margins of the crescents are produced downward into points and this gives a characteristic W-like cutting edge to these teeth, which shears down outside of the lower molars. The cross-crests, which characterize the upper molars of all other perissodactyl families, are lacking in all known brontotheres, even in the earliest, *Lambdotherium*, and the intermediate conules are reduced to vestiges, or are altogether wanting. In the later chalicotheres the transverse crests are much reduced and the brontothere condition is approximated, but, unless the supposed Uinta and Bridger Ancylopoda are mistakenly referred (e. g. *Eomoropus*) the reduced cross-crests are secondary, not primitive. In addition to the median enamel pocket, which the premolars possess, there is a posterior one, which is inside of the hinder external crescent and is smallest on m^3. Internally, there are two conical cusps, the proto- and hypocones, which have no connection with the outer wall and are entirely isolated. The upper molars of the Brontotheriidae are thus highly characteristic and almost unique among perissodactyls. In *Teleodus*, the two internal, conical cusps are of almost the same size in m^1 and m^2, but in m^3 the hypocone is very much reduced, hardly more than a remnant. The cingulum does not embrace three sides of the crown so distinctly as it does in the premolars and differs in prominence on the different sides and also according to the position of the tooth in the molar series; in m^1 the cingulum is prominent on the front and rear faces of the crown, very weak on the lingual side, though crossing the inner valleys and descending on the internal face of the antero-internal cone, somewhat as in p^2. In m^2 the outer part of the anterior cingulum is concealed by the close juxtaposition of m^1, but the inner half is high and, on the internal and posterior faces of the tooth, it is very much as in m^1. On the anterior face of m^3 the cingulum is low, but very distinct on the side of the antero-external crescent, while its inner half is relatively high, much higher than on any other tooth; internally also the cingulum is much more prominent, but is interrupted on the face of the protocone, or antero-internal cusp; otherwise, it is continuous around three sides of the crown.

The upper molars differ not only in size, but also in shape; in the outline of the crown, m^1 is symmetrical, rectangular and nearly square, with anterior and posterior transverse widths almost equal; m^2 is less symmetrical, owing to the outward projection of the anterior horn of the antero-external crescent, but the width is nearly the same before and behind. M^3 is decidedly asymmetrical, the anterior breadth exceeding the posterior, in the ratio of 69 to 56, because of which the buccal face seems to slope inward and backward and each crescent is inclined in similar fashion; the anterior crescent is rendered more deeply concave than in any of the other molars by the great prominence of the anterior horn.

There is every reason to suppose that all of the individuals found in the Quarry belong to the same species and yet, if so, surprising differences are found in the upper molar teeth. In No. 11,754, for instance, a young adult, in which m^3 was not completely erupted, the ribs on the buccal faces of the outer crescents are faintly marked, or entirely absent, whereas, in No. 11,866, these ribs are conspicuous.

Lower Teeth.—Peterson's account of the lower *incisors*, as already quoted, needs no addition. As he pointed out, there are three incisors in each ramus, the only known Oligocene genus of brontotheres of which that is true. The upper incisors, as before noted, are reduced to two and, in the White River, the number fluctuated in different individuals, or, perhaps species, from three to none. In *Teleodus*, the incisors are definitively of the useless, vestigial, Oligocene type as compared with the large, pointed, functional form, which is characteristic of the Eocene genera. All three of the mandibles, which the Duchesne River collection contains, are of old individuals, with much worn cheek teeth, but the incisors show no sign of abrasion.

The *canine* is small and inclined forward, much worn on the apex and posterior face, evidently by the abrasion of the upper canine, for the lower tusks project above the level of the grinding teeth.

The *premolars* number three in one of the mandibles; in one (C.M. No. 11,809) p_2 follows the canine without diastema and with only space sufficient for the occlusion of the upper tusk; there cannot have been a first premolar in this jaw. In a second specimen (C.M. No. 11,761) the number of premolars is four, and there is a considerable gap between the canine and p_2, in which is a very small remnant of a single-rooted p_1. This is the jaw figured by Peterson and, though decidedly old, with much worn teeth, it is somewhat younger than No. 11,809, which may have had an additional tooth, when young. Like the upper teeth, those of the lower jaw increase in size posteriorly, m_3 being much the largest of the grinding teeth; p_3 and p_4 are molariform, p_2 not quite so, for the anterior crescent is incomplete. All the other grinders are bicrescentic, the two crescents being placed one behind the other, and m_3 is enlarged by the addition of a basin-like heel; remnants of an external cingulum appear on several of the teeth, notably on m_3.

This bicrescentic pattern of the lower molars is very widely distributed among various groups of artiodactyls, perissodactyls, chalicotheres and the South American Litopterna, so that its presence is no proof of relationship. Originally, Leidy [44] was misled by it into referring the titanotheres (brontotheres) to the artiodactyls, family Anoplotheriidae; it was especially common in the Eocene and early Oligocene and, from the beginning to the end of their recorded history, the brontotheres displayed it. It is interesting to note that no hoofed mammals of Recent times possess it in unmodified form; it persists in the horse family, but is so masked by complications, as hardly to be recognizable.

SKULL

The Duchesne River Quarry, which has yielded so many skulls of *Teleodus*, does not contain a single uncrushed, undistorted specimen and the multitudinous, matrix-filled cracks have so obscured the sutures that but very few of the bones are distinguishable. This is true of even the youngest individuals, in which some of the milk-teeth are retained. Little can be done, therefore, in the way of description, more than to give an account of the various aspects of the skull, side, top, base and rear, but even in these general descriptions, it is often extremely difficult to make proper allowance for the effects of crushing. In most instances, the pressure has been downward and due, no doubt, to the weight of overlying sediments, for the formation has been subjected to very little diastrophic movement. The effect of vertical pressure has usually been to flatten the skull by reducing its dorso-ventral diameter, but there are some cases of telescoping, which has shortened and broadened the skull. Mr. Horsfall's drawings (pl. VII, figs. 1, 1*a*, 1*b*, 2, 2*a*) are, I believe, a very successful solution of the problem of distortion, but they leave much uncertainty concerning the limits of the bones.

Side View.—The skull shown in plate VII is the same individual as that figured by Peterson,[45] but in

[44] Leidy, J., *Jour. Acad. Nat. Sci. Phila.*, Ser. 2, **7**: 206, 1869.
[45] Peterson, O. A., *Ann. Carnegie Mus.* **20**: pl. XII, 1931.

his plates no attempt was made to correct the distortion. A comparison of Peterson's figures with those of plate VII in this paper will show the amount of change deemed necessary for this correction. The first glance suffices to show that this skull is Oligocene in character, in contrast with the Upper Eocene genera, such as *Diplacodon* and *Protitanotherium*. The small horns make it probable that this skull pertained to a female; other skulls, from the same quarry, which are interpreted as males, have larger, more prominent and rugose "horns" and more robust zygomatic arches. In this aspect of the skull, one is immediately struck by the large swelling, or convexity, in the parietal region, which so conspicuously characterizes this animal and is not known to occur in any other genus. I am at a loss to understand Peterson's meaning, when he says: "This convexity is observed in the crania of other Oligocene titanotheres; e. g. *Brontops*, *Megacerops*, *Brontotherium*, but not so general, or proportionately as large as in the species from the Upper Uinta." [46] Certainly, no other known member of this family has anything comparable to this hump. It is quite possible that the posterior part of the brain-case and the occiput are represented as too low in my plate and that they should be drawn more as in *Diplacodon*, though in lesser degree, for, even in the supposed males of *Teleodus*, the horns are much smaller than in the Uinta genus.

The extreme shortening of the face and elongation of the cranial region which characterize all of the White River brontotheres, despite the excessively small brain capacity, are equally marked in *Teleodus;* the anterior rim of the small orbit is above the front edge of m_1 and the postorbital processes of both frontal and jugal are more distinct than in most White River skulls, though there is great variation in regard to this. The orbit appears to have a higher position than in the latter, with narrower space above it, but, how far this is due to down-crushing, it is difficult to say. The infraorbital foramen is large and conspicuously open, placed very far forward, quite near to the hinder border of the narial incision and directly in front of the orbit. In this case, it would be more accurate to call the foramen preorbital rather than infraorbital. The very small premaxillae are concealed, in sideview, by the canine. The freely projecting portion of the nasals is very long and extends somewhat in front of the incisors.

As is usual in this family, the horns are formed chiefly by the nasals, capped by extensions of the frontals. In the White River, there is very great variability in the size and shape of these horn-like, bony outgrowths of the skull, but in all the skulls of *Teleodus uintensis*, the horns are very small, even in those which are believed to be males. This is the more remarkable, because in the Uinta genera, *Diplacodon* and *Protitanotherium*, the horns are much larger than they are in *Teleodus*, although the latter is White River and Duchesne River in date. *Menodus heloceras* (Cope), (to use Osborn's nomenclature) of the White River, has horns almost as small, proportionally, as those of *Teleodus*, far smaller than in other White River genera, in which the grotesque size of the horns is characteristic.

The zygomatic arch is long, laterally compressed, and relatively thin, usually a female characteristic; in one old male the arch is much stouter and the squamosal portion of it is rugose, but there is no approach to such massiveness, or roughness as occurs in many White River skulls, which, there can be little doubt, were derived from old males. This massiveness and rugosity reach a grotesque maximum in the species which Cope named *Symborodon bucco* (*Megacerops bucco* Osborn). The jugal is very long and has a long suture with the zygomatic process of the squamosal, beneath which it extends, but does not reach the glenoid cavity. No sutures are distinguishable in the temporal fossa, which is an elongate concavity with raised borders, made up of the parietal and frontal dorsally, the exoccipital behind and the squamosal below. The postglenoid and posttympanic processes are in sutural contact, converting the auditory meatus into a tube. The ventral part of the exoccipitals, with the condyles, projects very prominently backward, obviously a result of crushing, but, as none of the skulls is undistorted, no attempt has been made to show the normal position of these bones. The paroccipital process is relatively short, not projecting below the level of the postglenoid, from which it is widely separated—perhaps another result of crushing.

Top View.—The outline of the skull from above is wedge-shaped, the base of which is at the widest part of the zygomata, on a line passing just in front of the two glenoid cavities. Thence, the outline narrows forward to the premaxillaries, without interruption and with nearly straight sides. This figure (pl. VII, fig. 1*a*) is strikingly like the corresponding view of the skull of the White River rhinoceros, *Subhyracodon* [47] and differs as conspicuously from the two White River types of brontothere crania; in one of these types the skull widens gradually and regularly forward from the occiput to the base of the horns (*see* Osborn,[48] and in the second type the two sides of the skull are nearly parallel, producing a grotesque effect. The parallel-sided skull is usually associated with extravagantly large, antero-posteriorly compressed horns, making it one of the most bizarre of known mammalian skulls. Whichever of the two types of skull (Osborn, *A* or *B*) be selected for the comparison, the difference between the White River and the Duchesne River types and the

[46] *Ibid.:* 310.

[47] Scott, W. B., *Trans. Amer. Philos. Soc.* **28**, pt. 5: pl. LXXIV, fig. 1*a*, 1941.

[48] Osborn, H. F., The Titanotheres of Ancient Wyoming, etc. *U. S. Geol. Surv. Monograph* **55**: 463, fig. 393 *C* and *D*, 1929.

likeness of the latter to that of early Rhinocerotidae is not without interest.

Returning to a consideration of the top view, we may note that the zygomatic arches are thin transversely, the temporal openings narrow and elongate and that the sagittal crest, which is so conspicuous in the Middle Eocene genera of the family, has entirely disappeared and the cranial roof made flat by the development of sinuses within it, so that, except for the parietal bump, above mentioned, the cranium is flat transversely, concave antero-posteriorly, an obvious resemblance to the skull of the post-White River true rhinoceroses. The brontotheres are not nearly related to the Rhinocerotidae, despite the many remarkable resemblances between the two families, in their later members, and the recorded history of the two groups clearly shows that these resemblances form a remarkable example of convergent development. The likenesses may be traced, step by step, through the Lower, Middle, and Upper Eocene into the Oligocene and are obviously conditioned by steadily increasing bodily size and, more particularly, by constant increase of bulk and massiveness. The rhinoceros-like appearance of the skull is the mechanical response to large nasal weapons and the inverted arch of the skull top is a means of strengthening it against shock. The parietal hump and the naso-frontal horns are not conspicuous.

Base View.—The very small *premaxillae* are emarginated in the medium line between the two middle incisors and the incisive foramina are peculiar in not visibly perforating the premaxillaries, but are concealed by a shelf-like extension of the palatine plates of the maxillaries underneath them, which also hides the premaxillary spines from view. The hard palate is elongate, moderately vaulted transversely and widening posteriorly, as the tooth-rows diverge. The *palatine* bones contribute but a small share to the bony palate, for the posterior nares open far forward, opposite m^3, and the suture is on a line with the front edge of m^2, exposing two large vacuities, one on each side of the vomer. These are similar to, but much smaller than, the openings in *Menodus giganteus* or *Megacerops bucco* (to use Osborn's terms) as shown in that author's great monograph on the titanotheres.[49] The posterior narial canal is long and has rather low walls, which extend almost to the glenoid cavities, the pterygoid border sloping up to those cavities, no sutures between palatines and pterygoids are visible. In all the skulls in which the base has been exposed, the crushing obscures or obliterates the structure. No sign of tympanic bullae is to be seen and even the internal opening of the auditory tube is concealed. The cranial base behind the postglenoid processes is broad and flat, except for a medial keel; the paroccipital processes are widely removed from the postglenoids, to which they bear considerable resemblance in shape. The occipital condyles are broad transversely, but have relatively small dorso-ventral diameter.

Rear View.—The downcrushing, to which all the skulls of the "Titanothere Quarry" have been subjected, has so distorted the occiput, that none of them gives a clear picture. In its crushed condition, the occiput appears to be low and very broad and of rectangular outline, with a deep fossa above the condyles. In its original condition, however, it must have been much higher and narrower and probably shaped much as in the Uinta genus *Diplacodon*.

Mandible.—Attention has already been directed to the strange disproportion in numbers between the skulls and lower jaws occurring together in the "Quarry," 23 to 3, and not one of the skulls has its mandible attached to it. One of the lower jaws (C.M. No. 11,809) shown in plate VII, figure 1, has been much less crushed than the skulls, but has not entirely escaped distortion, for the left ramus has been pushed forward an inch or so beyond the right, but, otherwise, the shape would seem to have been little changed. Except for its small size, this jaw might belong to any one of the so-called species of the White River. The chin rises with a rather steep inclination, but is less abruptly rounded than is usual among the White River forms; there is a single large mental foramen beneath p_4. The horizontal ramus is moderately deep dorso-ventrally and deepens backward very gradually, the ventral border being nearly straight and becoming slightly concave just in advance of the angle. Transversely, the ramus is compressed and has almost flat sides, which is in marked contrast to the jaw-fragment, C.M. No. 11,996, which is so much thicker and dorso-ventrally shallower.

The front of the ascending ramus is so different on the two sides, that compression must have been involved; that on the right side must be the less altered. Behind m_3, the alveolar border becomes narrow and, after an angulation, passes into the inner border of the coronoid. The outer border of that process is continued down to a level below m_3 and bounds a shallow fossa. The ascending ramus of the mandible is broad and high, with the condyle much raised above the level of the teeth; the posterior border is straight and nearly vertical, the angle projecting hardly at all beyond the condyle. The masseteric fossa is shallow and not well defined, the angle rounded and descending somewhat below the ventral border of the horizontal ramus, from which it is separated by a shallow concavity; the angle itself is thickened, but not comparably to the jaw of a modern rhinoceros. The condyle is much extended transversely, projecting both outside and inside of the ramus. The coronoid is weak, but high and recurved and the sigmoid notch is broad and well defined.

Feet.—Associated with the skulls in and near the "Quarry" are several bones of the fore and hind feet, which are striking for their small size. The metacarpals are considerably longer than the metatarsals,

[49] *Ibid.*

but all of the metapodials are damaged. A number of astragali are the only tarsals preserved and these are of interest as being intermediate in form between those of such Eocene genera as *Palaeosyops* and *Telmatherium* and those of the White River type, as exemplified by *Brontops*. In the latter, the bone is conspicuously short proximo-distally and broad transversely; the neck has almost disappeared, the trochlea is much shortened and the cuboid facet is very large, relatively almost as broad as in a modern rhinoceros. In the Eocene *Palaeosyops*, although not believed to be in the direct line of brontotherian descent, is sufficiently near to be a suitable example, the astragalus, both trochlea and neck, is much more elongate proximo-distally and relatively narrow, while the facet for the cuboid is very small. As already noted, the astragalus of *Teleodus* is intermediate between these two extremes. It is shorter, neck and trochlea, than in *Palaeosyops*, longer than in *Brontops* and the cuboid facet is somewhat narrower than in the latter, but much broader than in the former. Altogether, this astragalus is somewhat nearer to the White River than to the Bridger type, as might have been expected, *a priori*.

MEASUREMENTS

	C.M. No. 11,754	*C.M. No. 11,759*
Skull, length occ. cond. to prmx.	523 mm.	508 mm.
Skull, length occ. crest to anterior end of nasals	513	521
Cranium, length occ. crest. to anterior border of orbit	410	368
Skull, width over horns	148	153
Skull, width over zygomatic arches (max.)	354	289
Skull, width of occiput	215	167
Face, length from orbit to tip of nasals	98	180
Nasals, width of free portion	92	91
Upper dentition, length i^2 to m^3 inc.	293	248
Upper incisor series, length	16	
Upper canine, antero-posterior diameter	22	19
Upper cheek-tooth series, length	268	232
Upper premolar series, length	94	95
Upper molar series, length	174	141
P^1, antero-posterior diameter	14	14.5
P^1, transverse diameter	13	14
P^2, antero-posterior diameter	21	20
P^2, transverse diameter	26.5	24
P^3, antero-posterior diameter	26	25
P^3, transverse diameter	32	30
P^4, antero-posterior diameter	35	28
P^4, transverse diameter	39	35
M^1, anterior-posterior diameter	46	34
M^1, transverse diameter	45	38
M^2, antero-posterior diameter	61	50
M^2, transverse diameter	53	52
M^3, antero-posterior diameter	65	56
M^3, transverse diameter ant. half of crown	61	56
M^3, transverse diameter post. half of crown	54	45

	C.M. No. 11,809
Lower dentition, length i_1–m_3	279
Lower cheek-tooth series, length	249
Lower premolar series, length	80
Lower molar series, length	172
P_2, antero-posterior diameter	23 mm.
P_2, transverse diameter	13
P_3, antero-posterior diameter	25
P_3, transverse diameter	19
P_4, antero-posterior diameter	33
P_4, transverse diameter	22
M_1, antero-posterior diameter	42
M_1, transverse anterior	23
M_2, antero-posterior diameter	49
M_2, transverse diameter	28
M_3, antero-posterior diameter	82
M_3, transverse diameter	31

	C.M. No. 11,809
Mandible, max. length from angle, right side	410
Mandible, max. length from angle, left side	390
Mandible, length of symphysis	115
Mandible, depth of ramus below p_4	70
Mandible, depth of ramus below m_3	68
Mandible, thickness of ramus below m_2	27
Mandible, breadth of angle diagonally from m_3	147
Mandible, height of condyle	182
Mandible, height of coronoid	233

	C.M. No. 11,756
Astragalus, prox.-dist. length, fib. side	56
Astragalus, prox.-dist. length, tib. side	54
Astragalus, width of trochlea	53
Astragalus, width of dist. end.	52
Astragalus, width of navic. facet.	33
Astragalus, width of cuboid facet	20
Metacarpal II, length	115
Metacarpal II, prox. width	34
Metacarpal III, length	123
Metacarpal, III prox. width	27
Metacarpal III, dist. width	31
Metatarsal II, length	97
Metatarsal II, prox. width	24
Metatarsal II, dist. width	29

Horizon: Duchesne River, Lapoint substage.
Locality: "Titanothere Quarry, 11 m. W. of Vernal, Utah."

?PROTITANOTHERIUM HATCHER

Protitanotherium Hatcher, *Amer. Naturalist* **29**: 1084, 1895.

In the Carnegie Museum there is a fragment of a lower jaw, containing the three molars (No. 11,996), which is so different from the corresponding part of the *Teleodus uintensis* mandible that it probably pertains to a different genus of the brontotheres. This is all the more likely because the fossil was derived from a lower and older horizon, the Halfway, while all the specimens of *Teleodus uintensis* were obtained in the Lapoint substage, which is at the summit of the Duchesne River formation in the Uinta Basin. It is entirely possible that this fragment should be referable to one of the characteristic Uinta genera, *Protitanotherium* Hatcher or *Diplacodon* Marsh. Between these two, it is not feasible to make a definite choice, because the fragment might equally well belong to either genus, but I have assigned it to *Protitanotherium*, since that is so very much more abundant individually. That the

Uinta genus should have persisted into the Duchesne River is not at all surprising, but, so far as is known, it did not extend up into the White River.

In this fragment, the first and second molars are very much abraded, so that little of the pattern is discernible; m_2 has a prominent cingulum on the hinder side of the posterior crescent, rising toward the lingual face and dying away upon the buccal. M_3 is very large and but little worn, the talon not at all; evidently, the tooth was erupted much later than m_2; it differs from that of *T. uintensis* in a number of particulars, as follows: (1) Larger size, in the proportion of 83: 76; (2) the two external valleys open more widely outward and in the posterior valley there is a basal pillar, or tubercle, of enamel; (3) the talon is larger and not basin-like, but merely slightly concave; in *Teleodus*, it is much more deeply concave and distinctly basin-like. Comparison of the other molars is impracticable.

The mandible is very little crushed, or deformed, and it is difficult to determine how far the very different appearance of the jaw in *Teleodus* and ?*Protitanotherium* has been produced by the crushing which the former has undergone. The most obvious difference is in the shape of the horizontal ramus; in the best preserved of the *Teleodus* mandibles, the right half of the jaw has suffered comparatively little deformation and may be used for the comparison. The striking contrast with the supposed *Protitanotherium* is in the dorso-ventral shallowness and transverse thickness of the latter; behind m_3, on the dorsal side of the ramus, is a rather deep fossa, bounded by the two *lineae asperae*, which converge upward and unite on the front of the coronoid. To a point beneath the talon of m_3, the ventral border is straight and horizontal; at that point, the border begins to curve downward to the angle and forms a concavity which is much more decided than in *Teleodus* and the angle itself projects father downward and has a thicker and more rugose margin. The masseteric fossa is curiously small and placed high upon the ascending ramus, but little below the level of the condyle. The latter is much extended transversely, but narrow antero-posteriorly; below the condyle, on the hinder side of the jaw, is a postcotyloid process, a knob-like, rugose prominence, which extends obliquely downward and inward. This process could not have had the functions which a somewhat similar projection has on the jaw of a modern rhinoceros, as is made clear by the shape of the postglenoid process, which is found in all the known genera of the brontotheres and is a transverse ridge that could have had no contact with the postcotyloid; furthermore, the latter has no articular surface. The coronoid has lost its proximal end, but was evidently recurved and probably shaped much as in *Teleodus*.

I feel confident that the two mandibles here compared belonged to species of different genera of the same family.

MEASUREMENTS

	C.M. No. 11,996
Lower molar series, length	179 mm.
M_1, antero-posterior diameter	?44
M_1, transverse diameter	31
M_2, antero-posterior diameter	55
M_2, transverse diameter	34
M_3, antero-posterior diameter	87
M_3, transverse diameter	36
Mandible, depth below m_1	58
Mandible, depth below talon of m_3	73
Mandible, thickness, below m_1	38

Horizon: Duchesne River, Halfway substage.
Locality: 12 m. S.W. of Vernal, Utah.

FAMILY **HELALETIDAE** OSBORN

HETERALETES PETERSON

(Pl. II, figs. 4, 4*a*)

Heteraletes Peterson, *Ann. Carnegie Mus.* **21**: 68, 1931.

Generic characters: I_3, C_1, $P?_4$, M_3.

P_2 with low tricuspid crown, P_3 submolariform; P_4 completely molariform; M_1 and M_2 with cross crests and anterior and posterior cingulae [*sic*] as in *Dilophodon* from the Bridger Eocene.

The incisors are fan-shaped, the first and second of subequal size, while the lateral incisor is reduced to almost half the size of those in front. The lower canine is rather low-crowned, but of considerable antero-posterior diameter at the base of the crown. The latter rises to a trenchant point. There is a considerable diastema between the canine and the cheek-dentition, which may possibly be slightly exaggerated [in the figure] due to the mending of the specimen. P_1 may or may not be present. P_2 a comparatively simple crown, consisting of the para-, proto- and metaconids, the protoconid the larger of the three. The crown of P_3 is almost completely molariform; besides the typical paraconid the crown of this tooth has the two complete crosscrests, as in the molars, and there is a well formed cingulum on the posterior face. P_4 has reached the complete molarization with the two cross-crests and well marked cingulum in front and back. The first and second molars are cross-crested with the cingulae [*sic*] in front and back, as in *Helaletes* or *Dilophodon* from the Bridger Eocene. M_3 is rather deeply buried in the jaw. Its detailed structure cannot be correctly described. The mandibular rami are quite heavy, deep, and have a strong symphysis, the posterior border of which is opposite the junction between P_3 and P_4.

Not knowing the detailed structure of M_3, the proposed genus is, with the advanced condition of P_3 and P_4 most nearly like *Dilophodon* of the Bridger Eocene, though smaller than the latter genus.[50]

Peterson's description is sufficient without adding anything further.

Heteraletes leotanus Peterson

Heteraletes leotanus Peterson, *Ann. Carnegie Mus.* **21**: 68, 1931.

As there is but one species assigned to this genus, the measurements are a sufficient description. In the following table, Peterson's figures are quoted (without verification).

[50] Peterson, O. A., *op. cit.* **21**: 68, 1932.

MEASUREMENTS

Length of jaw, incisors to angle of ascending ramus, approx.	95 mm.
Length of cheek-dentition p_2, m_2	33
P_2 length	4.5
P_2 breadth	5.2
P_3 length	6.5
P_3 breadth	4.
M_1 length	6.5
M_1 breadth	5
M_2 length	7
M_2 breadth	5

These dimensions show what a very small animal this type specimen was; it is much the smallest known representative of its family and one of the most diminutive of known perissodactyls; only some of the Eocene horses, *Hyracotherium-Epihippus* were as small.

Horizon: Duchesne River, Randlett substage: "Duchesne Oligocene, near base of series" (Peterson).
Locality: 3 m. N.W. of Leota Ranch, Uinta Basin, Utah.

FAMILY **HYRACODONTIDAE** COPE

MESAMYNODON PETERSON

(Pl. II, fig. 5)

Mesamynodon Peterson, *Ann. Carnegie Mus.* **21**: 71, 1931.
Hyracodon Peterson (*nec* Leidy), *Ann. Carnegie Mus.* **23**: 388, 1934.

I am convinced that Peterson's reference of the jaw fragment, which is the type of the species in this proposed genus, to the Amynodontidae is a mistaken one and, therefore, that the name *Mesamynodon* was not happily chosen; it is unfortunate that the law of priority forbids the substitution of some more appropriate term. I may add that Professor H. E. Wood, the leading authority on the American rhinoceroses, agrees with me in this opinion. I am, further, of the opinion that the specimen named *Hyracodon primus* in Peterson's posthumous paper of 1934, which did not have the benefit of the author's revision, should be placed in a less advanced genus.

The original description of *Mesamynodon* is as follows:

Generic Characters: P_2 vestigial; proto- and deuteroconid [of p_3], which form the main cross-crest, relatively low; paraconid and posterior cross-crest high and more completely developed, when compared with lower premolars of *Amynodon.* Molars as in *Amynodon*, but with cingulum better developed.

In excavating the alveolar border in front of P_3 of the type, a portion of a shallow alveolus with a small fragment of the root of P_2 was found. Judging from this very shallow alveole and the minuteness of the root-fragment this tooth is evidently reduced to a mere vestige in comparison to the already much reduced P_2 of *Amynodon.* In fact the tooth may be entirely wanting in some individuals, indicating a considerable step beyond *Amynodon* toward such forms as *Metamynodon* of the White River Oligocene. The prominence of the cingulum on both premolars and molars of the present specimen strongly suggests *Hyracodon* of the Nebraska-Dakota Oligocene, but P_3 is reduced too much in length, to say nothing of the vestigial P_2, to place the specimen with the *Hyracodonts.* Provisionally I therefore place *Mesamynodon* in the subfamily *Amynodontinae* pending the discovery of better diagnostic material.

I think that Peterson was mistaken in his interpretation of this fossil; its appearance has probably been changed since he wrote the description which is quoted above, for some plaster of Paris has been placed beneath p_3 and the alveole which he believed to be vestigial, has been excavated. *At present,* I can see no reason to regard p_2 as vestigial and the alveole might be that of the posterior root of a two-rooted p_2; p_3 and p_4 are less completely molariform than in *Hyracodon* and this feature, together with the upper premolars, which are not so completely molar-like as in the White River genus, justifies the recognition of *Mesamynodon* as a distinct genus. It is true that, in *Hyracodon*, the molarization of the premolars was subject to great variation in White River times and this has led to the proposal of several different species, but, in none of these variants, are the premolars so little complicated as in the unfortunately named Duchesne River genus.

In the posthumous paper, already referred to, Peterson proposed a new species of *Hyracodon*, naming it *H. primus*, the type of which is a fragment of an upper jaw, with p_4–m_3 in it, C.M. No. 11,914. In this maxillary, only m_2 is complete, the other teeth having lost more or less of their crowns and m^3 is not included in the figure (fig. 8) of Peterson's paper. A second, referred specimen, which Peterson probably never saw (C.M., No. 11,915) has enabled Mr. Horsfall to draw unbroken crowns of p^4 and m^1 and a nearly complete m^3, which lacks the postero-internal portion. The three specimens, upper teeth (No. 11,914 and 11,915) and lower jaw, type of *Mesamynodon medius* (No. 11,672) agree well in size and are probably referable to the same species. If so, the latter name must be used, as having priority.

Upper Teeth.—The two individuals in the Carnegie Museum, Nos. 11,914 (type) and 11,915 (referred) together supply p^4 to m^2 intact and m^3 substantially complete. Peterson describes his "*Hyracodon primus*" as follows:

In detailed structure the teeth agree most closely with those in *Hyracodon petersoni* Wood, except the less hypsodont crowns; the more prominent external cingulum of P^4, M^1 and M^2; and the greater convexity of the posterior portion of the ectoloph on P^4 and M^1. This convexity of M^2 is much reduced and more nearly approaches the concave external face of the metacone in the molars of *H. petersoni.* The crista of P^4 and the molars are of somewhat greater development than in the White River species.

Under the circumstances, a rather fuller and more systematic account than Peterson's preliminary description seems to be called for.

Upper Teeth.—P^4 is even more molariform than in most examples of *Hyracodon nebrascensis*, but is relatively smaller and lower crowned and the posterior crest is much farther forward, making the hinder valley and enamel pocket correspondingly larger. The crista likewise is much less prominent, being in an incipient stage; as in the White River species, the lingual end of the anterior cross-crest is recurved and reaches the posterior crest, obstructing the exit of the main valley; the cingulum is prominent on the anterior and posterior ends of the crown, but much less so on the lingual and buccal sides than in *H. nebrascensis*.

The *molars* are very much like those of *Hyracodon*, but with such differences as to make it inadvisable to refer them to that genus. There are also differences between the molars of the two individuals, which arouse some doubt as to whether they both belong to the same species, though I think that these differences are probably fluctuating variations rather than of specific value. In the type-specimen (No. 11,914) m^1, though much worn, shows a distinct antecrochet on the anterior cross-crest, while in No. 11,915 m^1 is without crochets of any sort. M^2 has a prominent and conspicuous crista, but no crochets and m^3 is without crista, or crochet, or antecrochet. In this tooth the external wall is extended behind the posterior crest much farther than in *Hyracodon*, making the crown much more quadrate and less triangular than in the genus last named.

Lower Teeth.—The type specimen of genus and species is a fragment of the left *ramus mandibuli*, in which four teeth are preserved, p_3, p_4, m_1 and m_2; the animal was past maturity, though hardly to be called old, as is indicated by the amount of wear of the teeth. No other specimen, referable to this species, has been found and, thus, nothing is known of incisors or canines in either jaw. The third *premolar*, p_3, is distinctly less molariform than in *Hyracodon*, the anterior crescent being less complete and almost straight and the anterior valley is broader and more shallow. P_4 is smaller than m_1 and the hinder horn of the anterior crescent is less inclined backward, but, otherwise, this tooth has acquired the molar pattern almost completely. The *molars* are very much the same as in *Hyracodon* and display no generic difference from it. As Peterson noted, the cingulum is prominent on both premolars and molars, but in the latter it is faint on the lingual side of the posterior lobe and the likeness, in these respects, to *Hyracodon* is so close, as almost to lead Peterson into referring these fossils to the White River genus.

Beside the teeth, Peterson described and figured a femur and an astragalus, which he provisionally referred to *Hyracodon*. So far as the femur is concerned, this reference is probably correct as to the family, probably not to the genus, which is not known to occur before the White River. As Peterson points out, the femur is like that of *Hyracodon*, but differs in several respects; the shaft is relatively more slender, the great trochanter rises higher above the level of the head and the third trochanter is larger and more prominent and has a rather more proximal position. The reference of the astragalus is very doubtful, but, almost certainly, is not to *Hyracodon* because of the shortness of the neck, which is almost as short as in the horses, the inner condyle extending nearly to the navicular facet, which is not produced beyond the trochlea toward the tibial side.

Mesamynodon medius Peterson

Mesamynodon medius Peterson, *Ann. Carnegie Mus.* **21**: 73, 1931.
Hyracodon primus Peterson, *Ann. Carnegie Mus.* **23**: 388, 1934.

As there would seem to be but a single species, the dimensions afford a sufficient diagnosis. Peterson's measurements of the type of *Mesamynodon medius* contain certain evident typographical errors, but are given here, with the measurements which I made at the Carnegie Museum in November 1941.

MEASUREMENTS

	Mesamynodon medius Peterson	*Mesamynodon medius* W. B. S.
Length of jaw fragment	66 mm.	72 mm.
Depth of ramus at m_1	31	30
Length of p_3.	13	14
Breadth of p_3 opposite posterior crest	95* [*sic*]	11
Length of p_4	15	15
Breadth of p_4 opposite post. crest	10.5	11
Length of m_1	17	17
Breadth of m_1 opposite post. crest	12	13
Length of m_2	20	21
Breadth of m_2 opposite post. crest	12.5	14

* Peterson probably wrote 9.5.

Horizon: Duchesne River, Lapoint substage.
Locality: "Titanothere Quarry," 11 m. W. of Vernal, Utah.

Peterson's measurements of the type of *Hyracodon primus* are repeated in the table below without verification.

MEASUREMENTS

	Hyracodon primus C.M. No. 11,914
Length of p^4–m^1 approx.	53 mm.
Length of p^4	12
Breadth of p^4	12
Length of m^1, ectoloph measurement	15
Breadth of m^1 opposite metaloph	15
Length of m^2 ectoloph measurement	17
Breadth of m^2 opposite protoloph	17
Breadth of m^2 opposite metaloph	14

It will be observed that the types of *Mesamynodon medius* and *Hyracodon primus* agree quite closely in size and may well belong to the same species.

Horizon: Duchesne River, Lapoint substage.
Locality: about ½ m. W. of Vernal, Utah.

FAMILY RHINOCEROTIDAE OWEN

EPITRIPLOPUS WOOD

Prothyracodon Peterson (nec. S. and O.), *Ann. Carnegie Mus.* **12**: 134, 1919.
Epitriplopus Wood, *Bull. Amer. Palaeontol.* **13**: 179, 1927.

In the Uinta stage of the Upper Eocene there occurs a group of fossils, evidently very closely allied, to which three generic terms have been applied, *Prothyracodon*,

Epitriplopus and *Eotrigonias*, all of which may eventually prove to be referable to the same genus. I am keeping them apart, provisionally, until more complete material shall have been found. All of the supposed genera agree in the form of the third upper molar, which differs from that of the hyracodonts and amynodonts and agrees with that of the true rhinoceroses in having the external wall confluent with the posterior crest; a vestige of the outer wall may or may not be present. In consequence, the crown has taken on a trigonal rather than a quadrilateral form. How far this exceptional and characteristic feature of the third upper molar is significant of relationship with the Rhinocerotidae, remains to be determined. For the present, I am including these forms, whether one genus, or two, or three genera, in the true rhinoceroses, though it is quite possible that the resemblance is due to convergence rather than to relationship. Wood, in his paper of 1927,[51] refers *Eotrigonias* to the Rhinocerotidae, *Prothyracodon* and *Epitriplopus* to the Hyracodontidae, a procedure which I cannot accept and I doubt that it represents his present opinion. Peterson overlooked the figure of the *Prothyracodon* dentition which I published in 1890,[52] and which showed that in this genus m^3 is of the trihedral, rhinocerotic type, with the outer wall and posterior crest confluent and the wall behind the crest reduced to a mere wrinkle of enamel. Nevertheless, I think it preferable to employ the three names, rather than to keep shuffling them back and forth, as better-preserved fossils are discovered. It is my present belief that at least two of the three alleged genera will eventually be merged into one, and that the undoubted differences between them will prove to be of no more than specific importance. *Prothyracodon* has a tridactyl manus and, if Wood is right in regarding *Eotrigonias* as ancestral to the White River *Trigonias*, it must have retained the fifth digit in the manus.

Wood's original definition of *Epitriplopus* is as follows:

The genoholotype is *Prothyracodon uintense* Peterson I_3, $C\frac{?}{1}$, $P\frac{4}{4-3}$, $M\frac{3}{3}$. P^2 has an ectoloph and a single large internal cusp connected with the paracone by a cross-crest. M^2 is very long in proportion to its width. The posterior buttress of M_3 is lost completely except for a trace near the base. The manus is tridactyl.[53]

Epitriplopus medius Peterson

Epitriplopus medius Peterson, *Ann. Carnegie Mus.* **23**: 387, 1934.

Peterson's diagnosis of this species is a comparison with *E. uintensis*, as described by Wood, though the comparison is implicit, not expressed as such.

Specific characters: P^3 with proto- and metalophs slightly separated on the inner face of the crown. P^4 with proto- and metalophs well separated internally; the base of the internal exit of the cross-valley is, however, closed up, forming a large and relatively deep median fossette. Furthermore, there is present on P^4 a very weak crista and a cingulum-like ridge on the posterior half of the ectoloph on M^1, of approximately the same diameter as in *Epitriplopus uintensis*, but with a heavier internal cingulum at exit of the cross-valley. There is a well marked cingulum on the posterior half of the ectoloph in M^1, absent on M^2, while in *E. uintensis* there are no external cingula on the molars. M^2 shorter than in *E. uintensis*.

The measurements which follow are Peterson's, given here without verification.

MEASUREMENTS

Length of cheek-teeth dentition p^3–m^2	59 mm.
Length of p^3, approximately	13
Transverse diameter opposite metaloph	12
Length of p^4, ectoloph measurement	13
Transverse diameter p^4, opposite protoloph	12
Length of m^1, ectoloph measurement	18
Transverse diameter of m^1, opposite protoloph	15
Length of m^2, ectoloph measurement	18
Transverse diameter of m^2, opposite protoloph	17

Horizon: Base of Duchesne River, Oligocene, Randlett horizon.
Locality: Two 2 m. N.E. of Randlett Point, Uinta County, Utah.

FAMILY **AMYNODONTIDAE** S. AND O.

HORACE ELMER WOOD, 2ND.

A new amynodont rhinoceros is one of the most interesting and instructive discoveries made by the Carnegie Museum expeditions in the Duchesne River formation. This long-sought link, now, fortunately, no longer missing, represents the logical continuation of the evolutionary trend visible in the successive stages of *Amynodon* from the North American Upper Eocene (Uintan), and foreshadows the additional specializations found in *Metamynodon*.[54]

It is a close decision whether to call this the most advanced species of *Amynodon*, thereby considerably extending the scope of the genus as previously known and distorting its verbal definition, or to make this a new, transitional genus; the least satisfactory treatment would be to include this form in *Metamynodon*, for which there would be little justification in logic or convenience. Since the decision must be based purely on general convenience, I am glad to defer to the judgment of my former teacher, the author of the rest of this study, and treat this form as a new genus.

It is a pleasure to acknowledge my obligations to Curator J. LeRoy Kay, of the Carnegie Museum, for his kindness in assigning this interesting material to me for study, as well as for numerous other courtesies over a long period; to Professor William B. Scott, for his sympathetic assistance in preparing this manuscript for publication, despite the complications occa-

[51] Wood, H. E., *Bull. Amer. Palaeontol.* **13**: 188, 1927.
[52] Scott, W. B., *Trans. Amer. Philos. Soc.* **16**: pl. XI, figs. 6, 7, 10, 1890.
[53] Wood, H. E., *loc. cit.* **13**: 179, 1927.
[54] Wood, H. E., *Trans. N. Y. Acad. Sci.*, ser. 2, **3**: 86–89, 1941.

sioned by my military service; and to the American Philosophical Society, in Philadelphia, for support of my studies of rhinoceros evolution.

Megalamynodon regalis, gen. & sp. nov.[55]

(Pl. VIII)

Taxonomy: Amynodon, n. sp. (Wood, *et al.*, *Bull. Geol. Soc. Amer.* **52**: 10, 1941; Wood, *Trans. N. Y. Acad. Sci.*, (2), **3**: 87 1941).

Types: type specimen, C.M. No. 11,953, consisting of both maxillae with the upper cheek teeth and the mandible with most of the lower teeth; the rear ends of the mandible with nearly all the molars: referred specimen, C.M. No. 9,961, upper and lower teeth in jaw fragments.

Locality: all three specimens (Field, No. 27/1933) were found in a quarry in the Halfway horizon (or local fauna) of the Duchesne River formation, Duchesnean [56] age (Eocene-Oligocene transition), one mile north and west of Twelve Mile Bridge on U. S. Highway No. 40, twelve miles southwest of Vernal, Uinta County, Utah, by the 1933 Carnegie Museum expedition, under the leadership of J. LeRoy Kay.

Generic and *Specific Diagnosis:* $i\frac{3??}{3-2}$, $c\frac{1}{1}$, $p\frac{2}{3}$, $m\frac{3}{3}$; two upper incisors preserved (probably three originally present), decreasing in size and complexity mesiodistally; upper canine suggests *Metamynodon* except in its much smaller size; buccal aspect of upper premolars suggests *Metamynodon*, but crown view retains most of the simplicity of the *Amynodon* pattern: p^2 is sharply triangular, with the protoloph slanting posteromedially at a small angle with the ectoloph, and a much less prominent metaconule, enclosing a medifossette; p^3 triangular to molariform; p^4 submolariform to molariform; upper molars intermediate in size and character between *Amynodon intermedius* and *Metamynodon planifrons* but nearer the former; i_3 was much larger than the other incisors, judging by the roots; lower canine relatively small, suggesting *Amynodon*, rather than *Metamynodon* or other advanced forms; no trace of p_1 or p_2; p_3 reduced as compared with *Amynodon*, but larger than in either *Paramynodon* or *M. chadronensis;* p_3 slightly simplified and shortened, antero-posteriorly, as in *Metamynodon* (as the anterior tooth of the functional series) but not as much so as in *M. chadronensis*, even; p_4 essentially molariform and not greatly shortened, antero-posteriorly; m_{1-3} much as in *A. intermedius*, i. e., not distorted out of all close resemblance to rhinoceros teeth, especially in the rather smooth (i. e., less swollen) buccal surfaces of m_2 and m_3, in particular, but nevertheless, intermediate between *A. intermedius* and *M. chadronensis.*

[55] *Megalamynodon regalis:* big + *Amynodon*; *regalis* refers both to the large size, compared with its Uintan predecessors, and to the leader of the field exploration through which this important material was secured, Mr. J. LeRoy Kay.

[56] Wood, H. E., *et al.*, *Bull. Geol. Soc. Amer.* **52**: 10, 19, pl. I, 1941.

DESCRIPTION OF INDIVIDUAL SPECIMENS

Since the left upper tooth row of C.M. No. 11,953, with p^2–m^3, occludes exactly with the lower tooth series bearing the same museum number, the association in a single individual seems indisputable, quite aside from the field label.

This is apparently a young adult, with all the permanent teeth fully erupted, but without serious attrition of the dental patterns. The upper diastema, as shown on the left side, is surprisingly short—only 10.5 mm. long. P^2 has a high protoloph and a low metaconule; its crista is beginning to form a sharp ridge; the valley is not quite open, internally. In p^3, a long crista first runs parallel to the metaconule, then joins it; the median valley is open, lingually. P^4 has a long crista which remains free at its internal end; it also has a metacrista which joins the metaconule, enclosing a pit (which lacks a special name) between metacrista, metaconule and metacone. (This peculiar arrangement is present, on one side only, in this specimen; it also appears on C.M. No. 9,961.)

Only the palatal aspect of the skull, C.M. No. 11,958, has been preserved. The individual was presumably considerably older than C.M. No. 11,953, since most of the distinctive characters of the tooth pattern had been ground away from the occlusal surfaces of the teeth. P^3 has a separate, somewhat puny, metaloph, with the median valley opening lingually. P^4 is fully molariform, or, rather, is like a stunted molar, with the metaloph thinner and lower than the protoloph; it is essentially like p^4 of *Metamynodon planifrons*, except that it is not bulbous. M^1 is wedge- or keystone-shaped, as in an unworn m^1 of *M. planifrons.* The proportions of m^{2-3}, also, are almost as in *M. planifrons.*

As he was unable to make more than a few measurements before leaving for induction in the Army, Professor Wood asked me (W.B.S.) to make them for him and I am, therefore, responsible in the matter. In the following table the term "dimensions" means the antero-posterior by the transverse diameter of a tooth-crown. All the specimens measured are in the Carnegie Museum.

MEASUREMENTS

	No. 9,961	*No. 11,958*	*No. 11,953*
Upper cheek-tooth series, length			204 mm.
Upper premolar series, length			.73
Upper molar series, length		147 mm.	144
P^2, dimensions	20×22 mm.		20×26
P^3, dimensions	25×28	26×38	26×31
P^4, dimensions	34×37	31×45	31×40
M^1, dimensions	48×41	50×49	46×42
M^2, dimensions	57×48	59×55	53×44
M^3, dimensions		51×44	49×46

Lower dentition, length i_3–m_3			290 mm.
I_3, dimensions			13×12.5
Lower canine, dimensions			24×16
Lower cheek-tooth series, length			217
Lower premolar series, length			88
Lower molar series, length		141 mm.	134
P_3, dimensions	21×14 mm.		22×17.5
P_4, dimensions	30×20		30×19
M_1, dimensions	39×23	49×21	39×23
M_2, dimensions	48×27	54×22	44×25
M_3, dimensions		53×27	51×26
Skull, breadth over zygomatic arches		298	
Palate, width at m_2		92	
Occiput, breadth over parocc. processes		141	
Occipital condyles, breadth		116	
Mandibular angle, breadth to m_3		170	
Mandible, height of condyle		181	
Mandible, height of coronoid		222	
Mandible, depth of ramus at m_2		68	

Horizon: Duchesne River, Randlett substage.
Locality: 1 M. N. & W. of 12 Mile Bridge, U. S. Highway No. 40, 12 M. S.W. of Vernal. Uinta Co., Utah.

CONCLUSIONS

The morphological and evolutionary position of this form is self-evident:[57] it falls into a "previously prepared position" in an unusually simple ("monophyletic") evolutionary sequence, intermediate between *Amynodon intermedius* Scott and Osborn, typical of Uinta C. and *Metamynodon chadronensis* Wood, from the Chadron formation, which latter species, unfortunately, is known only from the lower jaw. In a broader sense, it is an annectant form between the genus *Amynodon*, typical of the Upper Eocene (Uintan), and the Oligocene *Metamynodon*. All details of its structure fit into this generalization in an unusually harmonious evolutionary picture, without distortion of any kind—not even that of distorted coördinates.

ADDITIONAL NOTES ON *MEGALAMYNODON*

W. B. SCOTT

Professor Wood's induction into the Army has prevented him from preparing a more extended and detailed account of these most interesting fossils, which are surprisingly abundant for the Duchesne River; next to *Teleodus*, the individuals referable to this genus are more numerous than those of any other of the fauna. In the Carnegie Museum are two individuals which, considering the variety and generally fragmentary condition of the Duchesne River fossils, may fairly be called well preserved. The first of these (C.M. No. 11,953), which Wood has selected as the type of his proposed genus and species, *Megalamynodon regalis*, consists of the superior maxillaries, right and left, each of which retains all of the grinding teeth; two upper incisors, i^2 and i^3, of the left side, and the right upper canine were found detached, but evidently belonging to the same individual. In addition there is a mandible with symphysis and both horizontal rami, but lacking the ascending rami of both sides. On one side, or the other of this jaw are preserved the lateral incisor (i^3), the canine and all the cheek-teeth. The dentition is, thus, very completely represented.

The second specimen referable to this genus is C.M. No. 11,958, which Wood makes the paratype of his new genus and species. This second individual consists of the finely preserved (relatively speaking) *basis cranii*, with bony palate forward to p^3, and both zygomatic arches, undistorted; all of the cheek-teeth except p^2 of both sides, are retained in position. The anterior part of the lower jaw, including the symphysis and both horizontal rami in front of the molars, is missing, but the right and left ascending rami on both sides, with condyle and coronoid, are nearly complete. Several other more or less fragmentary specimens of jaws, with teeth, are in the collection, but add little to our knowledge of the genus.

Megalamynodon is decidedly larger than the Uinta species of *Amynodon*, but much smaller than the typical species, *Metamynodon planifrons*, of the White River, though the dental formula is the same, except for the incisors, as in the latter. Among the most characteristic features of the family, which, so far as North America is concerned, culminated and ended in the White River, are: (1) the great shortening of the facial region of the skull; (2) the shortening of the nasal bones, with long downward extension of the ventral borders *inside* of the premaxillaries; (3) the shortening of the premaxillae, which extend but little in front of the canines, but their anterior faces are much widened, which broadens the whole muzzle; (4) the shape of the anterior nares is greatly changed,[58] widening transversely and narrowing much dorsoventrally, thus giving a very different aspect to the view of the skull from in front; (5) the zygomatic arches are greatly expanded transversely and become broader dorso-ventrally and thicker transversely, giving to the skull, especially when seen from the front, much the appearance of a large carnivore; this resemblance is farther increased by the long and very prominent sagittal crest.

In the course of its known history, from the Washakie up through the Uinta, Duchesne River and White River, the teeth underwent considerable modification, each of the three genera in the ascending series, *Amynodon-Megalamynodon-Metamynodon*, displaying a characteristic stage of the change.

The successive steps of modification in the *dentition* are as follows: (1) The incisors, especially those of the upper series, grew steadily smaller and, in *Metamynodon*, they were reduced in number and would seem to have been functionless, which, apparently, was not

[57] Wood, H. E., *Trans. N. Y. Acad. Sci.*, Ser. 2, 3: 86–89, 1941.

[58] Scott, W. B., *Trans. Amer. Philos. Soc.* 28: 847, fig. 140, 1941.

true of *Megalamynodon*. At all events, the progressive reduction of the incisors had reached, in the latter, a stage intermediate between the Uinta and the White River representatives of the family.

The only incisors preserved in the Duchesne River material belong to C.M. No. 11,953 and are the left i^2 and i^3, the crown of i_3 and the roots of i_1 and i_2, also of the left side. The second upper incisor is very small and has a simple, antero-posteriorly compressed, conical and acutely pointed crown, without cingulum, and long, cylindrical root. I^3 is much larger and the compressed-conical crown is complicated by a prominent lingual cingulum, which encloses a groove between itself and the body of the tooth; the borders of the crown, mesial and lateral, are slightly, but distinctly raised and there is also, on the lingual side, a conspicuous, low ridge, which runs up from the cingulum to the mesial border, enclosing a shallow-enamel pocket. The root of i^3 has a diameter two or three times as great as that of i^2 and of a different shape, being strongly compressed laterally.

Of the lower incisors, only i_3 retains its crown; this is obliquely procumbent, simply conical in shape and antero-posteriorly thick and with a strong cingulum on the lingual face, which bears obscurely marked ridges; the root is stout and laterally compressed. As indicated by the alveoli, this is much the largest of the lower series. Of the other lower incisors, only the alveoli remain, but these suffice to show the relative size and shape of the roots; i_1 is much the smallest of the series and the cavity is very narrow transversely. The root of i_2 was evidently larger than that of i_1 and of cylindrical shape.

One of the most striking features of *Metamynodon* is to be seen in the large, hippopotamus-like tusks, upper and lower, which were obviously rootless and grew from permanent pulps. In *Amynodon*, the tusks, though effective and formidable weapons, were yet relatively much smaller than in the White River genus and are rooted. The Duchesne River genus, *Megalamynodon*, is intermediate in the character of the tusks between the Uinta and the White River genera; the upper canine is actually and relatively much larger than in the former, far smaller than in the latter, and almost certainly grew from a permanent pulp. The lower canine is proportionally smaller than the upper one, though larger than in *Amynodon*, and seems to have a root; at all events, the enamel ceases definitely at a level above the base.

To Wood's description of the grinding teeth, I need add only a comparison with those of the Uinta ancestor and the White River descendant of *Megalamynodon*. The *premolars* of the latter number two, which is one less than in the Uinta genus and the superior ones are larger and have a stronger cingulum; on both p^3 and p^4 the anterior cross-crest is complete, but the posterior one is divided in unworn teeth. In *Metamynodon* the premolars are still further reduced in size, but except for the convex external lobes, p^3 and ^{4} are more perfectly molariform and have complete crests. In the lower jaw, the premolars are still two in number and are even smaller proportionally. The progressive relative reduction of the premolars in this series is as below.

Like the premolars, the molars are intermediate in character between those of the Uinta and those of the White River genera; in the latter the molars have become very large and high-crowned, suggesting that this series would eventually have developed hypsodont teeth, had the family not become extinct.

EXPLANATIONS OF THE PLATES

PLATE I

FIG. 1. *Eosictis avinoffi* gen. et sp. nov. type; C.M. No. 11,847; right maxillary, with c–p^4, lingual side, slightly enlarged.

FIG. 2. *Eosictis avinoffi*; fragment of right squamosal, with glenoid cavity, from outer side. C.M. No. 11,347, slightly enlarged.

FIG. 3. *Hessolestes ultimus*, type; C.M. No. 11,763; fragment of right *ramus mandibuli*, lingual side, with m^2, × 9/10.

FIG. 4. *Poabromylus kayi*, type; lower dentition, buccal view, slightly enlarged.

FIG. 4*a*. The same, crown view.

FIG. 5. *Pentacemylus progressus* type; C.M. No. 11,865; upper molars, crown view, left side, × 5/2.

FIG. 6. *Pentacemylus progressus*, lower teeth, p_4–m_2, crown view, × 5/2.

FIG. 7. *Epihippus intermedius*, type; C.M. No. 11,845, right *ramus mandibuli*, with i_1–m_2, outer view, nearly natural size.

FIG. 7*a*. *Epihippus intermedius*, type; lower dentition, crown view.

PLATE II

FIG. 1. *Protoreodon tardus*, sp. nov., type; C.M. No. 12,049, skull top view, × 1/1 ?Duchesne River, Beaver Divide, Wyo.

FIG. 1*a*. *Protoreodon tardus*, the same, upper dentition, crown view, × 1/1.

FIG. 2. *Camelodon arapahovius*, type; upper cheek-teeth A.M.N.H. No. 14,604, Beaver Divide, Wyo. × 7/5.

FIG. 2*a*. *Camelodon arapahovius*, type; lower cheek-teeth, crown view, × 7/5.

FIG. 3. ?*Hypertragulid*, C.M. No. 10,199, p^3–m^1, crown view, × approximately 3/1.

FIG. 4. *Heteraletes leotanus*, type; C.M. No. 11,849; lower jaw left side, with i_1–m_3, × approximately 3/1.

FIG. 4*a*. The same, lower dentition, crown view.

FIG. 5. *Mesamynodon medius*; upper cheek-teeth, crown view, × 6/5.

PLATE III

Mesagriochoerus primus

FIG. 1. Skull, left side, C.M. No. 12,080.

FIG. 1*a*. Skull, base.

FIG. 1*b*. Lower dentition, crown view, i_1–m_3.

All figures natural size.

PLATE IV

FIG. 1. *Mesagriochoerus*, sp. A.M.N.H. No. 22,558, upper cheek-teeth, dp.^{3} and ^{4}, m^1, crown view, × 12/5.

FIG. 1*a*. The same, buccal side, × 12/5.

FIG. 2. *Mesagriochoerus* sp., lower teeth, dp_3, $_4$, m_1, crown view, × 12/5.

FIG. 3. *Diplobunops crassus* sp. nov. type; C.M. No. 2,967, skull, base, × 8/11.

FIG. 4. *Mesamynodon medius*; C.M. No. 11,762; fragment of mandible, left side, with p_3, $_4$, m_1, $_2$, × 8/7.

FIG. 4*a*. The same, crown view.

PLATE V

Diplobunops

FIG. 1. *Diplobunops crassus*, C.M. No. 11,301; skull, left side, × approximately 8/11.

FIG. 1*a*. The same, skull, top view, × approximately 8/11.

FIG. 1*b*. The same, skull, base, × approximately 8/11.

FIG. 2. *Diplobunops uintensis*, lower front teeth, ventral side, Uinta formation. P.M. No. 14,252.

PLATE VI

Brachyhyops wyomingensis

FIG. 1. *Brachyhyops wyomingensis*, type; C.M. No. 12,048; skull, side view.

FIG. 1*a*. The same, skull, base.

FIG. 1*b*. The same, top view.

FIG. 1*c*. The same, rear view.

All figures × 13/23

PLATE VII

Teleodus uintensis

FIG. 1. *Teleodus uintensis*, skull, left side, C.M. No. 11,759, teeth, No. 11,866 × approximately 3/10.

FIG. 1*a*. The same, base, × approximately 3/10.

FIG. 1*b*. The same, top view.

FIG. 2. The same, mandible, left side, C.M. Nos. 11,761 and 11,809, × approximately 3/10.

FIG. 2*a*. The same, crown view.

FIG. 3. The same, left astragalus, × approximately 1/2.

PLATE VIII

Megalamynodon regalis

FIG. 1. *Megalamynodon regalis*, type, C.M. No. 11,958, skull, base, × approximately 1/3.

FIG. 2. *Megalamynodon regalis*, mandible left side, C.M. Nos. 11,953 and 11,958.

FIG. 2*a*. *Megalamynodon regalis*; upper dentition, right side, crown view, × approximately 1/3. C.M. No. 11,953.

FIG. 2*b*. The same, lower dentition, crown view, × approximately 1/3.

FIG. 3. The same, upper incisors ^{2} and ^{3}, from behind.

FIG. 4. The same, left upper canine, lingual side.

EOSICTIS, HESSOLESTES, POABROMYLUS,
PENTACEMYLUS, EPIHIPPUS

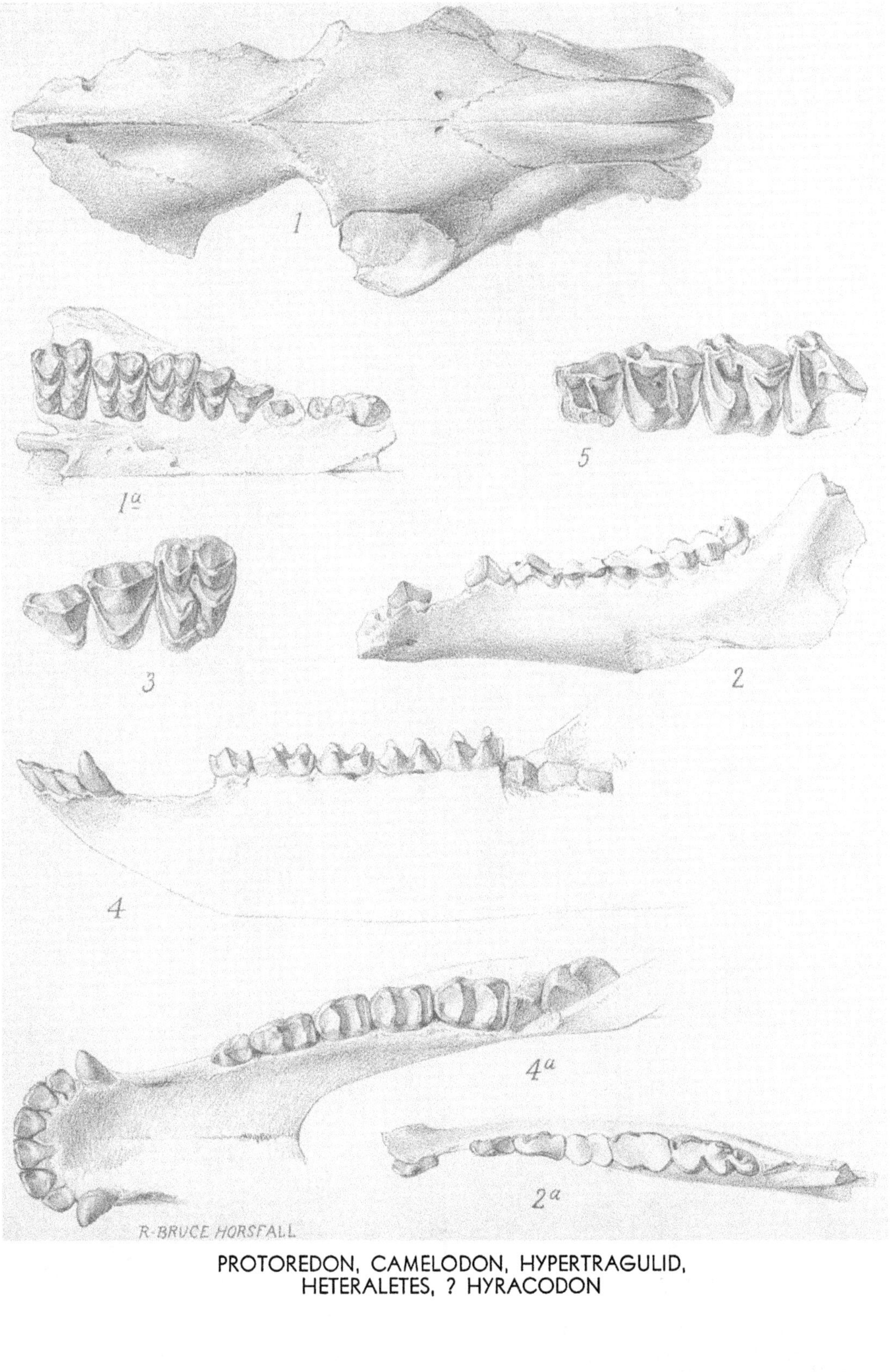

PROTOREDON, CAMELODON, HYPERTRAGULID,
HETERALETES, ? HYRACODON

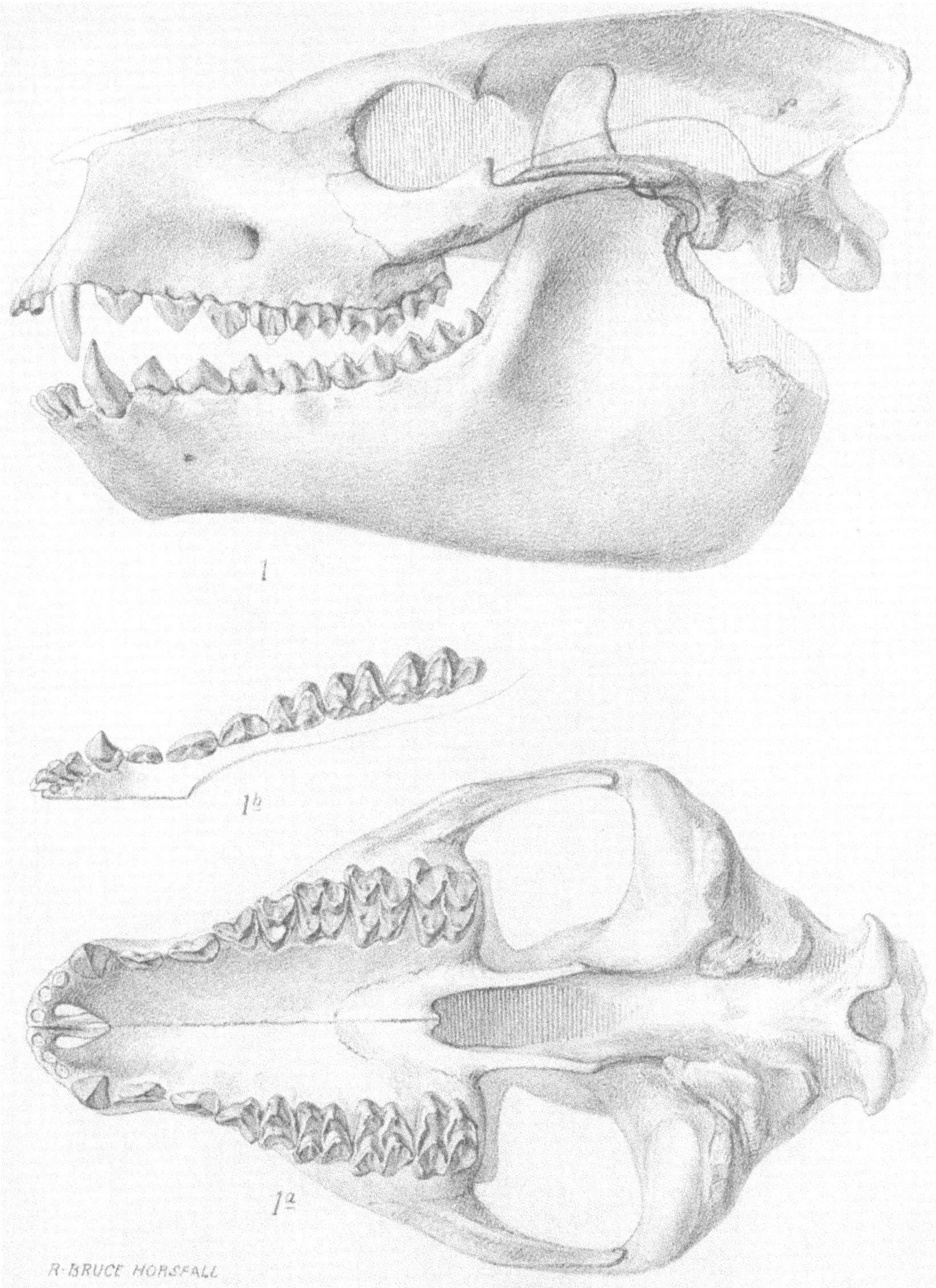

MESAGRIOCHOERUS

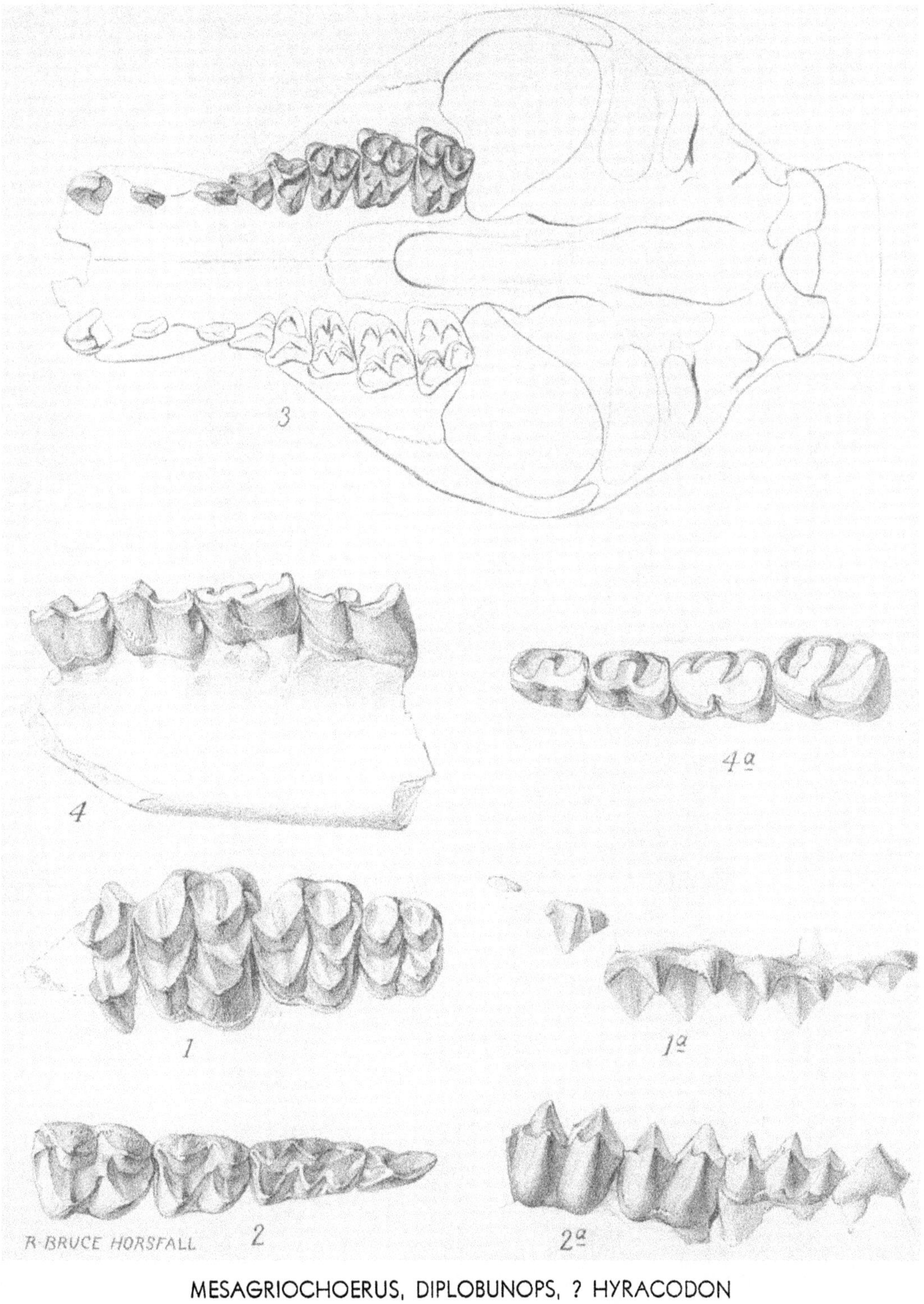

MESAGRIOCHOERUS, DIPLOBUNOPS, ? HYRACODON

PLATE V

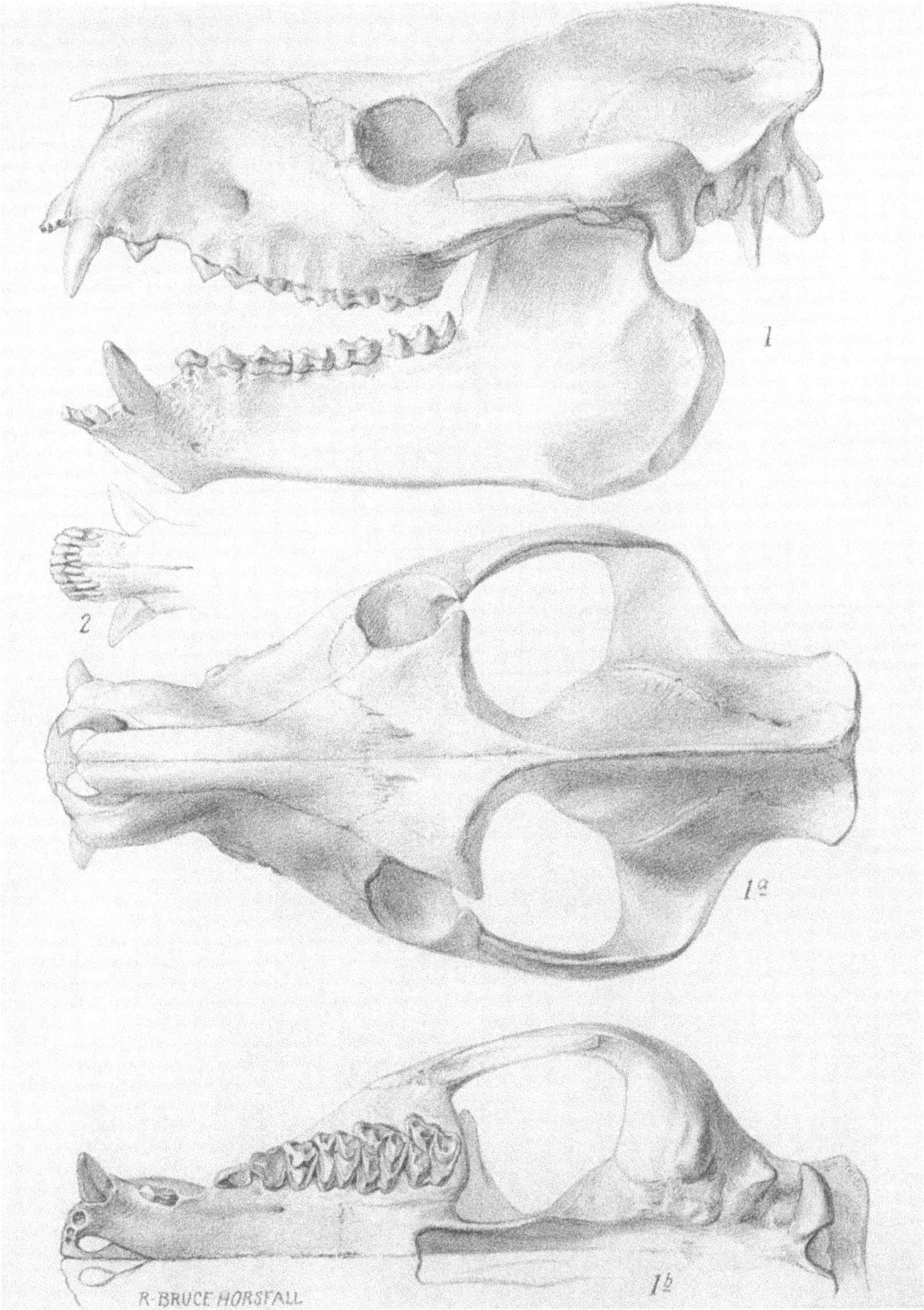

DIPLOBUNOPS

Plate VI

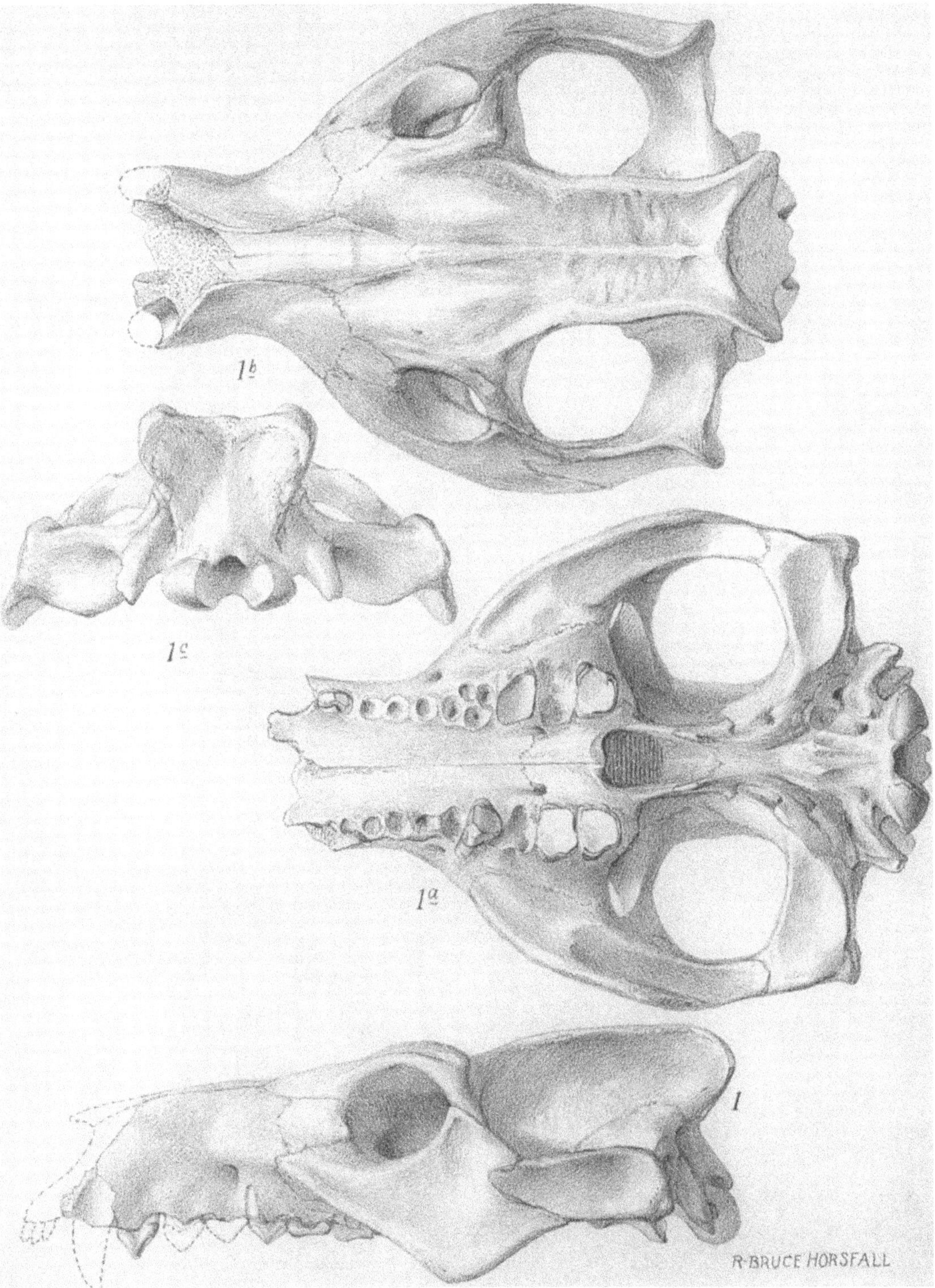

BRACHYHYOPS

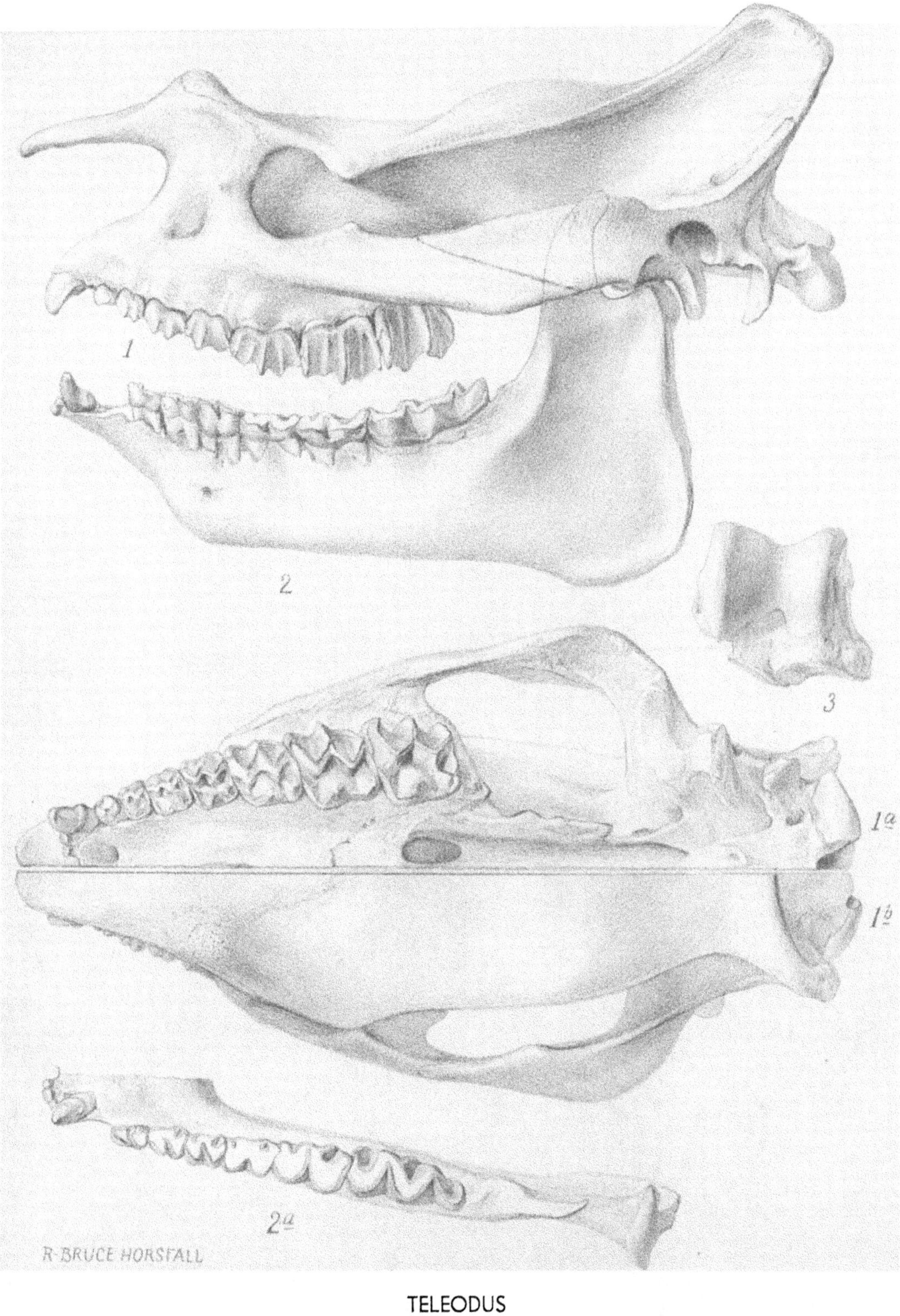

TELEODUS

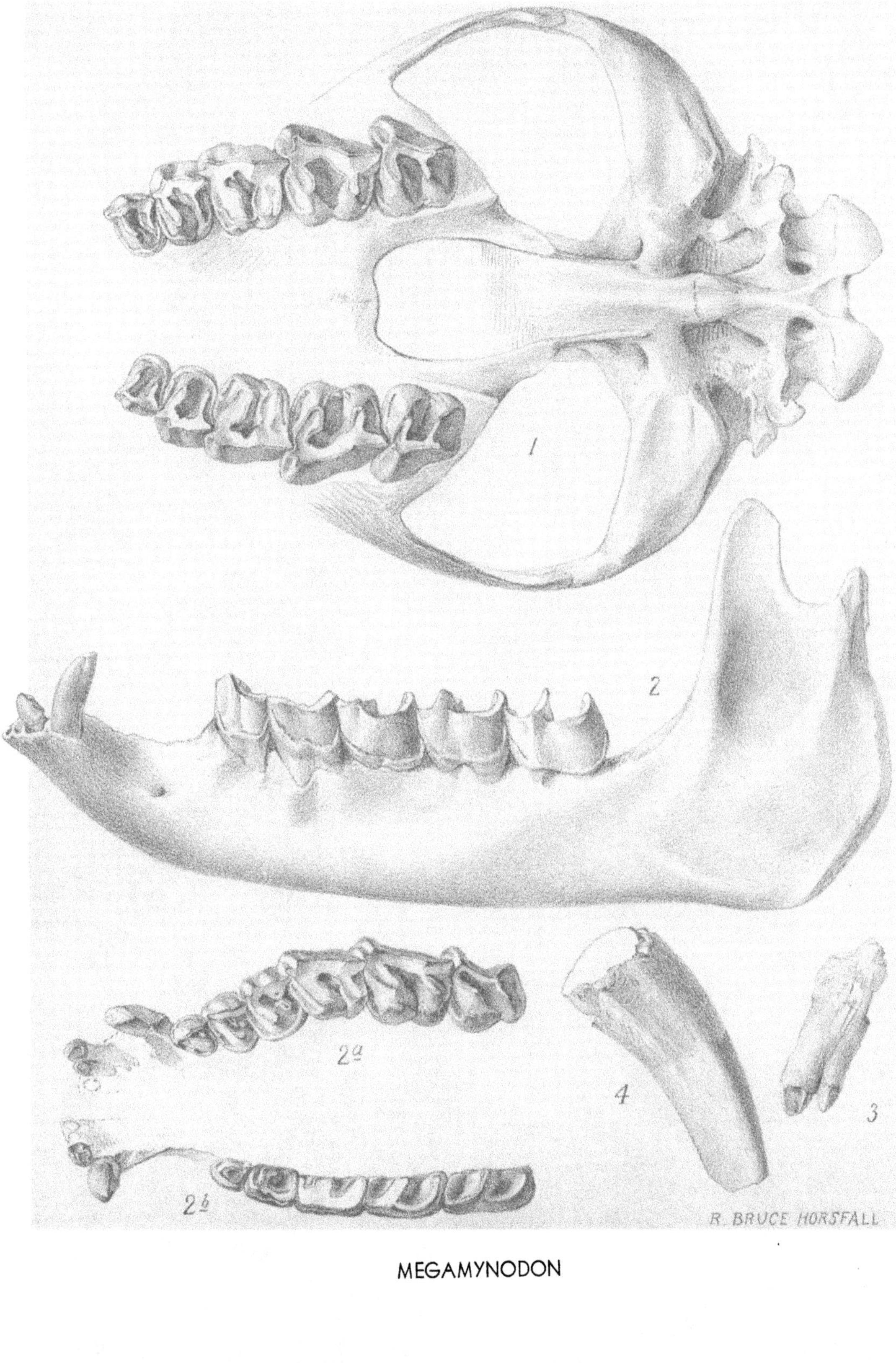

MEGAMYNODON

www.ingramcontent.com/pod-product-compliance
Lightning Source LLC
Chambersburg PA
CBHW081338190326
41458CB00018B/6039
* 9 7 8 1 4 2 2 3 7 7 2 4 6 *